Nutritional Ecology of the Ruminant

A gerenuk (photo courtesy of R. R. Hofmann).

NUTRITIONAL ECOLOGY OF THE RUMINANT SECOND EDITION

Peter J. Van Soest

CORNELL UNIVERSITY

COMSTOCK PUBLISHING ASSOCIATES A DIVISION OF

CORNELL UNIVERSITY PRESS | ITHACA AND LONDON

First edition published 1982.
Reissued 1987 by Cornell University Press.
Second edition published 1994 by Cornell University Press.

⊗ The paper in this book meets the minimum requirements
of the American National Standard for Information Sciences—
Permanence of Paper for Printed Library Materials, ANSI Z39.48-1984.

Library of Congress Cataloging-in-Publication Data

Van Soest, Peter J.
    Nutritional ecology of the ruminant / Peter J. Van Soest. — 2nd ed.
        p.   cm.
    Includes bibliographical references and index.
    ISBN 0-8014-2772-X (alk. paper)
    1. Ruminations—Feeding and feeds.   2. Animal nutrition.   3. Rumen
fermentation.   4. Feeds.   I. Title.
SF95.V36   1994   94-7001
636.2′0852—dc20

Cloth printing   10 9 8 7 6 5 4 3

To the memory of Dr. Lane A. Moore

Contents

Acknowledgments

The present book is a revision of the first edition and is based on my notes for the courses I teach at Cornell University on fiber and the rumen and tropical forages. I am indebted to all the students who have taken those courses since 1969. I am also grateful to the Agriculture and Food Research Council of Great Britain, whose Underwood Fund allowed me to spend 1988–89 at the Institute of Grassland and Animal Production at Hurley, Berks, where I made the initial preparations for this book. Particularly I thank my secretary, Terry Kinsman, who has faithfully typed this book and has endured my handwriting. I also thank J. B. Robertson for his suggestions and the reviewers who read individual chapters and provided many useful comments.

I am grateful to the following individuals and organizations for permission to reproduce, in original or redrawn form, figures and tables from their collections or publications:

Table 1.2, from R. E. McDowell, 1976, courtesy of R. E. McDowell

Table 1.3, from R. E. McDowell, 1977, courtesy of R. E. McDowell

Figures 2.2–2.5, from Fick and Mueller, 1989, courtesy of G. W. Fick

Table 2.4, from L. E. Harris et al., 1967, courtesy of L. E. Harris and the American Society of Animal Science, © *Journal of Animal Science* 26:97–105

Tables 2.5, 2.6, from Rohweder et al., 1978, courtesy of D. A. Rohweder and the American Society of Animal Science, © *Journal of Animal Science* 47:747–759

Figures 2.6, 7.6, 7.7, 7.10, Table 7.1, from R. E. McDowell, 1972, courtesy of R. E. McDowell

Tables 2.7, 11.3, from Kronfeld and Van Soest, 1976, courtesy of S. Karger AG, Basel

Figures 3.2, 4.1, 4.2, from Demment and Van Soest, 1983, courtesy of Winrock International

Figure 3.4, from Janis, 1986, courtesy of C. M. Janis and the Yale Peabody Museum of Natural History

Figure 3.5, from Moir, 1968, courtesy of R. J. Moir and the American Physiological Society

Figures 3.6, 3.8–3.10, from Hofmann, 1989, courtesy of R. R. Hofmann and Springer-Verlag

Figure 4.4, from Mertens and Ely, 1982, courtesy of D. R. Mertens and the American Society of Animal Science, © *Journal of Animal Science* 54:903

Table 4.6, from Udén and Van Soest, 1982b, courtesy of Elsevier Science Publishers

Figure 4.9, courtesy of Pamela Mueller

Figure 4.10, from Robert Grant et al., 1974a, courtesy of the American Dairy Science Association

Tables 5.2, 5.3, 5.8, from Parra, 1978, courtesy of R. Parra

Figures 5.3, 15.23, from C. E. Stevens, 1988, courtesy of C. E. Stevens and Cambridge University Press

Figure 5.5, courtesy of C. E. Stevens

Figure 5.7, from Bertone et al., 1989, reproduced with permission of the American Veterinary Medical Association

Figure 5.9, from Argenzio et al., 1974, courtesy of C. E. Stevens and the American Physiological Society

Figure 5.11, courtesy of B. D. H. van Niekerk and D. E. Hogue

Table 5.11, from Milton and Demment, 1988, courtesy of K. Milton and the American Institute of Nutrition, © *Journal of Nutrition* 118:1082–1088

Table 5.12, Figure 5.13, from Van Soest, 1993, © Springer-Verlag Heidelberg 1993

Table 6.1, from Robert Grant, 1973, courtesy of Robert Grant

Figures 6.3, 6.4, from Mowat et al., 1965, courtesy of D. N. Mowat and the *Canadian Journal of Plant Science* 45:322

Table 6.3, from Van Soest et al., 1978a, courtesy of the American Society of Animal Science, © *Journal of Animal Science* 47:712–720

Figure 6.6, from Minson and McLeod, 1970, courtesy of D. J. Minson

Table 6.7, Figure 6.13, from Capper, 1988, Crown copyright, courtesy of the Natural Resources Institute and Elsevier Science Publishers

Figure 6.8, from Kalu, 1976, courtesy of B. A. Kalu

Figure 6.9, Table 25.2, from Van Soest et al., 1978a, courtesy of the American Society of Animal Science, © *Journal of Animal Science* 47:712–720

Figure 6.10, from Minson and Wilson, 1980, courtesy of D. J. Minson

Figures 6.12, 16.8, courtesy of D. E. Akin

Table 7.2, from Mertens, 1983, courtesy of D. R. Mertens

Figure 7.4, from Riewe, 1976, courtesy of M. E. Riewe and Texas A & M University

Figure 7.13, from McCammon-Feldman, 1980, courtesy of B. McCammon-Feldman

Figure 7.14, courtesy of B. McCammon-Feldman

Figure 7.15, from Stanton et al., 1991, courtesy of T. L. Stanton and the American Dairy Science Association

Figure 7.16, from Mertens, 1983, courtesy of D. R. Mertens

Figure 8.1, courtesy of P. Udén

Figure 8.5, courtesy of P. Schofield and A. Pell

Table 9.4, from Conrad, 1983, courtesy of H. R. Conrad

Table 9.6, from Allen, 1982, courtesy of M. S. Allen

Tables 9.7, 10.8, from Van Soest and Jones, 1988, courtesy of Walter de Gruyter & Co.

Figure 9.8, from J. B. Russell, 1989, courtesy of J. B. Russell

Table 10.1, from Van Soest, 1967, courtesy of the American Society of Animal Science

Figure 10.2, from Van Soest, 1978, courtesy of the American Society of Clinical Nutrition, *American Journal of Clinical Nutrition* 31(Suppl.):S12–S20

Figure 10.4, from J. B. Robertson and Van Soest, 1981, p. 131, courtesy of Marcel Dekker, Inc.

Table 10.9, from McBurney et al., 1986, courtesy of Elsevier Science Publishers

Figure 10.11, from McBurney et al., 1983, courtesy of Elsevier Science Publishers

Figure 11.4, from French, 1979, courtesy of A. D. French and Siebel Publishing Co.

Table 11.6, from Englyst and Cummings, 1987, courtesy of H. N. Englyst

Figure 11.8, from Coughlan and Ljungdahl, 1988, courtesy of M. P. Coughlan and Academic Press

Figure 11.9, from Mayer et al., 1987, courtesy of M. P. Coughlan and the American Society for Microbiology

Table 11.11, Figures 11.13, 12.7, from Van Soest, 1965b, courtesy of the Association of Official Analytical Chemists

Figure 11.14, from Goering et al., 1973, courtesy of the American Society of Animal Science

Table 12.4, from Van Soest and McQueen, 1973, reprinted with the permission of Cambridge University Press

Figure 12.10, from Van Soest and Wine, 1967, courtesy of the Association of Official Analytical Chemists

Figure 12.12, from Lau and Van Soest, 1981, courtesy of Elsevier Science Publishers

Table 13.3, from Eglinton and Hamilton, 1967, copyright 1967 by the AAAS

Figure 13.5, from "Predicting Digestible Protein and Digestible Dry Matter in Tannin-Containing Forages Consumed by Ruminants" by T. A. Hanley, C. T. Robbins, and C. McArthur, *Ecology*, 1992, 73:537–541, copyright © 1992 by the Ecological Society of America, reprinted by permission

Figure 13.10, from Conklin, 1987, courtesy of N. L. Conklin

Figure 13.13, from Bush and Buckner, 1973, courtesy of L. P. Bush and the Crop Science Society of America

Figure 13.14, from Marten et al., 1981, courtesy of the Crop Science Society of America

Figure 13.15, from Barnes and Gustine, 1973, courtesy of R. F. Barnes and the Crop Science Society of America

Figures 14.2–14.4, 14.6, 14.9, from Pitt, 1990, reprinted with permission of The Northeast Regional Agricultural Engineering Service, Ithaca, N.Y. (607) 255–7654

Figure 15.1, courtesy of R. G. Warner

Figures 15.2, 15.3, 15.5–15.7A, 15.9, 17.1, from Hofmann, 1988, courtesy of R. R. Hofmann

Table 15.2, from Reed, 1983, courtesy of J. D. Reed

Table 15.3, from Rodrigue and Allen, 1960, courtesy of the Agricultural Institute of Canada

Figure 15.4, from Steven and Marshall, 1970, courtesy of D. H. Steven

Tables 15.4, 15.5, Figure 15.25, from Kay, 1960, courtesy of R. N. B. Kay and Cambridge University Press

Table 15.6, from C. B. Bailey, 1958, courtesy of Cambridge University Press

Figure 15.8, from Arnold and Dudzinski, 1978, courtesy of G. W. Arnold and Elsevier Science Publishers

Table 15.9, from Strobel and Russell, 1986, courtesy of the American Society for Microbiology

Figure 15.10, from C. E. Stevens and Sellers, 1968, courtesy of C. E. Stevens and the American Physiological Society

Figure 15.11, courtesy of J. G. Welch

Figure 15.13, from Welch and Smith, 1971, courtesy of J. G. Welch and the American Society of Animal Science

Figure 15.15, from Reid and Cornwall, 1959, courtesy of the New Zealand Society of Animal Production

Figure 15.20, from Welch, 1982, courtesy of the American Society of Animal Science, © *Journal of Animal Science* 54:885

Figure 15.22, from Briggs et al., 1957, courtesy of P. K. Briggs and CSIRO Editorial Services

Figure 15.24, from Leng, 1970, courtesy of R. A. Leng

Figure 16.2, from Chen et al., 1987a, courtesy of the American Dairy Science Association

Table 16.3, from J. B. Russell et al., 1988, courtesy of J. B. Russell and the American Society for Microbiology

Figure 16.4, courtesy of Carl Davis

Figure 16.5, courtesy of J. B. Russell

Table 16.5, from Bryant and Wolin, 1975, courtesy of M. P. Bryant

Table 16.7, from Baldwin, 1970, courtesy of R. L. Baldwin and the American Society for Clinical Nutrition, © *American Journal of Clinical Nutrition* 23:1508–1518

Figure 16.8, courtesy of D. E. Akin

Table 16.8, from Isaacson et al., 1975, courtesy of M. P. Bryant and the American Dairy Science Association

Figure 16.10, courtesy of P. H. Robinson

Table 17.1, Figure 23.8, from L. W. Smith, 1968, courtesy of L. W. Smith

Tables 17.2, 17.3, courtesy of T. R. Houpt

Table 17.5, from Kass et al., 1980, courtesy of the American Society of Animal Science, © *Journal of Animal Science* 50:175–197

Table 17.6, from Wanderley et al., 1985, courtesy of the American Society of Animal Science, © *Journal of Animal Science* 61:1550

Table 17.8, from Ørskov et al., 1972, courtesy of E. R. Ørskov and Cambridge University Press

Figure 18.2, from Van Soest and Mason, 1991, courtesy of *Animal Feed Science and Technology*

Figure 18.3, courtesy of J. W. Thomas

Figure 18.4, from Ellis and Lippke, 1976, courtesy of W. C. Ellis and Texas A & M University

Figure 18.6, from Satter and Slyter, 1974, courtesy of L. D. Satter and Cambridge University Press

Figure 18.7, from Van Soest et al., 1991, courtesy of the American Dairy Science Association

Table 18.8, from Mason, 1969, reprinted with the permission of Cambridge University Press

Table 19.1, from Renner, 1971, courtesy of R. Renner

Figure 19.2, from Nicolai and Stewart, 1965, courtesy of the American Dairy Science Association

Figure 19.4, Table 19.6, courtesy of D. E. Bauman

Table 19.5, courtesy of E. N. Bergman

Figure 19.7, from Bauman and Davis, 1975, courtesy of D. E. Bauman and the University of New England Publishing Unit

Table 19.7, from Peel and Bauman, 1987, courtesy of D. E. Bauman and the American Dairy Science Association

Figure 19.8, from Bauman, 1987, courtesy of D. E. Bauman

Figure 19.9, from J. W. Young et al., 1990, courtesy of J. W. Young

Tables 20.1, 20.3, 20.9, from Palmquist, 1988, courtesy of D. L. Palmquist and Elsevier Science Publishers

Table 20.6, from Katz and Keeney, 1966, courtesy of the American Dairy Science Association

Tables 20.7, 20.8, from Harfoot, 1981, courtesy of C. G. Harfoot

Table 20.11, from Christie, 1981, courtesy of W. W. Christie

Table 20.12, from Bauman, 1974, courtesy of D. E. Bauman

Figure 21.2, from Seoane et al., 1972, courtesy of J. R. Seoane and Brain Research Publications, Inc.

Figure 21.4, from Dinius and Baumgardt, 1970, courtesy of the American Dairy Science Association

Figure 21.6, from Van Soest, 1965a, courtesy of the American Society of Animal Science

Figure 21.7, from Milford and Minson, 1965, courtesy of D. J. Minson and Secretária da Agricultura do Estado de São Paulo

Figure 21.8, from Welch, 1967, courtesy of J. G. Welch and the American Society of Animal Science

Figure 21.9, from Bines et al., 1969, reprinted with the permission of Cambridge University Press

Figures 21.10, 21.15, from Forbes, 1986, courtesy of J. M. Forbes

Figure 21.11, from Osbourn et al., 1974, courtesy of D. F. Osbourn

Figures 21.12, 21.13, from Reid et al., 1988, courtesy of R. L. Reid and the American Society of Animal Science, © *Journal of Animal Science* 66:1275–1291

Figures 21.14, 22.12, from Mertens, 1973, courtesy of D. R. Mertens

Figure 21.19, courtesy of D. J. Thomson

Figures 22.4, 22.7, from Van Soest, 1967, courtesy of the American Society of Animal Science

Figures 22.13, 22.14, Table 25.5, from Van Soest, 1973b, courtesy of the Federation of American Societies for Experimental Biology

Table 23.1, from Van Soest et al., 1983b, courtesy of Butterworth-Heinemann Ltd.

Figure 23.11, from Allen and Mertens, 1988, courtesy of M. S. Allen, D. R. Mertens, and the American Institute of Nutrition, © *Journal of Nutrition* 118:261–270

Figure 24.3, from Pullar, 1964, courtesy of J. D. Pullar and Pergamon Press

Figure 24.4, from Flatt, 1964, courtesy of W. P. Flatt and Pergamon Press

Figure 24.5, courtesy of P. W. Moe and H. F. Tyrrell

Figure 24.6, from "A Technique for the Estimation of Energy Expenditure in Sheep" by B. A. Young and M. E. P. Webster, *Australian Journal of Agri-*

cultural Research, 1963, 14:867–873, courtesy of CSIRO

Figure 24.7, from Flatt and Coppock, 1965, courtesy of W. P. Flatt

Figure 24.11, from Bauman, 1979, courtesy of D. E. Bauman

Figure 25.1, from Van Soest, 1964, courtesy of the American Society of Animal Science

Figure 25.2, courtesy of M. S. Allen

Figure 25.3, from Moe et al., 1972, courtesy of P. W. Moe and the American Society of Animal Science

Table 25.3, from Van Soest and Robertson, 1980, courtesy of the International Development Research Centre

Figures 25.5, 25.6, courtesy of M. Barry and D. Fox

P. J. V. S.

1 Ruminants in the World

Grazing ruminants have a highly developed and specialized mode of digestion that allows them better access to energy in the form of fibrous feeds than most other herbivores. Study of their digestive system, which is characterized by pregastric retention and fermentation with symbiotic microorganisms, has led to the development of a field of nutrition that uniquely transcends the boundaries of several academic fields, namely, plant science, microbiology, animal science, and ecology.

Until the 1970s the study of ruminant nutrition stood apart from the rest of nutrition science, which had little appreciation for the role of dietary fiber and gut fermentations in nonruminant animals. But as nutritionists began to understand the importance of dietary fiber to human health and looked to rumen fermentation as a model of what normally occurs in the large bowel of humans, they developed a greater appreciation of the role of gut microorganisms in other nonruminant species.

Domesticated grazing ruminants (i.e., cattle and sheep) have evolved to maximize the utilization of cellulosic carbohydrates, yet even species that are not necessarily efficient utilizers of cellulose, such as goats, antelope, and deer, show adaptations to a diet high in cellulose. The appreciation of this diversity in nutritional ecology constitutes a further widening of the scope of ruminant nutrition to include other herbivorous species. In particular, the field of ruminant nutrition has provided the methodology for understanding fiber and microbial fermentations in all animal species.

The anatomical adaptations of their digestive system has allowed ruminants to exploit fibrous food resources and has rendered them relatively free from the need for external sources of B vitamins and essential amino acids. On the other hand, a price has been paid in the form of metabolic adaptations such as the need for gluconeogenesis to cover the loss of available carbohydrate.

1.1 Numbers of Ruminants

The enumeration of ruminants is important because this class of animals above all others has been important in the development of both hunting and agricultural societies. Most existing ruminants are domesticated, and wild or feral animals are uncommon. Of the 155 or so ruminant species, only about 6—cattle, sheep, goats, buffaloes, reindeer, and yaks—are domesticated, and many of the undomesticated species are in danger of extinction.

The three billion or so domestic ruminants outnumber wild ones by about ten to one (Table 1.1), but domestic ruminants belong to only 5 of the 78 genera of ruminants and tylopods.[1] Cattle and sheep are the two most numerous species. Almost all are domesticated (the few feral ones are derived largely from domestic sources),[2] and they have been genotypically and phenotypically altered during their long association with humans. Domestication may have occurred as long as 10,000 years ago, and world distribution undoubtedly reflects human migration. Most sheep and cattle occur in developing countries and in the tropics. The number of cattle in North America (70 million in the United States and Canada) is only 6% of the total in the world.

1.2 Wild Ruminants

Although less numerous, wild ruminants are nevertheless important to humans and to maintaining the world's ecological balance. Some species are endangered; their numbers are being reduced not only by poaching but also by loss of habitat. Farm fencing in East Africa, for example, is interfering in the natural migration of indigenous ruminants.

---

[1] Deer farming has become popular in New Zealand, supplying horns to East Asia and meat to Europe.

[2] Wild white cattle remain in certain parts of Great Britain.

Table 1.1.   Numbers of ruminants in North America and the world

Species	U.S. and Canada (millions)	World (millions)	Distribution
Sheep	12	1150[a]	Ubiquitous
Cattle	70	1291	Ubiquitous
Goats	2	460	Ubiquitous
Buffaloes	—	130	India, SE Asia
Reindeer	—	2.5	Arctic
Yaks	—	0.8	Himalaya
Camels	few	17	Asia, Africa
Camelids	few	4.8[b]	South America
Total domestic	84	3056	—
Wild ruminants	20.2	310	see Table 1.2
Total domestic and wild	104	3366	—

Sources: R. E. McDowell, 1976; D. E. Johnson et al., 1992.
[a]There are 200 million sheep in Australia and New Zealand alone.
[b]Inclusive of some wild camelids.

Estimates of the numbers of wild ruminants remaining in North America and the rest of the world are listed in Table 1.2. Wild ruminants are much more difficult to count than domestic ones, and the figure for Africa in Table 1.2 is especially doubtful. The most reliable estimates are from North America, and these are included for comparison, although wild sheep and goats are relatively few in North America compared with the rest of the world. There are 200 million sheep in Australia

Table 1.2.   Numbers of wild ruminants

Species[a]	Numbers (thousands)	Location
Bovids	100,000–500,000	Africa
White-tailed deer	10,000	Canada, U.S.
Mule deer	6250	Canada, U.S.
Roe deer	4000	Europe, former USSR
Black-tailed deer	1800	Canada, U.S.
Caribou	800	Arctic
Red deer	610	Europe, former USSR
Wapiti	600	Canada, U.S.
Pronghorn antelope	400	Canada, U.S.
Moose	250	Canada, U.S.
Bighorn sheep	30	Canada, U.S.
Bison	20	Canada, U.S.[b]
Mountain goat	17	Canada, U.S.
Feral goat and sheep	16	U.S.
Axis deer	5	U.S.
Total in U.S. and Canada	20,188	—
World total	130,000–500,000[c,d]	—

Source: R. E. McDowell, 1976.
[a]There are no data for giraffids.
[b]A small number of European bison remain in Poland.
[c]There are no estimates of the number of Asian and South American deer. There are another 30 million ruminants over which humans have partial control (R. E. McDowell, 1976).
[d]This total is seriously in doubt because the number of wild ruminants in Africa, which likely contains the largest populations, is unknown. If one uses the most extravagant estimates, however, the conclusion that there are fewer wild ruminants than domestic ones is not greatly altered.

and New Zealand and very many more in the developing parts of the world, so the ratio of wild to domestic species is likely highest in North America.

Deer, the most widely distributed of the wild ruminants, are found on all continents except Antarctica. They are restricted in Africa to regions north of the Sahara, whereas African antelope are sub-Saharan. Camelids (guanaco, llama, vicuña, and alpaca) are restricted to South America. Other groups are even more limited in their distributions.

Wild ruminants provide a small source of food and other products to societies devoted to hunting. Such animals need to remain in balance with their environments, and natural systems involve predators. Humans have disposed of many of the predators with the aim of conserving wild herbivores as their own resource. But if antihunting activists should be successful in reducing or eliminating hunting, herbivores would soon destroy their domain through overeating.

Not every wild animal species is in danger of extinction. White-tailed deer coexist quite well with humans, even in suburban areas, and there are probably more white-tailed deer in eastern North America now than at any other time. Hunting and exploitation have not at all reduced the numbers of this species. Yet others, such as the shy okapi, an African forest ruminant, cannot adapt to human settlement and clearing of forests and will require protection if they are to survive.

The extirpation of giraffes in southern Ethiopia emphasizes the interdependence of humans and animals. All the large animals were shot in the 1970s in the aftermath of the Somali war on the government's contention that wild herbivores compete with domestic herbivores. The government did not realize that a balance existed between the giraffes and the Burana people, who used the giraffe products in their everyday lives—making hides into water buckets, among other uses. Neither plastic nor other animal hides have successfully replaced giraffe hide. Further, giraffes and other antelope kept the browse trees in balance with the grassland that the cattle depend on. After the giraffes were eliminated, browse and bush encroached into grazing land already overpopulated by cattle.

1.3 Humans' Use of Ruminants

Domesticated ruminants have had a symbiotic relationship with humans since prehistoric times, and this association has resulted in many alterations in the species' characteristics. Specialized breeds probably depend on humans for their survival, since by now they must have lost aboriginal characteristics important to survival in the wild. At the same time many human societies could not be what they are without the association with ruminants.

Table 1.3. Classification of nonfood contributions of ruminants

Contribution	Main sources[a]
Fiber	
Wool	Sheep, camelids
Hair	Goats, yaks, sheep, camels
Skins	
Hides	All ruminants
Pelts	Sheep, camelids
Inedible products	
Inedible fats	Cattle, buffaloes, sheep
Horns, hooves, bones	Cattle, buffaloes
Tankage	Cattle, buffaloes, sheep
Endocrine extracts	Cattle, sheep
Traction	
Agriculture	Cattle, buffaloes, camels
Cartage	Cattle, buffaloes, yaks, camels
Packing	Camel, yaks, buffaloes, cattle, reindeer
Herding	Buffaloes, camels
Irrigation pumping	Buffaloes, cattle, camels
Threshing grains	Cattle, buffaloes
Passenger conveyance	Buffaloes, camels, yaks, cattle
Waste	
Fertilizer	Domestic ruminants
Fuel (dung)	Cattle, buffaloes, yaks, camels, sheep
Methane gas	Cattle, buffaloes
Construction (plaster)	Cattle, buffaloes
Feed (recycled)	Cattle
Storage	
Capital	Domestic ruminants
Grains	Cattle, buffaloes, sheep
Conservation	
Grazing	All ruminants
Seed distribution	All ruminants
Ecological	
Maintenance	All ruminants
Restoration	All ruminants
Pest control	
Plants in waterways	Buffaloes
Weeds between croppings	Domestic ruminants
Snails (irrigation canals)	Buffaloes
Cultural, including recreation	
Exhibitions, including rodeos	Cattle, sheep, goats, buffaloes
Fighting	Cattle, buffaloes
Hunting	Deer, elk, gazelles
Pets	Goats, sheep, deer
Racing	Buffaloes, cattle
Riding	Camels, buffaloes
Religious	
Instruments	Goats, buffaloes
Sacrificial	Buffaloes, sheep
Bride price	Cattle, sheep, goats
Social status	Cattle, sheep

Source: R. E. McDowell, 1977.
[a]Species listed in order of importance.

Not all societies possess or use domesticated ruminants. The relationship is maximized by agrarian societies in areas where arable land is limited and ruminants' harvesting of browse and forage grasses increases food resources. Even in some intensely cultivated regions—for example, the Gangetic plains of India and Pakistan—ruminants are crucial to the local economy. Existing on straw and waste cellulosic products, they supply the main power source for tillage,

transport, fuel (as dried feces), milk, and leather. This situation is also true in many areas of South America, Africa, and Asia, particularly those that are less industrially developed. Under these conditions ruminants do not compete with humans for food; in fact, they convert relatively unusable by-products (straw, browse, and weeds) to useful products. Other nonfood functions filled by ruminants are listed in Table 1.3.

The most important of these nonfood contributions is probably traction. No doubt farm machinery and tractors are energetically more efficient than animal traction, but large herbivorous animals are indispensable for hauling and plowing in undeveloped societies that cannot afford petrol and farm equipment—and given the lack of development capital in those countries, ruminants will continue to be important sources of power for some time to come. As already noted, ruminants also supply fuel energy (dried feces), which is derived from renewable resources. This is of major importance in developing countries but has not been greatly appreciated in developed ones until recently. Yet another nonfood use of ruminants is as a measure of wealth, as is the case, for example, in tribal African societies.

1.4 Economics and Animal Efficiency

A grazing ruminant optimizes its utilization of cellulosic carbohydrates by virtue of the arrangement of its digestive tract, in which the fermentation chamber (reticulorumen) precedes the main site of digestion. In this way, the fermentation products receive the most efficient use, unlike products of lower tract fermentation below the main site of digestion. Although large nonruminant herbivores such as horses compete with ruminants for food, they do not maximize their use of cellulosic matter for energy.

Pregastric fermentation is not without its disadvantages. More or less complete fermentation of dietary protein, starch, and soluble carbohydrates is offset for the host animal by the opportunity to digest the fermented products, microbial protein, and B vitamins. Only 50–70% of microbial nitrogen is in available protein, the remainder being bound in cell wall structures and nucleic acids. Ammonia is often a by-product. High-quality proteins are decreased in quality through fermentation. The fermentation of carbohydrates as an energy source results in the production of heat and methane. Hence, ruminants have an advantage over nonruminants because their digestive processes unlock otherwise unavailable food energy in cellulose. Postgastrically produced microbial protein may be lost to nonruminants because the fermentation occurs beyond the main sites of absorption. The volatile fatty acids (VFAs) are used by hindgut fermenters, but the microbial protein is lost in the feces unless coprophagy is practiced.

The relative efficiency (energy extracted from the food expressed as a ratio of net energy to available energy) of ruminants versus nonruminants may be related to diet quality. Ruminants adapted to poorer-quality fibrous diets perform as well as or better than nonruminants on the same diets. Size is also a factor in efficiency. Generally, larger animals of either class are better able to utilize nonprotein nitrogen than smaller animals. The ration quality at which animals might be equally efficient cannot be exactly stated unless the cell wall content (total dietary fiber) and its potential digestibility are known.

Grazing ruminants are adapted to consume foods with relatively high fiber content but sufficiently unlignified as to provide a large share of dietary energy in the form of cellulosic carbohydrates. It is the ability to utilize fiber that places the domesticated ruminant in its unique position in the world's economy.

1.4.1 Individual Efficiencies and Carrying Capacity

The efficiency with which animals utilize feed is a major area of research and application, for it involves not only the comparison of the productive efficiency of various animals but also the evaluation of diverse food resources. Because of this concern for efficiency, the production ability of individual animals has been emphasized in contrast to what animals might produce per unit of land (i.e., carrying capacity). The choice of whether production is evaluated per animal unit or per land unit may also depend on the relative cost of animals and land.

The need to evaluate the feed resource is a factor whether one deals with animal efficiency or with carrying capacity. The latter involves the forage yield, its quality, and the animal's efficiency in using it. The maximum production per unit of land brings into the discussion other ecologically interactive factors, such as the use of mixed grazing and browsing by complementary animal species. The increased productivity of land used by a combination of goats, sheep, and cattle or other species indicates the benefits of mixed systems. All these aspects have been ignored in the emphasis on individual animal efficiencies.

A second comparison is between animal species and their respective abilities to utilize feed resources. Mostly such studies compare stall feeding of controlled diets and involve ruminants such as beef or dairy cattle and nonruminants such as swine or poultry. The nonruminant species are often viewed as being more likely to compete with humans for food resources than ruminants, since some nonruminants are obligate consumers of concentrate diets. Farmers in many de-

Table 1.4. Feed efficiency of swine

Year	Conversion (feed/gain)	Daily weight gain (kg)	Days from 13–90 kg	Gross energy efficiency (%)
1910	6.3	.17	459	16
1930	5.2	.30	258	19
1945	4.0	.45	170	25
1969	3.0	.91	85	33
1989	2.5	—	—	—

Source: Modified from J. T. Reid, 1970.
Note: A digestibility of 80% is assumed.

veloping countries, however, feed monogastric animals (e.g., poultry and swine) on resources they do not consider suitable for human food.

The argument that humans and animals are competing for food is based on the assumption that food resources are in common between them and that land supplying feed resources for animals could be devoted to producing human food if the animals were not there. On the other hand, production of feed unusable by humans is inevitable even when land is used for human food production, and animals can contribute to the system by utilizing by-products inedible by humans.

Animal feed efficiencies have long been a means for evaluating individual animal performance as well as for comparing feeds and classes of livestock. Generally, nonruminants are more efficient utilizers of concentrates than ruminants, in which concentrate feeds are fermented by bacteria in the rumen before the main sequence of digestion, thus reducing efficiency.

Swine, for example, convert feed into meat with a higher efficiency than ruminants (Table 1.4). Although cows convert feed into milk more efficiently than into meat, the conversion is still not as efficient as meat production in pigs (Table 1.5). Cattle require more dietary fiber for normal function and therefore use the available cell contents less efficiently. Their lower efficiency is thus partly due to the lower caloric density of the fiber-rich diet. Also, not all dietary substances are subject to rumen fermentation.

Feed efficiency in all livestock has improved over the years as farmers have developed more concentrated diets and have increased the feed intake level so as to decrease the proportion of feed energy lost to maintenance. The time required to produce a given gain in efficiency is a prime function of this variable. Ruminants' less efficient use of high-fiber diets is another reason for the continued feeding of concentrates. The feasibility of feeding all-concentrate rations to ruminants was in doubt before 1950, but the fact that the cost per unit of net energy was less for corn grain than for forage pushed ruminant nutrition research to solve the problems of digestive disturbances that frequently resulted from concentrate feeding. (Most feedlot animals do not live long enough to experience the full toll of rumen acidosis, parakeratosis, and abscessed livers that are the result of overfeeding grains with too little dietary fiber.) The lower bulk of concentrate diets allows higher intakes and thus greater productivity.

1.4.2 Competition for Food

The feeding of concentrates to ruminants has evoked an economic argument about the competition between humans and animals for food. Because most Westerners see cows being fed grain, they erroneously assume that this occurs elsewhere and that animals are consuming food that could be given to people. Nothing could be further from the truth. In developing countries, as already noted, most domestic animals exist on products not edible by humans. Concentrates are fed to animals only in developed Western societies when the cost of food energy per unit is less for concentrates than for fibrous feed.

About 50% of the photosynthetic energy in cereal and seed crops is in the straw and stover portion, inedible by humans but consumable by ruminants and other herbivores. India (taken as an example here because it is a largely vegetarian society) has more than 60 million cattle. In the 1960s the U.S. Agency for Industrial Development (USAID) promoted a program that would have resulted in the elimination of sacred cows. This policy failed partly for religious reasons but mainly because of the overlooked economic contribution of these animals, which consume virtually no cereals. Cows, which subsist on straw, weeds, and garbage, provide dairy products and serve as a garbage collection system, and their feces are the main fuel resource for domestic cooking. These industries have been estimated by Indian economists to be worth billions of dollars annually.

Further, the complaint of Third World farmers needs to be heard. Their response to the shorter-stem high-yield varieties of cereals developed at CIMMYT (Centro Internacional de Mejoramiento de Maiz y Trigo) in Mexico and IRRI (International Rice Research Institute) in the Philippines was that they did not obtain enough straw to feed their oxen and thus could not plow the ground for next year's crop.

1.4.3 Feed Resources in the Developing World

As mentioned above, the feeding of concentrates is rarely economical in less developed countries because the cost of grains remains relatively high, and feed supplementation is often limited to by-products or other forages and browses. The challenge to ruminant nutritionists now is to improve forage and fiber utilization so that ruminants can exploit their true ecological and economic niches. This need is most keenly felt in the developing tropics.

Table 1.5. Feed efficiency of U.S. dairy cattle

	Milk/year (kg)	Feed/day (kg)	Digestibility (%)	Gross energy efficiency[a] (%)
1945	2173	10	56	10
1969	4159	13	58	14
1989	6500	18	60	18
Superior animals	13,600	28	62	22

Source: Modified from J. T. Reid, 1970.
[a]Includes energy cost of maintenance for 365 days.

The pressure for efficient human food production has been one of the arguments for reducing support of animal food agriculture. If all available land in the United States were converted into cereal production, however, three times the straw and stover necessary to support the present ruminant population would be produced (Moore et al., 1967). Needless to say, much of this forage resource is not utilized. Sixty-four percent of the world's land area is unsuited to tillage, and therefore for crop production. Such land could be used to produce animal protein through grazing and browsing ruminants. Cereal by-products and grazing lands are two major sources of feed energy that cannot provide food for humans without prior processing by ruminants, and ruminant's use of cereal by-products and grazing land is thus necessary for an efficient world agriculture.

Animal agriculture will always be an important complement to plant agriculture because animals can use by-products that people find unsuitable. Large land areas unsuitable for cultivation could be grazed by animals. Animal agriculture thus renders plant agriculture more efficient. If the world became entirely vegetarian, a major resource convertible to animal products (cellulosic residues) would perhaps be wasted.

1.5 Uses of Cellulose

It is possible that a fermentation technology could be developed to utilize waste cellulosic matter. The products might be sugars from enzymatic degradation or proteins and other products made by the rumen microbes such as alcohol or methane. The biological or enzymatic conversion of cellulosic matter into syrups is limited by lignification. Although it is a costly process, lignin can be removed, rendering the residual carbohydrate more digestible.

Cellulose could also be used to make paper products and in biogas (methane) production. The slower-digesting crystalline cellulose products might be more efficiently used by this method. Increased energy and fuel costs in the future could make biogas production a greater competitor.

Syrup produced from woody materials contains

much pentose from hemicellulose, which limits its use as feed for monogastric species. Such syrups could be fed to ruminants as concentrate substitutes; or they could be fermented with a nitrogen source to produce a single-cell protein, which would be suitable feed for monogastric animals. With more processing the protein could be prepared for human consumption, but our food preferences make it unlikely that single-cell protein will be used for human food.

Conversion ratios of cellulosic syrups into single-cell protein are often much greater than cellulose conversion in animal production; however, the cost of the protein's transformation into an edible product and the cost of recycling and utilizing the waste must be considered if the process is to be compared with a ruminant system. These problems also apply to large-scale animal farms, particularly those operated on a feedlot basis. It remains to be seen whether industrial conversion of cellulosic waste into food or feed is indeed more efficient and truly does not compete with humans for substrate and energy. A grazing system involves much less energy input and manpower than operating a plant to produce and process single-cell protein. The recycling problem of waste nutrients may also be greater than a well-integrated farming system. Thus the maintenance costs of the manufacturing plant versus the animal might be compared. The maintenance requirement and cost means that increasing ruminants'

efficiency in using grass and cellulosic matter is a prime challenge to ruminant nutritionists.

In the end, economics will determine the course of development and the cost of handling cellulosic matter. The cost of harvesting and processing the products must be balanced against the ruminant grazing systems in which the animal collects its own fodder. In this regard it is appropriate to recall what R. E. Hungate (1979:13) wrote:

An industrial cellulose fermentation might be profitable if the cost of collection of raw materials could be minimized through use of numerous small plants, if these small plants could be cheaply constructed, if operation could be made automatic to decrease necessary personnel, and if the concentration of cellulose fermented could be increased by continuous removal of fermentation products. Although such a situation is at present quite out of the question as an industrial process, it is an almost exact specification of the ruminant animal, a small fermentation unit which gathers the raw material, transfers it to the fermentation chamber, and regulates its further passage, continuously absorbs the fermentation products, and transforms them into a few valuable substances such as meat, milk, etc. To these advantages must be added also the crowning adaptation: the unit reduplicates itself.

2 Nutritional Concepts

2.1 Nutritive Value

Feeds and foods are not equal in their capacity to support the animal functions of maintenance, growth, reproduction, and lactation. They supply energy and essential nutrients in the form of protein, vitamins, and minerals. Energy and protein are often the most limiting factors for ruminants and have received the most attention in evaluation systems. Some feed or food characteristics are related to form (e.g., particle size) and have no relation to indigenous chemical composition. In short, animals' response to a feed depends on complex interactions among the diet's composition, its preparation, and the consequent nutritive value.

Domestic animals are fed under regimens that are intended to maximize their productivity, and feedstuffs are usually evaluated on the basis of their ability to elicit the desired response from the animal. Feeds are rated on their productive energy and protein content, and deficiencies in minerals and vitamins are usually compensated for by supplementation. Usually the assumption is made that the specific nutritive value quoted is operative in a ration balanced for other factors.

Wild animals and free-ranging domestic animals may not be treated so simply, and all the nutritive requirements necessary for their survival, growth, and reproduction may have to be considered comprehensively. In human nutrition and the nutrition of pet animals, adequate maintenance is of paramount importance and productivity is relatively meaningless. Indeed, diets that promote growth and obesity can be regarded as undesirable because they may shorten life. Food-producing animals generally do not live long enough to experience such consequences. The diet of the animals should provide all the essential nutrients, and the supply of vitamins, minerals, and protein is the criterion of evaluation.

Essential nutrients conventionally include water, energy, minerals, vitamins, and amino acids. For functioning ruminants, amino acids are usually considered under the general nitrogen requirement (as crude protein) because the rumen bacteria are capable of synthesizing them. The same applies to the water-soluble vitamins. The B complex and vitamin K are generally synthesized by rumen organisms and are thus related to microbial synthetic capability. Vitamin C is destroyed in the rumen, but the ruminant itself can synthesize this vitamin.

Mature ruminants require external sources for the fat-soluble vitamins A, D, and E, as well as essential fatty acids and minerals. Some wild herbivores may have higher requirements for these vitamins than do domesticated animals. Since their requirements are often based on National Research Council (NRC) requirements for domestic animals, wild ruminants may be underfed in zoos (Dierenfeld, 1989). Infant ruminants require the same amino acids and vitamins as nonruminants and are usually considered separately. Specific limiting amino acids or vitamins could be a problem for lactating adult females under the heavy stress of milk production. These factors interact strongly with rumen efficiency and proper feeding. For example, a high-producing, sensibly fed dairy cow might not respond to an exotic supplement—such as protected amino acids, vitamins, yeast cultures—but one on a less well balanced diet might respond by increasing productivity.

Nutritive value is conventionally classified by ruminant nutritionists and agronomists into three general components: digestibility, feed consumption, and energetic efficiency (Raymond, 1969). While feeds are individually evaluated in numerical terms for these categories, they vary in their amenability to such measurements.

The practical application of feed evaluation assumes that feeds are variable and animal responses are comparatively reproducible. Digestibility is more often measured than efficiency or intake, even though intake and efficiency are more responsible for total animal responses. Efficiency and intake offer more interanimal variation (Table 2.1), and thus the establishment of relative feed values for these components is more difficult than for digestibility. Often the assumption is made that intake and efficiency are related to digestibility; this is often, but not always, true. Animals' lack of response to differing digestibilities arises in part

Table 2.1. The approximate relative variation contributed by animal and diet (forages)

	Coefficient of variation (%)	
	Diet-related	Animal
Digestibility	30	3
Intake	50	30
Efficiency[a]	50	20

[a]Use of digested energy for productive purpose.

from their ability to compensate by eating more of a lower-quality feed; however, the amount of very poor quality forages that can be consumed is limited by their volume and by slow rates of digestion.

The quality of a feed is considerably modified by its physical characteristics, which may be relatively independent of its chemical composition. Factors such as caloric density, particle size, solubility in rumen liquor, buffering capacity, and the surface properties of the fiber particles (i.e., hydration capacity and cation exchange) influence the physiological effects of ingesta on the gastrointestinal tract. These factors are also likely to be changed by feed processing, further complicating evaluation.

2.2 Digestibility

The balance of matter lost in the passage through the digestive tract is the most reproducible measurement for a given feedstuff. The feces contain not only the undigested diet but also metabolic products including bacteria and endogenous wastes from animal metabolism. Consequently, *apparent digestibility* can be considered to be the balance of the feed less the feces, but *true digestibility* is the balance between the diet and the respective feed residues from the diet escaping digestion and arriving in feces exclusive of metabolic products. The coefficient of true digestibility is always higher than that of apparent digestibility if there is a metabolic loss in feces. In total diets, protein and lipids always have a fecal metabolic loss. For fiber and carbohydrates there is no metabolic fecal loss and apparent coefficients equal true digestibility. The relationships of apparent and true digestibility to metabolic products are explored further in Chapter 22. Food residues that survive the digestive tract are truly indigestible. Their identification is complicated by the fact that metabolic matter may be generated by fermentation of dietary matter that escaped digestion by the animal (Figure 2.1).

Microbial matter (Section 18.10) is derived by fermentation of both feed residues and secreted endogenous matter (such as urea and mucus in saliva), which cannot be easily distinguished in the microbial prod-

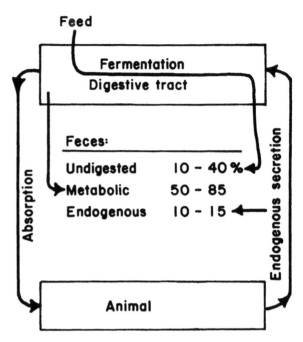

Figure 2.1. The origin of fecal matter and the relative proportions of undigested metabolic and endogenous matter.

ucts. In many animals some dietary starch escapes to the lower digestive tract where it is fermented by bacteria. This fermentation depends on urea and minerals secreted or sloughed off by the animal. Thus the products of this fermentation are produced from a combination of materials originating externally in the diet and internally in the animal. Gastrointestinal microbial activity is sufficiently high that the metabolic fecal component is largely microbial at the time of excretion in most animals (Mason, 1984). For ruminants and many other animals, the proportion of metabolic fecal matter that is microbial is about 85–90%.

The significance of true digestibility is that it represents that part of the feed available for digestion by animal *or* microbial enzymes. The availability of feed to rumen bacteria or animal digestive enzymes is the basic estimate made by in vitro laboratory procedures (Section 8.5). Thus, in vitro methods are related more to true digestibility than to apparent digestibility because in vitro methods are incapable of estimating fecal endogenous matter. Metabolic losses are also influenced by the physiological state and condition of the animal.

2.3 Measuring Feed Consumption

Ad libitum intake as a factor of feed quality is the most important factor affecting animal response, particularly efficiency. Intake measurement is complicated by animal variability, forage palatability, and forage selection. While intake of feed is demonstrably related to feed quality, the species of the animal, its status, its energy demand, and even its sex cause the voluntary intake of an individual animal on a given diet to vary. Thus, a growing animal consumes relatively more than a mature one, and a pregnant or lactating female consumes still more.

2.3.1 Intake Trials

The measurement of ad libitum consumption of a feed is usually conducted in a digestion stall, often in conjunction with a digestibility measurement. Feed is offered at 15–20% above the amount that will be consumed. Voluntary intake, or ad libitum intake, implies that a surplus of feed is offered, and consequently there will be a refusal. In a stall feeding situation this refused feed is termed *orts*. The composition of the orts may deviate from that of the total diet to the extent that the animal is able to practice selection. Selection can be restricted by reducing intake or by chopping, wafering, grinding or pelleting the diet. But this practice may cause the evaluation of the diet to deviate from conditions of practical feeding. Sheep and cattle are fickle and may alter their intake at any time for no apparent reason. Intake trials must therefore be conducted over several weeks so as to stabilize and even eliminate this variation.

Traditionally, feed acceptability and digestion trials have been conducted at a level of feeding below ad libitum intake. This controlled level, termed *restricted intake,* is not necessarily below the maintenance requirements of the animal but is often near it. The experimental level of feeding is then described on the basis of the plane of nutrition calculated as a proportion of maintenance. A problem with using this system is that it assumes that maintenance is constant, but in reality maintenance may vary depending on energy costs imposed by the environment. Some adjustments for maintenance depending on environment are now available (Robbins, 1993; Fox et al., 1990).

2.3.2 Expression of Intake

Feed intake has often been related to metabolic body weight on the assumption that intake is a function of metabolic requirements. Metabolic size is proportional to the three-fourths power of body weight across animal species. (This relationship does not apply to intraspecific relations.) For example, heat increment in cattle is related to the power 0.6 (Thonney et al., 1976). Data demonstrate that the fill-limiting aspects of intake are linearly related to body size (Chapter 4), so the amount of feed an animal consumes is often expressed as a percentage of its body weight. An adequate description of feed intake should include not only the amount of feed eaten but also body weight and other

information relative to nutritional status and energy demand.

2.3.3 Relative Intake

An experimental technique that seeks to correct for appetite differences among experimental animals measures the relative intake of a test forage. A standard forage (alfalfa) is fed to all the animals, and the intakes of the test forages are reported as a proportion of the standard. This procedure has practical problems. Fibrous plants vary in characteristics according to the environment in which they are grown, and thus the standard forage is generally not reproducible. One of the interesting results of relative intake trials is the disclosure of the eating capacity of individuals. A study of mature wethers by Osbourn et al. (1974) revealed that a large proportion of the variance in ad libitum intake is due to individual capacity for food consumption even in animals standardized for age, weight, and sex. This variation is considerably larger with higher-quality forages and much reduced when lower-quality forages are fed.

Nutritive value index (NVI), the product of relative intake and digestibility, was suggested by Crampton et al. (1960) as a measure of the productivity of a feed. The NVI was meant to estimate the intake of digestible nutrients on a basis that would give reproducible numbers of an intake-containing function and might relate feed quality more realistically to animal production. The problem of a universal standard feed remains unresolved, however. If such a system were evolved, its use would require the application of standard intakes characteristic of levels of animal function (lactation, growth, etc.).

2.4 Palatability

A major problem in evaluating the intake of a food lies in the reasons for which an animal may refuse the food. One reason might be palatability, the pleasing or satisfying aspect of a feed or food. Animals are unable to directly communicate their likes and dislikes, thus it is not easy to distinguish whether palatability or a physiological reason has caused feed rejection. *Voluntary intake* refers to the ad libitum intake of a single feed where choice and selection are eliminated to as great an extent as possible (Marten, 1985).

In ruminant feeding trials, *palatability* has come to mean the free-choice consumption of a feed when an animal is allowed to select among two or more diets. Feed choice is measured by offering the animal more than one feed simultaneously in compartmented mangers or allowing it to graze paddocks containing small plots of the test forages. Both methods are termed *cafe-*

teria trials. The intakes of the individual feeds are recorded, and the ranking of choice is based on the comparative intakes of the respective feeds. Experiments that unequivocally demonstrate choice on the part of animals are not easy to design, although it has been shown that sheep distinguish color and that dairy cattle choose according to smell and taste (Munkenbeck, 1988).

Palatability may also refer to free-choice consumption of some parts of a single feed over others when the feed is offered in a sufficient quantity and in a form which allows the animal to sort out the choicest bits. Such a situation often occurs under grazing and almost always under free-ranging and browsing conditions. Even in stall trials animals may find some parts of the feed more palatable than others, as in the selection of leaves over stems. The word *acceptability* is often used in grazing trials to describe aspects of quality that reflect palatability and voluntary intake and the degree to which individual kinds of plants are eaten relative to their abundance.

2.4.1 Selection

Selection is a source of variation when measuring intake, the essential problem being that more palatable portions are eaten first. Diet selection can cause the orts to have a different composition from the feed eaten.

Grazing studies are difficult to quantify because maturation of the sward parallels animal selection. For example, if animals are turned into a fresh paddock containing a single plant species, the more palatable portions are grazed first and the less desirable parts are eaten later. Thus, if a paddock is grazed for several days, selection and intake will alter as the forage supply decreases, and also if the stocking rate varies.

Selection of forage presumes morphological and nutritive differentiation in plants. Without differentiation, there could be no selective grazing. Forages vary in this parameter, which is influenced by individual plant species, number of plant species available, the environment for plant growth, and the age and maturity of the forage. The animal is another factor in the food selection process because an assumption is made that the animal has the ability and desire to select. Ability varies with animal species (Chapter 3), and desire may be regulated by hunger and the availability of feed. Generally, a hungrier animal is less selective.

2.5 Metabolizable Energy and Efficiency

The subtraction of urine and methane losses from total digestible energy is the classical way of calculating metabolizable energy. But metabolic losses are not

Table 2.2.   Digestibility and the fate of feed energy

Losses	Balance relative to dietary intake	
	Definition	Dietary energy disappearance (%)[a]
Fecal residues		
Undigested feed	True digestibility	72
Metabolic matter	Metabolic losses	−13
Net feces	Apparent digestibility	59
Urine	—	−4
Methane	—	−7
Net balance	Metabolizable energy	48
Heat increment		−9
Net balance	Net energy	39

Note: Example balance typical of a good-quality forage diet.
[a]Values are percentages of feed uptake.

confined to gases and urine. Metabolic wastes are excreted in the urine and feces, and as methane gas. Subtracting these losses from the intake yields the metabolizable substance and energy available to the animal. Metabolizable energy (ME), the quantity of metabolizable nutrients expressed as energy, is the most common basis for evaluating feed and expressing requirements in monogastric animal nutrition. Metabolizable energy as an evaluation presents several problems in view of the large and variable heat losses characteristic of ruminants. Table 2.2 describes the energy balance in an animal fed an average diet.

Fecal losses are divided among undigested feed and endogenous and microbial matter. Fecal metabolic losses are divided into microbial and endogenous substances; methane losses are entirely of microbial origin, primarily derived from feed substances, and urinary losses are largely of compounds of endogenous origin. Urine and feces do not represent a significant division of endogenous losses relative to function; rather, the division is based on the physiological properties of solubility and molecular size. Indigestible and unmetabolized substances of low molecular weight, such as dietary phenols and essential oils, may be absorbed and excreted in urine with little or no alteration. Thus both urine and feces can contain unmetabolized substances, although those that did not appear in feces would seem to have been "digested" when it reality they might have been passed on to the urine without energy being extracted.

In practical terms, accurate estimation of ME is limited by the analytical difficulty of calculating methane production, which is usually estimated rather than measured. The energy lost in urine, methane, and heat increment (defined below) is thought to detract from animal efficiency. The energy in the heat increment comprises the heat of fermentation, the work of digestion and rumination, and radiation loss from the body of the animal. This heat is theoretically partitioned into basal heat production, as sustained in a fasting animal, and the incremental heat from dietary intake, which can be considered a feed cost. This latter, the heat increment, is subtracted from the ME to yield net energy (NE), the energy content of the feed that is available for maintenance and production. Animal production can occur only when the consumption of dietary energy is sufficient to cover maintenance costs and allow a surplus. This presents a problem with regard to the NE evaluation of feeds because the value of feed to support maintenance is higher than the corresponding value for production. Heat increment may serve to warm the animal, an advantage under cold conditions, but it is a disadvantage under heat stress and thus changes the relative values of low-quality feeds that induce large heat increments.

2.6 Feeds

The many terms used to describe and name feeds refer to class, species, stage of growth, plant part, or products. Feeds are often divided into concentrates and roughages. Concentrates are high-quality, low-fiber feeds such as cereals and milling by-products that contain a high concentration of digestible energy per unit weight and volume. Very immature forages can have a quality equivalent to concentrates. Conventional classification (L. E. Harris et al., 1967) defines concentrates as feeds containing less than 18% crude fiber. Unfortunately, crude fiber does not recover lignin and hemicellulose, so this division is arbitrary and imperfect. For example, beet pulp with 17% crude fiber and 47% cell wall is considered a concentrate, while immature alfalfa with 24% crude fiber and 36% cell wall is classified as a roughage. There is no clear-cut division between concentrates and roughages. Roughage is characterized by the fraction of cell wall it contains, but the lesser amounts of cell wall in concentrates also contribute to the fibrous character of the diet. As the proportion of cell wall increases, feeds become bulkier, requiring a greater volume to hold an equal weight of substance.

Considerable effort has been expended to standardize feed terminology on a national and international basis so that every feed can be described in a form that will indicate its relative nutritive value.

Several systems in use today reflect different purposes and developments, although efforts are now being made to unify them. Older systems include the USDA grading systems, the U.S. Grain Standards Act, and the hay grading system, all of which emphasize botanical composition and appearance but are often poorly related to real nutritional value.

The feed nomenclature devised by the American Association of Feed Control Officials (AAFCO) is updated annually and is designed to set legal limits on the definition and composition of products. This organiza-

tion, which represents the state feed control agencies, sets labeling requirements that appear on the tags of commercial feeds. This system is closely associated with the feed industry and is indispensable in categorizing the many by-products generated in feed manufacturing and used in animal nutrition. Since its origin, the AAFCO system has been a practical one that emphasizes the description of products derived from a commodity.

A universal terminology was developed by L. E. Harris of Utah State University through his organization of the International Feed Institute (IFI; see Table 2.3). Until recently, the IFI was coordinated worldwide by the International Network of Feed Information Centers (INFIC).

The IFI and the INFIC were and are devoted to the worldwide collection of feed data. The USDA and the Food and Agriculture Organization of the United Na-

Table 2.3. Descriptive terms used in animal nutrition

Components[a]
Origin (species)
Variety
Part
Process
Stage of maturity
Cutting
Grade
Class
Form of feed[b]
Unknown
As harvested
Biscuits
Blocked
Caked
Chopped
Condensed
Cooked
Cracked
Crimped
Crumbled
Ensiled
Fermented
Flaked
Ground
Kibbled
Pearled
Pelleted
Pelleted, ground
Polished
Raw
Rolled
Type of feed[b]
Dry forages and roughages
Pasture, range plants, and green soiling crops
Silages
Energy feeds
Protein supplements
Minerals
Vitamins
Additives

[a]Proposed by L. E. Harris et al., 1967.
[b]From the IFI system for describing processed feeds.

Table 2.4. Stage-of-maturity terms for forage plants suggested by the National Research Council

Preferred term	Definition	Comparable term
Germinated	Resumption of growth by a seed embryo after a period of dormancy	Sprouted
Early leaf	Stage at which the plant reaches 1/3 of its growth	Fresh new growth, very immature
Immature	Period between 1/3 and 2/3 of the plant's growth (this may include fall aftermath)	Prebud stage, before boot, before heading out
Prebloom	Stage including the last 1/3 of growth before blooming	Bud, bud stage, budding plants, in bud, preflowering, before bloom, heading to bloom, boot heads just showing
Early bloom	Period between initiation of bloom and stage at which 1/10 of the plants are in bloom	Up to 1/10 bloom, initial bloom, heading out, in head
Midbloom	Period during which 1/10–2/3 of the plants are in bloom	Bloom, flowering plants, flowering, half bloom, in bloom
Full bloom	Two-thirds or more of the plants are in bloom	Three-fourths to full bloom
Late bloom	Blossoms begin to dry and fall, seeds begin to form	Seed developing, 15 days after silking and before milk, early pod
Milk stage	Seeds well formed but soft and immature	Postbloom to early seed, pod stage, early seed, in tassel, fruiting
Dough stage	Seeds are soft and immature	Seeds dough, seed well developed, nearly mature
Mature	Stage at which the plant would normally be harvested for seed	Fruiting plants, fruiting, in seed, well matured, dough to glazing, kernels ripe
Overripe	The plant is mature, seeds are ripe, and initial weathering has taken place (applies mostly to range plants)	Late seed, ripe, very mature, well matured
Dormant	Plants have cured on the stem, seeds, have been cast, and weathering has taken place (applies mostly to range plants)	Seeds cast, mature and weathered

Source: L. E. Harris et al., 1967.

tions (FAO) no longer support the IFI and INFIC programs, respectively, because of the cumbersome nature of their data collection systems, their inflexibility, their unselective appetite for data irrespective of its quality, and the lack of any meaningful editing of their files. The NRC has adopted parts of the nomenclature system of Harris et al. (1967). Table 2.4. shows the current NRC classification. A more accurate alternative for classifying feeds is the stage description (Section 2.8).

The historical issue of terminology may be observed by comparing the NRC's publications on beef cattle and dairy cattle over the years from about 1960 to 1985. Early adoption of the Harris system led to tables of composition based on the averages from the IFI data bank in Utah. For example, because of poor quality of inputs, mean values of total digestible nutrients (TDN) for alfalfa did not decline regularly with maturity. This problem was resolved later by editing of the files. The application of the Harris principles of data collection endeavored to recover the "lost" feed data around the world dealing particularly with the Third World and developing countries.

The failure of the IFI system to produce a meaningful scientific contribution to the worldwide description of feeds and forages was the result of several problems: the organizers did not foresee the consequences of overclassification and the limited availability of data (since resources for data collection were financially limited); scientists and their collaborators had to fill out complex forms and return the data to the Utah center, and limited time and resources on the part of the

scientists and laboratory personnel precluded cooperation.

2.7 Classification of Forage

A variety of systems exist for describing and classifying forages. Descriptions relative to forage quality are not yet wholly satisfactory, and the development and refinement of the descriptive systems is continuing. Description problems vary with plant species and their morphological characteristics. Most forages or feed sources for ruminants are angiosperms and are largely confined to two plant families: the Graminaceae and the Leguminosae. Angiospermous forage is commonly divided into grasses (grass and grass-like plants, including sedges), legumes (herbaceous legumes), forbs (nonlegume broadleafed herbs), and browse (woody plants, shrubs, and trees). This classification does not hold rigidly to taxonomic divisions; the terms refer in part to the morphological character of the plants. The most fundamental division is between legumes and grasses. Thus the most problematic classification deals with mixed forages containing both legume and grass. Grasses contain much lignified structural matter in their leaves, particularly the midrib, and in leaf sheaths. Legumes and some forbs, on the other hand, tend to be treelike on a miniature scale; that is, they have relatively unstructured leaves held on the ends of woody stems. The stems of herbaceous plants are not so woody that they cannot be eaten; usually they consist of annual growth. Consumption of

Table 2.5. Proposed market hay grades for legumes and legume-grass mixtures

Grade	Maturity stage	Definition	Physical description	Typical composition (%)[a]			Relative feed value[b] (%)
				CP	ADF	NDF	
1. Legume hay	Prebloom	Bud to first flower; stems beginning to elongate	40–50% leaves,[c] green; <5% foreign material[d]	>19	<31	<40	>140
2. Legume hay	Early bloom	Early to midbloom	35–45% leaves,[c] light green to green; <10% foreign material	17–19	31–35	40–46	124–140
3. Legume hay	Midbloom	Mid to full bloom	25–40% leaves,[c] yellow green to green; <15% foreign material	13–16	36–41	47–51	101–123
4. Legume	Full bloom	Full bloom and beyond	<30% leaves; brown to green; <20% foreign material	<13	>41	>51	100
5. Sample grade		Full bloom and beyond	Very poor quality;[e] >20% foreign material				

Sources: Hay Marketing Task Force (Rohweder et al., 1978).

Note: CP = crude protein; ADF = acid-detergent fiber; NDF = neutral-detergent fiber.

[a]Chemical analyses expressed on dry matter (DM) basis. Chemical concentrations are based on data from north-central and northeastern states and Florida. DM (moisture) concentration can affect market quality. Suggested moisture levels are Grades 1 and 2 <14%, Grade 3 <18%, and Grade 4 <20%.

[b]Nutritive value index.

[c]Proportion by weight.

[d]Slight evidence of any foreign material will lower hay by one grade, except grade 5.

[e]Hay that contains more than a trace of injurious foreign material (toxic or noxious weeds and hardware) or that has an objectionable odor, is undercured, heat-damaged, hot, wet, musty, moldy, caked, badly broken, badly weathered or stained, extremely overripe, dusty, is distinctly low quality, or contains 20% foreign material or 20% moisture.

browse is restricted to the leaves, shoots, smaller branches, and cambial layers of the larger branches. The leaf-stem ratio of herbaceous forage, often taken as an index of quality, assumes that leaves are of higher quality than stems. This index is more useful in describing the quality of legumes than grasses.

The USDA grading system for categorizing forage has been incorporated into the IFI system. The grading system is based on color, maturity, fineness, leafiness, and the presence of foreign material (Seiden and Pfander, 1957). Special grades describe leafiness, extra green, coarse, and fine. Hays are grouped into 11 categories based on species and admixture: (1) alfalfa and mixed alfalfa; (2) timothy and clover; (3) prairie; (4) johnsongrass and mixed johnsongrass; (5) grain, wild oat, and vetch; (6) lespedeza and mixed lespedeza; (7) soybean; (8) cowpea; (9) peanut; (10) grasses; and (11) mixed.

2.7.1 More Specific Agronomic Descriptions

It is difficult to devise terms that uniformly describe maturity among grasses, and broad-leaf plants may show no analogous development prior to formation of buds and flowering. Some vegetative plants such as Pangola grass rarely flower and can be described only in terms of age and leafiness. Some terms refer to the specific parts of plants that may be harvested. Usually "forage" is the aerial part cut a few centimeters above the ground. More comprehensive systems may be needed to account for differences in nutrient value among plant parts. The IFI-INFIC system attempts to

impose uniform descriptions for all forages (Tables 2.3 and 2.4), although obviously all terms cannot apply to all forages.

The unsatisfactory state of forage description and characterization has produced further efforts to refine hay-grading and classification systems. The proposed system (Table 2.5 and 2.6) adds relative composition and quality parameters to the maturity descriptions, and it also provides descriptive terms appropriate to grasses and legumes. This system is undoubtedly an improvement over older systems, although it does not resolve all the problems. Unfortunately, these systems described here, while more rational, have not been adopted, although they are included for their value.

2.8 Stages of Plant Development

The concept of providing accurate information to describe stage of plant development relative to nutritive value emerged from the efforts described in Section 2.7.1. Stage description tends to be specific for particular species, however, and different stage development criteria would have to be developed for all the important forage species. The beauty of stage description is that it recognizes that any standing vegetative plant exhibits varying stages of development within its structure that contribute to the net whole forage composition. This variety forms a basis for potential selective feeding or subsequent forage or feed fractionation in handling or processing.

A system of stage description for alfalfa has been

Table 2.6. Proposed market hay grades for grasses and grass-legume mixtures

Grade	Maturity stage	Definition	Physical description	Typical composition (%)[a]			Relative feed value (%)
				CP[b]	ADF	NDF[c]	
2. Grass hay	Prehead	Late vegetative to early boot; 2–3 weeks growth	50% or more leaves;[d] green; <5% foreign materials	>18	<33	<55	124–140
3. Grass hay	Early head	Boot to early head; emergence to ½ inflorescence in anthesis; 4–6 weeks growth[e]	40% or more leaves; light green to green; <10% foreign material	13–18	33–38	55–60	101–123
4. Grass	Head	Head to milk; ½ or more of inflorescence in anthesis to stage in which seeds are well formed; 7–9 weeks growth	30% or more leaves; yellow-green to green; <15% foreign material	8–12	39–41	61–65	83–100
5. Grass hay	Posthead	Dough to seed; >10 weeks growth[f]	20% or more leaves; brown to green; <20% foreign material	<8	>41	>65	<83
6. Sample grade		[g]	Very poor quality;[g] >20% foreign material				

Source: Hay Marketing Task Force (Rohweder et al., 1978).

Note: CP = crude protein; ADF = acid-detergent fiber; NDF = neutral-detergent fiber; relative feed value = digestible dry matter intake.

[a]Chemical analyses expressed on dry matter (DM) basis. Chemical concentrations based on research data from north-central and northeastern states and Florida. DM (moisture) concentration can affect market quality. Suggested moisture levels are Grade 2 <14%, Grade 3 <18%, and Grades 4 and 5 <20%.

[b]Fertilization with N may increase CP concentration in each grade by up to 40%.

[c]Tropical grasses may have higher NDF concentrations than indicated in this table.

[d]Proportion by weight.

[e]For grasses that do not flower or for which flowering is indeterminant.

[f]Slight evidence of any factor (see note g) lowers hay by one grade.

[g]Hay that contains more than a trace of injurious foreign material (toxic or noxious weeds and hardware) or that has an objectionable odor, is undercured, heat-damaged, hot, wet, musty, moldy, caked, badly broken, badly weathered or stained, overripe, dusty, is distinctly low quality, or contains >20% foreign material or >20% moisture.

developed at Cornell University (Fick and Mueller, 1989; Mueller and Fick, 1989). The model is presented here, and its anatomical descriptions are shown in Figure 2.2. The model depends on recognition of stages of development and of the quantitative contribution of parts that lie within the respective stages. The need to measure respective contributions has led to two descriptions: mean stage by count (MSC), which can be done in the field, and mean stage by weight (MSW), which is more accurate. Many ecological evaluations of plant material are based on count, so the comparison of the two methods is relevant. The relationships between count and weight are shown in Figure 2.3. Similar descriptions have been developed for red clover (Ohlsson and Wedin, 1989) and timothy (Fagerberg, 1988a, 1988b).

2.8.1 Descriptive Categories

The descriptive characteristics that identify the component plant parts are as follows.

Vegetative Stages

At early stages of development, reproductive structures are not visible on alfalfa stems. Leaf and stem formation characterizes vegetative growth.

Stage 0: Early Vegetative. Stem length ≤ 15 cm (6 in.). No visible buds, flowers, or seedpods. The junction between the main stem and a leaf or branch is called the axil. An axillary bud is present in each leaf axil; however, they are so small at this stage that they are not easily seen.

Stage 1: Mid-Vegetative. Stem length 16–30 cm (6–12 in.). No visible buds, flowers, or seedpods. As the stems continue to develop, axillary branch formation begins with the appearance of one or two leaves in the axil. Development of axillary leaves is more pronounced in the midportion of the stem than at the base or apex.

Stage 2: Late Vegetative. Stem length ≥ 31 cm (12 in.). No visible buds, flowers, or seedpods. Elongating branches are often found in the axils of the leaves at this stage. It may be possible to feel buds at the growing apex, but they are not visible unless one peels back the enclosing leaves. Stage 2 stems are rare in midsummer because of the rapid appearance of buds on shorter stems resulting from environmental conditions that hasten maturation.

Flower Bud Development

Flower buds first appear near the growing apex of a stem or an axillary branch. At the transition from the

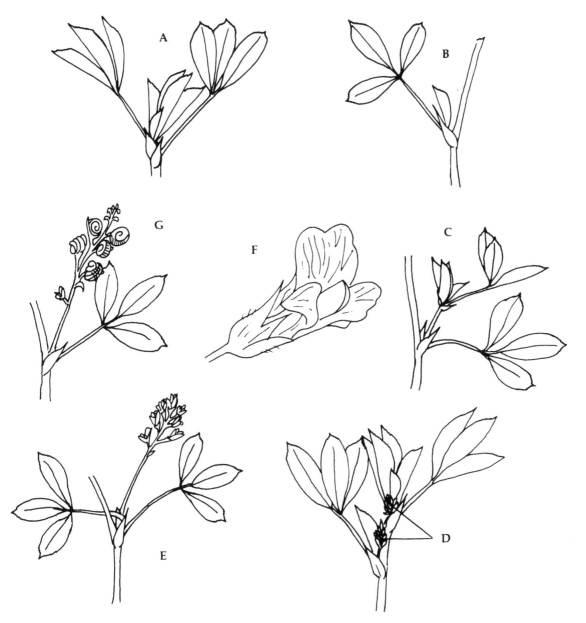

Figure 2.2. Development of a vegetative shoot (from Fick and Mueller, 1989). The growing tip is a cluster of folded and unfolding leaves (A). As the vegetative stem matures, it usually develops branches. The first sign of a branch is a folded leaf in the axil where a more mature leaf is attached to the stem (B). The branch develops with a typical clump of folded and unfolding leaves (C), and can also form flowers. Flower buds develop near the shoot tip, hidden in the cluster of unfolding leaves. They become visible (D) as their basal stalk elongates. At bloom, alfalfa flowers are clustered in a loose raceme at the end of a branch (E). The individual flower (F) has five petals, of which the standard is the largest and the first to unfold. Seedpods develop after pollination in a characteristic coil of up to four spirals (G). With maturity, seed color changes from green to brown. Because the shoot continues to grow and form new flowers after the first pollination, the first seedpods are found near the middle of the stem.

vegetative stages to the bud stages, flower buds can be difficult to identify. At first, buds are small, distinctly round, and hairy or fuzzy. In contrast, new leaves are flattened and oblong.

Stage 3: Early Bud. One or two nodes with visible buds; no flowers or seedpods. Flower buds appear clustered at the stem tip because of the closely spaced nodes on that part of the shoot. As the nodes elongate

during development into the next stage, it becomes easier to distinguish individual nodes.

Stage 4: Late Bud. Three or more nodes with visible buds; no flowers or seedpods. This stage differs from stage 3 only in the number of nodes with flower buds. The structure of the developing inflorescence becomes visible with elongation and clearer separation of individual flower buds in the raceme.

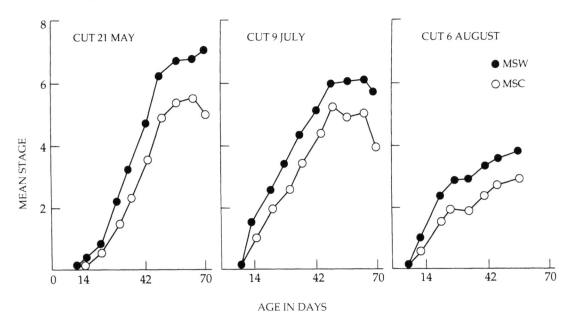

Figure 2.3. Comparison of mean stage of plant development by weight (MSW) and by count (MSC) (from Fick and Mueller, 1989). MSW changes faster than MSC, and MSW is less likely to decline in older canopies (Kalu and Fick, 1981, 1983). Regrowth occurred following harvest on the indicated dates. For the definition and method of calculation of MSW and MSC, see Sections 2.8.2 and 2.8.3.

Flowering

When environmental conditions meet specific requirements for temperature and photoperiod, flower buds develop into flowers (although in autumn, when there are fewer than 12 h of daylight, buds may abort without forming flowers). Flowers may be purple, blue, cream, yellow, white, or variegated combinations of those colors.

Stage 5: Early Flower. One node with one open flower and no seedpods; to be counted as an "open" flower, the standard petal of the flower must be unfolded. Because one raceme arises from each node, the number of racemes with an open flower or flowers is actually what is counted. Flowering usually begins near the apex of the stem while buds are still developing rapidly above and below the point of initial flower opening.

Stage 6: Late Flower. Two or more nodes with open flowers; no seedpods. This stage differs from stage 5 in that it has more racemes with open flowers. Nodes with flowers are spread throughout the midportion of the stem.

Seed Production

If flowers are pollinated, they usually develop seedpods. In some environments, pollination is poor and only a few flowers form seed. Typically, alfalfa is harvested for feed before the seed-bearing stages, when quality is lowest.

Stage 7: Early Seed Pod. One to three nodes with green seedpods. The spiral green seedpods are not always easy to see. Typically, seedpods first appear between the midportion and the base of the main stem while the upper nodes are still flowering. Again, one or more seedpods may be found on each raceme, but only one to three racemes with seedpods are present on stems in stage 7.

Stage 8: Late Seed Pod. Four or more nodes with green seedpods. At this stage the older stems are highly branched and may be lodged. Many leaves have fallen off the plant, and many of those that remain are yellow.

Stage 9: Ripe Seed Pod. Nodes with mature seedpods. As the seedpods ripen, they dry and turn brown. At this stage, most of the leaves on the lower portion of the stem have been lost and the stem is quite thick and fibrous. Alfalfa grown for seed is harvested at this stage, when most of the seed is mature.

2.8.2 Calculating the Mean Stage of Development

The evaluation of stage depends on the collection of a representative field sample. Two methods have been used to calculate the mean stage of development for alfalfa herbage: mean stage by weight (MSW), which

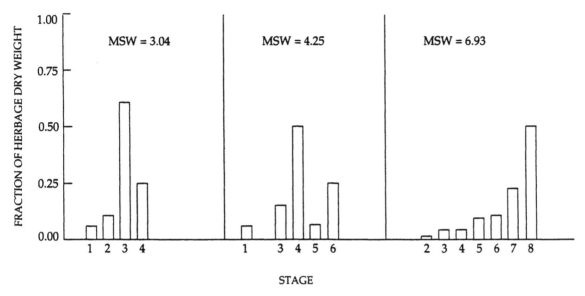

Figure 2.4. Each randomly selected forage sample has stems in several stage classes; MSW is the mean stage weight for the fraction of herbage dry matter in each stage. The samples described here were collected in central New York State. (From Fick and Mueller, 1989.)

is based on the dry weight of herbage in each stage, and mean stage by count (MSC), which uses the number of stems in each stage to quantify maturity. Each procedure requires a random sample of at least 40 alfalfa stems.

A representative sample of alfalfa stems can be collected from either a randomly selected square area (0.1 m², or 1 ft²) in the field or a specified distance along an obvious drill row. Cut the stems, leaving a 3-cm (1¼-in.) stubble. It is important to collect all the harvestable alfalfa stems in the sample area. Remove weeds and dead material from the alfalfa sample.

Samples can be stored temporarily in plastic bags in a cooler, but longer storage requires a freezer. If frozen, the sample should be allowed to thaw before the stems are examined and sorted. Otherwise, frozen tips will break off and cannot be associated with the original stems.

"Staging" the Stems

Separate the stems according to the nine stages of development. Experience with the system indicates that the most difficult aspects are the relatively subjective decisions involved in classifying individual stems. The most important concern is consistency. Do not count buds until they are visible. An elongated flowering branch is counted as a flower even if the flowers have fallen. Later in development, however, an elongated flowering branch without flowers is counted as a seedpod if seedpods are present on other stems. In New York, the most mature stems of an alfalfa stand usually pass through one stage per week during the first growth of the season.

If the shoot apex is removed or damaged, axillary buds below the point of injury form branches and development continues. The most mature characteristic of the shoot, whether on the main stem or on a branch, is used for classification. If the samples have been grazed by deer, discard the damaged stems if there are only a few in the entire herbage sample. If there are many, classify the damaged stems according to the most developed characteristic found on the axillary buds or lateral branches.

The stem classification system developed by Kalu and Fick does not depend entirely on morphological characteristics, because the three vegetative stages are distinguished by length. Stage 2 is sometimes skipped in the developmental sequences of summer, however, when stems are usually shorter than in the spring.

Counting the nodes for the vegetative stages would depend entirely on morphology, but it would also require a great deal more care and effort. Studies in Indiana by Volenec et al. (1987) showed that the number of nodes at a common stage depends on plant population, but stem length does not. Thus, counting nodes probably would not improve the precision of this system.

The time it takes to classify a herbage sample varies depending on the number and range of stages present. Most young samples have stems in only one or two categories. Older samples can have stems in each stage from vegetative through seedpod, thus taking longer to classify (Figure 2.4). On the average, 10–20 min per sample is required once the criteria for each stage are learned.

It is important to remember that all but the youngest herbage samples contain several stages. To avoid confusion, one should carefully distinguish developmental stage 3 (stems at the early bud stage) from mean stage 3.0 (the weighted average stage for all stems in a sample). It is recommended that a decimal point be included with mean stage values to avoid ambiguity.

2.8.3 Determining the Mean Stage

Stems from each stage should be counted to determine the mean stage by the MSC procedure. For MSW, stems from each stage should be dried in individual bags in a forced-draft oven at 65°C until they reach a constant weight. MSC is calculated as the average of the individual stage categories present in the herbage sample, weighted for the number of stems at each stage:

$$MSC = \sum_{0}^{9} \frac{(S_i N)}{C}$$

MSW is calculated similarly, except that the average of the individual stages is weighted for the dry weight of stems in each stage:

$$MSW = \sum_{0}^{9} \frac{(S_i D)}{W}$$

where S_i = stage number (0–9)
N = number of stems in stage S_i
C = total number of stems in herbage sample
D = dry weight of stems in stage S_i
W = total dry weight of stems in herbage sample

For example, if a sample of alfalfa had 10 stems in stage 3, 25 stems in stage 4, and 6 stems in stage 5, MSC would be calculated as follows:

$$MSC = \frac{(10 \cdot 3) + (25 \cdot 4) + (6 \cdot 5)}{(10 + 25 + 6)} = \frac{160}{41} = 3.90$$

Studies at Cornell University revealed that mean stage increases about 0.05 to 0.15 units per day while the crop is growing rapidly, with MSW increasing slightly faster than MSC. Four samples from a uniform area of alfalfa had MSC or MSW values with standard deviations of 0.1–0.4 units until flowers appeared. With more mature samples, the standard deviation tended to be higher, up to 0.6 units. When regrowth began before the herbage was harvested, MSC values declined and standard deviations for MSC became very large (up to 1.5 units). Standard deviations for MSW remained less than 0.8 units.

Both MSC and MSW quantify morphological development of alfalfa. Most users prefer MSC because it is less tedious to calculate; however, only MSW is closely related to forage quality once crown buds start to elongate into an older canopy. In New York, crown bud elongation usually begins after about seven to eight weeks of development. Therefore, older canopies contain both young and old stems. Because MSC is weighted for stem numbers, the increase in the number of younger stems leads to a leveling off and then a decline of MSC (Figure 2.3). The younger stems also enter into the calculation of MSW, but their small mass limits their impact. Therefore, MSW applies to alfalfa of all ages and is the preferred method for alfalfa grown for seed production. For practical hay crop management, MSC should be adequate. If it is necessary to convert MSC values to MSW, use the following equation:

$$MSW = 0.456 + 1.153MSC \qquad n = 569, r^2 = 0.98$$

This equation applies to samples up to eight weeks of age.

2.8.4 The Relationship between Mean Stage and Forage Quality

Because mean stage and quality are closely associated in alfalfa (Figure 2.5), the quality of the standing crop can be predicted by measuring mean stage. The following MSW prediction equations have been tested nationally:

$$CP = 36.15 - 6.09MSW + 0.48MSW^2$$

$$IVD(true) = 93.67 - 4.29MSW$$

$$NDF = 20.62 + 8.03MSW - 0.59MSW^2$$

$$ADF = 17.05 + 3.85MSW$$

$$ADL = 2.77 + 1.01MSW$$

Each equation predicts the percentage of a quality component in alfalfa dry matter. If necessary, MSC can be converted to MSW so the equations can be used.

Although the predictions are not perfect, mean stage appears to have a very important correlation to the quality of alfalfa. It is also an improvement over the old system. For example, in one study, samples at "first flower" at different times of year had MSW values ranging from 1.7 to 3.4. Thus, "first flower" in the old system corresponded to a range in estimated crude protein levels of 27.2–21.0%, respectively.

Mean stage (MSC or MSW) can be used to predict the quality of the standing herbage in the field. Many of the major changes in alfalfa quality occur after the herbage is cut, and thus mean stage is not a good predictor of hay or silage quality. Instead, it provides a starting-point prediction indicating what the quality might have been under ideal management without postcutting degradation. Predicting the quality of the standing crop can help one decide when to harvest and illustrates how much is lost through postcutting man-

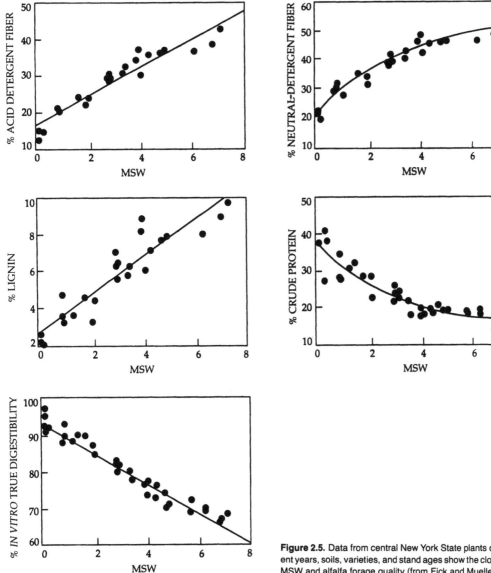

Figure 2.5. Data from central New York State plants collected in different years, soils, varieties, and stand ages show the close relationship of MSW and alfalfa forage quality (from Fick and Mueller, 1989).

agement. The system can be improved by developing quality prediction equations based on MSC.

2.9 Forage Quality

Forage quality is a complicated topic. It crosses academic disciplines, so the agronomist, who may understand little about animals, and the animal nutritionist, who may not understand forage plants, have different perspectives. Forage quality is perhaps the most important factor influencing ruminant productivity, whether grazing or feeding in a lot. The quality of a forage has a great deal to do with the amount of dietary fiber it contains. Fiber includes the bulk of the plant that has to be processed by the digestive tract, and it is also the

source of energy for rumen microbes and thus is important in promoting rumen function.

Forage quality is closely connected with that of fiber, which is needed in coarse form to maximize rumen function. The lignified part of fiber is indigestible, yet it is a requirement because unlignified material will not elicit adequate rumination. At the same time the forage must provide energy for microbial growth. Thus forage quality is indicative of several contrasting factors: the supply of plant cell wall, its optimal digestibility, and the rate of digestion. The rate of digestion is important because it sets the amount of food energy available per unit of time. Poor-quality forages that are too high in total fiber tend to have fermentation rates barely adequate to feed rumen bacteria near their maintenance requirements, thus setting severe limits on rumen output for the animal to use.

As plants age they generally decline in nutritive value as a result of increased lignification and a decreased proportion of leaves to stems. There are exceptions to this generalization. Not all leaves are more digestible than stems. Some crops, such as maize, do not decline in nutritive value with age and maturity although the stover may so decline. The quality of the whole plant is maintained by the formation and development of the seed head, which dilutes and compensates for the maturation processes in the stover. Parts of some plants may not change in quality with age; alfalfa leaves, for example, which have little structural function. In grasses the structural function of the leaves or the function of the stem as a storage organ can give stems a higher nutritive quality than leaves. This situation normally occurs in immature temperate grasses, although sugarcane is a tropical grass that falls into this category.

Factors influencing the quality of stems include diameter and whether they are hollow or filled with pith. If stems are large, lignified tissue may be more thinly distributed; consequently, the stem is more digestible. The pith is usually much less lignified than the cortex; therefore, hollow stems tend to be less digestible (e.g., alfalfa). The stems of immature grasses may hold reserve carbohydrates.

Generally speaking, cereals and seed crops at harvesting have reached ultimate maturity. Consequently, hulls, husks, bran, straw, and so on may be regarded as mature structures that have declined to their lowest quality. Nevertheless, there are important exceptions; soybean hulls and corn bran are relatively devoid of lignin and are highly digestible. Individual forages vary in the degree to which they decline in value with age, the effect being in part related to maturity (development, heading, seed production) and in part to the environment in which the plant is growing (Chapter 6).

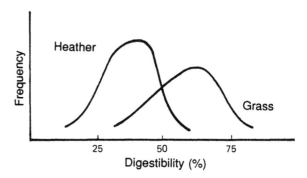

Figure 2.7. The concept of nutritive class is illustrated by the digestibility distribution ranges for heather and grass in the Scottish Highlands (see Sibbald et al., 1979). Frequency (distribution of total available forage) considers the disparate quality of leaves, stems, and branches of individual plants and the contrasting qualities of plant species. Sheep will graze out the higher-quality grass before consuming significant amounts of heather.

Plants that remain vegetative over a longer period may not decline in quality.

2.9.1 Nutritive Range and Class

The range in quality of standing herbage has led to the concept of *nutritive class* as the range and quantitative distribution of material of given digestibility and composition. A wide range in nutritive class may reflect the presence of plant species of different qualities, or it may be due to nutritive differentiation within a single plant species (Figure 2.6). Forage quality also varies with stratification relative to canopy (Buxton et al., 1985). The range in quality varies depending on species, maturity, and growth conditions. Immature grasses show little range in nutritive class, but the range widens with maturity. Browsing situations usually offer a wide range of plants in which recent annual growth is generally of a much higher quality than older woody material. Wide ranges in nutritive value may promote selective feeding, and narrow ranges inhibit it; for example, heather-grass combinations in Scotland (Figure 2.7). The grass will be overgrazed and killed before animals will consume significant amounts of heather.

2.9.2 Dietary Fiber

Fiber must be regarded as a biological unit, not as a chemical entity. The plant cell wall is a complex structure composed of lignin, cellulose and hemicellulose, pectin, some protein, lignified nitrogenous substances, waxes, cutin, and mineral components. This material divides into the insoluble matrix substances, including lignin, cellulose, and hemicellulose, which are covalently cross-linked, and the more soluble and extractable pectin, waxes, and protein. The cell wall contains most but not all of that portion of

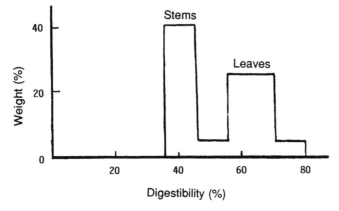

Figure 2.6. Digestibility of 30-day-old guineagrass (*Panicum maximum*) from Puerto Rico (from R. E. McDowell, 1972) showing the nutritive class range that can exist in "uniform" grasses. Some aspects of nutritive class are discussed in Section 2.8 relative to stage descriptions. Similar results exist for alfalfa (Collins, 1988).

plants which is resistant to enzymes secreted by the mammalian gastrointestinal tract. Although it may be considerably fermented by the intestinal microflora, the cell wall is rarely completely digestible. Complete digestion of the cell wall may occur when protective factors inhibiting cell wall degradation are absent. Fiber also imparts physical properties to food and feed and usually lowers caloric density. The composition of fiber is nutritionally significant and varies with the type of plant cell wall.

Fiber has long been recognized as important in human and nonruminant nutrition, and this recognition has resulted in further refinement of the dietary fiber concept. Human nutritionists define *dietary fiber* as the polysaccharides and associated plant cell wall substances that are resistant to mammalian digestive enzymes. This definition leads to the inclusion of substances that are not of plant cell wall origin and which can be scarcely called fibrous (Table 2.7). Because these other substances are water-soluble they behave differently in the digestive tract than does insoluble neutral-detergent fiber. These other substances are usually rapidly and completely fermented in the digestive tract. Pectin, although a cell wall substance, also falls into this class. Pectins are easily separated from the plant cell wall and are rapidly fermentable and virtually completely degradable by bacteria.

The total group of substances resistant to mammalian digestive enzymes is termed the *dietary fiber complex,* although the only true fiber is the relatively insoluble cell wall. A further distinction made regarding the carbohydrate components in nonruminant nutrition is that *unavailable carbohydrates* are those that give no sugar on digestion (Southgate, 1969, 1976). Carbohydrates of the dietary fiber complex are considered unavailable because when fermented the products are volatile fatty acids (VFAs) instead of sugars. In this sense, *unavailable* does not mean *indigestible.* This definition is useful in nonruminant nutrition, but in this context sucrose and possibly starch would have to be considered unavailable carbohydrates in ruminants. Indeed, the ruminant may receive no glucose from its diet and depends on gluconeogenesis for its sugar.

Not all indigestible cell wall structures are fibrous.

Table 2.7. Classification of plant substances relative to the concept of dietary fiber

Category	Function in plant	Degradability Mammalian gut	Degradability Ruminant gut organisms
Nonfiber			
Fructans	Storage	?	yes
Oligosaccharides	Storage	no	yes
Mucins	Connective tissue	no	yes
Tannins	Protective	no	no
Microbial cell wall	None	no	yes
Fiber complex			
Galactans	Storage	no	yes
Gums	?	no	yes
Pectin	Structural	no	yes
Insoluble fiber			
Hemicellulose	Structural	no	yes
Cellulose	Structural	no	yes
Lignin	Structural	no	no
Maillard products	None	no	no

Source: Kronfeld and Van Soest, 1976.

Cotton, which is nearly pure cellulose, is very fibrous and completely digestible, but the much more branched and less fibrous hemicellulose is more intimately linked with the three-dimensional lignin, which has the properties of a plastic polymer and is incompletely indigestible. Cellulose in fleshy vegetables and forage parenchyma tissue may not have much fibrous character and may be highly digestible.

The character and nutritive value of feeds and forage are determined by two factors: the proportion of plant cell wall and the degree of lignification. The amount of cellular contents in the dry matter of a feed determines the proportion of completely available nutrients present in the feed. The cellular contents comprise the bulk of the protein, starch, sugars, lipids, organic acids, and soluble ash. These are totally available to digesting organisms and are free from the effects of lignin or encrusted cell walls. The first evidence that protein and other cellular contents are available is that they may be extracted by suitable enzymes or by solutions of neutral detergent. Further evidence that in vivo true digestibility of protein and cellular contents is 90–100% is given in Section 22.3.1.

3 Feeding Strategies, Taxonomy, and Evolution

Herbivores depend on plants for food, and in turn plants depend on bacteria and herbivores for recycling their own nutrient requirements. This relationship is the result of interactions that were evolved over millions of years.

3.1 Plant Food Resources

The ultimate source of all food is solar energy, which photosynthesizing organisms store in their bodies. Green plants, which are autotrophic and can manufacture all the organic substances essential for life, supply the nutriment to animal herbivores and other organisms that feed on them.

Since the net process of photosynthesis is the fixation of carbon into organic substance, life on earth requires the recycling of fixed carbon through oxidation back to CO_2 from which it can be reduced again by photosynthesis. The oxidation of fixed carbon by nonphotosynthesizing organisms is necessary to maintain the CO_2 supply, without which photosynthesis would stop. Nonphotosynthesizing aerobic organisms, animals, and fungi require oxygen, which is provided by the photosynthesis process. A fourth, anaerobic, group is heterogeneous and syntrophic, and is comprised of eubacteria, some fungi and protozoa, and archaebacteria that ferment organic matter to a variety of products, including volatile fatty acids (VFAs) and methane.

The existence of biologically degraded carbon in peat bogs, coal beds, and petroleum suggests limits to biological degradation. Polyphenolic substrates, including lignin and condensed tannins, can be degraded only by aerobic systems (mainly fungal), and their accumulation in anaerobic environments promotes the deposition and protection of associated carbohydrate over long periods of time. These relationships in the food chain are outlined in Figure 3.1.

Like all living organisms plants have evolved protective systems to ensure their own continued existence and survival. Such protection includes physical and chemical structures that offer resistance to attack or ingestion by heterotrophic organisms (e.g., fungi, bacteria, and herbivores). Herbivores, on the other hand, have developed strategies to overcome these defenses.

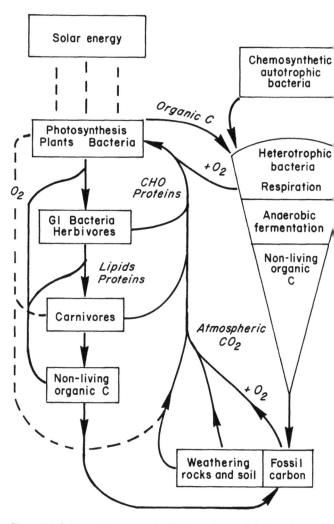

Figure 3.1. Carbon and oxygen cycles. The flow of oxygen is indicated by dashed lines; carbon by solid lines. The carbon that flows downward is generally organic carbon, and that which flows upward is CO_2. (Modified from W. C. Evans, 1977; see also Woodwell et al., 1978.)

Potentially degradable plant structures were present before parasitic organisms had enzymes to degrade them. Plants, in turn, developed (at a price) resistant protective structures, including structural celluloses and polyphenolic components such as lignin and tannin. A wide variety of secondary substances (tannins, alkaloids, terpenoids, etc.) also occur in plants. Some parasitic organisms, particularly fungi and bacteria, have enzymatic systems that degrade the more resistant cellulose and lignin. The cost to plants in elaborating these secondary substances is that the photosynthetic energy deposited in the secondary compound is lost to the plant, which has no metabolic system for retrieving the deposited energy.

Predatory organisms must have the requisite enzymes to break down the secondary substances (not all do), and their growth rate must be slow enough to be supported by slowly degrading substrates. The first limitation governs the extent of possible degradation; the second, the rate of degradation of insoluble material. A slow-degrading material means a slow release of food energy per unit of time, hence a low maintenance level for the organism subsisting on it.

3.2 The Fate of Cellulosic Matter

Cellulose is the most abundant carbohydrate in the world, and its recycling depends on microbial activity. It has been estimated that 85 billion metric tons of CO_2 are produced each year from the microbial degradation of cellulose. If such fermentation ceased and photosynthesis proceeded unabated, the earth would stagnate in about 20 years from lack of CO_2 (Cowling and Brown, 1969). But cellulose is not the only substance elaborated in such large quantities. Closely associated lignin and hemicellulose also add to the burden of carbon recycling. When one speaks of cellulosic matter, the associated lignin and hemicellulose are understood to be included.

Plant materials resistant to animal digestive systems are mainly found in the cell wall. Most animals lack enzymes capable of degrading cellulose, hemicellulose, and lignin. Although some snails and certain arthropods (Section 3.4.1) have intestinal cellulases, the organisms that attack these substances are mainly bacteria and fungi.

The ability of animals to utilize the carbohydrates cellulose, hemicellulose, and pectin as food depends on the presence of gastrointestinal organisms to degrade them and the capacity of the herbivore host to maintain these microorganisms and utilize their products. Use of the more slowly fermenting substrates is limited by retention time and the adaptation and evolution of their respective digestive tracts.

Ruminants are not the only organisms that use cellulose, and cellulose does not form the sole diet of ruminants. Cellulosic carbohydrate (including hemicellulose) may average about 50% of the metabolizable energy consumed by ruminants. Ruminants' contribution to the earth's CO_2 and methane levels is estimated in Table 3.1.

Gottschalk (1988) estimated total annual carbon flux from land and sea to the atmosphere at 200 gigatons (GT = 10^9 metric tons), of which about 90 GT arises from the ocean and 100 GT from the land. Cowling and Brown (1969) estimated annual recycled cellulosic matter to be 85 GT. CO_2 arising from this cellulose comes from both burning and fermentation. The yield of methane in peat bogs, paddies, and other longer-term anaerobic fermentations will be higher than in the rumen because the volatile fatty acids, acetate, propionate, and butyrate that are trapped by the animal as metabolizable energy are converted to methane and CO_2 in most other fermentations.

The most important sources of methane, in declining order, are natural swamps (26%); rice paddies (20%); oil, gas, and coal leaks (14%); biomass burning (10%); landfills (7%); and animal waste (3%). Other animal sources include termites (3.6%), equids (2%), pigs (1%), and humans (0.3%) (D. E. Johnson et al., 1990, 1992). These calculations place the contemporary concern over ruminants, methane, and the greenhouse effect in a new light. Because cattle and sheep are countable and their methane outputs are calculable, they are visible sources of production. But the great bulk of cellulosic matter is consumed by fermentation in oceans and the bottoms of lakes, ponds, and rice

Table 3.1. Contributions by world ruminants to net CO_2 and methane

	Domestic		Wild	Total
	Large	Small		
World population ($\times$ 10⁶)	1446	1610	310	3366
Body weight (kg)	300	40	100	—
Metabolic size (kg⁰·⁷⁵)	72.1	15.9	31.6	—
Metabolic mass (kg⁰·⁷⁵ $\times$ 10⁹)[a]	104.2	25.6	9.8	139.6
Net CO_2 (MT $\times$ 10⁶)[b]	2057	504	180	2741
Net CH_4 (MT $\times$ 10⁶)[b]	61	15	5.7	82
CO_2 as % net fermented carbon[c]	2.4	0.63	0.23	3.2
CH_4 as % potential CH_4[d]	11	2.7	1.0	14.7

Sources: R. E. McDowell, 1977; D. E. Johnson et al., 1992; see also Crutzen et al., 1986; Lerner et al., 1988.

[a] Product of world population and metabolic size.

[b] Yearly production from ruminants, in millions of metric tons (MT). An average of 80 g/kg⁰·⁷⁵ of feed intake and average methane yield of 2% of dry matter intake is assumed (D. E. Johnson et al., 1990).

[c] Cowling and Brown (1969) estimated annual recycled world cellulose equivalent to 85 billion metric tons (GT) of CO_2.

[d] Divided by an estimate of annual world methane output of 0.55 GT (D. E. Johnson et al., 1992).

paddies or consumed by soil and forest microorganisms, the net consumption of which constitutes about 90% of the total output of world CO_2 and 85% of methane arising from fermentations.

Domestic cattle in the United States and Canada constitute only about 6% of the total world cattle population. Of these, domestic dairy cattle, which produce the most methane per cow, number only 10 million, or 1% of the world total. Beef cattle (50 million) are usually kept in feedlots and are usually given methane-inhibiting substances. Another problem is that methane turnover (10–14 years) is more rapid than CO_2 turnover (50–200 years), and methane is produced at the expense of CO_2 and is in the natural cycle converted back to it. Methane from coal and gas production returns ancient fossil carbon to the atmosphere, whereas living animals (including bovines) merely recycle what has been recently photosynthetically fixed. Thus, as Drennen and Chapman (1992) have pointed out, the estimate that farm animals produce 15% of the world's methane overemphasizes the role of bovine emissions. In fact, manipulation of methane output in domestic cattle is likely to have little effect on net world methane output, and claims that reducing ruminant methane contributions can amount to a 20% reduction in upper atmospheric ozone is a gross overstatement.

3.3 Limits to Biodegradation

The physicochemical limits to biodegradation lie in the covalent bonding among the building units of the various macromolecular systems. All these systems involve an activating group acting on the α carbon, which allows the cleavage and utilization of the subcomponents, whether amino acid, sugar, or fatty acid. Thus, metabolic systems that allow synthesis and re-utilization make their possessors potential food for predators that have the same enzymatic systems.

The evolution of the polyphenolic polymers—most importantly lignins and tannins—represents a departure from the common mode in that interunit bondings are based on oxidative polyphenolic condensation in the form of ether or biphenyl linkages, which block hydrolytic activation and are generally resistant to the conventional modes of biological breakdown. Since laboratory analyses depend on the same chemical modes of attack, the peculiar refractoriness of lignins has retarded biochemical and biological understanding of these complicated polymers.

The unhydrolyzable linkages in lignin and other polyphenols have earned them the term *condensed substances.* Their nonhydrolytic character, however, has not prevented the evolution of novel means of breakdown, which seem to be possessed only by fungi and a few bacteria. Simple phenolic substances seem to be

utilized by anaerobic organisms, while the more condensed substances seem to be limited to those that can utilize oxygen. Hence the presence of anaerobic sinks, as attested by the existence of peat bogs, coal beds, and oil, which are the residues from materials either resistant to anaerobic degradation or that have fallen into environments of low oxygen availability, but which are potentially utilizable if oxygen becomes available. Most gastrointestinal digestive tracts are anaerobic and thus are limited by the extent to which anaerobically resistant substances exist within plants. Lignin sets a limit on the potential maximum degradability of plant cell wall. It protects about 2.4 times its own weight of cell wall matter that remains undegraded in a methane fermenter at 40 days (J. A. Chandler et al., 1980).

3.3.1 Anaerobic Metabolism

Metabolism by microorganisms in the absence of oxygen, generally termed *fermentation,* converts carbohydrate into organic products such as volatile fatty acids, lactic acid, and ethanol. These products retain much of the original energy of the substrate, a necessary consequence of the lack of available oxygen to oxidize them to CO_2 and water. The degree to which any substrate can be anaerobically metabolized depends largely on the ability of the metabolism involved to manipulate the oxygen contained within the molecules of substrate to produce CO_2 and the fermentive products.

The most efficient use of stored energy is through aerobic oxidation. Much plant organic matter ends up in places where oxygen is scarce, however, whether it be peat bog, rumen, cecum, colon, or sewer. Organisms adapted to these conditions can be classified into two groups: facultative anaerobes and obligate anaerobes. Facultative anaerobes can use free oxygen if it is available.

The metabolism of anaerobes parallels that of the photosynthetic process in green plants in that oxygen required for anaerobic metabolism must be stripped from the substrate—that is, from carbohydrate, CO_2 and other oxygen-containing substances. While photosynthesis reduces CO_2 to carbohydrate, further anaerobic metabolism results in the production of still more reduced organic substances enriched in carbon and hydrogen. Such reduced compounds include fatty acids, alcohols, and methane and must be regarded as excretory products of anaerobic systems. Dietary substances with low oxygen content will be metabolized very slowly or not at all by anaerobes. Thus fatty acids, waxes, and phenolic compounds do not provide energy for rumen fermentation, although they might be degraded to methane if the holding (turnover) time is sufficient. On the other hand, such reduced products may be directly incorporated into such organisms, saving them the energy expense of synthesis. This applies particularly to fatty acids and some amino acid skeletons. Longer-term fermentations in soil or in sewers may degrade substances not available to most animal gut fermentations. The time required (up to several weeks) is beyond the turnover capacity of the animal digestive tract (less than four days). Volatile fatty acids are degraded to methane and CO_2 in the sludge fermentation.

Other anaerobic systems may use photosynthesis to generate oxygen from CO_2 and water or may derive it by the reduction of sulfate or nitrate. Although light is excluded from the gastrointestinal tract, small amounts of sulfate and nitrate are ordinarily a normal part of the diet and can contribute some oxygen. The quantity of oxygen so derived is probably a minor part of the total oxygen budget in gut fermentation.

The potential energy available to herbivores from gastrointestinal fermentation includes that resident in the bodies of the microorganisms and in their reduced products such as VFAs (with the exception of methane). Reduced matter in the form of VFAs still contains the bulk of the stored energy from photosynthesis, and this energy can be made available through aerobic oxidation. Complete aerobic metabolism of 1 mole of glucose to CO_2 and water produces 38 moles of ATP, whereas anaerobic metabolism (in the absence of photosynthesis, nitrate, or sulfate) can produce only 2–6 moles of ATP depending on the fermenting organism and the ecological system. Therefore, the bulk of dietary energy is unused by rumen organisms and is passed on to their aerobic host. If rumen metabolism were aerobic, or if retention time were too long in an anaerobic system, no metabolizable products other than microbial cells would be available and the host would thus be wholly dependent on microbial cell yield. It must be understood, then, that microbial efficiency is antagonistic to the efficiency of the animal host. For example, an inefficient anaerobic system that produces a maximum of VFAs and a minimum of cells from a given amount of substrate yields a maximum of energy to the aerobic host, while an aerobic system that produces cells and CO_2 will yield protein and energy only in the form of cell yields.

The ruminant depends on prefermentation of its food by anaerobic microorganisms, and so the limitations of these anaerobic microorganisms also apply to the host. Energy expended for maintenance by the bacteria appears as heat and is lost to the host animal. Substrates with very low rates of degradation may be lost to both bacteria and host because they cannot be retained long enough to be fermented. This is the principle of rate limitation. Substances high in carbon and hydrogen are unavailable to rumen organisms and also unavailable to the host—unless it has the requisite enzymes. This latter limitation is utilized by plants in

Table 3.2. Occurrence of enzymatic systems for degradation

Material	Aerobic		Anaerobic (microbial)	
	Bacterial and fungal	Animal (mammalian)	GI tract	Sludge or soil
Sugar	+	+	+	+
Starch	+	+	+	+
Cellulose	+	−	+	+
Lignin	+	−	−	+
Keratin protein	+	−	−	+
Fats	+	+	−	+
Fossil				
Oil	+	−	−	−
Coal	−	−	−	−

their defensive compounds such as cutin and lignin, which are major factors influencing feed quality. These substances are resistant to anaerobic degradation because of their low oxygen content and condensed structure, which prohibit degradation and retard even aerobic catabolism.

The main microbial products unavailable to the host animal's digestive system are the microbial cell walls and methane, which the host animal's metabolism cannot utilize. In addition, animal digestive systems lack enzymes that can degrade very long chain paraffins and keratin proteins (Table 3.2). The generation of undegradable residues has significance not only with regard to the utilization of nutrients by ruminants but also as regards the disposal of manure. For any continuously running system to remain in a steady state it must ultimately recycle the residue, which will otherwise accumulate and unbalance the system. This recycling partly depends on aerobic metabolism by bacteria and fungi that can degrade lignin and thereby render residual cellulosic carbohydrate available for metabolism. The replenishment of the CO_2 supply is largely a function of aerobic heterotrophs. Almost all organic matter in nature is returned to CO_2 and water through bacteria and fungi; animals recycle only a very small part of it (Section 3.2.1).

3.4 Plant-Animal Interaction

Herbivores need plants for food, and in response plants have defenses against animal predators. But they also exploit animals and in their turn are dependent on animals for such things as seed dispersal and recycling of nutrients. Most of the food plants important to humans and their domesticated animals have been selectively bred during their long coassociation with man, but not all plants in the related ecosystems evolved in the presence of similar animal factors. So it is that to some degree plants follow alternative strategies. One strategy lies in allowing animals to eat them and then regenerating rapidly; another is based on total defense in order to avoid becoming food. Both gambits have their hazards. A plant cannot be completely eaten and still survive, but investing all energy reserves in protection sacrifices metabolic capacity and growth potential (Chapter 6). Protective compounds remove photosynthetic resources from potential metabolism and can be hazardous to the cells that produce them.

Plants adapted to grazing pressure tend to store energy reserves in roots and rhizomes that can regenerate rapidly under proper conditions of water and nutrients. A similar generalization applies to browse plants except that the older woody tissue may be heavily protected. The proportion of energy and nutrients that can be invested in growth is limited by environmental conditions of temperature, light, and supply of nutrients. These limiting factors impose adjustments on plants employing both strategies. A heavy investment in protection may be more limiting on potential growth under stressful conditions, which thus tend to promote higher nutritive values in plants that depend on regenerative potential. This aspect is explored in Chapter 6.

Plants have diversified their parts (leaf, stem, etc.) with respect to the distribution of nutrients. Leaves specialized for photosynthesis must necessarily contain the requisite metabolic apparatus in the form of enzymes, which are proteins and are liable to become animal food. Investment in protection prevents defoliation, but at the expense of plant reserves and photosynthetic capacity.

Grasses and dicotyledonous plants use different structural systems to cope with predators. Forage legumes retain a diminutive treelike structure in which leaves have little structural matter and lignin resides mostly in supporting stems and branches. Grasses, on the other hand, place their reserves in stems or roots, and their leaves have a midrib for support. Much of the lignin and resistant tissue is apt to be in this midrib and also in the leaf sheath that is wrapped around the stem (containing reserves in many cases) for protection. Thus a herbivore that approaches a grass plant may face siliceous sharp leaf edges at the front of the plant. Once the herbivore gets past these defenses, the good parts in the stems become vulnerable. Some grazed grasses have either hollow lignified stems (no doubt evolved under grazing pressure) and reserves underground or very thick cortex and cuticular stems for protection. Many tropical grasses are highly nutritively differentiated relative to available nutrients (see Figure 2.6), thereby offering grazing animals opportunities for selection. It is harder for animals to select upon grasses than upon legumes and other dicots because their outermost leaves and exposed parts tend to be more fibrous and protected.

Herbivores use various strategies to surmount plant defenses. For example, tropical grasses place high concentrations of lignin in their leaf parts. The herbivore responds by increasing consumption to overcome the limitation of low digestibility. Elephants, pandas, equids, and other perissodactyls use this stratagem, which sacrifices retention time, to utilize cell contents.

Alternatively, the grazing ruminants have specialized in extracting the more slowly degradable cell wall polysaccharides like cellulose. The price paid is gastrointestinal retention and, consequently, restricted forage intake. This presumes lower lignification, which is more likely to occur in plants in cooler and drier environments. The grazing ruminants certainly evolved under such conditions, and horses adapted to this strategy as well. Although ruminants do exist in humid tropical conditions, they are usually smaller and more selective and probably use the rumen fermentation more as a means of detoxifying low-molecular-weight secondary compounds. The latter strategy may give them an advantage over equids and other monogastric herbivores that need to be more careful in their choice of plants.

All herbivores, ruminants and nonruminants, have to deal with plants' secondary defensive compounds. One solution is cafeteria feeding: if no great quantity of any one plant is eaten, the relative amounts of poisons consumed across the spectrum of plant species will not result in a dangerous intake of any one. Thus, widening the range of choices, provided they are accessible, will reduce the hazard. This strategy is favored under tropical and temperate forest conditions that may present a wide variety of plant species. Grazing systems, which are relatively poor in plant diversity, may not present this opportunity.

The ability of the rumen microorganisms to adapt to secondary substances has been countered by plants that produce inhibitors that affect the rumen fermentation or are toxic to the animal; for example, tannins, estrogenic isoflavones, silica, and alkaloids. Some animals have a specialized liver metabolism that can detoxify specific toxic substances. For example, the koala bear feeds exclusively on eucalyptus, and various herbivorous insects are specialists that feed on toxic plants. Often this kind of evolutionary adaptation allows one species or group an exclusive resource unavailable to animals not possessing the requisite metabolic adaptation.

Rumen adaptation to secondary substances is a special topic relevant to bioengineering and its limits. Many secondary substances can be metabolized by gut microorganisms only after they have had time for metabolic adaptation—between 3 days and 3 weeks (Chapter 13). Other adaptations or adjustments require more extensive modifications. Two examples: ruminants cannot adapt to the mimosine in *Leucaena* without the introduction of requisite bacteria; and lignin cannot be metabolized at all because of the limitations of metabolic processes in anaerobic environments. Evolutionary adjustments like these are relevant to understanding the feeding behavior of wild and domesticated animal species as well as the feeding behavior of all herbivores.

Animal size relative to size of plants eaten is also a factor in animals' selection of plant foods (Owen-Smith, 1988). Herbivores that specialize on particular types or parts of plants frequently have specialized dentition and eating behaviors.

3.4.1 Invertebrate Herbivores

Probably the most important herbivores in the world in a quantitative sense are the insects and snails, which are often regarded as pests and competitors for man's crops. Their nutritive strategies are of interest, however. Except for the termites, most insects do not utilize fiber. Whether or not termites themselves digest fiber or even lignin remains controversial. Most wood-eating species probably depend on microorganisms for digestion (Breznak, 1982).

Some lower animal forms, including snails and certain arthropods, secrete cellulases (Dickenson and Pugh, 1974). These enzymes seem to be limited to the more available and uncrystalline sources of cellulose. Termites' ability to digest lignin and cellulose, if it exists, is likely related to the associated growth of symbiotic aerobic fungi (M. M. Martin, 1982). While termites have symbiotic cellulolytic organisms in their digestive tracts, they also consume fungal enzymes and achieve digestion through this means. Volatile fatty acids are produced in the hindgut of the termite (Odelson and Breznak, 1983), but there also seems to be conversion of methane and CO_2 to acetic acid (Breznak and Switzer, 1986).

3.5 Feeding Strategies and Plant Sources

The proteins and carbohydrates in plant cellular contents are completely available to all animals, but those in the cellulosic wall are available only to animals that harbor the requisite bacteria in their digestive tracts. The cell walls of many plants are lignified, and this protects a portion of the structural carbohydrates from microbial digestion. Animals that eat high volumes with low extraction of fermentable cellulose may adapt to consume more lignified sources. Secondary compounds require further adaptation. In fact, plants' structure and composition offer a wide variety of ecological niches to herbivores that can exploit them.

Plant families exhibit diverse morphological structures requiring herbivores to specialize or develop diversity in food choices.

Table 3.3. Classification of herbivores

Class	Ruminants	Nonruminants
Concentrate selectors		
Fruit and foliage selectors	Duikers, sunis	Rabbits
Tree and shrub browsers	Deer, giraffes, kudus	Sumatran and black rhinoceros
Intermediate feeders		
Prefer forbs or browsing	Moose, goats, elands	
Prefer grass	Sheep, impalas	
Bulk and roughage eaters		
Fresh grass grazers	Buffaloes, cattle, gnus, kobs, oribis	Hippopotamus
Roughage grazers	Hartebeests, topis	Horses, elephants, white and Indian rhinos, zebras[a]
Dry region gazers	Oryxes, camels, roan and sable antelopes	Kangaroos

Sources: Adapted and extended from Hofmann, 1973, 1989; Hansen et al., 1977; and Foose, 1982.

[a]Classification of zebra varies (see Foose, 1982). See bilateral classification (Figure 3.2).

3.5.1 Classification of Herbivores

Hofmann (1973, 1989) classified herbivorous mammals into three major classes based on their feeding preferences (Table 3.3). Langer (1988), on the other hand, classified these animals according to a herbivory rating from 1 to 6 which parallels that of Hofmann (Table 3.4). Langer's system presumes that grasses are more fibrous than nongrass forage. On the average this is true, but it is not a universal truth, since in browsing, fiber intake may depend on how much wood is eaten. Most woods contain more fiber than any grass. Another, but less satisfactory, system is that of Bodmer (1990). This system uses the proportion of grass in the diet as a criterion for classification, suggesting a con-

Table 3.4. Types of herbivores and their preferred foods

HR[a]	Preferred food	Feeding type[b]
1	Animal material, fruits, tubers, and other reserve organs; only occasionally buds and shoots	Omnivore
2	Occasional animal material; plant parts: fruits, tubers, seeds, buds, flowers, leaves, sap	Concentrate selector
3	Plant parts: fruits, tubers, seeds, buds, flowers, leaves sap; 30–40% is buds, blossoms, leaves, and young shoots	Concentrate selector
4	More than 60% is leaves, shoots, blossoms, and other plant parts; the rest is fruits, tubers, seeds, and buds	Intermediate feeder
5	Predominantly leaves buds, plant stems, considerable amount of grass	Bulk and roughage feeder
6	Specialized obligate grazers	Bulk and roughage feeder

Sources: Hofmann, 1973, 1989; Langer, 1988.

[a]Herbivory rating, according to Langer, 1988.

[b]After Hofmann, 1973, 1989.

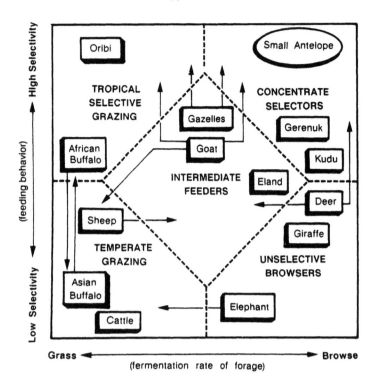

Figure 3.2. Herbivorous mammals classified by diet (from Van Soest and Demment, 1983; see also Van Soest, 1988b). The axes are the degree of feeding selectivity and the amount of grass versus browse in the diet. The arrows indicate mobility of a species with respect to these axes.

tinuum between grazing and browsing that may not be linear. The problem with one-way types of classification is that they assume that grazers are less selective than browsers; however, there can be very selective grazers (Hofmann, 1989) as well as less selective browsers, such as elephants. Concentrate selectors are unable to tolerate large amounts of fiber and are thus limited to feeding on low-fiber portions of plants. (Humans, if considered as herbivores, would fall into this group.) Some intermediate feeders pass relatively large volumes through the digestive tract and make limited use of the plant cell wall components in order to ingest sufficient amounts of the more readily available portions. They may be adapted to either browsing or grazing. Bulk and roughage feeders can digest the cell wall components. These include the grazing ruminants and some large nonruminant herbivores.

Intermediate feeders change their feeding behavior according to the availability of forage, and they are far more versatile than concentrate selectors or obligate grazers. Usually they eat only immature grasses; as forage matures animals move onto browses. Northern and Arctic species similarly need to adapt to summer and winter conditions and forage availability. Moose are classified as concentrate selectors, but they vary their diet seasonally to take advantage of available forage in winter. In Alaska considerable wood may be taken from willow species; but even then moose are

very selective, eating the cambial layers, buds, and the like. Other animal species adapted to tundra conditions include the camelids in the altiplano of South America and the yak in the Tibetan plateau, which selectively feed on sparsely distributed grasses, sedges, and forbs. These very large animals can cope on coarse forage as well as feed selectively (Cincotta et al., 1991).

Some of the smaller African intermediate feeders and grazers sometimes feed selectively on the nutritively differentiated tropical grasses. Thus intermediate feeders are sometimes selectors. Janis and Ehrhardt (1988) distinguished fresh grass eaters from dry grass eaters. Selective feeding is related to the range in nutritive classes available in a given habitat (see Figure 2.7). Tropical grasses in particular are more differentiated than are temperate ones, leading to the probability that tropical grazers will be more selective feeders than temperate ones are.

The larger the animal the less likely it is to be a selective feeder, if for no other reason than clumsiness in feeding on small things. Grazing is commonly associated with less selection and browsing with more; however, some large browsing ruminants such as the giraffe are probably less selective than the small grazing ruminants. The giraffe has a greater digestive capacity than the smaller selectors and for its size should be more tolerant of lower-quality browse. In contrast, the oribi (16-kg antelope) appears to be a highly selective grazer (Hofmann, 1989). Herbivores' feeding behavior can be illustrated with a two-dimensional classification in which ruminants and nonruminants may be compared (Figure 3.2). The system has a place for intermediate species such as the goat that show considerable versatility in their feeding behavior. The goat is a comparatively selective feeder and is inferior to cattle and sheep as a digester of fiber, despite its reputation for being able to digest almost anything.

The bulk and roughage eaters include (in decreasing order of their need for water) fresh grass eaters, roughage eaters, and dry region grazers. Temperate cattle are classified as fresh grass eaters mainly because of their need for water; actually, they are quite unselective. Unlike the African ruminants, cattle originated in temperate conditions, which favor less-differentiated plants.

3.6 Taxonomy

The ruminants constitute a suborder of the Artiodactyla (ungulates) and are divided into four families comprising about 155 species (Table 3.5). Camelids represent a sister group contained in a separate suborder, the Tylopoda, although some classification schemes include them in the Ruminantia. The tragulids tend to lack an omasum and may be more closely related to the

Table 3.5.   The suborders Tylopoda and Ruminantia

	No. of genera	No. of species	Examples
Tylopoda			
Camelidae	2	4	Alpaca, llama, camel
Ruminantia			
Tragulidae	2	4	Chevrotain, mouse deer
Giraffidae	2	2	Giraffe, okapi
Cervidae	17	38	Caribou, deer, elk, moose
Bovidae	55	121	Antelope, bison, buffalo, cattle, gazelle, eland, goat, sheep, yak

Sources: Simpson, 1945; Mochi and Carter, 1971; Hayssen et al., 1993.

tylopods (Gentry and Hooker, 1988). Only one or two genera of the tragulids, giraffids, and antilocaprids are extant today. Antilocaprids have recently been included in the bovids (Hayssen et al., 1993). All the African antelopes are bovids.

3.6.1 The Bovidae

The family Bovidae has the greatest number of genera and is the most diverse ruminant group; it includes the African antelopes and buffalo, cattle, sheep, and goats. The group includes the smallest ruminant, the royal antelope (*Neotragus pygmaeus*), 25 cm high at the shoulder (see Figure 4.9), and the largest, the cape buffalo. The grouping is somewhat problematic from the viewpoint of feeding habits since it includes both the tropical forest browsers and the most developed ruminants (temperate grazers). The modern bovid forest browsers probably represent an evolutionary radiation since the Miocene.

3.6.2 The Tylopoda

The tylopods consist of the Old World camels—the one-humped Arabian camel (dromedary) and the Bactrian two-humped camel—and the humpless South American camelids, of which there are four species: alpaca, llama, guanaco, and vicuña. Dromedaries and Bactrian camels are mostly domesticated, although small numbers of wild Bactrian camels may still survive in central Asia. Dromedaries are feral in the Outback of Australia. Alpacas are entirely domesticated and llamas largely so, while guanacos and vicuñas are mostly wild. All are adapted to arid or desert conditions, high mountain rangelands, or both. Thus their feeding strategies, which are not well described, are related to sparse vegetation of probably moderate nutritional quality (Franklin, 1983). Camelids can be considered as ranging from intermediate feeders to grazers. The altiplano of South America provides mainly grasses and forbs, and lower semidesert regions offer potential browse plants. The dromedary is a capable browser.

Table 3.6.   Classification of the family Bovidae

Subfamily and tribe	Example	Feeding habit
Cephalophinae		
Cephalophini (2)[a]	Duiker	Selector
Antelopinae		
Neotragini (8)	Klipspringer, suni, dik-dik, royal antelope	Selectors
	Oribi	Grazer (selective)
	Steenbok	Intermediate grazer
Antelopini (7)	Gerenuk	Selector
	Gazelles	Intermediate grazers
Hippotraginae		
Hippotragini (3)	Sable and roan antelopes, oryx	B + R[b]
Reduncini (5)	Waterbuck, kob, reedbuck	B + R
Alcephalini (5)	Topi, wildebeest, hartebeest	B + R
	Impala	Intermediate grazer
Bovinae		
Tragelaphini (4)	Kudu, eland, bushbuck	Selectors
Boselaphini (2)	Nilgai, 4-horned antelope	Selectors, intermediate
Bovini (6)	Cattle, yak, buffalo, bison	B + R
Caprinae		
Saigini (2)	Chiru, saiga	Intermediate browsers, or grazers
Rupicaprini (4)	Goral, chamois mountain goats[c]	Intermediate browers
Ovibovini (1)	Musk ox	Intermediate browser
Caprini (5)	Goats	Intermediate browsers
	Sheep, ibex	Grazers (selective)
Antilocaprinae	Pronghorn	Intermediate grazer

Sources: Simpson, 1945; Gentry, 1978.
[a]Number of genera is given in parentheses.
[b]B + R = bulk and roughage eaters.
[c]Mountain goats may belong in the Caprini (Gentry, 1978).

All tylopods chew their cud and have a three-compartmented stomach, lacking an omasum (Langer, 1988; C. E. Stevens, 1988). Reports of their digestive capacity are scarce. The llama is suggested to be a more efficient digestor of cellulose than the cow (Stevens, 1988). Foose (1982) indicated the same, but the feed intakes of the llamas he studied were submaintenance.

The camelids of South America apparently arrived in North America from Asia via the Bering Strait and then crossed into South America 2 million years ago when the isthmus of Panama formed and allowed the great interchange of land mammals in the Americas (Marshall et al., 1982). Camelids and many other species became extinct in North America in the Pleistocene.

3.7 Evolution of Herbivores and Gut Fermentation

Before presenting the main sequence of mammalian evolution, certain preliminary developments and their attendant problems should be described. The unit of 1000 million years (10^9 years) is termed an *eon*. The earth is about 4.6 eons old, and the oldest extant evidence of solid surface is about 4.2 eons. The oldest sedimentary rocks are about 3.8 eons, and the earliest evidence of life is found in such rocks about this age. The earliest organisms were prokaryotic and included the archea (archaebacteria) and bacteria. Photosynthetic bacteria appeared more than 3 eons ago. A eukaryotic oxygen-utilizing mitochondrion may have appeared about 1.5 eons ago (Woese, 1987). The eukaryotes appear to be descended from the archea (Woese, 1988). Multicellular animals appeared 800–600 million years ago when the oxygen level of the atmosphere had risen high enough to support them.

During this ancient Precambrian time, unicellular organisms were the main forms of life, and it was in the earlier part of this time that prokaryotic-type methanogens (archaebacteria) and eubacteria, including those that are represented by today's rumen bacteria (i.e., cellulolytics and other anaerobic carbohydrate fermenters), appeared. All existing large herbivores today depend on these organisms for digestion of cellulosic carbohydrates. This kind of symbiotic association may have evolved in Mesozoic times, possibly associated with the herbivorous dinosaurs. Mammals descended and evolved from the Therapsida during the Mesozoic but remained a minor group in this period.

The evolutionary history of the ruminants is essentially limited to fossil remains of bones, teeth, and horns. The phylogeny of the soft tissue reticulorumen is unknown because of the lack of a fossil record. Still, the fossil remains we do have indicate the nature of horns and the loss of the upper incisors, which are relevant. In any case, it is evident that nutritional strategy and adaptation to food resources have played a major role in the evolution of modern herbivores. Some idea of the course of evolutionary developments can be obtained from a survey of surviving forms (Hume and Warner, 1980; Janis and Ehrhardt, 1988).

The end of the Cretaceous and the beginning of the Paleocene period (65 million years ago) marked profound changes in the appearance of the earth and its plants and animals. Although these changes may not have occurred simultaneously, the preexisting dominant vegetation (cycads and gymnosperms) was replaced by the flowering plants (angiosperms; Regal, 1977), the dinosaurs became extinct, and the mammals emerged as a large and diverse group (Bakker, 1986).

3.7.1 The Herbivorous Dinosaurs

Since the 1980s our understanding of dinosaurs has undergone major revision (Bakker, 1986). Birds are now regarded as modern dinosaurs. It appears that the herbivorous dinosaurs, at least the more advanced ones (Farlow, 1987), had pregastric fermentation in the crop and were also probably warm-blooded (Robertshaw,

1984). Some of these animals were enormous (some species exceeded 50 tons), and the ruminant and herbivore models relative to size (Chapter 4) indicate their probable digestive strategy and feeding behavior. Larger animals have the potential capacity for greater retention and extraction of energy, but extrapolating retention to the capacity of a 50-ton dinosaur leads to an improbability. The extrapolated retention times are of such an order (i.e., a week or more) that methanogenic bacteria would have had time to degrade acetate and other VFAs to CO_2 and methane, just as they do in sewers and bogs today. This would result in a loss of more than 80% of the potential digestible energy from the food substrate as methane. It is not probable that any herbivores could exist under these conditions. Thus retention times must have been less than 4 days, and only the methanogens that convert CO_2, H_2, and formate to methane would have survived, as they do in the digestive tracts of most existing herbivores, including humans.

The strategy of fast passage and short retention is characteristic of some modern species, including the elephant, the largest living herbivore. The elephant has fast passage and lives on bulk, roughage, and low extraction, a strategy very like that of the large herbivorous dinosaurs. Vegetation at that time (fern trees, etc.) was likely lignified and not of the highest quality. A large intake and minimum chewing was handled by a huge crop that comminuted and crushed the woody tissue. The extraction rate (i.e., digestibility) was probably on the order of only 20%, as is that of modern elephants on poor forage (Hackenberger, 1987) and also of pandas on bamboo (Dierenfeld et al., 1982). Based on this theory, a homeothermic 50-ton herbivorous dinosaur would have ingested about 3 tons of fresh browse and voided about 2 tons of wet feces per day!

The dinosaurs probably ground their fodder in the muscular crop, which held cannon-sized stones (Bakker, 1986). This would give them an advantage in eating time over ruminants, which have to spend time chewing their cud.

3.7.2 The Paleocene and Eocene

The disappearance of the dinosaurs and many of the associated plants at the end of the Cretaceous eliminated groups occupying a variety of ecological niches. Mostly small species of plants (angiosperms) and animals were able to survive. The small mammals that survived radiated evolutionarily to fill the respective voids.[1] Most of the classes of modern mammals ap-

---

[1] Recent geological evidence suggests that a large (100 km) meteor crashed into the earth causing a blackout. Some contend that the atmospheric blackout was volcanic, and others that the dinosaurs were on their way out for other reasons; however, the volcanic outpour could have resulted from the meteor. Great lava lakes such as

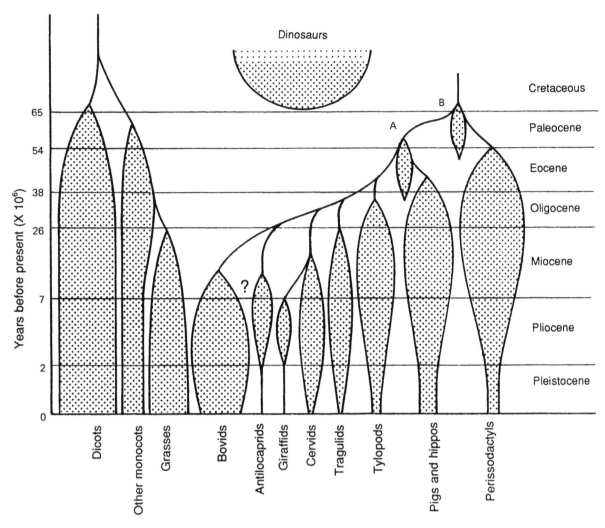

Figure 3.3. The appearance and development of ungulates and angiosperms (modified from Janis, 1976; Gentry, 1978; Hume, 1978; Hume and Warner, 1980; and B. A. Thomas and Spicer, 1986). All ruminant grazers are bovids and did not evolve until after the appearance of the grasses in the Miocene period. The American pronghorn antelope is the only extant antilocaprid; the Giraffidae are similarly reduced. (A) The ancestral artiodactyl common to both pigs and ruminants. (B) The common ancestor of Artiodactyla and Perissodactyla. The exact association of the various lines of ungulates is the subject of dispute. Recent cladistic studies suggest that the pronghorn antelope may be more closely related to bovids, and tragulids may be more related to tylopods. The existence of postulated ancestral groups A and B has been questioned (see Gentry and Hooker, 1988; and Janis and Scott, 1987, 1988). Furthermore, the Antilocaprinae have recently been included within the Bovidae (Hayssen et al., 1993), although they do not share a horn structure with the Bovidae (Figure 3.4). This association is indicated by a question mark.

peared during the Paleocene, including the Artiodactyla (pigs, hippos, and ruminants) and the Perissodactyla (horses and rhinos).

Ruminants and tylopods (camelids) appeared in the Eocene and represented an elaborate specialization within the Artiodactyla. It is debated whether camelids are true ruminants, but the two groups may have had a common ancestor (Figure 3.3).

Some interrelated problems are posed by the appearance of ruminants:

1. Early ruminants were small animals that weighed less than 18 kg and probably were adapted to forest conditions. They also did not have horns. No modern small mammal is an effective utilizer of cellulosic carbohydrate.

2. The appearance of pregastric fermentation and rumination probably did not coincide and likely developed for different advantages.

3. The early dominant herbivores were perissodactyls (horse-rhino types). These groups were largely replaced by ruminants after the Miocene and perhaps as late as the Pleistocene.

Modern perissodactyls (horse and zebra) are competitive grazers, although their digestive capacity is lower. These lines, which survived the competition of

---

the Deccan in India, Iceland, or the Columbia plateau in Washington might have been caused by meteor impact. Such basaltic lakes filling large craters are major features of the lunar landscape (Rampino, 1989).

Table 3.7. Possible sequences in herbivore evolution

Time and weather	Plants	Animals
Paleocene	Rise of angiosperms	
Eocene	Secondary compounds appear in plants for defense	Radiation of perissodactyls; pregastric fermentation (detoxification, amino acids, vitamins)
Miocene (colder and/or drier climate)	Rise of grasses; high reserve CHO and lower lignification	Grazing; less selective feeding; two groups of herbivores: ruminants (use cellulose) and nonruminant grazers (high intake–low cellulose utilization)
Quaternary (cyclic warming and cooling [Pleistocene glaciation])	Tropical plants; cold-adapted plants with large CHO reserves	Arctic ruminants; disappearance of many large monogastric herbivores; specialization in forest ruminants

Source: Based partly on Hume and Warner, 1980.

the ruminants, were also, since the Miocene, subjected to the same pressure to develop grazing behavior (Voorhies and Thomasson, 1979). Thus it is apparent that grazing developed independently in various lines.

Perissodactyls achieved dominance in the Eocene and Oligocene but declined after the Miocene period and were largely replaced by the ruminant in the Pleistocene. Grazing appeared in the Miocene at about the same time as grasslands.

The ancestral ruminants were probably tropical forest browsers. An old hypothesis is that the ruminant could better escape predators because it devoured its food quickly and masticated it later. An alternate, and much more likely, explanation is that predigestion detoxified secondary plant substances, allowing the pregastric fermenter a greater latitude in dietary choice and adaptation (Freeland and Janzen, 1974). The rise of angiosperms under tropical conditions was likely associated with the appearance of secondary compounds that were toxic to unadapted animals (Swain, 1977). Fermentation also eliminated the requirement for an external source of B vitamins and amino acids.

The modern small selector ruminants tend to choose a high-quality diet that supports high rates of VFA production (Section 4.4.4). Their rumens show increased surface area as an adaptation to the high rates of absorption necessary with high VFA production (Hofmann, 1973). They also seem to be able to bypass high-quality concentrate material to the abomasum (Hofmann, 1989). The ability to select seems to have reduced their need for digestive capacity, since all selector ruminants tend to have small rumens and no great capacity for digestion of fiber.

Regressive evolution has been suggested for the small tropical forest ruminants, most of which are comparatively poor digesters of fiber (Arman and Hopcraft, 1975), but this suggestion is controversial. The small African antelopes have undeveloped omasa, which are most unlikely to have regressed from a

grazer ancestor. Perhaps the ancestral bovid was a small-frame ruminant. The cladistic studies by Janis and Scott (1988) and Gentry and Hooker (1988), are based solely on bones, horns, and teeth. Small antelopes may ruminate for other reasons, perhaps to regulate rate of VFA production (Hoppe, 1977). The concept here is that the concentrate selector consumes such a high-quality diet that VFA production is potentially hazardous and needs to be regulated. Sunis and dik-diks consume high-energy food in a coarse form, and the rumination rate regulates the release of carbohydrates to the fermentation.

The development of rumination did not necessarily coincide with pregastric fermentation and was likely promoted to achieve a different advantage. For example, the modern small bovid antelope selectors are probably a modern reversal to browsing behavior and do not represent an ancestral form. On the other hand, the small tragulids (chevrotain and mouse deer) may be aboriginal types because they never left the forest. Table 3.7 shows possible sequences of ruminant evolution.

Modern grazing animals are less selective feeders. Being less selective is an advantage provided the plants are relatively undifferentiated (nutritionally) and contain sufficient available cellulosic carbohydrate. The animal must also be large enough so that rumen turnover time allows fermentation to take place. The remaining questions, then, is when did rumination evolve and why? Did the early pre-Miocene ruminants ruminate? And if they did, did they do it for reasons suggested above for the modern small forest ruminants?

3.7.3 The Appearance of Grazers

Grasslands appeared in the Miocene as the earth's climate cooled and semiarid zones developed. Apparently North America was a savannah during the Miocene. Grazing meant a less selective feeding pattern and a more developed cellulolytic rumen fermentation.

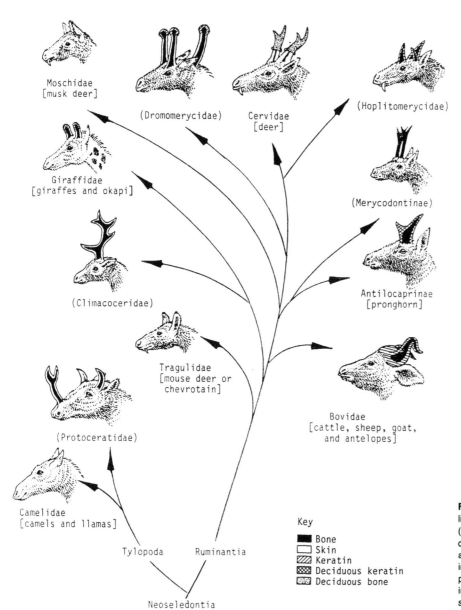

Moschidae
[musk deer]

(Dromomerycidae)   Cervidae
[deer]

(Hoplitomerycidae)

Giraffidae
[giraffes and okapi]

(Merycodontinae)

(Climacoceridae)

Antilocaprinae
[pronghorn]

Tragulidae
[mouse deer or
chevrotain]

(Protoceratidae)

Bovidae
[cattle, sheep, goat,
and antelopes]

Camelidae
[camels and llamas]

Tylopoda   Ruminantia

Neoseledontia

Key

■ Bone
□ Skin
▨ Keratin
▨ Deciduous keratin
▨ Deciduous bone

Figure 3.4. The interrelationships of living and extinct ruminant families (from Janis, 1986). Cervid antlers are deciduous bone, while bovid and antilocaprid horns are bone sheathed in keratin. Extinct families are in parentheses, and common names are in brackets. (The relative thickness of skin or keratin is not drawn to scale.)

Ruminating animals became larger and evolved horns independently in at least two lines, the cervids and the bovids (Figure 3.4; Janis, 1986). This adaptation coincided with a change to cooler and drier climates, which promoted less differentiated and developed forage plants of very great value to an animal that possessed adequate cellulolytic capacity in the rumen.

Extreme cooling of the earth's climate did not occur until the beginning of the Pleistocene, 2 million years ago, and widespread global glaciation lasted only until 10,000 years ago. The cooling occurred in several stages separated by relatively warm periods. These alternating weather cycles caused important changes in plants and the extinction of a number of large monogastric herbivores, particularly in the Americas. North and South America became joined at the isthmus of Panama near the beginning of the Pleistocene, and transmigrations led to competition and extinctions (Marshall et al., 1982; Owen-Smith, 1987). One may speculate on the interaction of climatic change and feeding strategies which overall, worldwide, favored ruminants over equids and other large monogastric herbivores.

Humans may well have been involved in the demise of some of these species, but herbivore competition and the climate should also be considered factors (Janis, 1984, 1987). The cold Pleistocene favored plants with higher nutritive value and sparse distributions, which should in turn have favored grazers. The equids' strategy of grazing for volume with low digestive extraction may have been advantageous in warmer conditions of lower-quality grass, but the situation may

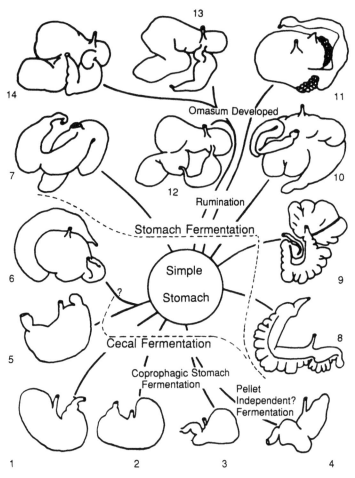

Figure 3.5. The radiation of stomach forms from the simple stomach (from Moir, 1968). (1) *Equus* (horse); (2) *Lepus* (rabbit); (3) *Dipodomys heermanni* (kangaroo rat); (4) *Dipodomys deserti* (desert kangaroo rat); (5) *Sus* (swine); (6) peccaries; (7) *Hippopotamus;*(8) *Macropus* (kangaroo, wallabies); (9) *Presbytis* (langurs); (10) *Bradypus* (three-toed sloth); (11) *Camelus* (camel); (12) *Hyemoschus* (chevrotain); (13) antelope; (14) *Ovis* (sheep). Evolutionary radiation has taken two directions: elongation of the body of the stomach with sacculation along greater curvature, as in the macropods and the colobine monkeys (not shown); and enlargement and modification of the cardiac region with constriction along the gastric canal. Ruminant development probably is a further modification of these latter features. Other pregastric fermenters include voles, hamsters, and some birds.

well have been reversed in the Pleistocene. The smaller ruminants had the advantage because they required less food and were more efficient at extracting energy.

Perissodactyls are represented by relatively few lines today, and the whole order, with perhaps the exception of the equids, seems to be headed for extinction except where protected by man (Janis, 1976). On the other hand, extant equids may represent a viable and continuing evolution of nonruminant grazing behavior (Foose, 1982).

It was in Africa that the most recent radiation of bovid ruminants occurred. More ruminant and non-ruminant herbivores coexist today in East Africa than anywhere else in the world. Their existence seems to belie the contention that coexisting herbivorous species are necessarily competitive, although over a much longer time competition might become evident and a few groups might emerge as survivors. Yet coexistence with collaborative feeding strategies is a major factor in maintenance of the existing grazing systems.[2]

3.7.4 Evolution of the Gastrointestinal Tract

Some idea of the evolutionary diversification of digestive tracts may be inferred from the anatomies of living animals, as well as from their respective embryologies (Langer, 1988). The reticulorumen and the omasum are elaborations of the nonsecretory tissue of the abomasum. This has given rise to criticism of prevailing terminology (e.g., monogastric) because the rumen system is an elaboration of a single organ. Complex and simple-gutted animals are alternative descriptions, but these terms have their own problems in that there are both parallel and diverse elaborations. Some of these have been described by Moir (1968) and are shown in Figure 3.5.

3.7.5 Pregastric Fermenters

Langer's work on the anatomies of extant herbivores (1988) makes it obvious that ruminants are only one subset of a larger group of animals with abomasal elaborations. Many nonruminant herbivores have complex forestomachs, but it is still not clear how many of them have true cellulolytic digestion. There is also elaboration in hindgut fermenters. While Langer describes this anatomy carefully, he gives no parallel descriptions of gastrointestinal contents and fiber-digesting bacteria, so the interpretation of these anatomies and their nutritive function remains difficult. Of the pregastric types there has been a growing list. Pregastric fermentation has recently been identified in the tropical hoatzin, a South America bird, and in some peccaries. Pregastric fermenters appear in so many taxonomic lines that parallel evolution of these forms is the only possible explanation.

Pregastric fermenters include some very small animals (e.g., the vole and hamster, which also practices coprophagy), some small to moderate-sized primates (e.g., colobine and langur monkeys), and a few very large species such as the hippopotamus. Ruminants seem to dominate the middle range in body sizes of herbivores, leading to speculation on the role of body size in the evolution of fermentation (Chapter 4).

---

[2] It is a feature of many evolutionary radiations that a variety of diverse forms appear but only a few groups survive.

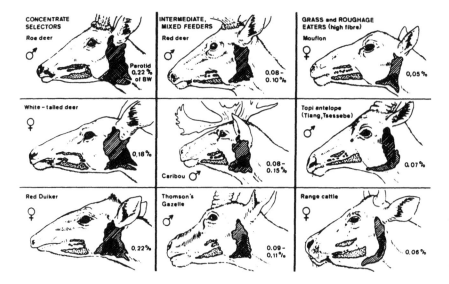

Figure 3.6. Comparative topographic representation of the salivary glands (from left to right: buccal, mandibular, and parotid) in 9 ruminant species of three feeding types, and the relative size of the parotid (from Hofmann, 1989). Values represent ranges as a percentage of body weight. All salivary glands (including the sublingual, not shown) regress with increasing adaptation to plant cell wall digestion, most distinctly the parotid, for functional reasons.

3.8 Ruminant Adaptations

Pregastric fermentation has had both positive and negative consequences. The latter include the need for gluconeogenesis and glucose sparing (Chapter 20), the conservation of essential fatty acids (Chapter 19), and the necessity of digesting protein arriving mainly in microbial form. Ruminant lysozymes differ from nonruminant types by being resistant to pepsin (D. E. Dobson et al., 1984). This allows the lysozymes in the ruminant abomasum to continue their action on bacteria even in the presence of pepsin. The browsing ruminants and some nonruminant herbivores have tannin-binding factors in their saliva (Austin et al., 1989), which temperate grazing cattle lack.

It is not known whether all enzymatic and metabolic specializations occur in all ruminants. So far, observations have been mostly limited to domesticated species.

3.8.1 Variation in Ruminant Anatomies

The spectrum of ruminant feeding behavior includes a variety of adaptations, and questions of adaptability and the anatomical and physiological aspects of these adaptations need to be examined. Selector animals dominate the smaller size range but are not restricted to that, while the true grazers, which use cellulosic matter for energy, are more likely limited to a larger size range (50+ kg). Further, many selectors feed on a wider range of plant species than do grazers, and secondary compounds (Chapter 13) are more abundant and variable in the diets of the selectors. This raises the question of how grazers are restricted from browses. Is it anatomy or some other adaptation that permits or restricts? The answer is probably both, as we shall see.

3.8.2 Salivary Glands

Hofmann (1989) noted that selector ruminants have comparatively large salivary glands. Both ruminants and nonruminants exhibit a proliferation of salivary secretion when offered tanniniferous feeds for sufficient time (Butler, 1989). Robbins et al. (1991) observed that grazer species (i.e., cattle and sheep) lack specific tannin-binding factors in their saliva. These binding factors are present in monogastric species, including humans and rats.

Salivary anatomy relative to feeding behavior in nine ruminant species is shown in Figure 3.6. The comparison probably combines both genetic differences among the species as well as the adaptive effects of the respective diets involved. Saliva production and probably size of salivary glands can be induced by dietary tannins but requires time and is seen only in species capable of the adaptation. Is this adaptive effect restricted to species that possess, for example, the salivary protection factor? This question is relevant to the position of cattle, which do not utilize tanniniferous plants and are not browsers, and which do not have the genetic capability to produce tannin-binding factors in their saliva.

3.8.3 Mouth Parts and Dentition

The evolution of different feeding behaviors has required the adaptation of mouth parts and teeth relative to the type of vegetation eaten and the digestive strategy. Janis and Ehrhardt (1988) measured the muzzle[3]

[3] Muzzle width index (MWI) is presented here as the ratio of the muzzle width to the palatal width (at the 5th molar). The original data of Janis and Erhardt are the inverse (reciprocal) of this function. Values here have been recalculated from their data in a consistent fashion and range positively with muzzle width.

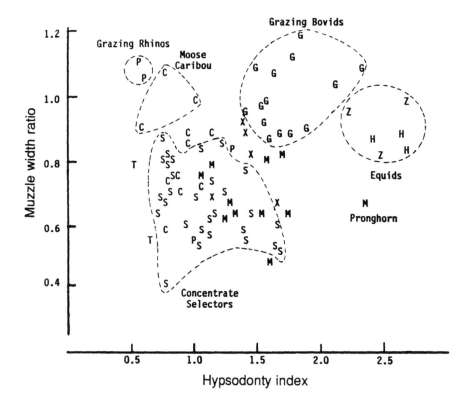

Figure 3.7. Muzzle width index versus hypsodonty index of herbivores classified according to feeding habit (modified from Janis and Ehrhardt, 1988). Equids (H, Z); grazing bovids, including antelope (G); selective grazers (X); mixed and intermediate feeders (M); caprids (C); ruminant concentrate selectors (S); rhinos (P); and tapirs (T). Classification system is modified from Janis and includes Hofmann's system; see also Janis and Fortelius, 1988.

width and the hypsodonty index[4] in 95 species of living ungulates and related these measurements to feeding behavior to examine the generally held hypothesis that grazers have wider muzzles (are less selective) and more selective species have narrower muzzles. In addition, the character of the molars in grinding and chewing is important in all grazers and ruminating types. The relationship between muzzle width index and hypsodonty index is shown in Figure 3.7. The data indicate that selector animals do indeed tend to have narrower muzzles, no doubt related to longer tongues and prehensile lips that may aid in selection.

Some larger selectors, such as eland and giraffe, have mouth parts suited to raking off and stripping branches. The giraffe tends to take the whole branch into its mouth. Hofmann claimed that they spit out the thorns, but as far as I know, no one has examined giraffe feces to verify that thorn remains are lacking.

The hypsodonty index is very high in equids, which need to chew grass on the first pass, since it is not regurgitated. The grazing ruminants, on the other hand, rechew after the ingesta have been softened in the rumen. Grazing bovids and equids have high muzzle width ratios and hypsodonty indexes, although the

equids have relatively higher hypsodonty indexes, and grazing bovids have wider muzzles. The pronghorn antelope has a high hypsodonty index with a relatively narrow muzzle, and Arctic ruminants such as moose and caribou have low hypsodonty indexes and wide muzzles. Perhaps crushing molars are not needed for the softer Arctic vegetation.

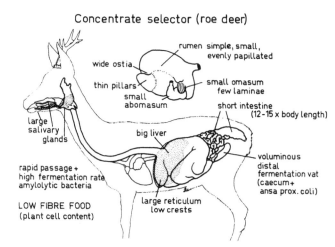

Figure 3.8. The roe deer, a concentrate selector, has morphophysiological characteristics common to all ruminants of this feeding type (from Hofmann, 1989).

[4] Hypsodonty index is calculated as the third lower molar height divided by the length of the second lower molar. It is a mean of the relative height of the molar teeth needed to masticate very fibrous matter.

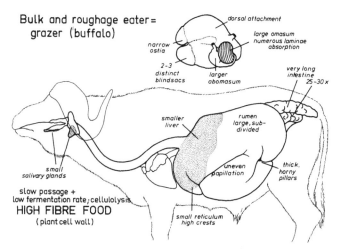

Bulk and roughage eater = grazer (buffalo)

dorsal attachment

large omasum numerous laminae absorption

narrow ostia

2-3 distinct blindsacs

larger abomasum

very long intestine 25-30 x

smaller liver

rumen large, sub-divided

small salivary glands

uneven papillation

thick, horny pillars

slow passage + low fermentation rate; cellulolysis
HIGH FIBRE FOOD
(plant cell wall)

small reticulum high crests

Figure 3.9. The African buffalo is a grass and roughage eater (from Hofmann, 1989).

3.8.4 Foregut and Hindgut

The gastrointestinal adaptations found in ruminants are so broad and overlap so much with parallel non-ruminant anatomies that Hofmann was led to remark that "the ruminant does not exist."

Within ruminants, selectors and mixed feeders tend to have larger hindguts relative to rumen size and smaller rumens relative to body size (see Chapter 4). These generalizations are shown in Figures 3.8 and 3.9, which compare a roe deer with a buffalo. Concentrate selectors have a relatively larger lower tract than grazers (see Figure 17.1).

Selectors typically have a small rumen, a less developed omasum, and a larger liver (no doubt needed for detoxification). The pillars and ruminal structure favor less selective retention and more passage of available carbohydrate and protein to a generally larger lower digestive tract (i.e., bypass).

The relative importance of pregastric fermentation differs among these ruminant species, and the lower tract is more important in selectors than in grazers. The observed anatomies relative to their function need further examination; however, the existing data very strongly suggest that bypass by selector ruminants is an important function largely lacking in grazing types.

Since most domesticated ruminants are grazers descended from nongrazers, it becomes interesting to relate the gastrointestinal capacities to feeding type and feeding strategies. It is the general practice to feed dairy cattle for the rumen, except for the emphasis on escaped protein (which may have been overvalued). Digestive efficiency is related to the rumen in animals whose lower tract digestive capacity is limited, particularly for carbohydrates like starch, which the ruminant grazer gut has not seen in 10 million years!

3.9 The Basis for Species Differences and Complementation

All animals are specialized in some way with respect to their nutritional strategies, and ruminants are no exception. While ruminants represent a particular gastrointestinal modification, there is within the group a wide range of adaptations relative to feeding strategies. The challenge is to understand the factors that promote advantages in a given situation (Figure 3.10).

Specialization is the result of evolutionary adaptation in response to a particular set of ecological circumstances. Specializations involve characteristics that may be disadvantageous outside this particular set of circumstances. For example, dairy cattle genetically bred for very high production probably could not survive in the wild. They suffer nutritive and other stresses when exported into unaccustomed environments such as the tropics, where mixed complementary animal-plant interactions have evolved. Similarly, today beef cattle can be maintained in feedlots, a management practice that was scarcely believed possible 50 years ago. This is not a natural situation for these animals, of course; they are not concentrate selectors by nature. Rumen acidosis, parakeratosis, liver abscesses, and other diet-related problems would follow in time, but animals in feedlots are slaughtered for meat and do not live long enough to experience all the problems of such a diet. The analogy of these problems to diet-related disease in humans is very evident (see Section 5.7).

The argument regarding efficiency in specialization can be carried too far, and the literature is full of claims about amazing efficiencies of overlooked species (e.g., the water buffalo, bison, goat, eland, etc.). Many of the claims relate to efficiencies of cellulolytic digestion that are not supported by rational biochemistry and kinetics. One can understand the relative abilities of animal species only on the basis of the real world of physical and biochemical limitations. There are no magic enzymes and no magic rumen bacteria that can exceed physicochemical limitations. These problems are considered in Chapter 4, but the overview and conclusions of that chapter are important for this discussion, and so I mention them below.

The constraints that food (substrate) composition places on the rate and extent of biodegradation are universal for all anaerobic systems. Moreover, the microbial population is determined first by the substrate offered and secondarily by the turnover and capacity of the fermentive compartment. The problems set by a particular diet offer various behavioral and morphological solutions. How will the animal cope with the problem of collecting the diet? Feeding behavior in conjunction with morphological adaptations of the mouth and digestive system constitute an animal's feeding strategy.

Ruminant morpho – physiological feeding types

(Co-evolution of plants and herbivores)

Plant cell   contents + dicot cell wall +monocot cell wall (fibre)

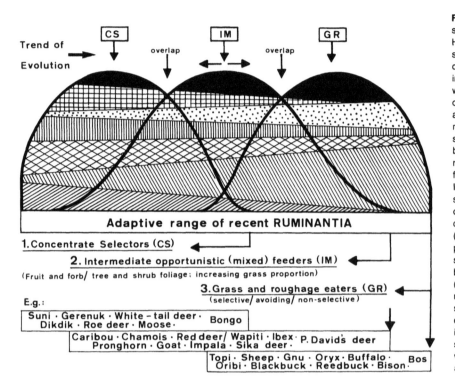

Figure 3.10. Ruminant diet specialization during evolution (from Hofmann, 1989). While the basic structural and functional design of the digestive system is retained, the increasing ability to digest plant cell wall carbohydrates (trend to the right) causes some components to regress and others to develop. Overlapping morphophysiological criteria ensure safeguards (e.g., nutritional bottlenecks) and a wide adaptive range. Separating limitations are found on both ends of the range. Intermediate species can switch seasonally from one strategy to the other. ▨, selectivity for plant cell contents; ▩, fiber digestion (cellulolysis) in rumen; ▦, HCl-producing tissue in abomasum, salivary gland tissue as proportion of body weight, and postruminal (cecocolic) fiber digestion; ▥, ruminal amylolysis, possibly bypass of solubles; ▤, effective food passage delay structures, mechanisms, total intestinal length; ▦, papillary surface enlargement on dorsal ruminal wall; ■, reticuloruminal capacity and weight relative to body weight.

Since the chemical composition and physical structure of the food set the rates of biodegradation, the digestive adaptation of the herbivores is limited to its capacity to retain ingesta for a sufficient time to extract nutrients. A low extraction rate may be counterbalanced to some extent by a higher food intake, in which case some of the matter digested at a slower rate will be lost in feces. The total nutrients utilized each day determine the plane of nutrition. Functionally, grazing ruminants are restricted by the sieving and sorting process of the reticulorumen and the omasum, which promote maximal retention and utilization of the plant cell wall substrate for available energy. Therefore, the adaptation of intake amount as a solution to feed quality problems will require regression of the selective retention function to either a tolerance of coarser particulate passage or a more efficient rumination process. This efficiency is related to body size. Other ways to reduce the dietary load of unavailable lignified residue are to feed selectively or to "open the pipe," reducing retention time and bypassing the indigestible load. This behavior is probably the major adaptation in temperate browsers and most tropical ruminants (Hofmann, 1986), while the increased intake of poorer-quality matter is more often characteristic of larger nonruminant herbivores (Foose, 1982). Washout of potentially digestible matter is a mechanism that may account for poor performance of animals with large appetites and small rumens that receive high-fiber diets.

The presence of adequate cellulolytic microorganisms depends on a supply of fermentable cellulose. The diet determines the microbes that are present in the rumen. Domesticated grazing species (cattle and sheep) are devoted to ruminal extraction, rumination, and passage, and thus depend on good-quality forage. The advantage of retentive rumination is lost if that which is retained has little potential fermentation energy.

Why do browsers have a lower digestive capacity in their rumens than do grazers? The eland, for example, has a cellulolytic capacity less than that of sheep (Arman and Hopcraft, 1975), and the same is true of deer in North America compared with sheep and cattle. Certain African ruminants cannot adapt to grass and consequently do not compete with grazing ruminants in the wild. Are browsers thus limited because of an inability to adapt to grass diets? Is the limitation set by rumination capacity? Many browsers eat some grass, but only the very immature growth.

The opposite adaptation—namely, that of grazers to concentrate diets—is clearly possible, as witnessed in the feedlot. It is doubtful, however, that cattle could adapt to a browsing strategy because of their size and mouth design (Owen-Smith, 1980). Sheep are more adaptable in this regard.

Any discussion of the relative efficiencies of herbivores must involve an understanding of their respective feeding strategies and digestive capacities. A species adapted to its environment may be more efficient in its own sphere of adaptation than an unadapted species. Efficient adaptation must involve interactions among diet characteristics, rumen organisms, and the characteristics of the reticulorumen and the omasum. Selective browsers are less efficient ruminators and have relatively smaller rumens and less-developed omasa (Hofmann, 1973), while grazers (particularly the large species) are unable to cope efficiently in situations requiring highly selective feeding, perhaps because of their size and mouth design. The intake of large grazers may be lower because they are unable to "pick" the proper forage according to the morphological parts of plants containing the higher concentrations of nutrients.

In this context a lower ability to harbor cellulolytic fermentation in the digestive tract is a sign not of lower efficiency but rather of specialization. Browsers consume the leafy and nutritious parts of woody plants that otherwise are far more lignified than any poor-quality forages encountered by grazers. Browsers seem to be adapted to faster passage of forage (Huston et al., 1986) and digest mainly the cell contents and easily digestible cell wall components. The majority of woody cell wall tissue is so lignified that no advantage is gained through any adaptation to its utilization. In contrast, grazers are adapted to grass-like plants of high fermentable cell wall content and derive considerable energy from cellulolytic fermentation. The limitations of grazer adaptations are therefore set by the availability of forage of requisite cell wall quality (Parra, 1978).

3.9.1 Energy Efficiency

A further question is whether metabolic efficiency varies among animals. Intermediary metabolism in homeotherms is quite uniform, and the view of biochemists and nutritionists is that the use of metabolizable energy (ME) is likely to be comparable for all homeotherms and is variable only in the contexts of nutrient imbalance and thermal stress. There could be significant variation in animals' ability to cope with the latter; examples include the recycling of nitrogen and gluconeogenesis in ruminants generally and, more specifically, the lowering of body temperature in deer as a response to winter cold stress.

Another basis for variation in efficiency might lie in feeding and gastrointestinal activity, factors that affect the heat increment (Webster, 1978a). Higher-fiber diets reduce efficiency through an increased heat increment that is probably a function of increased eating and rumination time. If browsers and concentrate selectors can avoid fiber—or alternatively, pass fiber as larger particles so as to avoid the work of digestion—higher efficiencies may result, although a lowered digestibility may be the penalty for the strategy of fast passage.

It is necessary to conclude that the basis for apparent interspecific differences in efficiency is gastrointestinal capacity for fermentation and the proportion of the diet that is catabolized in that fermentation. There is also the factor of animal size (discussed in Chapter 4). Generally, larger animals are more efficient in energy utilization but less efficient in food selection. Thus it can be expected that larger animals within a species (i.e., dairy cattle) will be more efficient (Blake and Custodio, 1984). The larger animal has the advantage because of its better digestive capacity (Korver, 1988).

The more time and energy that are spent on extracting energy from difficult substrates, the lower the overall efficiency due to intake limitations and the energy expended. Thus efficiency is a function of energy intake and the dilution of maintenance functions. Diet is determined by feeding behavior and forage availability, density, and morphology. Animal feeding behavior and the nature of the diet thus determines the rumen fermentation and its characteristics.

4 Body Size and the Limitations of Ruminants

Animal size has physical limits. Most very small living things are poikilothermic (e.g., bacteria and insects). Very small homeotherms have the problem of sufficient heat production to maintain their internal temperatures, and this factor presumably sets a lower limit to homeotherm size. Yet rumen bacteria are de facto homeotherms because of the regulation of their environment by the host animal. Thus the lower limit of size relative to homeothermal metabolism may be based on favorable thermal environments rather than size per se.

The opposite problem, large size, involves losing excess heat and the physical constraints of support and locomotion. The large terrestrial animals living today are probably not near the upper size limit, which may have been approached by the larger dinosaurs. The largest animal today (and larger than any dinosaur) is the blue whale, which weighs up to 150 tons.

Ruminants are neither the smallest nor the largest living animals, and they are subject to the same rules that govern all animal functions. This chapter offers perspective on the position of ruminants on the size scale and focuses on their unique digestive and feeding strategies. The overall topic of body size and allometry transcends the scope of this chapter, and the reader is referred to the literature (Calder, 1984; Schmidt-Nielsen, 1984; Martin et al., 1985; Illius and Gordon, 1987; Damuth and McFadden, 1990).

4.1 Problems of Animal Size

The potential limits to size arise because various animal functions are not directly proportional to the animal's size. Nonlinear functions tend to be geometric. Linear functions (e.g., length of limb bones; see Damuth and McFadden, 1990) tend to form a 0.33-power relationship with body mass. Heat production (metabolic rate), on the other hand, is two-dimensional and a function of surface area; thus its theoretical association with three-dimensional body mass is to the 0.66 power (although the practical value is nearer 0.75). Solid organs such as the digestive tract tend to scale

directly with body mass (Clutton-Brock and Harvey, 1983). Some organs, such as brain, heart, and liver, scale to the 0.66 power, like metabolic rate.

Functions proportional to metabolic rate are less likely to become limiting at small sizes than are functions proportional to body weight. This is particularly true with regard to functions such as digestive capacity that may be related to the difference between the respective powers of body weight and metabolic size.

Evolutionary pressure has promoted the elimination of expected limits through adaptation, so expected size limits for certain functions—grazing behavior, for example—must not be taken in a fixed literal sense. Suggested limits can be used to develop models, however, and the position of extant species can be discussed and explained in this context. The objective of such modeling is to achieve better understanding of specialized animals.

4.1.1 Energy Metabolism and Size

The conventional expression of animal requirements assumes that energy requirements are proportional to body weight raised to the 0.75 power, assuming heat loss from the body surface to be the principal limiting function. Thonney et al. (1976) collected and examined data regarding body size and metabolic heat production (Table 4.1). While body weight to the 0.75 power can be used to describe interspecific relationships, it may not account for differences in weight within species and sex categories, most of which exhibit power relationships significantly less than that. Thus body weight to the 0.75 power should not be used for intraspecific comparisons, although this is the common practice for expressing energy requirement and intake data. For more accurate and less presumptive treatment of data, the effect of weight on animal requirements should be accommodated through the use of a linear model that includes the logarithm of weight as a covariate.

The analysis of Thonney et al. (1976) suggests that log slopes are lower within species but show deviations associated with gender. Physical limitations set by

Table 4.1. Regressions obtained from plotting log heat production versus log body weight

Animal	Intercept	Power slope	Data source
Sheep (ewes)	2.04	0.64	Lines and Pierce, 1931
	1.86	0.71	Marston, 1948
	2.14	0.61	Pierce, 1934
Beef cattle	2.88	0.41	Benedict and Ritzman, 1927
(steers)	2.20	0.62	Mitchell and Hamilton, 1941
	2.38	0.57	Forbes et al., 1931
Dairy cows	2.25	0.61	Ritzman and Benedict, 1938
	1.76	0.80	Hashizume et al., 1962
	2.95	0.33	Forbes et al., 1926, 1927, 1928, 1930, 1931
	1.80	0.78	Flatt and Coppock, 1963
All species	1.82	0.76	—

Sources: Thonney et al., 1976, and sources therein.

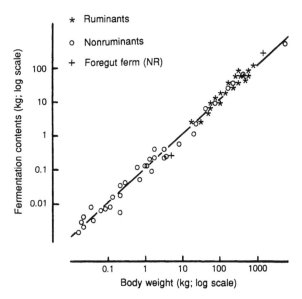

Figure 4.1. Wet fermentation contents plotted against log of body weight for African herbivores (from Demment and Van Soest, 1983; see also Demment and Van Soest, 1985). Nonruminants in the figure are hindgut fermenters. Foregut fermenters also do not ruminate. The regression equation for all herbivores is log Y = 1.032 log X − 0.936; r = 0.99, n = 59. The ruminant regression slope is 1.04, and slopes for ruminants and nonruminants are not significantly different. Regressions of this sort tend to overlook disparities characteristic of individual species (see Figure 4.3).

principles of physics and chemistry are vital in understanding biological functions but are not often used in the proper context. If a biologist rejects a generalization, he or she must seek to explain deviant biological phenomena in terms of relevant physical and chemical laws. Still, universality must not be allowed to override observations of real biological differences.

Analyses of the relation of heat production to body size indicate that the intraspecific power slopes are not often greater than 0.75, and are, in many instances, substantially less. Any power slope less than 1.0 means that small animals generally require more feed per unit of body weight for maintenance and general function than larger animals, and this effect is greater when young (smaller) animals are compared with larger adults within a species. The problem is relevant to ruminants since gastrointestinal size and capacity are limiting to the intake and utilization of forage diets. Therefore, one expects smaller animals to have a lower capacity to digest poor-quality forage than larger animals; and furthermore, there may be body size constraints on proper rumen function.

4.1.2 Gastrointestinal Capacity

Measurements of gastrointestinal capacity yield different results depending on what is measured and the method used. There are several kinds of measurements, volume or surface measurements of various organs and weight of organs versus volume among them. Surface measurements tend to be proportional to metabolic size (i.e., $kg^{0.75}$; see R. D. Martin et al., 1985). Indeed, small selector ruminants have greater papillar surface per unit area of rumen than larger ones (Hofmann, 1973), a feature possibly induced by the high rates of volatile fatty acid (VFA) production in these species (Section 4.4.4). The limitation of turnover and retention is the single most important factor setting limits to digestive capacity (see Chapter 23). For this reason

gastrointestinal volume, rumen volume, and measured rumen contents become the objects of interest.

The available data relating gastrointestinal contents to body weight were summarized by Demment and Van Soest (1985) and are shown in Figure 4.1. Values for ruminants and nonruminants fall on the same regression line. Nonruminants are classified as hindgut fermenters or foregut fermenters that do not ruminate. All are similar in gastrointestinal capacity.

Volume of gastrointestinal organs can be calculated by filling the organ with water after slaughter and fixing it in formalin to prevent stretching and then measuring the volume or weight of liquid; alternatively, the gut contents may be measured, usually by weighing the organ as excised and reweighing after washing out the contents. In the latter instance organ weight is usually recorded, but regrettably, the composition of the ingesta is rarely measured. A variable diet may induce variation in measurement.

Volume measurements provide consistently larger estimates than do ingesta weights because no allowance is made for gas space in the latter. On the other hand, measuring ingesta may not allow for variations due to time since the last meal, intake level, or dietary characteristics, all of which may vary considerably when interspecific comparisons are made. Both kinds of measurement are affected by individual animals' peculiarities. This is particularly a problem with many of the values for wild animals, for which data are often based on one or only a few individuals. Ingesta mea-

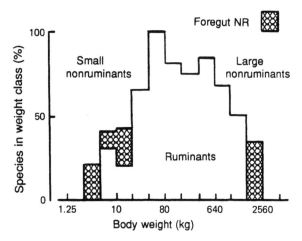

Figure 4.2. East African herbivores (186 species) that are ruminants, nonruminants (NR), or foregut fermenters (hatched histogram) are plotted in relation to body weight (from Demment and Van Soest, 1983; see also Demment and Van Soest, 1985). The weight scale is logarithmic; each bar represents a doubling in weight. The ruminant species dominate the middle body size range, while the nonruminants are primarily small or very large animals.

surements may be more prone to the influence of dietary variation in these cases. Although the two kinds of measurement cannot be combined unless interconverted by some method, either will disclose the relationship between gastrointestinal capacity and body size.

4.2 Sizes of Ruminants

Ruminants range in size from about 3 to 1000 kg, a range substantially narrower than that exhibited by mammalian herbivores generally (Figure 4.2). Ruminants as a group and their limitations relative to nonruminants must still be considered in relation to the problems of size and their respective feeding strategies. Small animals are better able to select food but lack digestion capacity. Larger animals may have better digestive capacity but are more limited in their selection abilities. As we shall see, there are exceptions to these generalizations. The problem with feeding behavior classifications is their linearity (see Chapter 3), the true radiation of evolutionary feeding behavior has many dimensions.

4.3 Modeling Limitations of Size

If it is accepted that the energy requirements of homeothermic animals are related to their metabolic size, a relationship generally assumed to be proportional to approximately the 0.75 power of body weight, a mathematical comparison of gastrointestinal capacities among species can be made. The relationship is described by the Brody equation (Kleiber, 1975):

Table 4.2. Regression values obtained by plotting log reticulorumen capacity (F and omasal capacity of African ruminants versus log body weight

Category	N	Body weights (kg)	RR power slope	Omasu power sl
Concentrate selectors	12	4–750	0.96	1.01
Intermediate feeders	5	10–63	0.83	0.98
Grazers	11	16–751	0.87	0.93
Total	28	4–751	0.94	0.99

Source: Calculated from data in Hofmann, 1973.

$$M = 70Wt^{0.75} \qquad (4.1)$$

where M is the fasting metabolism in kilocalories per day and wt is body weight in kilograms. The actual maintenance cost is roughly twice this factor.

The consequence of the 0.75-power relationship is that small homeothermic animals have a higher maintenance cost per unit of body weight. It is important to consider the relation of gastrointestinal capacity to body size. To maintain equal status, gastrointestinal capacity must be in proportion to metabolic size, and therefore proportionally greater in smaller animals. If not, nutritional status in small animals must be balanced by faster digestion and passage or a more concentrated diet. The relation of reticulorumen and omasal capacity to body size (Table 4.2) also varies with body weight. This implies that smaller ruminants are limited by ruminal and omasal capacity. Moreover, nonruminant herbivores fit the same pattern as do ruminants (Parra, 1978). Thus, the reticulorumen is replaced by a larger cecum or greater lower bowel volume in grazing nonruminants (see Table 5.3). The consequence of this relationship may set limits on the size for both ruminant and nonruminant herbivores. The linearity and high correlations of logarithmic plots of organ size versus body weight tend to overlook the obvious specialization in individual species (see Figure 4.3), which appear as deviations from the regression line in Figure 4.1.

If we overlook, for the moment, deviations in gastrointestinal size for the sake of explaining the general relationships and problems of animal size, mathematical modeling can be applied with interesting results. Assuming that metabolic rate determines the nutritional requirement and that gut size limits capacity to process food into nutrients, the nonlinear response of gut size produces higher rates of metabolism in small animals relative to large ones (Demment and Van Soest, 1985). Assuming that the gut is a space in which the ingesta mass flows like a liquid, the problem can be examined using single-pool kinetics (see Chapter 23). The time (T) a particle spends in the gastrointestinal mass (Q) can be expressed as T = Q/F, where F is the intake per unit of time and Q is the mass of ingesta as kilograms of dry matter. This equation ignores selec-

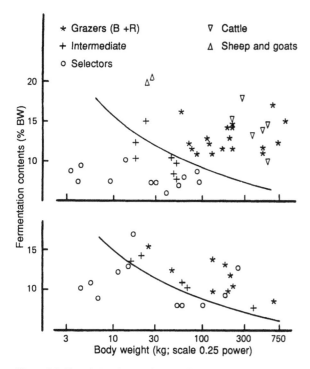

Figure 4.3. The relation of rumen fermentation contents (percentage of body weight) to body size in ruminants. The upper figure summarizes literature values of fermentation contents obtained by direct measurement (most of the values are from Hoppe, 1977; and Parra, 1978). The lower figure plots rumen volume measured by immersion volume (Hofmann, 1973). These data have been converted to estimated fermentation contents using a factor of 0.8. Fermentation contents are inherently more variable, but they show the same trends as the lower figure. The data show that concentrate selectors have small rumens and grazers and some intermediate feeders (notably sheep and goats) have comparatively larger rumens. Few grazers weigh less than 70 kg. The curved line in the figures is the limit to grazing activity and body size calculated with equation 4.3. A retention time of 17 h at 70% digestibility and 15% dry matter in fermentation contents is assumed. (The calculations in this figure, present in the first edition of this book, have been misinterpreted. Two erroneous inferences have been made: [1] there is a fixed limit for ruminants [and grazers], and [2] equation 4.5 and this figure are at variance with physiological intake models. To the first point the reader is referred to Sections 4.1 and 4.4.4; to the second, to Chapter 21. It is assumed in the calculations for this figure that the animal will consume feed to satiety, but retention time and digestive capacity become constraints to feasible solutions of equation 4.5. The calculations show that gastrointestinal expansion can account for some adjustment in digestive capacity in marginal grazers. It is significant that goats and sheep have the largest gastrointestinal contents.)

tive retention, however; an alternative is to consider it an expression of the cell wall volume and turnover. Cell wall content is the dominant dietary characteristic limiting intake (Section 21.4).

The relationship of fermentation contents (Q) to body weight is near unity, the power slope being 1.032 (Figure 4.1). (Hofmann [1973] obtained 0.94.) A significant factor contributing to variation in the slope is the distribution of browsers and small grazers relative to rumen size (Figure 4.3). Theoretical mathematical considerations predict a power slope of 1.0. The rela-

tionship between weight of dry ingesta (Q) and body weight (Wt) is:

$$Q = C_{dm}HWt \qquad (4.2)$$

where C_{dm} is dry matter content and H is the ratio of wet content to body weight.

The amount of dry matter intake (F) required to supply maintenance costs is obtained by dividing the turnover equation (T = Q/F) by digestibility (D) since M = FD. It is assumed that maintenance cost is twice the fasting metabolism (70) and that there are 4400 kcal/kg of dry matter. Substituting these equations and solving for intake yields:

$$F = \frac{140Wt^{0.75}}{4400D} \qquad (4.3)$$

$$T = \frac{Q}{F} = \frac{4400DC_{dm}HWt}{140Wt^{0.75}} \qquad (4.4)$$

$$T = 31.4DC_{dm}FHWt^{0.25} \qquad (4.5)$$

Turnover is the reciprocal of the total disappearance rate of food residues in the digestive tract (Chapter 23).

Equation 4.5 is pertinent to the discussion because it contains the variables that describe the evolutionary adaptation of gastrointestinal characteristics to diet. Turnover can be increased by speeding up either passage or digestion rate. Faster passage will result in loss of potential digestible matter, and increasing the digestion rate will require a high-quality diet. These aspects represent contrasting strategies that allow animal species of different sizes to adapt to their dietary situations. Increasing either fermentation content (H) or dry matter content (C_{dm}) effectively increases gastrointestinal capacity. Increase in the dry matter content works against the limits of osmotic pressure that cellulolytic bacteria can tolerate. The effect of body weight to the 0.25 power is to favor the larger animal with longer turnover and more efficient digestion and suggests a lower body weight limit for ruminants if extraction of energy from cellulose is the objective.

The advantage for larger animals is related to the difference between power slopes for gastrointestinal capacity and energy requirements (R. H. Peters, 1986). Thus, increase to a higher power for the relation of fermentation contents to body weight (i.e., >1) will increase the differential power in equation 4.5. Also, if the power slope of heat production to body weight is less than 0.75, the difference between the powers is increased. In either case, increasing the differential power in the equation will increase the disparity resulting from animal size and place greater limitations on smaller animals. This factor may be relevant in comparing young and older animals or males and females within a species, where a factor of body weight to the 0.25 power is probably a minimum value.

Illius and Gordon (1991) modeled the relationship between retention time of feed and body weight and arrived at a power function of 0.27. The power function of the time taken to comminute coarse fiber particles also scaled to a power of 0.27. They have applied their model to account for the behavioral and digestive differences between sheep and cattle.

These results have important implications for adaptive nutritional strategies as discussed in Chapter 3. The question is not how large or small an animal following a given feeding strategy can be, but rather, what are the adaptive options available to smaller herbivores that will help them avoid the constraints? A smaller animal may opt to feed selectively on more nutritious plant parts. Small ruminants and nonruminants alike do this. Alternatively, herbivores may eat for bulk and volume with fast passage and very low digestibility. The lesser and greater pandas do this to an extreme, but no ruminants fall into this class. Or, a bulk and roughage eater may retain food for maximal digestion at the expense of net intake. This group includes the grazing ruminants, which are able to digest fibrous grasses to a higher degree than most nonruminants of their size. Because of the energy spent on selective retention and rumination, these ruminants are more sensitive to low forage quality than the equids, elephants, and pandas of the bulk and roughage class.

Optimum digestion of cellulosic carbohydrates by ruminants depends on selective retention in the rumen. Failure of this selective retention will promote fecal loss of potentially digestible fiber in all animals, and proportionately greater losses in the smaller ones. With selective retention, potential loss of digestible nutrients is reduced, but at the cost of increased fill. The fill limitation is more severe in the case of lower-quality grasses. This presents a problem for small ruminants that have high energy and intake requirements relative to their gastrointestinal capacity.

4.3.1 Feeding Behavior and Size

The range of body weights in which the limitation of size appears to have its greatest effect on retention is less than 100 kg. A weight of about 40 kg represents the size below which unselective feeding ceases to be a viable strategy because the intake will be too great to allow sufficient digestion at the required faster rate of passage. Most, but not all, ruminants below this size are concentrate selectors (Figure 4.3). A number of factors—including gastrointestinal size, rumen fill rumination, tolerance to a larger particulate passage, digestive capacity, absorptive gut surface, feeding strategy, available forage quality, and gastrointestinal capacity—affect feeding strategy.

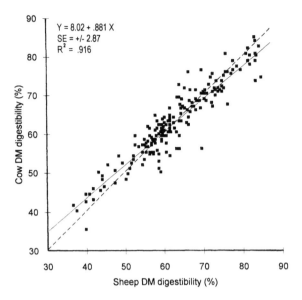

Figure 4.4. Correlation between digestion capabilities of cattle and sheep fed the same diets (from Mertens and Ely, 1982). The solid line is the regression line, the dashed line is the unitary equivalence. The slope of the regression differs significantly from unity. For feeds >66% digestible, sheep tend to have higher digestibility, probably reflecting their lower metabolic losses (Section 22.3.1). For feeds <66% digestible, cattle tend to have higher digestibility than sheep.

These variables can allow considerable latitude in size. For example, the oribi (16 kg), which is probably a selective grazer, is apparently the smallest grazer. The black Bedouin goat (30 kg) uses its rumen (like the camel) as a water reservoir. Infrequent drinking slows rate of passage and increases extent of digestion, allowing desert goats to compete with larger animals in digestive efficiency and dietary tolerance (Brosh et al., 1986).

Rumen turnover is a function of feed intake divided into net weight of rumen dry matter. Digestibility is limited by retention (turnover) time, so the constraint of 70% digestibility, 17 h retention, and 15% rumen dry matter (Figure 4.3) allows ruminal capacity to become the main free variable. It is assumed that intake will be driven by energy requirements. Small selector ruminants have relatively greater surface area for absorption to allow for the fermentation of higher-quality diets (Hofmann, 1973). Gastrointestinal volume is assumed to be more limiting than surface area of the digestive tract.

Selector ruminants may not have as great a need for a large rumen fill as grazers. This hypothesis is illustrated in Figure 4.3, which shows that grazer species tend toward a larger proportion of fermentation contents and, on the whole, remain above the line of the fill calculated for a 70% digestibility grass diet at a maintenance level. This observation confirms the speculation that grazers are limited in size by the size of their rumens and by diet quality.

The fermentation contents also vary within individuals and species depending on appetite and special physiological demands for energy (i.e., pregnancy, lactation, and growth). Fermentation contents values for cattle range from 10 to 17% of body weight, and it appears from Figure 4.3 that an upper limit of "stretch" is about 20% of body weight.

In general, almost all small ruminants are selector feeders and all bulk and roughage eaters are large ruminants. Hofmann noted the exception of the oribi, a 16-kg grazer (which is likely a highly selective grazer, however). On the other hand, there are also selectors in the larger body size range dominated by bulk and roughage eaters. The eland, for example, seems to be a specialist on browses and must cover a large area to meet its dietary needs. Large size and a small rumen are required for such behavior. The same argument may be made for the giraffe, a specialist in tree pruning. The length of the neck may be a factor limiting rumination capacity.

4.4 Digestive Capacity

Digestion trials were conducted with various species to compare coefficients of digestion, to demonstrate the superiority of one species over another, and (particularly in the case of sheep versus cattle) to demonstrate equal digestive ability among species so that sheep data could be substituted for cattle digestion coefficients.

There can be no question that the objectives of some digestion comparisons have biased the conclusions. The best documented case is a study by Swift and Bratzler (1959), who concluded that digestion values for sheep and cattle were identical. (Sheep digestion studies are more economical to perform than cattle studies.) Analysis of a larger data base, however, produced the equation shown in Figure 4.4. These data show that sheep may have higher digestion coefficients than cattle at high digestibilities but are inferior at lower digestibilities. The crossover point is at 66% digestibility.

Apparent digestibilities are calculated as the loss in dry matter between mouth and anus, but the feces contain both undigested feed and endogenous and metabolic products that are not a part of the feed. Mathematically, apparent digestibility equals true digestibility minus a metabolic endogenous component (see Chapter 22). Thus the difference in digestive capacity between cattle and sheep can be seen to involve two different biological factors: lower metabolic losses in sheep favoring higher digestibilities on better-

quality diets, and poorer ability to digest fiber, promoting lower digestibilities for sheep fed on lower-quality diets. Thus equal apparent digestibility in two species does not necessarily indicate identical metabolic and digestive capacities.

4.4.1 Comparative Digestion Studies

A number of comparative digestion trials have been conducted, the largest in scope being those of Foose (1982) and Prins et al. (1983). Other investigators have compared fewer species (see Table 4.3). The order of decreasing abilities of cattle, sheep, goats, and deer to digest fiber is the opposite of their respective abilities for selective feeding (Huston et al., 1986). Similar ranking probably relates African ruminants (Arman and Hopcraft, 1975) as well as native North American species (Kautz and Van Dyne, 1978). The more selective feeders tend to utilize fibrous carbohydrates less well. The selectors tend to utilize concentrate feeds better.

Smaller animals in any class are usually less able to digest forage. Concentrate selectors and browsers, regardless of size, seem less well able to utilize cellulose. Hofmann (1989) noted that cellulolytic activity in the rumens of selectors is lower than in other feeders. This also fits the pattern of smaller rumens, and therefore poorer retention, in these species (Figure 4.5).

A comparison of all these data involves the problem of varying diet quality. Prins et al. (1983) resolved this by measuring maximal extent of digestion and comparing species' ability to utilize the available part of the diet. The ranking order of animals according to protein digestibility is very often different from that of dry matter or fiber. This probably reflects differences in metabolic losses, but it might reflect smaller animals' relatively larger proportion of postruminal tract compared with reticulorumen (Hofmann, 1988).

4.4.2 Limitations of Digestion Trials

Comparative digestion trials such as those presented in Figure 4.4 involve feeding all the subjects identical diets. For grazing species such as cattle and sheep the disparity of the diet offered relative to the dietary preference is not large. As the disparity in foods eaten widens, however, choice of a common diet for comparison becomes somewhat problematic.

Foose (1982) presented comparative digestion data on 36 species of ruminants and nonruminants of widely differing feeding strategies. These data plus those from other sources are included in Figure 4.5. They show that extent of digestion of cellulose, the most slowly digested carbohydrate, is highly related to retention

Table 4.3. Literature comparing digestion and feeding behavior in ruminant species

Species compared	Parameter studied	Reference
Bos indicus > *B. taurus*	DM digestibility	Hunter and Siebert, 1985
Cattle > sheep	Fiber digestibility	Aerts et al., 1984
Sheep > cattle	Protein digestibility	Aerts et al., 1984
Cattle > sheep	DM digestibility	Playne, 1978
Cattle > sheep > horse	DM digestibility	Chenost and Martin-Rosset, 1985
B. idicus = *B. taurus* = sheep = bush duiker > hartebeest > Thomson's gazelle > eland	Fiber digestibility	Arman and Hopcraft, 1975
B. indicus > *B. taurus* = sheep > bush duiker = hartebeest = eland > Thomson's gazelle	Protein digestibility	Arman and Hopcraft, 1975
Cattle > sheep > goats > deer	DM digestibility	Huston et al., 1986
Deer > pronghorn > cattle = sheep > bison	Selectivity	Kautz and Van Dyne, 1978
Water buffalo > cattle	Fiber, DM digestibility	Sebastian et al., 1970
Carabao > cattle	Fiber, DM digestibility	W. L. Johnson, 1966
Carabao > cattle	Fiber, DM digestibility	Grant et al., 1974
Goats > sheep	DM digestibility	Masson et al., 1986
Goats = sheep	DM digestibility	Antoniou and Hadjipanayiotou, 1985
Sheep > red deer	Fiber, DM digestibility	J. A. Milne et al., 1978
Sheep > red deer	Fiber, DM digestibility	Maloiy and Kay, 1971

Note: DM = dry matter.

time and that this association holds true for ruminants and nonruminants alike.

Two standard diets were offered in Foose's study: timothy grass and alfalfa hays. The timothy diet offered the more stringent test for cellulose digestive capacity, but selector types did not eat enough of this forage to achieve maintenance (Figure 4.6). Higher intakes would have promoted faster passage rates and lower digestibilities. Retention is closely associated with digestibility (Figure 4.5), hence mere digestion coefficients viewed apart from other data may overvalue the capacity of smaller animals and selectors.

Another problem in dietary studies of selector feeders is that they may leave orts of lower digestibility and higher lignification, so that the diet actually eaten is better than that offered. This may explain some of the anomalous data in the literature comparing goats with sheep and cattle.

Results of studies of the digestive capacity of goats can be classified into two categories: one claiming that goats have a higher digestive capacity than sheep or cattle (Devendra, 1978); the other showing lower or equal digestive capacity of goats compared with sheep and cattle. Most of the trials claiming a higher digestive capacity for goats do not include adequate descriptions of the control of intake and the analysis of orts. Further, most have been conducted with tropical forages that offer an opportunity for selection because of the wide nutritive differentiation of these grasses (see Chapter 6). Thus the claim that goats digest lignin, for example, reflects overestimation of the lignin intake because the feed actually eaten was lower in lignin than the unanalyzed refusal.

Apparent digestibility balance does not determine whether the species difference seen in Figure 4.5 occurs because of metabolic losses or because of an inherent ability to digest fiber (cell wall). The results can vary depending on the quality of the diet. The sheep data (Figures 4.4 and 4.5) suggest that smaller animals are at a disadvantage in digesting fiber, even more so with lower-quality diets. The data offer the possibility that digestibilities by cattle and sheep are similar in a certain optimum size range and that the disadvantage of small size becomes apparent only at lower digestibilities. Most concentrate selectors have smaller metabolic losses than grazing species.

Studies comparing ruminant species other than sheep, goats, and cattle are much more limited in scope, often relying on one or a few diets. In sorting through this kind of data, it is important to examine digestion trial procedures, especially in the case of tropical forages, which have nutritively disparate parts that offer animals an opportunity to select. In the proper circumstances, selector species may exhibit higher digestibility than less selective species (e.g., cattle).

Olubajo et al. (1974) were unable to obtain complete consumption of tropical grasses by African sheep. Refusals were on the order of 60% of the forage offered, and the composition of the forage did not correlate with its in vivo digestibility. In such cases analyses of orts are obligatory for satisfactory results but unfortunately are not often done or reported. It is unsatisfactory to merely subtract the weight of the feed not eaten from that offered. The tendency in many digestion trials is to restrict intake to the point at which the animal cleans the manger. This procedure was not possible in Tes-

Table 4.4. Digestion capabilities of zoo ruminants

Species	R/Cª	NDFᵇ (%)	ANDFᶜ (%)	
Temperate grazers and mixed feeders				
American bison	0.6	63.5	85.6 ± 2.4	(3)
American bison	1.7	60.6	90.4	
American bison	3.5	59.7	92.0 ± 4.1	(3)
American bison	1.7	62.9	86.3 ± 2.9	(5)
European bison	3.0	63.0	86.0 ± 0.9	(6)
European bison	2.2	52.2	82.7 ± 6.1	(5)
European bison	2.7	56.4	86.6 ± 5.3	(3)
European bison	1.5	46.0	77.4 ± 0.9	(2)
Dwarf goat	0.7	53.6	75.4	
Tedal sheep	4.1	44.6	69.2	
European mouflon	6.4	45.7	69.0	
Mean			81.9 + 8.0	
Tropical mixed feeders				
Springbok	0.9	52.0	73.4 ± 9.4	(2)
Springbok	2.3	37.9	62.7	
Temperate mixed feeders				
Fallow deer	1.1	59.1	81.6	
Fallow deer	0.6	59.0	80.3	
Fallow deer	0.6	53.4	83.1	
Wapiti	2.5	57.9	79.2 ± 1.5	(6)
Père David's deer	1.0	53.0	74.5	
Red deer	0.3	41.3	56.1	
Mean			75.8 + 10.0	
Arctic ruminants				
Reindeer	4.4	68.3	88.4	
Reindeer	1.8	53.7	83.7	
Moose	0.4	41.8	78.3	
Moose	1.9	48.7	86.5 ± 3.4	(2)
Mean			84.2 + 4.4	
African concentrate selectors (except as noted)				
Giraffe	1.9	49.4	83.4 ± 2.2	(6)
Giraffe	1.3	34.8	66.7 ± 8.5	(6)
Giraffe	0.5	59.0	80.3	
Giraffe	0.8	61.7	87.9	
Pudu (S. Am.)	2.4	33.4	54.8 ± 13.5	(2)
Greater kudu	1.7	14.7	34.6 + 0.4	(2)
African grazers (except as noted)				
African buffalo	5.3	57.6	88.6 ± 2.2	(3)
Watusi	9.3	62.2	85.7 ± 0.2	(2)
Watusi	11.2	65.5	87.9	
Banteng (Asia)	6.3	62.2	84.3 ± 0.9	(4)
Blesbok	0.3	61.6	84.0	
Roan antelope	1.6	57.7	80.2 ± 8.3	(4)
Oryx	3.8	48.8	80.6	
Gayal (Asia)	2.9	50.8	81.4 ± 3.5	(4)
Gayal (Asia)	0.8	50.5	78.6 ± 2.4	(3)
Waterbuck	2.6	38.0	70.8 ± 0.7	(2)
Waterbuck	1.1	58.5	81.1 ± 2.6	(4)
Hartebeest	4.1	38.6	63.1	
Mean			82.1 ± 4.9	

Source: Prins et al., 1983.
ªRoughage-concentrate ratio.
ᵇObserved NDF (cell wall) digestibility in vivo.
ᶜANDF = available NDF, obtained by dividing animal digestible NDF by the residue after long-time in vitro fermentation with rumen microorganisms. Number of animals studied is given in parentheses.

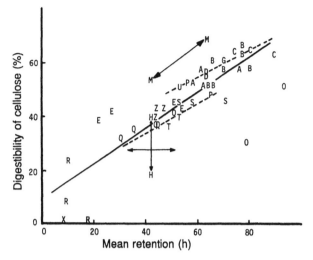

Figure 4.5. The relationship between cellulose digestion and mean retention in 46 species of herbivorous mammals fed grass-type fiber. Ruminants such as antelopes (A), grazing bovids (B), camelids (C), deer (D), giraffes (G), sheep, and goats (S) are better able to digest fiber than nonruminants, but smaller species of both groups are at a disadvantage (Van Soest et al., 1982). The ranges of values for humans (H) (Van Soest et al., 1978) and baboons (M) (Milton and Demment, 1988) are somewhat less than the figure for pigs (U) (Ehle et al., 1982) but more than that for rodents (R). Large animals such as hippos (O) and rhinos (P) may approach the capacity of ruminants, but elephants (E), equids (Q, Z), and tapirs (T) are less efficient (Foose, 1982; Hackenberger, 1987). Very small animals (rodents and lagomorphs [R]) have very low digestion capabilities (Udén et al., 1982). The panda (X) has the lowest of all (Dierenfeld et al., 1982). The overall correlation between retention and digestion is 0.86; the individual correlation for ruminants is 0.69; that for nonruminants is 0.78.

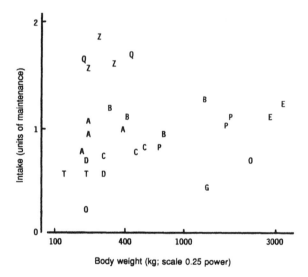

Figure 4.6. Intake of poor grass hay (intake of digestible dry matter divided by the estimate for maintenance) relative to body weight. Note the high intake by equids (Q, Z). Some selector ruminants could not achieve maintenance on this diet. (Data from Foose, 1982; see Figure 4.5 for identity of species.)

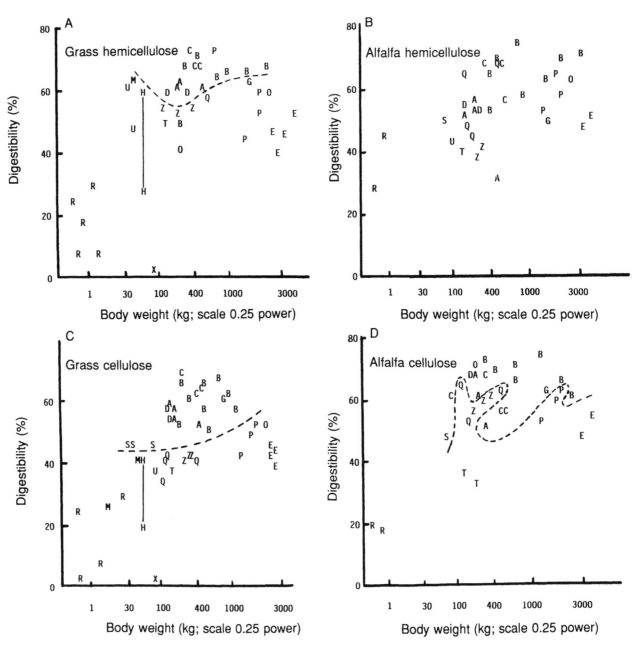

Figure 4.7. The relation between body weight and ability to digest hemicellulose and cellulose from alfalfa and grasses in diverse species of ruminants and nonruminants. (Data from Keys et al., 1969, 1970; Ehle et al., 1982; Foose, 1982; and Udén and Van Soest, 1982a; see Figure 4.5 for identity of species.)

sema's study (1972), in which sheep were fed low-quality guineagrass hay. The sheep could not be starved to the point where they would eat all of the coarse stems; however, complete consumption was achieved with cattle.

Selection leads to the consumption of the more nutritious portion, leaving orts of lower quality (Gutierrez-Vargas et al., 1978). Failure to correct for the composition of the refusal and to calculate the composition of the true intake can lead to a variety of possible anomalies, including increased apparent digestibility of dry

matter and fiber, better nitrogen balance, and faster in vitro fermentation rate (El Hag, 1976). If the intake of poorly digestible fractions is overestimated through failure to compensate for selection, digestibility will also be overestimated.

4.4.3 Body Size and Quality of Fiber

Foose (1982) calculated digestibilities of grass and alfalfa and rate of passage for 36 species of ruminants and nonruminants. To this information can be added the data of Keys et al. (1969), Udén and Van Soest (1982a), and Ehle et al. (1982). The results are particularly interesting in regard to use of cellulose and hemicellulose from grass and legumes (Figure 4.7). The data indicate that the slower-digesting cellulose from grass is limiting in all species, and that better-quality fiber from alfalfa is utilized by most species that weigh more than about 90–100 kg.

Retention time (the mean time feed residues remain in the digestive tract), although associated with body size in both ruminants and nonruminants, shows that small ruminants achieve more fiber digestion than non-ruminants of similar size. This is most apparent in the slower-digesting grass cellulose (Figure 4.7C) and less apparent in the digestibility of grass hemicellulose and alfalfa cellulose, which are generally digested faster. These carbohydrates seem to be more limiting to animals below 100 kg, arguing for a critical size cutoff for general use of structural carbohydrates (Demment and Van Soest, 1985). Alfalfa hemicellulose is likely an even more digestible fraction. Its quality is such that ruminants and nonruminants are not distinguishable.

The apparently inferior ability of smaller and selector animals to digest cellulosic carbohydrates might be related to inability to retain them, or perhaps to inability of cellulolytic microbes to adapt to their gut environments. Generally the survivability of a gut microorganism depends on its generation time, which must be less than gut turnover time in the fermentive chamber (see Chapter 16). Limited retention in small animals requires a higher relative intake, though the strategy of consuming a more lignified diet to obtain highly digestible cellular contents precludes exhaustive cellulose digestion. A final hypothesis is that there is a level of total diet digestibility that limits animal function because of the requirement for digestible energy, and that this point occurs at higher digestibilities for small animals and most selectors.

4.4.4 Small Ruminants

Volatile fatty acid data show higher rates of production in small concentrate selector species relative to larger ruminants (Figure 4.8). Although they may have higher rates of fermentation, the smaller selector species do not achieve as high a proportion of their requirements from VFAs as larger species do. Cecal production of VFAs is likely higher in nonruminants without pregastric fermentation, which tends to remove the more rapidly fermentable carbohydrates. The interpretation that fermentation characteristics are mostly representative of feeding behavior and type of diet assumes that rumen bacteria are mainly responsive to dietary input and that gastrointestinal architecture and ecology are less directly responsible for microbial selection and adaptation. This is consistent with the

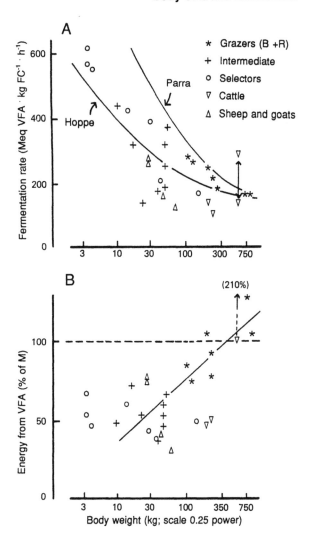

Figure 4.8. (A) The relationship between fermentation rate and body weight (summarized from Hungate et al., 1959, 1961; Allo et al., 1973; Giesecke and Van Gylswyck, 1975; Hoppe, 1977; and Parra, 1978; FC = fermentation contents). The two curves represent the VFA production required to support maintenance, assuming average gut size and dry matter content (Parra, 1978), and the regression of measured fermentation rates on body weights obtained by Hoppe (1977), who measured about 100 different animals of various species. Lower figure shows the percentage of maintenance (140 $W^{0.75}$) provided by estimated daily calories from VFA in relation to body weight. The divergence of Parra's line for maintenance from Hoppe's observed regression suggests that small ruminants may compensate by deriving energy from sources other than VFAs; namely, rumen escape, more efficient microbial yield through faster rumen turnover, or both. **(B)** The percentage of dietary energy from VFAs versus body weight (calculated from the data in A as the percentage of calories from VFAs that contribute to maintenance). The regression line is for grazers (bulk and roughage eaters) and intermediate feeders (r = 0.86). Selectors do not show any consistent trend, although their VFA yields are well below maintenance. This holds even for the small selectors such as the suni and dik-dik, which have very high fermentation rates **(A)**. Another possibility is that some of these species have low metabolic rates. Ranges in values for dairy cattle (Hungate et al., 1961) are indicated by arrows.

probability that high rates of VFA production reflect equivalent high quality of the selected diet.

Hofmann has suggested that the smaller selector ru-

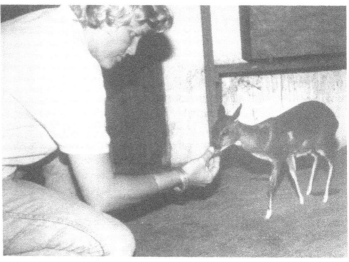

Figure 4.9. The royal antelope (*Neotragus pygmaeus*), the smallest ruminant. (Left) An ensnared individual with the hunter's hand on its side. (Right) A captive female with its keeper. Royal antelopes live in forests in central Africa. (Photos taken in Zaire; courtesy of Pamela Mueller, Cornell University.)

minants selectively bypass concentrate food items such as grains or fruits to compensate for the lower energy supply from VFAs. This behavior would allow small ruminants to avoid the energy losses involved in microbial fermentation, as well as losses of methane, some heat, and indigestible parts of microbes. Bypass would place these animals metabolically somewhere between nonruminants and true functioning ruminants. The smaller browsers are intolerant of higher-fiber diets, and forcing them to consume such feed induces omasal impaction, a suggestion of more limited ability to ruminate (Hofmann, 1973). Small ruminants compensate for their lesser digestive capacity by avoiding low-quality forage and browse.

Why are such small animals, which do not make use of the strategy of cellulose utilization, ruminants at all? One reason might be that they are adapting to secondary compounds via their microflora, which would allow them a wider selection of plant foods. Another might be so that they can utilize pectin, which occurs in high concentrations in many of the fruits eaten by small forest ruminants. The principal problem for small ruminants is rumination and particulate passage; this can be overcome by means of a special bypass mechanism (Hofmann, 1989). While the smallest living ruminants (*Neotragus*) weigh on the order of 1–3 kg (Figure 4.9), there are yet smaller pregastric fermenters, perhaps the smallest being the meadow vole (*Microtus pennsylvanicus*) at about 50 g (Keys and Van Soest, 1970).

4.4.5 Large Ruminants

Up to this point, the discussion of the problems of digestive efficiency has emphasized the limitations of small ruminants. The largest ruminants (cape buffalo and giraffe) weigh about 750 kg, although larger nonruminant herbivores exist (e.g., elephant). The advantages of rumination may decrease at large sizes because retentive mechanisms are not necessary to achieve an adequate energy extraction from the diet. This does not mean that a very large animal necessarily will be an efficient fiber digester, merely that a low extraction rate (i.e., low digestibility) is no longer limiting to animal function.

Water buffaloes are more efficient digesters of poor-quality and tropical forages than are cattle, which is to be expected on the basis of size. The data of Sebastian et al. (1970) comparing buffaloes with Sahiwal cows (Table 4.5), show that size is responsible for the better digestion in buffaloes. If the rate of total ration disappearance is about 20% per hour, the integrated depression in digestibility will be 2.6 units. Lignin's rate of passage is indicative of the passage rate of undigested fiber. If the rate of cell wall digestion is about 8% with a lag of 3 h, the integrated depression in cell wall digestibility will be about 4 units between 24 and 27 h, which are the estimated rumen turnover times for cattle and buffaloes, respectively. Assuming the ration is about 50% in cell wall constituents, this would correspond to a depression in digestibility of about 2.5 units of dry matter. This difference accounts for the slightly higher digestibilities reported for buffaloes.

Eating rate may also influence apparent efficiency. Grazers that bolt their food and later ruminate it by pulsative digestion and passage may be less efficient than species that eat at a more even rate. Pulsative consumption may be less disadvantageous to temperate grazers because cooler temperatures make the heat

Table 4.5. Digestion in water buffaloes and cows

	Buffalo	Cow
Body weight (kg)	587	357
Metabolic size (kg$^{0.75}$)	119	82
Dry matter intake (g·day^{-1}·MBS^{-1})	14.9	10.5
Intake/metabolic size (g)	87.1	86.5
Intake as % body weight	2.54	2.94
Lignin turnover (h)[a]	27.1	24.1
Estimated cell wall digestibility (%)[b]	77.1	73.6
Expected DM digestibility (%)[c]	67.4	65.0
In vivo digestibility (%)	69.7	67.1

Source: Data from Sebastian et al., 1970.

[a]Calculated from lignin rate of passage equation k_{pL} = 0.007 + 0.0196 CW intake/BW (Mertens, 1973) assuming 65% cell wall in forages.

[b]Assuming 85% potential digestibility of cell wall with a 3-h lag and k_d = 0.08 (k_d is the rate of digestion; see Chapter 22).

[c]Assuming metabolic loss (M_i) of 12.9%; calculated according to the summative equation.

increment more tolerable. Temperate breeds of cattle that have limited ability to select feed may compare unfavorably with native animals in tropical environments where selection of diet may be a matter of adaptation and survival. A selected diet will always appear to be more digestible than a diet of unselected feed. But this does not mean a more efficient rumen flora. Digestion rates are set by the physical and chemical composition of the plant's structural substances. The kinds of carbohydrates present in the diet determine the VFA distribution in the rumen. Claims for the superior digestive efficiency of an animal species must be interpreted in light of the ability of the animal either to select high-quality food or to more efficiently pass undigested matter. Removal of slower-digesting fiber will make the digestion rate a factor because rapidly digesting matter will form a larger part of the digestible energy, while passage of fiber will reduce efficient digestion of cellulosic matter.

4.4.6 Upper Limits to Size

While mathematical modeling has clarified the lower limits of size, the problem of upper limits remains. The very largest and very smallest herbivores are not ruminants (Figure 4.1). The largest ruminants are the giraffe and buffalo. Larger herbivores include the hippo, rhinos, and elephants (Owen-Smith, 1988). Retention times increase with body size, and digestibility is in turn a function of retention time. At large body size a point is reached where comparatively complete digestion will occur even without selective retention of ingesta. At still larger sizes, expected retention will be more than adequate for slow-digesting fiber. An upper limit to mean retention may be about 4 days, as this is the generation threshold for acetate-degrading methanogens. Conversion of VFAs to methane would represent a serious energy loss, and very large herbivores

thus have no need to reduce passage to optimize the yield of metabolizable energy.

Because large herbivores may have retention times sufficient for adequate extraction of slowly digestible nutrients and because their larger gastrointestinal volume is the best suited to ingesting bulky, fibrous forage, they are more tolerant to forages and browses of low quality. From equation 4.5 it is expected that animals that weigh more than 1000 kg will show equivalent digestibilities of coarse forages without the need for selective retention of fiber. The magnitude of the metabolic requirements of large herbivores and their mouth size relative to food resources available limit them to unselective feeding on high-fiber diets that are difficult to ruminate, and rumination is an essential process in the fermentative extraction of energy from retained fiber. If they are able to ingest a sufficiently large amount, low dietary quality and a low extraction rate become tolerable. Elephants, grazing rhinos, and hippos appear to fit into this category (Foose, 1982). The fermentation rate required for maintenance is low in large animals and can be achieved without special digestive processing of low-quality forage and browse. For these reasons an upper limit to ruminant size is postulated at about 1000 kg.

4.5 Measuring Rate of Digestion

Claims have been made for higher microbial efficiencies in certain ruminant species (e.g., goats: Devendra, 1978; and buffaloes: Robert Grant et al., 1974b) based on results obtained with the Hungate zero time method. This procedure measures not cellulolytic degradation rate but rather VFA production over shorter fermentation times. It is therefore more likely a measure of nonstructural carbohydrate digestion and, consequently, either recent feed ingestion or concentrate selection.

Measuring the rate of cell wall digestion allows comparison of the digestive capacities of inocula from various sources provided relevant forage standards are available. Extent of cell wall digestion is measured either in vivo or in vitro as residual cell wall remaining after varying times of incubation. Robert Grant et al. (1974a) compared in vitro cell wall digestion rates using European cattle and carabao inocula in the Philippines (Figure 4.10). Cattle and carabao fed on tropical forage grass showed similar rates and extents of digestion; however, digestion curves of these tropical forages used as standards with inocula from temperate cattle fed on timothy at Cornell University showed a lag effect. Digestive rates were similar, but presumably bacteria in the rumen of animals fed on timothy require time to adapt to a tropical substrate. The results indicate that digestion rates can be similar provided

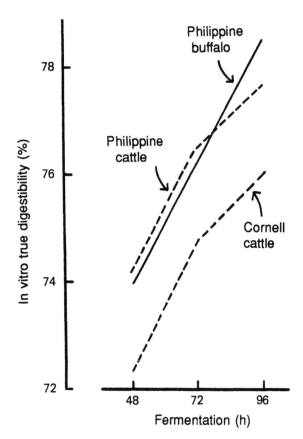

Figure 4.10. In vitro cell wall digestibility of tropical forage using rumen fluid from animals fed tropical grass or temperate grass (from Robert Grant et al., 1974a). The inocula from cattle and buffaloes fed on tropical forage achieved greater fermentation than the cattle at Cornell University fed on temperate forage.

animals of different species eat the same diet. This does not mean that dry matter digestibility will be the same, because rumen size and retention influence the extent of in vivo digestion. Carabao digested dry matter at a rate 2.6 units higher than European cattle fed on the same forage diets (Robert Grant et al., 1974a); however, they also probably retained ingesta in the rumen for a longer time.

Early digestion trials often showed that concentrate has a negative effect on fiber digestion (Wagner and Loosli, 1967). This depression was commonly assumed to have been the result of a depression of forage fiber digestibility since concentrates were overlooked as significant fiber sources. The experiments of Colucci et al. (1982) indicate that depression of digestibility in concentrate rations is due primarily to concentrate cell wall and secondarily to starch. Depression in digestibility of forage cell wall is a lesser effect than that contributed by concentrate. Digestion of both cell wall fractions is limited by competition between rates of passage and digestion (see Section 21.6). The microorganisms themselves appear to be affected by certain external factors. Often the inability of mixed cultures

of rumen organisms to degrade substrates is blamed on low pH, but this is not necessarily the case. It is possible that pH effects in the rumen may play a role in inducing lag effects in vivo. The observation that cell wall substrates are affected by inocula sources indicates a fundamental difference in the capacity of the inocula to digest cell wall that cannot be explained on the basis of substrate alone. Concentrate cell walls are less well digested than are forage cell walls by inocula derived from concentrate-fed animals. This kind of interaction is likely important in other situations in which diverse substrates are eaten. It may explain why animal species favor certain plants over others.

Inocula from deer fed on browse are equally or better able to digest maple cell wall than are cattle fed on grass hay, while the deer inocula are less able to handle grass cell wall (Robbins et al., 1975). Part of this interaction may be due to secondary compounds in the maple to which deer rumen bacteria may be better adapted. Deer also have the advantage of the salivary factor protecting them against tannins (see Chapter 13).

The character of the rumen microbial fermentation seems to be largely determined by the type of food eaten and generally agrees with the expected digestive capacity of selector animals such as deer (Short et al., 1975). When secondary compounds are present in browse, an assay for digestibility by inocula from forage-fed animals is apt to underestimate the digestibility that can be achieved by adapted animals. Fermentation rates and the characteristics of the rumen microbes do not explain why certain animal species are specialists in their diets, because the microbes found in the rumens of animals that eat such diets appear to be there as a result of dietary adaptations, not vice versa.

4.6 Rumination Capacity

The time spent ruminating is generally proportional to cell wall intake. Animals with greater appetites ruminate less per gram of cell wall, resulting in a greater fecal particle size. A parallel response that may be an adaptation allows concentrate selectors to consume more feed and avoid the rumen fill limit by passing larger and less-processed ingesta. This strategy inevitably results in lower fiber digestion. Nonruminant herbivores certainly take advantage of this factor. The question then becomes: How much does the reticuloruminal and omasal sorting and filtration system limit this adaptive capacity?

The possibility that low rumination capacity and fill prevent ruminants from taking advantage of fibrous diets needs further examination. Quantitative data for ruminants are largely limited to sheep, cattle, and goats. Udén and Van Soest (1982b) fed sheep, goats, cattle, horses, and rabbits on grass hay diets and found fecal particle size to be larger in nonruminants of comparable size (Table 4.6); yet regardless of class, the smaller animals tended to pass smaller fecal particles. That goats pass larger particles than sheep may be partial evidence of adaptation in a browsing ruminant species.

On the other hand, Parra (1978) observed that the fecal particle size of capybara (a sheep-sized rodent) is the same as sheep particle size. The capybara obviously chews its food exceptionally well. Both species achieve comparable fiber digestion. Production of finer fecal particles in small animals is likely the result of the smaller animal needing to comminute particles to pass through a smaller orifice.

Welch and his students (Bae, 1978; Hooper and Welch, 1983; Welch, 1982) examined rumination rates in goats, sheep, and cattle on similar diets in which the main variable within each species was whole animal size. Their results show a linear relationship between log of rumination rate and body size (Figure 4.11). The power slope across species is essentially unity, illustrating that rumination rate parallels the relation of digestive tract or rumen size to body weight. The slope of log rumination rate plotted versus body size of adult dairy breeds is 0.94. Rumination rate is thought to be a factor limiting feed intake. Taylor et al. (1986) found a slope of 0.81 among 25 breeds of cattle, a value intermediate between those for metabolic size and gut function.

Slopes for immature animals of the same species (goats or cattle) approach 1.5, indicating that young animals are more limited than adults. Thus it appears that rumination rate is also a possible limiting factor in small ruminants. The unity slope is comparable to

Table 4.6. Mean particle size of fiber in the gastrointestinal contents of herbivores

Animal	Particle size (μm)				
	Reticulorumen	Omasum	Abomasum	Cecum	Feces
Large heifer	2290				890
Small heifer	1670				640
Goat	1470	530	570	520[a]	460
Sheep	1290	550	530	490[a]	460
Pony					1600
Horse					1630
Rabbit				450	520

Source: Udén and Van Soest, 1982b.

Note: Timothy hay was the sole feed for all animals but rabbits, which were fed 60% hay and 40% concentrate. The hay was fed long to the horses; chopped to 5 cm for the heifers, goats, sheep and ponies; and ground to 2 mm for the rabbits.

[a]Cecum-proximal colon.

that observed between gastrointestinal size and body weight (Figure 4.1) and imposes a similar penalty relative to energy requirements as had been calculated from rumen and gastrointestinal size. Small ruminants ruminate more per gram of feed but also have a higher energy requirement per unit of body size. They probably have to grind fiber to a smaller size than larger ruminants do in order to get the dietary fiber past their relatively smaller sorting and filtering systems.

Allowing for some adaptive ability, browsing ruminants appear to be more limited than nonruminant herbivores in their capacity to exploit the strategy of high forage intake and rapid transit to achieve adequate intake of highly fibrous diets. This limitation likely involves the anatomical features required for adequate rumination: the arrangement of teeth and the omasum. Concentrate selectors' and smaller animals' smaller ru-

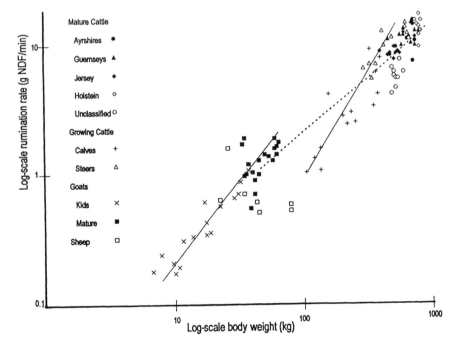

Figure 4.11. Body size and rumination rate of mature animals (goats, sheep, and cattle), plotted on a logarithmic scale give a slope of 0.90. Mature cattle have a similar slope (0.94); however, immature animals (goat kids and bovine calves) show slopes considerably greater than 1.0 (kids, 1.21; calves and steers, 1.53). Data illustrate the greater chewing efficiency of large and mature animals. (Recalculated from the data of Bae, 1978; Welch, 1982; and Hooper and Welch, 1983.)

Table 4.7. Feeding activity of male (m) and female (f) East African herbivores

Species	Body weight (kg)	Feeding time (%)	Rumination time (%)	Feeding activity (%)	Rumination-feeding ratio
Yellow baboon (f)	~13	24.4	—	24.4	—
Yellow baboon (m)	~22	21.8	—	21.8	—
Gerenuk (f)	?	60.3	19.2	79.5	0.32
Gerenuk (m)	31	47.0	17.2	64.2	0.36
Springbok (m)	32	67.0	11.4	78.4	0.17
Impala (f)	42	40.0	12.4	52.4	0.31
Impala (m)	56	34.0	13.8	47.8	0.40
Warthog (f)	57	38.3	—	38.3	—
Warthog (m)	85	27.5	—	27.5	—
Waterbuck (f)	181	64.1	19.5	83.6	0.30
Waterbuck (m)	238	43.7	21.5	65.2	0.49
Cape buffalo (f)	447	38.0	27.0	65.0	0.71
Cape buffalo (m)	751				
Cape buffalo (f)	424	41.8	35.4	77.2	0.85
Cape buffalo (m)					
Giraffe (f)	793	53.1	15.0	68.1	0.28
Giraffe (m)	1122	26.9	24.9	51.8	0.53
Black rhino (f)	1179	33.3	—	33.3	—
Black rhino (m)		32.4	—	32.4	—
White rhino (f)	1600	48.8	—	—	
White rhino (m)	2150				
African elephant (f)	2800	74.2	—	74.2	—
African elephant (m)	5000				

Sources: See Demment and Van Soest, 1983, for sources.

mination capacities are likely compensated by factors that allow ruminal escape and bypass. Hofmann's view is that the reticulo-omasal opening is the bottleneck for passage and outflow from the reticulorumen. This orifice and all other stomach openings appear to be proportionally larger in selector ruminants than in grazers; this applies to even large selectors like the moose and eland (Hofmann and Nygren, 1992). Ingesta escape through the omasal orifice may vary with dietary situation, allowing the moose, for example, to be a kind of intermediate feeder capable of using a wider range of foods in response to climatic effects on food availability. Such a view would be consistent with the discussion in Section 4.4.

4.6.1 Rumination and Feeding Activity in Wild Species

Feeding and rumination times have been reported for various African ruminants (Table 4.7). These data agree with the observation of Welch (1982) that rumination does not exceed 10 h/day. Sheep and cattle cannot be forced beyond this limit. The proportion of time spent ruminating is probably larger for grazers than for browsers that may spend more time collecting a higher-quality diet.

4.7 Integrative Assessment

Ruminants range in body weight from about 3 to 1000 kg. Limits to effective grazing, presuming that energy is derived from the processing of such forages, is 30–50 kg, and fermentation capacity is 20–30 kg. The ability of ruminants to handle fiber as an energy source is in the same range (Table 4.8). Newer information indicates that the smaller species practice some monogastric behavior (Figure 4.8; Hofmann, 1988) and only a part of their energy derives from the rumen fermentation process. Further, they do not derive important amounts of energy from cellulose and cannot tolerate low-quality fiber. The significance of the modeled limits discussed above should not be misunderstood. I do not mean that there cannot be ruminants beyond these specified limits. Rather, one should ask: How, taking into consideration all the physical and biological limits, does a species outside the capability of utilization of cellulose operate as a ruminant?

4.7.1 Fermentative Capacity and Digestibility

Limitation of fermentative capacity could be compensated for by a higher dry matter content in the diet; this would allow a smaller gastrointestinal capacity to contain more dry matter. Higher amounts of rumen dry matter could be one means of adaptation. The limits to dry matter content are established by osmotic pressure and pH as a result of fermentation products. These, in turn, could be obviated by more epithelial surface and absorptive capacity. The smaller selector ruminants seem to have achieved higher rates of fermentation, but this fails to compensate for their energy needs (see Figure 4.8A). Lower-quality diets lead to relatively

Table 4.8. Body size limits for a diet of grass forage without selective feeding behavior

	Limits of body weight (range, kg)
Rumination[a]	100–500
Digestive capability[a,b]	20–105
Fermentation volume[a]	16–100
Fermentation rate	30–40

[a]Values for grass at 50% digestibility and 70% cell wall.

[b]Size range in which rumen capacity limits adequate retention for utilizing cell wall in grass.

faster passage and thus require a reduced mean retention time and higher intake in order to meet metabolic requirements. This could be offset by eating a calorically denser diet.

The body weight limit for grazers (i.e., consumers of grass) will be higher than that for animals that feed on dicotyledonous herbs because of the nutritive differentiation among plant parts in dicots as opposed to grasses. The smaller grazer will need to feed selectively—a feeding process that is more difficult with nutritively less-differentiated grasses. One alternative is to postulate that small grazers may follow equids' strategy of higher consumption and lower extraction, although in that case selective retention would be severely diminished. Another alternative is selective grazing. There are probably ruminant and nonruminant species that exploit both alternatives. Similar adaptations could occur within the browsers. The cell wall contents of browse are often less than grasses or forbs but contain little digestible cell wall, so retention (except as it limits intake) ceases to be an important factor in the utilization of woody browses.

Variable tolerance to passage could be one factor in sexual dimorphism. Females have higher energy requirements during pregnancy and lactation that may lead toward selective feeding and, perhaps, faster passage (Demment and Van Soest 1985).

The slower-digesting tropical grasses are not conducive to high intake, and larger bodies or, alternatively, selective feeding on grasses (e.g., the oribi) are necessary strategies. Tropical grasses are rarely more than 70% digestible and decline to 40% or less at maturity, but they show a wide nutritional range in the standing forage. Smaller ruminants could feed unselectively on young tropical grasses but would have to become increasingly selective as the grasses matured. This feeding behavior would be favored by small tropical grazers since the grasses also become nutritively differentiated in response to environmental temperature. Alternatively, selectors might consume grass only when it is immature and of high quality and then move to browses and forbs after the grass quality declined. This behavior is characteristic of deer in temperate regions.

4.7.2 Fermentation Rate

The rate at which metabolizable energy must be absorbed per unit of body weight for maintenance increases as animal size decreases. Therefore, fermentation rate and production of VFAs could be possible limits to adaptable ruminant function. Fermentation rate is determined to a considerable degree by dietary composition and therefore is greatly influenced by feeding behavior. The interpretation of Figure 4.8A is that small antelope select better-quality diets, allowing a higher fermentation rate. The quantity of VFAs is inadequate as an energy resource for these small species, however, and bypass is almost certainly necessary (Hofmann, 1989). VFA production is less efficient than direct nonfermentative digestion of carbohydrate. It also likely bypasses high-energy sources such as fats because of infermentability due to the higher carbon-hydrogen density. Thus the small antelope may be somewhere between monogastrics and ruminants in net function (Hofmann, 1989). On the basis of Hoppe's data, Demment and Van Soest (1985) estimated that fermentation rate would become limiting at body weights less than 14 kg. The data in Figure 4.8 suggest that this limitation could be as high as 100 kg, depending on the diet.

The smallest concentrate selector ruminants (royal antelope, dik-dik, suni, and duikers) fall within the weight range of 3–6 kg. The smaller tragulids are also in this range. The majority of herbivores smaller than 3 kg are rodents or lagomorphs, some of which have pregastric fermentation but not rumination. These animals tend to practice coprophagy, an obvious strategy to overcome the limits of short retention and fermentation times. Small animals constrained by body size to eating concentrate foods have evolved gastric digestion and hindgut fermentation. Their passage characteristics might follow plug flow rather than competitive turnover in a mixing organ (see Section 23.5.3; Udén, 1989). Direct ingestion provides the energy of the cell contents without losses to microbial fermentation while permitting the cell wall to be fermented secondarily in the hindgut.

Energy density of the diet sets a theoretical limit to homeothermic function because the high maintenance cost for very small animals sets the demand for a high-energy diet. The effective limiting density is the caloric value of fat. A small species may have a low metabolic rate to reduce this problem, but low metabolic rates are associated with low reproduction rates, which may be disadvantageous in interspecific competition (McNab, 1980). The smallest organisms that live in a homeothermic environment are the rumen bacteria, and their metabolic temperature is set by the host. The problem of maintenance energy cost for organisms in the micrometer size range is solved by adopting growth rates and generation times that depend on the intrinsic rate at which the substrate can be catabolized.

4.7.3 Rumination Rates

The time ruminants spend chewing their cud is proportional to the amount of cell wall in their diet. Rumination rate (g cell wall/min) is directly related to animal size. More forage with lower cell wall content can be ruminated in less time (Welch and Smith, 1969). Welch (1982) found that ruminants spend no more than 9–10 h/day ruminating. From these factors rumination rate as a limitation to intake can be calculated. The average rumination rate (Figure 4.11) is about 0.02 $g \cdot kg$ body weight$^{-1} \cdot min^{-1}$, and the maximum rate is twice this rate. Assuming a maximum rumination time of 10 h daily, the average rate will be 12 $g \cdot kg^{-1} \cdot day^{-1}$, cell wall and the maximum rate 24 $g \cdot kg^{-1} \cdot day^{-1}$. Since smaller animals require more feed per unit of body weight, knowledge of the cell wall content of the diet allows calculation of the intake limit for an animal of a known size.

The limiting range for poor-quality hay (50% digestibility at 70% cell wall) sets the upper body weight limit at 490 kg. On such a diet even large ruminants will need to increase rumination rate if they are to achieve adequate intake for functions above maintenance. Higher-quality forages will promote adequate consumption at lower body weights. Small ruminants must either expand their rumination capacity to the limit or adopt selective feeding habits to avoid intake of cell wall.

Table 4.8 compares the size limits imposed by the various factors under consideration, and it appears that rumination rate is more limiting than digestive capacity, fermentation rate, or gastrointestinal volume (the latter three factors are closely interrelated). Thus, it appears that the advantage of the ruminant in being able to digest fiber exists at the penalty of the rumination limit.

The only way to relieve pressure of fill is to allow passage or oral ejection of coarser material. There is undoubtedly such specialization in selector ruminants, including both small ones and large ones such as moose (Nygren and Hofmann, 1990). Udén and Van Soest (1982b) found larger fecal particles in goats compared with sheep. This is consistent with Hofmann's suggestion that small selector ruminants can pass coarser matter and even bypass the rumen. This passage suggests a less-developed omasum and less-selective retention.

5 Nonruminant Herbivores

Ruminants are by no means the only animals that exploit fibrous plants; they represent but one of a number of digestive strategies. Nonruminant herbivores use many of the same feeding strategies as ruminants do, albeit with somewhat different anatomies and in some cases with anatomical adaptations in the form of pregastric fermentation. Another group of nonruminant herbivores does not have much fermentation and consumes plants for their protein, sugar, and starch. The taxonomy and gut morphology of these animals (e.g., pandas) link them to carnivores. How this adaptation occurred or why other carnivores seem unable to exploit plants is not well understood.[1]

5.1 Digestion Sequences

Fibrous carbohydrates must be digested symbiotically by gut microorganisms in all higher animals that have not evolved cellulases, hemicellulases, or pectinases. If the fermentation chamber is postgastric, the animal host has first chance at the available carbohydrates and protein in the food but loses the chance to capture microbially synthesized proteins and vitamins, which can be very important if the dietary sources are poor in these nutrients.

The simplest sequence is that exemplified by humans, dogs, and other carnivores (Figure 5.1A), which lack a cecum as a separate compartment. Many herbivorous animals, including humans, have sacculated colons. Sacculation probably helps slow the passage of fibrous solids and leads to more efficient extraction of fermentable energy; it indicates a herbivorous evolutionary ancestry (C. E. Stevens, 1988). Nonruminants such as the pig and horse possess a sizable cecum (though smaller than the colon), in contrast to rodents and lagomorphs, whose cecum is proportionally larger than the colon.

The main site of fermentation in many rodents and lagomorphs (Figure 5.1B) lies in the cecum. Many of

these animals practice coprophagy to capture microbial protein and vitamins. Within this group there are even more specialized species, such as rabbits and lemmings, whose cecum selectively admits only fine matter; coarser fiber is excluded and excreted in day feces (Bjornhag, 1972; Udén et al., 1982). Night feces are reingested, allowing the animal to recapture microbial protein and vitamins derived from the most fermentable carbohydrates. Because the coarse fiber is rejected, fiber utilization is very low. Rabbits and lemmings probably exploit vegetative tissues containing pectin and other rapidly fermentable unlignified carbohydrates. Coprophagy can be viewed as an adaptation of small herbivores, in which the limiting effect of rate of passage is a special problem. This strategy allows these small herbivores to consume fiber without the penalty of energy intake restriction, although many potentially digestible cellulosic carbohydrates are lost in the feces.

Some animals possess pregastric fermentation without rumination (Figure 5.2A). This group includes a wide spectrum of mammals, including some kangaroos (Hume, 1982), hamsters (Ehle and Warner, 1978), voles (Keys and Van Soest, 1970), colobine (C. E. Stevens, 1988) and langur monkeys (Bauchop and Martucci, 1968), and hippopotamus (Moir, 1968). At least one bird with pregastric fermentation has been discovered, the hoatzin (Grajal et al., 1989), and the phenomenon probably exists in other species as well. The microbial contents of the hoatzin's crop have been shown to contain anaerobic bacteria similar to those of the rumen, in addition to ciliate protozoa (Dominguez-Bello et al., 1993). Pregastric fermentation was discovered in the hamster when it was noted that animals showed no negative response to amino acid–deficient diets (Banta et al., 1975). Subsequent examination revealed cellulolytic digestion in the abomasum (Ehle and Warner, 1978). Large herbivorous dinosaurs may have had a pregastric fermentation chamber in the crop (Farlow, 1987). Cannonball-sized stones plus heavy musculature let the organ to do the equivalent of rumination while the animal was allowed greater freedom to feed in contrast to modern ruminants (Bakker,

---

[1] "And the cow and the bear shall feed: their young shall lie down together: and the lion shall eat straw like the ox" (Isaiah 11:7).

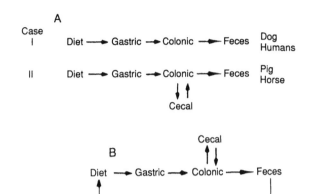

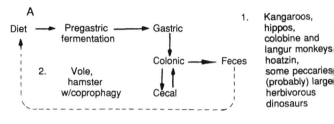

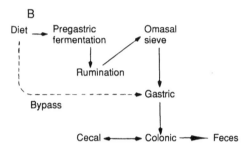

Figure 5.1. **(A)** Sequence of digestion in nonruminants. The simplest case (little or no cecal digestion) occurs in most carnivores and in humans. Cecal digestion occurs in large herbivores but is secondary to colon digestion (C. E. Stevens, 1988). **(B)** Sequence of digestion in rodents, most of which practice coprophagy; cecal fermentation is dominant over secondary colonic fermentation. Some rodents (hamster and vole) exhibit pregastric digestion (see Figure 5.2).

Figure 5.2. **(A)** Sequence of digestion in nonruminants with pregastric digestion. **(B)** Sequence of digestion in ruminants. The sieving system of the omasum is much more developed in grazing species.

1986). The definition of *dietary fiber* for these animals would follow that used for ruminants (Figure 5.2B) rather than nonruminants, because soluble unavailable gums (see Chapter 11) would ferment along with other available carbohydrates in the pregastric fermentation. Thus, classifying animals as ruminants and nonruminants is an oversimplification. Ruminant-like capacities (e.g., pregastric fermentation) exist in animals with grazing and selector types of feeding and in true ruminant and nonruminant groups.

5.2 Nonruminant Animal Anatomies

The earlier discussion of ruminant gastrointestinal architecture invites comparison with nonruminant anatomy, particularly the lower digestive tract, of which more is known. On the other hand, much microbial physiology and gut ecology gleaned from rumen studies can be applied to nonruminant lower tract fermentation. There are several habitual misconceptions regarding lower tract fermentation. Nonruminant nutritionists tend to ignore or underestimate its significance, and ruminant nutritionists think of most nonruminants as "simple gutted" or "monogastric." Some digestive tracts of nonruminant mammals are quite complex; others are comparatively simple. There is a wide variation among species representing different dietary adaptations. Gut fermentation as a source of energy is important for most herbivores that consume fiber.

Mammalian species can be classified into several groups based on the characteristics of their digestive tracts (Table 5.1). This classification may be compared with that based on dietary specialization in Chapter 3.

The ruminants represent the most developed and specialized herbivores in view of their ability to use fiber and other carbohydrates unavailable to animal digestion. Nonruminants exhibit a variety of adaptations relative to herbivore specializations and gut architecture (Moir, 1968; see Figure 5.3).

Feeding strategy does not follow gastrointestinal anatomy. Grazers (bulk and roughage eaters), selectors, and even some omnivores can be found within any class based on gastrointestinal anatomy. For example, the ruminants include grazers, selective folivores, and frugivores (very small antelope), and many of the latter have relatively small rumens (Hofmann, 1988).

Table 5.1.   Mammals classified by gastrointestinal anatomy

Class	Species	Dietary habit
Pregastric fermenters		
Ruminants	Cattle, sheep	Grazing herbivores
	Deer, antelope, camel	Selective herbivores, ing folivores and frug
Nonruminants	Colobine monkey	Selective herbivore
	Hamster, vole	Selective herbivores
	Kangaroo, hippopotamus	Grazing and selective vores
	Hoatzin	Folivore
Hindgut fermenters		
Cecal digesters	Capybara	Grazer
	Rabbit (lemming)	Selective herbivores
	Rat, mice	Omnivores
Colonic digesters		
Sacculated	Elephant, horse, zebra	Grazers
	New World monkeys	Folivores
	Pig, human	Omnivores
Unsacculated	Panda	Herbivore
	Dog, cat	Carnivore

Sources: Parra, 1978; Hume, 1982; C. E. Stevens, 1988.

Table 5.2. Gastrointestinal volume of mammal species

Species	Total contents (%)	Reticulorumen (%)	Omasum (%)	Abomasum (%)	Small intestine (%)	Cecum (%)	Colon and rectum (%)
Ruminant							
Cattle	13–18	9–13	1.1–2.8	0.5	0.9–2.3	0.8	0.8–1.5
Sheep	12–19	9–13	0.1–0.3	0.7–1.6	1.0–1.6	0.9–1.6	0.5–0.7
Camel	—	10–17	—	—	—	0.1–0.3	1.0–2.2
Nonruminant							
Elephant	—	—	—	—	—	3.2	8.8
Horse	16.4	—	—	1.3	2.6	2.4	8.8
Pig	10.4	—	—	3.6	1.9	1.6	3.4
Rabbit	7–18	—	—	2–7	0.6–1.8	2.5–7.8	0.7–1.3
Rodent							
Capybara	16.5	—	—	1.8	1.0	11.7	1.3
Guinea pig	8.0	—	—	1.9	0.9	3.7	1.6
Rat	3.1	—	—	0.4	0.9	1.0	0.9
Vole	—	—	—	—	—	5.9	—

Source: Parra, 1978.
Note: Volume is weight of liquid contents as percentage of body weight.

Pregastric fermentation occurs in widely differing taxonomic groups—including rodents, primates, ungulates, birds, and perhaps dinosaurs (Figure 5.2A)—of diverse sizes and feeding behaviors.

Hindgut fermenters can be divided into those that are mainly cecal (mostly rodents and lagomorphs) and the colonic fermenters, many of which have some cecal capacity.

In the large nonruminant herbivores (horse, rhinocerus, elephant, etc.) fiber digestion is more important in the large bowel compared with the cecum. All these animals have sacculated colons. The colon is also sacculated in humans and other primates, although the cecum is much reduced. The sacculated colon and the relatively small cecum may represent special adaptations of the lower bowel to fermentations, and therefore a history of evolutionary adaptation to some degree of fiber in the diet (C. E. Stevens, 1980).

Omnivores and carnivores are classes in which the importance of fiber and gastrointestinal fermentation has been largely ignored. These animals have an unsacculated colon and little or no cecal capacity, although cats and dogs are known to eat some grass. A digestion balance study in which cereal bran was fed to dogs indicated about 20% digestibility of cereal fiber (Visek and Robertson, 1973). It is apparent that some fermentation can occur in animals lacking a cecum and sacculated colon, provided the substrates are rapidly degradable. Anatomically, the pandas fall into this group, but they use the extreme strategy of bulk and roughage consumption without much fermentation.

5.2.1 Fermentation Capacity

The capacity of nonruminant herbivores to digest cellulosic carbohydrates is related to body size just as it is in ruminants, because average gut capacity is related

to the 0.25 power of body weight (Chapter 4). Within any size range, however, there is a large variation in fermentative capacity in nonruminants that generally exceeds that of ruminants. This occurs because of the wide range in the proportion of gut volume devoted to fermentation, which can be viewed on the basis of the proportion that the respective digestive chambers form with body weight (Table 5.2) and as a proportion of the digestive tract (Table 5.3).

The comparative importance of fermentation as a means of digestion is demonstrated by the proportion of digesta residing in fermentative compartments relative to the whole digestive tract. Ruminants are not the only animals with a large proportion of the digestive tract devoted to fermentation. The capybara, a large South American rodent, has the digestive capacity of a sheep and is a true grazer (Parra, 1978). Pigs, rabbits, and rats have a lesser capacity. The latter two are actually inferior to the pig in fiber digestion because of their small size. Humans and dogs devote substantially

Table 5.3. Fermentative capacity expressed as percentage of the total digestive tract for mammal species

Species	Reticulorumen (%)	Cecum (%)	Colon and rectum (%)	Total fermentative capacity (%)
Sheep	71	8	4	83
Capybara	—	71	9	80
Cattle	64	5	5–8	75
Horse	—	15	54	69
Guinea pig	—	46	20	66
Rat	—	32	29	61
Rabbit	—	43	8	51
Pig	—	15	33	48
Human	—	—	17	17
Cat	—	—	16	16
Dog	—	1	13	14

Source: Parra, 1978.

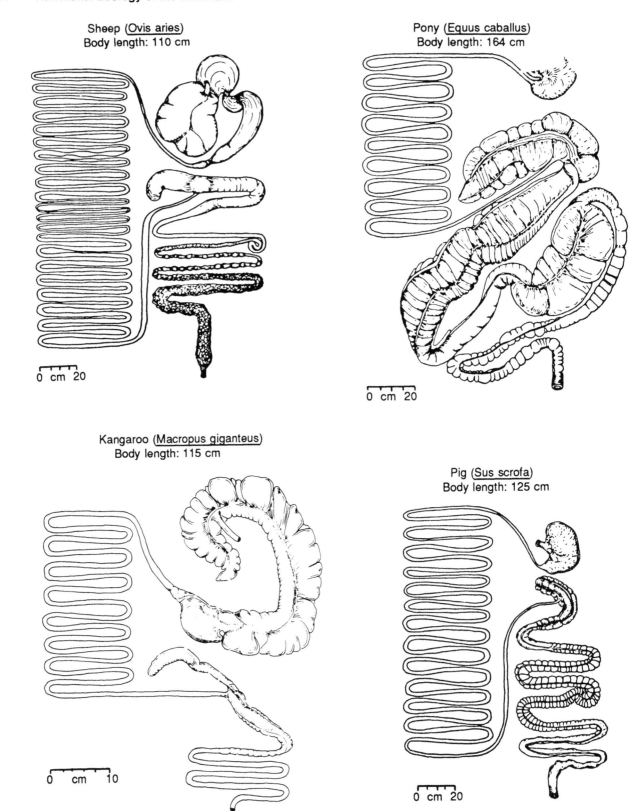

Figure 5.3. Eight mammalian digestive tracts, including pregastric digesters (sheep and kangaroo) (from C. E. Stevens, 1988). Note the similarity between the sacculated kangaroo stomach and the hindgut fermenters (pig and pony). The fermentative compartments of herbivores tend to be sacculated. In carnivores such as the mink and dog sacculation is largely absent. In the cecal digesters (rabbit and rat) the cecum is sacculated. The cecum is very reduced in the dog and absent in the mink. The human colon (not shown) is also sacculated, probably indicating herbivore ancestry.

Rabbit (<u>Oryctolagus cuniculus</u>)
Body length: 48 cm

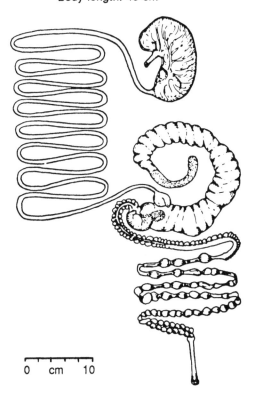

Dog (<u>Canis famillaris</u>)
Body length: 90 cm

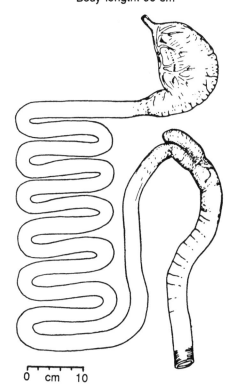

Rat (<u>Rattus norvegicus</u>)
Body length: 17 cm

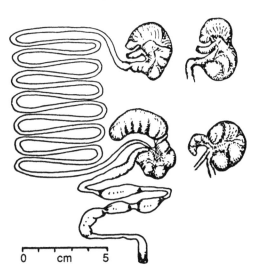

Mink (<u>Mustela vison</u>)
Body length: 42 cm

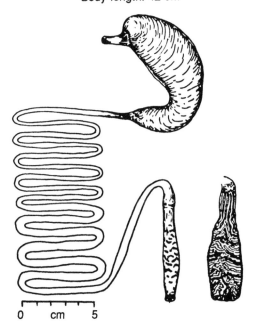

smaller proportions of their tracts to microbial fermentation, a feature generally characteristic of omnivores and carnivores.

Although data are limited, several generalizations can be made. Fermentative compartments tend to form equivalent proportions of the digestive tract in both ruminants and nonruminants. A similar compensation occurs between colonic and cecal fermenters. A smaller proportion of the gut devoted to fermentation is probably reflected in a more limited tolerance to dietary fiber or, alternatively, a depression in digestibility of fiber if more fiber is eaten. A larger intake of fiber relative to a smaller fermentation chamber will induce a more rapid turnover, although this can be modified somewhat by a sacculated gut and plug flow mechanics (see Section 23.5.3). Thus humans probably have a larger digestive capacity for fiber than dogs, which do not have sacculated colons (C. E. Stevens, 1988).

5.3 Comparative Digestion: Nonruminants

Although nonruminants' ability to digest fibrous feeds is often lower than ruminants', there is considerable variation in the capacity to utilize fibrous carbohydrates among animal species. Some reported digestibilities of various fiber sources are shown in Table 5.4. Retention times are unknown, but the data show less difference among animal species when they are fed high-quality unlignified fibers, for example, vegetables, fruit, or leaves. Bran is fairly equally digested by all species, but cellulose from forages—particularly grass fiber—is poorly digested by smaller species and immature animals.

It is important to consider the factors responsible for this variation and to compare individual species and their ability to utilize fibrous carbohydrate with respect to their dietary adaptations and ecological niches. This consideration is important for human nutrition because it can help identify animal species that are suitable models for studies of dietary fiber and human diseases, and which animal species convert feed and forage resources into food for humans most efficiently.

Cellulolytic capacity can be measured by placing animals on the same diet and conducting digestion balances, although interpretation of results is difficult when the species being compared have different feeding habits (Arman and Hopcraft, 1975). Another way of studying the problem is to compare in vitro (rumen or cecal) digestion of standard substrates using inocula from the various animals.

Nonruminants' digestion of cellulose and hemicellulose indicates varying capacities to digest these carbohydrates. Generally, smaller animals are less able to handle cellulose. Higher digestibilities are seen in

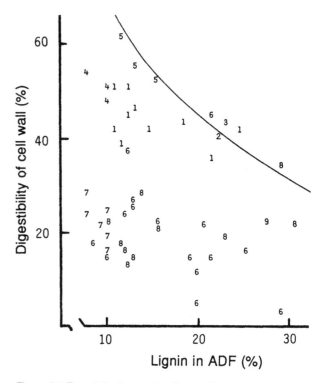

Figure 5.4. The relation between lignification (lignin content of acid-detergent fiber) and nonruminant species' ability to digest fiber (modified from Parra, 1978). Species: 1 = horse; 2 = zebra; 3 = Przewalski's horse; 4 = pig; 5 = capybara; 6 = rabbit; 7 = rat; 8 = vole; 9 = tortoise. The curved line is the regression line for domestic ruminants (Goering and Van Soest, 1970). Note that smaller species are concentrated at the bottom of the figure, indicating lesser ability to digest fiber. See Section 4.3 and Demment and Van Soest, 1985, for a discussion of animal size.

certain small herbivores; for example, the pigmy hippo, hamster, and the vole. This exceptional capability might be the result of their ability to select feed or of some specialization of the gastrointestinal tract selective retention of the available cellulose. Humans' ability to digest alfalfa is apparently low (Williams and Olmstead, 1936), but their ability to digest wheat bran is only slightly less than ruminants'. Humans digest the more rapidly fermentable vegetables easily.

Another way of looking at nonruminant digestibilities is to compare them on the basis of the lignin-to-cellulose ratio, the main factor setting a limit on extent of digestion. The curve in Figure 5.4 represents the average digestion for ruminants (sheep and cattle) as a general upper limit to digestion of fibers with varying potential digestibilities. Some nonruminants, such as the capybara, are very close to this line, but values for most nonruminants are scattered below the curve, indicating the inadequacy of the smaller ruminant species to harbor and contain a fermentation that will extract all the available cellulosic carbohydrate from the fibrous feed. Some potentially digestible cellulosic carbohydrate is lost in the feces, the quantity being larger in animals with lesser retention and gut capacity.

Table 5.4. Digestibilities of fiber components reported in nonruminants

Species	Body weight (kg)	Diet	Digestibility Hemicellulose (%)	Cellulose (%)	Reference
Vole	0.05	Alfalfa	39	34	Keys and Van Soest, 1970
		Grasses	34–38	18–29	Keys and Van Soest, 1970
Rat	0.2	Alfalfa	47	21	Keys et al., 1970
		Grasses	6–11	1–2	Keys et al., 1970
		Microcrystalline cellulose	—	9	Hsu and Penner, 1989
		Vegetables	50–95	38–63	Nyman et al., 1990
Guinea pig	0.26	Alfalfa	—	49	Fahey et al., 1979
		Cellulose		84	Fahey et al., 1979
Rabbit	3	Timothy	11–13	7–9	Udén, 1978
		Alfalfa	47	21	Udén, 1978
Turkey	4–6	Cellulose	—	3	Duke et al., 1984
Howler monkey	5.2–8.4	Fruit	16–21	20–33	Milton et al., 1980
		Leaves	57–69	66–69	Milton et al., 1980
Dog	10	Cellulose[a]	—	5–10	Burrows et al., 1982
	c	Brewers' grains	30–60	7–25	Visek and Robertson, 1973
Beaver	14–22	Poplar	—	30(ADF)	Hoover and Clarke, 1972
Kangaroos					
red	27–33	Alfalfa	33[b]	36[b]	Hume, 1974
		Oat straw	17[b]	17[b]	Hume, 1974
		Wheat straw	43[b]	33[b]	Hume, 1974
euro	23–31	Alfalfa	38	46	Hume, 1974
		Oat straw	27	21	Hume, 1974
		Wheat straw	40	39	Hume, 1974
Capybara	41	Tropical grasses	—	52–61	Gonzalez and Parra (unpublished)
Pig	44	Alfalfa	18–28	28–33	Kuan et al., 1983
	48–90	Alfalfa	23	7	Kass et al., 1980
obese	83–100	Alfalfa	34–65	14–64	Varel et al., 1988
lean	92–110	Alfalfa	44–84	30–78	Varel et al., 1988
	120	Alfalfa	49	47	Ehle et al., 1982
		Wheat bran	72	21	Ehle et al., 1982
		Cellulose[a]	88	48	Ehle et al., 1982
	120	Rutabaga	93	93	J. A. Robertson et al., 1987
		Wheat bran	43	43	J. A. Robertson et al., 1987
Onager	c	Pellet diet	47	59	Hintz et al., 1976
Ponies	115–147	Alfalfa	60	51	Hintz et al., 1973
		Timothy	45	35	Hintz et al., 1973
Horse	450	Alfalfa	55–72	45–66	Hintz et al., 1971
		Alfalfa	33	45	Hintz et al., 1971
		Grasses	42–53	42–49	Fonnesbeck, 1967
		Timothy	37–45	33–43	Udén, 1978
Przewalski's horse	c	Pellet diet	42	46	Hintz et al., 1976
Zebra	250	Pellet diet	39	50	Hintz et al., 1976

Note: This table reports values for species in which retention time has not been measured. Values for other species are in Tables 5.7, 5.10, 5.11, and Figure 5.12.

[a]Solka floc (commercial wood cellulose).

[b]Digestibility of acid-detergent fiber. Control sheep digested hemicellulose in alfalfa at 43%, oat straw at 23%, and wheat straw at 43%; ADF digestibilities were 44%, 21%, and 39%, respectively.

[c]No values given.

The lower fermentative capacity of many nonruminant herbivores may not limit their ability to operate as grazers, because their digestive tracts do not contain filtering mechanisms that promote retention of fiber at the expense of intake. The larger herbivores (e.g., horse and elephant) resolve the problem by higher intake, which offsets the lower degree of digestive extraction. Fecal particle size may offer a clue to the nutritive strategy of a species.

5.4 Gut Fermentation in Nonruminants

Fermentation in the gut of nonruminants parallels the range found in ruminants. There are pregastric nonruminant fermenters as well as hindgut fermenters. As expected from their anatomies, the latter are further subdivided into colonic and cecal fermenters. Fermentation in all these animals has many similarities to fermentation in the rumen with respect to microbial species and microbial products. The lower tract fermentations have been regarded as a sort of "displaced rumen" (Argenzio and Stevens, 1984).

5.4.1 Pregastric Fermentation

Pregastric fermentation may require only hypo-acidity and buffering in the non-acid-secreting portion of the abomasum. The fundamental requirements for such a condition include sufficient volume relative to intake so that turnover (i.e., retention) can be long enough to harbor carbohydrate-fermenting bacteria. The most rapidly fermenting carbohydrates are sucrose, pectins, and some starches, the sugar and starch leading to lactic acid. This kind of fermentation could exist in some herbivores without cellulose-digesting capability, although digestion would require longer retention. The rule involving survivability of gut microflora is that retention (turnover) time must exceed generation time. Many lactic acid–producing organisms in the mouth and upper tracts of monogastric herbivores overcome this limit by attaching themselves to the mucosal lining.

Langer (1988) described sacculation in the foregut of various herbivorous nonruminants. Unfortunately there is little data on foregut microorganisms, and details of fermentation are often lacking. There probably are more foregut fermenters that can handle cellulose than have been identified. Known cellulolytic foregut fermenters are listed in Figure 5.2A and Table 5.1. These fermenters are found in both very small and very large weight classes.

5.4.2 Lower Tract Fermentation

The sources of fermentable substrate in the large bowel and cecum include any carbohydrate escaping or arriving from the upper tract. In the case of ruminants, the fermentable substrate for the lower tract is limited to the slower-digesting carbohydrates escaping the rumen plus some secreted mucins. Crystalline starches are probably an important source of fermentable carbohydrate. In addition to dietary sources, saliva and mucus add mucopolysaccharides as a fermentation substrate. In nonruminants that consume a low-fiber diet, such as humans, the mucopolysaccharides may constitute the main substrate for colonic fermentation (Salyers et al., 1982). The ruminant lower tract receives generally less fermentable carbohydrate than the nonruminant tract because the rumen will have removed the more available fraction of the dietary fiber.

While some bacteria occur in the small intestine, the major sites of fermentation and VFA production are the cecum and large intestine, which exhibit higher VFA concentrations (Figure 5.5). The major organisms in the lower tract in humans include the genera found in

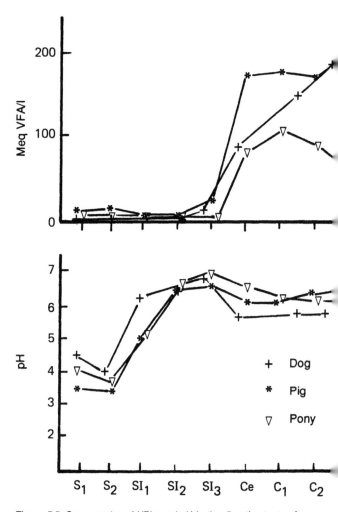

Figure 5.5. Concentration of VFAs and pH in the digestive tracts of nonruminants (C. E. Stevens, 1978). Stomach (S) and small intestine (SI) are divided into segments (S₁, S₂; SI₁, SI₂, and SI₃, respectively). Ce is cecum, and colon segments are C₁, C₂, and C₃.

ruminants, although the relative numbers may differ and there are some unique species that occupy comparable niches (Holdeman et al., 1976). Nutritional requirements for most of these bacteria are similar to those of rumen bacteria (Bryant, 1974). This similarity is emphasized by the fact that in vitro cultures of colonic and fecal organisms from humans are optimally grown on rumen fluid media (Bryant, 1978).

The environment of the lower tract fermentation is more constant and less influenced by dietary pulses than is the rumen. Food proteins and other easily digestible dietary components do not ordinarily reach this section of the digestive tract. The pH of the cecum and the large bowel remains essentially neutral, even in the face of considerable VFA production, although, like the rumen, cecal pH drops if an excess of rapidly fermentable carbohydrate is supplied that causes the buffering capacity to be overwhelmed. The carbohydrate source has to be resistant to upper tract digestion

Table 5.5. Influence of diet on VFA Proportions in the lower tracts of pigs, steers, and ponies

Species	Diet	Rumen		Stomach		Small intestine		Cecum		Colon	
Pig[a]	High cellulose	—			(10)		(<5)	75:25:5	(80)	73:24:4	(120)
	Concentrate	—		53:38:9	(20)	85:10:5	(25)	57:34:9	(200)	54:35:11	(200)
Steer[b]	Timothy	74:15:6	(59)	92:6:1	(12)	85:10:2	(6)	79:13:3	(34)	73:13:5	(17)[c]
	Timothy + oats	69:20:11	(91)	—		—		—		—	
	Clover	75:15:9	(110)	—		—		—		—	
	Clover + oats	74:17:9	(86)	—		—		—		—	
Pony[b]	Timothy	—		82:6:2	(14)[c]	65:10:6	(7)[c]	75:18:6	(56)[d]	80:11:5	(26)
	Timothy + oats	—		—		—		70:22:8	(50)[d]	—	
	Clover	—		—		—		75:19:5	(64)[d]	—	
	Clover + oats	—		—		—		69:24:7	(74)[d]	—	
Pony[e]	Timothy	—		—		—		71:19:9	(41)	—	
	Timothy:grain 1:1	—		—		—		69:18:11	(41)	—	
	Timothy:grain 1:4	—		—		—		57:24:13	(37)	—	
	Alfalfa dehy	—		—		—		76:15:8	(57)	70:16:8	(42)[c]
	Alfalfa:grain 3:2	—		—		—		70:21:7	(52)	68:15:9	(48)[c]
	Alfalfa:grain 1:4	—		—		—		61:26:10	(48)	67:17:9	(28)[c]

Note: VFAs are given in molar percentages of acetate:propionate:butyrate, respectively; values in parentheses are total VFAs, in meq/l.
[a]Read from graphs in Argenzio and Southworth, 1974.
[b]Data from Kern et al., 1973, 1974.
[c]Contains significant molar percentage of isobutyrate.
[d]Diets of ponies and steers are comparable and indicate less butyrate in lower tract relative to rumens of timothy-fed steers.
[e]Data from Hintz et al., 1971.

and yet rapidly fermentable. Galactans (in beans) are such carbohydrates. Lactose intolerance in humans is another factor that can upset the pH of the lower tract. Lactase deficiency allows lactose to reach the lower tract, where it ferments to the stronger lactic acid and causes diarrhea. Large changes in flora and inhibition of cellulose digestion must be accompanied by a drop in pH as occurs in the rumen. Mechanisms are discussed in Chapter 15. A similar problem can exist relative to sucrose in preruminant calves, which lack sucrase.

When the salivary secretion that aids in rumen buffering is absent, pH regulation depends on the transit of free acid across the gut wall and the secretion of bases into the intestinal lumen. As in the rumen, fatty acids are absorbed as free acids, although at cecal pH only a very small proportion is in the free form. The diffusion of sodium ions and urea into the bowel offer buffering and a nitrogen source for the fermentation, respectively. Urea is rapidly hydrolyzed to ammonium bicarbonate to support its utilization (Visek, 1978). Fermentable carbohydrate promotes microbial growth, and the nitrogen requirement of the microbes provides an ammonia sink in the form of microbial cells. It is in this context that the role of fiber in increasing the metabolic fecal nitrogen should be understood (see Section 18.10).

Mucopolysaccharides are secreted at various sites in the gut and are fermented in the cecum and colon. This material contains glucosamine, which could liberate some ammonia on fermentation as well as contribute to the microbial component of metabolic fecal nitrogen.

Hydrolysis of urea produces ammonium bicarbo-nate, which can act as buffer. The liver's uptake and conversion of ammonia to urea, and then its excretion via the kidneys, is an energy-expensive process associated with high-protein diets. The potential toxicity of ammonia in this situation constitutes one theory of carcinogenesis in the lower bowel in humans that can be connected with the overconsumption of protein and the lack of fiber or other fermentable carbohydrate in the diet (Visek, 1978).

While the lower tract is relatively cushioned from dietary variations, it nevertheless shows responses comparable to those of the rumen. The ratios of VFAs are similar to those in the rumen and, in pigs and horses, show shifts toward high propionate when high starch levels are part of the diet (Table 5.5). This indicates that some starch reaches the lower tract of non-ruminants.

5.4.3 Passage

As in the rumen, several factors affect the ecology of the lower tract; for example, particle size of dietary fiber influences passage rate and microbial turnover. In contrast with the rumen, more rapid turnover is promoted by coarse fiber; the consequent increment in fecal nitrogen is due to enhanced microbial yield (Table 5.6). Fine particle size reduces the bulk of ingesta, thereby slowing the passage rate. Note that equal intakes of coarse and fine bran induce disparate fecal nitrogen excretion. Unlike the rumen, the lower tract of the nonruminant (and many other species) has no filters, so fine grinding of the food increases the density of the ingesta and the retention, just the opposite of

Table 5.6. Wheat bran fiber particle size and fiber digestion and passage in male students

	Coarse bran	Fine bran	Low-fiber basal
Fiber particle size (μm)	744[a]	173[b]	—
Fiber intake (g NDF/day)	22[a]	22[a]	4[b]
Digestibility			
Cellulose	41[a]	53[a]	86[b]
Hemicellulose	58[a]	56[a]	68[b]
Passage			
Mean retention (h)	41[a]	57[b]	62[b]
Turnover, lower tract (h)	12	15	18
Bulk density–hydration capacity			
Stool water (g/week)	846[a]	665[b]	545[c]
Water (% feces)	77.3[a]	73.6[b]	74.7[c]
Total metabolic fecal N (g/day)[d]	2.0[a]	1.4[b]	0.8[c]
Increase in metabolic N/kg fiber	57[a]	37[b]	0[c]
Metabolic N (g/kg fiber digested)[e]	175[a]	120[b]	270[c]

Source: Van Soest et al., 1978b.

Note: NDF = neutral-detergent fiber.

[a,b,c]Numbers with different superscripts are statistically different.

[d]Total fecal nitrogen corrected with nitrogen in fecal acid-detergent fiber.

[e]The metabolic nitrogen associated with the low-fiber basal is probably microbial cells produced from the fermentation of secreted mucopolysaccharides. See Section 18.9.3.

Table 5.7. Comparative passage rates of liquid and particles

Species	Body weight (kg)	Whole tract retention[a] Particles[c] (h)	Whole tract retention[a] Liquid (h)	Retention in fermentation compartment[b] Particles (h)	Liq...
Ruminants					
Large heifers	555	79	29	47	
Small heifers	243	62	30	38	
Sheep	30	70	38	35	
Goats	29	52	39	28	
Nonruminants					
Horses	388	29	29	10	
Ponies	132	34	26	10	
Human	70	41	39	12	
Rabbit	3	9	193[d]	4	1

Sources: Udén, 1978; Van Soest et al., 1978b.

Note: Heifers, sheep, goats, horses, ponies, and rabbits were fed a standard ti... thy diet; human subjects were fed a standard diet including 20 g of dietary fiber ... wheat bran.

[a]Mean retention according to Faichney, 1975.

[b]First pool turnover (k_1) according to Grovum and Williams, 1973. The ferme... tion compartment is the rumen in ruminants and the cecum and large bowel in ...ruminants (Grovum and Williams, 1977).

[c]Particulate passage based on chromium mordant of the dietary fiber.

[d]Coprophagy and recycling of indicator are responsible for the high value.

what one sees in ruminants. Sorting mechanisms are not important in humans and other species that have relatively equal passage of liquid and fiber in the lower tract. Some coprophagous nonruminants (e.g., vole, rabbit, hare, and lemming) have selectively more rapid passage of fiber compared with liquid (Bjornhag, 1972; Udén et al., 1982; Hume et al., 1993).

Fine grinding promotes more rapid passage of fiber from the rumen, but slower passage in the lower tract of man. In humans the water content of feces from finely ground diets is reduced; this points out the failure of finely ground fiber to alleviate constipation (Van Soest et al., 1978b; van Dokkum et al., 1982).

The relative passage rates of liquid and particles vary among animal species. In general, ruminants tend to have the slowest rates of particulate passage relative to liquid (Table 5.7). Nonruminants present varying situations. The rabbit passes particles selectively faster than liquid, but there is little separation of liquid and particles by humans and other nonruminants. There is a trend toward longer retention of particles in larger animals.

Transit also varies among animal species relative to the proportions of the digestive tract. For example, carnivores typically have short intestines and very reduced fermentative capacity. Tract length is proportionally larger relative to volume in many herbivores. Thus transits through the stomach, intestines, cecum, and colon can vary relative to the mean retention time. Detailed flow characteristics can be obtained by radio-

opaque markers (Clemens and Stevens, 1980) or by slaughter and measurement of each segment of the digestive tract (Vidal et al., 1969).

5.4.4 Cecal Fermenters

The cecum has fermentation characteristics similar to the colon but differs in that ingesta must pass out of

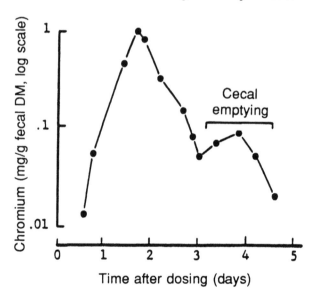

Figure 5.6. Chromium mordant excretion in a 90-kg pig fed fine bran (from Van Soest et al., 1983a). This is an example of a biphasal excretion pattern involving cecal emptying.

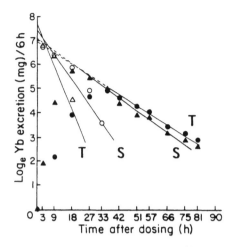

Figure 5.7. Ytterbium excretion in sham-operated (S) and transection colonectomized (T) horses 3 wk after surgery (from Bertone et al., 1989). Transected horses showed a significant increase in K_2 passage kinetics but did not differ in K_1 values.

the blind sac by the route of entry. This provides the possibility of the cecum being a special environment with unique retention of selected feed fractions. The cecum tends to empty pulsatively every day or two, leading it to be more characteristic of a batch culture. The passage rate is difficult to calculate under these conditions, since the cecal emptying can prevent kinetic marker calculation (Figure 5.6).

A curious situation is the consumption by Central American horses of certain large tree seeds, which, eaten whole, appear to enter the cecum and can have very long residence times (Janzen et al., 1985). Even more curious are the cecotrophs (rabbit and lemming), which selectively retain fine ingesta in the cecum that is passed out as night feces and reingested, the day feces being passed directly out and not recycled. These animals retain liquid for very long periods. Because of the selective reingestion it is not possible to calculate the actual retention on one pass from mouth to anus.

Table 5.2 shows that the capacity of the cecum relative to that of the colon varies among animal groups. The larger hindgut fermenters have a reduced cecum relative to the smaller rodents. Nevertheless, the cecum forms a sizable portion of the total hindgut.

Bertone et al. (1989) showed that when the colon was transected as completely as possible in horses, the solid passage rate for the slower pool (K_1) did not change, but there was a significant decrease in the speed of the faster-moving pool (K_2; Figure 5.7). This indicates that the K_1 pool is associated with the cecum and the K_2 pool with the colon. Digestibilities of fiber and total retention were reduced by the transection but tended to recover 6 months after surgery.

5.5 Utilization of Products of Lower Tract Fermentation

Products of fermentation include VFAs, particularly acetic, propionic, butyric, isobutyric, and isovaleric acids. These are produced along with some gas, CO_2, methane, hydrogen, and the normal gut bacteria. The process can be expressed in an equation:

$$\begin{array}{c}\text{Hemicellulose, cellulose, pectin, or lactose}\\+ \text{ } NH_3 \text{ or urea} \rightarrow\\\text{Microbes, protein, and lipid}\\+ \text{ VFA or lactate}\\+ \text{ gas}\end{array} \quad (5.1)$$

The equation, which must balance according to rules of chemical stoichiometry (Wolin, 1960), is written in a form that indicates alternative substrates and products that can affect the ecology and health of the colon. The equation is essentially the same in the rumen; however, the nutritional difference is that the bacteria containing protein are unavailable in the lower tract because the fermentation site is past the point of gastric digestion. Thus, in nonruminants, hindgut fermentation of fiber promotes fecal nitrogen loss in the form of microbial cells. This phenomenon should not be confused with digestion and utilization of amino acids, since the nitrogen sources for the hindgut fermentation are unspecific and include urea from the blood diffusing across the colon/cecal wall and secreted mucins.

The VFAs are normally absorbed into the bloodstream directly across the colon wall in the form of free acid, thus relieving the acidity and maintaining the pH of the colon above 6, as the normal fiber-digesting bacteria require. This process also is known to occur in humans, pigs, horses, and dogs (Argenzio and Southworth, 1974; Argenzio et al., 1974; McNeil et al., 1978; C. E. Stevens, 1988).

Fecal water is held by the water-holding capacity of the solids and by osmotic pressure. Although VFAs are small molecules capable of osmotic pressure, they are probably not a major factor in promoting fecal water because they are largely absorbed (Argenzio et al., 1974; McNeil et al., 1978; C. E. Stevens, 1988). Volatile acid concentrations vary in human feces (Ehle et al., 1982) and probably reflect the balance between rates of production and absorption. Generally, a higher concentration indicates a higher net absorption (Leng, 1970).

Urea and bicarbonate tend to flow into the colon, where the urea is hydrolyzed to ammonia and CO_2, the ammonia supplying the nitrogen requirement for the growth of the fiber-digesting bacteria. Imbalances in equation 5.1 can lead to problems. Too rapid a fermentation (which could be from lactose or from galactans) leads to rapid gas and acid production. Too much acid, particularly lactic acid, which can come from lactose or starch, can overpower the buffering mechanism and the pH of the colon will drop below 6, causing diarrhea and great discomfort. The normal bacteria are also affected because fiber-digesting bacteria cannot tolerate low pH.

If the fiber intake is at a very low concentration, the balance of equation 5.1 becomes altered relative to the substrate, as shown in equation 5.2:

$$\begin{aligned} \text{Protein, mucins} &\rightarrow \\ \text{VFA} & \\ + \text{ NH}_3 \text{ and amines} & \\ + \text{ small amount of microbes} & \\ + \text{ gas} & \end{aligned}$$

(5.2)

Because carbohydrate is deficient, the facultative bacteria switch to proteins, which may be from the gut lining and the mucins secreted by it. The bacteria no longer need urea because the nitrogen supply is now in excess of their needs. Putrefaction, with the production of deaminated products and excess ammonia, results.[2] This process is also associated with long retention and transit times. The excess amines and ammonia are absorbed and must be detoxified by the liver and excreted. This phenomenon constitutes one of the postulated mechanisms of colon cancer because of the stress and increase in cellular turnover involved in the process (Visek, 1978).

Estimates of the energy contribution of VFAs in nonruminant species indicate that VFAs form a significant part of the dietary energy (Table 5.8). Such averages depend on the diet, of course; if more fermentable carbohydrate reaches the lower tract, the contribution of VFAs will be larger.

---

[2] Fermentation of proteins is generally regarded as putrefaction and occurs in systems with a high protein-to-carbohydrate ratio. Microbial growth is inefficient and putrefactive end products accumulate, thereby significantly affecting the nature of feces. This phenomenon has been succinctly described in the following poem (Anonymous, 1963):

Protein Foundation

How inoffensive are the feces
Of all the graminivorous species
That grind on grain and graze on grasses,
Like sheep and horses, mules and asses,
Or, practiced in regurgitation,
Spend idle hours in rumination.
Such are the cows, the goats, the camels
And other ungulated mammals.
But ah, how offal they excrete
Who pry their protein needs from meat
From chops and steaks and yet, from cheeses
And pork and everything that pleases
From sulfurous eggs and oily fishes
And all the highly seasoned dishes;
Such is the ordurous part of man
Devoted to his frying pan.

Table 5.8. Estimated energy obtained from volatile fatty acids

Animal	Fermentation site	
	Foregut (kcal/BW$^{0.75}$)	Hindgut (kcal/BW$^{0.75}$)
Cattle	70–80	0–15
Sheep	57–79	—
Goat	37–46	—
Deer	25	—
Langur	>100	—
Rabbit	—	8–12
Porcupine	—	6–39
Beaver	—	19
Rat	—	9
Pig	—	5–28
Humans	—	0.7–20[a]

Source: Parra, 1978.
[a]Values vary depending on the fermentable fiber intake.

The major site of disappearance of plant cell wall carbohydrates in the horse and most large ungulates is in the large bowel (Figure 5.8). Volatile fatty acid absorption from the large intestine of the horse is similar to absorption in the rumen, although the rumen appears to absorb acetate and propionate faster (Figure 5.9). There appears to be some metabolism of VFAs, particularly butyrate, in the intestinal mucosa.

The microbial fermentation is an important facet of nitrogen balance in all animals possessing gut fermentation, ruminants and nonruminants alike. A model of nitrogen metabolism in nonruminant herbivores is shown in Figure 5.10. Feeding animals fermentable fibrous carbohydrate raises the ammonia requirement of the microorganisms to support their growth. This ammonia is supplied to a major degree by urea secreted across the gut wall, whether colon, cecum, or rumen. This increases fecal loss of microbial matter at the expense of urinary urea and is the reason that amino acid balances are of little significance. The overall effect of fermentation on nitrogen balance is greatest in species with large fermentation capacity and is less

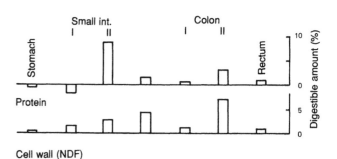

Cell wall (NDF)

Figure 5.8. The disappearance of protein and plant cell wall in the digestive tract of the horse (modified from Hintz et al., 1971). The disappearance of cell wall in the stomach and small intestine may represent a gastric attack on cell wall protein and, perhaps, associated hemicelluloses. Negative values for protein represent endogenous secretion. Upper and lower parts of the small intestine and colon are indicated by I and II, respectively.

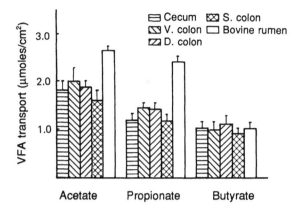

Figure 5.9. Transport of volatile fatty acids (acetate, propionate, and butyrate) by the equine large intestine (which is partitioned into the cecum, ventral colon, distal colon, and small colon) and bovine rumen epithelium (from Argenzio et al., 1974). Intestinal mucosa exhibit significant differences in relative rate of fatty acid transport compared with rumen mucosa.

important in species with lower fermentation capacity and faster transits.

There is also considerable disappearance of nitrogen in the lower tract. Very little available true protein is voided in the feces of ruminants; its quantity in nonruminants is possibly greater. Probably most microbial protein is lost in the feces, and coprophagy is the main means of recapturing it. The utilization of lower tract

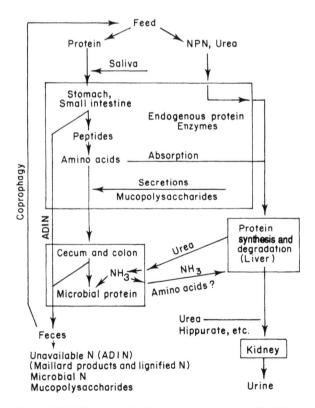

Figure 5.10. Nitrogen metabolism in nonruminants (modified from Slade, 1970).

microbial protein as well as nonprotein nitrogen (NPN) through this lower tract absorption is probably more important at low protein intakes (Houpt and Houpt, 1971; Glade, 1984).

One way to consider the question of amino acid absorption in the colon is to measure the proportion of microbial matter contaminating the metabolic fecal nitrogen in response to added dietary fiber (Mason, 1984). Such a response can be compared with microbial yields from fermented carbohydrate. The expected bowel loss of microbial cell mass is about equal to that expected from fermentation balance if dietary sources are the only substrate. Some utilization of nitrogen could be argued on the basis that endogenous mucopolysaccharides contribute significantly to the fermentation (Vercellotti et al., 1978) and that observed yields are therefore lower than expected from fermentation balance.

Horses survive in the wild on low-quality forages with protein contents less than 6% of the dry matter. Efficient nitrogen use is expected in this situation. Perhaps the horses practice coprophagy (Francis-Smith and Wood-Gush, 1977) or have higher intakes and passage rates. It should be noted that nonruminants digest and use essential amino acids in the diet, and lower tract fermentation is supported entirely by nonspecific nitrogen sources such as urea and mucus.

5.6 Nonruminant Grazing Strategies

Grazing herbivores are not a uniform group. They use at least three nutritional strategies. One, of course, is exemplified by the grazing large ruminant that has evolved a large retentive pregastric fermentation designed for maximum energy extraction from cellulosic carbohydrates. Species that obtain most of their dietary energy from grasses include, besides a few smaller specialized antelopes, nonruminants that tend toward consumption of bulk at the expense of efficient fermentative action. Their relative intakes are much higher than intakes of grazing ruminants. An example is the white rhino,[3] a true grazer with a digestive capacity equal to that of the large grazing ruminants. Other grazers, in order of decreasing capacity, are the hippos, the equids and elephants, and the panda. There are apparently no ruminants that exploit grasses as inefficiently as elephants and pandas. Thus there is a major group of nonruminant grazers (bulk and roughage eaters) that eat for volume and low extraction for which there is no ruminant counterpart. It has been suggested that dinosaurs probably belonged to this class, as do certain grass-eating birds such as geese (Buchsbaum et

[3] The name "white rhino" is derived from a corruption of the Afrikaans *weit rhino*, meaning the rhino with the wide muzzle.

Table 5.9. Digestion and passage in African (*Loxodonta africana*) and Asian (*Elephas maximus*) elephants fed on grass hays

	African	Asian
Number of observations	39	42
Body weight (kg)	2805 ± 506	2623 ± 813
Intake per day (% body weight)	1.7 ± 0.3	1.3 ± 0.3
Mean retention (h)	22 ± 2.0	26 ± 2.5
Digestibility		
Dry matter (%)	38.3 ± 3.0	43.1 ± 3.2
Neutral-detergent fiber (%)	38.1 ± 3.2	42.5 ± 3.0
Cellulose (%)	40.4 ± 3.1	44.3 ± 2.9
Hemicellulose (%)	41.2 ± 1.4	46.0 ± 3.0
Protein (%)	44.8 ± 6.8	54.0 ± 5.7
Fecal nitrogen (FN) (%)	0.89 ± 0.11	0.65 ± 0.08
Net metabolic loss (M_i) (%)	11.1 ± 1.7	13.3 ± 1.7
FN/M_i	0.080	0.049

Source: Data from Hackenberger, 1987.

Table 5.10. Digestion balance of the giant panda

	Male	Female
Body weight (kg)	123	114
Total daily food intake (% BW)	5.6	2.9
Bamboo intake per day (% BW)	4.2	1.6
NDF intake per day (% BW)	3.2	1.2
Proportion from bamboo (%)	90	54
Mean retention (h)	8	8
Digestibility		
Dry matter (%)	36	45
NDF (%)	19	25
Cellulose (%)	14	2
Hemicellulose (%)	25	29
Protein (%)	30	64
Digestible protein (g/day)	187	189
Digestible energy (Mcal/day)	11.9	7.7

Source: Data from Dierenfeld et al., 1982.

al., 1986). The advantage for these grazers is that their digestive parts have no filters to hinder passage and they can spend time eating bulk forage while their ruminant counterparts must stand aside to ruminate. The equids have received considerable attention in this regard, but two other cases (elephants and pandas) have been only recently accurately categorized.

5.6.1 Elephants

The two species of living elephants are the African (*Loxodonta africana*) and the Asian (*Elephas maximus*). There are two sets of elephant digestion data available: those of Foose (1982) on a few animals, and the more extensive studies of Hackenberger (1987), who found faster passage and lower digestibilities in both species than did Foose. Hackenberger also showed the African species, although larger, to have shorter retention times and lower digestion abilities than the Asian species (Table 5.9). Fecal nitrogen losses were larger in the African elephant, but net metabolic loss (M_i) was greater in the Asian species, with larger microbial losses as well. Perhaps the nitrogen losses in the African species are from unfermented endogenous matter. The data support the view that both species tend toward higher intakes, faster passage, and lower extraction relative to other herbivores.

5.6.2 Pandas

Pandas are remarkable for their specialized feeding on bamboo, a treelike grass. Of the two extant species, both of which are exclusive to Southeast Asia, the giant panda has been classified with the bears (Ursidae) and the lesser red panda with the raccoons. These classifications are controversial, however, and some have suggested that the two pandas are more closely related to each other and fall into a group intermediate between raccoons and bears (Schaller et al., 1985). Alter-

natively, O'Brien et al. (1985) suggested that there has been convergent evolution to similar herbivory in the two pandas.

The red panda, native to the Himalayan region from Nepal eastward into China, weighs 4–5 kg; males are larger than females (Roberts and Kessler, 1979). Like the giant panda the red panda eats bamboo and probably lesser amounts of small mammals, bird eggs, blossoms, and berries and other fruit (Roberts and Gittleman, 1984). Also like the giant panda the red panda has a short carnivore-like digestive tract and a modified thumb that enables it to grasp bamboo while feeding (Holmgren, 1972). It is likely that the red panda eats large amounts of bamboo with rapid passage, low digestibility, and little fiber utilization, although there are no data to substantiate this.

Most of the detailed panda studies have been conducted on the giant panda in captivity (Dierenfeld et al., 1982; Hirayama et al., 1989) and in the wild (Schaller et al., 1985). The digestion studies performed by Dierenfeld et al. (1982) indicate high neutral-detergent fiber (NDF) intake—higher in males than in females (Table 5.10). Digestibility of fiber is low, and hemicellulose is the only structural carbohydrate that contributes to the panda's energy budget. Despite the low digestibilities, the daily intake of protein and energy are above maintenance level. Bamboo is unique in the grass family in having relatively high crude protein in leaves (up to 15%) even though the NDF is about 80%.

Schaller et al. (1985) quoted the seasonal variation in dry matter digestibility in wild pandas as 12% in spring, 23% in summer, and 29% in winter, with an overall average of 17%. The average digestibility of hemicellulose was 22%. This indicates that digestibility in the wild is lower than in zoos, but it may reflect the fact that pandas in the wild consume more bamboo. Mean retention time in the wild of 5–13 h is similar to the values in Table 5.10.

The giant panda has a typical carnivore's digestive

Figure 5.11. Baboon and impala feeding on the same acacia tree in Kruger National Park, South Africa, an example of convergent feeding behavior. (Photo courtesy of B. D. H. van Niekerk and D. E. Hogue.)

tract, shorter and unsacculated as compared with most other herbivores. The rapid passage rate is not conducive to fermentation, and cellulose digestibility is not significantly above zero. This agrees with the observation of Hirayama et al. (1989), who found low counts of obligate anaerobic bacteria in panda feces.

Most studies conclude that the giant panda represents the most extreme adaptation to high intake and low extraction, tendencies also seen in equids and elephants. While pandas seem to feed almost exclusively on bamboo, they have been observed to catch rodents and will eat some meat if offered (Schaller et al., 1985).

5.7 Primate Feeding Strategies

The mammalian order Primates is of interest in this discussion of ruminants because it features evolutionary developments in herbivory parallel to those of the ungulates. Further, the role of fiber in human nutrition has become an area of major interest in relation to diet and disease. The application of the principles of rumen fermentation to the human colon is particularly relevant.

The feeding strategies of primates range from insectivory to omnivory, frugivory, and folivory. Setting aside the insectivores, omnivores (e.g., humans) clearly have a herbivorous ancestry. Humans' social evolution has proceeded faster than that of their digestive tract and metabolism, leading to the dietary epidemiology of low fiber and high meat and fat diets in modern civilization.

Man's closest relatives, the pongid apes, have diverged in dietary habits. Gorillas are exclusively vegetarian and lean toward folivory. Orangutans are largely frugivorous; chimpanzees are more versatile, leaning toward frugivory, but they also catch and eat an antelope now and then. Other groups such as baboons (cercopithecids) are omnivorous and eat a wide variety of plants. Much of their natural diet is highly fibrous. Male baboons eat 50% or more NDF; females, which are smaller, are more selective (Demment and Van Soest, 1985). Other groups, such as New World monkeys and colobines, are folivores. Colobines in Africa and langurs in India and Southeast Asia have rumen-like pregastric fermentation. Fermentation capacity may supply most of their energy in the form of VFAs, but no passage or digestion data are available to confirm this.

Thus it seems that some of the more advanced primates have converged with small ruminants in their feeding strategies (Figure 5.11). Small antelopes in Africa also have diversified into frugivory and folivory as concentrate selectors, but they are at present not well described.

5.7.1 Primate Digestion Studies

Digestion studies are available for only two nonhuman primate species: howler monkeys (Milton et al., 1980) and chimpanzees (Milton and Demment, 1988). Digestion balances were conducted on howler monkeys fed their natural diets, fruits and leaves from *Ficus* and *Cecropia*. Animals weighing 3–8 kg digested 75% cellulose and 81% hemicellulose in fruit, and 33% cellulose and 36% hemicellulose in leaves. There was no association between body size and digestive capaci-

Table 5.11. Fiber digestion by chimpanzees fed wheat bran in comparison with humans and pigs

Species and diet	NDF		Digestibility	
	Dietary content (%)	Digestibility (%)	Cellulose (%)	Hemicellulose (%)
Chimpanzees				
Low fiber	15	71	68	77
High fiber	34	54	38	63
Human				
High fiber	10–15	51	41	58
Pig				
High fiber	17	65	24	74

Source: Milton and Demment, 1988.

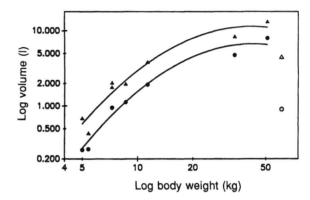

Figure 5.12. Volume of hindgut (cecum + colon) and total gut of hominids plotted against body weight (from data summarized by Milton and Demment, 1988). The hindgut volume (○) and total gut volume (△) of humans are plotted with those of six ape species (●, ▲). Humans have smaller hindgut and total gut volumes than other hominids relative to body weight. Listed in order of ascending body weight in the figure, ape species represented are *Hylobates lar, Hylobates pileatus, Symphalangus syndactylus, Pongo pygmaeus* (immature specimen), *Pan troglodytes,* and *Gorilla gorilla.*

ty, yet fiber consumption (NDF) was 3–7% of body weight.

The chimpanzee studies were conducted with standard wheat bran as a fiber source so that results on digestion and passage could be compared with studies of humans and pigs (Table 5.11). The data show that chimpanzees digest relatively more fiber than humans or pigs. Humans digest more fiber than pigs, but their intake is much lower. Cellulose and hemicellulose digestibilities are significantly correlated with mean retention time in chimpanzees (Milton and Demment, 1988) and in humans (Van Soest, 1977). Larger chimpanzees had higher fiber digestion and longer retention times.

Although the digestive patterns of chimpanzees and humans have similarities, the proportion of the gut devoted to fermentation is smaller relative to body size in humans than in other hominids (Figure 5.12). The hindgut volume in apes is about 52%; in humans, about 17–20%. The fact that all humanoid apes have relatively more of the digestive tract devoted to fermentation indicates that humans' ancestors were decidedly herbivorous. Man escaped the constraint of body size and digestive capacity through technological and social innovations that permitted improved net return from food gathering and a reduction in dietary bulk (Milton and Demment, 1988).

5.7.2 Dietary Fiber and Humans

Fiber has come into the limelight in popular nutrition largely because of hypotheses advanced by Burkitt (1973) and Trowell (1975) regarding the relationship between lack of fiber in the diet and human disease. The traditional attitude among monogastric and human nutritionists had been to treat fiber as a negative index of quality. Although nutritionists recognized its value as a natural remedy for constipation, it was otherwise ignored. The hypotheses advanced by Burkitt and Trowell have resulted in an enormous rise in dietary

fiber research, which is confirming, to a considerable extent, the fiber-disease association.

Burkitt, a surgeon who worked for many years in African hospitals, was struck by the almost total lack of coronary heart disease, colon cancer, diabetes, and diverticular disease in native Africans, while Europeans living in the same countries have relatively high incidences of these diseases. Africans who adopt Western diets and ways of living tend to develop the same diseases. The differences persist when the epidemiological data are adjusted for age and mortality differences between the populations. Historical evidence indicates that these diseases were less frequent in Europe in past centuries, before cereals were refined to produce white flour and sugar and animal product consumption increased.

Associations between diet and heart disease, cancer, diabetes, and so on are not new; during the 1960s cholesterol and protein from animal fats were implicated as causative agents for these diseases. The evidence for associations between diet (fiber included) and disease rests on epidemiological surveys. The relationships are correlations only; they provide no actual support for cause and effect.

The hypothesis that lack of fiber results in certain diseases stands as an alternative to the animal fat, cholesterol, and protein hypotheses, but it is difficult to separate these alternatives because human diets high in fiber tend to be low in animal products, and vice versa. There is even some evidence that fiber protects against heart disease associated with animal product intake (Morris et al., 1977). The negative correlation between fiber intake and consumption of animal products leaves a chicken-and-egg argument as to whether animal products or fiber deficiencies are the cause of disease. Because humans tend to eat to satiety, the increased intake of one component signifies that the intake of some other dietary component must necessarily decrease.

Theories Regarding Fiber and Disease in Humans

If fiber *is* a preventive factor in human disease, why is this so? Fiber has been proposed to exert beneficial effects on (1) passage of digesta, (2) colon bacteria, (3) physical binding by indigestible fiber fractions, and (4) fecal composition. From these effects can be derived properties of dietary fiber that would be desirable and which could provide a working basis for evaluating fiber quality in human diets.

Dietary fiber is said to benefit human health because it increases the frequency of bowel movement and the softness of the stool, the latter being associated with water content, which reduces intracolonic pressure and therefore colon stress and disease (Burkitt, 1973). These concepts emphasize the physical property of water-holding capacity of fiber and overlook fiber's role in supporting normal gut microorganisms. Another, not mutually exclusive, hypothesis relates colon health to metabolism of normal gut microflora, whose growth is supported by fibrous carbohydrate. Visek (1978) suggested that potentially toxic ammonia and amines are scavenged by fiber-digesting bacteria that need these nitrogen sources for cellular growth. A survey of recent advances in the understanding of how fiber is related to certain diseases follows.

Diabetes. J. W. Anderson (1985) treated diabetes with dietary fiber and was able to reduce or eliminate insulin dependency in many patients, although this depended on the individual having a pancreas with some insulin capability. Viscous types of fiber are the most effective in this treatment and appear to work in part by reducing the rate of absorption of sugar after a meal (D. J. A. Jenkins et al., 1977), thus placing less urgent demands on insulin. Another factor reducing the need for insulin is the fermentation of fiber in the colon. Propionic and isobutyric acids are among the major products of this fermentation. These acids are efficiently absorbed (McNeil et al., 1978; Fleming et al., 1983), enter the bloodstream, and are metabolized by the liver into glycogenic products that do not require insulin for metabolic induction (J. W. Anderson, 1985).

Cancer. The hypothesis that fiber offers some protection against colon cancer is supported by studies on rats and a limited number of epidemiological surveys in humans. Several additional hypotheses have been formulated concerning the protective effects of fiber:

1. Lignin fractions and perhaps pectins and gums sequester cholesterol and other substances potentially convertible into carcinogens and cause their loss in feces.
2. Transit time is reduced, thus contact with mutagens is decreased.
3. Fiber reduces stress and cell turnover in the colon wall. Ammonia, amines, and perhaps lipid compounds are used by growing bacteria, which are overfed on fibrous carbohydrates but deficient in nitrogen and other cofactors, which are scavenged and incorporated into the cellular mass, where they become harmless (Visek, 1978).
4. Fermentable fiber promotes growth of normal safe bacteria over potentially pathogenic types that may convert compounds into carcinogens.
5. There are special phenolic compounds in the lignin fraction that protect DNA from alteration.

The above hypotheses cannot be regarded ·as established—these are ideas under active investigation at the present time. Many researchers feel that the first two mechanisms are too simplistic and that the microbes and their fermentation are somehow intimately involved. This is supported by observations on inducible carcinogens that show that the more fermentable types of fiber are the most protective. Some fibers do not protect, yet they may be fermentable.

Heart Disease. The case for dietary fiber protecting against heart disease is similar to that for colon cancer. The hypothesis involving the binding of cholesterol and lipid is prominent; it is less clear how fermentation might be involved. The best epidemiological information involving fiber is from Morris et al., 1977, a study showing that higher fiber intake reduced the risk of disease from consumption of fat. Fiber and fat are often associated in dietary intakes: fiber negatively with calories and fat positively so.

Other Diseases. Fiber seems to act to prevent diverticulitis, a disorder that may be preliminary to colon cancer. The lower the amount of fiber in the diet, the higher the intracolonic pressure. Dietary fiber may increase water content and reduce firmness of feces. Here the hypothesis of Stephen and Cummings (1980) is relevant. Essentially the hypothesis states that the action of fiber on the large bowel is to a considerable extent mediated though the anaerobic microorganisms that grow on the fermenting fiber. Further, there is a difference between brans (less fermentable) and vegetables, fruits, and soluble types of dietary fiber in regard to the microbial effect. Stephen and Cummings (1980) measured the microbial and residual dietary fiber components in feces. Their data show that cabbage contributes less fiber to the feces and more micro-

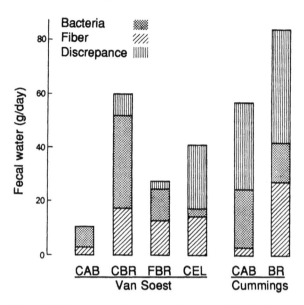

Table 5.12. Influence of type of fiber on gastrointestinal responses in humans

Fiber type	Gastric emptying	Pushes transit	Promotes colon fermentation
Insoluble lignified (coarse brans)	+	+ + + +	+ +
Vegetable or fruit fiber (pectinaceous)	+ +	+	+ + + +
Gums	+ + +	+	+ + +
Processed celluloses	+	+ +	− a

Source: Van Soest, 1993.
a Decreases fermentation relative to control.

Figure 5.13. The calculated increments of fecal water elicited by dietary fiber over controls based on the expected hydration capacities of residual dietary fiber and bacteria (from Van Soest, 1993). Microbial mass was estimated by metabolic fecal nitrogen in the Cornell University study; Stephen and Cummings (1980) used a different procedure. Diets: CAB = cabbage-based; CBR = coarse wheat bran; FBR = fine wheat bran; CEL = wood cellulose; BR = wheat bran (fed only in Stephen and Cumming's study).

bial matter, while bran does the reverse—fewer microbes and more fecal fiber.

Fecal Water

Stephen and Cummings (1980) calculated the positive responses of microbial matter and water to cabbage and bran diets and postulated the mechanisms by which the two fiber sources promote fecal water differ: microbes dominate in the case of cabbage, and undigested fiber is more important in bran. Their data are compared with those of a Cornell University study in Figure 5.13. The data agree that cabbage is very fermentable; however, the Cornell experiment showed a much smaller, though still significant, response. Finely ground wood cellulose and fine bran produced less water and bacteria but relatively equal fibrous residues. The bacterial response was least in the case of wood cellulose, and water response was poorest in the case of fine bran.

These results do not support the broad generalization that insoluble lignified sources of dietary fiber uniformly promote fecal water through indigestible residue. Also, while microbial responses can be highly variable, the generally poorer microbial responses to fine bran and wood cellulose may be the result of chemostatic effects in which increased passage and colonic turnover influence microbial efficiency (see Chapter 16). The efficiency rises because faster transit causes the voiding of a younger microbial population

with less death and recycling of nutrients, resulting in more VFAs and fewer microbes.

Coarse bran fiber also promotes the loss of starch from the terminal ileum, leading to increased fermentation of the escaped starch (McBurney et al., 1988). Insoluble coarse fiber is more effective than fine bran in promoting transit and alleviating constipation, as Table 5.6 has already shown (Van Soest et al., 1978b; Van Dokkum et al., 1982).

These observations do not diminish the importance of the physical properties of dietary fiber in eliciting colonic responses. It has been suggested that water-holding capacity promotes fecal response (McConnell et al., 1974), but Stephen and Cummings (1980) pointed out that a high degree of fermentation may ruin this capacity. Hydration capacity and cation exchange are possible factors promoting fermentibility (McBurney et al., 1983), and if one considers the replacement of fermented fiber with produced microbes, that original hypothesis may still stand (McBurney et al., 1985).

A comparison of wheat bran and psyllium gave similar results: both are only partially fermentable, and the unfermentable fraction pushes transit (J. S. Stevens et al., 1987b). This is in contrast with pectin, which is completely fermentable and pushes transit much less. Thus it appears that even within the gums the same variability may be seen that we saw in less soluble sources of dietary fiber. The considerable water-holding contribution is due to bacteria that have a high exchange and water-holding capacity.

Table 5.12 surveys the effects of various kinds of fiber in human diets. Generally, fermentable sources do not push transit, while those with an insoluble core do promote transit.

Weight Control

Unsubstantiated claims have been made for fiber as a weight-reducing aid. While animal studies do support the claims, the relative level of fiber feeding that caused lower caloric efficiency may be beyond what is practical in human diets. Optimization of fiber intake in pigs, above a very low dietary fiber control, increased growth efficiency and weight gain (Kornegay,

1981). There is some evidence to suggest that increasing the amount of fiber in the diet results in reduced intake, weight loss, or both over a period of days (Evans and Miller, 1975; J. S. Stevens et al., 1987a), and that fiber helps people consciously restrict their intake to eat less and lose weight (Dodson et al., 1981). A recent Cornell study with women (J. S. Stevens et al., 1987a) indicates that consumption of psyllium reduces caloric intake and wheat bran has no effect. In the case of bran the reduction in intake of digestible calories was offset by increased food intake.

One problem in determining the influence of fiber on energy balance is estimating fiber's contribution in the form of VFAs to the energy of the diet. Atwater and others calculated the caloric value of many foods containing fiber in actual balance studies (for review, see Merrill and Watt, 1973; Miles et al., 1988). The results of these studies were generalized for food classes, and coefficients were developed to estimate the available calories of the foods in each class, as listed in the old USDA *Agricultural Handbook,* no. 8. American food companies calculate the caloric value of specific high-fiber foods by measuring the gross energy value and the crude fiber content of the food. They estimate the available calories in the food with the following formula: (Gross energy of intact food [kcal]) − (crude fiber [g] × estimated gross energy of fiber [kcal/g]). This method is based on the assumption that fiber yields no calories from fermentation. Since crude fiber underestimates dietary fiber, the values obtained in this manner are similar to those published in the USDA handbook. The underestimation of the fiber content of the food was largely counterbalanced by the contribution of calories via VFA from the fermentation of fiber, however, resulting in caloric estimates similar to those obtained from metabolic studies. Miles et al. (1988) found that the energy from VFAs tends to be offset by increases in fecal fat and nitrogenous losses so that at higher fiber intakes total calories may be somewhat overestimated by the values in the handbook.

Recently the Food and Drug Administration (FDA) suggested substituting dietary fiber in the same calculation. This change led to a considerable underestimation of the caloric value of high-fiber foods and a flourishing dietary fiber food industry based on faulty claims for weight loss.

The caloric value of the fiber in a food depends on its fermentability. Highly fermentable (digestible) fibers like the pectin in fruits and vegetables yield more calories than cereal brans, which are more lignified and less fermentable. The caloric values of wheat bran and psyllium are each about 1 kcal/g according to balance measurements in women (J. S. Stevens et al., 1987a). By comparison, the value of highly fermentable vegetable fiber and pectin is about 3 kcal/g. The real absorption of calories in the form of VFAs in the colon is really several times larger than observed, but this effect is counterbalanced by large endogenous fecal losses, which vary considerably with the type of dietary fiber. Again, not all fibers are equal.

5.8 Fiber Requirements

Ruminants generally require adequate dietary fiber for normal rumen function. Positive effects of fiber in the diet also have been observed in several nonruminant species, however, including pigs, guinea pigs, and humans. The optimum fiber for humans has been estimated at 40 g/day, which may correspond to about 10% of dietary dry matter intake. Feeding alfalfa to growing pigs affects gross feed efficiency and body composition. Up to a certain level (6–12% NDF in total diet) fiber does not alter use of digested energy, and may even improve it (Kornegay, 1981). Feeding more fiber elicits some loss in overall efficiency in most species (Figure 5.14).

Dietary fiber acts to alter body composition. Fiber-fed pigs are leaner, have less fat, and have a larger gut fill and a heavier gut mucosa, probably stimulated by enhanced VFA production (Kass et al., 1980; Pond et al., 1988). Butyrate is metabolized by the colonic and cecal mucosa just as it is in the rumen wall and is quite stimulatory of mucosal growth.

An increased intake of fermentable fiber leads to a greater portion of the energy being derived from VFAs, leading in turn to greater gluconeogenesis from

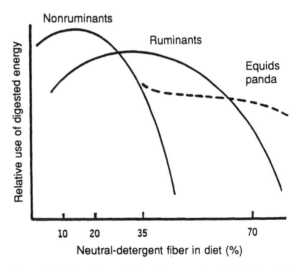

Figure 5.14. The optimum fiber level for nonruminants is less than that for ruminants because microbial protein is lost in the feces of nonruminants and their digestive capacity for fibrous carbohydrates is less than ruminants' capacity. A few nonruminants tolerate very high fiber diets by increasing intake to achieve their requirements from nonfiber components. This results in fast passage and low fiber digestibility. Ruminants fail on low-quality very high fiber diets because of the cost of rumination and other digestive work required to eliminate lignified fiber from their complex gut. (From Van Soest, 1985.)

propionate and increased metabolism of butyrate by the colonic mucosa. The gluconeogenic effect of propionate may contribute to the beneficial effect of high-fiber diets for diabetics. All animals derive some of their energy from VFAs, the proportion being associated with greater intake of fermentable fiber. Caloric inefficiency in ruminants has long been associated with high-fiber diets and high ratios of acetate to propionate among the fermentation products.

A final question: Does increased intake of fiber and associated VFAs induce caloric inefficiency and provide a basis for weight control in humans? Most animals attempt to eat to their level of energy requirement (i.e., satiety), and the declining phases in Figure 5.14 may represent failure to achieve optimal intake due to dietary bulk (see Chapter 21). In the aforementioned pig studies, fiber increased leanness but did not decrease body weight because of compensatory increases in gut weight. An argument based on the model in Figure 5.14 is that a moderate increase in fiber intake will (at least up to the optimal level) be compensated for by increased food intake, so no reduction in net caloric intake will occur unless gut fill becomes limiting. This probably occurs past the optimum point and involves fiber intakes that most people will not accept. On the other hand, the argument from rumen metabolism is that acetate production is inefficient relative to ATP use, so the heat increment is elevated even though caloric intake may not have changed. Fiber may also promote loss of endogenous lipid. This is an aspect of ruminant studies that needs application and understanding in nonruminant and human nutrition.

6 Plant, Animal, and Environment

Soil, weather, animals, and disease all influence plant growth and composition. Plants derive their energy from the sun and use it to fix carbon into their cellular structures. The distribution of this carbon and energy within plant parts is greatly affected by the external factors of the environment, and nutritive value and forage quality are consequences of these conditions.

6.1 Factors Affecting Plants

Two major strategies employed by plants for survival are relevant to the nutritive quality of forage: storage of nutrients and defense against external threats. Nutrient reserves are essential for survival during cold or dry periods and to support regrowth following adverse weather, defoliation, grazing, or cutting. Reserve substances are generally highly digestible. On the other hand, defensive compounds—including lignin, cutin, phenols, terpenes, and alkaloids—are required for resistance to wind, disease, and defoliation, and their presence generally reduces the nutritive value of the forage plant. Resistant substances are often unavailable to the plant itself and thus are synthesized at the expense of reserves and the metabolic pool (Figure 6.1). Stress, weather, disease, and herbivorous animals restrict the deposition of reserves and promote their mobilization. At the same time, deposition of resistant structures like lignin and cell wall is also restricted.

Environmental factors may be divided into those that alter reserves and those that promote the development of resistant structures (Figure 6.2). The nutritive value of forage is primarily determined by its composition; consequently, a sequence of cause-effect relationships exists among environment, plant response, composition, and nutritive value.

6.1.1 Physiological Factors

A plant must have reserves if it is to survive periods of stress. If the aerial part remains vegetative, the re-

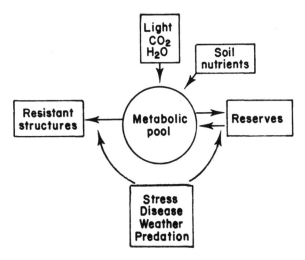

Figure 6.1. The relation of environmental factors to plant metabolic components. The weather and disease and predation stress modify the distribution of photosynthetic products and mineral nutrients from soil among the metabolic pool, plant reserves, and protective substances. Arrows indicate the direction of possible mobilizations. Resistant structures (lignin and cellulose) are irretrievable sinks. Stress inhibits deposition of energy in resistant structures and tends to deplete resources. The size of the metabolic pool is diminished by high metabolic rate and environmental temperature.

serves may be used to maintain the quality of that tissue. If tissue death occurs through senescence, reserves are often moved into storage organs or seeds, leaving behind depleted dead material of very high cell wall content. Annual plants store most of their available energy in seeds, and perennial plants deposit much of it in roots, lower stems, or cambial layers. The digestibility of the cell wall may remain high if reserves are formed at the expense of cell wall development and lignification, often a feature of standing hay that develops under arid conditions. Aridity also retards deterioration of dead matter through biotic processes.

Environmental factors that stimulate plant growth promote the use of reserves and the development of aerial tissue. Plant development involves maturation, and a plant's nutritive value eventually declines through deposition of photosynthetic carbon in the ir-

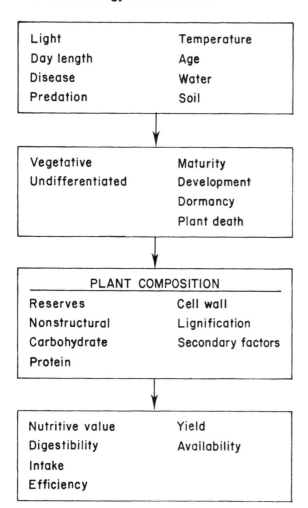

Light	Temperature
Day length	Age
Disease	Water
Predation	Soil

Vegetative	Maturity
Undifferentiated	Development
	Dormancy
	Plant death

PLANT COMPOSITION

Reserves	Cell wall
Nonstructural	Lignification
Carbohydrate	Secondary factors
Protein	

Nutritive value	Yield
Digestibility	Availability
Intake	
Efficiency	

Figure 6.2. The cause-effect relationships between environmental factors and nutritive value. Factors generating positive effects on quality are on the left; negative factors are on the right.

retrievable sink of structural matter. As forage plants mature, the accumulation of structural cell wall dilutes the metabolic pool as represented by cell contents (see Figure 6.1).

6.1.2 Genetic and Evolutionary Factors

Plants have adapted to specific environments through selective processes that can be reduced to the common factor of species survival under a particular set of ecological conditions. Most domesticated forages originated in conditions involving interaction with grazing animals. Plants and animals in grazing and browsing systems are interdependent. The nutritive value of the plants is essential for the survival of grazing animals. The plants, in turn, depend on the animals to maintain the grazing environment, disperse their seeds, and recycle the system's nutrients. Plants exist not to be food for animals but rather to survive, and the evolution of a grazing system involves selec-

tion for plants of higher quality as an advantage for their own survival. Grazing originated in the semiarid grasslands of Asia and Africa.

Plants that adopted the strategy of symbiosis with animals as opposed to heavy defenses against them (Chapter 6) needed to maintain reserves for regrowth after grazing and to avoid excessive energy deposition in the sink of lignification. The need for regrowth led to a further strategy of spreading into disturbed areas.

6.2 Species Differences and Plant Morphology

Figure 6.3 shows that not all forage species have the same digestibility when grown under identical conditions. The digestibility of legume stems is lower than that of most grasses at any stage of growth; and the digestibilities of respective grass species also vary. For example, orchardgrass shows lower stem digestibilities and lower digestibility overall than timothy, brome, or ryegrass. Although orchardgrass is a productive grass, this feature has led to its virtual abandonment in western Europe, although it is still grown in North America.

Similar differences are apparent among tropical species: Pangola grass (*Digitaria decumbens*; a C_4 grass) declines in digestibility less than most other tropical grasses and tends to remain vegetative, in contrast to others such as *Panicum maximum*. Cogon grass (*Im-*

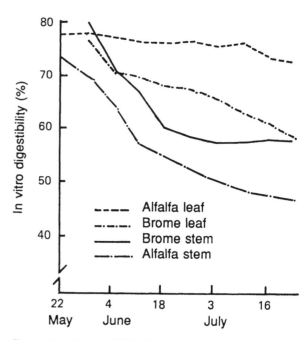

Figure 6.3. In vitro digestibility of leaf and stem portions of alfalfa and bromegrass averaged over three seasons in Ontario (from Mowat et al., 1965). Note the relative constancy of alfalfa leaf quality relative to age while grass leaves deteriorate with age. Compare with Figure 6.4.

Table 6.1. Composition of Philippine forages

Grass	Age (days)	TDN (%)	NDF (%)	ADF (%)	Lignin (%)	Crude protein (%)	Silica (%)	Ash (%)
Pennisetaum purpureum	20	59	54	36	3.3	7.3	7.0	14.9
(napiergrass)	45	54	62	43	6.3	7.5	5.9	15.3
	60	51	66	47	8.8	6.7	5.6	13.8
Panicum maximum	20	63	62	41	5.6	11.2	6.2	12.3
(guineagrass)	40	55	65	45	5.6	8.5	6.0	12.0
	60	47	72	51	8.0	5.5	5.9	11.5
Brachiaria mutica	20	57	60	34	4.2	11.0	5.6	13.3
(paragrass)	40	49	65	39	5.4	6.5	6.1	11.5
	60	51	67	38	5.3	4.2	4.7	9.7
Cynodon plectostachyus	20	65	69	39	4.9	10.8	3.5	9.9
(stargrass)	40	57	71	42	6.7	9.1	3.8	9.1
	60	45	74	47	8.9	5.5	4.9	9.0
Dicanthium aristatum	20	70	63	41	3.6	10.2	5.7	11.4
(alabang X)	40	63	68	47	5.7	6.4	5.8	10.4
	60	57	73	51	6.8	4.0	5.6	9.1
Imperata cylindrica	20	57	77	46	5.6	8.5	3.9	6.7
(cogon grass)	40	50	75	47	5.8	6.7	5.0	7.8
	60	41	74	46	8.5	5.2	4.7	7.7
	90	36	73	50	8.7	4.3	6.7	8.7
Setaria sphacelata	20	61	56	33	2.7	11.1	4.9	14.1
(golden timothy)	40	55	63	42	6.6	8.5	3.5	11.6
	60	51	71	47	8.6	5.3	4.9	11.7
Stylosanthes gracilis (Townsville Lucerne)	75	55	57	44	13.8[a]	9.7	0.6	5.2
Calopoqonium mucunoides (calopo)	75	47	54	41	13.4[a]	16.8	0.3	7.4
Centrosema pubescens (centro)	75	38	63	45	15.6[a]	17.0	1.2	7.0
Macroptilium atropurpureus (siratro)	75	46	55	44	12.3[a]	13.9	0.7	7.0

Source: Robert Grant, 1973.
Note: TDN = total digestible nutrients; NDF = neutral-detergent fiber; ADF = acid-detergent fiber.
[a]Lignin values probably reflect the presence of tannins.

perata cylindrica, see Table 6.1) is poorer in quality than other forage species grown in the same environments. Tropical legumes tend to be higher in crude lignin and protein and lower in cell wall than tropical grasses, and higher in cell wall and lignin than most temperate legumes. The crude lignin value is elevated by the presence of tannins in most tropical legumes.

6.2.1 Leaves and Stems

Forage frequently shows a decrease in leafiness and an increase in the stem-to-leaf ratio with age. Stems are often of a lower quality than leaves in mature forage. This generalization, however, is not universal; there are important exceptions. The quality of the stems compared with the leaves depends on the function of these structures in the particular plant species. Decline in quality is usually associated with an increase in the proportion of lignified structural tissue. In alfalfa and browse species the stems are structural organs and the leaves are metabolic organs. In grasses, on the other hand, the leaves have an important structural function through the lignified midrib. The result, in terms of nutritive value, is that alfalfa leaves maintain their

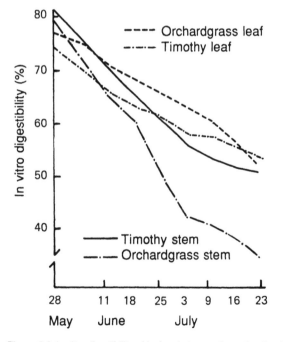

Figure 6.4. In vitro digestibility of leaf and stem portions of orchardgrass and timothy (from Mowat et al., 1965).

quality as they age (Figure 6.3), while grass leaves decline in quality, though not as rapidly as stems.

In some grasses (e.g., timothy and sugarcane) the stem is a reserve organ. This leads to the anomaly of stems having a higher quality than leaves, particularly at early stages of growth (Figure 6.4). Stem quality, furthermore, varies significantly among plant species. The leaf-stem ratio must therefore be used cautiously as an index of quality. It is more valuable in legumes than in grasses. If leaf digestibility is equal to or lower than stem digestibility, the ratio is useless.

6.3 Environment and Forage Composition

The chemical composition of plants, and consequently their nutritive value, is a result of the distribution of photosynthetic resources into the various plant tissues (Figure 6.1). This distribution into metabolic pool, reserves, and structural parts is relevant to vegetative forages. Generally, the lignified structural portion is unavailable to mobilization and thus is regarded as an irretrievable energy sink. Seeds produced at maturity likewise become unavailable energy. Nutritive value is usually assigned to the aereal part of the plant, which is classified according to the distribution of resources into cell contents (including the metabolic pool, reserves, and seed storage), which are potentially completely available nutrients, and the cell wall structure, which is incompletely and variably available depending on lignification.

Thus distribution of resources involves (1) the dilution of the aerial cell wall structure with metabolic reserves and seed storage, (2) the distribution of reserves between roots and aerial parts, and (3) the degree of lignification of the cell wall structure. In the simplest terms for vegetative forages, dilution of the cell wall structure with metabolic reserves and the lignification of cell wall are the two overriding forces.

The above concepts are required to explain the effects of climate and season on plants. Climate refers to the weather patterns peculiar to a geographical region or locality, and its effects on nutritive value of forage account for regional differences in composition. Seasonal changes in the weather pattern in a locality account for seasonal variation in forage composition and nutritive value. This latter association—composition and nutritive value—is the basis for various systems for predicting digestibility from composition. General equations based on one fiber value do not account for regional differences or the differences in forage quality in different regions.

In order to explain the effects of climate and season on forage quality it is necessary to factor out the environmental variables that affect forage composition one by one. These are, in declining order of importance,

temperature, light, water, fertilization, and soil. Disease and other plant stresses also influence composition. Studies of forage grown in growth chambers and greenhouses have allowed some assessment of the specific effects of individual environmental factors. Field studies have also contributed valuable information, particularly comparisons of the same or similar forage species grown in different climates.

6.3.1 Temperature

Lower digestibility at higher temperatures is the result of the combination of two main effects. High environmental temperatures result in the increased lignification of plant cell wall. High temperatures also promote more rapid metabolic activity, which decreases the pool of metabolites in the cellular contents. Photosynthetic products are thus more rapidly converted to structural components. This activity decreases nitrate, protein, and soluble carbohydrate, and increases the structural cell wall components. Enzymatic activities associated with lignin biosynthesis are also enhanced by increased temperature.

The general effects of temperature appear uniform in all species studies, although the quantitative effects of temperature on forage quality vary with plant parts and plant species (Figure 6.5). Temperature's greatest effect on plant development is through promoting the accumulation of structural matter. For example, plant species that remain vegetative, whether because of low environmental temperature during growth or genetic character, are almost always less lignified than plants that develop to the flowering stage under similar environmental conditions.

The behavior of alfalfa is characteristic of plants in which the leaves have no structural function. The leaves show little change in digestibility in relation to environmental temperature; however, lignification of stems and consequent lower digestibility occurs with hotter temperatures. This is offset by the unchanging digestibility of leaves and an increase in the leaf-stem ratio. Thus, a warmer environment results in a widening of the range in quality between the more and the less digestible parts of the same alfalfa plant. The digestibility of alfalfa declines with increasing temperature because the maturation process is more rapid. Digestibility also declines at an equal stage of growth unless counterbalanced by an increase in the proportion of leaves relative to stems. Under field conditions, where both day length (Section 6.3.2) and temperature vary, the digestibility of flowering alfalfa declines in midsummer and recovers in autumn (Kalu, 1976).

In grasses, both leaf and stem qualities decline with increasing temperature, the effect being more pronounced in tropical grasses. Leaf quality declines particularly as a result of lignification of the midrib, which

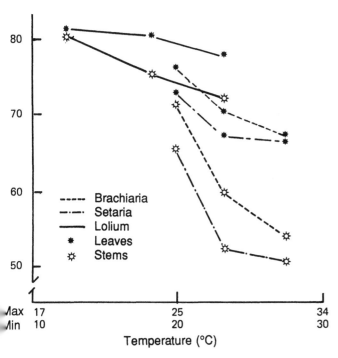

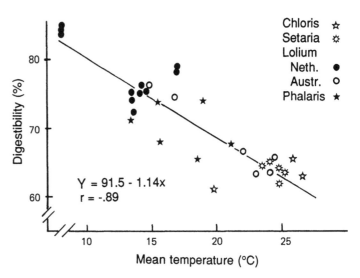

Figure 6.6. Relation of environmental temperature to digestibility of temperate and tropical grasses, summarized from world literature (from Minson and McLeod, 1970). Although digestibility is plotted versus mean temperature of growing period, data were also affected by day length and total light through latitudinal association. Compare with Figure 6.7, which uses some of the same data and presents an alternative view.

Figure 6.5. The effect of temperature on digestibility of leaves and stems of three grasses at tillering stage. Digestibility declines with temperature more severely in the tropical grasses (*Brachiaria* and *Setaria*) than in the temperate grass (*Lolium*). Maximum and minimum temperatures (day and night) are included in the X = axis. "Stems" include leaf sheath (Deinum and Dirven, 1975, 1976). Note the poorer quality of *Setaria* compared with *Brachiaria*.

6.3.2 Light and Day Length

The effect of light, the energy source for most plants, is exerted directly on metabolism through photosynthesis. Several parameters are involved, including total light received, light intensity, and day length. Total light available sets the upper limit of energy for plant use. Photosynthetic efficiency is low; only 1–3% of the total light received is actually fixed by the photosynthetic process. The end product of photosynthesis is glucose, and additional light promotes the accumulation of sugar and the general metabolism of nitrogen. Nitrate requires photosynthetic energy for its reduction to ammonia and synthesis into amino acids. Thus greater light promotes reduction in nitrate level. Cell wall components decrease with increasing light, probably through dilution by the nonstructural carbohydrate, amino acids, and organic acids formed.

Light intensity is influenced by the angular incidence of the sun, which decreases with latitude; light intensity under clear skies is greatest at the equator and least at the poles. Net total light is a product of the day length and the solar incidence.

Cloud cover and shade, which affect the amount of light plants receive, tend to decrease the nutritive value of forage. Nitrate accumulation in forage is maximum under cool, cloudy conditions, which reduce photosynthesis and the reduction of nitrate to amino acids. Moisture per se (Section 6.3.4) promotes plant development and lowers forage quality. Cloudy weather and moisture thus interact to produce the lower forage quality associated with wet climates.

contains the major portion of the lignin in grass leaves. Since stems also decline in quality at an equivalent stage, an increase in temperature usually causes an overall decline in grass quality. High temperatures promote a greater disparity in quality among plant parts, a factor leading to the greater advantage of selective feeding under tropical conditions (Deinum, 1976; Struik et al., 1985). There is also a tendency for warm season forage to have a slower fermentation rate (Fales, 1985).

The quantitative effects of temperature on the digestibility of grass are calculated by plotting digestibility versus temperature (all other factors are controlled). A partial regression obtained by Deinum et al. (1968) showed a decline of half a unit of digestibility per degree Celsius increase in temperature when light, age, maturity, and fertilization were controlled. Minson and McLeod (1970), however, obtained a corresponding value of 1.14 when they compared forages grown in different environments (Figure 6.6). The latter value may be of practical significance in comparing temperate and tropical forages. It includes the associative effect of day length because the data were derived from experiment stations at different latitudes (see Figure 6.6).

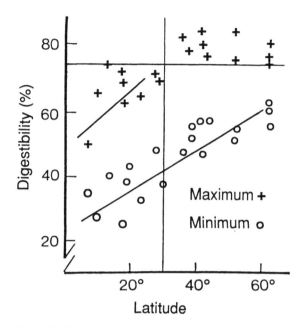

Figure 6.7. The relation between digestibilities of perennial grasses and the latitude at which they were grown. First-growth grasses usually have maximum digestibility (+), whereas grasses at maximum maturity have minimum digestibility (o). Digestibility of early first cuttings is constant in regions north of about 30° latitude, since first growth occurs after frost. Below 30° latitude, digestibilities of mature grasses decline progressively toward the equator (Van Soest et al., 1978a). Vertical bar at 30° latitude is the approximate northern/southern limit of the semitropics.

Few studies have been conducted to assess the effects of photoperiod on forage quality and digestibility. Longer dark periods probably reduce forage quality because nutrients are metabolized but none are produced. A changing photoperiod is characteristic of high latitudes, so interaction of photoperiod and with other factors of the climatic cycle is likely. Since growth occurs only in the summer, photosynthesizing plants are primarily adapted to long days and short nights. Tropical plants are subjected to longer and less variable dark periods. They may be sensitive, however, to small changes in day length, which may trigger physiological and adaptive responses such as flowering. Tropical forages tend to be of lower quality in part because of the metabolic adaptations associated with the longer dark periods and the higher temperatures during growth.

6.3.3 Latitude

Figure 6.7 compares reported digestibilities of forage grasses grown at experiment stations at different latitudes. The figure shows the maximal digestibilities at the earliest cutting and the lowest digestibilities at maturity. The minimum digestibilities at plant maturity result from cumulative environmental effects during growth and maturation. Level of digestibility is related to latitude and reflects an inverse relationship to environmental temperature and day length. Longer days and lower temperatures are associated with higher latitudes and probably interact to increase or decrease forage value with age. Since these factors have counterbalancing effects, temperature will be dominant in tropical and temperate latitudes and generally in regions with continental climates, and day length will be the more important factor at high latitudes and in regions with very tempered maritime climates.

Maximum digestibilities of temperate grasses show little change with latitude because once frost has ended, uninterrupted growth can begin. Tropical forages decline in digestibility with decreasing latitude because of the warmer conditions at the inception of growth, which is usually after a dry period. The same factor promotes lower digestibility of second cuttings in temperate regions.

6.3.4 Water

Lack of water tends to retard plant development and thus to slow maturity, with the result that digestibility is somewhat increased and dry matter yields are reduced. Various studies have shown that lack of water increases digestibility and irrigation tends to decrease it (J. R. Wilson, 1981, 1983; T. R. Evans and Wilson, 1984; Collins, 1985b; Dias Filho et al., 1991).

Some of these studies separated water's effects from light and temperature effects, which rain or the addition of water can alter through cloud cover and evapotranspiration. Cloud cover particularly interacts with moisture since it reduces light under conditions of more available water, with both combining to lower forage quality.

Desert-adapted perennial plants may revert to dormancy in the dry season by transporting nutrients to the roots and leaving an aerial part of decreased nutritive value. Many plants adapted to arid conditions are annuals, however, and depend on seeds to survive the dry season. The leaves and stems of grasses tend not to be very lignified because they must grow rapidly and reproduce in a short period—a factor that precludes much investment in irretrievable sinks. Cereals tend to mobilize reserves into the grain, leaving a depleted straw of lower value. Lack of water retards seed development and lignification, leading to higher-value straw.

Desert browse plants often maintain their foliage, particularly in tropical areas with long dry seasons. Many of these species are legumes and are very deep rooted, allowing them to maintain green leaves during prolonged dry spells. Browse plants often have defenses against animal herbivory and defoliation. These may be in the form of spines or, in a few species, silicification, but more commonly they are secondary compounds such as tannins and alkaloids, substances that are energetically cheaper than heavy lignification or cutinization. They also do not reduce digestibility as much for ruminants, since they are often more toxic to the animal than to gut microorganisms, an advantage to selector ruminants that can modify secondary substances through saliva or rumen fermentation.

6.3.5 Fertilization

Nitrogen has the greatest effect on plant composition relative to other mineral elements in fertilizers; it increases the protein content and the yield. Amino acids and protein are synthesized from sugar, and thus an increased nitrogen supply reduces the sugar content. This effect is promoted at higher temperatures and retarded at low temperatures, where nitrate can accumulate instead, in which case sugar remains. Protein and nitrogenous products accumulate mostly in the cell contents, thus diluting the cell wall and increasing digestibility. This can be offset by the increased lignification of the cell wall when adequate nitrogen is present for plant growth and development. Any increase in the nitrogen component thus requires a compensatory depression in non-nitrogenous components, especially sugar. Changes in digestibility depend on the balance of compensatory factors. Reduction in cell wall (fiber) is positive, while lignification is negative but can be canceled by the cell wall change. The balance of these factors is undoubtedly influenced by temperature, light, and water supply. Water stress, as mentioned, increases digestibility but reduces the efficiency of mineral use (Dias Filho et al., 1992). Positive, negative, and insignificant changes in digestibility have been reported in response to nitrogen fertilization. On the average, however, nitrogen fertilization tends to reduce digestibility slightly.

Some fertilizers stimulate more rapid development and increase the plant's yield at the potential expense of quality. This reduction in quality, predicted by theory, is not certain because most of the macronutrients (N, P, K, Ca, Mg, and S) are retained within the cell contents of the plant, leading to higher digestibility. Not many forage quality studies have been conducted on elements other than nitrogen; however, some potential effects of specific elements can be pointed out.

The cationic elements (K, Mg, Ca) increase the buffering capacity of the forage and are almost always associated with some anion (organic acid or nitrate) to maintain ionic balance in the plant cell. Excessive potassium, antagonistic to magnesium, is associated in grasses with excessive levels of the anions aconitate or nitrate. Magnesium is a part of the chlorophyll molecule needed for photosynthesis, and conditions in which magnesium is a limiting nutrient could conceiv-

Table 6.2. Influence of soil on composition and digestibility of Pangola grass (*Digitaria decumbens*)

Component	Crude protein		Lignin		Silica		In vitro digestibility	
	ult	inc	ult	inc	ult	inc	ult	inc
Whole plant	10.1	8.4	6.9	8.3	1.2	3.3	55.6	49.6
Young leaves	14.9	13.3	6.3	5.5	1.7	3.4	61.8	60.7
Mature leaves	13.0	11.8	5.4	5.1	1.9	4.8	68.0	64.1
Old leaves	9.3	7.2	9.0	9.6	2.2	7.1	59.5	53.0
Upper stems	10.9	9.1	7.8	8.5	1.1	2.4	55.3	49.3
Lower stems	7.8	6.0	10.3	10.7	0.7	2.3	43.0	41.8

Source: Data are from the Vicente Chandler farm at Aguas Buenas, Puerto Rico.
Note: All sampes of 60-day-old forage were taken from the same field. ult = ultisolic bench; inc = inceptisol.

ably limit photosynthesis and sugar production and lower digestibility (R. L. Reid et al., 1984).

6.3.6 Soil

Plants grown on different soils are offered a different balance of mineral elements, which influences their growth and composition. Thus soil's effect on plants will be similar to that of fertilization. Soil effects can be viewed from two points: the accumulation in the plant of minerals and the influence of the minerals in the plant on its organic matter yield, composition, and digestibility (Metson, 1978).

The nutritive elements in soils depend on the rocks and minerals from which the soils are derived and the degree of weathering to which they have been subjected. Very old soils become depleted in the more soluble elements and become acidic and rich in iron and aluminum oxides that can be toxic to plants. This process is accelerated in warmer and more humid climates. In very wet environments, mineral nutrients may have been almost completely leached out of soils and remain largely in the tissues of living plants, so that continued growth depends on death and recycling. Man has altered this situation through fertilization and cultivation, particularly with the addition of potassium, nitrogen, phosphorus, calcium, and other elements known to be necessary to growth.

Forage plants grown in the same general climate on different soils may have different compositions at the same age (Table 6.2). Identical fertilization does not guarantee equalization of plant nutrition because microclimate, leaching, and available silica can have a considerable effect on grasses and some other plants (see Section 9.7). Highly weathered soils low in silica and other minerals (oxisols and ultisols) tend to produce forage plants of higher digestibility than unweathered soils. Often, lower yield is associated with higher digestibility because the deficiencies in soil nutrients tend to restrict plant development.

6.3.7 Defoliation and Disease

The physical loss of leaves, stems, or both represents a major stress that puts pressure on the plant to mobilize its reserves and put forth new leaves to restore its photosynthetic capability (Parsons et al., 1988; Parsons and Penning, 1988). Since this process precludes formation of lignified tissue, the effect of defoliation on quality is always positive. From the point of view of the plant, the loss of tissue, whether due to a mowing machine, an animal, fire, or insects, makes relatively little difference, except that smaller herbivores might be more selective in removing the more nutritious parts. It should be understood that improved forage quality as a result of any predation, including by insects, is a function of the regrowth, not of the remaining untouched material. Harvesting forage at the optimal growth stage is a managed form of predation and needs no further comment here, except to say that if overdone, yields and plant survival will be adversely affected; the same applies to overgrazing and insect damage, although Liu and Fick (1975) showed a substantial increase in digestibility of regrowth alfalfa infested with weevils.

Diseased plants that have impaired growth and thus a lack of development and lignification may also be more digestible. On the other hand, isoflavones (Chapter 13), and probably tannins, are reported to increase in diseased or overgrazed legumes. Some of these compounds are inhibitory to fungi, microorganisms, and animals alike. Malechek and Balph (1987) reported increases in secondary compounds in overgrazed browse plants.

6.4 Environmental Interactions and Plants

Among the climatic variables, light and temperature are the most important, followed by moisture (Table 6.3). This is most apparent in the seasonal cycles characteristic of temperate regions. The growing season commences in spring with slowly increasing tempera-

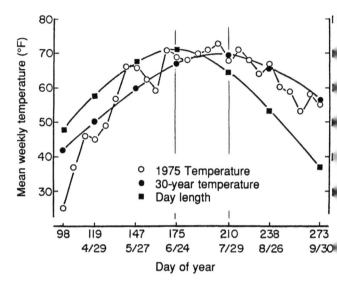

Figure 6.8. Seasonal changes in temperature and day length during the growing season in a temperate environment (Ithaca, New York) (from Kalu, 1976). The growing period is divided into spring (up to June 21); midsummer, during which temperature is still increasing but light is declining; and late summer, when both temperature and light are declining.

tures and rapidly increasing amounts of light until the summer equinox. The maximum temperatures are reached in the summer after day length has begun to shorten (Figure 6.8).

The growing season in temperate regions can be divided into three periods: spring, when light and temperature increases are positively associated with plant age; midsummer, when temperature is comparatively constant, the light level decreases, and plants mature; and autumn, when both light and temperature decrease as plants age. Light, temperature, and plant maturity all have distinct effects on plant composition, and these effects vary and interact according to season. In addition, fertilization, water, and predation must be considered (Table 6.3)

One consequence of climatic interactions through the various seasons is changes in chemical constituents as the forage matures. In temperate spring, light and temperature on average increase each day, resulting in positive associations between daily mean temperature and maturity of first cuttings and consequently between carbohydrate and lignin production. These associations become altered in midsummer when light and temperature patterns become inverted. The most crucial change is the negative pattern between lignin and cellulose in second cuttings (Figure 6.9). Second cuttings show a different compositional behavior from first cuttings. The amount of cellulose in a forage is related to digestibility secondarily through its correlation with lignification. This leads to a variation in the degree of association of fiber (predominantly cellulose) with digestibility.

Table 6.3. Influence of environmental factors on composition and nutritive value of a forage

	Temperature	Light	Nitrogen fertilizer	Water supply	Defoliation harvest
Yield	+	+	+	+	−
Water-soluble carbohydrate	−	+	−	−	+
Nitrate	−	−	+	NA[a]	NA
Cell wall	+	−	±	+	−
Lignin	+	−	+	+	−
Digestion	−	+	±	−	+

Source: Van Soest et al., 1978a.

Note: Positive direct effect is indicated by +, negative association is indicated by −, and variable association by ±. NA = not available.

In late summer and autumn, temperature decreases, as does day length and total light. The effect of temperature at mid-temperate latitudes is sufficient to override the negative effects of declining light, promoting an increase in forage quality with age (Van Soest et al., 1978a). The increased metabolic pool and cell contents diluting the cell wall are partly responsible for the improvement; in addition, lignification of new growth diminishes with the cooler temperatures of autumn.

Fall cuttings may vary with latitude and climate. Although these factors are insufficiently studied, experience in Britain suggests that fall cuttings are inferior and decrease in quality with age. This observation may be rationalized by the fact that the British climate is very mild and cool year-round with smaller temperature changes in the fall season, while at the higher latitudes the daily decrease in daylength may be the overriding factor.

6.4.1 Age and Maturity

Stage of growth in terms of plant development is a common means of describing forage quality (Section 2.8), and plant age (date of cutting) has been utilized for the same purpose. The general assumption that age and maturity are synonymous, however, is not true. In plants, maturity means morphological development culminating in the appearance of the reproductive cycle: tillering, flowering, pollination, and seed formation. This sequence in many plants depends on specific signals; for example, a particular day length (photoperiod) or temperature. Plant age is generally defined as the period since the beginning of regrowth in spring following winter, or growth of aftermath following cutting. Forage plants that remain vegetative can be described only in terms of age or the height of the sward, and any meaningful distinction between age and maturity becomes difficult.

Temperature, light, and water accelerate the maturation process; clipping, grazing, and disease retard it. These factors, positive and negative, can be separated into those that can cause variation in plant response at a given site (weather, water, temperature, and management) as opposed to those that can vary geographically (light, day length, soil, and climate). These latter factors can account for variation in forage at different locations.

Agronomic generalization tends to associate decline in forage quality with plant maturity. Maturation is undeniably a major feature of the aging of forage; however, the relationship can be greatly modified by individual plant responses and environmental factors. Variation in plants' composition at equal age and physiological stage of maturity may occur as a result of genotypic differences among forage plants or species as well as physiological responses of individual plants

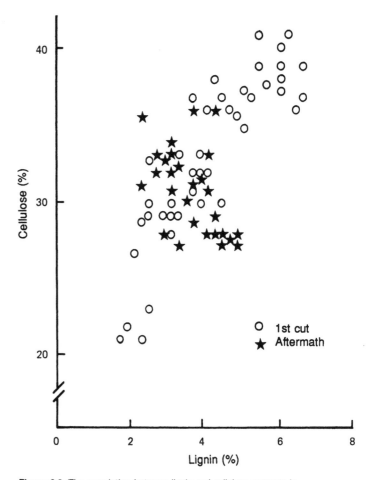

Figure 6.9. The correlation between lignin and cellulose contents in West Virginia grasses. Note that the relation in aftermath cuttings (after June 21) is inverse (Van Soest et al., 1978a). Compare with Figure 6.10, which depicts the contrasting situation in tropical forages.

as a result of environmental factors that influence composition but not physiological stage of development.

The overall remaining problem is to build an adequate model that integrates environmental and climatic factors into a model for composition and digestibility. A partially successful model is that of Fick and Onstad (1988).

6.4.2 Date of Cutting

The association of age with maturity has led to the application of cutting date as a criterion of forage quality. The general decline in nutritive value with age is best exemplified in first cuttings of hay in temperate regions. The date for obtaining an optimum yield of digestible matter is relatively later in more northern regions and in areas of higher elevation. A major factor is the starting date for spring growth. J. T. Reid (1961) constructed a regional map comparing optimal first cutting dates in the United States that shows the later dates for mountains and northerly latitudes but does

Table 6.4. The effect of season and year on alfalfa quality

Harvest date	5/30	6/10	6/20	6/30	7/12[a]	7/19[a]
1983		Cold spring			Hot summer	
Degree-days	444	616	913	1167	—	—
CP (%)	22	21	20	19	19	19
NDF (%)	31	35	41	45	43	46
ADF (%)	24	28	34	38	36	39
TDN (%)	69	65	59	55	56	53
1986		Warm spring			Cool summer	
Degree-days	828	1061	1274	1505[b]	—	—
CP (%)	19	19	18	19	21	20
NDF (%)	39	43	46	48	35	41
ADF (%)	33	36	39	40	28	34
TDN (%)	60	56	53	51	65	59

Source: Data are from G. Fick, Cornell University; values are based on the model equations of Fick and Onstad, 1988.

Note: CP = crude protein; NDF = neutral-detergent fiber; ADF = acid-detergent fiber; TDN = total digestible nutrients.

[a]Second cuttings, previously harvested June 10.

[b]Tillering.

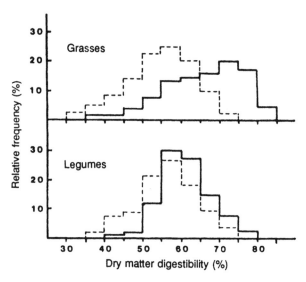

Figure 6.10. Range in digestibility of tropical forages (dashed lines) and temperate forages (solid lines) (from Minson and Wilson, 1980). The mean difference in total digestible nutrients (TDN) between tropical and temperate grasses is about 15 units (upper graph). The differences between tropical and temperate legumes is smaller (lower graph). Fifty-two percent of tropical grasses are below 55% TDN, as contrasted with only 4% of temperate grasses.

not reflect the generally lower quality of forages grown at lower latitudes and in warmer climates (Section 6.4.3).

Date of cutting has been advocated as a means of predicting digestibility, and indeed it would be a simple means of estimating the value of first cuttings. But such a system does not reflect latitudinal and temperature differences and is utterly inadequate for predicting digestibility of second or third cuttings because the starting dates for initial growth (date of previous cuttings) vary relative to the climatic cycle. Second cuttings have a lower digestibility than first cuttings of the same chronological and physiological ages. Higher temperatures promote lignification and more rapid physiological development, and aftermath forage is usually less nutritive at a younger age. In autumn, forage can actually increase in digestibility with increasing age provided the environmental factors are favorable.

The years 1983 and 1986 were unusual in the northeastern United States in that the spring of 1983 was cold and the ensuing summer hot; in 1986, a warm spring was followed by one of the coolest summers ever recorded. The effects on alfalfa quality of these unusual conditions are listed in Table 6.4. The fiber level was so low in the second cutting of the cold year, 1986, that many dairy cattle produced low milk fat due to unbalanced and acidotic rumen conditions.

6.4.3 The Tropics

Most of the common generalizations about forage quality and composition derive from studies in temperate areas with a typical four-season pattern. The tropical pattern is one of relatively unchanging day length, high temperatures, and the absence of winter. Regions near the equator tend to exhibit two dry periods and two rainy periods (the short rains and the long rains) that result from the sun crossing and recrossing the equator twice annually; these features are absent in temperate zones.

"The tropics" is usually taken to mean the geographical regions of the world that are free from frost. Consequently, in these portions of the world forage can exhibit a more or less continuous growth provided sufficient moisture is available. In temperate latitudes (above 30 degrees) growth begins with the cessation of frost. In tropical areas, however, growth begins at relatively higher temperatures, usually after cutting or when rains end a dry spell. Maximum digestibility declines with latitude below 30 degrees, where interruption of growth is caused not by cold but by lack of water. Tropical forages have little need for coldhardiness, but they have greater problems with disease and predation than temperate plants. Tropical climates would thus be expected to have forages of low nutritive value and a high proportion of protective structures to help prevent predation. The additional factors of long warm nights promoting respiration and warmer growth temperatures increasing lignification, and the fact that most of the cultivated grasses are C_4 plants (see Section 6.5), combine to lower nutritive quality in tropical plants.

Summaries of reported digestibilities show that tropical forages average about 15 units of digestibility lower than temperate forages (Figure 6.10). The lower quality of tropical forages is due to the generally higher proportion of cell wall and its greater lignification. The

Table 6.5. Digestibility and components of tropical and temperate forages

	Dry matter basis								
	Digestibility (%)	Crude protein (%)	Crude fiber (%)	Cell wall (%)	Cellulose (%)	Hemicellulose (%)	Lignin (%)	NFE[a] (%)	Soluble components[b] (%)
Temperate									
Alfalfa	60	17	30	40	24	8	7.5	43	33
Corn silage	70	9	24	45	26	16	3.0	61	40
Orchardgrass, young	70	15	27	55	26	25	4.3	49	21
Timothy, mature	52	7	34	68	31	29	7.3	54	20
Tropical (60 days)									
Pangola grass	54	11	30	70	34	29	7.0	50	10
Guineagrass	54	9	34	70	35	26	8.0	49	9
Bermudagrass	50	9	30	77	32	38	7.0	56	8
Napiergrass	50	9	31	72	36	28	8.0	50	9

Sources: Riewe and Lippke, 1970; Van Soest, 1973a, 1973b.
[a]Nitrogen-free extract.
[b]Soluble components not accounted for by cell wall, protein, or ash.

availability of protein and soluble fractions appears equal to temperate forages (Combellas et al., 1971).

There is also important variation in quality among tropical forage plant species. For example, Pangola grass, which remains largely vegetative, declines more slowly in digestibility than the more intensely flowering species such as guineagrass (Gutierrez-Vargas et al., 1978). Under similar growth conditions, the quality of Pangola was initially the lowest but also declined the least with age. The declining order of quality was napier > congo > star > guinea > Pangola at 30–60 days of age (Arroyo-Aguilu et al., 1975).

Tropical forages are characterized by lower soluble carbohydrate and higher cell wall and lignin contents (Table 6.5). The more productive grasses are highly responsive to nitrogen fertilization, which tends to decrease digestibility and soluble carbohydrate content and increase crude protein content. Many tropical grasses are high in hemicellulose, which elevates values for nitrogen-free extract (NFE), the use of which in the past has given a false impression as to the quality of tropical grasses. The digestibility of tropical grasses is difficult to predict from their fiber composition. In tropical environments, where temperature is likely to be seasonally more constant and day length is less variable, there tends to be a lack of association between cellulose and lignin in grass species (Figure 6.11). The inverse relationship between crude fiber (largely cellulose) and digestibility found in temperate grasses simply does not hold for tropical plants. The situation is comparable with that of temperate forage grown in midsummer. Butterworth and Diaz (1970) found a correlation of −0.37 between crude fiber and digestibility for 270 tropical grasses. The comparable relationship in temperate grasses is about −0.7.

Generally, all fiber values are poor predictors of the nutritive value of tropical forages. The proximate system (see Section 10.2), the use of which is supposedly justified by its economy, is a particularly poor predictor.

Another characteristic of tropical grasses is the wide range in quality possible within the same standing plant (Figure 6.5, Table 6.2). Temperature is a likely influence here, but soil type is also a factor. The wide range in the quality of available herbage offers opportunity for selection by grazing and browsing animals. Stall-fed animals must be allowed a larger refusal—up to 60% for practical feeding (Olubajo et al., 1974)—to compensate for the low quality of the poorest parts. The need for selection (and hence larger orts) is greater in small ruminants such as goats and sheep that cannot be made to eat all the offered forage under practical

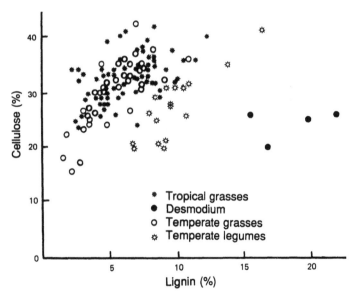

Figure 6.11. Lignin and cellulose contents of tropical forages from Puerto Rico and Florida (continuous aftermath). The correlation between lignin and cellulose is insignificant in the tropical grasses and +0.7 in the temperate first cuttings.

conditions; even a small refusal can result in under-nutrition. This creates a problem in evaluations of trop-ical forages (Zemmelink, 1986), since a standard orts cannot be set, as has been done with temperate forages. In a Cornell University study in which sheep were fed guineagrass hay from Puerto Rico, the animals could not be forced to consume the stems (Tessema, 1972).

Domestic legumes in the tropics have not been im-portant forages despite their wide distribution. Tropi-cal legumes, like temperate ones, are higher in protein and lower in cell wall, but do not have higher di-gestibility compared with tropical grasses (Milford and Minson, 1965). Further, tropical legumes are higher in lignin than temperate legumes, and many contain tan-nins and alkaloids. Still, the differences between tem-perate and tropical legumes are not as marked as they are for grasses. All tropical legumes are C_3 plants, whereas most tropical domestic grasses are C_4 plants.

It should be clear that animal production in tropical countries is handicapped by the low quality of the available forage. The greatest limitations are evident in introduced European dairy cattle, which are the least selective in their feeding habits and have high nutrient requirements. It was once hoped that the humid tropics would help solve the world's food problem, but the lack of high-quality forage remains an obstacle. There are three approaches to a solution: management, chem-ical treatment, and plant breeding. Management can help a great deal in solving production problems of quality forage, and the judicious use of local concen-trate by-products, including molasses, can help animal production. Chemical treatment is expensive and re-sults are variable. Genetic improvement would be cheaper in the long term but has received less attention.

6.4.4 The Arctic

The polar environments are characterized by short summers with relatively continuous light. In the Arctic these summers can be relatively hot in some inland regions, but the permafrost is not far underground and there may be frost almost any day of the summer. There are a few areas amenable to agriculture, includ-ing the inland valleys of Alaska and some parts of northern Scandinavia. Barley is the main cereal, and smooth brome grass is the main forage; there are essen-tially no legumes.

Perennial plants that grow under these conditions must store reserves for the long, dark winter. Woody plants store reserves in buds and cambial layers to resist very low winter temperatures.

The three native Arctic ruminant groups—moose, reindeer or caribou, and musk ox—are all relatively large animals and can be classed as intermediate feed-ers since they must shift their feeding from herbivorous material in summer to lichens or wood in winter.

The Arctic forage consists mainly of grasses, forbs, browse, and lichens. As might be expected, forage quality is high but of low availability in the agronomic sense. Arctic conifers, which are extremely slow grow-ing and very heavily defended with secondary com-pounds (Chapin et al., 1986), do not figure signifi-cantly in the browse budget. Arctic grasses do not decline below 60% digestibility, and even willow twigs remain relatively digestible as woody tissue. The wil-lows are capable of secondary compound production if overgrazed.

Lichens play a major role in the nutrition of Arctic ruminants, although rumen bacteria must adapt before they can digest the carbohydrates in lichens. Adapta-tion takes about a week and appears to involve the elaboration of enzymes capable of breaking down lichenin, a type of beta glucan. There do not seem to be any secondary compounds involved. Digestibility by unadapted rumen fluid is about 15%; however, diges-tion trials with adapted fluid and caribou indicated about 80% digestibility.

6.5 C_3 and C_4 Plants

The first stable products of photosynthesis in C_4 grasses are four-carbon compounds, while those of C_3 grasses are three-carbon compounds, hence, their des-ignation as C_4 and C_3 species. The difference is signifi-cant because C_4 plants are photosynthetically more ef-ficient. Tropical C_4 plants tend to exhibit high dry weight accumulations that are often low in nutritive value. C_4 plants have only a few mesophyll cells be-tween vascular bundles, compared with 10–15 cells in C_3 plants. Mesophyll cells are unlignified in temperate grasses, and therefore their proportion influences qual-ity (Figure 6.12). Differences between C_3 and C_4 plants are listed in Table 6.6.

The generalization that C_4 plants have a lower nutri-tive value than C_3 plants is not universally true, and a few exceptions deserve mention. Corn and sorghum are C_4 plants derived from tropical ancestors. Studies show that corn grown under tropical conditions has a much lower nutritive value than corn grown in a tem-perate environment (Deinum, 1976). The low quality of C_4 grasses presents a challenge to plant breeders, but there are solutions. After all, corn is a C_4 plant that has been greatly altered by genetic manipulation.

A few C_4 plant species are adapted to cold condi-tions (Waller and Lewis, 1979; Bae et al., 1984), main-ly in semiarid areas of the temperate zone that have high total light. Some are found as far north as the Canadian plains. Phosphoenolpyruvic acid carboxy-lase, the CO_2-fixing enzyme in C_4 plants, requires temperatures of 30–35°C; while diphosphoribulose carboxylase, the major analogous enzyme in C_3 plants,

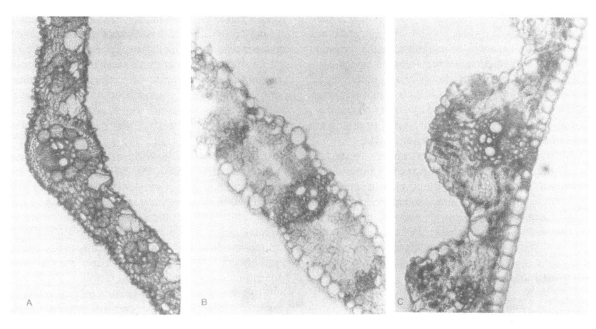

Figure 6.12. Electron micrographs of leaf sections of some C_3 and C_4 grasses (courtesy of D. E. Akin). (A) Coastal bermudagrass (C_4). Cross section of leaf blade shows that vascular bundles are closely spaced and have a thick-walled, rigid parenchyma bundle sheath. Mesophyll radiates from vascular bundles and has few air spaces. ×98. (B) Orchardgrass (C_3). Cross section of leaf blade shows more widely spaced vascular bundles and small, thin-walled parenchyma bundle sheaths. Mesophyll is loosely arranged, and air spaces are prevalent in this tissue. ×98. (C) Annual ryegrass (C_3). Cross section of leaf blade shows that vascular bundles are widely spaced and the parenchyma bundle sheaths are small and thin-walled. The mesophyll is loosely arranged, with air spaces prevalent in this tissue. ×98. Compare A with B and C to illustrate the differences between semitropical bermudagrass and the more digestible temperate grasses; see also J. R. Wilson and Hacker (1987).

le 6.6.   Characteristics of C_3 and C_4 Plants

C_3	C_4
ompounds are the first stable notosynthetic products	C_4 compounds are the first stable photosynthetic products
ılose diphosphate carboxylase arboxydismutase) is the major nzyme of the first step of CO_2 fix- ion.	Phosphoenolpyruvate (PEP) carboxy- lase is the major carboxylating en- zyme
temperature optimum (20–25°C) r the major carboxylating enzyme	High temperature optimum (30–35°C) for the major carboxylating enzyme
iest photosynthetic accumulation distributed throughout mesophyll ılls	Earliest photosynthetic accumulation is concentrated in the bundle sheath cells
ally 10–15 mesophyll cells be- ween vascular bundles	Only 2–3 mesophyll cells between vascular bundles
v translocation of photosynthate ɔm the leaf	Rapid translocation of photosynthate from the leaf
starch storage (grasses only); no norphic chloroplasts	Storage of starch in large dimorphic chloroplasts
atively low dry weight accumula- ɔn	Relatively high dry weight accumula- tion
ı transpiration	Low transpiration
ı photorespiration (1–3 times the ırk respiration)	Low photorespiration (essentially zero)
net CO_2 exchange rates (15–30 g·dm⁻²·h⁻¹)	High net CO_2 exchange rate (40–50 mg·dm⁻²·h⁻¹)
light saturation (7000 foot- ındles)	High light saturation (10,000 foot- candles)

ource: Information from C. A. Jones, 1985.

requires temperatures of 20–25°C. C_4 plants have adapted to tropical conditions, where low respiration at high temperatures and in long dark periods conserves energy. This advantage may be lost at lower tempera- tures and with longer day lengths and lower light inten- sities (C. A. Jones, 1985).

Most domestic tropical grasses are C_4 plants, and all legumes—including tropical ones—are C_3 plants, leading to problems of balance in pastures where le- gumes and grasses are combined. The C_4 grasses may outgrow the legumes, and if intensive grazing is prac- ticed to improve grass quality, the legumes may not survive. Most temperate perennial grasses are C_3 plants.

6.6 Genetic Improvement

The search by plant breeders for forages and cereals of superior nutritive value is a recent development made possible by laboratory methods suitable for eval- uating small amounts of plant material. Plant breeders seek to produce higher-yielding and higher-quality forages, with the largest effort being made to foster adaptability, yield, and resistance to disease. Much plant breeding research necessarily involves the pro- duction of disease-resistant varieties of high yield. But selection for high yield or disease resistance may inadvertently be selection against nutritive value. Plants that have a high food value may have low survivability, so that more care will be required for their maintenance and cultivation. Selecting for resistant and high-yielding strains does not necessarily produce nutritionally inferior plants. On a statistical basis these qualities have a significant but low-order negative association (Coors et al., 1986), and it may be possible to find plants that have both quality and yield.

Nutritive value is a somewhat elusive factor. Attempts to improve it include increasing digestibility, reducing lignification, modifying plant morphology, and reducing potentially toxic secondary compounds. A plant's amenability to genetic manipulation requires a detectable range of inheritable characteristics and reproducibility of those factors phenotypically. Nutritive value is a complex phenomenon not likely to be accounted for by a few genes.

6.6.1 Legumes

Legumes are characterized by high protein and lignin and low cell wall relative to grasses. The amounts consumed by herbivores are large compared with their digestibility, provided secondary compounds are not a factor. Tannins, alkaloids, estrogenic compounds, and factors promoting bloat are prominent secondary factors, although they may not be present in all leguminous species.

Moderate levels of tannins may improve rumen protein output, although they become toxic at higher levels and are convenient for stressed plants to produce. Tannins also inhibit bloat, and there has been an attempt to genetically engineer tannin-producing genes into white clover in New Zealand. Not all alkaloids and tannins are toxic, and biological assays are needed to examine the effects of each substance.

The search for low-lignin cultivars of alfalfa has led to the proliferation of weak or prostrate plants because lignification is related to stem strength; thus, in the end, a compromise has to be made. It seems that there is a lower-order negative correlation between digestibility and yield in alfalfa (Coors et al., 1986), and it might be possible to find more digestible varieties that do not necessarily suffer from being less adaptable to the environment.

Attempts to increase protein content in legumes by increasing leaf proportion have been less successful. Ruminants do not need increased protein content in legumes, which may be overoptimal for rumen organisms; however, such efforts reflect attempts to provide protein for nonruminants.

6.6.2 Grasses

Grasses as a group tend to be relatively high in cell wall and low in lignin, leading to lower voluntary feed intake relative to digestibility. The lignification affects a larger proportion of the available digestible matter in grasses than in legumes, however, because of the high cell wall content. Grasses may contain secondary compounds such as cyanides (in sorghums and sudangrasses), endophyte alkaloids (in fescue), and indolealkylamine alkaloids (in *Phalaris*). These have been studied (Marten et al., 1981) and are discussed in Section 13.7.4.

Although less NDF would be desirable in grasses, scientists at the Welsh Plant Breeding Station have been unable to reduce the content. Attempts to lower lignification or increase rate of digestion may be more successful. An example is the coast-cross variety of bermudagrass, which has lower yield but gives higher animal productivity than the standard variety (Burton and Hanna, 1985). The same is true for the brown midrib mutant of corn (Section 6.6.5).

"Improved" varieties of grasses may be more susceptible to disease and environmental stresses, leading to lower yield. This is true of the brown midrib corn mutant and endophyte-free fescue. Casler et al. (1987) and Godshalk et al. (1988) discuss genetic variation in smooth bromegrass and switchgrass, respectively.

6.6.3 Straw Quality

Cereal straws have been greatly modified by breeders' development of short-stemmed varieties that increase the proportion of photosynthetic energy in the form of grain at the expense of stalk. Straw is a major animal food resource in the Third World, however, and reception of the new varieties has been mixed. Some farmers complained that there was not enough straw to feed their animals, particularly the draft animals used in plowing (R. E. McDowell, 1986).

Shortening a plant reduces its stem area, thus increasing the leaf-stem ratio (Figure 6.13). The effect on nutritive value depends on the relative quality of leaves and stems, which varies among the cereal species. Barley stems tend to be of lower quality than the leaves, so digestibility has been improved in the shorter varieties (Capper, 1988). In rice, leaves exhibit more silicification than stems, so the reverse may be the case (Table 6.7). Wheat and triticale are similar to barley. Even though barley is taller than wheat, its straw is on the average more digestible (National Research Council, 1984, U.S.-Canadian tables).

The quality of leaf and straw, as well as their proportions, is affected by environmental factors, the most important of which is moisture. Aridity increases the leaf-stem ratio through favoring shorter, less-developed plants that may retain soluble carbohydrate in stems or leaves because they do not form a fully developed seed head.

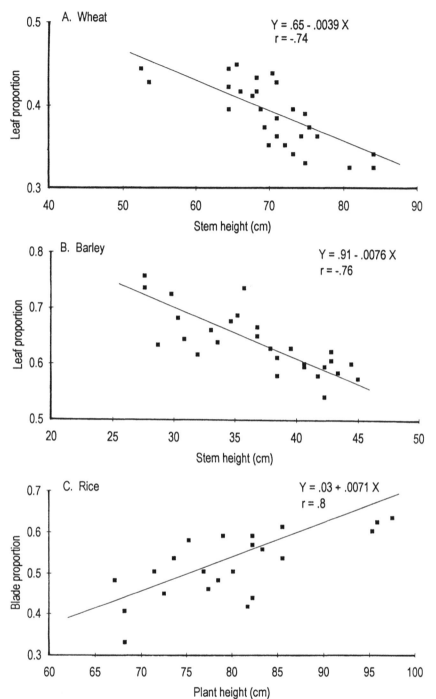

Figure 6.13. The proportion of leaf to stem in wheat **(A)**, barley **(B)**, and rice **(C)** samples from the Plant Breeding Institute, Great Britain (barley); Syria (wheat); and the International Rice Research Institute, Los Baños, Philippines (rice) (from Capper, 1988).

6.6.4 Maize and Sorghum

Maize, sorghum, and related millets usually differ from other cereals in having a filled pithy stem that serves as a reserve for soluble carbohydrates. These C_4 plants have been modified from their tropical ancestors so that they can survive in temperate climates. The quality of maize leaf and stalk is greatly influenced by temperature (Deinum, 1976).

Stem diameter, which influences the ratio of carbohydrate-containing parenchymous cells to the lignified vascular cortex, is associated with digestibility. This may be particularly important in sorghums. The leaf sheath, which contains lignin and silica, is a protective layer. For example, silicification of the leaf sheath appears to be a more important factor than lignin in protecting the plant from the corn borer. The sheath protects the stalk, and thus the composition of a relatively

Table 6.7.   Digestibility of leaf blade, leaf sheath, and stem in barley and rice straw

Crop	Method of digestibility determination	Digestibility (%)					
		Leaf blade		Leaf sheath		Stem	
		Mean	SE	Mean	SE	Mean	SE
Barley (ICARDA, 1984)	Tilley-Terry[a]	48	0.5	42	0.5	35	0.6
Barley (ICARDA, 1985)	Tilley-Terry	58	0.4	49	0.4	32	0.5
Barley (PBI, 1984)	Cellulase[b]	51	0.7	30	0.4	20	0.3
Barley (PBI, 1985)	Cellulase	44	0.6	27	0.5	19	0.5
Rice var. (IRRI, 1985)	Cellulase	30	0.6	25	0.6	44	0.6
Rice lines (IRRI, 1985)	Cellulase	31	0.6	26	0.5	43	0.4

Source: Capper, 1988.
Note: ICARDA = International Center for Agricultural Research in the Dry Areas; PBI = Plant Breeding Institute; IRRI = International Rice Research Institute. Years in parentheses are years of harvest.
[a]Tilley and Terry, 1963.
[b]Goto and Minson, 1977.

small plant part, weightwise, may allow a higher nutritive value in the bulkier stem. Resistance to lodging and stem diseases and infestation is not necessarily related to lignification.

6.6.5 Brown Midrib Corn

Brown midrib was originally discovered in a collection of Mexican maize at Purdue University. The significance of soluble colored matter in the midrib of leaves (the main reservoir of lignin in maize) was not appreciated until the character became associated with low lignin contents and high soluble polyphenolic matter. This type of mutant has since been found or induced in sorghum and millets (Cherney et al., 1988). Because these cereals (particularly maize) have been used primarily for their grain, leaf and stalk quality

have been overlooked by commercial North American corn seed producers. The brown midrib gene may be of value in warmer and tropical countries where lignification of forage is a greater problem. Plants bearing the mutation appear to contain both a less polymerized lignin and a considerable amount of soluble polyphenolic substances in the midribs that do not affect digestibility like normal lignin does. Cell walls are more digestible and ferment at a faster rate (Wedig et al., 1988). On the other hand, the plants may be shorter and more prone to lodging.

The brown midrib gene introduced into European hybrids is responsible for a significant improvement in the nutritive value of European corn silage (Deinum, 1987). Differences between European and North American hybrids average as much as 12 units of digestibility.

7 The Free-ranging Animal

The survival, growth, and productivity of free-ranging animals depend on external factors such as forage nutrient supply, composition, spatial distribution, and temporal characteristics, and on the animal itself—its behavior, requirements, and interaction with the forage. In this context the animal is free to express its choices and behavior. Although the free-ranging animal can select food according to its preferences, it must at the same time deal with a changing environment. Pasture research and management has been reviewed by J. L. Wheeler (1987).

The plant-animal relationship is part of an open ecological system (Figure 7.1) in which the flow of photosynthetic energy moves from plant to herbivore and is returned to the soil as feces and urine (Barnes and Taylor, 1985). Decomposition of organic matter by fungi and bacteria releases mineral nutrients for reuse by plants. Nutrients are also supplied by the breakdown of rocks in the weathering process. Plants supply energy, protein, vitamins, and minerals to animals, but require only mineral sources and CO_2 for their own photosynthesis and growth. The recycling of mineral sources is vital to the system, the balance of which may be altered if plants or animal products are removed from the cycle. The basis for fertilization and mineral supplementation is the need to correct this upset balance and overcome inherent nutrient deficiencies within the system. Failure to replenish these resources reduces productivity to the limits allowed by geochemical weathering and biological recycling.

Farmers try to remove the secondary converters (insects, predators) to maximize the recovery of animal or plant products. They also tend to restrict the system to a single grazing species. Under range conditions, some regulation of the herbivore density is necessary to balance forage yield; otherwise, overgrazing will result. The restriction of herbivore species may mean that forage is incompletely utilized.

The main problem in grazing studies is change, for nothing remains the same from day to day. In feeding experiments diet is kept as constant as possible in order to monitor the animal's status. The changing sward and the response of the animal to environmental conditions further complicate this fluctuating situation. It is therefore not surprising that a degree of imprecision creeps into grazing measurements, or that grazing and browsing have become the subjects of modeling and simulation studies. Since so many variables impinge on both forage and animal, it is not possible to state any general policy or management system. Solutions are only possible given a set of conditions. In the case of animal production, economic constraints must also be considered (Morley, 1987).

The free-ranging herbivore exists in an environment that imposes varying degrees of constraint on its choice of forage. Such factors as the number of plant species available, the morphological differentiation in quality within species, and the plant density determine the amount of nutritive matter per unit area. An additional component is animal density, which determines grazing pressure. As grazing pressure increases, animal selectivity and the amount of food per animal decreases (J. E. Ellis et al., 1976).

Grazing and browsing cannot be entirely differentiated from one another as there is a continuum of feeding behavior between the two extremes. Under intensive conditions of cultivated temperate pastures only 1 or 2 plant species are usually present; on natural rangeland, however, 30 or more species may be present. Improved pasture imposes a limit on the grazing animal in the sense that forage choice is very limited, although availability and quality are high. Range conditions may offer both differentiated plants and a variety of plant species. Arboreal browsers have many possibilities to select from, although availability and overall nutritive quality may be low.

Computer models can integrate these complex factors to allow a balanced judgment of the composite effects of several parameters. Models are valuable for assessing the total effect of a complex set of factors and for evaluating the limitations of existing information. It must be remembered, however, that while models can be very helpful in predicting effects within complex systems, they present the hazard of losing sight of cause-effect relationships and alternative theories (Chapter 23). Most methods of research applicable to

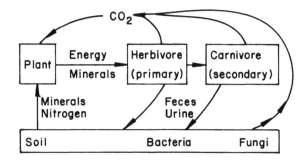

Figure 7.1. Open ecological system showing interdependent plant-animal relationships.

grazing or browsing are limited in scope relative to the range of possible situations considered and their ability to describe observed effects.

7.1 Carrying Capacity

An alternative to the complex techniques of measuring forage digestibility and intake is to settle for a measurement of animal productivity under grazing conditions. Since the pasture responds to grazing pressure, measuring carrying capacity becomes essentially a titration of pasture capacity by varying the number of animals. *Grazing pressure* or *intensity* is defined as the number of animals per unit of area. Animals' response to the intensity of grazing pressure is complex and includes a number of factors, some of which are described in Figure 7.2. Low grazing pressures, or *undergrazing,* allows forage to develop and mature such that individual animal responses may be reduced by low forage quality and reduced intake of digestible nutrients. An increase in grazing pressure results in increased animal yield per area up to a point. Beyond this point the addition of animals causes a sharp decline. The point of maximum yield of animal product is the *carrying capacity.* This level is generally beyond the most economic grazing pressures, indicated by the brackets in Figure 7.2. Note that the maximum total animal yield occurs at a grazing pressure well above that which produces maximum yield per animal. These levels of grazing pressure are characteristic of many situations in Africa and Asia, and less characteristic of South America. Carrying capacity varies continually throughout the season, so the manager must decide the proper balance between animals and pasture. This operation requires careful management, because judgments must be made about the effect of current weather and other contemporary events on future digestible nutrient availability. Management also has to consider the optimum level of economic output and the likelihood of risk (White and Morley, 1977; Morley, 1987).

Maintaining a balance at the maximum carrying capacity carries the risk of shortfall in forage if lack of

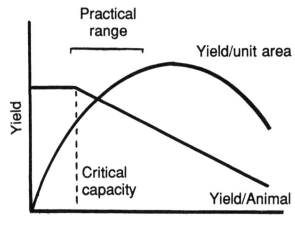

Figure 7.2. Animal production as a function of grazing pressure. The yield per unit area peaks after the point where individual animal yield begins to decline (Raymond, 1969). Undergrazing may reduce the yield per animal because plants are allowed to mature and decline in nutritional quality. Severe overgrazing destroys the desirable plant species. The range of practical grazing is indicated by the bracket. There is controversy with regard to the shapes of the declining animal yield (see Figure 7.3).

rainfall or other causes prevent normal growth. As a result, most managers attempt to keep a feed reserve and operate below maximum carrying capacity and optimal animal output.

Optimal economic return and reduced risk occur in the range of practical grazing as indicated in Figure 7.2 (Bransby et al., 1988; Hart et al., 1988). If grazing pressure is too low, the result is overmature wasted forage. There is some evidence that undergrazing in humid tropical conditions reduces individual animal output. Grazing too near the maximum capacity causes reductions in individual animal yields.

The management of grazing pressure varies considerably around the world. The most studied systems are those of New Zealand and Australia. In Latin America management systems tend toward undergrazing, whereas those in Africa are characterized by overgrazing. In North America, Europe, and parts of Asia forage is often cut and carried to the animal—a zero grazing system. Sub-Saharan Africa probably presents the highest pressures in that ruminants are valued more as a bank than for their productivity.

Figure 7.3 presents several possible relationships between grazing pressure and yield. If pasture yield and quality are unaffected by animal pressure, the function will follow that shown by curve A. This curve may be representative of instant points in time, but it fails to account for longer-term effects on forage composition and quality caused by grazing. Some results suggest a linear decline, as shown by curve B (Hart, 1978), and a curvilinear decline (curve C) was suggested by Connolly (1976). A sigmoid shape (not

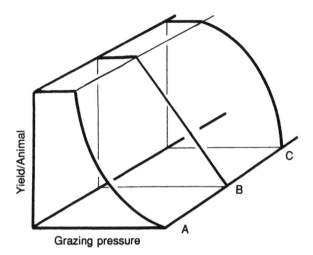

Figure 7.3. Proposed relationships between animals' weight gain and the stocking rate (modified from Hart, 1978). All three theories recognize a critical capacity below which animal yield is independent of stocking rate.

shown) has also been suggested (Conniffe et al., 1970).

7.1.1 Pasture Productivity

There is no really satisfactory way of determining the amount of forage actually available for grazing, but fairly reliable estimates have been derived from agronomic measures, simulated grazing, and chemical analysis. Chapter 8 discusses the matter in more detail; here the discussion is devoted to animal output in terms of carrying capacity.

One method of measuring carrying capacity is the put-and-take system. Animals are divided into two groups: (1) test animals that remain permanently on pasture and have their productive performance recorded and (2) animals that are added or removed as needed to maintain forage in an optimal state. The carrying capacity is the number of animals per unit area that can be maintained in a state of good nutrition at any time. Nutritive yield of pasture is expressed in terms of animal product. A problem with this system is that it allows no flexibility in situations in which a constant number of animals has to be maintained or supplementation is required. Removal and return of animals presumes external sources of feed or an ability to sell and buy at inconvenient times. The put-and-take system is thus a good research tool but not a system of management.

An alternative method, less convenient for experimentation, is to supplement feed as needed to a fixed number of animals on a fixed area. This is how many animal management systems operate (particularly dairy farms). A difficulty with this system is that grazing pressure is reduced by supplementation, partic-

ularly in the case of poor-quality pastures in which animals cease to graze and eat the supplement instead. In such cases animal response varies directly with the level of supplementation rather than according to pasture quality.

The object of both systems is to obtain some estimate of nutrients received by the pastured animals and convert it to animal product. The put-and-take system has the advantage of simplicity, but supplementation becomes necessary in cases of high animal output or relatively low quality pasture. The problem becomes more complicated and the results are more tenuous when attempts are made to assign a nutritive value to the pasture in the form of total digestible nutrients (TDN) or net energy (NE). In this situation, energy for maintenance has to be estimated, and an arbitrary value is assigned to supplements and their efficiencies.

In the evaluation of managed pastures it is important to restrict the study to the limiting nutrients. Usually, this means energy and protein. Assessment of protein levels for ruminants is complicated because the nitrogen level of the forage often changes more or less along with energy content and interacts with rumen function. Maintenance of nitrogen and other nutrients at a level above that required (via supplement or fertilization) tends to place the limits of animal production on forage supply and quality. An adequate nitrogen level is more easily achieved when animals are growing and being fattened than during gestation and lactation. If nitrogen is supplied in the supplemental diet, animals may tend to eat more supplement and less forage, particularly if the latter is low in energy.

7.2 Animal Productivity

The ultimate objective in grazing management is to discover the optimum balance between animals and pasture. Forage yield may be affected by general plant density, and shading by overlapping material may reduce photosynthetic capacity. Such situations may be associated with undergrazing and can be controlled by clipping, although forage may be wasted.

Leaf-area index, the total leaf area per unit land area, is a major factor influencing how forage responds to grazing. Undergrazing allows overgrowth and shading by senescent foliage, which reduces photosynthesis and increases respiration. Optimum grazing pressure improves the effective leaf-area index, and higher pressures that result in excessive defoliation and a resultant decrease in forage yield diminish it. Higher forage yield is not necessarily consistent with the maximum feed energy per unit area. Extra predation often increases digestibility because plants stressed in this way limit lignification and investment in plant cell wall structures (Chapter 6).

The response in total animal output per unit area undoubtedly reflects the yield of forage digestible matter. Predation on forage generally increases its nutritive value through inhibition of the maturation process. The point of maximum forage yield is not identical with maximum animal yield, since production of a smaller yield of higher-quality forage often produces a greater animal response. Part of the forage response is due to recycled nutrients from the urine and manure returned to the field, but these are often distributed unevenly in the pasture, and the recycling of nutrients to plants may be inefficient.

The reduced animal production at high grazing pressure (overgrazing) results from excessive defoliation, reduction in photosynthetic capacity (leaf-area index), and, in the extreme, the death of desirable plant species may also occur at lower grazing pressure if swards are mixtures of plants of disparate quality. If animals have to travel further to obtain feed, maintenance cost increases and intake is reduced (Hogan et al., 1987).

7.2.1 Grazing versus Crop Production

In many parts of the developed world, northern Europe and North America in particular, grazing has been replaced by "zero grazing," which means not allowing animal access to fields; instead feed is cut and brought to animals or stored. This management system allows better control of animal intake and feed quality. The animal's maintenance cost under conditions of zero grazing is substantially less than its cost under grazing (Waldo et al., 1961), and there is the added advantage that animals do not tread on and damage the plants. Total mixed rations (TMR) have become popular in North America as a means of regulating diet composition for high-producing animals. Sophisticated programs for designing such diets balance the input of carbohydrate and protein to match rates of fermentation for rumen bacteria and optimize rumen output (Fox et al., 1990). Application of these principles under supplemented grazing is more uncertain.

On the other hand, zero grazing has produced negative results in the humid tropics. For example, in one case chopped napiergrass was fed to corraled cattle. The grass had been managed at a high yield at 60–90 days of age such that stem material was of low quality. Animals that were expected to eat the total chopped material did not do as well as free grazers, who avoided the stem material.

7.3 Grazing Behavior

Animals exert their preferences and a certain capricious behavior in their choice of feed. What they choose to eat, however, depends on the diversity of the

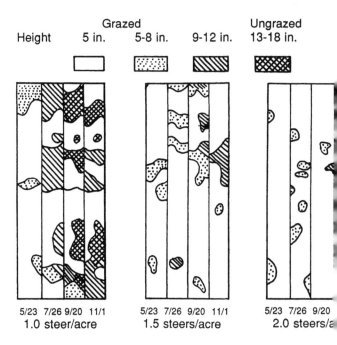

Figure 7.4. Typical grazing patterns showing the effect of stocking rate on degree and nature of strip-grazing for steers on dallisgrass–white clover pastures at different seasonal dates in Angelton, Texas (from Riewe, 1976). Undergrazing results in zones of overmature forage.

forage. Selection is diminished by high grazing pressure and uniformity of the sward (Black and Kenney, 1984). Vegetative temperate grasses are the least differentiated; leaves and stems often have the same digestibility. A single-species stand strip-grazed at a high stocking rate for a day or two per strip also allows no selection. Tropical forages, mature forages, and species mixtures offer a range of expression for selection (e.g., choice of leaves, stems, and plant species of disparate quality or palatability) and do not lend themselves to strip grazing. Extreme situations offer special problems. For example, grazing studies in the Scottish Highlands show that sheep will graze grass to its elimination and death before significant amounts of the much lower quality heather are grazed (Sibbald et al., 1979).

Capricious feeding behavior is usually more evident at low stocking rates. Animals may graze an area to the ground and allow adjacent areas of the same forage to grow to maturity (Figure 7.4). Manure spots are generally avoided even though the initial nutritive quality (particularly nitrogen) of the forage in these spots may be better than that on adjacent ground. Rejection is presumably a palatability factor based on smell or taste and perhaps related to the avoidance of recycling internal parasites.

Plant morphology influences grazing behavior and intake. Leaf size and distribution are important in plants with highly lignified stems. Such morphology discourages random biting, and intake may be reduced if the animal has to select each leaf and avoid stems. Bite size is then a factor in grazing behavior, since it is affected by plant size and morphology and the ability of the animal to cope with selection and still achieve adequate intake (Stobbs, 1973a, 1973b).

Another factor influencing behavior is forage density or availability. Under range conditions where forage may be sparse, it is more economical energetically to graze the available forage before moving on to other areas. The grazing pattern is also dictated by herd movement, which in turn may depend on water supply. Consequently, overgrazing may occur near water sources.

7.3.1 Diurnal Patterns

Ruminants show definite diurnal eating and rumination patterns (Figure 7.5). Generally, cattle and sheep graze in the morning and evening and ruminate mainly

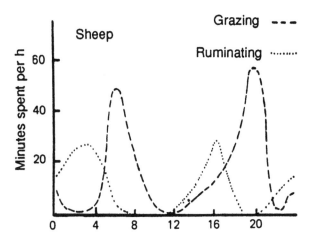

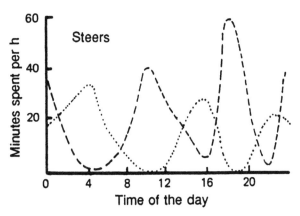

Figure 7.5. Time spent grazing and ruminating by sheep and cattle on legume pasture. Although sheep and cattle diurnal patterns differ, both species tend to graze in the morning and evening and rest during the hotter part of the day. There is some night grazing. Intake patterns are also influenced by diurnal factors (see Figure 15.8). Rumination and grazing are generally exclusive activities, although there are a few short periods devoted to both. (For a comparison with behavior of dairy cattle in the tropics, see R. E. McDowell, 1972:497.)

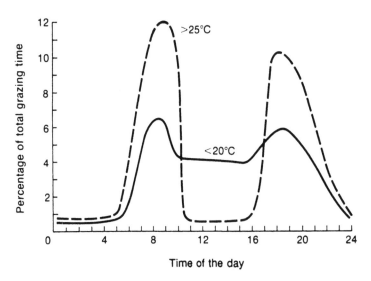

Figure 7.6. Contrasts in grazing patterns in lactating dairy cows on good pasture at two temperature regimes: maximum daily temperature <20°C and >25°C (from R. E. McDowell, 1972).

at night, though there is some rumination at midday. These pulsative eating patterns may be more characteristic of temperate grazing species (R. E. McDowell, 1972). Figure 7.6 shows the expected grazing patterns for lactating cows on good pasture when the maximum daily temperature is 20°C. The preferred grazing times at 25°C strongly suggest that maximum intakes can be expected if the milking operation is completed before 6 A.M. The afternoon milking should likewise be early, 2–4 P.M., in order to get the cows back on pasture for two or more hours before complete darkness. Poor-quality pastures maximize the time spent eating and ruminating, and this may limit intake.

7.4 Variation among Animal Species

Feeding behavior varies even among grazing types. The grazing preferences of cattle and sheep overlap by perhaps 70%. Cattle's preferences overlap much less with goats, and only about 5% with selector species such as deer. Species with large overlaps tend to compete for food, especially if overgrazing occurs. On the other hand, the more disparate the feeding behavior the more complementary the two species are, and combining them in the same pasture balances and optimizes forage utilization. If overgrazing is allowed to occur under these conditions, however, undesirable plant species can become dominant.

7.4.1 Mixed Grazing

Several animal species grazing in the same area usually results in increased animal output per unit area compared with grazing a single animal species (T.

Nolan and Connolly, 1977). The animal species involved do not eat the same forage, and the complementation results in better forage utilization. There is likely also a predation effect on forage and browse whereby nutritive value is improved. The increased output is particularly striking in mixtures of African ungulates (McNaughton, 1979), in which there is maximal interspecific complementation and less competition for food. Less selective feeders are a necessary part of the group to eliminate coarse forage or browse and render the range more acceptable to smaller, more selective species (Reed, 1983).

Ranges are often characterized by diverse plant species and herbivore types. Compared with nonruminants, ruminants are less discriminate consumers of plant species. The rumen bacteria may act as a buffer by detoxifying many potentially harmful substances; they also provide a more constant supply of amino acids and vitamins to the animal. Nonruminant herbivores may need to select more carefully from a wider range of plants in a given day to achieve the same results. Problems with potentially toxic plants usually arise under conditions of nutritional stress and forage scarcity when less desirable plants are consumed in sufficient quantities to induce problems (Culvenor, 1987).

The ability of animals as closely related as sheep and cattle to avoid competition has been underestimated. Unpublished data from Ireland (Fermoy Station) show that the addition of up to two sheep per hectare in a managed grazing of dairy cattle on pure ryegrass pastures at a fixed optimum stocking rate does not reduce milk production. The sheep eat the manure spots, and the grazed ryegrass responds in nutritive value. Pasturing goats with cattle is a common way of controlling weeds and incipient browses that are eaten by goats and not cattle. Such a system may reduce internal parasites.

7.5 Grazing Management

Optimized animal production on improved pastures depends on maintaining an optimum carrying capacity and altering animal density to maintain this balance. This means making day-to-day decisions based on forage status, number of animals, and other factors (Riewe, 1976). It may also mean harvesting excess herbage at times of peak growth and storing it to use during periods of feed deficiency or, alternatively, selling animals before periods of pasture decline or diet supplementation during periods of shortage. Managers seek to minimize periods of excessive or scarce forage (Figure 7.7). The object of this management is to reduce carrying costs, that is, to reduce the total grazing

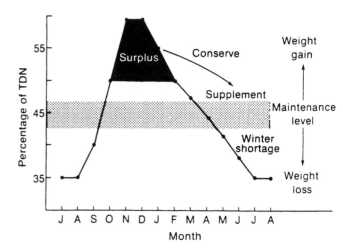

Figure 7.7. Principles of fodder conservation (from R. E. McDowell, 1972).

days per unit of product. Productivity is measured as the productive response above animal maintenance costs.

The grazing intensity that produces the maximum economic return is not identical with that which produces the maximum yield per unit area. Generally, high input costs make gain or production per animal a more important factor than production per unit of land. Fewer animals means less risk of a shortfall in feed supply. Pasture improvement or supplemental feeding must be paid for through increased individual animal productivity to be economical.

Dairying is more labor- and cost-intensive than meat or wool production. It demands higher animal intake and higher-quality forage. Feed quality is a major reason why dairying is so successful in cool-temperate zones. In addition to protein, legume forages in cooler areas are superior to grass because they usually contain a lower content of plant cell wall, a primary factor in feed efficiency and voluntary intake. An exception is the perennial ryegrasses adapted to Britain, northwestern Europe, and New Zealand that are low in cell wall and high in soluble carbohydrates. In Europe, the choice of forage has been set by the economics of nitrogen fertilization necessary for high-quality grass; in New Zealand, ryegrass–white clover mixtures reduce the need for added nitrogen. The problem of high nitrates in drainage water is changing European practices, however, particularly in the Netherlands.

Rotational grazing has often been promoted over continuous grazing, partly because grazing pressure is easier to control and less forage is wasted; however, continuous grazing may be as efficient or more efficient if properly managed.

It is important to consider the kind and amount of forage needed to supply the nutrient requirements

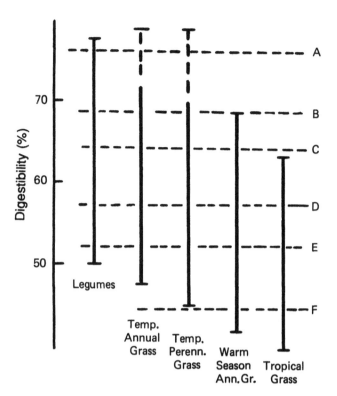

Figure 7.8. Relation between digestibility of forage classes and nutrient requirements of cattle and sheep (modified from Riewe, 1976). (A) Dairy cow producing 22 kg milk/day. (B) 200-kg steer at 0.7 kg daily gain, ewe with twins, or young lambs. (C) Lactating ewe or heifer with first calf. (D) Beef cow with calf, weaned at 225 kg. (E) Dry cows gaining condition, or sheep at maintenance. (F) Cattle at maintenance. The highest digestibilities for temperate grasses (dashed bars) are achievable only in temperate zones in spring.

of a particular class of animal. Dairy cattle have the most rigid requirements, followed closely by pregnant sheep; lactating beef cattle have lesser requirements, and dry mature stock have the fewest requirements. The level of digestibility that becomes limiting to production decreases in that same order.

Lactating dairy cattle and pregnant sheep may not be able to derive all their energy requirements from consumed feed and are prone to ketosis and nutritional stress. The relative capabilities of various forages to supply animals' requirements are shown in Figure 7.8. Legumes are superior to grasses by virtue of their higher protein content and intake by animals, and any forage grown in a cooler season or climate is generally superior to that grown under warmer conditions. Some warm-season grasses not noted for exceptional yield—for example, Kleingrass (*Panicum coloratum*), dallisgrass (*Paspalum dilatatum*), and some C_3 grasses such molasses grass (*Melinis multiflora*)—provide better acceptability, intake, and digestibility than other semitropical grasses and are more responsive to fertilization but require more intensive management to achieve acceptable animal gains (Riewe, 1976).

7.5.1 Strip Grazing and Rotational Grazing

Animals confined to paddocks are forced to consume all available forage before they are transferred to the next paddock. Strip grazing can be accomplished by a movable boundary, with similar results. The grazed area is allowed to recuperate and regrow before the next cycle of grazing. This method of management favors tall, erect, easily defoliated plants. Low-growing plants such as white clover tend to be shaded out by taller plants during regrowth. Consumption of most of the aerial growth severely reduces the leaf-area index. Therefore, maintenance of optimum forage productivity requires plants (e.g., alfalfa) that can tolerate periodic defoliation.

Whether animal production is favored by strip grazing or rotational grazing depends on total yield and the quality of all the pasture plants. High-quality material is eaten first, leaving material of lesser nutritive value for later. This produces a cyclical pattern as animals are rotated from paddock to paddock. Rotating on tropical and warm-season forages produces poorer results than continuous open grazing at a properly managed stocking rate. Rotational and strip grazing are most successful for intensive management of nutritionally uniform high-quality forages under cool, temperate conditions.

7.5.2 Alternatives

A simpler system developed in Ireland offering some flexibility is to combine continuous open grazing with variable restriction of the open grazing area to balance stocking rate with carrying capacity. This technique requires movable fencing to regulate the grazing pressure as forage production varies. Ungrazed areas are saved for harvest as hay or silage to provide feed in periods of low forage productivity.

7.5.3 Fertilization

Improving pastures to maximize their yields requires the addition of the mineral elements that are limiting to plant and animal production. The most important supplements for plants are nitrogen, phosphorus, potassium, and lime (to correct soil acidity). Of these nutrient factors nitrogen is most relevant here. The other elements are discussed in Chapter 9.

Nitrogen promotes yield and increases the crude protein content of the sward. In many grass species total growth outweighs the increase in protein content because growth is so great that the dilution of organic matter limits crude protein levels in mature forage to low levels, not greatly different from unfertilized forage. Thus the sequential response to nitrogen is very

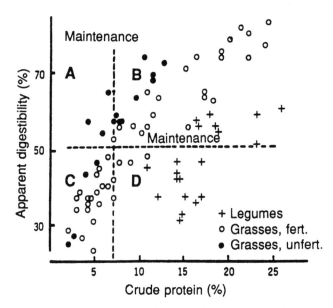

Figure 7.9. Digestibility and crude protein content of tropical forages. Horizontal line (signifying energy level of maintenance) and vertical line (protein level of maintenance) divide the figure into quadrants. Forages in quadrant A are deficient in protein; in quadrant C, deficient in both energy and protein. Forages in quadrant D are deficient in energy, and available protein will be unused and wasted as urea in urine. Only quadrant B has an adequate balance. (From an unpublished summary of tropical forages analyzed at Cornell University by P. Van Soest, 1970–1974.)

high initial crude protein levels that drop to low levels as the plant matures.

The amount of nitrogen required by the rumen bacteria for their own growth and reproduction is generally less than the nitrogen level in grass that will produce maximum forage yield. Raising the levels of nitrogen through fertilization increases the soluble crude protein, particularly nonprotein nitrogen (NPN). At the same time soluble carbohydrate levels are depressed because the promoted protein synthesis occurs at the expense of glucose, the primary product of photosynthesis. As a result, nitrogen-fertilized pastures tend to be too high in soluble nitrogen and too low in rapidly fermentable carbohydrate to make for efficient rumen fermentation. This problem is more severe in warm-season and tropical grasses, which are inherently low in soluble carbohydrate. Forage nitrogen (see Section 18.2) consists of NPN and protein that are relatively soluble in the fresh state, promoting rapid fermentation. Pasture-fed ruminants probably depend on microbial protein as their main nitrogen source since rumen escape is likely to be low. Therefore, nitrogen is wasted when administered at levels that maximize grass yield because the level of nitrogen in the forage is in excess of microbial requirements. In view of energy costs, an alternative is to use legumes to maximize the utilization of fixed nitrogen. The use of legumes will not overcome the protein solubility prob-

lem, however, unless tanniniferous species are employed. Tannins are double-edged swords that may be most advantageous for animal species with salivary adaptation.

Because of the limited rates of fermentation of cell wall and the shortage of carbohydrate in straws and tropical forages, microbial yields are not high and the rumen requirement is about 12 units of crude protein, of which about 5 can be obtained from saliva and ruminal diffusion. The ratio of digestible carbohydrate relative to crude protein is critical, and imbalances leading to high rumen ammonia and wasted nitrogen are easily produced by fertilization and the introduction of legumes (Figure 7.9).

7.6 Tropical Conditions

Grazing and animal management in tropical environments present a special set of problems. Tropical climates, although seemingly desirable, do not support the quality of agriculture production in proportion to land area that temperate regions do (Whyte, 1962). This is particularly true for net animal production, although the tropical environment is harsh and restrictive to plants and animals alike (Zemmelink, 1986) and does not respond to many of the improvement practices that work in temperate regions (R. E. McDowell, 1972).

Animal production in the tropics is frequently based on unsupplemented grazing, particularly in Latin America, Australia, and some parts of Africa. Eastern Asians rely more on feeding animals straw from cereal production. The grazing areas suffer from lack of conservation methods and from the nutritive class problems discussed in Chapter 2 and illustrated in Figure 7.10.

7.6.1 Seasonal Variation

Forage yield and quality vary throughout the year. The prime variables are water, plant nutrients, temperature, and light, which in turn determine plant response and maturity. As a consequence, carrying capacity is not constant, regardless of management. Seasonal curves vary with climatic geographic regions (Figure 7.11).

Pasture management in temperate zones takes for granted the necessity of conserving forage for winter. Tropical climates have dry periods, which contribute to seasonal low production of forage; however, conservation of forage is rare. As a result, animal growth and production tend toward a strict seasonality. Animals must rely on energy stored in adipose tissue to get through lean periods. Growth under such conditions is characterized by large variations in body weight (Figure 7.12). This pattern is also exhibited by wild grazing

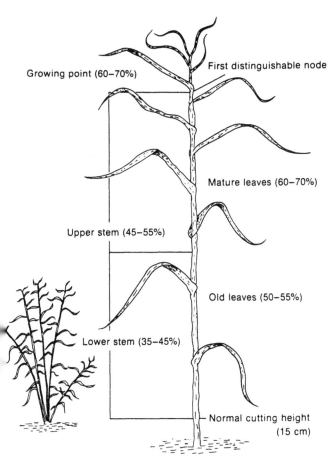

Figure 7.10. The variation in digestibility of portions of the stems and leaves of tropical grasses (from R. E. McDowell, 1972).

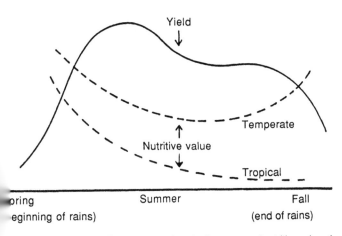

Figure 7.11. Typical seasonal production curve and nutritive value of temperate and tropical forages. Curves characteristic of monsoon or seasonal rainy periods in the tropics are similar except for the response in nutritive value. Yields and nutritive value are often influenced by peculiar weather in individual seasons; moisture and temperature are probably the most important factors.

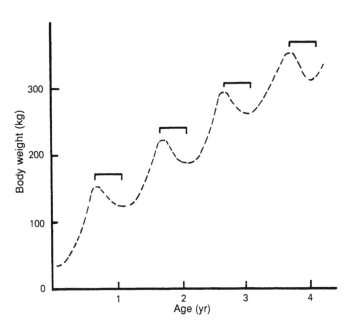

Figure 7.12. Growth patterns of cattle on unimproved tropical pastures without supplementation. The pattern (dashed line) is also characteristic of wild ruminants living in temperate and Arctic zones, with winter corresponding to the dry season (brackets), during which animals lose weight, and summer to the wet period (R. E. McDowell, 1972; Moen, 1973). Selectors may not show this pattern (see Figure 7.13).

herbivores (Moen, 1973) in temperate regions. Under conditions of seasonal food shortages, animal numbers are balanced against forage availability through winter die-off, migration, and predation.

Some adaptable browsers or intermediate feeders in the tropics may avoid seasonal weight losses. Goats observed in Nicaragua, for example, did not lose weight (Figure 7.13) in the dry winter season because they ate deep-rooted shrubs and trees that remained green throughout the dry season. Goats can live on this type of food because of their dexterity in picking leaves from among the thorny branches of leguminous shrubs and even, on occasion, climbing small trees (Figure 7.14).

7.6.2 Traditional Practices

Ethnic social traditions and diverse environments allow a wide range of techniques in tropical land and animal management. The small-farm mixed agriculture typical of tropical Africa contrasts with the cereal-based agriculture of Southeast Asia and India, but both systems depend on animals for tillage and for a variety of products. Often there is little or no grazing, and animals subsist on plant residues and other wastes.

The seminomadic herders in drier regions (e.g., the Sahel) have their own set of problems related to overgrazing. Water often is the limiting factor, and the grazing pressure (overgrazing) is related to the distances animals have to walk to water holes and back to

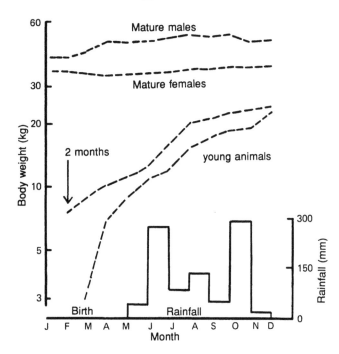

Figure 7.13. Seasonal body weight changes in mature and growing goats in Nicaragua (from McCammon-Feldman, 1980). Goats do not decline in weight during the dry winter periods because they collect an adequate diet by browsing on shrubs and small trees (see Figure 7.14).

Table 7.1.   The influence of mowing on the carrying capacity of four grasses in Colombia

Grass	Carrying capacity (steers/ha)	
	Unmowed	Mowed bimonthly
Pangola	1.56	2.50
Para	1.60	2.50
Guinea	1.45	2.00
Puntero	1.50	2.50

Source: R. E. McDowell, 1972.

grazing areas. Three days between waterings is the limit for cattle.[1] This leads to overgrazed areas surrounding available water and underutilization of forage in regions farther from water.

Slash-and-burn agriculture has been a practice in forested regions from ancient times. It was practiced by Indians in northeastern North America before Europeans came. A patch of forest is burned, the area is farmed until soil nutrients are depleted, then a new patch is burned and the old patch is abandoned. The practice is easy to condemn, especially when it gets out of hand, as in the Amazonian rain forest and in Madagascar, but management is possible. A continuing slash-and-burn agriculture depends on recycling land and a reforestation program. An efficient system demands narrow strips of farmed land lying in the lee of the prevailing wind. The width limit is set by the need for new forest to reseed from adjacent trees.

The still more extensive cattle ranch systems in South and Central America exploit C_4 grasses imported from Africa. This single-species forage tends to limit production. Great improvements in production would arise from using by-products indigenous to the region, including molasses, cassava, tree legumes, and

waste bananas, as food supplements. On the other hand, all these products introduce problems in nutrient balancing, particularly in nitrogen-carbohydrate balances with regard to protein.

Since the grazing pressures applied in Latin America are generally somewhat below carrying capacity, mature herbage is likely to accumulate (Figure 7.4). Good management requires mowing to eliminate the undergrazed material, including weeds and overmature forage. R. E. McDowell mentioned that mowing is controversial and results may vary according to plant species and the situation. It may do more harm than good for some grasses and may seriously damage trailing or climbing legumes. In Puerto Rico a mixture of tropical kudzu and napiergrass provided 880 kg of animal gain per hectare when mowed once a year compared with 650 kg when mowed bimonthly. But later experiments in Puerto Rico showed no significant benefits from periodic clipping of stands of guinea, Congo, star, and Pangola grasses. On the other hand, in Colombia mowing at bimonthly intervals increased the carrying capacity per hectare for Pangola, para, and guinea by 60% over no mowing (Table 7.1).

Overgrown pastures of Pangola, star, molasses grass, or kudzu and molasses grass mixtures do not require mowing because animals trample them down while grazing on the newer growth. Even though much of the forage is wasted, the pasture recovers quickly if there is sufficient moisture. Nonetheless, on lands where mowing by machinery is possible, it may be desirable to cut back these pastures occasionally as an aid to weed control and to cut back undergrazed spots where the forage has become less palatable.

7.6.3 Grasses and Legumes

Tropical environments are characterized by higher temperatures, longer nights in the growing season, and C_4 grasses. These factors lead to generally lower nutritive value, greater lignification, low soluble carbohydrate, and high cell wall levels. C_3 forage plants such as legumes do not compete well with C_4 grasses. Grazing pressures that optimize the grass may cause loss of the legume, and reducing grazing pressure to maintain

[1] Brosh et al. (1988) showed that black Bedouin goats can go up to 4 days without access to water. They report that Bedouin goats may increase their body weight 35–45% in 2 min when offered water. These animals can store water in the rumen and have low rates of liquid passage.

Figure 7.14. Goats are better adapted to browsing than are cattle or sheep. They are very active animals with dextrous tongues and mouth parts that allow them to efficiently select their diet. They use their forefeet (like deer) to pull down lower branches to reach leaves (left). Smaller goats even climb trees to gain access to higher foliage (right). This feeding behavior is advantageous in the drier tropical and semitropical regions of the world. Some of the selector African antelope—the gerenuk, for example—also exhibit this ability. (Photos courtesy of Beth McCammon-Feldman.)

the legume leads to overmaturation of the grasses (D. Thomas and de Andrade, 1986).

The alternative is to grow legumes separately and provide them as supplementary feed. Legumes can be grown in enclosed fields or as trees, many of which are deep-rooted and remain green in the dry season and can serve as boundaries or fences. Branches are lopped and offered as supplementary feed. Sometimes the branches are shredded and the leaves beaten off to produce a protein supplement. The practice is so common in some parts of the world (e.g., India) that it is part of the deforestation problem.

Another practice is alley farming, a mixed system of animals and crops. Legume fodder trees are planted in rows about 4 m apart and interplanted with crops that benefit from the nutrient exchange from the trees, which have to be pruned to avoid overshading the smaller plants. The loppings and the crop residues become the basis of animal feeding.

7.6.4 Forage Conservation in the Tropics

Techniques for preserving forage (hay and silage) are largely temperate developments based on forage responses in temperate environments. The basic aspects of hay and silage are discussed in Chapter 14. Here I present the special problems of warm climates. The objective of conservation is to preserve quality feed for times of shortage—in temperate zones the cold winter, in the tropics a relatively warm dry period. Preservation in the tropics must overcome problems of fermentation, molding, and Maillard reactions (Section 11.7) promoted by the warm temperatures and intense solar radiation. Season rainfall can be so intense as to prevent any kind of wilting, much less drying, so that hay production is possible only at the end of the rainy period when the forage is overmature and of low value.

The tropical grasses themselves offer some problems. The more productive species (e.g., guinea, napier, etc.) have thick culms with low dry matter content (10–15%) and are not amenable to wilting. If they have been fertilized with nitrogen, soluble carbohydrate is very low and normal lactic fermentation is doubtful. This problem might be correctable with appropriate silage additives, but in most developing countries the necessary chemicals, enzymes, and microbial cultures are too expensive. Molasses remains the main additive, and it tends to give a very acidic product with low-dry-matter forage. Silos in developing countries tend to be very simple, usually trenches or bunkers. Sunny tropical conditions require a roof or other cover to block solar radiation, which, beating down on black plastic, is quite capable of setting off a Maillard reaction, leading to possible conflagration.[2]

---

[2] I actually saw this happen at the National Dairy Research Institute in Karnal, India, where sorghum silage had been put up in typical midwestern fashion in a bunker covered by black plastic.

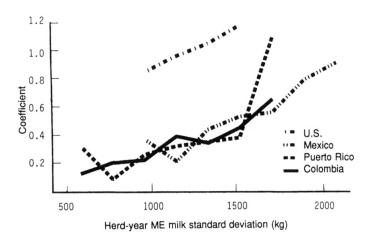

Figure 7.15. Milk response per unit of sire-predicted differences for milk (1982 basic year) by herd-year standard deviation class in Colombia, Mexico, Puerto Rico, and the United States (from Stanton et al., 1991). Note the similarity to Figure 21.11, which depicts the relation between cell contents and voluntary feed intake. ME = mature equivalent.

7.6.5 Australian Systems

The Australian subcontinent has a considerable tropical area and also large desert regions that provoke many of the same ecological problems seen in the less-developed tropical regions. However, intense agricultural research there has made the Australian systems potential models for less-developed parts of the world. Australian researchers are noted for their development of tropical legumes, many of them collected in South America. Australian scientists genetically improved the species and designed management systems for them, and then reintroduced them in South America.

7.6.6 Tropical Dairying

The literature abounds with data from short-term grazing trials indicating that grazing alone will support average daily gains of 1.6 kg or higher in cattle and daily milk yields of 18 kg. But on a year-round basis, about the best performance that can be expected from grazing alone is average daily gains of 0.5–0.7 kg and average daily milk yields of approximately 12–14 kg, with total yields of 2000–3000 kg in a lactation period lasting about 7–8 months. These levels of performance may be near the maximum genetic potential for some stock, but more energy input is required for maximum performance from European-zebu crosses or high-grade European breeds. Since there is a ceiling on intake—determined by digestibility of forage, dry matter content, bulkiness of the grass, protein content, and type of forage (soft or hard leaf)—a more concentrated supply of energy is needed to achieve maximum performance.

The problems of dairying in tropical environments have been further exacerbated by the introduction of temperate genetic stock of very high productive capacity, particularly Holstein-Frisian. Generally there is insufficient feed to support such high levels of production, and the result is nutritional stress that is often blamed on high temperatures. Temperate cattle are adapted to comparatively unselective feeding on a range of forages but tropical grasses are uniformly high in cell wall contents, a factor that is not entirely overcome by supplementation. Under such conditions phenotypic expression becomes repressed and animals of superior capacity are not able to consume to their capacity and requirement. The result is reduced milk production (Figure 7.15).

7.7 Supplemental Feeding

Providing energy or protein supplements to pastured animals is one means of overcoming shortage of grazable forage. Supplementation reduces grazing pressure since animals obtain a portion of their requirements from the supplement. To avoid undergrazing, which leads to overmaturity of the forage, an increase in stocking rate is required when supplements are provided. If the pasture quality is poor (high maturity), animals reject more of the forage in favor of supplements. This is particularly a problem in poorer-quality tropical grass pastures. Concentrate supplementation in the tropics is controversial, especially if the sources are imported, and therefore probably expensive. Alternatives are to use high-quality by-products, if available. Some nutritionists think that grazing ruminants should utilize fibrous sources of feed, but such forages by themselves lead to mediocre production.

The interaction between forage quality and concentrate supplementation is often overlooked. Because of this interaction, animals fed poorer-quality forages will never produce at the level reached by animals fed better-quality forage with less concentrate, even when fiber levels are optimized (Table 7.2 and Figure 7.16). The data in Table 7.2 show that bermudagrass plus 60% soy-corn concentrate yielded less milk than alfalfa plus 30% concentrate. The implication for tropical forage is that concentrate supplementation is more expensive because relatively more concentrate is needed to optimize fiber levels while milk production remains inferior. Supplementation is not a way to overcome the limitations of low-quality forages, which are best resolved by managing for better-quality forage.

Experiments in tropical areas have shown little increase in milk yield following the supplementation of forage with concentrates. A number of these tests used 2–3 kg of molasses per day as the supplement, without significant results. No doubt the cows reduced their intake of forage, thus failing to increase their total

Table 7.2. The productivity of alfalfa, bermudagrass, and corn silage–based rations fed at equal (36%) NDF content

	Alfalfa[a]	Corn silage[b]	Bermudagrass[c]
Concentrate[d] (% of whole diet)	30	45	60
Intake (kg/day)	24	20	19
Estimated TDN (mixed ration)	65	72	71
Discounted to 3M[e]	61	65	62
Net energy (kJ/kg)	6.28	6.74	6.40
Intake of net energy (kJ/day)	148	131	122
Milk (FCM) (kg/day)	23	20	18

Source: Mertens, 1983.
Note: TDN = total digestible nutrients; NE = net energy; FCM = fat-corrected milk.
[a]Alfalfa at 46% NDF and 58% TDN.
[b]Corn silage at 55% NDF and 65% TDN.
[c]Bermudagrass at 70% NDF and 55% TDN.
[d]Concentrate balanced for protein, made of corn grain (12% NDF) and soybean meal (14% NDF).
[e]A 4.5% discount was used on concentrate and corn silage; a 2.5% discount was used for southern alfalfa; and 10% was used for bermudagrass per unit of maintenance intake. (See Chapter 25 for an explanation of discounts.)

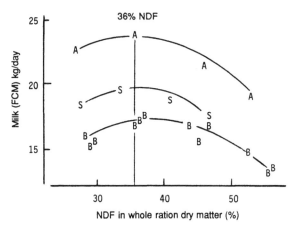

Figure 7.16. The relationship between milk yield and neutral-detergent fiber (NDF) contents of whole diets fed to dairy cattle (from Mertens, 1983). Optimum production is obtained at 34–36% NDF in all diets. Diets are made isonitrogenous by exchanging corn for soybean meal of about the same NDF contents. A = alfalfa; S = corn silage; B = bermudagrass; FCM = fat-corrected milk.

intake of energy and actually decreasing their protein intake (Meijs, 1986). If the animal is grazing forage of 5–7% digestible protein on a dry weight basis, it obtains a submarginal to marginal level of protein. The level of nitrogen (as urea or ammonia) that must be added to molasses to make it self-sufficient in rumen fermentation is about 16% crude protein equivalent. Sulfur will also be required at a sulfur-to-nitrogen ratio of about 1:12 (Section 9.5).

7.7.1 Methods of Supplementation

Supplementing animals on pasture raises several problems, including the regulation of both the intake of the supplements and the consumption of the supplement relative to optimal rumen fermentation and use of forage. Milk animals may be fed supplements during milking; although this allows easy regulation of amounts, optimal ingestion relative to rumen fermentation is probably not achieved because animals return to the pasture after milking and fail to eat. This problem is more severe with lower-quality pastures, resulting in a "titration" of forage intake and leading to less than optimal results of the supplementation (R. E. McDowell, 1972).

Part of the problem with supplementation is that carbohydrate and protein supplies are mismatched. Nitrogen-fertilized pastures are low in available soluble carbohydrate, particularly in warm or tropical weather, and so are forage legumes used as an alternative nitrogen source. Unfertilized or overmature tropical grasses are also deficient in both nitrogen and soluble carbohydrates.

These difficulties can be surmounted by providing licks and feeding blocks in the grazing areas. The licks are often composed of molasses and minerals and per-

haps a nitrogen source. Unlimited intake at the licks can be controlled by adding salt or using hardened blocks.

One of the more clever ideas is a supplement in the form of blocks that are sufficiently crystallized so as to limit intake yet still provide the nutrient requirements for tropical pastures. Blocks, however, have their own set of problems. They must be designed so that they can be fed on at reasonable rates, they do not disintegrate in rain, and they are not so solid as to resist feeding. Cement is sometimes combined with the molasses for durability, but the amount must be carefully regulated. Additional components of blocks are of a supplemental nature and include protected proteins, isoacids, and trace minerals.

Because these effects are not adequately addressed by the quality control of the system, results are mixed. Heating to form solid blocks can damage protein and produce toxic Maillard products such as 4-methyl imidazole. There are no reports assaying the problem of compositional quality and contribution from this variability of the ingredient components. The development of blocks has provided little research on the factors affecting quality (in the engineering aspects) and is badly needed.

Use of supplementation systems such as molasses blocks and forage evaluation systems employing nylon bags has been promoted in Australia and is being fostered in other tropical regions with varying success. The attitude of the consultant Australian scientists is that more intensive studies of forage composition are unnecessary since the problems have been solved by the use of nylon bags or the studies are too expensive. The limitation of this attitude is that it tends to import foreign ecological experience into environments that may need more study for their own sake. For example, should one import developed tropical species or ignore evaluation of plants native to a particular region? The problem with molasses blocks, for example, is that the technique presumes available sources of molasses, additives, and insoluble bypass protein. If these need to be imported, the process can be too expensive. There is also the danger that techniques, plants, and animals can be promoted in cultural environments where they are not the most acceptable.

7.7.2 By-product Feeds

Vegetable by-products from the food industry are extremely important feed supplements. Often they are cheaper sources of protein and energy than intact cereal sources, although they reflect the shortcomings of their plant sources as well as problems that arise in their processing. In many cases nutrients have been removed to leave an altered product, as in extracted oil seeds or sugar beet and cane pulps. For example, the quality of molasses—the uncrystallizable residue from sugar production—varies inversely with the efficiency of sugar extraction. Table 7.3 lists by-product feeds and their nutritive value.

Feed products in North America are legally defined relative to composition by the American Association of Feed Control Officials (AAFCO), and feed regulation guarantees quality in most developed countries. The Third World generally lacks such regulation, however, and the quality of concentrates for ruminants has been known to fall below that of forage because of failure to control ingredients. Particular cases in Latin America include rice bran adulterated with rice hulls and substitution of similar-looking coffee hulls for soybean hulls. Heat damage, tannins, and silica are special problems that can be solved through suitable analysis.

Many by-product feedstuffs do not contribute sufficient amounts of effective digestible fiber, although they may be important sources of protein and energy and useful in balancing dairy rations; citrus pulp, beet pulp, and brewers' grains are important feeds in this group. Also included are products of low or no feeding value. Some of these are put into rations based on their proximate analyses, which tends to overvalue them, because no account of lignin, tannins, or Maillard products that they may contain is reflected in proximate analysis. Examples of such products are coffee wastes and grape wastes (see also Table 7.3).

Products from the essential oil and spice industries are a peculiar and variable group. They include such things as cocoa wastes, nut hulls, and aromatic seeds from various plants (Wohlt et al., 1981; see Table 7.3). Generally they have low digestibility and not much available protein. The compounds that inhibit the di-

Table 7.3. Composition and nutritive value of some by-product feeds

	NDF (%)	Lignin[a] (%)	Available protein (%)	TDN (%)	Other compounds
Acorns	34	13	0	60	Tannins
Almond hulls	31	10	0	56	Tannins
Apple pomace	14	5	<3	50	Tannins
Bakery waste	18	1	9	89	—
Black pepper	52	18	8	43	Unknown
Brewers' grain	46	6	24	67	Maillard prod
Celery seed	56	22	15	44	Oils
Cinnamon bark	81	20	0	13	Phenolics
Citrus pulp	23	3	6	82	—
Cocoa bean	30	25	16	57	Tannins[a]
Coffee bran (hulls)	93	20	0	0	—
Coffee grounds	74	15	0	0	Maillard prod
Corn bran	60	1	11	83	—
Corncob	90	7	3	50	—
Cottonseed hulls	90	24	4	45	—
Cottonseed meal	28	6	44	78	—
Distillers' grains	50	5	22	86	Maillard prod
Fruit pits	—	40	—	0	Cyanogens
Ginger root	48	6	10	61	Unknown
Grape pomace	51	17	0	35	Tannins
Grape stems	67	12	<3	53	—
Linseed meal	25	7	38	82	—
Nutmeg hulls	79	18	1	22	Oils
Oat hulls	78	8	4	35	—
Olive waste	56	28	0	36	Tannins[a]
Peanut hulls	74	23	7	22	Pesticides
Peanut meal	14	—	52	83	—
Peanut skin	36	7	<3	24	Tannins[a]
Pineapple bran	73	7	4	68	—
Potatoes	6	1	5	81	—
Rice bran	25	—	13	70	Silica
Rice hulls	82	17	3	11	Silica
Sesame seed	17	2	—	83	—
Soy hulls	67	2	12	80	—
Sugar beet pulp	54	5	7	74	—
Sunflower hulls	71	19	—	31	—
Sunflower meal	40	12	26	60	—
Tomato pomace	55	11	16	58	Maillard prod
Tomato vines	43	—	—	64	—
Vanilla bean	70	14	3	36	Oils
Wheat bran	51	3	17	78	—

Sources: Van Soest and Jones, 1968; Van Soest and Robertson, 1976a; Wo[h]l[t et] al., 1981; Larwence et al., 1984; Van Horn et al., 1984; Van Soest et al., 1984; [ ] Smith and Adegbola, 1985; G. M. Hill et al., 1986; Coppock et al., 1987.

[a]Many of the tannin-bearing by-products cannot be analyzed by convent[ional] methods for analyzing tannin. Polymerized insoluble tannins are recovered in [ ] lignin.

gestibility of energy and protein are not entirely known but probably include the volatile oils contributing to flavor.

Pomaces

Fruit residues are a potential source of feed. Generally high in pectin and sugar, they can have very high digestibilities provided they do not contain high levels of tannin. (Examples of tanniniferous sources are apple pomace [Nikolic and Jovanovic, 1986], grape pomace [Rebolé and Alvira, 1986], and olive waste [Nafzaoui and Vanbelle, 1986].) Generally they are low in available protein. Sometimes fruits are pressed with a filter substance to obtain juice. The filter may be grain hulls, such as rice hulls, or paper pulp. Although rice hulls are essentially indigestible, other filters may contain some digestible energy.

Citrus pulp, beet pulp, and tomato waste are all fairly digestible, but if they are heat-damaged, the protein may not be available. Materials of this sort require evaluation on a batch basis for protein availability through a determination of the nitrogen content of acid-detergent fiber.

The by-products of the wine industry include grape pomace and raisin stems. These products tend to have a low feeding value because of their lignin and tannin contents. The analysis of grape waste and browses in general is complex because they contain, in addition to tannin, both lignin and cutin in the seed coats and stems.

Tannin-bearing feeds tend to be undervalued when analyzed by unmodified in vitro rumen methods. Sup-plementation with adequate nitrogen or protein may overcome this limitation. Tannins cause nitrogen deficiency in nonadapted bacteria and thus inhibit cellu-lolytic digestion. Adaptation apparently depends on the sacrifice of some cells or protein to inactivate the tannin. Mucins in saliva may play a similar role in browsing ruminants that consume significant amounts of tannin. These animals selectively regulate their rate of intake to avoid excessive tannins (Section 13.3.5).

A further limitation in the practical feeding of tan-niniferous feeds is the amount that can be fed, since there is a tolerance limit for tannin. Most low-quality by-products can be fed at a low level (5–10% of the ration) without much observable effect, but larger amounts become counterproductive.

Unusual feeds involve a broader group of plants and technical processes than large-scale cattle farmers are accustomed to using. If they are to come into more common use, these by-products will have to be evaluated in terms of rumen fermentation and the animal.

8 Forage Evaluation Techniques

A variety of indirect techniques have been developed to measure the nutritive value of pastures and the amounts eaten by animals in them. These techniques are complex because the control of animal intake and the collection of fecal output are much more difficult in pastures (and in situations such as feedlots, farms, and zoos) than in stalls.

8.1 Pasture Evaluation

The total yield of the range or pasture is not available to the grazing herbivore. Indigenous (and often unwanted) herbivores such as wild game, insects, and rodents compete with domestic ruminants for forage. The measurement of total consumption by primary users is difficult. In addition, unused dead forage is decomposed by microbial action in the soil. This balanced system makes as little as 50% of the total yield available to the browsing or grazing domestic animal. Competing or complementary species using forage not eaten by domestic animals may promote forage quality because they add to the net grazing pressure, but this is a controversial topic. The total yield of forage is thus more than that assayed by measuring the animal product.

Forage quality and yield can be estimated by observing plant density, sampling, and laboratory analysis of plants, and by measuring carrying capacity (number of animals that can be supported on a given area; see Table 8.1). No single type of measurement is capable of a quantitative description of the nutritional relationship between plants and animals. Carrying capacity and animal productivity, for example, both depend on the intake of digestible plant matter and animal efficiency.

Assessing the net energy of available forage is more difficult on free range than under stall feeding or even grazing conditions. The more varied environment ensures that many more factors influence the efficiency of feed used on free range. Heat, light, and moisture affect both grazing animals and forage plants. The animal's energy use is affected by its grazing activities and

Table 8.1. Methods for evaluating pasture

	Factor	Method
Forage yield		Cages, clipping, animal off-take/ha
Forage quality	Composition	Chemical analysis
	Digestibility	Esophageal fistula, in vitro rumen, in situ bag digestion
	Intake	Fecal bag, grab-sampling, markers
Animal yield	Carrying capacity	Stocking rate, put-and-take, yield/ha
	Efficiency	Portable respiration apparatus

Sources: Raymond, 1969; Streeter, 1969; Langlands, 1975; Cordova et al., 1978; Holechek et al., 1982a, 1982b; Cook and Stubbendieck, 1986.

its need for thermal control. No effective method for separating and evaluating these factors has been developed. Measurements of forage intake and digestibility leave the remaining variation in animal performance to be counted under the heading of efficiency. In general, maintenance requirements of free-ranging animals are estimated to be 140–170% of the requirements for stall-fed animals. Direct determination of maintenance costs of range animals via a portable respiration apparatus attached to the animal may overestimate values because of the energy cost of carrying the apparatus. The principal criticism of systems that attempt to estimate total digestible nutrients (TDN) or net energy production per acre through animal production involves the inherent error in converting an animal function involving efficiency to a number (e.g., TDN) that is a function of the nutrient availability of the plant.

8.2 Sampling

8.2.1 Cages

The simplest assay of forage production is dry matter yield, but measuring yield under grazing conditions is complex. Pasture yield can be estimated by clipping patches of randomly selected areas protected by the use of movable cages. Some procedures cited by McDowell (1972) are the following:

1. Forage dry matter available (lb/acre) = 85 (H) − 190, where H is the sum of four height measurements in inches taken at the four corners of a piece of light plywood measuring approximately 25 in.^{2} dropped into a stand of grass. Assuming the four height measurements total 50 in., $85 \times 50 - 190 = 4060$ lb dry matter per acre (4540 kg/ha). If the animals that will be grazing the stand require 11 kg of dry matter per day, the stand will in theory supply their needs for 413 days; however, the calculated requirements of the animals should be increased by 30% for waste (e.g., damage caused by trampling). Hence, for a 28-day period the stand would be expected to carry 11 head per hectare, or about 0.86 animals on a yearly basis.

2. Available forage can also be estimated as dry matter per unit of land = weight of fresh forage from quadrants × dry matter content. This system involves weighing samples cut from measured areas randomly selected throughout the stand.

3. Simulated grazing is estimated by hand plucking materials an animal might consume. The extent of the plucking on an individual plant is determined by the height the operator expects the animal to graze the stand. If a designated area is used, such as a 1-m^2 caged area, an estimate of the dry matter the animals may consume can be made as in item 2. Simulated grazing is somewhat superior to the other methods for estimating what the animals are likely to consume.

For these estimates to be of practical value there must be corresponding estimates of the quality of the forage. Forage quality varies tremendously with age of the plant, season, and the portion of the plant being consumed. From a practical standpoint, the best means of determining how well the animals are fed is to observe which portions of the plants the animals eat at a given time. When animals are turned onto a young, growing stand of grass, they first consume the growing tips, the youngest and most tender portion of the plant. Next, they graze the mature leaves, and then, if the stocking rate is high enough that the animals are grazing faster than the growth rate, the upper stem and old leaves (those which have lost their bright green color). If the animals are left on a stand of guineagrass, a typical tropical grass, until the plants are grazed down to 30 cm height, the average digestibility of the consumed material will range from 50 to 55%, which is marginal for animals expected to make good gains. In contrast, in the early days of grazing this stand the digestibility of the ingested forage was 65% or better.

Cage systems are not amenable to browse or mixed browse systems. Clipping only protected areas eliminates any effects the animal has on the plants. Such effects may include trampling and manure and urine spots, all of which influence plant growth. Further, cages also may produce errors such as those due to border effects, in which taller plants at the boundary receive more light than the more shaded plants in the interior. Clipping forage at a relatively uniform height may not reflect the selective feeding practiced by animals. Although selection in improved temperate pastures is largely eliminated by high stocking rates, this is not true in browsing situations and under some tropical grazing conditions where highly differentiated forage contains parts with low value that are not ordinarily eaten. Intensive strip grazing may eliminate the difference in measurement due to protected and unprotected forage by the cage system, but it also represents a kind of management that may not be the most appropriate or efficient from a practical point of view.

8.2.2 Quadrats

Quadrats are similar to cages except that square areas are surveyed on a statistically random basis relative to field area. This method has advantages where grassland is intermixed with trees and shrubs and a larger area is needed for an accurate survey.

8.2.3 Transect Survey

Ecologists developed the transect survey to catalog the available plant species based on plant density and occurrence. Random lines are drawn across the terrain to be surveyed, and all plants occurring on the lines of the survey and within some meters to either side are recorded. This procedure, although particularly adaptable to browse situations, is apt to overestimate available forage in native mixed situations, since plant species may be present that are not eaten. Those eaten may also be of widely differing nutritive value.

8.2.4 Clipping

The assay of actual forage intake requires a sampling procedure that mimics animal selectivity. Various techniques have been employed. One method involves following the grazing animal and clipping material physically comparable to that selected by the animal. This is the least intrusive procedure relative to the animal, but it requires that the animal be sufficiently tame that the operator can observe the parts selected at close range. This is perhaps the best system for identifying plant species and parts chosen (Cook and Stubbendieck, 1986). One alternative to hand clipping is to mow a strip that represents a cross section of the field, but this procedure is apt to represent what is eaten only when applied to uniform pastures.

Selective hand clipping is useful for surveying mixed vegetation, including browse, and is in some ways less labor-intensive than other procedures.

McCammon-Feldman (1980) compared a transect survey of the vegetation with clipping that mimicked the selective feeding of goats in Nicaragua. The transect showed the existence of more than 100 species of plants, of which 30 were occasionally taken by the animals, but only 3 species constituted more than 80% of the feed intake. The selective clipping method, while simple, gives accurate data provided animal observation covers sufficient time and numbers of animals. Actual intake can be estimated from total fecal collection combined with the analysis of an internal marker (lignin, etc.) on the clipped material as representing the diet and also the feces (Section 8.4.1).

8.2.5 Fistulation

A more direct approach to measuring feed intake is to assay the plant material that the animal actually consumes. This may be accomplished via esophageal or rumen fistula or by fecal analysis (McManus, 1981; Chenost, 1986). Fistulation techniques are much more intrusive for the animal and can affect results. These procedures are labor-intensive and require considerable expertise.

An esophageal fistula is formed by surgically transecting the esophagus and inserting a cannula. During sampling the cannula is replaced by a collection device while the animal grazes. Sampling via a rumen fistula requires removal of the rumen contents into a container before the animal begins to graze. After grazing, samples of ingesta are collected and the rumen contents are returned. This method is less sophisticated but requires more labor. Esophageal fistulae require constant care and maintenance.

Samples obtained by either esophageal fistula or by rumen emptying and sampling are contaminated with saliva, which contains both mineral and organic components, so the chemical composition of forage samples does not exactly reflect that of the forage eaten. Correction for its contamination is not simple and is often ignored. The mucopolysaccharides in saliva also interfere in lignin determination, by causing elevated values (Theurer, 1970). Generally, it is more difficult to identify plant species and parts after they have been eaten.

8.2.6 Fecal Analysis

It is possible to identify plant species and parts in the fecal residues. This procedure is very time-consuming and requires expert training. The method can provide valuable data and can also be applied to rumen contents and esophageal material. It is, however, difficult to relate results to the chemical composition of the original forage ingested. Fecal composition will reflect the amounts of lignified tissue present, but it tells nothing about the amount and quality of the more digestible components that are not represented in the feces.

8.3 Estimating Digestibility and Intake

The intake of the animals in pasture trials and the digestibility of the forage eaten are unknown and must be estimated. The relationship used for estimation is:

$$F_i = \frac{P_r}{R_a} \qquad (8.1)$$

where intake (F_i) equals fecal residue (P_r) divided by apparent indigestibility (R_a).

Apparent indigestibility can be estimated from internal markers such as fecal nitrogen or chromogens. Chromogens work well only in situations of uniform green swards, where pigment concentrations are high and samples representative of selective intake are obtained. Fecal nitrogen is mainly of microbial origin. Its amount depends on the animal's nutritional status, and it requires that a standard be set by feeding cut forage in stall trials and collecting all feces; usually, a regression equation is developed to relate digestibility to fecal nitrogen. (Works that discuss indirect methods for determining digestibility include Bartiaux-Thill and Oger [1986], Wofford et al. [1985], and Bruckental et al. [1987].)

An alternative is the use of esophageal fistulae (McManus, 1981; Chenost, 1986). The use of the ratio of lignin in esophageal samples to that in feces may not work well because saliva interferes in the determination of lignin in esophageal samples (Theurer, 1970). If digestibility is determined by in vitro methods, internal indicators (e.g., fecal nitrogen) can be used to estimate intake (Chenost, 1985). For this purpose it is necessary to know or to be able to estimate total fecal nitrogen output, which is a function of intake. Usually, the relationship is calculated by means of a regression equation. Fecal nitrogen is probably a more accurate estimation of digestibility than of intake (Cordova et al., 1978).

Total fecal output can be measured by using bags attached to the animal. The bag technique has the disadvantage of a high labor requirement as well as the possible detrimental effect of the load on the animal and its grazing habits. The latter might be more serious in low-density swards where distance traveled may affect intake. It is much easier to use the marker techniques that have been developed for estimating fecal output.

External markers are most commonly used for estimating fecal excretion (Section 8.4.2). Daily administration of chromic oxide (Cr_2O_3) in a form that has a controlled rate of release is one such method. Assuming total recovery of the marker in the feces, knowledge of its amount and concentration in a representative sample of feces will allow estimation of total fecal output. This technique depends on sampling in such a way that diurnal variation in fecal concentration does not contribute to error. Measuring the diurnal excretion pattern can provide information for establishing an adequate sampling schedule.

Various techniques exist for fecal sampling. One method is to grab-sample fecal matter from the rectum. This method does not work well when chromic oxide powder is the marker because the powder tends to move with the liquid rather than with the particulate phase. This method has been improved by the use of chromic oxide–impregnated paper or mordanted preparations, which give a more even excretion pattern. Alternatives are preparing chromic oxide bullets that disintegrate at a constant rate in the rumen and sampling the feces at times that overcome the variations in the diurnal fecal excretion curve (Langlands, 1975; Raleigh et al., 1980).

Another procedure is to sample droppings in the field. Individual animal droppings can be identified if colored plastic pellets are administered with the chromic oxide (Minson et al., 1960). This method works best in a dry environment where leaching of the droppings is minimal and a relatively complete recovery of feces is obtained.

Telemetric devices have been used to record numbers of bites and chews (Penning, 1983). Penning and Hooper (1985) measured short-term forage intake by weighing the animals.

8.4 Markers

Markers are used when total fecal collection is inconvenient or the direct measurement of intake is difficult, as, for example, in grazing studies. The marker is usually maintained at a certain level such that fecal output ratio becomes an estimate of indigestibility. There are three types of markers in this context: internal, that is, already contained in the diet, like lignin; generated mathematically, as with fecal nitrogen; and external markers (Section 8.4.2).

The mathematical relationship of indigestible and unabsorbable indicators is essentially one of relative concentration according to the degree of digestion; indigestible residues become proportionally concentrated in the feces. From equation 8.1, if X is indigestible and R_{xa} is unity, then the concentrations of X in feed (C_{xi}) and feces (C_{xr}) are:[1]

$$R_a = \frac{C_{xi}}{C_{xr}} \qquad (8.2)$$

[1] See Section 22.1.2 for definition of terms.

The same is true for metabolic matter since $M_i = R_a C_{mr}$:

$$R_a = \frac{M_i}{C_{mi}} \qquad (8.3)$$

In the case of an indigestible component such as lignin or added marker, the concentrations of the reference substance in feed and feces are needed. In the case of a metabolic constituent such as fecal nitrogen, however, the M_i constant must be derived from total collection digestion trials.

Digestibility is a linear function of the reciprocal of fecal concentration. An equation for estimating digestibility is obtained by plotting digestibility versus fecal content, a process applicable to any indigestible fecal constituent, metabolic or otherwise. The direct relationship between digestibility and fecal concentration is curvilinear. In practice, the calibration of fecal nitrogen is a regression of digestibility on a function of fecal nitrogen concentrations.

Precision and accuracy in estimating digestibility are diminished by variability in measurements of indicators, but such variation will not create a bias toward high or low values unless there is a problem in recovery. Failure to recover the marker results in an underestimation of the digestion coefficient. This might be adjusted by introducing a correction for the digestibility of the marker as determined in a total collection trial. This can be justified in cases of natural feed ingredients such as lignin but not in cases of metabolic substances or added heavy metal markers such as chromic oxide, where it becomes a cover for faulty technique.

It is essential that the marker be recoverable, since passage is a function of the flow of undigested residues. If all the marker is not recovered, disappearance rates are assayed instead, and disappearance is the sum of passage plus apparent digestion. The apparent digestion arises from the error resulting from failure to recover the marker. The second requirement of a marker is that it must flow with the feed or the portion of the feed that is being investigated. Feed residues do not flow at equal rates; fine, soluble, and liquid matter escapes the rumen more rapidly than coarse, light, solid matter. Hence there is no single rate of passage. Differential rates can be measured with multiple markers as long as it is known which marker moves with which fraction and there is no migration of marker from one fraction to another.

8.4.1 Internal Markers

Recoverable indigestible feed fractions are the basis for internal markers, which are convenient in pasture and other balance studies in which an estimate of digestibility is required (Mayes et al., 1986). They are not useful for pulse dose administration for rate of passage, but they are useful for turnover measurements based on gut contents such as obtained by rumen emptying (Chapter 23). Other internal markers are fecal nitrogen and chlorophyll-related pigments.

Lignin

While the acid-detergent sulfuric lignin behaves in the Lucas test (Sections 22.1–4) more or less as an ideal indigestible fraction, immature grasses and other forages of low lignin content often show apparent digestibilities on the order of 20–40%. The lack of recovery may be due to a number of factors: contamination of the crude lignin with nonlignin matter from the feed, some loss of immature lignin (see Chapter 12), formation of soluble phenolic matter, failure to recover finely divided lignin in feces, and overdry feed, which is more sensitive to heat damage. The problem is complicated by the lack of a clear definition of lignin (Chapter 12). Generally speaking, lignin is a better marker in rations with high lignin content, especially above 5% of feed dry matter. The standard error of the lignin determination is about 0.4 lignin unit as a percentage of dry matter under the best conditions. This error tends to be independent of the lignin content; the error in estimating digestibility is about 10% at 4% lignin in the diet and about 20% at 2% lignin in the diet. As a marker, sulfuric lignin is better than permanganate lignin because the latter gives low values in samples of high lignin content. Recovery of permanganate lignin from the feces can be improved by increasing the time of oxidation of fecal samples to three hours or by performing sequential analysis with 72% sulfuric acid followed by permanganate. Fahey and Jung (1983) reviewed the use of lignin as a marker.

Acid-Insoluble Ash and Silica

Silica and acid-insoluble ash (AIA) have been used as internal markers, with variable success. It is not always possible to determine the nature and source of siliceous ash. Insoluble minerals in the diet arise from two sources: biogenic mineral fractions in the forage and contamination from soil and dust. Native biogenic insoluble ash is probably an acceptable internal marker because it is truly a part of the feed. Soil or dust contamination does not satisfy the condition of a true internal marker because technically it is not a part of the plant matter. Soil minerals have a higher density than feed and are apt to have a different rate of passage. Very fine material may have the characteristics of chromium oxide powder in that it is neither a true particulate nor a liquid marker. Animals that have ingested soil during grazing tend to retain the larger stones in the rumen,

and these continue to erode and leach silica. One study found seven times more insoluble ash in the feces than that consumed (R. R. Parra, pers. comm., 1980). Thus the proper use of AIA and silica seems limited to clean animals and to conditions in which sufficient biogenic silica-containing forage is eaten.

Grasses and sedges characteristically contain high levels of biogenic silica, but legumes usually contain very low levels (Section 9.7). Many cereal by-products contain biogenic silica, and it can be a marker for concentrate-forage combinations involving graminaceous plants. The level of silica is influenced by soil availability of orthosilicic acid.

AIA measurement can also be difficult. A renewed interest in AIA emerged after the publication of a procedure by Van Keulen and Young (1977). This procedure, however, does not result in the total recovery of silica in plant ash. The method fails for two reasons: first, the whole sample is ashed and the alkalinity of sodium, potassium, and calcium can create soluble glasses from originally insoluble matter; second, the steps for acid dehydration of soluble silica to convert it back to the insoluble form are grossly inadequate (Van Soest and Robertson, 1985).

The acid-detergent insoluble ash procedure overcomes this problem and recovers all silica regardless of its origin or form. Several studies have pointed out the superiority of the acid-detergent fiber–AIA measurement over Van Keulen and Young's technique (Porter, 1987; Giner-Chavez et al., 1990).

Indigestible NDF or ADF

Lignin and AIA may occur in such low concentrations in the consumed herbage that their usefulness as internal markers is precluded. The standard error of both determinations is about 0.3% on a forage dry matter basis. The sample error can be reduced if an indigestible component of higher dry matter percentage can be found, and indigestible neutral-detergent fiber (NDF) and acid-detergent fiber (ADF) have been proposed for this purpose (Lippke et al., 1985; Clar et al., 1988). The measurements require long digestion times (up to two weeks of digestion with nylon bags). There is a danger of bag contamination in such long in situ incubations, and estimation may be more accurate by in vitro batch fermentation.

Chromogens

Chlorophyll and its degradation products, chromogens, are acetone-soluble pigments whose use as internal indicators is based on their relative indigestibility (Kotb and Luckey, 1972). Although not utilized by rumen bacteria, chlorophyll is degraded in the animal into colored products with altered spectra. The principal changes are loss of magnesium and the formation of the yellow pigment pheophytin, which has its maximum absorption at 415 nm. Chlorophyll may be chemically converted into pheophytin by treatment with acid or chelating agents. Chlorophylls and their degradation products are light-sensitive substances. Chlorophyll degradation in hay and other exposed dead plant material results in bleaching and the formation of substances with a ruptured porphyrin ring.

The practical use of chromogen as a marker is essentially limited to forage with high chlorophyll content, such as fresh green herbage. The method is empirical and requires standards set through stall-feeding digestion trials on the particular type of forage in question. The wavelength at which density is read is arbitrarily chosen so as to give apparent recovery of pigment.

Waxes

Long-chain hydrocarbons are a part of the cuticular surface waxes in forages and browses. Chain length varies from 21 to 37 carbons, (waxes more than about 33–35 carbons long are reasonably indigestible and recoverable in feces. Shorter chains appear to be absorbed and at least partly metabolized (Mayes et al., 1986). Since hydrocarbon distribution is a characteristic of plant families, it has become possible to apply long-chain hydrocarbon analysis as an indicator of grazing intakes with reference to selectivity toward legumes, grasses, and so on. A solution of long-chain waxes sprayed on forage can serve as an external marker in the diet.

Fecal Nitrogen

Fecal nitrogen content is one of the most common methods for estimating digestibility of pasture (Chenost, 1985; Holechek et al., 1986). Total collection trials with clipped herbage must first be carried out to establish the fecal metabolic constant for the experimental trial. Fecal metabolic fractions vary with the intake and quality of the diet and the animal species. Although researchers have attempted to improve the use of fecal nitrogen as an indicator by fractionating the total fecal nitrogen, success has been very limited.

The truly indigestible nitrogen in feces can be separated by means of acid- or neutral-detergent solutions, allowing an estimate of the total metabolic nitrogen fraction by difference. Such corrected values, however, show no less variation than uncorrected values. Bacterial matter contributes more than 80% of the total fecal nitrogen and is thus responsible for the variation. Fecal nitrogen gives an estimate of digestibility and intake with an error of about 10–15%. This error may be diminished somewhat by calibration with stall-fed animals given the same forage as that grazed in the total collection trials.

Diaminopimelic Acid (DAPA)

This amino acid, unique to bacteria, has been used as a marker to estimate the microbial output from rumen fermentation and the proportion of microbial matter in feces. As a component of the fecal metabolic microbial matter (and the fecal nitrogen) it shares the characteristic behavior of other metabolic fecal fractions. A particular difficulty with using DAPA as a marker is that its content in individual strains of bacteria can vary. This variability precludes the use of DAPA to obtain anything other than general estimates. Some correction can be made by culturing the microbial species to characterize its particular level of DAPA.

8.4.2 External Markers

A substance added to the diet as a marker is known as an external marker. The substance chosen will vary with the individual requirements and convenience of the experimenter. Markers are used to save labor—as, for example, to avoid total fecal collection in a digestion trial—or to obtain information otherwise difficult to acquire, such as rumen volume, rates of passage, or yield of rumen fermentation products. External markers can be used in two basic ways: a constant level is added for digestibility studies, and pulse doses are used to study rates of passage and digesta flow. The behavior of markers in the digestive tract is more critical in experiments on rates of passage and digestive flow than in digestion studies.

For most applications an external marker needs to be recoverable, indigestible, and nonadsorbed to the walls or the lining of the digestive tract. An external marker should have no effect on the animal or on digestibility and should not occur in the diet or soil. Other requirements may be added according to specific need and use. Liquid markers must not associate with solids, and particulate markers must remain associated with the fraction intended to be labeled.

Markers may be grouped according to their composition and properties, which are related to their use. Kotb and Luckey (1972) provided a lengthy and detailed list. An abridged list of commonly used markers is given in the following sections.

Stains and Dyes

Stained particles have the advantage that they are feed particles, whereas chromic oxide and plastic particles may move independently of feed particles and liquid. Feeds stained with suitable organic dyes are used as particulate markers in rate-of-passage studies (Balch and Campling, 1965) since the dyes are more or less indelibly bound to the surface of the feed particles. This binding prevents quantitative measurement, and passage studies are usually based on direct counting of the dyed particles that appear in the feces. Rumination and digestion result in the formation of very finely ground particles, which are difficult to measure in the feces. The stained particle method gives relative rates of passage since an absolute measurement would depend on the recovery of this fine material.

Plastic and Rubber

Synthetic organic substances such as beads, particles, and plastic ribbon have been used as markers for particulate passage studies. These substances have an advantage over stained particles in that they can be easily separated from feces and are therefore potentially completely recoverable. If they are impregnated with barium or chromium compounds they are radio-opaque and can be counted by x-ray examination. Such plastic products are available in a variety of identifiable shapes and are used in human studies as well. Polyethylene or polypropylene plastics end up in the lignin fraction on treatment with 72% sulfuric acid. Permanganate oxidation will isolate them from most organic feed substances. Plastic ribbon simulates the action of coarse hay and is ruminated into fine particles that pass down the digestive tract (J. G. Welch and Smith, 1971) provided the size of the ribbon is small enough to be prehended by the bolus-forming process. Plastics do not have the same physical properties as feed particles (e.g., density, ease of rumination, etc.) and therefore yield only relative data. Small plastic particles can be counted and therefore are quantifiable in monogastric studies provided they are not chewed into smaller pieces.

Metal Oxides

Insoluble inorganic substances have been widely used for digestibility and marker studies. They have the advantage of being quantitatively determinable according to the chemical properties of the individual element involved. Only substances that do not occur naturally in feed and soil can be used, of course. This eliminates iron, titanium, and silicon compounds, which are widely distributed in feed and soil. Heavy elements that do not occur in significant concentrations in soil and plants make suitable external markers.

The most commonly used material is chromic oxide (Cr_2O_3), a very fine, heavy, insoluble powder. Because of its density and particle size it, like barium sulfate, tends to behave as a heavy liquid when in water suspension. These two substances pass more rapidly from the

rumen than coarse fiber and tend to be associated with the movement of the liquid fraction (Kotb and Luckey, 1972). Chromic oxide is more suitable as a digestibility marker than as a passage marker, provided a constant fecal output can be achieved. Digestibility measurements depend on even distribution of the Cr_2O_3 in the feed and its constant passage down the digestive tract. Various sampling techniques are used to overcome diurnal variation in fecal chromium concentrations. The grab-sampling methods (random sampling of feces to avoid total collection) depend particularly on the control of diurnal variation.

The heavy metal or oxide is bound to an organic matrix or made in the form of a bullet that will release the respective heavy element at a constant rate. Chromium-impregnated paper is no longer available and has been replaced by chromium mordants in which the chromium is indelibly attached to fiber (Udén et al., 1980). This is a preferred method, although the level of chromium should not be so high as to affect particle density (Ehle et al., 1984). Oxide bullets are an alternative to chromium mordant if they can achieve constant release. This marking technique is an art, but it can be workable with efficient handling.

Rare Earths and Other Mordants

A variety of rare earth and heavy metal compounds have been used as markers (Mader et al., 1984; K. R. Pond et al., 1985). Of particular interest are those that can be chemically bound to the particulate fraction, called *mordants* after the textile-dying technology employing the system. Basically, a mordant is formed when a coordinate or covalent bond is induced between the organic matrix and the respective heavy element. The type of bonding depends on the element. Chromium forms coordinate bonds via hydroxyl groups, and rare earths bond with the organic matrix through cation exchange. The first requirement in choosing elements is that there be no indigenous background; this precludes elements like calcium, aluminum, titanium, and probably scandium. Rare earths and chromium are not as common in soil and feed. All these elements, when bound to the cell wall, are inhibitory; that is, they reduce digestibility (Chapter 9) and form ligands resistant to digestive action. This seems to be the price of an indelible bond.

To prepare mordanted fiber one must first isolate the plant cell wall, because chromium and rare earth elements are generally reactive with the carbohydrates, phenols, and phosphate in the soluble fractions. If a rare earth is indiscriminantly applied, finer matter will be selectively labeled (Erdman and Smith, 1985). Recovery of a rare earth in a centrifuged pellet does not necessarily indicate binding to plant cell wall, since rare earths form insoluble salts with phosphate present in rumen fluid and buffers. Soluble rare earth salts added to the rumen have slower passage than liquid markers and faster passage than true particulate markers (Bernal-Santos, 1989). Gold is deposited as the metal but digests off in about the same proportion as the digestibility, while chromium and rare earths are more tightly bound and reduce digestibility of the plant cell wall particles to which they are attached. Rare earth elements are not identical in physical properties, and caution is necessary in choosing an element (see Section 9.10.1).

Ruthenium Phenanthroline

Ruthenium as a phenanthroline complex is sometimes used as a heavy metal marker (Faichney, 1984). Its use is based on its observed affinity for particulate matter as a complex-bound lipid in histological studies. Ruthenium, a very expensive element, is a homologue of iron, hence its affinity for phenanthroline. The valence form for the complex is critical and like the dyes does not seem to have a specific attraction for fiber. As a particulate marker it thus appears to move with other lipophilic complexes in the digestion process (R. M. Dixon et al., 1983).

Isotopes

Organic substances in the feed can be labeled with specific isotopes such as ^{14}C, ^{35}S, and ^{15}N, and rates of disintegration of the substances via fermentation or metabolism in the animal can then be measured. Labeled acetate, propionate, and butyrate (^{14}C) have been used to measure the rumen pool sizes and flux of these acids. Radioactive ^{14}C-labeled carbohydrate or plant cell wall has been used in rate-of-digestion and passage studies. The use of the radioactive isotopes has become limited, however, because of safety restrictions. The stable rare isotopes ^{15}N and ^{13}C are safer to use but require experimental mass spectrophotometry for measurement. The isotopes ^{12}C and ^{13}C differ in their metabolism in C_3 and C_4 plants, and the result is differing ratios in photosynthesis, which thus can be a basis for distinguishing sources of forage—whether C_3 or C_4 plant. Labeled sulfur and nitrogen are used to measure the incorporation and synthesis of feed amino acids into microbial protein.

^{14}C-labeled plant cell wall can provide an absolute method of measuring passage, since the labeled material has the same properties as the diet (provided they are of the same plant type). It is necessary to grow the labeled forage in a light chamber, however, which perhaps limits more widespread use of this method. Undigestible plant cell wall (neutral-detergent fiber) must first be isolated from ingesta and feed to eliminate contamination from digested and recycled ^{14}C, which

becomes widely distributed in the animal. Only the undigested labeled isotopes are of interest in the biological measurements. This method has given some insight into the problems of particle reduction and passage in the tract as well as a critical evaluation of other passage markers.

Soluble Markers and Metal Chelates

Soluble markers are intended to measure liquid flow. They should be ideal solutes, yet possess high enough molecular mass (500+ daltons) to be unabsorbable (and thus recoverable in feces). Larger molecules tend to be more adsorbable to surfaces, and smaller ones are more likely to be absorbed. The soluble marker must not be reactive with any other component in the diet.

Metal chelates with ethylenediaminetetraacetic acid (EDTA) or similar chelates are anionic complexes of relatively strong acids. Trivalent metals with tetravalent EDTA produce monovalent anions. Trivalent metal ions with pentavalent diethylenetriaminepentaacetic acid (DTPA) produce divalent anions, and so on. Generally, the greater the number of ionic groups or valence, the more stable the complex.

Cobalt and chromium EDTA are the easiest to determine and have more or less replaced polyethylene glycol (PEG). Cobalt EDTA is favored if chromium is used as a mordant to the fiber. At very low levels EDTA complexes adsorb to protein and particles. EDTA chelates of trivalent ions are monobasic anions of a relatively strong acid and at very low levels are probably adsorbed by anion exchange to feed particles and protein. This has been a problem with the use of low levels of ^{51}Cr-EDTA (A. C. I. Warner, 1969), which can be overcome by dilution with sufficient cold marker. About 4–5% of Cr-EDTA is absorbed by ruminants and excreted in the urine. This quantity increases at higher osmotic pressures obtained by feeding salt or more rapidly fermentable matter (A. Dobson et al., 1976). Rare earth complexes with EDTA are unstable to phosphate or oxalate in the rumen and form insoluble complexes (Bernal-Santos, 1989).

Polyethlyene glycol of molecular weight 1000 or greater is not absorbed; however, there have been some difficulties in achieving complete recovery. PEG appears to be precipitated by tannin and capable of associating with particulate matter of ingesta if samples are frozen (A. C. I. Warner, 1969).

8.5 Rumen Fermentation Techniques

Digestibility is often estimated by in vitro rumen systems that simulate the digestion process. In vitro systems can be more accurate, because in vivo microorganisms and enzymes are sensitive to undetermined factors that influence rate and extent of digestion. Chemical systems are faster and offer better replication; however, they do not reflect the biological process of digestion that occurs in the rumen environment. The cloth bag technique may provide a better indication of digestion in the rumen, but it has its own set of problems. The success of any rumen in vitro system depends on the degree to which it reflects rumen events and the sequential processes of the ruminant digestive tract. The ultimate superiority of any in vitro system over that of any purely chemical analysis rests on the accuracy of its biological response.

8.5.1 In Vitro Techniques

The sequence of all in vitro rumen procedures is an anaerobic fermentation of a sample substrate with medium and filtered rumen liquor followed by an end point measurement. The medium is usually a buffer solution simulating ruminant saliva. It is important to observe careful anaerobic technique and to supply all possible nutrients, particularly ammonia, that might become limiting in poor-quality forages. Unlike the rumen, in vitro systems do not have a continual supply of saliva, which might supply nitrogen. The time of batch fermentation is commonly 48 h for digestibility estimation, although other time periods from 3 h to several hundred hours have been used to estimate rate of fermentation. Voluntary intake is best related to a 6-h value, and digestibility is better associated to a 36–48-h value. Longer times are needed for maximum extent.

8.5.2 Cloth Bag Methods

The intraruminal incubation of samples in cloth bags is preferred by those who wish to avoid the details of anaerobic technique. Cloth bags are inserted into the rumen through a fistula. On the whole, this method has given more variable results than the Tilley-Terry system (Section 8.5.5). The site of incubation in the rumen must be controlled and is best near the bottom. A major problem has been the integrity of the cloth bag as an analytical filter. Lignified matter can enter and accumulate in the bags, causing low or even negative results (Figure 8.1). Improved methods utilize cloths of specific pore size and control the ratio of sample weight to surface area of the bag (Udén et al., 1974; Van Hellen and Ellis, 1977). Bags with a large ratio of surface area to sample size minimize the error. The optimum pore size is about 30 μm. Smaller pore sizes retard the entry of organisms and thus inhibit optimum fermentation, while larger ones permit transit of lignified particles.

Although cloth bags of controlled pore size can be

very useful in measuring in vivo rates of digestion (Nocek, 1985), digestibilities determined by bag techniques are subject to the same end point measurement problems as those found in in vitro methods. Dry matter disappearance is the most common measurement for digestion studies, but this fails to distinguish bacteria from undigested substrates and leads to low results, especially in higher-fiber feeds (Sauvant et al., 1985; Varvikko, 1986). Neutral-detergent extraction has given more repeatable and biologically relevant results for digestibility of cell wall; however, protein correction for DAPA may be required if true protein digestibility is to be measured. Small, mobile bags that pass through the gut of the animal and are recovered in the feces have been used to estimate lower tract protein digestion (de Boer et al., 1987).

8.5.3 End Points

Various end point procedures are used to ascertain the extent of digestion or substrate utilization. The end point measurements are relevant to both in vitro Tilley-Terry and cloth bag digestions in situ. End point measurements may be of cellulose disappearance, residual dry matter, residue after pepsin digestion, or neutral-detergent residue. Gas production (Menke et al., 1979) and volatile fatty acid (VFA) production are exclusively in vitro measurements. These end points, although correlated, are not equivalent, and their respective utility depends on the purpose to which they are being applied. They may be grouped according to whether the products of fermentation (VFA and gas) or the residual unused substrate are being measured.

The products of microbial fermentation are cells, fermentation acids, and gases. Net gas production is correlated with the extent of digestion, although it does not give a direct value, and so gas production is a convenient assay of metabolic activity. As a measure of the rate of fermentation it represents the whole substrate, the soluble portion of which ferments faster than the cell wall part, leading to a complex of multiple rates that are difficult to sort out. Use of either gas or VFAs as end points requires an inoculum and medium of low fermentable energy so that these products are kept low in the blank control fermentations.

VFA measurement is subject to the same limitations as gas measurement but has the additional problem of the distribution of products between microbial cells and VFAs. Early in the fermentation the rapid growth of cells causes more energy to accumulate in cells relative to that in acids (Figure 8.2). After the cells reach their maximum number (stationary phase), VFAs become the major product. Later, when the culture becomes senescent, cell death, lysis, and refermentation of cellular matter lead to a further increase in VFAs and a reduction of the insoluble residue. As a result, VFAs

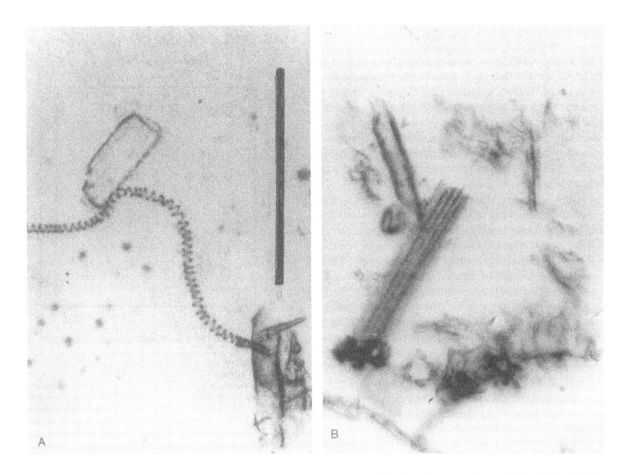

Figure 8.1. Particulate fractions obtained by means of 70-μm pore size synthetic cloth bag from the rumen of a cow fed timothy hay. The bag contained a plastic hair curler to provide the overfilling effect that enhances particle movement into nylon bags. Photograph A shows a long tracheal element that entered endwise through the pore. The bar at right is 200 μm long. Photograph B (same magnification: ×200) shows other fibers and cellular fragments. The chemical analysis of these particles revealed a lignin content of 52%. (Photos courtesy of P. Udén, Agricultural College of Sweden, Uppsala. Experiments performed by P. Udén at Cornell University.)

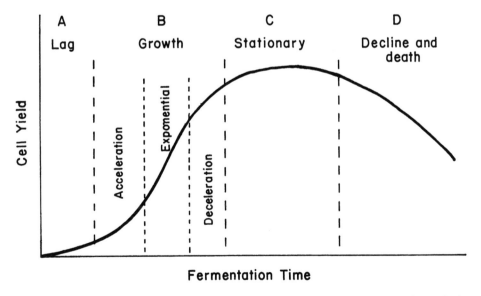

Figure 8.2. The stages of microbial growth in batch fermentations. During initial lag and acceleration phases bacteria are adapting to the substrate and are the limiting factor in fermentation rate. In the exponential and deceleration growth phases bacteria are in equilibrium with the substrate, which is now the limiting factor. During the stationary phase cell yield peaks and substrate becomes exhausted. This is followed by a declining phase accompanied by cell death and secondary fermentation of cell products. The time scale on the bottom axis is relative to the potential fermentation rate of the substrate and the generation time of the bacteria.

are a poor indicator of fermentation rate and extent of digestion, particularly in shorter-term incubations. The value of VFAs as an estimate of feed efficiency is questionable since the batch culture departs from the natural ecology of the rumen. Microbial yield and microbial products are better measured in a continuous fermentation system; however, continuous systems are impractical for measuring extent of digestion because they are so labor-intensive.

The total organic residue obtained by the filtration or centrifugation of the sample at the end of fermentation will be a measure of the sum of undigested substrate plus microbial cells. Therefore, the numerical value combines undigested matter and digested products. Apparent digestibility values calculated in this way are too low if obtained in times less than 50 h. Later (96 h), after much cell lysis and refermentation, digestion values approach those obtained by the Tilley-Terry method (48 h fermentation followed by a pepsin–hydrochloric acid digestion). The long-time fermentation method (den Braver and Eriksson, 1967), like the unmodified Tilley-Terry procedure, is good only for measuring digestibility and is not useful for rate measurements because disappearance of substrate is confounded with cellular production.

8.5.4 Cellulose

The desire to separate substrate and products in fermentation residues led to the development of methods that assay substrate only. Cellulose analysis was the earliest assay to be applied. Since cellulose is only a part of the cell wall complex, it is necessary to convert the cellulose digestion values back to dry matter digestibilities using a regression calculation. The value of cellulose digestibility depends on its correlation with total dry matter digestibility, and regression equations assume that the relation between the two is consistent. Unfortunately, this is not the case. The cellular contents are entirely available, but cellulose forms a variably digestible part of the cell wall. The most successful in vitro assays are those in which indigestible cell wall components are quantitatively recovered.

8.5.5 Pepsin-Insoluble Residue and the Tilley-Terry Procedure

Pepsin-insoluble residue measurement directly applied to forage samples gives values closely related to NDF, although its net value may be somewhat higher. Application of pepsin-acid digestion to the fermented residue after in vitro rumen fermentation is important in removing microbial protein from the residue.

The Tilley-Terry system has largely replaced other fermentation systems for estimating digestibility. The method involves two stages: 48-h digestion with rumen

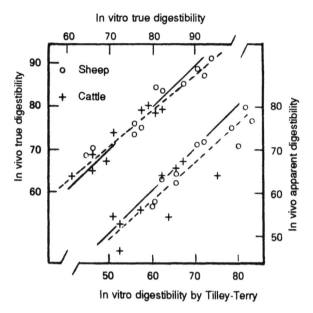

Figure 8.3. True and apparent in vivo and in vitro digestibilities (from Van Soest et al., 1966). Solid lines are the locus of 1:1 equivalence; dashed lines, regression. In vivo true digestibility was determined by analysis of feces with neutral detergent, and in vitro apparent digestibility was obtained using the unmodified procedure of Tilley and Terry (1963). In vitro true digestibility was obtained by substituting neutral-detergent extraction for the second-stage digestion with pepsin. The mean difference between true and apparent digestibility in vitro was 11.9 units of digestibility, while the difference for in vivo digestibilities was 13.9 units.

organisms followed by 48-h digestion with pepsin in weak acid (about pH 2). The residue is composed of undigested plant cell wall and bacterial debris and yields values comparable to in vivo apparent digestibility. The success of the method is related to the recovery of indigestible cell wall matter and its similarity to the ruminant digestion sequence. Rumen bacteria are partially digestible, and the comparability of in vivo and in vitro digestibility values rests on the fact that ruminant feces are composed of roughly equivalent amounts of undigested cell wall and bacterial remnants similar to those obtained in the in vitro situation (Figure 8.3).

The main disadvantage of the Tilley-Terry system is the long time required to do the analysis and the number of steps. Modifications and alternatives developed to shorten and simplify the procedure include the substitution of McDougall's solution with a high-phosphate buffer, which allows direct acidification and addition of pepsin at the second stage, and shortening of the pepsin digestion to one day or substituting a neutral-detergent extraction for the second stage.

There have been various attempts to standardize Tilley-Terry (1963) procedures without much avail since the proponents did not understand analytical principles relative to reagent saturation and to end points (Van Soest and Robertson, 1985). As a result, the application of in vitro procedures to poor-quality forage and straw has been relatively unsatisfactory. Overall, the Tilley-Terry procedure applied to reasonable-quality forages and feeds without supplements has been a relatively accurate mode of measuring digestibility in the laboratory.

8.5.6 Neutral-Detergent Extraction

This treatment extracts all indigestible microbial matter and leaves a residue of undigested plant cell wall. Fermentation of residues followed by neutral-detergent extraction yields values that are estimates of true digestibility; apparent digestibility must be estimated by subtracting a metabolic value of 11.9 (for cattle and sheep). The method is just as precise as the original Tilley-Terry procedure and requires half the time to complete. Results of the neutral-detergent modification should be expressed on a silica-free or organic matter basis to correct for the neutral-detergent solubility of silica in siliceous forages.

The difference between apparent digestibility and true digestibility for any feed is in fecal metabolic matter, which includes animal endogenous matter. Undigested plant cell wall can be separated from metabolic matter by neutral-detergent extraction. The neutral-detergent extraction of the in vitro residue similarly removes metabolic matter, which in this case is entirely microbial, since animal endogenous matter is not present. It is therefore a true indigestible residue. The quantitative similarity between Terry-Tilley digestibilities and apparent in vivo values leads to the conclusion that the metabolic amounts from in vitro and animal origin are of similar constitution, a fact that has been verified by analysis (Mason, 1979). The quantity of metabolic matter in vitro is slightly less (85%) than that in vivo, thus the endogenous fraction in the feces is small. The neutral-detergent modification also allows estimation of cell wall digestibility provided the NDF content of the diet is known.

Digestibility and voluntary intake are correlated and vary according to fermentation time. (Figure 8.4). Maximum correlation with digestibility of temperate forages occurs at about 36 h, perhaps reflecting mean in vivo retention. Longer times of digestion give somewhat lower correlations. The maximum correlation between digestibility and voluntary intake occurs at 6–12 h, which corresponds approximately to a neutral-detergent extraction since protein and other easily degradable matter of the cell contents are the main components digesting at these early times. The high correlation at 6–12 hours can also be viewed as a measure of disappearance at a time when the animal may eat a second meal, and supports the idea that available rumen space created by digestion is a model of feed intake (Chapter 21).

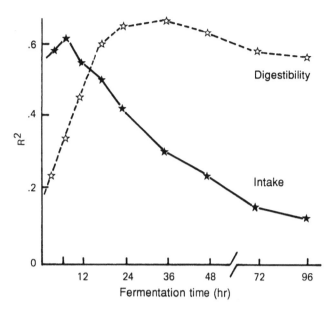

Figure 8.4. The correlation (R^2) between in vitro rumen digestion and in vivo digestibility and intake of 187 forage samples, including 61 legumes and 126 grasses of diverse species. Maximum correlation is at 6 h for intake (r = 0.79) and 36 h for digestibility (r = 0.83).

8.5.7 Gas Production Systems

Gas produced in fermentation is generally proportional to net microbial metabolism and is therefore a possible end point for estimating digestibility. Its use as such has lagged behind gravimetric procedures because of the lack of convenient, accurate measurement techniques. Early work was limited to manometers, which are affected by temperature and barometric pressure. Nevertheless, a competitive system was developed by Menke et al. (1979) and Menke and Steingass (1988) using large-bore syringes as the measuring device. This system has been successful in predicting digestibility and metabolizable energy (ME) by relating expected gas production to organic matter fermented.

More recently small-scale pressure transducers have been used to overcome many measurement problems (Theodorou et al., 1992). The combination of pressure transducers and computer software allows continuous recording of gas production (Pell and Schofield, 1993). This advance allows measurement of rate and extent of fermentation in a single fermentation flask, in contrast to batch fermentors and nylon bags that require the collection of replicate samples at specified time intervals to describe rate and extent.

Gas production can be related to substrate via fermentation of isolated subcomponents and subtraction of component curves. This mathematical procedure is valid if the preparations of the subfraction have not been altered. Figure 8.5 shows the results of fermentation of brome grass and its water-extracted residue.

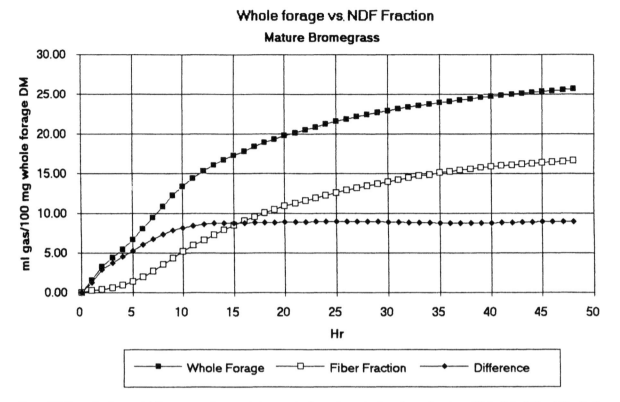

Figure 8.5. Example of a gas yield from a computer-recorded gas production system using transducers (courtesy of Peter Schofield and Alice Pell). The curve for soluble matter was obtained by subtracting the curve for insoluble residue from that for the whole forage.

Because the gas production system can utilize small amounts of substrate (on the order of 100 mg or less), fermentation patterns of small preparations and anatomical fractions of forage can also be studied.

8.6 Enzymatic Cellulase Procedures

Systems that use cellulases to measure digestibility appeared when fungal enzymes became commercially available. Cellulase systems avoid the problems of anaerobic technique; however, their precision tends to be less than systems that use rumen organisms. Enzymes lack the ability of living organisms to adapt to a substrate, and the quality of commercial cellulases has been variable in the past (Gabrielsen, 1986). Enzymatic systems are also limited by the completeness of the enzyme component. For example, there is evidence that proteins interfere, thus requiring deproteinizing steps. The same problem exists if starch is present. As a result, sequential procedures involving more than one enzyme have been developed (Aufrére, 1982). A more serious problem is that most enzyme sources are deficient in hemicellulolytic activity (McQueen and Van Soest, 1975; Roughan and Holland, 1977). The

ratio of cellulase to substrate is also critical. Sequential treatment with proteases or neutral detergent prior to cellulase treatment results in procedures that take just as long as the Tilley-Terry system. Rumen bacteria are more efficient at digesting structural carbohydrates than are purified fungal enzymes (as presently available) and yield a greater extent of digestion of forage cell walls. All enzymatic systems fail if the feed is high in tannins.

The end point limit of digestion in a cellulase system is a true digestibility since the medium is sterile and no microbial matter is formed. The metabolic microbial matter normally accounts for 11–12% of the substrate; therefore, the in vitro digestibility value should exceed the in vivo apparent digestibility by that amount. Most published equations for cellulase digestibility do not reach this limit, indicating a lesser degree of in vitro cell wall digestion than is observed in vivo (Dowman and Collins, 1982). If a crude filtrate from a fresh *Trichoderma* culture adapted and grown on a forage substrate is used as the enzyme source, however, the rate and limit of digestion can be comparable with rumen bacterial digestion (Roughan and Holland, 1977). Complex enzymatic series involving proteases and amylases have also been developed for use with concentrate feedstuffs (De Boever et al., 1988).

8.7 Near Infrared Reflectance Spectroscopy and Nuclear Magnetic Resonance

Rapid automatic nondestructive analysis of feeds and forages is possible by means of near infrared reflectance spectrophotometry (NIRS; Bertrand and Demarquilly, 1986; Robert et al., 1986; Abrams et al., 1987; Coelho et al., 1988). The procedure involves combining an infrared reflectance spectrophotometer with a programmed computer. The system scans the reflectance spectrum of a feed and correlates the resulting spectra with those of standard samples of known composition that have also been scanned by the system. The computer calculates the chemical composition of the unknown samples for which data input and standards have been provided.

Since the spectra as currently used do not have any absolute significance with regard to specific chemical structures or components, the system is liable to all the problems inherent in predictive regression techniques. Consequently, it is very important to calibrate with standards that are as similar as possible to those that are to be evaluated and which contain the kind of compositional variation expected in the unknown samples. The system does not represent a breakthrough in relating physicochemical properties to digestibility, but it is a shortcut for estimating the chemical fractions known to be involved in the availability of nutrients in forages and feeds, namely, cell wall, lignin, protein, in vitro digestibility, and so on. Future workers should develop associated adsorptions with chemical components to provide a more scientific basis (Downey et al., 1987)

The NIRS system is not practical for basic work on the composition of forages but can be useful for evaluating large numbers of similar samples, as in farmers' cooperatives or plant-breeding programs. About 50 samples of known composition are needed to establish an adequate calibration.

Nuclear magnetic resonance (NMR) is a potential means of analyzing feeds and forages that may well be more sensitive and accurate than NIRS. Like NIRS, NMR is safe and nondestructive and analyzes solid materials. Also like NIRS, the equipment is expensive and practical only for large-scale analyses. It is already used in medical diagnoses. NMR operates on the principle that atomic nuclei with unpaired protons or neutrons selectively absorb very high frequency radio waves in the presence of a strong magnetic field. ^{13}C is the principal isotope providing information. The absorptions are modified by the immediate chemical environment, from which information on organic structures can be ascertained. Himmelsbach et al. (1983) reported on carbohydrate lignin and protein in grasses using NMR, and Elofson et al. (1984) estimated nutritive value.

9 Minerals

Although minerals are essential for the nutrition of all living things, mineral nutrition is often treated fragmentarily, with various plants, animals, and elements all described in separate treatises. The topic is thus not peculiar to ruminants, but what *is* unique about mineral nutrition in ruminants is that microbial requirements and interactions need to be considered; otherwise ruminant mineral nutrition coincides with that of most other herbivores.

This chapter gives an overview and a supplementary coverage of areas of mineral nutrition often overlooked by animal nutritionists. Relevant areas include geography, geological provinces, geochemistry, the periodic table from a biological point of view, and the contrast between plant and animal requirements and elemental interactions.[1] The coverage of the mineral elements here is biased toward those that have been overlooked. Elements that are well covered in current literature are treated here only summarily.

Mineral nutrition involves the physicochemical attributes of the biologically important mineral elements. These attributes affect the interactions among soil, plant, microbe, and animal, including problems of supply relative to requirements and the availability of requisite elements from the feed sources.

Many of the better chemistry texts relevant to nutrition and functional biology were written in the 1940s, when biochemists were discovering vitamins and essential mineral elements. These texts offer some understanding of aspects that have become obscured as biochemists have turned away from nutrition. Modern developments in biochemistry are important to nutrition, but the problem of essential factors has become of secondary interest in that field.

9.1 Geography and Geology

The availability of minerals to plants, and consequently to animals, is greatly affected by geochemical processes on rocks and soil. The surface of the earth, with its turbulent oxidizing wet atmosphere, is continually eroding, turning material over and releasing its elements. The composition of the rocks that are parent to soil is variable, and the effects of weathering and leaching are not the same for all elements. Weather and climate vary over the globe, leading to a multiplicity of interactions among elements, rocks, soil, erosion, and weather.

Generally, the processes of weathering and erosion are promoted by water and its reactions with minerals, and by temperature, which influences the rate of chemical reactions. Thus surfaces in cold and dry environments weather more slowly than those in warm, humid areas. Pleistocene glaciation produced much of the topography characteristic of the Northern Hemisphere. The power of ice to grind rocks to powder and move large amounts of till over wide distances is responsible for many of the geographical characteristics in the northern temperate regions.

The first problem of mineral availability is the collective rock structure of the geographical neighborhood. Soluble minerals leach down from mountains and concentrate in lower reaches (or the sea). If they were not present in the parent rock, they will not occur in the deposited soils. Thus characteristics of regional soils are set by the parent geological structure from which they were formed. The renewal of the earth's mineral resources is undoubtedly maintained by the combined growth and erosion of the land surface. The rate of these processes varies such that one can speak of old soils and young soils. Older soils are more likely to be depleted in mineral nutrients.

The features of world geography vary in age from the geologically young to very old. One can begin with the ancient shields—the Canadian, the Ghanian of West Africa, or the Fennoscandian in northern Europe. The shields are characterized by poverty of soil re-

---

[1] Perhaps the most significant publication on mineral nutrition is the excellent treatise by the late Eric Underwood (1977). In 1987 the text was reissued under the editorship of Walter Mertz with contributed chapters on the respective elements. The reader is referred to this book for a discussion of important elements in nutrition. Another useful review is by Ammerman and Goodrich (1983).

Table 9.1. Geochemical classification of the elements, according to their distribution in iron, sulphides, silicates, atmosphere, and organisms

Iron (siderophil)	Sulphide (chalcophil)	Silicate (lithophil)	Gases (atmophil)	Organisms (biophil)
Fe, Ni, Co, P, As, C, Ru, Rh, Pd, Os, Ir, Pt, Au, Ge, Sn, Mo, W, Nb, Ta, Se, Te	S, Se, Te, As, Sb, Bi, Ga, In, Tl, Ge, Sn, Pb, Zn, Cd, Hg, Cu, Ag, Au, Ni, Pd, Pt, Co, Rh, Ir, Fe, Ru, Os, Mo	O, SO$_4$, P, H, Si, Ti, Zr, Hf, Th, F, Cl, Br, I, Sn, rare earths B, Al, Ga, Sc, Li, Na, K, Rb, Cs, Be, Mg, Ca, Sr, Ba, Fe V, Cr, Mn, Ni, Co Nb, Ta, W, U, C	O$_2$, N$_2$, CO$_2$, H$_2$, Ar, He, Ne, Kr, Xe	C, H, O, N, P, S, Cl, I, B, Ca, Mg, K, Na, V, Mn, Fe, Cu

Source: Modified from Goldschmidt, 1958.
Note: Elements may be found in more than one classification.

sources since they are ancient platforms from which most of the important nutrients have been eroded and lost. The foldings and mountains that produced these sheets of rock exceed 2×10^9 years, and, in the Northern Hemisphere at least, glacial ice removed most of the surface soil, pushing it southward. Thus the existing soils developed from a limited mineral base, the result being fragile acidic lakes and marginal soils.

In the tropics, rates of weathering are higher. Because of the intensive leaching from the ancient rock soil base, most available nutrients reside in living systems or have otherwise been leached away. In these environments nutritional poverty occurs not so much as a result of poor soil, but from plant defense. Old tropical soils are likely to limit plant production and animal utilization in that order. Where there is adequate water, competition between plants and herbivores is severe and leads to problems of nutrient distribution and plant defense (Chapter 6).

9.1.1 Geochemistry

One of the most relevant aspects of geochemistry pertinent to plant and animal nutrition is the mobility of the essential chemical elements. The elements sorted themselves out relative to density and volatility so that the solid surface (lithosphere) is comparatively low in heavy and volatile elements and rich in comparatively nonvolatile light ones.

The distribution of elements in the earth has led to their classification into categories that illustrate their affinities (Table 9.1). Siderophils are heavy and may occur as free metals in the nickel-iron core of the earth, but metallic nickel and iron hardly ever occur on the surface, which is exposed to oxidation. The more inert platinum metals and gold commonly occur in the elemental form, however. Chalcophilic elements include many transition metals important to metabolism in living organisms. Chalcophilic behavior signifies affinity toward sulfhydryl and amino groups and involves the interaction of metal ions, prosthetic groups, and enzymes. Other categories are lithophilic elements that associate in silicate rocks, the gaseous atmophilic ele-

ments, and the biophilic elements that form the bulk of organic living matter.

Many of the nutritional problems of living organisms can be related to the relative mobility of the essential elements. Animals derive minerals from plants, and plants get their minerals from soil, which is derived from eroded rock. Each turnover involves some loss, with the more soluble and mobile elements tending to become washed downward and eventually out to sea. They may be redeposited somewhere down the line if solubility limits are breached or new insoluble mineral combinations can be formed. The relative order of leaching is as follows: $Cl, Na > K, Mg > Ca$, $SO_4 > PO_4 > SiO_2 >$ trace elements $> Mn, Fe, Al$, and rare earths. Generally the parent materials tend to become depleted in basic elements, leaving behind the less basic and more acidic, polyvalent ones, ultimately leading to a rise in soil acidity.

9.2 Biological Requirements

Many elements not required or required only at low levels by plants are necessary to animals in more substantial amounts. Conversely, plants have higher needs for some elements than animals. These differing requirements lead to situations in which plants grow normally but animal production is limited and, alternatively, plant growth is limited by soil nutrients and animal response is limited by available plant organic matter and not by plant nutrients per se.

Plants' intake of essential elements is a function of their availability in the soil, hence geographical areas of deficiency are related to geology and soil availability. Toxicity can be a problem when certain elements that occur in high concentrations in the geochemical environment are taken up by plants.

Whether nutritional problems occur in free-ranging herbivorous animals depends on the supply of nutrients from the pasture or range. Feeding supplements grown in geochemically different locations may modify or obliterate nutritional problems resulting from mineral deficiencies. An example is the selenium deficiencies

Table 9.2. mineral requirements of plants and animals listed in decreasing order of likely animal deficiency

	Plants		Animals		
Element	Marginal tissue level (ppm)	Range (ppm)	Rumen microbes (ppm/diet)	Marginal level (ppm/diet)	Field observations
Na	NR	100–high	<2000	500–1800	Def
Cl	<100	100–high	<2000	500–3000	Def
P	1000–2000	1400–3000	1000	2000	Def
Mg	2000	—	<75	2000	Def
Ca	400–1000	300–30,000	<250	4000	Def
S	250–1000	1200–3000	75	2000–6000	Def
Zn	10–50	30–100	3–10	20–30	Def
Cu	5–10	—	0–0.1	5–20	Def
Co	0.08	—	0.07–0.20	0.0004 as B_{12}[a]	Def
I	NR	—	20	0.08	Def
Se	NR	Wide	0–0.1	0.1	Def-Tox
Fe	100	Wide	2–5	30–60	Rare
K	10,000	15,000–30,000	<2000	6000	?
Mn	10–20	50–150	10–20	10	—
Mo	0.2–0.5	3–100	10–20	<0.1	Tox
Si	?(low)	300–200,000	—	50–100	Tox
B	10–50		NR	NR	?
Ni	?		2–5	0.05	—
Rb	?		20–70	?	

Sources: Summarized from Hodgson et al., 1962; L. H. P. Jones and Handreck, 1967; Martinez and Church, 1970; Reisenauer, 1976; Underwood, 1977; and Spears, 1984.

Note: Elements are listed in decreasing order of likely deficiency in animals. NR = not required; Def = deficiency; Tox = toxicity.

[a]The cobalt content of vitamin B_{12} is 4.3%. The B_{12} requirement is 0.01–0.015 ppm.

in eastern North American farm animals that subsist on local feed. These deficiencies are absent from the human population because they eat wheat grown on seleniferous soils of the Midwest that is shipped into the region.

Nutritional problems due to minerals, other nutrients, or toxicants are related to the diversity of the plant species available for grazing or browsing. Generally, nutritional problems involving deficiency or excess are accentuated when the range is limited to a single plant species. Potentially toxic plants are often avoided by animals or ingested at levels that do not cause problems. Overgrazing and the concomitant loss of desirable forage plants tend to result in nutritional problems caused by the intake of less desirable plants.

Unless they are fed supplements, grazing animals must derive their nutrients from the plants they consume. Thus plants' mineral requirements are relevant both to their own growth and to animals' mineral requirements. The question is, Which requirement is lower: the plant's or the animal's?

Table 9.2 classifies minerals into two categories: those that occur in plants at levels often inadequate to support animals, and those that occur at levels usually adequate for animal functions. The first group includes sodium, chlorine, phosphorus, magnesium, copper, zinc, cobalt, chromium, and, perhaps, calcium in the

case of animals on grass pastures and lactating animals. Some of these elements have specific functions in rumen microbes or animals but no function in plants. These include sodium, cobalt (as vitamin B_{12}), chromium, iodine, and selenium. Plants may absorb sufficient amounts of the elements to sustain animals if the elements are available in the geochemical environment. Areas of plant deficiency or excess are thus geographically associated with corresponding soil levels (J. K. Thompson and Warren, 1979).

Microbial mineral requirements for adequate rumen fermentation are generally lower than those for the host's requirements. Microbes do not have a structural requirement for calcium or phosphorus, as do vertebrates generally for bone formation, but they do need these in the context of cellular metabolism. Hydrogen, carbon, oxygen, nitrogen, phosphorus, and sulfur are the major elements in organic cell composition. Proteins are the major sources of nitrogen and sulfur. Ratios of nitrogen to sulfur are about 12:1 relative to the contents of essential amino acids. Microbes, and consequently their ruminant hosts, may require some trace elements in greater quantities than are needed by other animals and plants, for example, cobalt, which is used in pseudo B_{12} forms in microbes (Elliot, 1980), and nickel, which is required by microorganisms and also is a cofactor in urease (Spears, 1984).

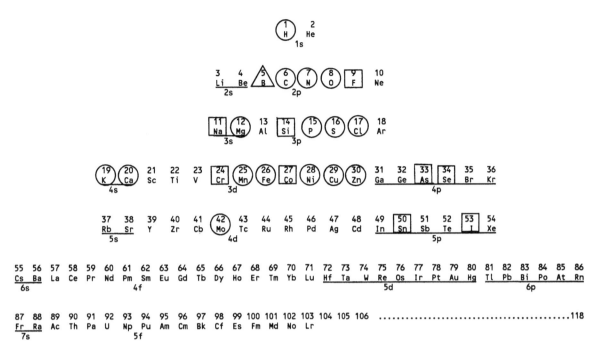

Figure 9.1. Periodic table showing the proper relationship of subshells and electronic filling. The third and fifth periods are repetitive because the third suborbit cannot fill until after 4s is filled, and the 4f suborbit fills only after 5d and 6s. The maximum number of electrons in a shell is 2n²: thus 2 for the first shell, 8 for the second, 18 for the third, and 32 for the fourth. Because of the different energies of the subshells, the filling within the fourth and higher periods involves suborbits of respective shells. In rare earth elements (lanthanides) the 4f subshell is filled. Essential elements are mainly light elements. Only a few such as molybdenum and iodine occur above number 30. Essential elements for plants and animals are in circles; essential to plants only, in triangles; and to animals only in squares. The transuranic elements are synthetic. Curium (96) is the heaviest naturally occurring element.

9.3 A Biological View of the Periodic Table

The tables of chemical elements perpetrated on chemistry students are usually pushed far back into the recesses of the mind of the average student of biology. The apparent lack of relevance to biologically oriented individuals has tended to relegate the whole of inorganic chemistry into the same pit of unusable information. Inorganic chemistry's relevance to biology is seldom taught by teachers of inorganic chemistry and only belatedly realized by biologically oriented students who pass through the college system. An example of a periodic table that indicates essential elements for plants and animals is given in Figure 9.1. A biological classification of elements is shown in Table 9.3.

Periodicity arises because of the arrangement of electrons around the atomic nucleus in concentric shells. The capacity of any shell is 2n², where n is the number of the shell. Thus the first shell has a capacity of 2 electrons, the second 8, the third 18, and the fourth 32. The differences between these numbers give the series 6, 10, and 14, which represent the expansion of new elements in periods II, IV, and VI of the table. Each expansion is repeated because the filling of s and p electrons in outer shells precedes that of d and f. The order of filling follows energy level, with the lowest levels filled first.

The chemical properties of an element are determined by its atomic number, which sets the configuration of elections around the atomic nucleus. The atomic number represents the net positive charge on the atomic nucleus and allows a possible element for each positive integer beginning with 1 (hydrogen) and proceeding up the numerical scale to bismuth (83), above which there are no stable nuclei. The existence of heav-

Table 9.3. Biological classification of the elements

Functional role	
Structural	H, C, O, S, Ca, P
Cell environment	H_2O, CO_2, Na, K, Mg, Ca, PO_4, Cl
Metabolic (coenzymes)	
Plant	B, Mo, Cl
Plant-animal	Mo, Mn, Fe, N, Cu, Zn
Animal	Co, Cr, Se, I
Toxic-protective	
Secondary	S, F, rare earths, Se, As, Sn, Ba
Inimical	Be, Cd, Sb, As?
Excluded	Ru, Rh, Pd, Ag, W, Os, Ir, Pt, Au, Bi, Ba
Inert	
Gases	He, Ne, Ar, Kr, Xe
Questionable	Al, Ga, Ge, Br, Rb, Y, Zr, Hf, Nb, Ta, In
Essential in some species	V, Si
Substitutes for K, Ca	Rb, Sr

Sources: Bowen, 1979; Mertz, 1981.

ier elements is allowed by the primordial survival of isotopes of long life (e.g., thorium [90] and uranium [92]) whose complex disintegrations provide sources for adjacent radiogenic elements (84 to about 96) that could not otherwise exist due to the short half-lives of their isotopes. There are stable isotopes for all elements up to bismuth with the exception of technetium (43) and promethium (61).

Radioactivity is generally inimical to life, and the heavy elements have no particular relation to nutrition other than their high potential toxicity; most nutritionally important elements are light ones. But radioactivity is not exclusive to heavy elements of high atomic number. Radioactive light elements exist, particularly long-lived ^{40}K and the unusual isotopes, ^{14}C and ^3H among others, that are cosmically produced and rain down from the sun to end up in living matter in very low concentrations.

A number of nonnaturally occurring radioactive substances have been introduced to the world since the discovery of atomic fusion and the atomic bomb. These include a wide range of atomic numbers, and the most dangerous are those that have biological affinities and long enough half-lives to persist in the environment.

Many of the synthetic radioactive substances with short half-lives (but long enough for consequent assay) are used as tracers to study the metabolism of the respective elements. Application of radioactive tracers is limited by the requirements of safety, and alternatives are offered by stable isotopes that can be assayed by neutron activation of samples. This removes much of the problem of environmental contamination.

The successive filling of orbital shells leads to the repetitive characteristics of the periodic table first noticed by Dmitry Mendeleyev. Repetition—that is, the vertical occurrence of families of elements having similar valence and often other properties—results from similar configurations of their outer electron shells.

The sequential filling of the shells results not only in repetition of outer electronic configurations (promoting repetition of chemical properties) but also in expansion of atomic size. The larger atoms lose electrons more easily because their electrons are farther from the nucleus, and they are generally more stable in higher positive valences. The result is a remarkable variety in physical and chemical properties.

Valence results from the acceptance or loss of electrons creating a net charge. The most stable valences are those that achieve a noble gas configuration, such as that of helium, neon, or argon. For example, the sodium ion has a neon configuration, and the chloride ion an argon one. It is more difficult for elements in the middle of a period to achieve this, since many electrons may have to be involved. The maximum negative valence is 4 (a net gain), achievable in only group IVb

elements; and the maximum positive valence of 8 (net loss) is achieved only in ruthenium, osmium, and xenon. Many elements in the middle of the longer periods are unable to achieve noble gas configurations and exhibit characteristics that make their positions in the table harder to comprehend. This is particularly the case with transition elements such as iron, cobalt, nickel, and most of the rare earths.

The relevance of the periodic table to mineral nutrition lies in the understanding of biological interactions and antagonisms among elements it can offer. The vertical columns of family groups list elements of similar valence, but they also represent sequences of shifting properties. Ionic size increases as one goes down the chart, basic properties increase, and acidic properties usually (but not always) decrease. Substitution of an analogue from the column may result in loss of biological function and promotion of toxicity. This is particularly noteworthy in such sequences as K-Rb-Cs, Ca-Sr-Ba and Zn-Cd-Hg. On the other hand, the sequence F-Cl-Br-I proceeds from toxicity with fluorine (Bunce, 1985), probably an essential trace element, to chlorine, an essential macroelement (Coppock, 1986), to apparently nonessential bromine, to iodine essential for animals.

9.3.1 Group I and II Elements

Elements that have positive monovalence or divalence due to loss of s electrons tend to be strong bases and perform the function of cations in biological systems. The B families, which include copper and silver, and zinc and cadmium, are less typical and really do not belong in the group. They have closer affinities to the transition metals (Section 9.8). There are similarities between some IIA and IIB elements such as magnesium and zinc.

The lightest metallic elements (lithium and beryllium) are not essential and may be toxic. Lithium affects the central nervous system and is used to treat schizophrenia. The next two rows (Figure 9.2) contain some of the most important nutritional elements, several of which (sodium, potassium, magnesium, and calcium) can cause problems in grazing animals and ruminants due partly to disparate availabilities and partly to antagonistic behavior in metabolism.

Potassium is the important cation in plants, which generally seem to have no use for sodium. As a result, many herbivores eat a K:Na ratio as high as 15:1 when their tissue contains only a 1:2 ratio. Wild herbivores often seek out salt licks to obtain sodium. Potassium, magnesium, and calcium are necessary for muscle activity; imbalances can lead to tetany. The calcium-to-phosphorus ratio is also critical (the ratio is 2:1 in bone). These problems are discussed further in Section 9.11.2.

<table>
<tr><td>Group</td><td></td><td>Group</td><td></td></tr>
<tr><td colspan="2" align="center">I</td><td colspan="2" align="center">II</td></tr>
<tr><td colspan="2" align="center">Li</td><td colspan="2" align="center">Be</td></tr>
<tr><td colspan="2" align="center">Na</td><td colspan="2" align="center">Mg</td></tr>
<tr><td>IA</td><td>IB</td><td>IIA</td><td>IIB</td></tr>
<tr><td>K</td><td>Cu</td><td>Ca</td><td>Zn</td></tr>
<tr><td>Rb</td><td>Ag</td><td>Sr</td><td>Cd</td></tr>
<tr><td>Cs</td><td>Au</td><td>Ba</td><td>Hg</td></tr>
<tr><td>Fr</td><td>(Tl)</td><td>Ra</td><td>(Pb)</td></tr>
</table>

Figure 9.2. Elements with only s electrons in the outermost subshell. Many of these are strong alkalis, particularly those in the A group. Beryllium is a weak base and toxic. In the B group, involvement with d electrons causes copper and gold to behave as transition metals and not strong bases, in contrast with silver and zinc, which are stronger bases than copper but weaker than sodium, potassium, and calcium. Thallium and lead are not properly members of this group; however, the peculiar effects of the lanthanide contraction create difficulty in the loss of 6s electrons, and this causes them to behave as alkaline elements. Thallium chemically is somewhere between potassium and silver, and lead resembles barium. Thallium and lead are toxic, while silver and gold are benign because they have no easy entry into metabolism. Once inside the cell, they are toxic.

The disparate behavior of potassium over sodium, or calcium over magnesium is due to the expansion in atomic size resulting from the filling of 4s before 3d in potassium and calcium. Potassium is far more resistant to leaching than sodium and is the dominant cation in soils and plants, while sodium is dominant in seawater and potassium is present in low concentrations. A similar parallel exists between magnesium and calcium. Seawater is a source of magnesium but very low in calcium, which occurs in bone, while most of the magnesium exists in soft tissues. Potassium is in cells while sodium is dominant in extracellular fluids of animals.

Rubidium, in the fourth row, is very similar to potassium and can partly, but not completely, replace it. It also appears to stimulate microbial digestion (Martinez and Church, 1970). Likewise, strontium can partly replace calcium. Neither rubidium nor strontium is very toxic, and an essential role has been suggested for strontium. In contrast, the still larger cesium and barium ions (row 5) are comparatively toxic. Barium and lead have similarities: both form insoluble sulfates and concentrate in bone. Curiously, barium is accumulated by certain plants (e.g., the brazil nut contains up to 3000 ppm of barium!). This might be a secondary defense mechanism.

9.3.2 Nonmetals and Metalloids

This group features the most diverse behavior, largely associated with nonmetallic characteristics that diminish as one proceeds through the periodic table. These elements all involve p electrons, which fill to culminate in a noble gas.

The nonmetallic elements located between the metals and the gases in the periodic table provide the standard building blocks for most things related to biology —carbon for organic forms and silicon for mineral structure.

There is an enormous contrast between carbon as a representative of the C-N-O group and silicon representing the Si-P-S group and the other members of groups IV, V, and VI. The smaller, lighter carbon, nitrogen, and oxygen can form double bonds much more easily than the heavier elements, a characteristic that results in stronger acids with the lighter elements. Strong acidity in part depends on the double liganding with oxygen, and an acid is greatly weakened when the respective bonds are substituted by single bonds. Thus, for example, sulfuric and selenic acids are strong, and hexahydroxytelluric acid is very weak.

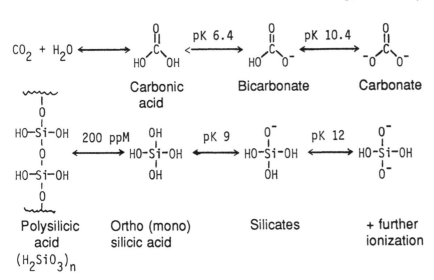

Figure 9.3. Comparison of carbonic and silicic acids. Carbonic acid is the stronger acid. Both are unstable and undergo dehydration; however, carbonic acid decomposes to gaseous CO_2 while silicic acid polymerizes into polysilicic acids. The water content of polysilicic acid (silica gel) is variable. Salts of polysilicic acids can also be formed at high pH and with alkaline fusion (Iler, 1955).

Similarly, carbonic acid is stronger than silicic acid (Figure 9.3).

The sequence N-P-As-Sb-Bi shows the family relationship. All possess negative trivalence and positive pentavalence. But only the first members of the group are of major biological importance; indeed, all metallic members are nonessential elements, and the essential ones near the nonmetal-metal boundary are toxic (e.g., arsenic), but also potentially essential at a trace level. The sequence from fluorine to iodine is exceptional in that while fluorine is toxic, iodine is the heaviest element required by animals.

9.4 Phosphorus

Phosphorus lies in group V beneath nitrogen, and it is no surprise that it has many similarities to nitrogen in its biological behavior. In contrast to nitrogen, valence +5 is dominant, and −3 is too reducing to become an article of commerce in biological systems, although trivalent compounds can be extraordinarily toxic. Phosphorus is ubiquitous in living cells and is a part of the molecular systems that involve the genetic code (DNA, RNA) and energy storage and transmission (ATP). In animal bone it assumes a structural role. Like nitrogen, phosphorus is an integral part of all metabolic tissues.

Rumen microbes contain a high proportion of RNA and DNA relative to most other cells and thus have a high phosphorus requirement. Like nitrogen (as urea), phosphate is recycled through saliva to the rumen and incorporated into rumen organisms that are separately digested in the lower digestive tract; and also like nitrogen, phosphorus is a component of the microbial cell wall, which is incompletely degradable and thus appears as a part of the metabolic fraction in feces. This creates the same relation between metabolic losses and true digestibility as is the case with nitrogen. As a result, Lucas analysis describes the digestion balance for phosphorus just as it does for nitrogen (Table 9.4). The analysis shows that the apparent balance is always less than the true utilization, because of endogenous fecal losses. The main repository for phosphorus is in

bone along with calcium, with which it interacts in nutrition (see Section 9.11.2).

Phosphorus deficiency in ruminants manifests itself in peculiar symptoms. Its lack induces pica, a peculiar feeding behavior involving the consumption of materials not ordinarily eaten. In grazing systems this has manifested itself in the consumption of the tissue and bones of dead animals; thus an early report of phosphorus deficiency was associated with botulism from the consumption of rotten meat.

9.5 Sulfur

Sulfur is just beneath oxygen in group VI. While oxygen is largely confined to −2 valence, sulfur exhibits −2, +4, and +6 valences, represented by sulfides, sulfites, and sulfates. While plants take up sulfur mainly as mineral sulfate, the natural form of sulfur found in ordinary forages and feeds is principally as sulfur amino acids, although small amounts of other forms do occur. Some plants form thioglycosides (e.g., *Brassica*). Sulfates may be found in hard drinking water or soil, often as gypsum.

Most forms of sulfur can be used by bacteria because sulfate is reduced through sulfite to sulfide, the article of commerce in the rumen. Elemental sulfur is also utilizable. Rumen sulfide occurs as hydrogen sulfide (H_2S) and hydrosulfide ion (HS^-), the proportion depending on rumen pH. The nonionized compound is generally the most abundant because the first pK (6.7) of H_2S is higher than the usual rumen pH.

Nonionized H_2S is absorbed about four times faster than HS^-. Consequently, more HS^- is absorbed as pH decreases. The half-life of absorption is relatively short and ranges from 8 to 18 min. H_2S, like ammonia, appears as a waste product when protein is fermented in excess of microbial needs. Therefore, rumen sulfur concentrations often depend on the supply of sulfur from the fermentation of protein in excess of microbial growth requirements.

The concentration of sulfur in the rumen that appears to limit growth of rumen organisms is about 1 mg/1 (Bray and Till, 1975). The output of microbial protein to the lower tract is closely related to the non-sulfate sulfur leaving the rumen, indicating that sulfur, in addition to nitrogen, can be a limiting factor in protein synthesis by bacteria. The N:S ratio of ingesta leaving the rumen is 14±3:1.

The metabolism of sulfur in ruminants parallels that of nitrogen, the common factor being their role as essential components of protein. Sulfur, however, occurs in mainly two amino acids, cystine and methionine, whereas nitrogen occurs in all amino acids. Methionine is an essential amino acid for the animal, but the rumen fermentation requires only a source of sulfur.

Table 9.4. Calcium and phosphorus absorption and losses in cows weighing 450 kg

| Major source | Endogenous losses | | Fecal bacteria (% of intake) | Digestibility (true) (%) |
	Urine (g/day)	Feces (g/day)		
Calcium				
Legumes and grasses	nil	8.4	—	38
Bone meal and grain	nil	8.6	—	55
Phosphorus				
Legumes and grasses	0.27	3.9	14.1	79
Bone meal and grain	0.37	3.7	12.3	74

Source: Conrad, 1983.

Sulfur also has a role as a mineral element in rumen metabolism and has significant interactions with copper and molybdenum (Section 9.11.3). The specific content of sulfur-containing amino acids in proteins establishes the dietary nitrogen-to-sulfur ratio. The critical ratio of nitrogen to sulfur varies relative to specific situations. For examples, in the case of wool growth, the high content of sulfur in keratin sets a lower required N:S ratio. Microbial proteins have less sulfur than wool proteins, thus a higher ratio in rumen output sets a limit on rate of wool growth (Barry and Andrews, 1973). The same question of dietary supply via the rumen may be addressed relative to tissue synthesis or milk production. Also, the sulfur content of rumen organisms is not constant and probably varies directly with the true protein content and inversely with cell wall content of the organisms.

Isotopic sulfur (^{35}S) has been used to label rumen microbial protein. Its use depends on the assumption of a fixed N:S ratio in microbes. This value is about 13:1, but there is a considerable range. Part of the problem is that true protein averages only about two-thirds of the total microbial nitrogen, with microbial cell wall and nucleic acids forming a substantial variable portion of nonproteinaceous substances that are likely to be lower in their inherent sulfur content. The tendency to express protein fractions in the rumen and GI tract contents on a total nitrogen basis does not resolve this problem, because large proportions of nonammonia nitrogen are included in the denominator.

9.5.1 Animal Requirements

The need for sulfur is primarily related to the metabolic requirements for methionine specifically, and for cystine as a sulfur source; therefore, the sulfur requirement is related to the protein requirement. The requirement is fixed by the proportion of sulfur in the proteins of the animal body. Sulfur requirements are highest in lactating or rapidly growing animals; that is, in proportion to their protein requirements. Sulfur deficiency is not likely to be a problem provided protein requirements are met and the sulfur content of the animal protein synthesized is within the range of that provided by the diet and the rumen microbial synthesis (Kandylis, 1984).

Lactation and heavy wool growth may present problems because the required N:S ratios are lower than those provided by the bacteria. Microbial synthesis of sulfur-containing amino acids may limit animal function, and the rumen escape of dietary protein of suitable amino acid composition is needed to maintain a high production level. This rationale provides the basis for response to administration of methionine hydroxy analogue in a form designed to escape the rumen. Wool growth generally responds to such administration, as, in some instances (but not universally), does milk production.

9.5.2 Recycling of Sulfur

Hydrogen sulfide is the microbial end product of excess sulfur supplied to the rumen. It is absorbed by the liver and converted to sulfate, which is also the product of catabolized sulfur-containing amino acids. Sulfate is secreted in urine and saliva. Recycling of sulfur through saliva is relatively independent of dietary intake and less important than in the case of nitrogen. Amounts of sulfur available for salivary secretion are apt to be minimal in low-protein diets, and the N:S ratio may be as high as 80:1 (Bray and Till, 1975). As a result, recycled sulfur is likely to be inadequate for rumen organisms if the diet does not contain the sulfur necessary to meet the N:S ratio required by rumen organisms (c. 12:1).

Sulfur should be considered when nonprotein nitrogen (NPN) supplements the diet, since the addition of NPN usually increases the N:S ratio (and increases the microbial requirement for sulfur if the NPN is used for microbial synthesis). A combination of sulfur and nitrogen in the NPN supplement at a ratio of 1:12 may be required.

9.6 Selenium

Selenium is the element in group VI directly below sulfur, and like sulfur it exhibits valences of -2, $+4$, and $+6$. It is less volatile, more metallic, and potentially much more toxic. Paradoxically, selenite ($+6$) is more oxidizing and less stable than sulfate. Selenium can substitute for sulfur in most organic substances, including sulfur-containing amino acids, albeit with altered physical and biological properties.

Selenium is a component of the glutathione peroxidase molecule, which explains its interactive role with vitamin E and the sulfur-containing amino acids. The major role of the enzyme may be destruction of peroxides, while vitamin E prevents formation of peroxides by scavenging free radicals. Selenium deficiencies are well known in the form of white muscle disease in ruminants and other animals.

Selenium deficiency shows a definite geographical pattern. The glaciated areas of the northeastern United States are particularly low in selenium. White muscle disease is present in the human population and is associated with cancer in parts of China and Asia. In the United States, white muscle disease is seen only in the animal population. The pattern of selenium availability also includes areas of excess selenium, where the mineral is taken up by certain accumulator plants (e.g., *Astralagus* sp.) to levels that are toxic to animals that graze on them.[2] Toxicity is generally related to lack of variety and unavailability of less toxic plants. Excess selenium can be alleviated with arsenic, another element that is probably essential in trace amounts but toxic at higher levels.

9.7 Silicon

Falling under carbon in group IV, silicon might be expected to possess similar properties of forming structural compounds, and indeed the solid surface of the earth is composed mainly of silicate polymers. Silicon is the second most abundant element (27%) in the earth's crust after oxygen. It occurs as the oxide (silica) in the forms of quartz, opal, and amorphous silica and as complex silicates in rocks, sand, and clays. Its abundance has overshadowed any nutritional importance, and it has long been regarded as inert and passive in biological systems (L. H. P. Jones and Handreck, 1967).

The silicon atom has a larger radius than carbon, leading to less stable atom-to-atom bonding and the inability to form stable double bonds with either itself or with other important divalent species such as oxygen. The inability to form stable double bonds eliminates the possibility of silicon compounds equivalent to aldehydes, ketones, and unsaturated compounds, features providing the basis for the flexible elastomers in carbon compounds. The flexible silicon-based polymers that do exist (e.g., silicone rubber, etc.) are formed by bonding with carbon.

Most attempts to prepare analogues of ketones and aldehydes result in polysiloxane polymers known as silicates, which are comparatively rigid structures. In the case of the respective acids the difference between the two elements is particularly striking. Silicon cannot form the equivalent of a carboxyl group (Figure 9.3), and as a result, all silicic acids exist as polyols. Alcohol (silanol) groups ionize in water only at pH 9 or higher, with the result that all silicic acids are nonionized at physiological pH.

Silicon is an essential nutrient for rats, chickens, diatoms, *Equisetum,* and rice. It is not known whether all plants require it (Raven, 1983). Many grasses and some other plants accumulate silicon, causing problems for herbivores through excessive intake or other effects of silica on nutritive quality. The main form in grasses is opaline silica. Alfalfa and other temperate legumes restrict their absorption and never contain more than a few hundred parts per million in their tissues.

Silica is used by some plants as a structural element, complementing lignin, to strengthen and rigidify cell

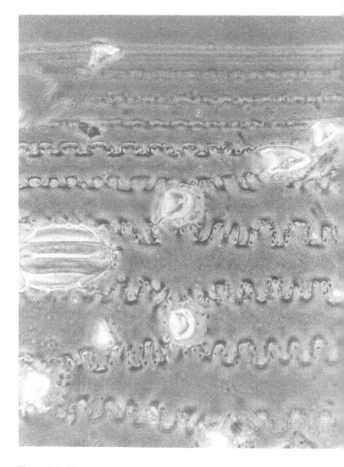

Figure 9.4. The opaline skeletal structure of an oat leaf blade prepared by wet ashing with mineral acid (from L. H. P. Jones and Handreck, 1967). The mineralogical structure is composed of hydrated silica, which has the crystalline pattern of opal. Upper boundary is the leaf edge.

walls (Figure 9.4). It appears to influence carbohydrate metabolism (mainly studied in sugarcane) by promoting the accumulation of sucrose and lowering protein and lignin contents. These effects have been exploited in the use of silicates to fertilize sugarcane, where they also cause the release of phosphates in soil and detoxify excess iron, aluminum, and manganese. Silica is deposited in hairs on the plant surface and cuticular edges and contributes to defense mechanisms in certain plants.

The consequences of silica in forage are complex and varied. The silica level in grasses is highly dependent on soil type, the availability of silica, transpiration, and the nature of the plant species. Many soils, particularly the products of weathering such as clays and soils high in aluminum or iron oxides, contain low concentrations of silica. Consequently, there is a wide variation in silica content among forages of the same species grown on different soils, with resultant differences in quality. Availability of soil silica is certainly the largest single variable influencing the silica content

---

[2] *Astralagus* may still be toxic when grown on low-selenium soils, in this instance due to alkaloids.

of any plant (L. H. P. Jones and Handreck, 1967). These differences in quality may be due to lowered digestibility or to palatability problems caused by sharp siliceous projections on the edges of leaves. Quality differences are apparent among the same forages grown in different soils, but it is more difficult to distinguish such differences in forage from a single place. This may account in part for the contradictions in the literature regarding whether silica influences digestibility (Minson, 1971).

In vitro studies have provided direct evidence that silica reduces cell wall digestibility. Three kinds of experiments have provided positive evidence that allows some speculation as to the mechanisms of inhibition. One type of study is hydroponic culture with and without silica added to the nutrient solution. Hartley (1981) showed about a unit of depression in organic matter digestibility per unit of deposited silica in ryegrass, with cellulases as the digesting medium. Shimojo and Goto (1989) obtained similar results. Soluble silica added to Tilley-Terry in vitro fermentation also reduces digestibility (G. S. Smith and Nelson, 1975). In another study silica was selectively removed from natural silicified forages with a concomitant increase of cell wall digestibility of about 1–2 units of organic matter digestibility per unit of silica removed (Van Soest, 1981). M. G. Jackson (1977) found increased digestibility when silica was removed from rice straw.

The nature of the observed depression in digestibility remains puzzling, since at least two other hydroponic experiments with silica fertilization of rice (Balasta et al., 1987) or reed canarygrass (Van Soest and Grunes, unpublished 1980 data) showed no effect. Other suggestions (G. S. Smith and Nelson, 1975) are that silica promotes sucrose formation, which may affect decline in cell wall digestibility, or that silica promotes trace element deficiencies that are overcome by mineral supplementation. The matter is unresolved.

9.7.1 Effects of Silicate Minerals in Feed

Grazing animals may ingest large amounts of soil when available forage is sparse. Most soils are composed largely of clays, silicate rocks, and quartz plus an organic humus component. It is often assumed that the effect of soil minerals on digestion is largely dilatory since most of the minerals are highly insoluble. Clays and related silicate minerals have substantial cation exchange capacity, however, and in this role selectively absorb certain ions and have buffering capacity and effects on attachment and turnover of rumen bacteria.

Other effects of silica in forage are still to be explored. Siliceous forages are more abrasive than less silicified ones; thus silica could affect palatability and

forage selection. Silicified plant fragments may have exceptional ability to stimulate rumination (Baker et al., 1961), as, for example, do rice hulls added to high-concentrate feedlot rations.

Silica is relatively soluble under rumen conditions. It can be absorbed by the animal and is normally excreted in the urine. Ruminants grazing in arid regions with limited access to water where grasses are high in silica may suffer from siliceous kidney stones. Consuming large quantities of potentially soluble silica relative to available water causes urinary concentrations to exceed the saturation of silicic acid, leading to precipitation and the formation of siliceous urinary calculi (C. B. Bailey, 1981). Phytoliths have been found as nuclei in siliceous stones (Jones and Handreck, 1967). Siliceous phytoliths are often needle-like and may be widely distributed in the tissues of most herbivores (L. H. P. Jones, 1978). One hypothesis for their formation is that a phytolithic crystal in the kidney is the seed for the formation of a mineral stone. Large amounts of silica and silicates ingested in soil provide for some cation exchange and may influence the utilization of other elements (L. H. P. Jones, 1978). The solubility of silica at rumen pH is relatively high, while gastric acidity promotes the conversion of plant silica to a relatively insoluble form. Perhaps this is the reason for the comparatively unique occurrence of siliceous stones in ruminants (Hopps et al., 1977).

Feeding silica or silicates to animals induces definite but as yet unexplained effects on growth (G. S. Smith et al., 1971). Silicon is required for collagen synthesis and bone formation on the order of 50 ppm in the diet (Carlisle, 1974). Although there are highly depleted tropical soils, clinical silicon deficiencies are unknown in the field.

9.8 Transition Metals

The expansion of the third period with the filling of the 3d suborbital results in 10 new positions; all these elements are metals, but in their higher valences (up to manganese) they mimic the behavior of the respective family groups in the periodic table. The biologically important valences are not the maximum ones but the intermediate ones, which often involve oxidation-reduction capabilities. As one proceeds to the right in the period (beyond chromium), chalcophilic affinities increase, but these diminish from gallium onward.

Most of the biologically essential elements are in period 4, there is only one, molybdenum, in period 5. Scandium is relatively devoid of biological affinities. Titanium is sufficiently inert so as to be used as an index of soil contamination and has been suggested as a gastrointestinal marker. Vanadium may be an essential trace element (Nielsen, 1991), and it is a constituent of the respiratory pigments in some of the lower forms of marine life.

The major elements manganese, iron, copper and zinc are required by all animal species and are not unique to ruminants. These elements are extensively discussed in the literature; however, cobalt poses some unique problems for ruminants.

9.8.1 Cobalt

All animals require cobalt in the form of vitamin B_{12}. For animals unable to manufacture the vitamin, cobalt has no other known function besides the B_{12} cofactor. Vitamin B_{12} synthesis seems to be limited to microorganisms, including those in the gut. Among plants, legumes appear to have a requirement for cobalt, but plants themselves apparently do not synthesize vitamin B_{12}. Plants are able to absorb cobalt from the soil according to its availability (C. F. Mills, 1987).

Ruminants' cobalt requirement is unique among species in that the element is used and required by rumen microbes that convert cobalt into vitamin B_{12} and its analogues. Since the animal requirement is specifically for B_{12}, the much higher apparent requirement for cobalt in ruminants is inversely related to the efficiency (only about 3%) of its conversion to vitamin B_{12} (Table 9.2). Rumen microbes appear to synthesize analogues of cobalamine, which may explain the higher cobalt requirement of the ruminant. Forms of pseudo B_{12} are synthesized in greater quantity in the rumens of animals fed on high-concentrate diets. Although the analogues appear to be absorbed by the animal, their effects on animal metabolism are unclear (Elliot, 1980).

The cobalt requirement in ruminants is higher than in nonruminants. The reason seems to be that the microbial requirement is higher and that cobalt is used in factors relative to vitamin B_{12} that are not useful in the metabolism of the ruminant animal. Indeed, it has been speculated that B_{12} and its cofactors could be involved in ketosis, milk fat depression, or both (Elliot, 1980).

Cobalt deficiency is geographically and geologically dependent. In grazing animals it manifests itself in unthriftiness, listlessness, and emaciation. Poor appetite due to insufficient cobalt exacerbates the cobalt undernutrition. Cobalt supplementation is accomplished better through a salt or other mineral supplement than by fertilization. Staggers, a nervous degeneration that may be caused by a plant toxin found in *Phalaris* pastures (Section 13.7.4), also responds to cobalt.

9.9 Nickel

Nickel has been suggested as essential for animals because very low dietary levels are associated with reduced growth in rats and chicks (Nielsen et al., 1976). Nickel has not been proven to be required by plants, although it is present in jackbean urease (Nielsen et al., 1974). Since nickel seems to be a ubiquitous cofactor in urease enzymes, it is likely to be universally essential. Nickel also appears to be important in nitrogen metabolism in the rumen (Oscar and Spears, 1988). Methanogens and nitrogen fixers (both microbes) require nickel in coenzymes that metabolize hydrogen. Certain nickel-binding compounds (e.g., hydroxamic acids) are antimethanogenic.

Nickel may play a special role in the ruminant because it is required by the urea-degrading bacteria in the rumen. The level that produces optimum effects in sheep and cattle (ca. 5 ppm) is higher than that observed for chickens (0.05 ppm) and is more critical on low-protein diets. Ureolytic activity is reduced in the rumen on low-nickel (0.03–0.07 ppm) diets (Spears, 1984) and is associated with the region near the rumen wall; presumably, ureolytic organisms are attached to the papillar mucosa, a region favorable to urea diffusion across the rumen wall.

9.10 Rare Earths

The rare earths have been largely overlooked in the general assumption that they are strange, rare, and of no biological relevance, although the latter belief has promoted their use as biological markers. This use and its associated problems justify the following discussion. Rather than being rare, these elements and their relations represent a quarter of the earth's elements (Muecke and Moller, 1988). Their abundance is nearly equivalent to that of zinc, lead, and copper, and they are hundreds of times more abundant than silver and gold (Figure 9.5). Although they are widely distributed in soils (J. B. Dixon et al., 1977), they are "rare" in the sense that they do not often occur as concentrated ores.

The rare earths and some closely associated elements lie beneath aluminum and silicon in four families (IIIA: Sc, Y, La, IIIB: Ga, In; IVA: Ti, Zr, Hf; and IVB: Ge; see Figure 9.6). Elements in the B families tend to emulate their "parent"; thus germanium is the closest analogue to silicon, and gallium to aluminum, respectively. The A families tend to be stronger bases and have larger ionic radii. Some of the rare earths (cerium, praseodymium, terbium, and thorium) can be tetravalent, and in this state they resemble group IV elements. Others (europium, ytterbium, and samarium) can be divalent, and then resemble the alkaline earth, barium, with characteristic insoluble sulfates. These differences could be important when considering the geological and biological fate of these elements (Goldschmidt, 1958). Europium is sufficiently stable in the divalent state to mimic barium, and the geochemical distribution of europium is demonstrably af-

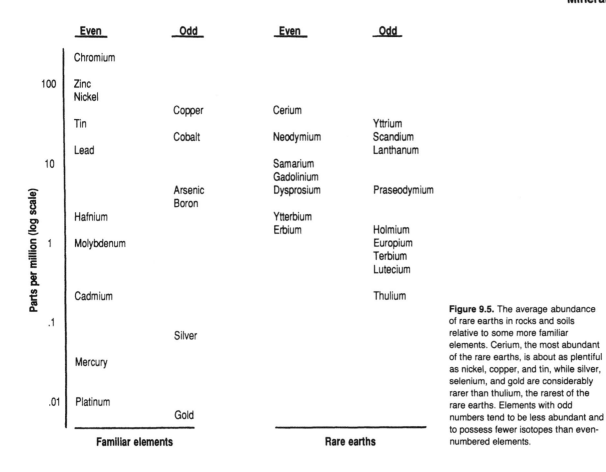

Figure 9.5. The average abundance of rare earths in rocks and soils relative to some more familiar elements. Cerium, the most abundant of the rare earths, is about as plentiful as nickel, copper, and tin, while silver, selenium, and gold are considerably rarer than thulium, the rarest of the rare earths. Elements with odd numbers tend to be less abundant and to possess fewer isotopes than even-numbered elements.

fected by this property. Strongly reducing conditions can also produce divalence in ytterbium. It is a moot point whether this occurs in rumen contents when ytterbium is used as a marker. If it did, the migration of the rare earth would follow barium, strontium, and possibly calcium rather than the expected patterns of trivalent elements.

Rare earths form complexes with many anions, among them volatile fatty acids (VFAs) (particularly acetate) and acidic sugars. The complexes with mu-

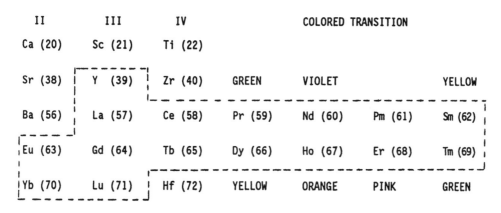

Figure 9.6. Arrangement of group III and IV elements into a periodic classification that includes rare earths (enclosed by the dashed line) and related elements (see Cotton and Wilkinson, 1966). Elements in periods III and IV (within dashed line) have dominant stable valences lower or higher than 3 or 4 and chemical characteristics of toxic heavy metals. Promethium (Pm) does not have any long-lived stable isotopes and is irrelevant to the discussion. While the true rare earths (lanthanons) include cerium (Ce) through lutecium (Lu), yttrium (Y) and lanthanum (La) are often included because of their similarity. Scandium (Sc), sometimes included, is a weaker base and has somewhat different behavior. Trivalent lanthanum and cerium are colorless, but praseodymium (Pr), neodymium (Nd), and samarium (Sm) form a colored transition series followed by colorless europium (Eu), gadolinium (Gd), and terbium (Tb) under periods II, III, and IV respectively. Dysprosium (Dy), holmium (Ho), erbium (Er), and thulium (Tm) form a second colored transition followed by colorless ytterbium (Yb) and lutecium. Cerium, dysprosium, gadolinium, terbium, and ytterbium absorb in the ultraviolet. Spectra are due to the incomplete filling of the 4f electronic shell. The greatest stabilities are a 0, 7 (half-filled), or 14 (completely filled) 4f subshell, which explains the two colored transitions and the divergent valences of cesium, europium, terbium, and ytterbium.

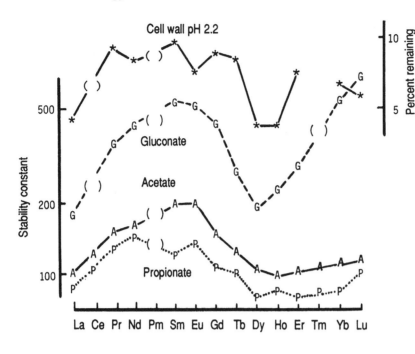

Figure 9.7. The stability of rare earth chelates (Sinha, 1966) and their retention on acid-washed forage cell wall preparations (Allen, 1982). Parentheses indicate missing data. Rare earths are significantly chelated by rumen VFAs. Rare earth complexes of acetate are unstable relative to deposition on particulate matter. Stronger chelation by gluconate is indicative of the probable binding by uronic acids in pectin and hemicellulose. Polyphenolic substances (e.g., tannins) form even stronger complexes. All chelates show a peak of stability at about samarium and a trough at dysprosium and holmium. Reliable data for thulium are not available.

cins, bacterial cell wall, phenolic acids, oxalate, and phosphate are insoluble and are probably among the particulate matter labeled by rare earth when solutions of their salts are indiscriminately added to rumen contents. The complexing abilities of rare earth ions are not equal. The relative stability of acetate, gluconate, and cell wall complexes is shown in Figure 9.7. Stabilities generally increase with atomic numbers up to samarium, diminish to about holmium, and then increase again. Rare earths form more stable complexes with cell wall and gluconates than with acetate or propionate. The insoluble phosphates are more stable than any of these, and phosphate will decompose even EDTA complexes of the rare earths.

9.10.1 Biological Activity of Rare Earths

Measurable levels of rare earth elements in plants and animals (Table 9.5) raise the question of how they got there and whether they have any essential function. Aluminum, chromium, and rare earths were reported to activate the succinic dehydrogenase cytochrome system (Horecker et al., 1939). The order of enzymatic activation had chromium being greater than lanthanum and samarium, which in turn were greater than neodymium and aluminum. Other reports, less well documented, indicate that rare earths stimulate plant growth (Vickery, 1953; Tang et al., 1985), with the absorption mechanism unknown. Their incompatibility in physiological solutions (rare earths [and aluminum] precipitate phosphate as insoluble complexes) would require complexation in a soluble form as a means of transport.

The best evidence for a light rare earth requirement (i.e., Ce and its relatives) is the extraordinary concentrations of light rare earths in hickory (W. O. Robin-

son, 1938; W. O. Robinson et al., 1958) and the fact that these trees spend ATP on mobilizing rare earths (W. A. Thomas, 1975). Plants' use for these elements is not known; perhaps they are chemical defenses.

Data on rare earths indicate variable toxicity in animals if given in excess. Injected light rare earths appear to be more toxic and produce different symptoms (including liver necrosis) than heavier elements, which are probably less well absorbed (Magnusson, 1963). Very high levels occur in certain accumulator plants grown on soils with high rare earth content. Some species accumulate relatively more lanthanum and neodymium (hickory), while others favor cerium and samarium (pond weeds). Plants that accumulate rare earths do not necessarily accumulate aluminum, and vice versa (Hutchinson, 1945). High oxalate content may be associated with rare earth accumulation—as, for example, in rhubarb.

9.10.2 Rare Earths in Feed Analyses

Rare earth ions are useful for measuring the cation exchange of fiber and for isolating tannins and other phenolic complexes from forage. They form strong complexes with pectins and other uronic acid–containing carbohydrates. The complexes formed resist rumen bacterias' efforts to remove them, and tend to depress digestibility of treated cell walls (Table 9.6). Heavier rare earths tend to bind more strongly and depress digestibility more than lighter, or group IV, elements; this effect probably arises from their selective attachment to the cation exchange sites. Rare earths could compete with microbes for attachment sites on the cell wall. Results with tetravalent elements (silicon, zirconium, and hafnium) indicate that the mode of rare

Table 9.5. Rare earths (RE) in animal and plant tissues

Material tested	Total RE (ppm)	Elements measured (ppm)	Elements detected	Reference
Bone meal	8000	—	Y, La, Ce, Nd, Sc, Gd, Dy	W. O. Robinson, 1938 Borneman-Storenkevitch et al., 1941
Human diets	—	Y 0.02, La 0.13, Ce 0.2, Sm 0.01	Ge, Ti, Sc	H. J. M. Bowen, 1979
Cadavers (ash)	—	Y 30–90[a]	—	Erämetsä, 1968
Rhubarb	100	—	—	W. O. Robinson et al., 1958
Pine	149	—	—	W. O. Robinson et al., 1958
Hickory leaves	3–2300[b]	Y 42, La 16,[c] Ce 6, Nd 18	—	W. O. Robinson et al., 1958
Hickory leaves	149	Y 33, La 40, Sm 3, Ce 10, Nd 37	Pr, Eu, Gd, Tb, Dy, Ho, Er, Tm, Yb, Lu	W. A. Thomas, 1975
Beet leaves, ash	31[a]	Y, La, Ce > Nd	Pr, Eu, Gd, Tb, Dy, Er, Yb	Borneman-Storenkevitch et al., 1941
Lupin, ash	29[a]	Y, La, Ce > Nd	Pr, Eu, Gd, Tb, Dy, Er, Yb	Borneman-Storenkevitch et al., 1941
Atriplex, ash	8[a]	Y, La, Ce > Nd	Pr, Eu, Gd, Tb, Dy, Er, Yb	Borneman-Storenkevitch et al., 1941
Red maple, oak	3–4	—	—	W. A. Thomas, 1975
Pond weeds[d]	—	Y 62–80, La 32–64, Ce 70–125, Nd 9–20, Sm 9–23	Pr, Gd, Dy, Er, Yb, Sc, Zr, Hf	Cowgill, 1973

[a]Content of the ash only.
[b]Variable according to soil.
[c]Molar percentage.
[d]*Decadon verticulatus* and *Nuphor advena* from Connecticut.

earth attachment varies, since variable effects on digestibility were observed. Tetravalent elements (particularly silicon) are much less ionic in character and probably are more attracted by the nonionic physical surface, in contrast with the ionic preference of the rare earths.

9.11 Inorganic Antagonisms

Some of the nutritional problems that concern minerals involve antagonisms between elements, as might be predicted from the arrangements of elements in vertical columns and within families; for example, potassium versus sodium, or barium versus strontium and calcium, or cadmium versus zinc. Antagonisms arise

Table 9.6. Effect on digestibility of forage cell walls mordanted with rare earths and group IV elements

Element	Fiber source	Deposition[a]	Digestibility[b]
Lanthanum	Orchardgrass	4	75
	Alfalfa	5	66
Samarium	Orchardgrass	10	67
	Alfalfa	14	68
Ytterbium	Orchardgrass	14	34
	Alfalfa	14	47
Silicon	Ryegrass	128	89
	Wheat straw	143	93
	Filter paper	189	70
Zirconium	Orchardgrass	13	93
	Wheat straw	32	90
	Guineagrass	31	92
Hafnium	Orchardgrass	11	98
	Wheat straw	16	72
	Guineagrass	48	98

Source: Allen, 1982.
[a]Millimoles of the element deposited (on the cell wall) per 100 g cell wall dry matter.
[b]True organic matter digestibility (48 h) as percentage of the digestibility of an unlabeled control of the respective forage cell wall.

for other reasons as well; for example, minerals may combine to form very insoluble substances such as $MgNH_4PO_4$, and $CuMoS_4$, and insolubility renders the component elements unavailable. Some antagonisms involve diagonal relationships within the periodic table, such as selenium and arsenic, and potassium and magnesium. Some of these antagonisms are discussed below because they involve grazing or other problems peculiar to ruminants.

9.11.1 Grass Tetany

Low blood magnesium is characteristic of grass tetany, which is not, however, a simple magnesium deficiency, even though the classic description of magnesium deficiency includes tetany. Tetany is characteristic of animals kept in grass pastures during cool periods of cloudy weather, usually during spring or fall periods of lush growth. Symptoms of grass tetany are nervousness, ears pricked, head held high, staring eyes, stiff and stilted movement, staggering, muscle twitching, and extreme excitement. Later developments (a few hours to days) include violent convulsions frequently, but not always, followed by coma and death. Chronic forms also occur and last for several weeks. Animals either recover or pass into acute tetany (Underwood, 1977). Grass tetany in lactating cattle may be related to milk fever with low blood calcium or to nervous ketosis. Treatment involves the administration of magnesium as salts or as the oxide. Prevention is through magnesium fertilization of pastures, provided soils are acid and magnesium is available. Unless animals are receiving other supplements, direct feeding of magnesium is inconvenient (Care, 1988).

The causes of the disorder involve unavailable or low forage magnesium and perhaps depleted reserves in the animal. Magnesium availability seems to be af-

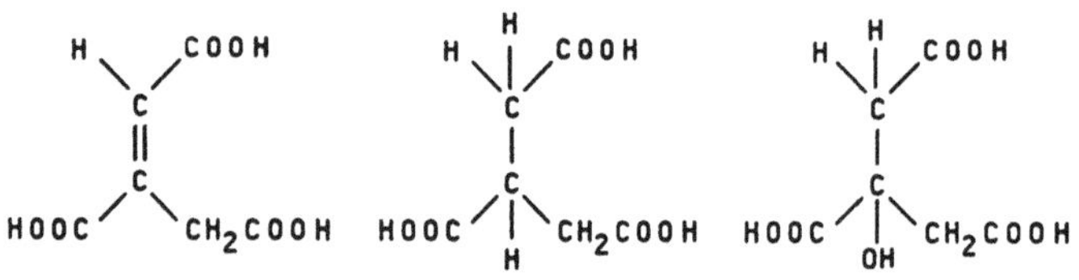

Figure 9.8. Structures of *trans*-aconitic, tricarballylic, and citric acids (from J. B. Russell, 1989).

fected by several factors that may interact. Unavailable forms of magnesium may reside in the forage cell wall. Nitrogen fertilization and possibly potassium appear to increase the incidence of tetany. Nitrogen fertilizer increases NPN content in forage, causing a rise in rumen NH_4^+, especially under cooler conditions that limit conversion of nitrate. Nitrate uptake seems to stimulate magnesium uptake (D. L. Robinson et al., 1989).

The cool, cloudy weather conditions associated with the tetany cause other changes in grass composition, including a rise in organic acid and potassium concentrations. Alternative hypotheses, all of which have some viability though none are proven, take these latter factors into account and suggest $MgNH_4SO_4$, potassium-magnesium antagonism, *trans*-aconitic acid, and tricarballylic acid as causes (Grunes and Welch, 1989).

The formation of the very insoluble magnesium ammonium phosphate ($MgNH_4PO_4$) occurs in the presence of NH_4^+, phosphate, and relatively high pH (above 7). Such conditions might be favored by high-nitrogen–low-soluble-carbohydrate forages. Potassium salts of organic acids would also have an alkalizing effect in the rumen through fermentation of the anions to bicarbonate; however, it is doubtful that rumen pH is ever high enough for precipitation of the complex. The higher pH of blood (7.4) might provide more favorable conditions if ammonia levels in blood were significant. Ammonia is extraordinarily toxic but is rapidly converted to urea by the liver.

The argument for potassium-magnesium antagonism is on a surer base, since potassium inhibits magnesium absorption. Tetany can be produced by the administration of potassium salts in combination with citric acid or *trans*-aconitic acid. Studies in Holland related the incidence of tetany to high dietary ratios of potassium concentration in relation to the sum of magnesium and calcium concentrations (K:Mg + Ca).

The most recent hypothesis involves *trans*-aconitate. *Cis*-aconitate is the normal metabolite in the citric acid cycle of aerobic organisms. For some unexplained reason grasses can accumulate up to 7% of their dry weight as the *trans* isomer. At first this fact seemed to explain tetany (Stout et al., 1967) because the unnatural *trans* acid is a good chelator of magne-

sium. *Trans*-aconitate is rapidly fermented by rumen bacteria, however, and enthusiasm for the hypothesis faded. J. B. Russell and Van Soest (1984) later found that rumen bacteria metabolize a considerable part of *trans*-aconitate to tricarballylic acid (Figure 9.8), which is not metabolized by either rumen bacteria or the animal and is also a strong chelator of magnesium. Tricarballylate is absorbed into blood and (J. B. Russell and Mayland, 1987) appears to inhibit aconitase (J. B. Russell and Forsberg, 1986). Thus a secondary toxic role exists for *trans*-aconitate in addition to that of binding magnesium.

9.11.2 Calcium and Phosphorus

The calcium content in grasses is often less than the amount animals require, while legumes and many dicots are high in calcium. Liming the soil produces forage of higher calcium content (although mature grasses may still remain low). Animal responses are influenced by the calcium-to-phosphorus ratio, so that exclusive feeding of one or the other causes problems. A low Ca:P ratio leads to osteopenia; yet the consumption of alfalfa (Ca:P ratio = 6:1) alone leads to a pathology characterized by excessive calcification. It is much less of a problem in lactating females, which can dispose of excess calcium through their milk.

Under conditions of optimum pasture management phosphorus occurs at levels sufficient for grazing animals (except for lactating females, which may have high requirements) because it usually is added as fertilizer to supply plant needs. Under range conditions or on unfertilized pastures, however, the phosphorus level in grasses is well below that needed by animals (D. Scott, 1986).

9.11.3 Copper, Molybdenum, and Sulfur

Molybdenum's suppression of copper in ruminants has been known for a long time. Molybdenum toxicity has all the characteristics of copper deficiency. The symptoms of copper deficiency include unthriftiness, anemia, and bleaching of the hair coat (Suttle, 1978).

The discovery of an extremely insoluble compound of copper, molybdenum, and sulfur—copper thiomolybdate ($CuMoS_4$)—in rumen contents and plasma (J. Price et al., 1988) implicated sulfur as well and explained some anomalies relevant to the simple Cu:Mo ratio, since an adequate or excessive sulfide supply in the rumen fermentation is essential for the formation of the complex. Sulfate, sulfite, and sulfur-containing amino acids are generally degraded in the rumen to hydrogen sulfide, which is free to combine with chalcophilic chemical elements such as copper and molybdenum. The hexavalent molybdenum binds sulfur to form the thiomolybdate anion, which precipitates an insoluble salt of copper.

9.12 Organic Antagonisms

Many organic substances complex or chelate metal ions, leading to problems of potential mineral availability; however, most of the essential metal ions are also complexed in living systems. As I noted above, elements involved in plant structure may not be available to animal digestion. This unavailability is likely the result of the strength of the particular complex or chelate relative to the eluting power of gastric acid or other complexing systems in the digestive system. Tricarballylic acid is one example of an insoluble complex; another is the calcium salts of long-chain saturated fatty acids. Others, discussed below, are fiber and phytate.

9.12.1 Chelating Agents and Ionophores

Chelating agents are the organic compounds that bind metal ions to form organo-metallic complexes; in this form the metal atom ceases to express its ionic character. Many chelating agents are derivatives of ethylenediamine or glycine and have affinity for cations in order of their valences. Thus the binding strength is weakest for monovalent ions such as sodium and potassium and strongest for polyvalent atoms such as rare earths. Ionophores are a special group that have affinities for monovalent ions. Chelating agents with affinity for polyvalent cations contain carboxyls, amino groups, and phenolic hydroxyls. The ionophores' attachment to monovalent ions depends on the presence of polyether or diketone structures in the ionophore. The order of binding within a family of the same valence is usually related to atomic weight: thus $Ba > Sr > Ca > Mg$, or $Cs > Rb > K > Na > Li$. Ionic radius also decreases in the same order, however, and certain organic structures can bind a specific element because its radius fits the hole in the chelating structure. Thus the order for Monensin is $Na > K > Li > Rb > Cs$, and for Lasolacid, $K > Rb > Na > Ca > Li$.

The biological effects of chelating agents and ionophores depend on the strength of binding of a respective ion relative to the biological binding in the gut and in the animal cells. If the chelator is too strong, unavailability will result—for example, calcium oxalate is unavailable. Also, EDTA binds calcium more strongly than any gastrointestinal or cellular device, and hypocalcemia may result. On the other hand, the affinity of EDTA for such chalcophilic elements as copper, zinc, and iron, while considerable, is less than the affinity of those elements for sulfide groups and

other structures in proteins, so the chelated EDTA form, which is soluble, makes the chalcophilic elements more available.

The ionophores that bind sodium and potassium interfere with hydrogen transport because the alkali metal ions (particularly Na) are involved in transporting of hydrogen ions through cell membranes. The methanogens in the rumen may be the most affected. The inhibitor is sensitive to the supply of sodium and potassium. For example, high dietary potassium decreases the activity of Lasolacid, while high sodium increases its antimicrobial activity. The ionophores can also affect the balance of divalent ions. Generally ionophores improve absorption of calcium, magnesium, and zinc (M. W. Smith, 1990).

9.12.2 Soaps

The fatty acids liberated in lipolysis can chelate most divalent and trivalent cations. The most sensitive cation is calcium, which forms very insoluble complexes with long-chain saturated fatty acids. Unsaturated fatty acids do not seem to have the same effect. There is thus a dietary interaction between calcium and fat that is probably more evident in ruminants than in other species because of the ruminal biohydrogenation of unsaturated fatty acids. Excess calcium tends to depress absorption of saturated fat, and vice versa. Calcium soaps are excreted in feces, where they confound traditional proximate analysis because they are ether-insoluble.

Saturated fatty acid chelation may be limited to alkaline earth ions, since it appears that soaps of chalcophilic elements such as zinc are quite available (Dell'Orto et al., 1990). This difference probably reflects the stronger ligands of chalcophilic elements for the sulfhydryl and amino groups in proteins that are likely the means of their selective absorption.

9.12.3 Fiber

Plant fibers have the ability to bind and hold metal ions on their surfaces in the same way that soil minerals and soil organic matter hold cations in soil. This load of metal ions is exchangeable in the same manner as in soil. The main functional groups involved in the binding are phenols and carboxylic acids. The cation exchange capacity of fibers is an important physical property that affects hydration of fiber and attachment of gut microbes.

Recognition of this binding led to the speculation that excess fiber might lead to fecal losses of trace metals. This concern came largely from human nutritionists in response to the surge of interest in dietary fiber during the 1970s, but the expectation has not been supported by experimental studies. The balances from

Table 9.7. Cation exchange and fermentability[a] effects on mineral balances (amount/day) for dietary fiber sources in 12 male students

	Diet			
	Cabbage	Wheat bran[b]	Wood cellulose	Low fiber control
Fe (mg)	9.9**	8.0***	10.4	11.3
Ca (g)	0.26**	0.35	0.45**	0.35
Mg (g)	0.21**	0.23***	0.06**	0.14
Zn (mg)	6.8**	2.5	−1.0	4.4
Cu (mg)	0.45*	−0.07	−0.35	0.06
Mn (mg)	−1.6	−1.8	−4.4	−1.5

Source: Van Soest and Jones, 1988.
[a]Extent of colonic fermentation.
[b]Coarse and fine bran combined.
*$P < .1$.
**$P < .05$.
***$P < .01$.

a Cornell University study in which cabbage, wheat bran, and wood cellulose were added to a low-fiber control diet showed variable responses (Table 9.7).

The addition of some fibers, particularly cabbage, improved the balances of magnesium, zinc, and copper but reduced calcium and iron. On the other hand, lignified wood cellulose reduced the mineral balances of zinc, copper, magnesium, and manganese. The behavior of calcium and iron appears to support the hypothesis that fiber can cause loss of certain elements; however, considering the amounts of fiber fed (<40 g/day), these small negative balances should not cause alarm that higher-fiber diets will induce serious deficiencies.

Other studies also indicate that dietary fiber affects mineral excretion in various ways, some positive and some negative (Van Soest and Jones, 1988). Ismail-Beigi et al. (1977) showed that cellulose reduced zinc levels in diets containing red wheat, which is high in polyphenolics that bind zinc. The curious effect of the respective dietary fibers in the Cornell study was that balances of zinc, magnesium, calcium, and manganese were inversely related to cation exchange capacities (Table 9.7) but were also related to fermentability (see Section 5.7.2).

When one considers that bacteria require trace elements and that fermentable fiber promotes bacterial growth, a microbial mechanism for fecal mineral loss is probable in monogastric animals, since the site of fermentation is below the sequence of gastric and intestinal digestion and absorption. The bacteria also produce VFAs, which are absorbed efficiently by the colonic mucosa. The absorption of the VFAs is accompanied by mineral ion exchanges, both into the colonic lumen and across to the blood. Two mechanisms have been suggested: one for mineral loss by bacterial feeding, and the other for absorption of available quantities of colonic mineral ions that may be swept across the colonic membrane by VFAs.

9.12.4 Phytate

The hexaphosphate ester of inositol (phytic acid) occurs in cereal bran. Although it is not classed as dietary fiber, it is indigestible by mammalian enzymes, including phosphatases, although it is highly fermentable. Phytases also occur in the brans themselves, particularly in wheat. Phytic acid can complex most di- and trivalent metal ions as phytates, which are unavailable to monogastric digestion but are decomposed in the rumen and other pregastric fermentations. Phytate is not known to have any negative effects on mineral absorption in ruminants, but some accounts in the literature suggest that it may have that effect in simple-gutted animals, particularly rats and chickens. Phytate's effects in human nutrition are less than might be expected. Indigenous phytases in the respective brans have a significant effect in human and animal nutrition. The effect of fiber and colonic absorption have been largely overlooked in monogastric animals, and most of the negative effects have been seen in young animals whose gut and colonic fermentation may be less developed. It might be predicted that in mature simple-gutted animals of a sufficient size for adequate retention that the negative effects of phytates are unimportant.

9.13 Mineral Availability in Forages

The mineral content of forages changes with season because the availability of nutrients in the soil and the capacity of the root system to absorb them are affected by weather patterns. This variability can produce situations in which feeds fill the mineral requirements of herbivores in some months of the year and fail to do so the rest of the year. Generally, important nutrients parallel the digestibility of a given forage, and they decline as cellular contents and metabolic tissue diminish with maturity (Whitehead et al., 1985; Minson, 1990).

Comparatively little is known about the true availability of mineral elements from plants. There is considerable speculation concerning the effects of fiber and lignin in promoting fecal loss of magnesium, zinc, and iron either through binding via cation exchange or by the presence of unavailable forms in the fiber matrix. Phytates found in the matrix between plant cells are believed to bind minerals through chelation. Some iron is present in the cell wall of legumes, probably as a matrix, and is unavailable in this form. Forage cell walls contain some nonexchangeable iron and zinc. The silica fraction may be responsible for binding between these two minerals. One theory explains silica's inhibition of cellulolytic digestion by postulating that high silica intakes create trace metal deficiencies in rumen bacteria (G. S. Smith and Nelson, 1975).

Some studies have attempted to relate extractability of minerals to their availability and physical properties in soil. One method is to equilibrate plant cell wall with an isotope of the element and then examine the isotopic ratio of the residual cell wall versus metabolized element. One study found that resistant pools of iron, zinc, and perhaps magnesium exist in the cell wall matrix (Dennis Miller, pers. comm., 1978).

Animal mineral balances are difficult to interpret because of the excretion of the respective elements in feces. Attempts to apply Lucas analysis to this situation are not very successful because physiological regulation at excessive intakes and consequent modulation of absorption lead to large variation in estimates of true availability.

10 Fiber and Physicochemical Properties of Feeds

The availability of nutrients in a feed is essentially determined by the chemical constitution of the feed: first, with respect to the concentrations of available and unavailable components, and, second, taking into account organic structures and inhibitors that may limit the availability of the components with which they are associated. From a physicochemical point of view, digestibility is a function of cumulative availability of net nutrients. Nutrient availability is further limited by the extent of obligately unavailable matter and by competition between the rates of digestion and passage, which results in some potentially digestible matter remaining undigested.

Characterizing the availability of energy and protein in a feed requires analyses that estimate digestibility and other parameters of nutritive value. Laboratory analyses are a rapid, economical means of quality control of feeds and of predicting animal response in different feeding situations.

10.1 Systems of Analysis

Laboratory analyses involve chemical evaluation and in vitro digestion with rumen bacteria or enzymes as well as adjunct methods such as in situ cloth bags and near infrared reflectance spectrophotometry (NIRS) (Chapter 8). In vitro digestion gives a direct and realistic estimation of digestibility but is lengthier, more expensive, and less reproducible than the general analysis for fiber. Enzymatic analyses are limited by the quality of commercially available enzymes. The cheaper, faster chemical analyses do not give a direct estimate of nutritive value but rather depend on statistical associations between the content of analyzed components and quality. More expensive component chemical analyses can provide important biochemical information but only estimate digestibility and nutrient availability. Lignin is the most important single fiber component limiting nutrient availability; however, its effects are not uniform. NIRS, nuclear magnetic resonance (NMR), and other instrumented methods have considerable promise but also have their own peculiar limitations.

Much effort has been expended in developing regression equations that relate various compositional parameters to digestibility. Unfortunately, interspecific variation among plants and environmental effects invalidate many of these efforts, particularly so when inadequate mathematical models are involved (Chapters 22 and 23, and Section 25.5). This does not nullify the premise that composition determines availability but rather indicates that there are several compositional factors involved and that their influence varies depending on conditions. An adequate predictive model must consider these variables.

Some feed components (e.g., protein and cellulose), while important in themselves, have little direct influence on digestibility. Any predictions based on their content in a feed must reflect their secondary associations with lignification and other protective factors. Nutritionists need to identify causative primary factors. It is possible to identify conditions in which almost any component—even lignin—appears to have no consistent association with digestibility. As a consequence of this complexity, it is difficult to find any single compositional parameter that will adequately indicate nutritive value over a range of feedstuffs. In a way this is connected with the inertia of overcoming the archaic proximate analysis system (Section 10.2), since it has been difficult to show in many cases any improvement in predictive accuracy with prospective new methods of analysis. Proximate analysis is outdated, as is the mode of its application. Substitution of ADF or NDF into a crude fiber model is a misuse of these methods and does not lead to a real advance. Scientifically the analyses and models employed should provide understanding of the balance of limiting factors in a particular feed or forage. The purely empirical approach may gloss over this in favor of an expedient answer that may not be very accurate, although it appears to give respectable correlations with the standard feeds used in calibration. The problem has to do with reproducibility of the regression when applied to sets of forages foreign to the one from which the relationship was derived.

The need for an adequate evaluation system has led to a search for the causal relationships between feed composition and nutritive value. The analytic and mathematical approach developed in Chapter 22 shows that the problem resides in the components and structure of the plant cell wall. The problem is (apart from lignification) that digestibility of the cell wall is regulated more by the intrinsic character of cell wall components than by proportions of the components.

The biochemistry of cell wall availability is complex and not completely understood. This much is known: no single chemical analysis can describe the biodegradability of cell wall matter by rumen bacteria, although such a description may be possible by combining the results of a sufficient number of analyses. There is no fractionation system that can separate the available from the unavailable. Such separation is possible only with live rumen bacteria or with adequate cell wall–degrading enzymes. Commercial cellulases are not as efficient as intact bacteria. In the end, the only completely accurate way to evaluate a diet is to feed it to an animal.

The unsatisfactory state of knowledge about the ultimate mechanisms regulating biodegradability should not prevent nutritionists from seeking a system of analysis that is consistent with current information. The discussion in Chapter 22 makes it clear that a satisfactory fiber analysis must recover from the feces the unavailable residues from the diet, and the fiber residue must not be confounded with metabolic matter. For example, enzymatic methods for total dietary fiber will isolate some microbial matter when applied to feces. A practical method must be economically competitive with the proximate system of analysis and must be consistent with the general fractionation schemes for structural analyses of plant components, so that when further advances are made, the applied system is more easily adjusted to the new information.

10.1.1 Toward a Rational System of Feed Analysis

Correct analysis depends on identifying the physical and biochemical factors that influence the biological availability of various feed fractions and knowing which of those fractions are influenced by a common factor so that they can be grouped in a general classification. The statistical test to solve this problem was devised by H. L. Lucas (Sections 22.1–4).

Basically, the challenge is to understand the cause-and-effect mechanisms behind digestibility. The results of the Lucas test and allied biochemistry reveal that biological availability (Table 10.1) can be divided

Table 10.1. Bioavailability of forage components

Component	True digestibility (%)	Limiting factor[a]
Class 1		
Soluble carbohydrate	100	Intake
Starch	90+	Passage with fecal loss
Organic acids	100	Intake and/or toxicity
Protein	90+	Fermentation[b]
Pectin	98	Fermentation[c]
Class 2		
Cellulose	Variable[d]	Lignification, silicification,
Hemicellulose	Variable[d]	and cutinization
Class 3		
Lignin	Indigestible	Limit use of cell wall
Cutin	Indigestible	Limit use of cell wall
Silica	Indigestible	Limit use of cell wall
Tannins, essential oils, and polyphenols	Not available[e]	Inhibit proteases and cellulases

Source: Van Soest, 1967.

Note: Class 1 = completely available; class 2 = partly unavailable due to lignification; class 3 = unavailable.

[a]First limiting factor relative to animal utilization and response.

[b]Fermentation may waste valuable protein by catabolism to VFAs and ammonia.

[c]Pectin can be utilized only via microbial fermentation to VFAs and other microbial products. This characteristic is shared with cellulose and hemicellulose.

[d]Fermentability of cellulose and hemicellulose is limited by lignification.

[e]Low-molecular-weight components may be absorbed but excreted in urine without being used.

into three classes: (1) total availability, actual extent of digestion determined by the competition between rates of digestion and passage; (2) incomplete availability, a refractory entity with enzymatically unhydrolyzable bonds is associated with the available portion, subject to limitations of rates as in class 1; and (3) total unavailability, the lignified fraction.

In class 1 are the entire cellular contents, including sugars, starch, protein, organic acids, and lipids. Pectin, normally a cell wall component, is also included because it has a very high nutritive availability. In the case of nonruminants, a division between soluble and insoluble components resistant to animal digestive enzymes is recognized; however, the soluble components, pectin and gum, contribute little to fecal residue. Class 2 includes the structural carbohydrates, cellulose and hemicellulose. The availability of these fractions varies widely among plant sources. Also, there is much evidence of biochemical nonuniformity. Class 3 includes lignins, cutin, Maillard products (i.e., protein damaged by heat in a Maillard reaction; see Section 11.7), and other indigestible substances.

A sequence of chemical analyses devised to divide feed dry matter according to the above classification will not necessarily estimate digestibility because of the variable character of structural carbohydrates in class 2. Unfortunately, there is no chemical method to divide class 2 material into digestible and indigestible fractions; such separation is most conveniently obtained by using rumen bacteria or the requisite enzymes. Digestibility can be estimated by regressions based on lignification (summative equation) or by in vitro digestion with rumen fluid or cellulases.

A final problem involves all systems that attempt feed fractionation based on these described principles: the statistical relationships among the fiber fractions recovered in the indigestible components may be poorly correlated to digestibility. This failure may reflect the statistical systems used as much as the chemical analysis (Chapter 22).

10.2 The Proximate Analysis System

The proximate system, now in use for more than 100 years, consists of the following steps:
1. Dry matter at 100°C.
2. Ether extraction of the dry residue to estimate lipids.
3. Reflux of the fat-extracted residue 30 min with 1.25% sulfuric acid followed by 30 min with 1.25% sodium hydroxide. The insoluble residues are dried, weighed, and ashed and the insoluble organic matter reported as crude fiber.
4. Determinations of nitrogen and ash on separate portions of samples.
5. The calculation of nitrogen-free extract (NFE) as the dry matter not accounted for by the sum of ether extract, crude fiber, ash, and crude protein (calculated as nitrogen × 6.25).

This system is the basis on which TDN (total digestible nutrients) is calculated, using the following assumptions:
1. Ether extract recovers lipids and fats, which contain 2.25 times the energy of carbohydrates.
2. All nitrogen is in protein, which is 16% nitrogen.
3. Crude fiber recovers the least digestible fibrous and structural matter of the feed.
4. The NFE represents highly digestible carbohydrates.

None of these assumptions is true, and the degree of error varies greatly. Ether extraction includes waxes and pigments of little value and does not recover soaps in feces, which are the main form in which undigested fatty acids are excreted. Forages contain no triglycerides, and leaf galactolipids contain less energy than the factor 2.25 would imply. The error involved with ether extraction is a relatively minor one unless lipid is a large component of the dry matter. Ether extraction can be ignored in the analysis of most forages and many other ruminant feeds.

Plant tissue contains a variety of nitrogenous constituents, which may be divided into proteins, nucleic acids, water-soluble nonprotein nitrogen (NPN), and

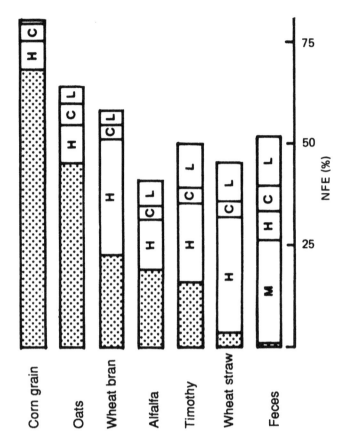

Figure 10.1. The composition of nitrogen-free extract (NFE). Stippled bars show content of water-soluble carbohydrate and starch. Contaminating components are hemicellulose (H), cellulose (C), and lignin (L). Feces contain a large metabolic component (M) composed largely of microbial debris.

very insoluble fractions associated with the crude lignin. The nitrogen content of plant proteins varies from 15 to 16% (Section 18.1). True protein, however, accounts for only about 70% of the forage nitrogen and little or none of the fecal nitrogen, so the application of the 6.25 factor to all feed nitrogen constitutes an error that is reflected mainly in the NFE calculation. The magnitude of this error depends on the nitrogen content of the diet. The error is most serious in fecal analysis. Ordinarily, little true protein is found in feces, and the main nitrogenous constituents are microbial substances or Maillard products with only 7–11% nitrogen. Most feces yield considerable NFE on analysis but do not ordinarily contain water-soluble carbohydrate (Figure 10.1). Insoluble starch is the only nonstructural carbohydrate likely to appear in feces, and only at high intake does it appear in substantial amounts.

The NFE contains the cumulative errors of all the other determinations. The largest of these errors is due to the solubilization and loss of lignin and hemicellulose in the preparation of crude fiber. Even cellulose is not wholly recovered, and the behavior of these

Table 10.2. Percentages of original feed lignin, pentosans, and cellulose dissolved in the crude fiber determination

Class	Lignin	Pentosans	Cellulose
Legumes			
Range	8–62	21–86	12–30
Average	30	63	28
Grasses			
Range	53–90	64–89	5–29
Average	82	76	21
Other[a]			
Range	10–84	43–84	7–32
Average	52	64	22

Source: From the summary in Van Soest, 1977.
[a]Gymnosperms and angiosperms exclusive of legumes and grasses.

materials in different plants is quite variable (Table 10.2). Generally, lignin in grasses is more soluble than that in legumes. The error caused by the inclusion of cell wall fractions in the NFE is lowest in the case of concentrate foods, in which about three quarters of the NFE is starch and soluble carbohydrates. In alfalfa, available carbohydrate and organic acids make up about 50% of the NFE, whereas in mature grasses and straws very little of the NFE is available carbohydrate.

Including cell wall fractions in the analysis causes the apparent digestibility of the NFE to be less than that of the crude fiber in a significant number of cases (Table 10.3). The presence of a prominent metabolic fraction in the fecal NFE contributes greatly to this effect. The digestibility of the crude fiber equals or exceeds the digestibility of the NFE in about 30% of all feedstuffs, but the error is greater in grasses, which contain more hemicellulose and soluble lignin (Table 10.3). The error is largest in the case of tropical grasses and straws.

The greatest and most fundamental error of the prox-

imate system of analysis lies in its division of the carbohydrates into crude fiber and NFE. All attempts to unseat and replace crude fiber as a system of analysis have attacked in one way or another the problems of carbohydrate fractionation and analysis. The unequal recovery of lignin, cellulose, and hemicellulose in the crude fiber fraction leads to a variable relationship between crude fiber and plant cell wall. The wide scattering of points in Figure 10.2 points out the futility of trying to predict the net insoluble fiber content from crude fiber. There is, however, more consistency within plant groups in which the ratios of cellulose to hemicellulose and lignin are less variable. Grasses, which have much hemicellulose and a moderate lignin content, provide the poorest recovery of cell wall components in crude fiber, while legumes, lower in hemicellulose but higher in lignin, are intermediate. Nonlegume dicots (principally umbellifers and composites in Figure 10.2) are chiefly unlignified vegetables and have a high rate of cell wall recovery as crude fiber because the principal component of their insoluble cell wall is cellulose. These plants often contain considerable pectin and other water-soluble components that are not recovered in the neutral-detergent fiber (NDF) or crude fiber.

10.2.1 NFE Calculation Utilizing Other Fibers

Replacing crude fiber with acid-detergent fiber (ADF) or neutral-detergent fiber analyses does not make the use of NFE less objectionable. The problem resides in its composition. In the case of ADF, hemicellulose is included in the NFE and in the feces, which contains virtually no soluble carbohydrate; the fecal fraction is made up of hemicellulose, microbial cell walls, and mucopolysaccharides. In the case of NDF,

Table 10.3. Relative digestibilities of crude fiber (CF) and nitrogen-free extract (NFE) with the proportion of cell wall components

	CF digestibility ≥ NFE		Average composition				
Feed type	N[a]	Digestibility (%)	N[a]	Cell wall (%)	Hemicellulose (%)	Cellulose (%)	Lignin (%)
Concentrate							
Whole seeds	55	11	5	21	12	8	2
Oil meals	24	10	5	33	9	18	6
Brans	4	0	3	43	28	13	2
By-products	36	6	6	48	21	24	3
Hulls	9	44	4	79	21	47	11
Forage							
Temperate legumes	104	9	39	43	10	24	9
Tropical legumes	18	28	—	—	—	—	—
Nongrass nonlegumes	49	18	—	—	—	—	—
Annual grasses	71	34	5	59	23	30	6
Temperate grasses	117	62	26	59	26	29	4
Tropical grasses	266	74	51	69	28	33	8
Straws	9	100	5	74	21	42	10

Source: Van Soest, 1975.
[a]N = number of samples.

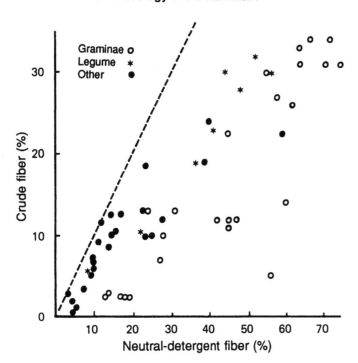

Figure 10.2. The relation between crude fiber and insoluble dietary fiber determined by the neutral-detergent method (from Van Soest, 1978). The dashed line represents equivalence of the two methods. Cereals and grasses yield relatively less crude fiber; legumes are intermediate. The "other" category is composed mainly of dicotyledonous nonlegume vegetables. Grasses are high in hemicelluloses and intermediate in lignin, and legumes are low in hemicellulose and higher in lignin. Vegetables are higher in cellulose and contain very little lignin.

calculation of NFE in feed might be correct, but fecal calculation will be confounded by the metabolic component (M_i; see Chapter 22). The defect is in the concept, not the analytical method. Metabolic fecal nitrogen is combined with organic matter at a ratio of 14:1 instead of 6.25:1 (Section 18.10), and using the 6.25 conversion factor produces an NFE component as a mathematical artifact when applied in the proximate system. This kind of error will also occur in feeds such as silages and heat-damaged feeds that include large NPN components.

10.2.2 Replacement of Crude Fiber

The problem of finding a practical and rational replacement for the crude fiber method transcends the mere chemistry of the problem. It involves the definition of fiber in the nutritional sense and the problem of the relation of fiber to nutritive value. These two aspects—definition (Section 2.9.2) and relation—present contrasting and, to a degree, conflicting philosophies. There is no guarantee that a division of plant carbohydrates to conform to an acceptable definition of the term *dietary fiber* will yield a more realistic relation to nutritive value.

The definition of *dietary fiber* is based on resistance to digestion by mammalian enzymes, leading to a division of plant substances based on chemical linkage and structure. A consequence is the requirement that the fiber residue must recover all the truly indigestible matter in the fiber preparation. This is not compatible with the desire for a fiber residue that will give a consistent relation with digestibility and also reflects practical usage in the animal feed industry, in which fiber content is a negative index of quality. Fiber's relation to any parameter of nutritional quality is a purely statistical one and depends on the association of the major components—cellulose and hemicellulose—with the primary factors such as lignin that control nutrient availability. That association is controlled by environmental factors that affect plant growth (Chapter 6), and with suitable environmental manipulation it is possible to destroy the correlation of fiber with digestibility in any crop. As a result, the search for an ideal fiber that will always relate to nutritive value is unrealistic.

Since the time of Heinrich Einhof (1778–1808), a number of other criteria for fiber definition have surfaced, including indigestibility of the fiber fraction (which has turned out to be unrealistic) and the concept of fiber as a pure compound of definite composition. While the latter is biologically unrealistic, the notion is not entirely dead. Crude fiber was used for analyses instead of Einhof's macerated fiber because crude fiber (as cellulose) was presumed to be a uniform chemical entity representing glucan, and the character of cellulose was thought to be representative of other unavailable carbohydrates. Unfortunately, this approach ignored lignin and hemicellulose and did not satisfy the requirement that dietary fiber recover the truly indigestible components of the diet. Further, cellulose is not a uniform material from either a nutritional or a biochemical point of view (Section 11.5.2).

Another approach has been to use lignin as a replacement for crude fiber, this based on the observed effect of lignin as a primary limiting factor. Lignin, although it is indigestible and affects the availability of cell wall carbohydrates, nevertheless cannot be absolutely defined chemically or nutritionally and does not always account for all the truly indigestible matter present in the structural carbohydrates. The accepted definition of the term *dietary fiber* is "polymers that are unavailable to animal digestive enzymes." For nonruminant nutrition, equating fiber with unavailability is the most significant aspect, although many nonruminants ferment fiber in the hindgut (Chapter 5). For ruminants, fiber represents the fraction that has largely lost its chance of being used once it is past the rumen. There is thus a tendency to use fiber as a negative index of quality by regressing fiber on digestibility. This last use is difficult to reconcile with the definition as stated for nonruminants because of the highly variable digestibility of plant cell wall. The total fraction satisfying quantitative recovery of truly undigested residues is the cell wall, but it should be obvious that cell wall composition influences fiber quality. The proportions of lignin and hemicellulose present are the most important quality factors, but these are often poorly related to fiber content.

There can be no question that the survival of crude fiber as a measure of fiber content has been related in part to the simplicity of the procedure as well as to the large reservoir of analytical data. A further problem is that any method that gives correct values for total fiber (e.g., plant cell wall) will involve inherently different and unfamiliar numbers. Proponents of the continued use of crude fiber depend on its ability to correlate with nutritive value and are hopeful of a relationship among crude fiber, cell wall, and digestibility. These expected associations are often ruined by variability in cell wall composition and the environmental factors affecting lignification. Now, after a half century of forage work to replace crude fiber, better analytical methods are generally applied—although not always intelligently —in developed countries. The Third World countries, however, continue to use proximate analyses, usually on the basis of cost and expediency. Ironically, it is in developing countries in the tropics that the association of fiber and digestibility is the poorest (because of the large lignin-temperature interaction), so the promotion of cheap proximate analysis there has resulted in a compilation of useless analyses.

Accurate prediction of nutritive value from any single chemical analysis is difficult because of the complex nature of the nutritive quality variables. Digestibility, the most reliably measured aspect of nutritive value, is not accurately estimated by any single chemical analysis over a range of feeds. The application of regression equations assumes that a high correlation denotes a controlling influence of the measured parameter on nutritive value. The usefulness of fiber-based regressions involves two further assumptions: that the same factors influencing the nutritive availability of fiber also influence the nonfiber fraction, and that the content of a fiber is related to its digestibility. These assumptions are, of course, false. A regression of digestibility on lignin content assumes that lignification influences the availability of the whole diet, including non–cell wall matter. The Lucas analysis for uniform feed fractions (Chapter 22) disproves all these hypotheses.

10.3 The Detergent System

The detergent system was devised to provide a rapid procedure for determining the insoluble cell wall matrix and estimating its major subcomponents: hemicellulose, cellulose, and lignin. Subsequent develop-

ments in the technique allow the partition of feed nitrogen and protein and measurement of heat-damaged protein as representing the nitrogen content of acid-detergent fiber. The use of neutral detergent on fecal and rumen contents is the most convenient method for separating undigested matter from microbial and metabolic contaminants.

The principal obstacle in preparing plant cell wall residues in which indigestible components are recovered is removing the contaminating protein. This is one reason why sodium hydroxide is employed in the preparation of crude fiber. Unfortunately, when the protein is removed, most of the hemicellulose and lignin goes as well. Many cell wall and lignin procedures use proteases to degrade the protein. Another alternative is to employ detergents that can form soluble protein complexes (Figure 10.3).

10.3.1 Neutral-Detergent Fiber

Anionic detergents form polyanionic complexes, the sodium salts of which are soluble above pH 6. Interference by heavy metal or alkaline earth metal ions is prevented by ethylenediaminetetraacetic acid (EDTA), which chelates them. Extraction of forage with a neutral (pH 7) solution of sodium lauryl sulfate and EDTA allows the preparation of a fiber residue that recovers the major cell wall components: lignin, cellulose, and hemicellulose (Figure 10.3). The residue also contains minor cell wall components, including some protein and bound nitrogen, minerals, and cuticle. Pectins are removed, although they are a cell wall component. The neutral-detergent extraction is nonhydrolytic and recovers the insoluble matrix. Pectin's loss in this procedure and its relatively complete nutritive availability are evidence of its lack of covalent bonding with the lignified matrix.

Common contaminants of the neutral-detergent residue include starch, animal keratin, and soil minerals. The starch can be eliminated either through pretreatment or by concomitant treatment with amylases (Van Soest et al., 1991). Failure to remove starch leads to difficulties in filtering the fiber and increases the analytical error.

There have been remarkable improvements in the quality of commercial amylases. A heat-stable amylase that is specific and will hydrolyze starch to oligosaccharides, in combination with urea, is capable of attacking the most resistant and insoluble starches. This availability has led to new procedures for determining NDF and total dietary fiber (Van Soest et al., 1991).

Insoluble indigestible animal proteins (keratins) occur in mixed feeds through the inclusion of animal products and in feces through hair and epithelial sloughing. Keratins can be eliminated from cell wall

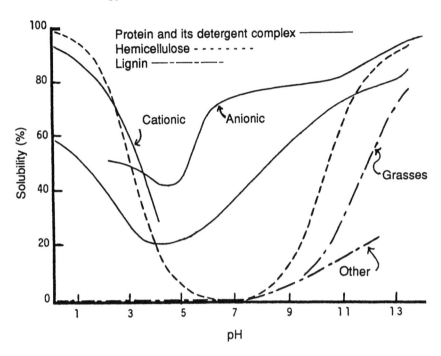

Figure 10.3. The relative solubility of forage hemicellulose, lignin, protein, and protein-detergent complexes in boiling aqueous medium at different pH's (see Van Soest and Robertson, 1977). Lignin is soluble at pH > 7, and hemicellulose is soluble in either strong acid or strong alkaline conditions. Cellulose (not shown) is generally insoluble over the pH range shown. Other = lignin from nonlegume, nongrass forage such as brassicas and composites.

with sodium sulfite, which cleaves disulfide bridges between peptides and renders keratinous proteins soluble. Unfortunately, the sulfite also attacks lignin, thereby reducing its recovery. Sulfite-treated cell walls show an increased in vitro digestibility. Sulfite must not be used if the cell wall preparation is to be used for any further purpose.

The natural mineral components of the cell wall are poorly recovered in the neutral-detergent residue because of the necessary chelation of metal ions needed to prevent their interference. A high sodium concentration results in the solubilization and loss of much of the biogenic opaline silica. This loss is measurable if the amount of opaline silica in the sample is less than that required to reach the saturation limit in hot neutral-detergent solution (430 ppm SiO_2). In a 0.5-g sample, this is roughly equivalent to 1% of SiO_2 per unit of dry matter. On the other hand, most soil mineral silica is insoluble in neutral detergent. The residual insoluble silica is thus an estimate of the soil contamination of the sample provided the above-stated saturation limit has not been exceeded. Other major minerals that are relatively insoluble in neutral-detergent are ferric iron, alumina, other aluminum-containing minerals such as clay, and probably naturally occurring rare earths.

10.3.2 Acid-Detergent Fiber

A low-nitrogen residue that recovers lignin and cellulose can be prepared by extracting plant tissue with strong acid solutions of quaternary detergents (Figure 10.3). This procedure is essentially a modification of normal-acid fiber, which has too high a nitrogen content. The residue does not represent a fiber residue that

can fill the ideal position of an estimate of dietary fiber; rather it is a fraction of the cell wall useful in the partitioning of the major cell wall components. The truly indigestible components of feed are recovered in the neutral-detergent residue; acid detergent divides this residue into fractions soluble and insoluble in 1 N acid. The acid-soluble fraction includes primarily hemicelluloses and cell wall proteins, while the residue recovers lignin, cellulose, and the least digestible noncarbohydrate fractions. Acid detergent has the advantage of removing substances that interfere with the estimation of the refractory components, so the ADF residue is useful for the sequential estimations of lignin, cutin, cellulose, indigestible nitrogen, and silica. In contrast with neutral-detergent extraction, silica is quantitatively recovered in the ADF residue.

Acid-detergent fiber is widely used as a quick method for estimating the fiber in feeds, often substituting for crude fiber as part of a proximate analysis. The use of ADF as a predictor of digestibility is not founded on any solid theoretical basis other than statistical association. It is influenced by the environmental associations described in Chapter 6. Nevertheless, there have been attempts to improve predictability by modifying the ADF procedure to give values that correlate better with a set of forages of known digestibility. The modified acid-detergent fiber (MADF) of Clancy and Wilson (1966) is such a development.

10.3.3 Modified Acid-Detergent Fiber

The MADF, developed in Ireland, used as its standard forage dried at a high temperature in the process of sample preparation. The forage actually fed to ani-

Table 10.4. Compounds that interfere in the estimation of hemicellulose
as the difference between neutral-detergent fiber (NDF) and acid-detergent fiber (ADF)

Fraction	NDF	ADF	Effect on estimate	Reference
Cell wall protein	Recovered	Largely dissolved	Increase	Keys et al., 1969
Biogenic silica	Considerable solution	Quantitative recovery	Decrease	Van Soest and Jones, 1968
Pectin	Dissolved	Partial precipitation	Decrease	Bailey and Ulyatt, 1970
Tannin	Precipitation as protein complex	Partly dissolved	Increase	Van Soest and Robertson, 1985

mals, however, did not exhibit the heat damage imposed in the standard's preparation. The study found that prolonged boiling with acid of higher concentration reduces the bound nitrogen and improves the association of fiber with animal digestibility. A subsequent study in Britain using more carefully prepared standards does not support the original observation. The

MADF procedure includes oven drying at 95°C as a preliminary step. Unfortunately, this treatment prevents the use of MADF as a means of assaying for heat damage and unavailable protein, one of the more valuable ADF applications. The interactive combination of ADF and acid-detergent insoluble nitrogen (ADIN) improves estimates of digestibility of silages and hays

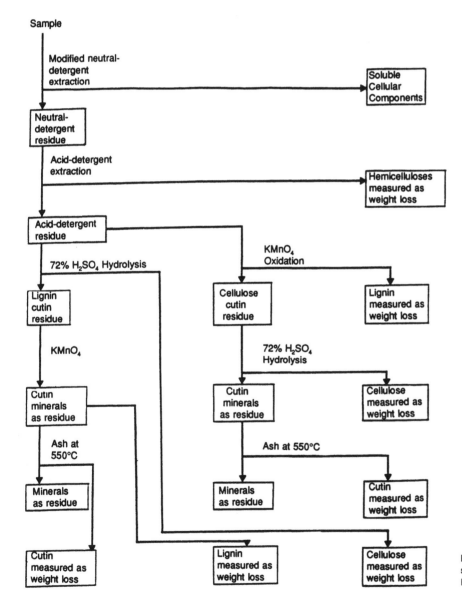

Figure 10.4. Flow diagram for sequential analysis of fiber (from Robertson and Van Soest, 1981).

(Yu and Thomas, 1976). In these instances ADIN can be substituted for a lignin value with which it is highly correlated.

10.3.4 Sequential Analyses

Neutral detergent dissolves pectin and opaline silica, while acid detergent recovers silica and dissolves some tannin-protein complexes. Galactouronan is precipitated into the ADF as the quaternary detergent salt. Acid-detergent residues are usually lower in protein than neutral-detergent residues but recover silica with no loss. The influence of these effects on estimates of hemicelluulose are shown in Table 10.4. Some of these errors are partly self-canceling when used statistically. R. W. Bailey and Ulyatt (1970) recommended that neutral-detergent extraction precede acid detergent if purity of the acid-detergent fiber is sought (Figure 10.4). Preextraction will eliminate interference by pectin, tannins, and silica, although silica will no longer be measurable. The silica content of ADF minus that of NDF can be used as an estimate of soluble opaline silica. Subtracting NDF from ADF tends to overestimate hemicellulose in grasses relative to figures obtained by measuring its component sugars (Theander and Westerlund, 1993).

Tannin content may be estimated by measuring lignin in a double-sequential manner. However, tannins in dried plant materials are impossible to remove from lignin, and thus the measures of crude lignin will include the contaminating tannin.

These problems illustrate the difficulty of designing a single system to analyze all conditions. For example, sequential treatment removes the pectin interference but sacrifices the determination of total biogenic silica. As I said earlier, no single analytical protocol can satisfy all conditions.

10.4 Alternative Systems for Fractionating Feed

The detergent system of analysis is not the only one that can accomplish the desired partitioning of feed components according to nutritional criteria. Systems of fractionation that antedate the detergent system include that described by Paloheimo (1953), Gaillard's modification of Harwood's procedure (1958), and the methodology of Waite and Gorrod (1959) and Southgate (1969) as applied to human foods. More recently, improved methods for determining the sugar components of plant cell walls have been devised (Åman, 1993). The relative advantages of the various strategies of forage analysis depend on their intended use and application.

Many systems for fractionating plant cell wall have

Table 10.5. Forage dry matter (DM) composition and digestion coefficients (components

	Fresh grass		Fresh alfalfa		Fresh clo
	DM (%)	DC	DM (%)	DC	DM (%)
Sugars (mono- & disaccharides)	4.1	100	2.7	100	4.9
Fructosan	5.7	100	—	—	—
Starch	—	—	0.4	100	1.2
Polysaccharides					
Soluble in 0.5% ammonium oxalate (pectin)					
Anhydrous galacturonic acid	2.6	95	5.0	95	5.0
Anhydrous arabinose	1.8	99	0.6	97	0.5
Soluble in 5% and 24% KOH					
Anhydrous galactose	1.3	85	0.5	82	0.6
Anhydrous glucose	1.9	88	0.9	91	0.8
Anhydrous arabinose	3.5	86	0.8	85	0.6
Anhydrous xylose	8.5	72	5.8	33	3.6
Cellulose residue					
Anhydrous arabinose	0.3	60	0.4	79	0.4
Anhydrous xylose	0.6	38	0.3	29	0.1
Pure cellulose	21.1	79	22.2	41	18.2
Aldobiuronic acid of xylan	7.3	79	7.3	52	7.1
Lignin	6.2	0.2	9.1	0.4	7.0
Organic matter	90.0	71.7	89.4	59.0	88.8

Source: Modified from Gaillard, 1962.

been devised; the detergent system is one. The older systems are tedious and not biologically realistic. Gaillard (1962) demonstrated the complete availability of pectin, the complexity of pentose digestibility, and the nonuniformity of the lignified plant cell wall component. This classic study also demonstrated the futility of gross component analyses. For example, the total analysis of a food or feed for net glucose is meaningless unless the polymer in which it occurs is specified: starch and cellulose have utterly different nutritional implications. For pentoses, arabinose is more available than xylose in the same fractions (Table 10.5). Xylose in crude cellulose is less digestible than xylose extractable into a hemicellulosic fraction. This observation considerably weakens criticism that ADF contains residual pentosan, since these residual pentosans have a lower availability than extractable ones and contribute to the correlation of ADF and indigestibility (Gaillard, 1962; Bittner and Street, 1983).

Analyses for sugar components give valuable basic information, but the quality of the preparatory steps needs to be emphasized. Too much elegant chemistry has been applied to poor preparations, with confusing results. The advent of good-quality commercial cellulases and hemicellulases, as well as rumen organisms and pectinases, should improve this situation and provide a new direction to plant cell wall research.

Most early procedures were restricted to relatively few samples, which were examined and analyzed in great detail, often ending with a structural analysis of the sugar components of the fractions. The advent of automated gas-liquid chromatography (GLC) has somewhat alleviated the problem of time constraints (Englyst and Hudson 1987). The advantage of component analysis is that the structural carbohydrates are determined directly, before sugar composition is determined, which is valuable for extending knowledge about the inherent composition and variability of cellulosic and hemicellulosic fractions. The detergent system does not determine sugar components, which represents a sacrifice for the sake of utility. On the other hand, hybrid systems linking the detergent system and component fractionation have been developed (R. W. Bailey et al., 1978). Newer methods seek milder means of extracting cell wall that utilize enzymatic degradation, avoiding the bond fractures that occur with harsh acidic or alkaline extraction and the resultant artifacts.

Many methods of isolating specific polysaccharides are based on their unique solubility characteristics. Methods for estimating the amounts of these materials are termed definitive in that the material obtained is essentially defined by the procedure and experimental conditions. This leads to the problem of specificity of methods and interference from undesired components. Starch, pectin, and hemicelluloses may be difficult to obtain free of one another because their solubility characteristics overlap. Accurate determinations require specific enzymes and perhaps component sugar analyses.

Hemicelluloses and pectins are more difficult to work with because they are less well defined than starch. The tendency to regard net hemicellulose as pentosan and pectin as polyuronic acid (galacturonan) is an oversimplification leading to errors when methods based on these principles are applied. Pentoses occur in both pectin (as side chains) and hemicellulose, and also in RNA as ribose. Thus, measuring the total pentosan may overestimate hemicellulose and its digestibility since pectins and ribose are unlignified and highly digestible. In rumen cultures, RNA synthesis may confound the measured utilization of hemicellulose based on pentose. This is overcome if ribose is distinguished from xylose and arabinose.

Hemicellulose is particularly susceptible to alteration and damage. For this reason, cell wall analysis is better applied to a fresh preparation than to the residue left after the analysis of nonstructural carbohydrates. Cooking and heating cause protein to become refractory to enzymatic extraction, and protein and amino acids catalyze the destruction of sugars, among which pentoses are the more sensitive (Section 11.7).

Conventional systems for fractionating carbohydrates from forages often involve preliminary extraction of lipid from the dried tissue with ether, chloroform, or alcohol-benzene, followed by sequential extractions of the carbohydrates based on solubility.

The disadvantages of preliminary extraction lie in the loss of simple sugars in the solvent and potential heat damage to the residue. Extraction with 80% aqueous alcohol will remove soluble sugars and most of the lipid. Many old-fashioned methods employ heating to gelatinize starch and solubilize pectin, thus denaturing protein, and then employ inefficient methods to deal with this now-refractory protein and its stubborn contamination of subsequent residues.

Hot-water extraction removes the rest of the soluble carbohydrates, excluding most of the pectin and starch, which require separate steps for their division. Extraction of starch with acidic reagents or impure amylases results in loss of some pectin and hemicellulose. Therefore, starch determination requires a glucose-specific assay. Glucose oxidase or GLC can be used for this purpose; however, all starch must be reduced to glucose if this is to be effective.

Deproteinization is required for any realistic measure of dietary fiber and may be partly accomplished with neutral detergent or proteases. Complete removal of protein is not possible, and a fraction remains that is probably truly indigestible. This fraction averages 7% N in forages and more in heated feeds and foods. Neutral proteases may be more desirable than acid pepsin because available protein is more exhaustively extracted without danger of hydrolyzing the more acid-labile glycosidic bonds. On the other hand, acid pepsin more closely mimics the sequences and effects of digestion in nonruminant animals. The removal of extractable proteins should precede any step that involves heat. All forage cell wall preparations contain residual nitrogen; a portion of it is available protein, and the remainder is an unavailable fraction associated with the crude lignin.

Isolating the pectin fraction requires chelating agents in neutral solution to remove the calcium and convert the pectate to a water-soluble form. Unfortunately, this results in some alteration and debranching of pectin, so it is not possible to isolate the less-soluble pectins without some alteration (Bucher, 1984). Some pectins (in cabbage, for example) possess significant cold-water solubility without any pretreatment, while alfalfa pectins require a chelating agent and heat. As much as 40–50% may dissolve in the presence of phosphate, a chelating agent. General schemes cannot be indiscriminately applied without knowing the peculiarities of the plant material at hand. Hot neutral detergent dissolves most pectins, leaving a cell wall residue of hemicellulose, cellulose, and lignin. This can be an advantage or a disadvantage, depending on the objective.

The fact that pectins can be removed without hydrolytic cleavage (although some β elimination of side groups may occur; see Sections 11.5.4 and 11.6) at pH 7 indicates that they are not likely to be interlinked with

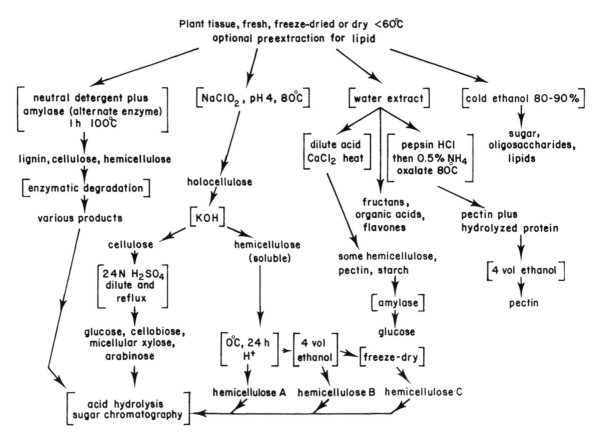

Figure 10.5. Some common sequences of carbohydrate fractionation and separation.

10.5 Physical Properties of Fiber

the cell wall matrix. This is in agreement with the complete availability of pectin to fermentation, indicating freedom from the effects of lignification.

The partition of hemicellulose has been traditionally accomplished with alkaline extraction, which cleaves all ester bonds and probably some glycosidic linkages as well (particularly β 1–3). Lignin is also largely dissolved in this procedure, and the carbohydrate is partitioned according to solubility in weak acid and alcohol (Figure 10.5). Cellulose is defined as the carbohydrate residue remaining insoluble after treatment with strong alkali. This residue still contains some arabinose and xylose, which exhibit lower digestibilities than true cellulose (Table 10.5), digestibility of arabinose being higher than xylose in all fractions. Since both sugars may be a part of the same arabinoxylan polymers, the different digestibilities may involve the labile linkage of arabinose, a close association of xylose with lignin, or both. Xyloglucans associated with cellulose are also quite insoluble and could be part of the latter fraction.

The extraction of plant cell wall by chemical means to obtain carbohydrate fractions relevant to nutritional quality is largely an unsolved problem. Partitioning hemicellulose and cellulose fails to provide any meaningful interpretation of sugar composition in relation to digestibility of structural polysaccharides.

The quality of feeds and forages is greatly affected by physical attributes that may not be at all associated with chemical fractions or obvious from chemical analysis. These properties include physical density, hydration capacity, cation exchange, and fermentation rate.

10.5.1 Density

Physical density, or the concentration of digestible energy and nutrients per unit volume, is related to composition of the plant as harvested but can be greatly altered by processing. The maturation of plant cell walls involves thickening of the secondary layers with concomitant lignification. Cell size is fixed at an early stage, and wall thickening occurs at the expense of the intracellular space. This results in an increase in density of the wall with physiological maturity (Figure 10.6). Young plant cells have a high water content, but this declines with maturation. Nutritional effects attributed to water content could be the result of cell volume, which must contain all the water of the living plant. While the bulk volume of young cell walls is high, young cells also are more digestible and more easily ruminated. Nevertheless, young plant cell walls

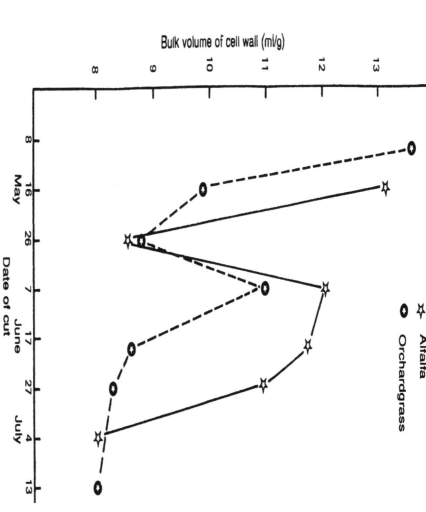

☆ Alfalfa

Ⓦ Orchardgrass

Bulk volume of cell wall (ml/g)

Date of cut — May 8, 16, 26 / June 7, 17, 27 / July 4, 13

Figure 10.6. Change in bulk volume of forage cell wall with age and maturity in forage grown in West Virginia (van der Aar and Van Soest, unpublished data, 1978). Volume declines with age as lignification and cell wall thickening proceed. The abrupt rise in volume after May 26 is associated with tillering and flowering and involves the dilution effect of new tissue and organs.

are consumed in the same quantity by weight as are those of more mature forage. Bulk volume is less well related to voluntary intake than is cell wall content. This may reflect either a digestibility factor or a rumination factor. The latter seems more likely on the basis of rumination studies. Possibly the higher digestibility and easier communition compensate for the increase in bulk volume.

The physical volume of a plant is a property of the structures that house the cellular contents. If the soluble matter and cellular contents are removed, the hollow cell wall structure remains filled with gas or water. Consequently, the absolute density of the cell wall

structure may be of little significance. Instead, its bulk volume and hydration capacity determine its effective volume in the rumen (Hooper and Welch, 1985). R. E. Miller et al. (1984) developed a procedure for measuring ruptured cells in macerated forages.

Isolated forage cell wall possesses a volume similar to an equivalent weight of whole forage of the same particle size. The cell wall itself weighs less and may not pack as well (Figure 10.7). Removing the lignin and hemicellulose leaves the structural volume of the cellular structure intact, so that although as much as 70% of the forage weight is removed, the effective volume has not been diminished. Communiting the fiber through grinding, pelleting, or rumination will increase its effective density because the cellular structure collapses. The effect of the collapse can be compared to the demolition of a skyscraper: the contents of the rooms—nonbearing walls, plaster, doors, and furniture—can be expelled without altering the volume of the building. Only when the wrecking ball smashes the structure does its effective volume decrease.

10.5.2 Particle Size

Animals ruminate in proportion to the cell wall content of their diet. Communition is the physical reduction of particulate matter in size and volume. Communition of cell wall structure by rumination is an important process assisting digestion and passage of matter from the rumen; rumination is greatly suppressed in animals fed pelleted or ground diets. Chewing is probably the major force in reducing particle size

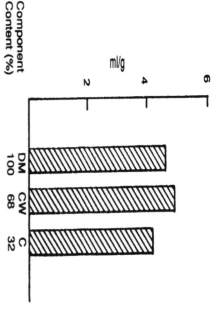

Component Content (%)

ml/g

DM 100 / CW 68 / C 32

Figure 10.7. The relation between volume of milled (20 mesh) forage dry matter (DM), cell wall (CW), and cellulose (C) in orchardgrass (modified from Van Soest, 1975). Volume was measured for air-dry samples in graduated cylinders. Extraction removes much intracellular and secondary wall matter without disturbing the bulk volume.

Fiber **Fiber 151**

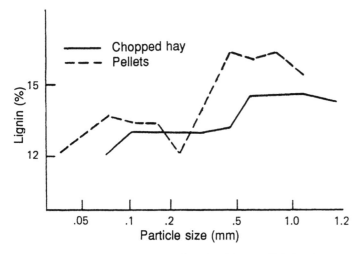

Figure 10.8. Lignin content of milled and chopped alfalfa hay of various particle sizes.

Table 10.6. Median particle size of three feeds sieved by wet and dry procedures

Feed	Particle size (μm)		
	Wet	Dry	Difference (dry − wet)
Wheat bran	858	895	−37
Alfalfa	530	434	96
Timothy	645	478	167

Source: Modified from Allen et al., 1984.
Note: Data are the result of eight replicated sievings. Median values were determined by fitting gamma distributions.

of lignified materials that are cross-linked; however, microbial digestion may contribute to the breakup of unlignified tissues that fall apart on degradation.

The physical and milling properties of cell wall are greatly influenced by their chemical composition. Cellulose, a long, linear chain, is flexible. Lignin, a three-dimensional plastic-like polymer, is rigid and inflexible. Woody tissue with a low lignin-to-cellulose ratio tends to bend rather than break, while tissue with a high lignin-to-cellulose ratio tends to break rather than bend. Consequently, the less lignified grasses tend to mill in long, thin, fibrous particles while more lignified alfalfa fragments into shorter, fatter pieces.

Lignification increases the force required to shear fibers. In milling, lignin becomes selectively distributed among the larger particles (Figure 10.8); however, this does not occur in the rumination process.

Particle size is not often measured as a forage characteristic, probably because of the labor involved in measuring it. Particle size may be measured in two ways: by dry-sieving or by slurrying wet suspensions through calibrated screens arranged by decreasing pore opening from top to bottom. Wet-sieving tends to sort particles according to length, while the dry method tends to sort according to cross-sectional diameter. Both characteristics are important in classifying fibers. Since the rumen is a wet system, wet methods of sizing particles in feeds and rumen contents have come to be favored. Soaking the feed in water will dissolve much of the finely divided non–cell wall matter; however, in ruminant feeds it is more relevant to measure the insoluble fibrous matter. Table 10.6 compares particle sizes determined by wet and dry sieving.

Legumes shatter into short particles, while grass particles are more needle-like. The dry method has problems with the electrostatic charge on particles, which leads to aggregates, particularly in smaller parti-

cles. Sample size must be small, and the ratio of sample size to screen area controlled.

The expression of particle size as a single number offers special problems. The distribution of dry matter per screen size classification is logarithmic, not linear, giving an arithmetic mean biased to low values. Valid statistical treatment requires converting to the logarithm of particle size and calculating the mean logarithmic particle size. The standard deviation allows an estimate of the size range within a sample. The logarithmic uniformity can be tested by plotting the cumulative weight that is less than the stated size against log size. A straight line results in the case of uniformity. A problem may occur if there happens to be an overlapping distribution of two particulate populations. This would be evident by inflection or nonlinearity in such a plot. An alternative is to normalize the data via a gamma distribution, which gives standardized statistics (Allen et al., 1984). The gamma distribution may deviate from mechanistic reality, however; that is, it satisfies statistical requirements but does not resolve questions of physical relationships.

An older and somewhat simpler system of expressing particle size is by modulus of fineness (American Society of Agricultural Engineers, 1961), in which the problem of logarithmic distribution is resolved by using standard screens that form an approximate logarithmic series. The modulus is expressed as an average screen size number that, although related to size, has no dimensional units. This system is somewhat arbitrary and is less satisfactory in detecting nonuniformity in a particle population.

10.5.3 Hydration

Hydration involves the ability of feed particles to adsorb and hold water, ions, and other soluble substances. Fractions of the diet that are important contributors to hydration effects form gels or are insoluble and have sufficiently slow digestion rates that they persist in the digestive tract and continue to exert their effects for some time.

The cell wall fraction is the most significant contributor to hydration because it has the slowest digestion

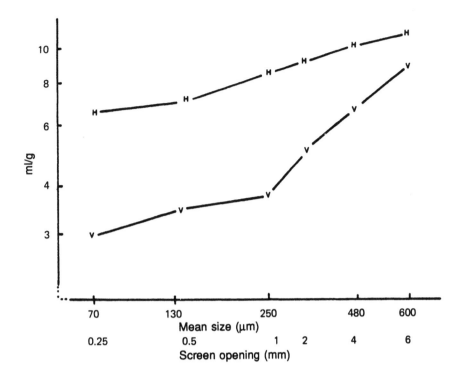

Figure 10.9. The effect of grinding on the dry bulk volume (V) and hydration capacity (H) of wheat straw (Sequeira and Van Soest, unpublished data, 1981).

rates and contains undigested components that survive to reach the feces. Particle size is inversely related to surface area per unit weight; however, reduction in feed particle size also reduces the internal cellular space in cell walls. These two aspects affect the water-holding capacity of cell walls. The alteration in hydration capacity caused by grinding (Figure 10.9) and rumination involves the interaction of increased surface area and decreased intracellular space. Bulk vol-

ume includes the space trapped within the cellular matter. Grinding destroys much of this space. The interior space of the cellular structure is an important factor influencing bulk volume and hydration capacity.

The ability of the surface of a solid to hold water and other molecules varies with the type of chemical substituent group (e.g., hydroxyl, carboxyl, amino) on that surface. The alcohol groups in cellulose are extensively interhydrogen-bonded, leading to a low swell-

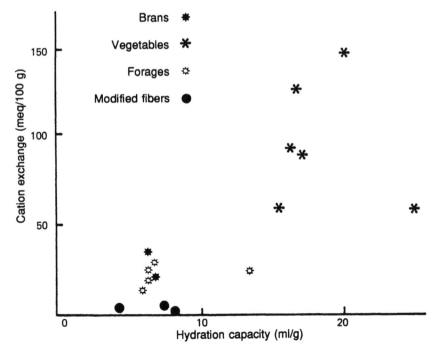

Figure 10.10. Cation exchange and hydration capacity in plant cell walls (from Van Soest and Robertson, 1976b). Modified fibers, including purified celluloses and partially delignified pulp, have relatively low exchange and hydration capacities. Vegetable fibers have the highest exchange and hydration capacities, and forages are intermediate.

Table 10.7. Functional groups and their contributions to adsorption and exchange capacity

Group	Source	Effect
Alcohol	Carbohydrates	Hydrogen bonding
Phenol	Lignin	Chelation and cation exchange
Carboxyl	Pectin, hemicellulose	Chelation and ion exchange
Amino	Protein, Maillard products	Anion exchange, chelation
Peptide	Protein	Lipophilic adsorption
Phenyl	Lignin	Lipophilic adsorption

Sources: McConnell et al., 1974; Eastwood et al., 1983.

ing capacity and minimal water adsorption. The presence of free carboxyl, amino, or other hydrophilic substituents increases hydration capacity and ion exchange (Table 10.7).

Pectin and crude lignin fractions appear to have a major role in binding bile acids and fats. The Maillard component may be of particular importance in binding lipids since it is high in nitrogen that may provide anion exchange capacity. Proteins form well-known lipophilic complexes with fatty acids and detergents in which the hydrophobic ends of the fatty acid molecules are oriented into the peptide structure and the hydrophilic or ionic groups point outward (Table 10.7). When proteins form complexes with ionic detergents their solubility increases, a feature that is the basis for the detergent system for fractionating forages. The ability of starch to form complexes with lipid is related to its helical configuration because fatty acids fill the core of the coil. Pectin promotes the loss of bile acids and cholesterol in feces, but the mechanisms are not well understood.

10.5.4 Buffering and Cation Exchange

The cation exchange capacity of cell walls has been implicated as a possible factor affecting the availability

Table 10.8. Cation exchange of forage fibers (meq/100 g)

Forage	Mechanical pulp[a]	NDF[b]	Pepsin digested, residue	Rumen digested, residue
Fescue	59	111	43	32
Timothy	68	132	51	39
Ryegrass	97	143	68	32
Orchardgrass	72	120	44	40
Rice straw	43	57	25	32
Horsetail	171	162	70	181
Alfalfa	152	104	68	31
Red clover	169	139	85	47
White clover	294	249	204	65

Source: Van Soest and Jones, 1988.
[a]Prepared by macerating fresh forage with water in a Waring blender.
[b]Neutral-detergent fiber.

Table 10.9. Correlation coefficients (r) of cation exchange capacities determined with copper (valence II) and praseodymium (valence III) at specified pH with composition and fermentation kinetics of batch-isolated neutral-detergent fibers

	Cu (II) pH 3.5	Pr (III) pH 3.5	Pr (III) pH 7.0
Lignin (%)	0.763***	0.693***	0.843***
Hemicellulose (%)	0.486	0.563*	0.481
Cellulose (%)	0.114	0.032	0.155
Nitrogen (%)	0.699**	0.500*	0.579*
Cu at pH 3.5		0.963***	0.949***
Pr at pH 3.5			0.941***

Source: McBurney et al., 1986.
Note: Regression equations are significant at:
*P < .10.
**P < .05.
***P < .01.

of zinc, iron, and copper to nonruminants. This aspect of fiber quality has been ignored in ruminants, although the exchange properties of fiber are probably of considerable importance in buffering the rumen and the lower digestive tract. The evidence for this is largely indirect and involves the general beneficial effects of feeding buffers to ruminants on low-fiber diets. Sodium bicarbonate and similar salts are ephemeral in that once the CO_2 is expended, their buffering capacity is dissipated. Other buffers (e.g., clays and bentonites containing insoluble and undegradable anions) may be more effective because they not only service the rumen but also can carry buffer and exchange capacities to the lower tract since they become recharged when the pH rises. The same thing happens to lignified fiber as it passes down the tract. The absorption of mineral ions involves replacement with hydrogen ions (Figure 10.11).

Cation exchange capacities for a number of forages are compared in Table 10.8. The value for purified cellulose is very low, and that of graminaceous plant fiber is less than that of legumes. Silicified cell walls like those of *Equisetum* have a high cation exchange capacity that survives cellulolytic digestion, while the high value for white clover is largely due to pectins that are fermented away. Nevertheless, cation exchange remains substantial in the indigestible fraction. In most common forages lignin and other polyphenolics are the most important sources for cation exchange (Table 10.9).

The affinity for cation exchange is greatest for heavy metals and least for alkali metals. Ions of higher affinity tend to displace those of lower affinity. The affinity tends to increase with valence charge and atomic weight. The factors influencing affinity are essentially those controlling adsorption of ions to soil colloids (McConnell et al., 1974). Of the macro elements, calcium, magnesium, and to a lesser extent potassium constitute the major ions carried on the exchange.

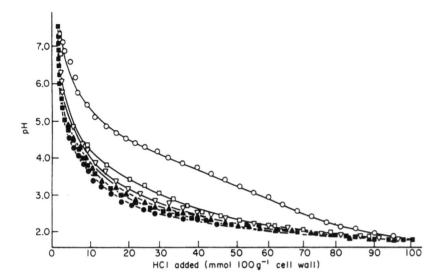

Figure 10.11. Titratable acidity of plant cell walls from pH 7 to pH 2 with 0.1N HCl (from McBurney et al., 1983). Note the greater buffering capacity of alfalfa. Symbols: ●, timothy hay; ■, oats; ▲, maize silage; □, wheat middlings; ○, alfalfa hay.

The ion exchange in pure nonioinic polysaccharides (e.g., cellulose and starch) is essentially zero, unlike that in intact cell walls, which varies widely. The components that contribute to the exchange are probably lignin, pectin, and to a lesser extent hemicellulose, which contain the requisite functional groups.

11 Carbohydrates

Carbohydrates are the main repository of photosynthetic energy in plants. They constitute roughly 50–80% of the dry matter of forages and cereals. The nutritive characteristics of carbohydrates depend on their sugar components and linkages with polyphenolic lignin and other physicochemical factors. Generally speaking, plants have a much greater variety of sugars and linkages than animal tissues have. Plant carbohydrates contain sugars and linkages not found in animal systems, and the animal digestive systems lack the respective carbohydrases required for digesting some of them. Nutritive availability depends on the animal's ability to cleave glycosidic bonds in plant carbohydrates and between the carbohydrates and other substances. The nutritional chemistry of carbohydrates, then, is largely a description of the degradation of structural and nonstructural carbohydrates and the factors influencing their availability to animal and microbial digestion.

11.1 Sugars and Linkages

Most sugars in plants are combined by glycosidic linkages as disaccharides, oligosaccharides, and polysaccharides and carry noncarbohydrate moieties. A *glycoside* is any compound having a hemiacetal linkage between two sugars or between a sugar and a noncarbohydrate component. The noncarbohydrate moieties are termed *aglycones* and may include phenols, lipids, alkaloids, nucleic acids, and peptides. The lower-molecular-weight glycosides formed from sugars and phenols known as terpenoids or alkaloids are of interest in nutrition not because of their carbohydrate content but rather for their potential toxicity. These aspects are discussed in Chapter 13.

The most stable forms of sugars are cyclic hemiacetals either as 5-membered (furanose) or 6-membered (pyranose) rings (Figure 11.1). Sugar residues that are glycosidically linked are exclusively cyclic. Small amounts of the aldehydic (open chain) form exist in equilibrium in solution, as evidenced by the ability to reduce mild oxidizing agents like copper or silver.

Fructose, arabinose, and ribose invariably occur as furanosides in glycosidic linkage. Five-carbon sugars are not all furanosidic, nor are all 6-carbon sugars pyranosidic. Note that fructose, a 6-carbon sugar, is furanosidic, while the 5-carbon xylose is pyranosidic. The ring form is favored by the maximum equatorial positioning that favors stability (Figure 11.2). The furanosidic linkage is considerably weaker than the pyranose one, with the result that furanosides are hydrolyzed by relatively weak acid and represent labile sites in the polysaccharides that contain them.

The formation of a ring through the first carbon (the second in the case of fructose) forms a new asymmetric center, the resulting hydroxyl being labeled α or β. The conformation of a pyranosidic ring is not planar but in a "chair" form (Figure 11.2). Substituent groups that project into the plane of the ring are termed *equatorial,* and those that project out of the plane are termed *axial.* The steric effects of axial groups tend to reduce molecular stability of their form such that the equatorial conformation is favored. All substituent groups in β-glucose are equatorial, which is also true for β-xylose, but all other sugars and all α forms contain one or more axial groups that project out of the plane of the ring. That β-glucose is more stable than α-glucose is shown by the mutatory equilibrium of glucose in water solution, which favors the β over the α form by a ratio of 2 to 1. The pyranose forms of sugars such as fructose result in 2 or more axial groups lending instability. All natural occurrences of fructose in its respective oligomers and polymers are exclusively furanosidic. The 5-membered ring is more planar than the pyranosidic one and is known as the "envelope" form because the ring oxygen lies slightly below the plane of the four carbons. Sugars glycosidically linked in furanosidic configuration are more labile to acid and other destructive effects, including that of heat in the Maillard reaction.

Greater molecular stability implies a greater energy cost (activation) for cleavage or degradation and is a negative factor relative to nutritive availability. The relevance of this concept may be seen in the contrasting properties of starch (α 1–4 glucan) and cellulose (β 1–

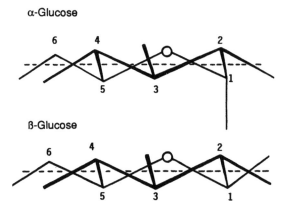

α-Glucose

β-Glucose

Figure 11.2. The spatial configuration of α- and β-glucose showing the "chair" form, with the equatorial plane approximately 90° to the page. Compare with Figure 11.3, in which the angle to the page is 60°, the conventional presentation. As in Figure 11.3, the oxygen and carbons 1, 5, and 6 are at the rear. The equatorial plane is established by carbons 2, 3, and 5 and the ring oxygen. Dashed line shows the axis of the molecule as it occurs in cellobiose and cellulose (Figure 11.3). Three carbons (1, 3, 5) lie below the axis, and three carbons (2, 4, 6) and the ring oxygen lie above it. The hydroxyls all point into the plane of the molecule in the case of β-glucose (lower figure); however, the hydroxyl on carbon 1 of α-glucose points down perpendicular to the plane and is termed *axial*. Equilibrium in solution favors the β form (66%). Generally, equatorial positioning favors molecule stability. Compare with linear and Haworth formulas of glucose in Figure 11.1.

4 glucan), which are essentially isomers. Starch occurs in many living systems as a reserve carbohydrate, whereas cellulose occurs only as a structural element that is irretrievable by the plant as a source of energy. The refractoriness of cellulose is not entirely explainable by the β 1–4 linkage, which in itself is somewhat more stable than α 1–4, but depends to a major degree on the intermolecular association and conformation of the macromolecule.

The strength of intermolecular associations such as hydrogen bonding is inversely related to the intermolecular distance; molecular shapes closely packed in a crystal lattice are more stable than shapes that do not allow close packing. Biological regulation of molecular shape is extremely important in all classes of substances. The branching in polysaccharides tends to decrease their crystallinity and increase solubility because it is difficult to closely pack branched molecules. The extended straight conformation of cellulose contrasted with the coiled form of amylose is illustrative. Solubility generally decreases with increasing molecu-

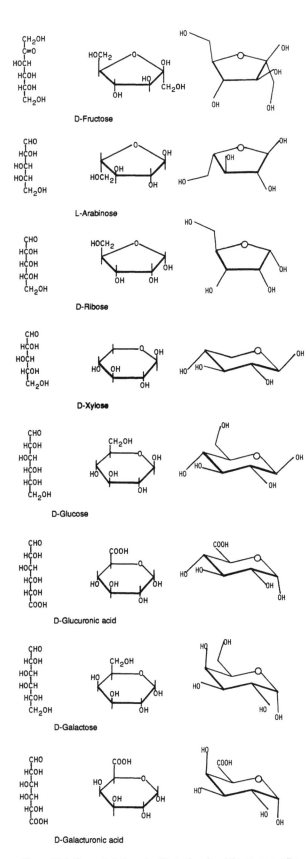

D-Fructose

L-Arabinose

D-Ribose

D-Xylose

D-Glucose

D-Glucuronic acid

D-Galactose

D-Galacturonic acid

Figure 11.1. Open-chain formulas (first column) and the corresponding ring forms of the more common sugars in forages. Only glucose and

fructose are encountered in significant amounts in the free form. Glucuronic and galacturonic acids are important in hemicellulose and pectins, respectively, and have the same configuration as their parent sugars. The projection of the ring forms according to the Haworth convention does not reveal actual ring geometry. Spatial presentation at the right shows the three-dimensional molecular geometry at a plane of 60° to the plane of the page. Heavy lines indicate the near side. The furanosidic rings (right side of figure) have an "envelope" form in which carbons 1, 2, 3, and 4 form a plane while the oxygen at the rear dips below the page.

Table 11.1.   Carbohydrates in feedstuffs and forages not recovered in neutral-detergent fiber

Carbohydrate	Source	Solubility	Plant reserve	Availability[a]	Fermentable
Sucrose	All plants	+ +	+	+	+ + +
Oligosaccharides	Legumes[b]	+ +	+	0	+ + +
Galactans	Legumes	+	+	0	+ +
Fructans	Grasses, composites	±	+	?	+ + +
β-glucans	Cereals	+ +	?	0	+ +
Pectins	Legumes, fruits	+	0	0	+ +
Starch	Diverse	0	+ +	±	±
Soluble hemi-cellulose	Grasses and treated forages	+	0	0	±

Source: Data from D. Smith, 1973.
[a]Availability to digestion by animal digestive enzymes.
[b]Legumes are a major source, but transglucosylation can produce oligosaccharides in molasses and other by-products.

lar size and is increased by hydrophilic or potentially ionizable groups. Suppression of ionization in pectic acid by the addition of mineral acid, for example, decreases its solubility. While branching of the molecular structure increases solubility, interchain cross-linking decreases it. The presence of esters either as methyl or acetyl groups decreases the hydrophilic character of the molecule and therefore its solubility. Acid or alkaline treatments readily remove ester groups and increase the solubility of the exposed acidic molecules.

11.1.1 Classification of Sugars

The biochemistry of organic substances in forage plants can be viewed from two aspects: the plant physiologist emphasizes biosynthesis, and the nutritionist emphasizes biodegradation. From the point of view of plant physiology, carbohydrates can be classified into three distinct categories: (a) simple sugars and their conjugates active in intermediary plant metabolism; (b) storage reserve compounds, for example, starch, sucrose, and fructans; and (c) structural polysaccharides, principally pectins, hemicelluloses, and cellulose, which are generally irretrievable (Tables 11.1 and 11.2).

Plants have enzymes both to synthesize and to degrade the compounds they use in metabolism or store. The degradation of the structural polysaccharides, however, is largely a function of microbial and fungal activity. The enzymes that synthesize the structural carbohydrates are prevented from degrading them by the complex intermolecular hydrogen bonding and cross-linking.

The above classification essentially divides plant carbohydrates into those available for metabolism and structural compounds. From a functional point of view, cellulose content is more constant among plants than any other carbohydrate. Fibrous cellulose is combined in varying proportions with lignin and noncellulosic carbohydrates. A minimum cellulose content is probably essential for proper construction of the cell walls. Thus, it is not surprising that the cellulose content of cabbages and oak trees does not differ that much.

On the other hand, the composition of storage polysaccharides differs greatly among plant species and in some cases in respect to plant parts. In temperate grasses starches occur only in the seed, and fructosans are the storage form in leaves and stems. Tropical grasses and all legumes store starch in leaves and stems. Fructans are also characteristic of the tubers of composites such as Jerusalem artichoke. In addition, the pectin content of legumes is much higher than that of grasses, whose level is so low that it is often ignored in analysis.

Carbohydrates serving as storage and energy reserves in plants are categorized as total nonstructural carbohydrate (D. Smith, 1973). This term includes insoluble starch, galactans, water-soluble carbohydrates, sucrose, oligosaccharides, and fructans, and is an expression having more meaning for the plant than for the animal that has no digestive enzymes to degrade galactans and certain oligosaccharides. Table 11.3 gives a general summary of carbohydrates that animals can digest.

Animal digestive enzymes can hydrolyze α 1–4 and α 1–6 linkages in starch but not the α 1–4 linkages in

Table 11.2.   Storage carbohydrates in common foodplants

	Sucrose	Galactans	Starch	Fructosan
Cereals	+	0	+	0
Forage grasses				
Tropical	+	0	+	0
Temperate	+	0	0	+
Legumes				
Soybeans	+	+	(+)	0
Alfalfa	+	(+)	+	0
Composites[a]	+	0	?	+

Source: Data from D. Smith, 1973.
Note: + = present; 0 = not present. Parentheses indicate only low amount present.
[a]Including safflower, sunflower, and Jerusalem artichokes.

Table 11.3. Summary of common dietary carbohydrates digested by animal enzymes

	Simple sugar components	Digestion	Digestibility	Digestive products	Linkage
Maltose	Glucose	Maltase[a]	Complete	Glucose	α 1–4
Sucrose	Glucose, fructose	Sucrase[a]	Complete	Glucose, fructose	α 1–2 β
Lactose	Glucose, galactose	Lactase[b]	Complete[c]	Glucose, galactose	β 1–4
Starch	Glucose	Amylases[a]	High	Glucose	α 1–4, α 1–6
Fructans (grass)	Fructose	Gastric acid	High	Fructose	β 2–6
Galactans	Galactose	Fermentative	High	VFAs and bacteria	α 1–6
Cereal gums	Glucose	Fermentative	?	?	β 1–3
Pectin	Arabinose, galactose	Fermentative	High	VFAs and bacteria	α 1–4
Cellulose	Glucose	Fermentative	Variable	VFAs and bacteria	β 1–4
Hemicellulose	Arabinose, xylose, mannose, galactose, glucuronic acids	Fermentative	Variable	VFAs and bacteria	Mixed
Mannan	Mannose	Fermentative		VFAs and bacteria	β 1–4

Source: Kronfeld and Van Soest, 1976.

[a]In some species (e.g., ruminants) fermentative digestion of these soluble carbohydrates yields acetic, propionic, and butyric acids.

[b]Lactase activity diminishes after weaning in some species, and digestion of lactose may then depend on fermentation.

[c]Volatile fatty acids include acetic, propionic, and butyric acids; smaller amounts of lactic and other acids may be produced.

pectins and galactans. The glucose-galactose linkage in lactose is β 1–4, which is hydrolyzable by lactase. The fourth hydroxyl on galactose is axial, however, leading to the different molecular configuration of lactose compared with cellobiose.

11.1.2 Solubility, Hydrolysis, and Molecular Conformation

The ease with which acids and probably also enzymes hydrolyze polysaccharides is influenced by several factors. The nature of the glycosidic linkage, intermolecular forces and linkage between chains, molecular size, and the affinity for the medium are important in this regard. Polysaccharides' susceptibility to acid hydrolysis is described in Table 11.4. Polymers involving chains of furanosidic sugar units are so susceptible to hydrolysis by very weak acid that polysaccharides containing susceptible groups are difficult to isolate without altering and to some degree degrading them. In these polymers it is possible that nonenzymatic cleavage occurs during gastric digestion. The pyranosidic polymers are much more resistant to acid hydrolysis. Starch polymers and xylan are comparable in their susceptibility to hydrolysis, but cellulose is more resistant, and polyuronides are still more so. Only the very strongest mineral acid dissolves cellulose. Hydrolysis of the dissolved cellulodextrins is still slower than

Table 11.4. Solubility and hydrolysis of polysaccharides

Polymer[a]	Ring form	Acid normality (HCl)	Time and temperature	Reaction	Products
Arabinan	Furan	0.1	3 h, 24°C	Partly hydrolyzed	Arabinose
Fructan	Furan	0.1	3 h, 25°C	Partly hydrolyzed	—
Fructan	Furan	0.01	5 h, 70°C	Hydrolyzed	Ketoses and homologues
Xylan	Pyran	0.1	3 h, 25°C	Partly soluble	Xylose, xylobiose and xylotriose
		1.0	3 h, 25°C	Soluble	
		0.33	8 h, 100°C	Hydrolyzed	
Amylose	Pyran	0.1	1 h, 100°C	Soluble	
		1.0	1 h, 100°C	Soluble	
		1.0	1 h, 100°C	Partly hydrolyzed	Glucose
		24.0	1 h, 100°C	Hydrolyzed	Dextrins
Cellulose	Pyran	1.0	3 h, 100°C	Insoluble	Glucose
		24.0	3 h, 24°C	Partly hydrolyzed	Cellodextrins
Galactouronan	Pyran	0.1	3 h, 25°C	Soluble	
		1.0	3 h, 25°C	Soluble	Poorly hydrolyzed
		24.0	3 h, 25°C	Partly hydrolyzed	

Source: Data from R. W. Bailey, 1964.

[a]See Tables 11.1 and 11.2 for the occurrences of these polysaccharides in plants and cereal foods.

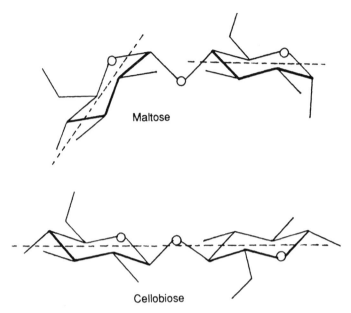

Maltose

Cellobiose

Figure 11.3. The conformations of maltose and cellobiose. Planes of the rings are turned 60° into the page. Heavy lines indicate near side. The oxygen bonds are at a 108° angle to one another. This leads to the feature in maltose (and in starch) that the substituent glycosidic group cannot lie in the same plane as the first sugar unit. Dashed lines indicate axes of the respective rings. Even though the second ring can rotate 360°, there is no position that will allow it to achieve a "straight" extended conformation. In cellobiose, the β linkage (equatorial) allows the molecule to assume a position in which the rings lie on the same axis, thus forming the extended conformation observed in cellulose.

that of comparable oligomers of starch and xylan (Southgate, 1976a). Since all three polymers are 1–4 linked, the resistance of cellulose must be attributed in part to the conformation allowed by the β 1–4 linkage.

The difference between starch and cellulose may be regarded as the consequence of the glycosidic linkages. In maltose (Figure 11.3) the angle of the anomeric bond prevents the second pyranosidic group from lying in the same plane as the first, with the result that the molecule cannot conform to a linear axis, even though the substituent group can rotate 360°. This feature leads to coiling in the amylose chain (Figure 11.4). In cellulose the alternate glucosidic groups are rotated 180° with a *trans* configuration relative to one another (Figure 11.5), and the molecule can assume a straight extended conformation in which the respective rings lie in the same plane.

Hydrogen bonding between the ring oxygen and the third hydroxyl of the adjacent pyranose ring renders the linear ribbon form of cellulose stable. Cellulose is soluble in cupriethylenediamine as a complex (pH > 13) and with hydrolysis in 24 N H_2SO_4. Acetylation promotes solubility in organic solvents. Introduction of a methyl or carboxymethyl group on carbon 6 or randomly on other positions promotes water solubility.

The stereo conformations of the respective polymers (Figures 11.4 and 11.5) show the difference between

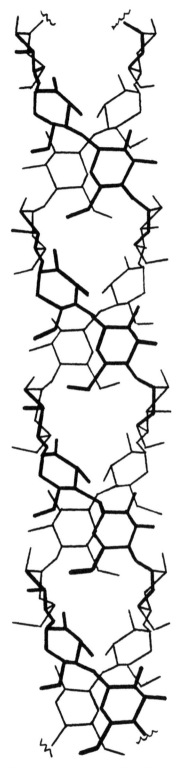

Figure 11.4. The probable structure (double helix) of native amylose (from French, 1979; see also French, 1973): spirals normally contain six sugar units per turn; however, expansion can occur with swelling to seven or eight units per turn. The coil is arranged such that all hydroxyls face outward, leaving the inner hollow relatively hydrophobic. Complexes with fatty acids or iodine are formed when the foreign unit resides within the hollow of the coil. The tightness of the coil also varies. More stretched and disordered forms probably occur in solution.

A

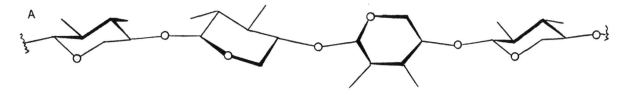

B

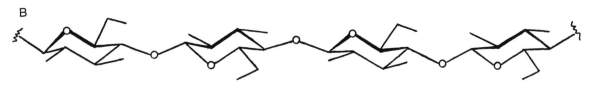

Figure 11.5. Xylan (A) and cellulose (B) structures. The cellulose chain is a ribbon in which glucose rings are *trans* to one another in a two-unit repeating segment. The hydroxyl on carbon 6 is important in the cross-chain hydrogen bonding that helps stabilize this conformation. While xylan is also linear, the chain twists 60° counterclockwise so that a 180° twist is accomplished in three units. This causes the fourth ring to be in the same configuration as the first in the figure. The repeating unit is therefore three units of xylose (Nieduszynski and Marchessault, 1971; Southgate, 1976a). Compounds attached to xylan, including glucuronic acid or its 4-methyl ether, are joined α 1–2 or 1–3 (see Figure 12.5). Uronic acid and arabinose substitutions may not occur in the same xylan chain (R. W. Bailey, 1973).

α- and β-linked polymers. Chains in which the 6 carbon of glucose is in a *cis* arrangement result from α linkage, while β linkage causes the *trans* configuration of cellulose to be the most stable form. Xylan, the β 1–4 xylopyranoside, is linear like cellulose; however, the chain twists counterclockwise on a threefold screw axis (Figure 11.5). The resistance of cellulose to solution and hydrolysis may be involved with the arrangement of the 6 carbon and the cross-chain hydrogen bonding through the alcohol group on the 6 carbon. Cellulose chains are relatively straight and highly polymerized. Xylan is just as soluble and hydrolyzable as starch, although the native form is insoluble because of cobonding with lignin. Xylans containing branching linkages with arabinose and glucuronic acid are more soluble than those that do not.

In pyranosidic polymers with 1–4 ring linkage, branching may occur at 1–2, 1–3, or 1–6 sites, decreasing the ease of acid hydrolysis in that order. The 1–3 linkage is unstable to alkali and undergoes a cleavage (β elimination), leaving an unsaturated bond in the 2–3 position. Variety in branching points is characteristic of hemicelluloses.

The solubility of polymers is influenced by a variety of factors. Generally, ionized forms such as pectic acid are soluble unless they form insoluble salts, as in calcium pectate. The competitive forces promoting hydrogen bonding between chains and the molecules of the water medium are important in nonionizable polymers. Long, linear polymers with strong intermolecular association are less soluble; branching increases solubility. Consequently, the relatively branched amylopectins are more soluble in hot water than unbranched amyloses. The binding forces between molecules are strengthened by an increased degree of polymerization.

11.2 Water-Soluble Carbohydrates

The water-soluble carbohydrate in forage represents the more rapidly digestible part of the nonstructural or storage carbohydrate of the plant. The two terms are not synonymous because *storage* includes starches that are generally insoluble in cold water and cannot be removed entirely with hot water. The term *water-soluble carbohydrate* is generally intended to mean compounds soluble in cold water or in gastrointestinal contents and includes monosaccharides, disaccharides, oligosaccharides, and some polysaccharides (Table 11.5).

Many free sugars have been detected in forages in small amounts, but most occur in sufficiently low concentrations to be unimportant in nutrition. Sucrose, the main sugar in the sap of plants, serves as the primary vehicle for energy transport and, in some plants, energy storage. Many plants convert sucrose into polymeric forms for storage. Temperate grasses store fructans in leaves and stems and starch in the seed. In

Table 11.5. Composition and linkage of some oligosaccharides and glycosides

	Sugar components	Linkage
Sucrose	Glucose, fructose	α 1–2 β
Melibiose	Galactose, glucose	α 1–6
Raffinose	Galactose, glucose, fructose	α 1–6, 1–2 β
Stachyose	Galactose, galactose, glucose, fructose	α 1–6, α 1–6, α 1–2 β
Kestose	Glucose, fructose, fructose	α 1–2 β, 1–2 β
Isokestose	Glucose, fructose, fructose	α 1–2 β, β 2–6
Neokestose	Fructose, glucose, fructose	β 2–6, α 1–2 β
RNA	Ribose	α 1,6
DNA	2-Deoxyribose	α 1,6
Rutin	Rhamnose, glucose	α 1–6, β 1–(quercitin)
Amygdalin	Glucose, glucose	β 1–6, β 1–(mandelic nitrile)

Source: Data from D. Smith, 1973.

contrast, tropical grasses and most broad-leaved plants store only starch.

11.3 Fructans

There are two types of fructans: the grass-type levan with β 2–6 linkage and the inulin type characteristic of composites, with β 2–1 linkage (Figure 11.6). The grass levans are quite soluble; the inulins are less so. A considerable part of the soluble polysaccharide in onions and garlic consists of fructans, but some linkages of the inulin type do occur in grasses as a system of branching in fructosan chains. All three trisaccharides —kestose, isokestose, and neokestose—have been found in grasses (Table 11.5). Fructans in grasses are probably predominantly linear molecules that can vary extensively in degree of polymerization. All are water-soluble, though solubility in alcohol-water mixtures decreases as chain length increases. Fructan content is enhanced by low environmental temperature and may increase to as much as 30% of dry matter in cool-season perennial ryegrass as grown in Britain.

The sugar content of forages, which is important to their palatability and suitability as silage, is markedly affected by environmental growth conditions. High light intensity and photosynthetic rate increase sugar content, and high temperatures promote increased metabolic rate and lower sugar content. Consequently, marked diurnal variations in sugar content occur in living plants. Respiration in cut and drying forage may reduce the sugar content substantially. Determining the physiological levels of sugar in plants requires careful sampling and handling. Plants may be collected and placed in liquid nitrogen or directly extracted with 95% ethanol, or rapid heating to 100°C for a brief period followed by 60°C may be satisfactory. Prolonged heating at high temperatures promotes loss of sugar through the Maillard reaction.

A variety of sugars are bound in glycosides and oligosaccharides (Table 11.5). These include ribose in RNA and deoxyribose in DNA. Ribose and deoxyribose are totally hydrolyzable and digestible in the rumen, although the aglycone portions, if phenolic, may resist degradation. Quantitatively, oligosaccharides are not important as energy sources, and sugars of this group may be responsible for flatus or diarrhea in nonruminants. They contain some linkages not hydrolyzable by animal enzymes, so they pass to the lower tract for rapid fermentation. Young calves do not produce sucrase, and the ingestion of sucrose will generate scouring (diarrhea). Lactose intolerance in humans may be a similar problem.

11.4 Starch

Starch is the most important of the storage carbohydrates in plants. It is sometimes classified with the soluble carbohydrates because of its gelatinization and partial solubility in hot water. More properly it can be regarded as a reserve carbohydrate. Some forms of starches (i.e., uncooked potatoes, stale bread, etc.) can be extraordinarily insoluble and resistant. Sol-

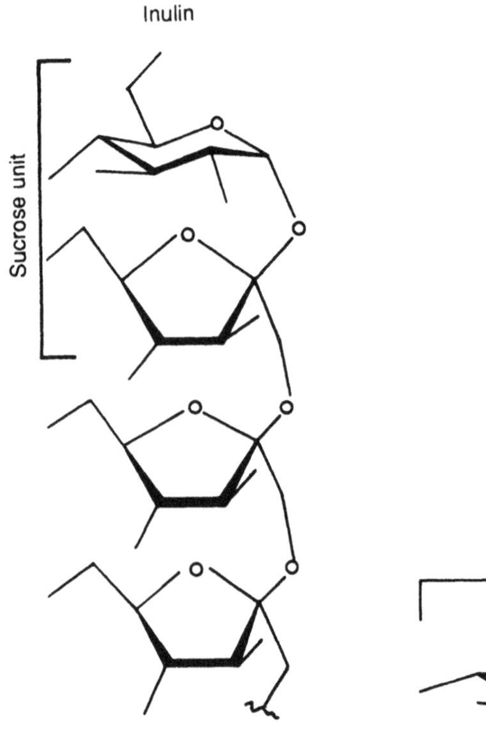

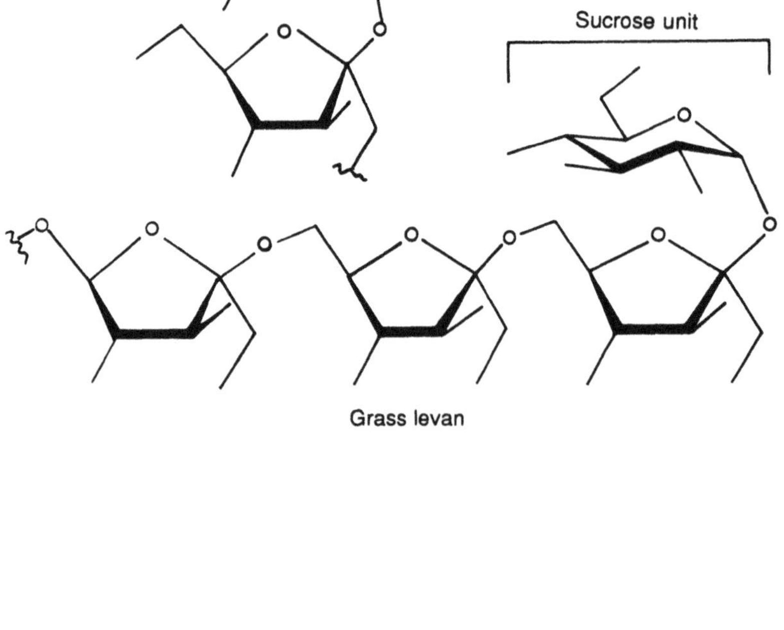

Figure 11.6. Structural formulas of fructans showing the two main types: grass levan with β 2–6 linkages and inulin with β 2–1 linkages (see D. Smith, 1973). Both polymers begin with a sucrose unit. Inulin is usually larger and less soluble than grass levan.

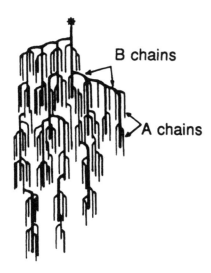

B chains

A chains

Figure 11.7. Conjectural arrangement of the amylopectin fraction of waxy corn. Reducing end (asterisk) of the chain is at top. Each chain consists of helical coils (see Figure 11.4). The linear A chains are associated with more crystalline regions of the macromolecule; B chains (heavy lines) are about 45 units long and are associated with branching and with less dense and less crystalline regions. The pattern (descending) is alternating bands of less crystalline (branching) regions and more crystalline (linear) regions. The banding pattern has a frequency of about 60 angstroms (15–20 sugar units) (French, 1973; Robin et al., 1975).

Table 11.6. Proposed nutritional classification of starch

Starch	Examples	Digestion in intestine
Readily available		
Freshly gelatinized	Fresh-cooked food	Complete
Amorphous crystalline	Most raw cereals	Slow but complete
Partially resistant		
Physically inaccessible	Unmilled or coarsely milled grains	Partial
Resistant granules	Raw potato, banana	Partial
Retrograded amylopectin	Cooled, cooked potato	Partial
Resistant		
Retrograded amylose	White bread, corn flakes	Relatively indigestible[a]

Source: Englyst and Cummings, 1987.
[a]But slowly fermentable in colon or rumen fermentation.

ubility but not hydrolysis is assisted by weak acids and chaotropic reagents that undo hydrogen bonding. Examples of chaotropic reagents are 8M urea, guanidine thiocyanate, strong solutions of calcium chloride, and some other salts.

Two types of polymers occur in starch: linear amylose, consisting of α 1–4 glucopyranosidic chains, and branched amylopectin. Branching occurs at the sixth carbon (α linkage) to form side chains of α 1–4 units.

The proportions of amylose and amylopectin vary in cereals and are to a considerable extent genetically controlled (Theurer, 1986). The amylose content of maize increases as the plant matures. The quality of the starch granule is affected by the type of starch it contains. Floury endosperm in maize characteristically has a high proportion of amylose, and flinty or horny characteristics are associated with amylopectin. The chains of amylopectins are arranged as in Figure 11.7.

All the chains indicated in Figure 11.7 represent double helical arrangements like those in Figure 11.4, in which the hydroxyls face outward and the hydrophobic hydrogens face inward. This arrangement increases reactivity with fats to form inclusion complexes in cooking and with iodine to form blue inclusion complexes. When starch is heated in 60°C water it tends to unwind, and a viscous solution results. The uncoiling may be only partial, and cooling may result in reassociation; branched systems break down more easily than unbranched ones. The reassocia-

tion is termed *retrogradation* and can lead to forms of starch resistant to amylases (Table 11.6).

Resistant starches escape amylolytic digestion to ferment slowly in the lower tract of nonruminants (Gee et al., 1991). Because of the slow rates starch fermentation is similar to that of cellulose. Natural variation in degradability of starches also occurs; for example, sorghum is less easily degraded than corn, wheat, and barley (Hibberd et al., 1982). Retrogradation can occur in any of these cereals after cooking, baking, or steaming, however, and it is responsible for staleness in bread (Berry, 1986); it is probably aided by drying or cooling if the starch is in a hydrated state.

The two enzymes that hydrolyze starch are α and β amylase. Alpha amylase cleaves starch chains randomly and will degrade both amylose and amylopectin. Beta amylase cleaves units from the ends of chains; it degrades amylose, but its activity is limited to the peripheral parts of amylopectin. Starch isolated mechanically from many plants cannot be degraded by amylase until the starch grains have been ruptured.

Crystallinity occurs in both amylose and amylopectin; however, hydrogen bonding is stronger in amylose chains with a high degree of polymerization (1000–2000 units). Crystallinity is noted by birefringence of starch granules in polarized light with the appearance in the granules of dark Maltese crosses. Amorphous areas are isotropic and transmit polarized light in all directions. Heating starch granules with water disrupts hydrogen bonding and results in loss of crystallinity and birefringence. The temperature at which this occurs is called the *gelatinization temperature*. It varies with the source of starch and the proportion of amylose (Table 11.7). Amylose is responsible for greater crystal strength and requires a higher temperature for gelatinization of the granules compared with amylopectin. There are crystalline regions in amylopectin; however, these are associated with less branched regions of the molecular structure (Figure 11.7). Dissolved starch polymers tend to reassociate on cooling. Amylose can

Table 11.7. Amylose contents and gelatinization temperature range of whole granular cereal starches

Cereal	Amylose in starch (%)	Gelatinization range (°C)
Barley	22	59–64
Oats	27	
Wheat	26	65–67
Sorghum	25	67–77
Maize, dent[a]	28	62–72
Maize, flint[a]	1	66–69
Maize gene combinations		
ae (amylomaize)	61	92[b]
du, wx	0	74[b]
wx	71	71[b]

Source: Armstrong, 1972.

[a]Genetic forms of maize are available that contain various ratios of amylose to amylopectin.

[b]These birefringence end point temperatures are the higher of the two temperatures describing the gelatinization temperature.

Table 11.8. Animal responses to treated cereal grains

Cereal	Treatment	Digestibility change	Intake	Gain
Barley	Steam flaked	↓	↑	↑
	Pressure steam	↑ ?	↓	↔
Corn	Steam, cracked	—	—	↑
	Cold grinding	↑	—	—
	Flaked maize	↑	—	—
	Heat rolled	↔	—	—
	Expanded gelatinized	—	—	↔
	Complete gelatinized	? (cattle)	↓	↓
	Complete gelatinized	↓ (sheep)	—	↔
Sorghum	Cold ground	↑ (cattle)	—	—
	Cold ground	↔ (sheep)	—	—
	Steam flaked	↑ (cattle)	—	↑
	Steam flaked	↑ (sheep)	—	—
	Steam processed	↑	↔	↑

Source: Summarized from the literature by Armstrong, 1972.

recrystallize (retrogradation) and become even more refractory to digestion. Amylopectin tends to gel, with limited reassociation of molecular chains. The temperatures at which reassociation occurs are specific with respect to different starches.

The way starch reacts to moist heat is partly responsible for the improved utilization and feeding efficiency resulting from steaming, flaking, micronizing, and pelleting, although denaturation of the protein and protein-carbohydrate condensations (Maillard reaction) are involved also. The Maillard reaction decreases availability. The latter associations along with retrogradation may lead to lower nutritive availabilities, but this depends on the associated molecular system, which can impede reassociation. Gelatinization requires water, as does the Maillard reaction, although the latter also requires greater heat input. Excess heat will caramelize carbohydrates, a process that involves chain fracture, migration and racemization of linkages, and dehydration to form unsaturated carbonyls, which interact with proteins (Section 11.7).

Thus the effects of hydrothermal treatment of starch (see below) can be either positive or negative. Since the negative effects require rather large inputs of heat, a diminishing returns model of heat input on feed availability can be postulated. Practical feeding studies of processed grains sometimes offer conflicting results (Table 11.8), which may reflect too little or too much heat and failure to find the optimum, or, alternatively, retrogradation on cooling and drying.

Methods for processing cereals can be divided into cold, dry heat, and hydrothermal. Cold processing—including grinding, cracking, rolling and crimping, extrusion, and pelleting—is designed to break the pericarp and expose the endosperm to digestive attack. Some heat may be produced through friction in these processes, particularly in the cases of extrusion and pelleting, which are sometimes considered dry heat

processes. Another cold process is the ensiling of corn silage or high-moisture grain. Fermentation and acidity develop, which tend to preserve the residual carbohydrate. Organic acids such as propionic acid and acetic acid may be used as preservatives (Chapter 14).

Dry heat processes include popping and micronizing. Expansion by popping includes exposure to 230–240°C for 30 sec. Grains expand on the order of 1.5–2 times their original volume with rupture of the pericarp. Micronizing consists of exposure to infrared radiation at 150–180°C for 30–60 sec followed by flaking through rollers.

The hydrothermal method uses moist heat applied for 8–25 min with or without pressure, and drying after flaking or rolling. This process is characteristic of American flaked corn or sorghum. In Britain a more extensive process involves cracking, dampening with water, leaving for 24 h, redampening, and leaving for another 24 h. Grain is then steamed under pressure for 10–15 min and flaked and dried. Most starch gelatinization occurs after the steamed grain is flaked. Steaming causes 2–16% gelatinization in maize and sorghum, and flaking increases gelatinization to 40–48%.

Animals do not respond equally to hydrothermal processing effects, which are probably related to the character of individual starches. In general, increased animal efficiency is obtained with hydrothermally treated sorghum, maize, and barley, decreasing in that order. The responses to wheat treated in this way are more marginal and can be negative.

The improved utilization of cereal starch involves several factors. Crushing the seed coat improves digestibility in cattle but not in sheep because sheep chew the seeds into smaller particles. The most marked improvement is associated with the gelatinization and rupture of the starch granules in a process requiring flaking in addition to steaming. Presumably reassociation via retrogradation is limited by the branched structure of the polysaccharide. Increased availability increases the rate of digestion, which may increase the efficiency of use of nonprotein nitrogen. In addition, gelatinization promotes starch digestion in the rumen and increases liability to lactic acid fermentation and acidosis. Escape of fermentable starch to the abomasum may be a factor in displaced abomasa in dairy cattle. Amylolytic digestion of starch in the duodenum and small intestine may be very efficient provided there is amylolytic capacity; however, fermentation in the lower tract may mean loss of microbial nitrogen in the feces. If the ruminant has relatively weak amylolytic capacity, the fate of bypassed or escaped starch may be mainly inefficient lower tract fermentation. In general, the rumen escape of starch results in lower efficiency in contrast to the escape of protein. Indirectly, the gelatinization of starch and the heat denaturation of protein can diminish rumen escape of starch and increase the passage of unfermented protein.

Inefficient starch utilization in ruminants may be related to the pH effects of lactic acid fermentation; for example, it appears that the pH optimum for bovine amylase is critical (W. E. Wheeler, 1977). Starch digestion in cattle has been improved by feeding them buffers. The negative effect of dietary starch is related to its more rapid fermentation and the development of large amounts of lactic acid as a primary product. The pH of the rumen and the lower tract declines because of inadequate buffering capacity relative to high concentrations of a nutrient to which the system is constitutionally unadapted. These effects of starch are related to gastrointestinal disturbances.

11.5 Plant Cell Wall

The covalent linkages among the respective cell wall carbohydrates fail to account for the overall properties of the plant cell wall. Rather, the large-scale organization of the cell wall matrix with its cross-linking appears to be the overriding factor. Acetyl and methyl groups promote hydrophobicity.

An important concept of cell wall structure considers the cell wall as a giant macromolecule with covalent bonds running through β-glucan, xylan, and araban to peptides (extensin), with cross-linking involving extensin, and dimers of ferulic and *p*-coumaric acids as cross-linking agents (Fry and Miller, 1989; Iiyama et al., 1993). Presumably lignin has some role in this (Chapter 12).

These models emphasize the laying down of the cell wall in the growing plant. It is probable, however, that the mechanisms of degradation and digestion differ from those of development and growth. Chesson (1993) presented a model of the degradation process that is related to the structural model of Iiyama et al. (1993). Cell wall degradation likely differs from synthesis because the digestive enzymes must deal with the cross-linking and lignification intended to protect the plant.

Circumstantial evidence indicates that hemicellulosic polysaccharides are held in place by cross-links to lignin, because on delignification much of the hemicellulose becomes soluble. Also, the treatment of cell wall with acid pepsin dissolves both protein and some hemicellulose (Ely et al., 1956). Lignified forage cell walls also contain unavailable arabinoxylan as well as peptides, much of which is probably recovered as nitrogen insoluble in acid detergent. Harkin (1973) suggested that the distance from the lignin bond was a factor in the availability of hemicellulose chains, but it is not known why the lignin bonding protects other organic components of the cell wall from degradation.

11.5.1 Structural Polysaccharides

The polysaccharides associated with the plant cell wall can be divided into two classes based on biological associations and nutritional availability: those that lack covalent links to the lignified core and are more soluble and completely fermentable in the rumen, and those that have some covalent linkages to the lignified core and thus are incompletely or only partially digestible. Generally, the former are dissolved in neutral detergent, and the latter are recovered as neutral-detergent fiber. Delignification promotes solubility in the case of hemicelluloses, but delignified cellulose remains insoluble, although its fermentability is increased.

The overall classification of the cell wall polysaccharides is problematic in that modern biological understanding does not coincide with the conventional historical classification imposed by paper and cellulose chemists. Many fractionation systems for structural polysaccharides depend on rigid classifications based on β-glucosan (as cellulose) and pentose and uronic acid as hemicelluloses. No fractionation system is "clean" in this context. For example, all prepared cellulose from lignified tissues contains some pentose, and some glucose will be found in hemicellulosic preparations, although no formal structural understanding allows it to be a part of the hemicellulose system.

The fundamental problem confronting any nutritionally meaningful fractionation of cell wall polysaccharides was exposed by Blanche Gaillard (1962) in a classic paper that is often overlooked. Gaillard observed that total glucose, xylose, or arabinose is not meaningful nutritionally because each of these sugars is incorporated into structures of diverse availability (her data are summarized in Table 10.5). Pectin contains galacturonic acid with arabinan side chains which have virtually complete availability—which verifies the lack of lignification of these structures. The lignified fractions of arabinoxylans soluble in alkali have variably different digestibilities for the respective sugars; for example, arabinose is always more digestible than xylose within any fraction. These sugars are less digestible in the more lignified legumes compared with grass. The arabinose and xylose residual in the crude cellulose are less digestible than the glucose of the same structure, and less digestible than the arabinose and xylose that have been extracted by alkali. Thus Gaillard's data expose the problem that total arabinose, xylose calculated to xylan, or total glucose calculated to glucan are not nutritionally meaningful partitions.

These observations have stimulated some researchers to examine the physicochemical reasons for the respective diversity. Bittner and Street (1983) suggested that linearity of xylans may promote their inclusion into cellulose, but this does not explain its lower digestibility. After all, xylan is a more destructible molecule because it lacks the 6-carbinol that limits cross-chain hydrogen bonding. The hemicellulosic fractions are the only ones that have demonstrated linkages with lignin, yet the digestibilities of hemicellulose and cellulose are inversely associated with degree of lignification. Could the lower digestibility of xylose in the residual cellulose reflect interlinkages between lignin, pentose, and cellulose? Arabinose appears to be the sugar directly involved in the lignin linkage in grasses (Mueller-Harvey et al., 1986).

From this overview it seems that the physicochemical factors influencing nutritive availability of most plant polysaccharides are related to linkage, but association factors at the macromolecular level remain unelucidated. Merely analyzing for component sugars sheds little light on the problem.

11.5.2 Cellulose

Cellulose, the most abundant carbohydrate in the world, amounts to 20–40% of the dry matter of all higher plants. While it is a major (and variable) portion of the plant cell wall structure and the principal fibrous substance, its amount or concentration is not a good measure of fibrousness or total fiber (plant cell wall), although many nutritionists have used it for this purpose. Cellulose, as isolated from forages, usually contains, in addition to β 1–4 glucan, about 15% of the pentosans (mainly xylose and some arabinose) and most of the cutin and silica present in the entire plant tissue (I. M. Morrison, 1980). The nonglucan substances are commonly regarded as contaminants, but they are impossible to remove without destructive degradation of cellulose. Treating celluloses as pure β-glucans does not provide a rational means of comprehending their biological character because this does not provide any accounting of the widely differing characteristics of naturally occurring celluloses, most of which are combined with hemicellulose and lignin.

In nature, pure cellulose—in the sense of insoluble crystalline β-glucan—is a biological rarity confined to specialized hairs; as, for example, in cotton. All structural celluloses are combined to some degree with lignin, hemicellulose, cutin, and minerals in the plant cell wall structure. Total cell wall glucose obtained through hydrolysis of cell wall and followed by chromatography has been recommended as a method for determining cellulose; however, this assumes that all cell wall glucose is bound in cellulose. Direct determination of cellulose as glucose has merit; however, noncellulosic glucose would have to be distinguished as well. The simplicity of gravimetric methods for determining cellulose is partly responsible for their continued use, although gravimetric methods tell nothing about the isolated material. This situation may be tolerable in the case of well-understood fibers (i.e., those from common legumes and grass forages), but for unfamiliar cases it is unsatisfactory. This criticism may be applied to fiber determinations that include the use of NDF or ADF.

Isolated cellulose does not include all the β 1–4 glucan present in a plant's tissues. Water and alkali dissolve β-glucans, which are regarded as distinct from cellulose and are included in the hemicellulose fraction (R. W. Bailey, 1973). The view of cellulose as a refractory, insoluble carbohydrate has been dictated by the practical aspects of paper chemistry that have been applied to the analysis of forages. Cellulose, practically determined, represents the insoluble core left after harsh conditions for hydrolysis and oxidative delignification. It should be no surprise, then, that this residue has no biochemical or nutritional uniformity.

The nutritional availability of cellulose varies from total indigestibility to complete digestibility, depending largely on lignification (Table 11.9), although there are other inhibitors and limiting factors, including silicification, cutinization, and intrinsic properties of the cellulose itself. Generally, delignification increases digestibility.

Alternative views have been developed to account for the variable nutritive character of cellulose. One is that the nutritional value of cellulose is regulated by its ratio to lignin. This concept leads to the expectation that digestion rate and extent of digestion are related to lignin content. In fact, this appears to be the case, but rate of digestion may not be related to lignification at all since the available cellulose may have different physicochemical associations with lignin. Nonruminants digest more hemicellulose from the same forage, indicating that, on the whole, hemicellulose is more rapidly degradable than cellulose (see Figure 4.7).

The other alternative view is that there are two cellu-

11.9. Digestibility of cellulose from various materials

Material	Nonruminant digestibility (%)	Ruminant digestibility (%)	Lignin/cellulose ratio[a]
▮	20–30	40–60	0.18–0.30
▮erate grass	0–20	48–90	0.08–0.20
▮al grass	0–20	30–60	0.11–0.24
▮s	Negligible	40–60	0.10–0.26
▮an hulls	40	94	0–0.03
▮nseed hulls	Very low	50	0.55
▮ulls	0	0	0.45
▮on newsprint	0	23–27	0.34–0.43
▮pers	Low	20–99	0–0.50
▮	0	0–40	0.30–0.60
▮ables	40–80	90–100	0–0.05

▮urces: Van Soest 1973b, 1976a.
▮cid-detergent sulfuric acid lignin, incuding cutin fractions expressed as a ratio to
▮se. A high ratio is associated with low digestibility.

loses: one lignified and protected, the other unaffected by lignin. This concept explains the kinetic behavior of cellulose in rate-of-digestion studies. The existence of available and unavailable cellulose reinforces the view of nonuniformity of cellulose since unlignified fractions show much diversity in digestibility. The greatest contrast is between cotton and vegetable celluloses, both of which are unlignified and ultimately completely digestible but have very different rates of digestion. Alfalfa cellulose prepared by chlorite delignification and alkaline extraction of hemicellulose shows a slower digestion rate and greater digestibility than cellulose in intact alfalfa cell wall. Possibly the removal of lignin and hemicellulose allows cellulose chains to become more closely aligned and thus results in greater crystallinity. The fermentation of prepared isolated celluloses tends to have a long lag time followed by slower digestion than is normally seen in good forage even though the total digestibility of delignified tissues is higher (Van Soest, 1973b). This alteration is seen in many delignified celluloses and is an obstacle to their efficient use as ruminant feed.

Cellulose possesses variability in nutritive quality quite independent of lignification, and such behavior must be ascribed to intrinsic properties. It has been tempting to attribute nutritive differences in forages to crystallinity, although all evidence of cellulose crystallinity has been obtained solely with cotton and purified cellulose prepared from wood. Crystallinity is observed by the passage of x-rays at different angles through crystalline tissue causing diffraction and the production of regular patterns, which are absent in amorphous tissue and difficult to measure in heterogeneous cell wall residues. This does not mean that crystallinity is not a factor in intact forage cell walls; it merely has not been observed. In forage cell walls, cellulose is much diluted by lignin, hemicellulose, and other substances. Removal of these may alter the resid-

ual cellulosic matter, as suggested by cellulolytic rates of digestion.

Cellulose crystallinity is diminished by certain treatments such as alkali swelling and ball milling. The latter depolymerizes cellulose and is a physical method of removing cellulose from lignin. These treatments also render cellulose susceptible to enzymes that would not otherwise attack the substrate. R. W. Bailey (1973) distinguished cellulosic fibrous matter from matrix substances containing hemicellulose and lignin. Bailey characterized the cellulose as crystalline and the matrix substances (lignin and hemicellulose) as amorphous. If this is the case, crystallinity has little to do with digestibility, since some of the xylan and uronide fractions of hemicellulose are less digestible. Perhaps these fractions are also crystalline minor components (Bittner and Street, 1983); or, alternatively, perhaps they are closer to the linkage with lignin (Chapter 12). Conceivably, crystallinity limits rate but not extent of digestion.

11.5.3 Cellulases

Enzymes' ability to attack cellulose involves much more than their capacity to cleave β 1–4 linkages. A sharp distinction must be made between true cellulases and β 1–4 glucosidases, which are capable of hydrolyzing cellobiose and related oligocellulosic fragments but cannot attack native cellulose. This distinction has led to the postulation of enzymatic sequences in which hydrogen bonds are first cleaved with possibly some endocleavage (amorphogenesis), rendering chains susceptible to sequential hydrolysis yielding cellobiose and ultimately glucose (Coughlan, 1991). Most commercial cellulases possess some hemicellulolytic activity but are comparatively deficient, so the largest undigested fraction of cellulase-prepared residues is apt to contain enzymatically available hemicellulose. The problem may be well one of enzymatic purity; that is, too pure systems are enzymatically inefficient. Introduction of carboxymethyl groups promotes water solubility and increases susceptibility to enzymes that are unable to attack native cellulose. Enzymes that can attack this carboxymethylcellulose may be utterly unable to attack an insoluble substrate. Other agents that bind (perhaps indiscriminately) with hydroxyl groups in cellulose may protect cellulose fibers. Mordanting the fiber with silica and chromic salts can cause cellulose to become more resistant or wholly refractory to cellulases. Treating forage cell walls with cellulase reduces NDF contents but also "cracks" the remaining structure so that it is more rapidly fermentable in ensuing rumen fermentation.

Much of what is known about cellulases is derived from organisms that secrete extracellular enzymes, and

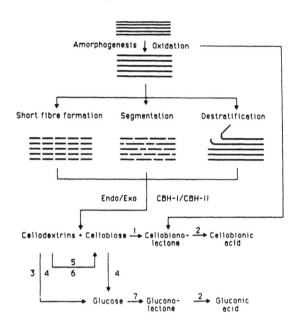

Figure 11.8. Proposed mechanism of cellulose degradation by fungi (from Coughlan and Ljungdahl, 1988; see also Coughlan, 1991). The numbered enzymes are (1) cellobiose oxidase/dehydrogenase, (2) lactonase, (3) exo-glucohydrolase, (4) β-glucosidase, (5) endo-glucanase, (6) exo-cellobiohydrolase, and (7) glucose oxidase. CBH-I = cellobiohydrolase I; CBH-II = cellobihydroolase II.

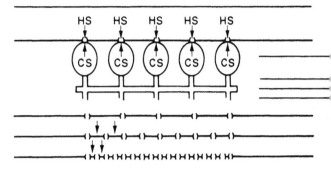

Figure 11.9. The *Clostridium thermocellum* cellulosome and its inherent structure-function relationships (from Mayer et al., 1987; see also Coughlan, 1991). CS = catalytic site; HS = hydrolysis site (i.e., the site of the cellulose chain at which a glycosidic bond is cleaved); LF = central string of unknown material; UF = ultrathin fibrils; C_n, C_4, C_2, C_1 = cellodextrins of various lengths (C_n being cellulose and C_1 being cellobiose). See text for details.

these generally prefer to grow noncompetitively in pure culture. Scientists' understanding of the enzymatic activities that degrade plant cell wall in the rumen has been retarded because many of the organisms in the rumen do not secrete enzymes extracellularly. This is no doubt an ecological adaptation to interspecies competition on the part of strains adapted to slow-digesting substrates. Active true cellulases do not exist in filtered rumen liquor. Cellulolytic rumen bacteria attach themselves to fibers and etch pits (Latham et al., 1978) so that cellulases are retained in the filtered fiber mat. Cellulases can be obtained by agitating rumen solids with phosphate buffer; the phosphate appears to enhance activity (Francis et al., 1978). In contrast, filtered rumen fluid can have considerable hemicellulase activity (Dekker, 1976). Thus, the limited information available indicates that rumen cellulolytic activity differs greatly from that in aerobic fungi.

Coughlan (1991) summarized the current models for cellulytic activity in fungi and bacteria (Figures 11.8 and 11.9). Initially there seems to be some amorphogenesis (breaking of hydrogen bonds) followed or in conjunction with interaction of endo-glucanase and cellobiohydrolases. The sequences in Figure 11.8 are not mutually exclusive relative to order of attack.

The cell wall–bound model developed for *Clostridium thermocellum* probably applies to rumen organisms such as *Fibrobacter* (Figure 11.9) as well (McGavin and Forsberg, 1989). The term *cellulosome* has been applied to the multicomponent cellulolytic

complex produced. Aggregates of these cellusomes mediate attachment of the cell to cellulose. Endoglucanase activity is associated with these aggregates arranged in rows, such that multicutting takes place with the release of oligosaccharides of about four cellobiose units. Further hydrolysis to cellobiose allows transfer into the cell. The cellulase complex of the anaerobic rumen fungus *Neocallimastix* may also be cell-bound under some growth conditions (T. M. Wood, 1989).

11.5.4 Hemicellulose

Hemicellulose is a heterogeneous collection of polysaccharides, and its collective composition varies greatly from one plant species to another. The common characteristic of the hemicellulose fraction is that it represents native insoluble polysaccharides that are soluble in acid or alkali and are associated with lignin. The current tendency to report the compositional values of component sugars as a simple inclusive fraction tells little of hemicellulose's internal complexity and nonuniformity (Wilkie, 1979). Digestibility of hemicellulose is directly related to that of cellulose and inversely related to lignification. Hemicellulose is more closely associated with lignin than any other polysaccharide (Sullivan, 1966). Evidence of direct bonding to phenolic constituents includes an ester linkage to arabinoxylan and possibly other glycosidic linkages. Hemicellulose and lignin together form the encrusting material of the secondary wall thickening. In forage plants hemicellulose is found mostly in lignified walls and is generally insoluble, but it does become water-soluble when delignified. Its isolation depends on solution or destruction of lignin with alkali or oxidizing agents. These treatments likely cause considerable alteration and degradation of the original molecular structure, which remains obscure.

After delignification, hemicelluloses are extracted from the holocellulose (the net cell wall carbohydrate) with hot water and alkali of varying strengths. The more soluble matter is richer in arabinans; some hexosans are dissolved along with xylan and a little araban with very strong (17–24%) alkali (R.W. Bailey, 1973). Acidification with acetic acid followed by refrigeration causes hemicellulose A to precipitate while hemicellulose B remains in solution. Subfractionation of hemicellulose A or B can be accomplished by precipitation with copper, quaternary ammonium salts, or iodine. The variation in hemicellulose composition among plant parts and species was summarized by R. W. Bailey (1973) and Åman (1993). Hemicellulose A is less bonded, often higher in xylan, probably more linear, and more common in stems than in leaves. Hemicellulose B contains branched fractions often high in arabinose or uronic acid, but not both in the same polymer chain. There is variation in the degree of branching, and some units have a very high substitution of arabinose units (Åman, 1993).

Hemicellulose is a mixture of polysaccharides, often with the common factor of β 1–4 linkage in the main xylan core polymer (Figure 11.5), although branching with a variety of other glycosidic linkages occurs. The hemicellulose in grass and legume leaves and stems seems to be largely arabinoxylan. Linkages between arabinose and xylose are 1–3, while uronic acid may be 1–2, 1–3, or 1–4 linked. While bonding to lignin occurs in both plant families, however, the nature of the linkage is very different. A suggested model of linkage in grasses is shown in Figure 12.5.

Some of the linkages in hemicellulose are susceptible to alkaline attack; these include ester linkages and 1–3 glycosides. Ester linkages are hydrolyzed through saponification, and 1–3 glycosides are cleaved through β elimination and the formation of an unsaturated bond (the latter are particularly vulnerable to even mild alkali). Both ester linkages and 1–3 glycoside linkages are seen in the branching of pectins and hemicelluloses. Consequently, the traditional method of alkaline extraction of hemicelluloses must produce extensive degradation.

The digestion of hemicellulose by ruminants and nonruminants is not in proportion to their digestion of cellulose. Nonruminants digest relatively more hemicellulose than cellulose, and ruminants digest about equal amounts of both carbohydrates. Nonruminants are better at utilizing legume compared with grass hemicellulose than ruminants. In ruminants most cellulose is digested in the rumen, but a substantial portion of hemicellulose escapes the rumen to be fermented in the lower tract. Perhaps xylan cannot be attacked until the arabinosyl side chains are removed; or perhaps its digestion depends on removal of some encrusting cellulose (Francis et al., 1978). Arabinofu-

ranosyl linkages should be sensitive to gastric acid, thus exposing xylan to further digestion in the lower tract. It is not understood why arabinose groups survive the rumen. It may be that some hemicellulosic carbohydrates occur as a glycoprotein releasable by acid pepsin. Ely et al. (1956) noted that hemicellulose could not be recovered from grasses pretreated with acid pepsin.

Cellulases release hemicellulose–ferulic acid esters, which may be related to Neilson and Richards's curious observation (1978) of soluble lignin-hemicellulose complexes in rumen fluid that are resistant to degradation and survive to reach the feces. Apparently, covalent linkage of carbohydrate to lignin can protect the carbohydrate from digestion, even in solution (Jung, 1988). These complexes are soluble in neutral and alkaline solution but are precipitated by weak acid.

11.5.5 Hemicellulases

Digestion of hemicellulose is a complex affair because hemicellulose is a composite of various sugars and glycosidic linkages. Moreover, the character of hemicellulose differs in various forages and types of plant cell wall. Hemicellulolytic enzymes present in filtered rumen fluid have the ability to cleave a variety of linkages. Because rumen hemicellulases are more soluble than cellulases, somewhat more information is available concerning them (Dekker, 1976). Hemicellulases are produced by some rumen bacteria and ciliate protozoa. All enzymes so far identified appear to be of the endo type, which attack the glycosidic chain linkages randomly. Most of these enzyme systems have been only partially purified and characterized. Enzymes that attack side chains include α-D-glucosiduronidase, which attacks the α-D-(1–2) linkage in glucoronoxylans, and α-L-arabinofuranosidases, which hydrolyze the 1–3 linkage branch points in arabinoxylan. The xylanases include β-D-xylanase (endoenzyme) as well as components that attack oligomers but not xylan or xylobiose. These enzymes efficiently degrade linear xylan; branched molecules are more slowly or incompletely degraded. The process of xylan hydrolysis at least partly depends on the action of arabinosidase to remove branching groups.

11.5.6 Micellular Pentosans

Almost all crude celluloses prepared from delignification of woody tissues contain residual hemicellulose that is not easily removed. These residual hemicelluloses are less digestible than the more extractable pentosan (Lyford et al., 1963; Table 10.5). They are intimately associated with cellulose, but their function is not well understood (Wilkie, 1979). Their presence in ADF probably improves the negative association with digestibility (Bittner and Street, 1983).

11.6 Soluble Fiber

The components of the plant cell wall that have no covalent linkage with lignin appear to be completely available to fermentation and are comparatively soluble, some sufficiently so that they are classified in the general category "soluble carbohydrates." The most important fractions included in this classification are pectins and β-glucans. Gums, also soluble carbohydrates, occur in various vegetables and legumes. They are highly fermentable and available but could have special significance to ruminants because they also have attributes of cellulose: that is, they do not give rise to lactic acid, and their fermentation is inhibited by low pH.

11.6.1 Pectin

Pectin is a polysaccharide rich in galacturonic acid that occurs in the middle lamella and other cell wall layers. It is the cement in plant cell walls. Hemicellulose is a lignified fraction in the secondary wall thickening that is not ordinarily soluble but may contain some of the same sugars. Unlignified hemicelluloses occur in vegetables and are water-soluble. Thus, the distinction between pectin and hemicellulose is by no means clear. Chemical classification based on solubility characterizes pectins as soluble in hot neutral solutions of ammonium oxalate or EDTA, while hemicellulose requires acid or alkali. Some hemicellulose in delignified tissue is hot-water-soluble, however, and some galacturonic acid is trapped in cell walls of grasses in a form not accessible to hot EDTA (Wilkie, 1979). Doner (1986) noted a pectin–ferulic acid (ester) linkage. Pectins are much more abundant in dicots than in monocots. Commercial sources are derived mainly from apples or citrus fruits (Comstock, 1986). Most of the information relevant to forages is from legumes.

Some chemists tend to avoid the terms *pectin* and *hemicellulose* and classify all noncellulosic cell wall polysaccharides in one group (Southgate, 1976b), but this view results in an inconvenient nutritional classification of carbohydrates. A small amount of galacturonic acid in the insoluble cell wall seems less digestible, like the glucuronide counterparts, and has been used to support an argument that pectin is lignified (Hatfield, 1993), but the research methods are suspect and the results do not agree with results from other studies suggesting that pectin is highly digestible (Chesson and Monro, 1982). The problem seems to be one of classification and definition. One view is that pectin is defined by its galacturonic acid or even its total uronic acid and is thus a nutritionally heterogeneous entity. This view has arisen from the studies of Albersheim's group and Fry (cited in Hatfield, 1993), who are not concerned with degradability, but rather with cell wall synthesis and growth. They have also largely confined their studies to vegetative unlignified plant cell walls.

The other, older view states that extractable galacturonan is the pectic fraction, and nonextractable uronic acids constitute another entity within hemicellulose and may be quite lignified. The analogy is to glucose, which can appear in the form of sucrose, starch, β-glucans, and cellulose, all of which have unique and intrinsic nutritional value and biochemical behavior. Galacturonic acid as well as the pentoses, arabinose, and xylose, surely occur in various polysaccharides of varying availability. Of these two contrasting views, I favor the latter, because only it is in agreement with nutritional conceptual principles. The following discussion assumes this classification.

Methods of preparing pectin that use acid risk including some hemicellulose. Forage hemicelluloses are attacked by acid as weak as 0.01 N (Sullivan, 1966). Some galactan and araban may be associated with cell wall protein and may be difficult to extract if cell walls are heated. Predigestion with pepsin causes loss of pectic or hemicellulosic carbohydrate from holocellulose prepared by the chlorite procedure (Ely et al., 1956).

On the other hand, there is evidence that heating alone is sufficient to debranch and thus partially degrade pectin and is a factor promoting its solubility. Temperatures greater than 80°C appear necessary for relatively complete pectin extraction; near 40°C at neutral pH only half as much will be extracted (Bucher, 1984). Weak acids increase pectin's apparent solubility under physiological conditions. Most methods for examining dietary fiber avoid the use of any acid, and thus whether pectin is a soluble fiber and whether it should be recovered in NDF are still unclear.

The solubility of pectins is relevant to their nutritional properties. Phosphate is a sufficiently strong chelating agent to remove calcium from cross-links and promote solution of pectin in the digestive tract. Pectic acid along with some phosphate likely promotes solution as well. For these reasons pectin may be treated as a soluble dietary component even though it is not soluble in the native unaltered cell wall, being held there by calcium cross-links.

Pectin is largely dissolved by neutral detergent and thus not recovered with other cell wall components (R. W. Bailey and Ulyatt, 1970). This loss is not a serious defect, however, because the dissolved fractions have an essentially complete availability. In human and nonruminant nutrition, pectin and other soluble gums are recognized as a part of the dietary fiber

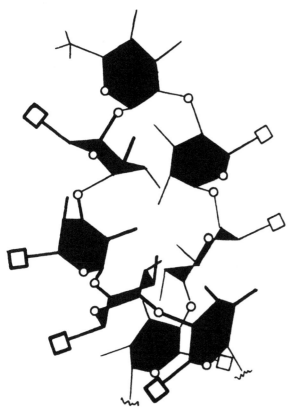

Figure 11.10. The polygalacturonic acid backbone of pectin is shown in a configuration in which carboxyl groups (squares) face outward (from Theander, 1976). The axial linkages cause the chains to kink or coil. Pectin differs from starch because of the axial bond on carbon 4 (equatorial in starch). The rhamnose unit (topmost sugar unit in the figure) linked α 1–2 in the chain causes a sharp turn so that the tails fall back on one another. The association between the chains is unknown, but a helical structure is possible. The formation of helical complexes with lipids could be important in accounting for pectin's promotion of fecal lipid and bile salt excretion (B. A. Lewis, 1978).

Table 11.10. Oligosaccharides formed on partial hydrolysis of pectins

Oligosaccharides	Alfalfa	Citrus	Soybean hulls	Soybean meal
GalA 1–4 GalA	+	+	+	+
GalA 1–4 GalA 1–4 GalA	+	+	+	+
GalA 1–2 Rha	+	+	+	+
GA 1–6 Gal	0	0	0	+
GA 1–4 Gal	0	0	0	+
GA 1–? Fuc	0	0	(+)	+
Gal 1–4 Gal	0	+	0	+
Gal 1–4 Gal 1–4 Gal	0	+	0	+
Arabinose (furan)	+	+	+	+

Source: Aspinall, 1965.

Note: + = present; 0 = not present. Parentheses indicate only low amount present.

complex. Separate determination apart from insoluble fiber is recommended, and methods are available for doing so (Van Soest et al., 1991).

Pectin Structure

Pectin consists essentially of a galacturonic acid (galactouronan) chain substituted with arabinan and perhaps galactan side chains. The acid groups are combined with calcium ions or as methyl esters. The galactouronan backbone has interpolations of rhamnose in the chain, causing an abrupt turn in the molecular structure (Figure 11.10). Like starch, pectin is linked α 1–4, leading to coiling of the polygalacturonic acid chain. Pectin differs from starch, however, in the axial position of the linkage on carbon 4, and pectins are not attacked by amylases. A variety of saccharidic linkages that vary with the plant species has been isolated from pectin (Table 11.10). Legume and citrus pectins are insoluble in the native form, in con-

trast with those of *Brassica* (cabbage group), which are relatively soluble in cold water. The insolubility depends on calcium cross-chain bonding or methylation. Ester links to ferulic acid are also known. Many of these would be exposed and broken in boiling with acid detergent, leading to a general precipitation of pectic acid into ADF as the quaternary detergent salt.

Pectin offers some contrasts with cellulose, starch, and xylan in its physical properties. While galactouronan (the backbone of pectins) is linear, its axial linkages force the molecule to coil or kink; rhamnose units force an abrupt turn (Figure 11.10). Polygalacturonic acid is soluble in neutral or alkaline solutions through ionization of the carboxyl groups. The greater solution achieved in hot alkaline solutions may be due to β elimination reactions. Glycosidic linkages involving the third (beta) hydroxyl are prone to cleavage with heating in neutral systems, with the formation of an anhydrosugar with a double bond between the second and third carbons. The polyuronide chain is, however, very resistant to hydrolysis. Pectins in the cell wall are soluble and are easily removed with hot solutions of chelating agents that convert calcium pectate to sodium or ammonium pectates, which are soluble. In pectin, many carboxyl groups may be bound as methyl esters, which also decreases solubility.

Figure 11.11 is a hypothetical model of pectin structure. Although no helical arrangements of galactouronan or galactan have been proposed in the literature, such a possibility can be inferred from the model. The angular turn of the chain attached 1–2 (vicinal) to rhamnose forces a near 180° turn in the chain. Because galacturonic acid and galactose are linked axially, the sugar rings cannot lie in the same plane and are forced to turn or kink into coils. Thus the two chains attached to rhamnose can be expected to interact. Figure 11.11 also shows the araban side chains and a "hairy" region in the middle of the molecule.

Methods of isolation tend to alter pectin through loss of methyl groups and arabinosyl side chains that are very susceptible to weak acid. Removal of methyl groups tends to increase the solubility of the pectic

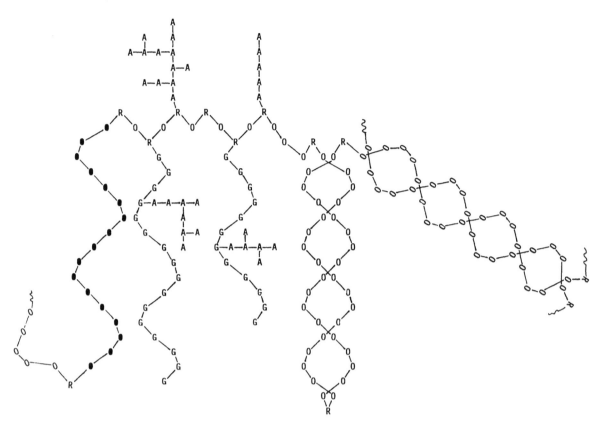

Figure 11.11. Structure of pectin (modified from Jarvis, 1984). The original reference shows the block chains of galactouronan as straight and linear, which is not possible according to the structural model shown in Figure 11.10. Also, Jarvis's rhamnose linkages show too wide a turn, and the twin chains do not interact in his figure. Symbols: 0 = galacturonic acid; ● = methyl ester of galacturonic acid; R = rhamnose; G = galactose; A = arabinose.

acid, as does the removal of calcium with chelating agents. Weak acids in combination with sugar or alcohol precipitate pectin in the form of a gel. This interaction is important in the preparation of jellies and jams.

11.6.2 β-Glucan

These relatives of cellulose have been observed in the cell walls of grasses, usually in very small amounts, but in oats and barley they occur in much larger quantities, generally associated with the bran. Thus these substances are important fractions of the cereal grains that contain them. Nomenclature in this class of celluloses is potentially confusing since all celluloses are β-glucans. The term has come to mean β-glucans that are soluble and not isolated with cellulose, however. Beta-glucans are usually partially soluble in water, forming viscous solutions, although complete extraction may require sodium hydroxide (Jeraci and Lewis, 1989). The tendency to gel and form solutions appears to result from random β 1–3 linkages in the otherwise cellulosic chain that cause a right-angle turn that may occur every four or five sugar units (Figure 11.12). Thus the soluble β-glucan mole-

cule appears to have a zigzag conformation. Its lack of packing ability leads to solubility and gel formation.

Beta-glucans are unavailable to mammalian diges-

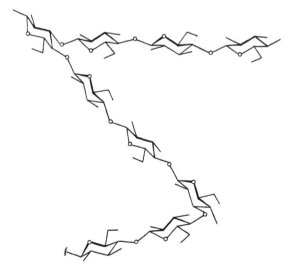

Figure 11.12. The structure of β-glucan with β 1–3 and β 1–4 linkages. The β 1–3 linkages that occur about every three to five sugar units cause the molecule to fall back on itself, forming angular coils.

tive enzymes, but they are highly fermentable and have been noted as antinutritional factors in poultry. They are prominent in beer and contribute to flatus. From the point of view of ruminant nutrition these nonstarch polysacchrides, like pectin, should be used efficiently by fiber-digesting bacteria.

There are a variety of other gums besides the β-glucans; these include xyloglucans, mannoglucans, and glucan gums in many plant seeds. The gumlike mannans and xyloglucans and β-glucan (Figure 11.12) are unlignified.

The structure of the glucomannans includes backbone chains of mannose and glucose in an alternating pattern, while galactose may form side chains. These polymers seem to be characteristic of legumes. Xyloglucans with similar properties occur in rapeseed meals and other dicotyledonous plants.

11.7 The Maillard Reaction

Products of the Maillard reaction occur widely in foods and feedstuffs that have been cooked or heated. The reaction is important in developing flavor because its products are responsible for producing caramel, brown crispiness, and, if overdone, a burnt, acrid taste. Its value as an assay of the quality of processed feeds cannot be underestimated or overlooked. While the reaction is important in carbohydrate chemistry, the information it provides about protein availability in processed animal feeds is even more significant (see Chapters 14 and 18).

Carbohydrates are susceptible to thermal degradation (heat damage) in the presence of amines or amino acids and water. This reaction, called the Maillard, or nonenzymatic browning, reaction, involves the condensation of sugar residues with amino acids followed by polymerization to form a brown substance consisting of about 11% nitrogen that possesses many of the physical properties of lignin. Other feed substances, including unsaturated oils and phenols, may participate by copolymerization. The most important reaction is the condensation with amino acids through which proteins are rendered indigestible, particularly those with a free nucleophilic group such as the amine or sulfur in lysine, cystine, and methionine, and, perhaps, the phenolic group in tyrosine.

Because water increases the rate of the reaction, laboratory samples must be dried carefully to prevent the loss of carbohydrate in Maillard products. The relation between damage and temperature (Figures 11.13 and 11.14) indicates that drying at 60°C or below can be relatively safe provided it is accomplished rapidly, although small amounts of artifact will be produced. The effect of water is variable, the maximum catalytic activity of water being at 30% of the sample weight.

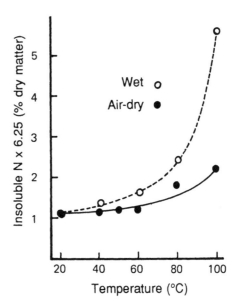

Figure 11.13. Effect of drying on the bound nitrogen in wet and air-dry ladino clover (from Van Soest, 1965b). Samples were heated in open dishes in a forced-draft oven and bound nitrogen was determined as that residual in acid-detergent fiber. Results illustrate the greater sensitivity of wet material to the Maillard reaction, which is further enhanced if moisture loss is restricted (see Table 11.11).

Heating dry material produces much less damage (Table 11.11). Different forages heated at constant temperature, time, and moisture show greatly different susceptibilities to heat damage that cannot be accounted for by their carbohydrate composition or protein content (Goering et al., 1973). The reaction rate may be affected by other chemical agents; for example,

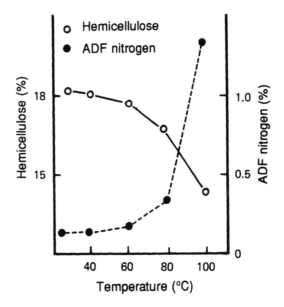

Figure 11.14. The hemicellulose content of orchardgrass declines with increasing temperature as it is used in the Maillard reaction (from Goering et al., 1973). Damaged carbohydrates no longer show up as such in analyses and appear instead in the lignin fraction.

Table 11.11. Effect of various methods of heating on yield of acid-detergent fiber and lignin

Initial moisture content (%)	Treatment	Acid-detergent fiber[a,b] (%)	Residual insoluble protein[a] (%)	Apparent lignin (%)	Corrected lignin[b] (%)
80	Dried 20°C, 72 h	30.3	1.1	5.5	4.4
9[c]	Dried 100°C, 20 h	32.4	2.2	7.0	4.8
80	Dried 100°C, 20 h	36.8	4.3	9.4	5.1
9[c]	Dried 100°C, 120 h	34.6	3.4	9.3	5.9
40[d]	Stoppered flask 100°C, 20 h	41.8	8.3	14.4	6.1
9[c]	Boiled 20 h	41.1	9.3	15.9	6.6
40	Autoclaved 127°C, 4 h	48.0	11.5	21.6	10.1

Source: Van Soest, 1965b.

[a]Calculated as N × 6.25 and expressed as a percentage of whole forage dry matter. Residual nitrogen is quantitatively recovered in the crude lignin.

[b]Calculated as −(N × 6.25).

[c]Air-dry material from 20°C treatment.

[d]Water added to air-dry material.

sulfite is a very powerful inhibitor of the Maillard reaction.

11.7.1 Pathways for the Maillard Reaction

The Maillard reaction consists of the formation of active carbohydrate degradation products that condense with any available amino groups. Amino acids such as lysine with a free amino group in the peptide form are very reactive. All the protein in a feed can become bound in this process. The degradation of carbohydrates is shown in Figure 11.15. Sugars change configuration and dehydrate after condensation with amines. Furfurals are extremely reactive, but their reactivity can be blocked by bisulfite (Van Soest, 1965b). Presumably, the bisulfite reacts with carbonyls to prevent their further polymerization but does not limit thermal breakdown of the carbohydrates. Furfurals are undetectable in heat-damaged forage, however; possibly they are labile intermediates. Alternatively, other reaction mechanisms may occur (Figure 11.15).

Carbohydrates' variation in susceptibility to the Maillard reaction renders it impossible to predict heat damage on the basis of drying temperatures or moisture content. The most reactive carbohydrates are the hemicelluloses and soluble carbohydrates, whose concentrations decrease on drying. Cellulose and starch are comparatively more stable. Figure 11.14 shows examples of the loss in hemicellulose due to Maillard reactions. Sugar residues are aromatized and the carbohydrate becomes an indigestible part of the lignin complex. The digestibility of the unreacted polysaccharide is not changed, however. Peptides and hemicellulosic polymers are so reactive that the reaction must involve relatively easy chain cleavage. The most reactive units are sugars with furanosidic rings (5-membered). Arabinose (in pectin and hemicellulose) and fructose (in sucrose and fructans) are furanosidic, and this may help to explain the comparative reactivity of soluble carbohydrate and hemicellulose. These carbohydrates are most easily converted to furfurals.

Basically, amino acids catalyze the destruction of sugars by converting them into phenols. The amino acid can be regenerated from this reaction. Following this, however, the phenol can catalytically destroy amino acids with a free amino group nitrogen in a Strecker degradation (Figure 11.16). The amino group enters the phenolic molecule, which is highly reactive toward condensation and polymerization (Figure 11.15), resulting in a brown to black plastic substance (Theander, 1980).

The Maillard polymer has most of the physical and chemical properties of lignin. It is soluble in alkali and insoluble in acid. Indigestible protein formed through the Maillard reaction is recoverable in lignin and acid-detergent fiber (Table 11.11). The proportions of nitrogen to artifact indicate a 1:1 adduct between sugar and amino acid, so that a N × 6.25 correction will not account for the increase in lignin that results from the addition of damaged carbohydrate (Van Soest, 1965b). Moreover, the Maillard product also possesses the phenolic absorbance characteristic of lignin. Some interference might be due to the phenolic group in tyrosine. Theander and his group in Sweden discovered the aromatiziation of sugars through the Maillard process occurring at a relatively neutral pH (Olsson et al., 1978; Theander, 1980). Sugar units were converted to 5- and 6-membered ring compounds (Figure 11.17). Phenols, once formed, may participate readily in condensations with aldehydes and amines, particularly through the bakelite reaction; these are mechanisms common to the polymerization of phenylpropanoid units in the formation of higher-molecular-weight lignin.

The Maillard reaction appears to produce both low- and high-molecular-weight products. At the present

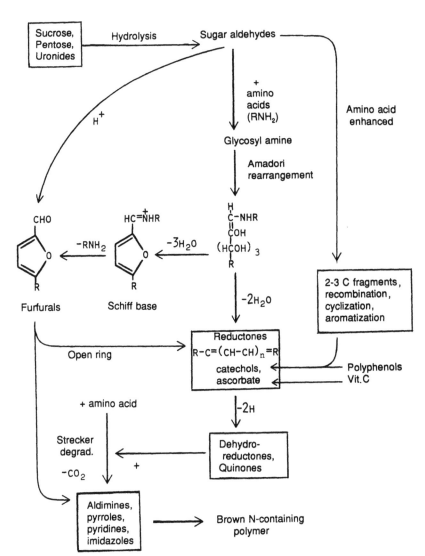

Figure 11.15. Some pathways in the Maillard reaction. Furfural formation direct from sugars is characteristic of relatively acidic systems; however, furfural formation at mild pH is catalyzed by amino acids that participate in the Amadori rearrangement. The formation of reductones (where n = 0 or 2) can arise from the Amadori rearrangement products, furfurals, or through recombinant aromatization reactions. Dehydroreductones are very reactive and promote Strecker degradation of amino acids (Figure 11.16) as well as copolymerization to the ultimate Maillard polymer. For a more complete discussion, see Hodge, 1953; Theander, 1980; and Dworschak, 1980.

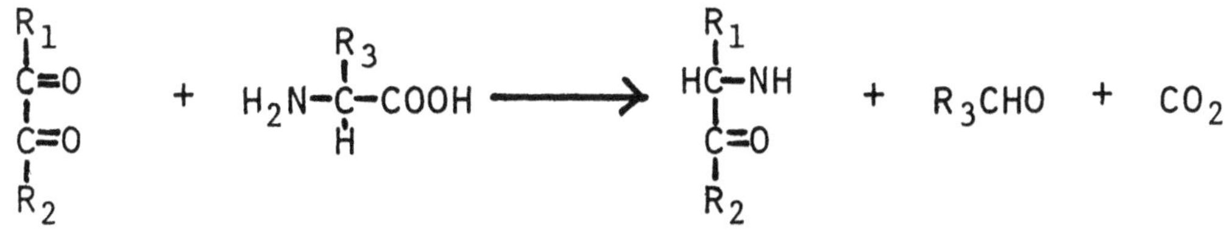

$$\begin{array}{c} R_1 \\ | \\ C=O \\ | \\ C=O \\ | \\ R_2 \end{array} \quad + \quad \begin{array}{c} R_3 \\ | \\ H_2N-C-COOH \\ | \\ H \end{array} \quad \longrightarrow \quad \begin{array}{c} R_1 \\ | \\ HC-NH \\ | \\ C=O \\ | \\ R_2 \end{array} \quad + \quad R_3CHO \quad + \quad CO_2$$

Figure 11.16. The Strecker degradation of amino acids is a transamination reaction whereby the amine group is transferred to dehydroreductones derived from sugars. The remainder of the amino acid carbon structure is separated as an aldehyde and CO_2. This reaction is dominant under neutral and slightly acidic conditions. The rearrangement of these transamination compounds and their subsequent polymerization may account for the lignin-like character of the final Maillard product produced in heated forages. The final product has a high nitrogen content (ca. 11%) that might be explained by the loss of the nonnitrogenous moiety of the amino acids.

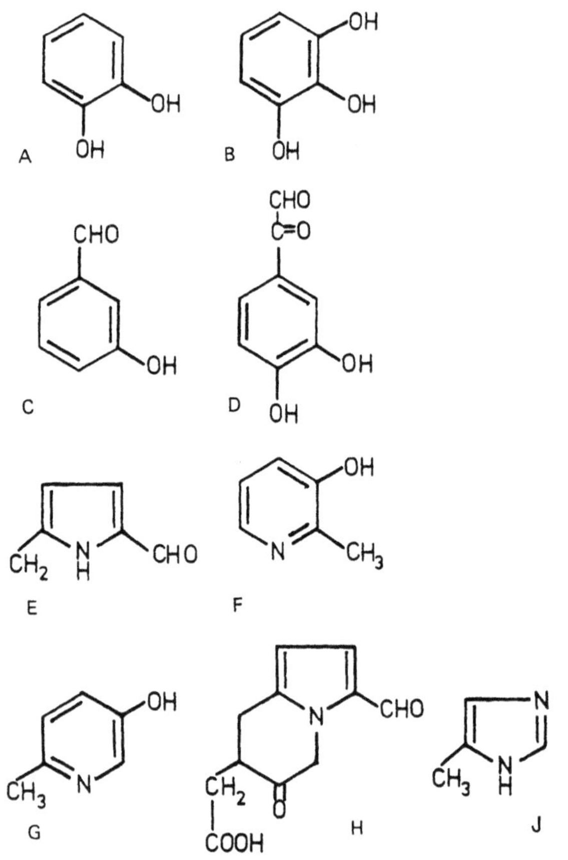

Figure 11.17. Some aromatic and heterocyclic products that result when glucose and glycine are boiled. Substantial amounts of high-molecular-weight products (brown insoluble matter) are also formed but have not been characterized (Olsson et al., 1978). A–D are nitrogen-free structures. Structures containing the catechol groupings (A, D) or pyrogallol (B) are extremely susceptible to oxidative polymerization. E–H are nitrogen-containing forms. J is 4-methylimidzole, the toxic product formed when molasses and other sugar-containing feeds treated with ammonia are heated (Perdock and Leng, 1987).

time only the monomeric, or low-molecular-weight, products have been characterized. Low-molecular-weight condensates between sugar degradation products and amino acids, if absorbed, are not utilized by the animal and may be excreted in urine. Some, however, may be of high enough molecular weight to contribute to water-soluble nitrogen in feces (Van Soest and Mason, 1991). The soluble Maillard nitrogen not recovered in fecal acid-detergent insoluble nitrogen does not appear to be used by the animal (Nakamura et al., 1991). The very toxic 4-methylimidazole formed in heated ammoniated molasses and in ammoniated forages is known to cause bovine "bonkers" (Perdock and Leng, 1987). These substances, which may be responsible for the dark color of aqueous extracts of damaged forage, are not assayed (as artifact lignin) in any of the present measures of the Maillard reaction for heat damage in feeds.

12 Lignin

Lignin is the most significant factor limiting the availability of plant cell wall material to animal herbivores and anaerobic digestion systems, yet it remains an enigma because of its intractability to biochemical elucidation. It is the only major plant polymer whose subcomponents are not clearly known, as amino acids are for proteins and sugars for polysaccharides. The condensed structure of lignin forbids hydrolysis in the ordinary sense, and the models that have been synthesized are based on in vitro studies, not on natural material. Such models are often based on wood, such as spruce, but it is becoming increasingly evident that lignins vary taxonomically, especially among monocots, dicots (angiosperms), and gymnosperms.

Animal nutritionists need to understand the integrated role of lignin and other protective agents that give rigidity to plant cell walls and provide resistance to biodegradation. The consistency of lignin's association with indigestibility is the basis for a biological view of it as an indegradable entity that evolved as part of the system plants use to protect themselves against herbivores. Lignin is a perfect protective material. Yet progress in natural polymer chemistry depends on the isolation and identification of its building units.

12.1 Definition

The definition of *lignin* can vary depending on the discipline and the viewpoint. Biochemists and plant physiologists tend to emphasize the role of plant cell wall in plant development and may not appreciate the problems of biodegradation and digestion in the nutritional sense. Wood chemists' definition is peculiar to the nature of the material and the object of its use. The nutritionist's view of lignin is also narrowly defined in the context of lignin's application to nutritive availability. To this list must be added the attitude of botanists and some lignin biochemists who tend to view lignin legitimately as a unique plant product and look askance at the various contaminating products included in the Klason residue used as its measure (see Section

12.1.2). It is hardly surprising that there is a certain lack of harmony in the sum total of all of these concepts.

12.1.1 Nomenclature: Core Lignin

The problem of defining *lignin* is linked with its nomenclature, which at the present time is not consistent in the literature on lignin and related phenolics. *Core lignin* is a term introduced to distinguish the principal lignin polymer from the extractable phenolics associated with samples. The extractable phenolics then represent non–core lignin. Proponents of the core concept include all cell wall phenolics in lignin proper. One definition of *non–core lignin* is that it comprises the coumaric acid fraction. If this definition is taken to its logical conclusion, there is no core lignin in grasses. Another view is that acid-detergent lignins represent core lignin. These concepts are incompatible.

Another class of phenolics extracted from wood are the low-molecular-weight relatives of lignin that are optically active and have differing linkages. The cellulase-releasable coumaric acid esters from grasses do not fall into this class.

Jung and Fahey (1983) suggested that acid detergent and neutral detergent remove some true lignin. This view considers all cell wall phenolics, including ferulic and *p*-coumaric acid esters as lignins (Jung and Deetz, 1993). This concept of core lignin is not supported by other lignin chemists (Ralph and Helm, 1993), who regard the extractable components as low-molecular-weight phenolics. This argument as to whether the extractable phenolics are true lignin has affected the methodology for studying lignin. The Klason method (Section 12.5) can recover most of these fractions, plus other interfering compounds as well. The use of Klason lignin can increase values in grasses up to threefold over acid-detergent lignin, but has less effect on legume lignin. This interaction would destroy the modeled relationships in which lignin appears to have uniform effects on cell wall digestibility (see Section 12.4).

The attempt to include in lignin the naturally soluble

phenolics that probably have some digestive disappearance causes the crude lignins obtained that way to depart from the nutritional objective of a truly indigestible residue. If lignin becomes nutritively heterogeneous through this means, it becomes less useful.

For example, using sugar component analysis to determine total glucose and glucosans in a feed is useless because glucose occurs in several substances (sucrose, starch, and cellulose) that are recognized for their nutritional and biochemical individuality. By the same token, phenolic acids occur in more than one place and therefore may have more than one biological effect. The fact that the monomeric cinnamyl esters might be precursors of grass lignin does not justify their inclusion in total lignin. Thus it is necessary to distinguish all the phenolic components, their locations, and their biological and nutritional effects. The phenolics, like the carbohydrates, should be classified according to their biological availability.

12.1.2 Klason Lignin and Its Contaminants

Wood chemists first characterized lignin as a polyphenolic insoluble indigestible component that is more or less isolated by sequential removal of extractable compounds with solvent followed by treatment with 72% sulfuric acid, dilution with water, and more refluxing. The insoluble organic residue is called Klason lignin. The analytical variations of this extraction method are discussed in Section 12.5.1. The application of the unmodified method to forages and feeds suffered from serious limitations because protein, which is negligible in wood, caused serious interference. Protein removal is essential because if it is not removed, it will be measured as lignin. Even so, the protein-depleted residue from forages and feeds is still apt to contain nonlignin components that have the mutual characteristics of being indigestible and insoluble in 72% sulfuric acid. Because wood is very low in nitrogen, the original Klason residues from wood also contained little nitrogen, promoting the concept of lignin as a nonnitrogenous polymer. Crude lignin prepared from forages or feeds, no matter how carefully prepared, always contains significant amounts of nitrogen (Section 12.2.3).

Crude Klason lignin in feeds can include a variety of nonlignin components; namely, phenol aldehyde polymers, tannin and tannin-protein complexes, Maillard polymer from heat damage, cutin (nonphenolic), and plastic (additive contamination). These subcomponents are treated below under their respective headings. Their common denominator is indigestibility and unfermentability. They do not occur in all feeds; tannins and the protein complexes are absent from many domesticated forage and feed species, and Maillard products occur only in excessively heated products.

Klason lignin has been much criticized for its impurities, its inability to recover phenols, and the alteration of lignophenols resulting from the use of strong sulfuric acid. Methods more productive of "true" or "pure" lignin have been developed to satisfy the criteria of chemical purity and integrity, although the notion that a measurement of the true and purest lignin would give an improved estimation of digestibility is incompatible with the observation that indigestible associations tend to be additive, as evidenced in the Lucas analysis and the summative equation (Chapter 22). For example, cutin, silica, and other nonlignin products are also associated with indigestibility. The value of acid-detergent lignin in feed and forage analyses is that it passes the Lucas test of nutritive unavailability and acts equally in all feedstuffs. On the other hand, soluble lignin-like complexes can have negative effects on digestion through toxicity to rumen bacteria. These latter aspects are controversial (Section 12.3).

12.1.3 Integrative Assessment

Lignin is a condensed product (along with Maillard polymers, leather, and plastics), but chemical analysis depends on characterizable residues that can be extracted and separated. Since most of these products are insoluble and cannot be extracted without destruction, their characterization remains problematic. Perhaps they may be amenable to some future solid-state method such as NMR or NIRS (Himmelsbach and Barton, 1980; Cyr et al., 1988). From a nutritional standpoint, lignin's greatest relevance is as a signal of indigestible residues, which are the dietary source of indigestible bulk.

A more specific lignin analysis method may exclude associated indigestible and inhibitory matter, whereby the net indigestible fraction would be underestimated. It depends on whether one wants a specific value for polymeric cell wall phenolics or an account of unavailable residues. It could be that a too puristic view of lignin is counterproductive to assessing cell wall indegradability. What may be in order is a clarification and nutritional classification of categories in the crude Klason lignin; that is, true lignin, as well as the nonlignin components contained in a net noncarbohydrate fraction in dietary fiber.

12.2 Biosynthesis

All lignin compounds and their phenolic relatives—flavones, isoflavones, and so on (Chapter 13)—arise from the shikimic acid pathway of phenol synthesis (Figure 12.1), which yields all phenylpropanoid products. The shikimic acid pathway is a synthetic route that requires boron for some steps in phenol synthesis,

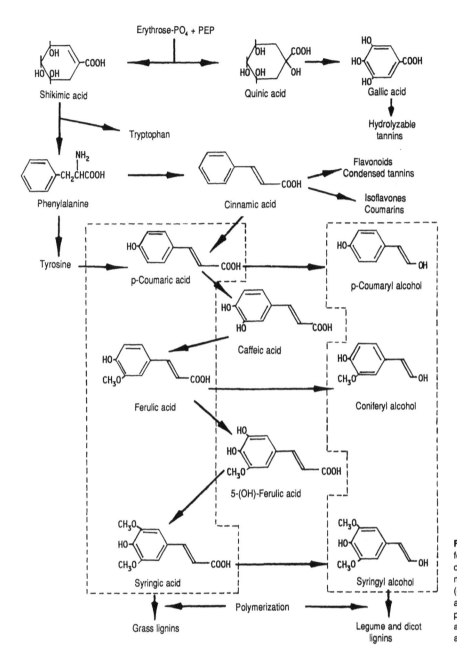

Figure 12.1. Shikimic acid pathway for the formation of aromatic compounds in plants. The routes to monomeric precursors of lignin (cinnamic acids and cinnamyl alcohols) are shown, as are the probable routes to tannins, coumarins, and isoflavones. Intermediate steps and coenzymes are not shown.

and thus boron deficiency in plants has been related to inefficient lignification (D. H. Lewis, 1980). The variability of lignification in response to nutrition and stress can be considerable (Chapter 6). The alternative end products of cinnamic acid metabolism are also of interest in this context, since flavonoids (including tannins) are responsive to stress and have significant nutritional interactions.

The biosynthesis of aromatic compounds in plants can occur in three ways: some phenylpropanoid compounds are formed through the shikimic acid pathway (Figure 12.1), others are formed via gallic acid from quinic acid (Mueller-Harvey and McAllan, 1989), and terpenoid compounds are formed from the polymeriza-

tion of isoprene (3-methylbutene). A fourth, non-biological, source is the Maillard reaction, in which catechols are formed from sugars under the influence of heat, moisture, and amino acids (Section 11.7). Plant phenolic compounds are often combined with substances from other classes, such as the terpenoids, or are conjugated with sugars as glycosides. Lignin itself is covalently linked to hemicellulose, and although cellulose linkages are unknown, they could exist undetected because they are destroyed by the method of isolation.

The most important building block for polyphenols is the phenylpropanoid group, which includes the aromatic amino acids phenylalanine, tyrosine, and tryp-

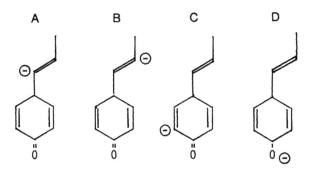

Figure 12.2. Free radical structures believed to be intermediates in lignin polymerization. A–C are quinone methides formed by loss of a proton at the indicated position. D is the radical formed by loss of a proton at the hydroxyl position. Note that formula C is not possible with a syringyl component because the position is blocked by a methoxyl group.

involves the formation of free radicals resulting from reaction with oxygen or peroxide. Peroxidases as well as ultraviolet light can induce dimerization and polymerization. The principal free radical forms, which can exist for any of the *p*-hydroxyl-substituted compounds, are shown in Figure 12.2. Condensation of the quinone methides can induce carbon-to-carbon bonds between any of the indicated sites. Formula D, which involves the phenolic oxygen, leads to ethers. Not all of the possible combinations have been found in lignins, although this does not mean that they do not exist. Some have been found in vitro but not in vivo. Figures 12.3 and 12.4 show naturally occurring structures that are products of cinnamyl alcohols and cinnamic acid esters.

The oxidative polymerization of the respective monomeric phenylpropanoids results in loss of identity of their precursors. The polymerization is irreversible, and this makes it very difficult to identify the components of the structure. Another, and more profound, problem is that of the random character of the polymerization reaction, at least as seen in vitro. Does the plant have any control over it? It is an unanswered question. Harkin (1973) suggested that random oxidative polymerization of lignin could occur when woody tissue is dried and aired. This possibility relates to the polymerization steps of the Maillard reaction (Section 11.7) in which free radical condensation from catechols and nitrogen-substituted catechols leads to the artificial lignin-like polymer when proteins and carbohydrates are heated. The parallel between phenolic polymerization in these dead systems leads one to wonder where life begins and ends.

The products of polymerization have a condensed structure containing primarily ether and C—C linkages between phenylpropanoids in a three-dimensional structure, which explains why lignin is so refractory to hydrolysis. Structures that incorporate these features have been proposed for softwood lignin (Harkin,

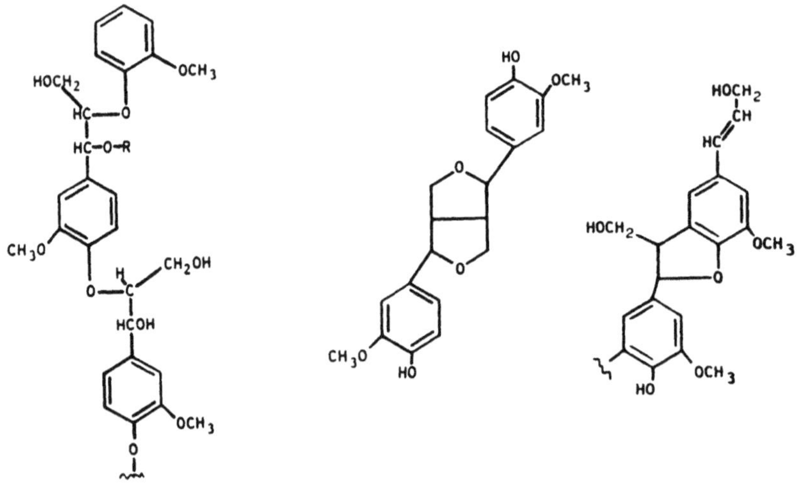

tophan. Polyphenolic products include the flavones, condensed tannins, lignans, lignins, and isoflavones. The cinnamic acids are produced from phenylalanine and tyrosine. Figure 12.1 shows the sequential introduction of phenolic and methoxy groups. The corresponding alcohols are produced from the acids.

12.2.1 Lignin Structures

Lignin arises from the polymerization of cinnamic acids or their corresponding alcohols. Conventional lignin as seen in gymnosperms and dicot angiosperms tends to be of the polymerized alcohol type; the corresponding acids that seem to be important in grasses, esters of ferulic and *p*-coumaric acids (but not syringic acid), can be obtained from cellulase-treated cell walls. Although these substances are not considered lignin, they may be precursors of it in grasses. The condensed lignin of grasses contains a significant proportion of syringyl groups, even though syringic esters are lacking.

The polymerization of phenylpropanoid monomers

Figure 12.3. Dimeric and trimeric products of cinnamyl alcohols thought to be components of spruce lignins and possibly dicot lignins (see N. G. Lewis, 1988). Formulas show condensation with coniferyl units, possible also with *p*-coumaryl units and with syringyl units.

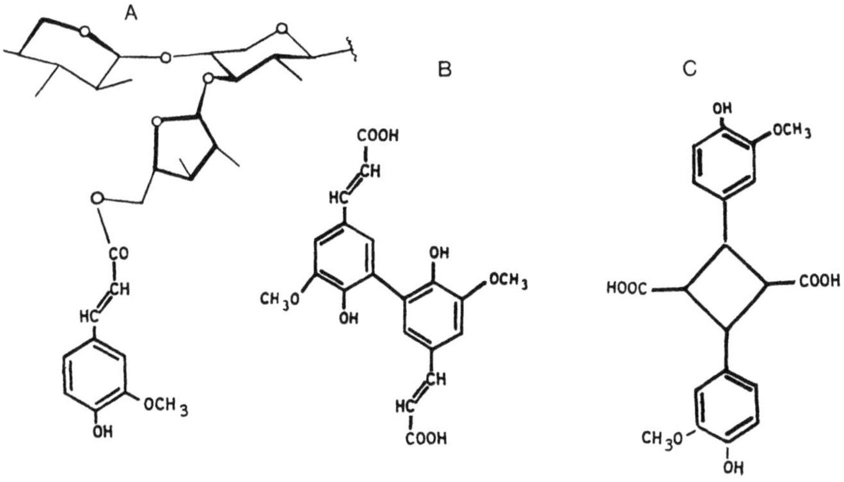

Ferulic acid ester of   Diferulic acid   Truxillic acid
arabinoxylan

Figure 12.4. Dimeric structures of cinnamic acid esters that have been found in grasses. Arabinoxylan ester groups could be substituted at the carboxyl positions in B and C. Diferulic acid is a biphenyl condensation product (Hartley and Jones, 1976), and truxillic acid involves the condensation of the side chain (Eraso and Hartley, 1990; Ford and Hartley, 1990).

1969), but their relevance to forages is unknown. Some of the basic condensation products between phenylpropanoid units are shown in Figures 12.3 and 12.4. The structures proposed for gymnosperms and dicots (Figure 12.3) could be present in legumes; structures found in grasses are shown in Figure 12.4.

12.2.2 Models of Lignin Synthesis

The polymerization of the phenylpropanoid components in lignin synthesis results in a loss of chemical identity. None of the phenylpropanoid precursors can be retrieved as such from the lignin product by hydrolysis or other means. Much of what is known of lignin structure has been obtained by the synthetic combination of probable building units or by following labeled compounds that are metabolized into lignin. Only in the grasses have the coumaric acid esters and their dimers been found. The cinnamic alcohols have not been isolated as such from native lignins.

Nevertheless, several models of the formation of true lignin in the young secondary plant cell wall have been proposed (Ralph and Helm, 1993). The first, which could only occur in grasses, is the dimerization of feruloyl or p-coumaryl groups attached to arabinoxylan. This results in cross-linking of adjacent chains (Figure 12.5). The dimers thus formed are not large enough to account for mature lignin in grasses, and a second, more speculative, suggestion is that the feruloyl- and p-coumaryl-substituted arabinoxylan becomes a template for deposition and polymerization of phenylpropanoid units, and the lignin polymer grows from the template surface. This is where syringyl groups must enter, since syringic acid does not occur in the arabinoxylan esters. The model composed of these steps accounts for the known facts about grass lignin.

All linkages to carbohydrate are saponifiable, and an unhydrolyzable phenolic polymer with syringyl groups also exists.

Another proposed template involves a tyrosine-rich peptide chain. The peptide is presumably a glycoprotein in which the peptide serves as a linking unit between carbohydrate and lignin. Such a structure would be more resistant to hydrolysis by alkali than the arabinoxylan esters are and might account for the properties of lignin in legumes. Legume lignins generally contain more nitrogen than grass lignins and are resistant to mild alkaline treatments that break the ester bonds in grasses; however high-temperature and pressure treatment of legume lignin will result in its dissolution with alkali.

12.2.3 Composition of Forage Lignins

The composition of lignin preparations varies according to the method of isolation. Sulfur-containing reagents (i.e., sulfite and sulfuric acid) add sulfur to the lignin. The physical properties of lignin are greatly altered by strong acids, which promote further polymerization and condensation and may cause originally soluble matter to become insoluble in the products.

The alternative to crude isolates is to attempt isolation of fragments that will identify the building components, although as I mentioned, they tend to lose their identity through oxidative polymerization. Lignin, like all polyphenols, is labile to oxidation. Oxidative treatments such as alkaline fusion or oxidation with nitrobenzene (Reeves, 1985) yield aromatic products that may be classified into three groups: p-hydroxybenzyl, guaiacyl, and syringyl (Figure 12.6, Table 12.1). The aldehydes are degradation products of the parent compounds from which lignin was presumably synthesized biologically. Grass lignins are quite vari-

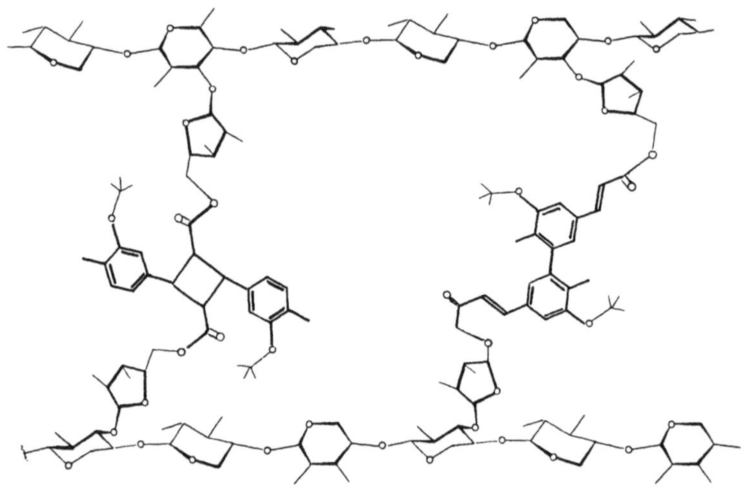

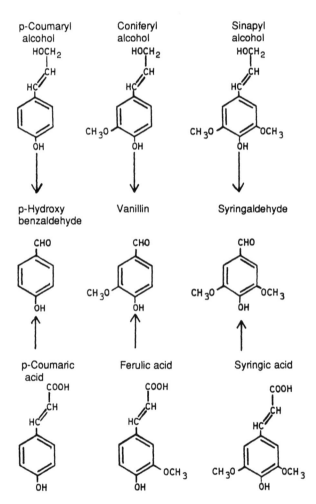

Figure 12.5. Possible structure of cross-linked arabinoxylan chains connected by diferulic acid (right side of figure) and by truxillic acid (left) (based on Hartley et al., 1990b). The linkage involving diferulic acid allows more space between the chains, and other cinnamyl groups could enter the matrix.

Figure 12.6. Relation between cinnamyl alcohols, acids, and their nitrobenzene oxidation products. The acids can also arise from nitrobenzene oxidation (Reeves, 1985). Vanillin (guaiacyl), *p*-hydroxybenzaldehyde, and syringaldehyde are produced from the degradation of the corresponding phenylpropanoid alcohols or acids. *Guaiacyl* is a collective term for the functional group common to coniferyl alcohol, ferulic acid, and vanillin.

able in composition and contain larger amounts of *p*-coumaric acid than most other types of plants.

In grasses, however, the corresponding substituted cinnamic acids may represent the major portion of lignin. These acids, present as esters in grasses, also form the corresponding derivatives on alkaline oxidation (Figure 12.6), and the yield of the respective aldehydes gives no information as to the properties of ester-linked components in grasses. Histochemical tests may give some idea of the relative distribution of guaiacyl (vanillin) and syringyl groups.

Ferulic and *p*-coumaric acids, but not syringic acid, have been isolated in the monomeric state in the form of esters to hemicellulosic side chains. These complexes are released when grass cell walls are treated with cellulase (Hartley et al., 1974). Grass lignins contain *p*-coumaric acid in amounts sufficient to account for most of the *p*-hydroxybenzaldehyde and enough ferulic acid to account for about a quarter of the vanillin yielded on nitrobenzene oxidation, with the acidic components being preferentially extracted. The acidic components are not uniformly distributed in the lignified tissue of grasses.

A number of histochemical tests indicate the presence of the various constituent compounds in lignin. The Maule reaction gives a yellow color if ferulic acid or coniferyl alcohol is present. The phloroglucinol-HCl test is sensitive to vanillin and ferulic acid and turns the sample red (see Akin et al., 1990). The inverse of this test is to apply vanillin-HCl to condensed tannins that contain 1,3,5-benzyl structures: vanillin in HCl gives products containing phloroglucinol groups a red color (see Figure 13.4). Results with tested grass cell walls show vanillin groups present in parenchyma and mesophyll cells, while phloroglucinol stains the vascular tissue, indicating vanillin (guaiacyl) groups.

12.1. The apparent proportions of *p*-hydroxybenzyl, guaiacyl, and syringyl ~s obtained by oxidation of lignins

Source	*p*-Hydroxybenzyl (%)	Guaiacyl (vanillin) (%)	Syringyl (%)
ood (gymnosperms)[a]	14	80	6
voods (dicotyledonous iosperms)[b]	4	56	40
i[c]	7	39	54
ut hulls[c]	3	90	7
ses[b]	22	44	34
stover[c]	30	34	36
it internode[d]	21	27	52
t straw[c]	11	47	42

stimates from Harkin, 1973.
veraged from the data of Higuchi et al. (see Harkin, 1973).
eeves, 1985.
am et al., 1990.

Methoxy Groups

The methoxy groups in vanillin and syringaldehyde have been used as a criterion of the presence lignin and even for estimating amounts of lignin. Methoxyl content varies among lignins and appears to be partially lost from fecal lignin. On the other hand, not all methyl groups in cell wall are in lignin. Methyl ester groups occur in hemicelluloses and pectins, and the methodology now in use cannot distinguish these from methoxy groups in lignin. Determination of methoxy groups is based on distillation with hydriodic acid, which liberates methyl groups as the volatile methyliodide, the basis of measurement.

Nitrogen Content

The lignins of grasses and legumes always contain some nitrogen. Under the best conditions of isolation, with proteases to remove contamination with protein, this nitrogen content is about 1.5–2%. This quantity also seems to be bound in lignin of fresh forage and is of very low digestibility; it occurs in larger amounts in legumes than in grasses. There is less indigestible nitrogen in acid-detergent fiber in fresh forages than in heated forages (Yu and Thomas, 1976). The lignins of woody trees, on the other hand, contain no nitrogen, a factor that led to the old view that forage lignin preparations are contaminated with protein because they cannot be isolated free of nitrogen. The presence of amino acids in lignin preparations has been used to support this hypothesis (Czerkawski, 1967). On the other hand, the template for lignin biosynthesis could involve peptides.

The presence of nutritionally unavailable amino acids in lignin may indicate a close association of lignin and hydroxyproline peptides of cell wall proteins (Fry, 1988) or, alternatively, a condensation of lignin with tyrosinyl groups in protein at the benzyl position of a cinnamyl group (Harkin, 1973). The theory that residual nitrogenous components are present in forage lignins is consistent with observations on nitrogen availability. The nitrogen content of acid-detergent fiber of forages has been correlated positively with lignin content and negatively with digestibility. Crude lignin in unheated forages has low amounts of nutritionally unavailable nitrogen (4–15% of forage total nitrogen). Much higher amounts of indigestible nitrogen result when the same forages are heated (Van Soest, 1965b). Thus unavailable nitrogen in feeds and forages has at least four potential sources (not seen in all products), including native indigestible nitrogen, Maillard nitrogen, tannin-bound nitrogen, and keratins of animal origin. All these compounds are common in feces and in mixed prepared feeds. Thus, it is difficult to distinguish the sources of nitrogen in crude lignin.

Nitrogen Correction of Lignin Preparations

Protein constitutes a serious interference in lignin determination. It is impossible to completely remove it; therefore the expedient of subtracting N × 6.25 has been applied. This factor assumes that all nitrogen in lignin is in protein, but insoluble nonprotein heterocyclic substances have also been reported. These and the Maillard polymers found after heat damage require a factor greater than 6.25. The data in Figure 12.7 indicate that the indigenous nitrogen content of un-

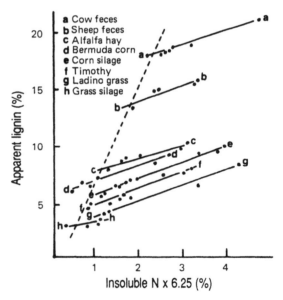

Figure 12.7. The nitrogen content of heated and unheated lignins in relation to crude lignin content (from Van Soest, 1965b). The dashed line represents the regression from unheated samples. The solid lines represent deviation in composition as a result of varied degrees of heating of materials. The solid lines have a common slope of 8.9, which corresponds to a polymer containing 11% nitrogen.

heated feeds is about 2%, whereas the nitrogen content of the Maillard polymer is about 11%. All the nitrogen fractions shown in the figure are associated with indigestibility if the fractions are present in feed.

Laboratory artifacts formed through the Maillard reaction can be avoided by drying at temperatures below 65°C and using a ventilated oven to ensure rapid moisture removal. Acid detergent removes carbohydrate and protein components active in the production of Maillard polymers. Therefore the temperature limitation does not apply to the prepared fiber and lignin. Further, one must distinguish between Maillard polymers already present in processed foods and those induced during sample preparation. The former are important in the nutritive characterization of the material, while the latter should be avoided.

12.3 Forage Lignins

Most forage plants belong to one of two families: the Graminae (grasses) and the Leguminosae (legumes). The grasses are monocots and the legumes are dicots. Together the monocots and dicots comprise the angiosperms. Yet the emerging view of the biological diversity of lignins seems to indicate that grass lignins are divergent in having many ester linkages that are largely absent in dicot angiosperms (P. J. Harris and Hartley, 1980; Hartley and Harris, 1981). Monocots (and grasses) probably evolved during the Jurassic period from herbaceous nonlignified water plants (Nymphaceae). The evolution of lignifying graminaceous species has obviously led to a novel kind of lignified structure. The alternative hypothesis—that monocots are misclassified in the angiosperms—is less acceptable, although ester-linked phenolics ought to be looked for in cycads. In the context of this discussion it is necessary to appreciate the consequences of the apparent differences in structure as they affect rumen digestion.

The principal differences between grass and legume lignins are listed in Table 12.2. Treating grass cell walls with fungal cellulases releases the components by which ferulic or p-coumaric acids are ester-linked to a hemicellulosic side chain of xylose and arabinose (Mueller-Harvey et al., 1986). These soluble phenolics are positively related to digestibility (Figure 12.8).

On the other hand, Gaillard and Richards (1975) isolated a soluble hemicellulosic complex, presumably also containing lignin, from rumen contents of cattle fed on speargrass. These soluble complexes do not appear to be digestible; they undergo precipitation on reaching gastric acidity in the lower tract and are recoverable in feces (Neilson and Richards, 1978). The substances involve protein as well as phenolics (Nordkvist et al., 1989). Many studies have indicated that ferulic

Table 12.2. Comparison of grasses and legumes relative to cell wall characteristics

Grasses	Legumes
Cellulase treatment releases soluble ester-linked phenolic acid carbohydrate complexes	No release of soluble complexes
Alkaline saponification accounts for major effects on digestibility	Alkaline saponification has little effect on digestibility
Alkaline saponification results in considerable solubilization of lignin	Alkaline treatment at mild temperatures results in little solubilization of lignin
Ferulic and p-coumaric acids are present as esters with arabinoxylans	Ferulic acid is present in small amounts, but not as ester with hemicellulose

acid and p-coumaric acid inhibit rumen bacteria (Jung 1988), although the studies did not consider the possibility that rumen bacteria might adapt, and metabolism of monomeric units has been reported (Fukushima et all, 1991). Anaerobic fungi possess esterases capable of hydrolyzing the phenolic ester bond (Borneman et al., 1990); indeed, soluble phenolics are directly related to digestibility in ryegrass (Figure 12.8). On the other hand, Hartley and Akin (1989) found that toxicity to rumen bacteria decreases with dimerization of phenol. High-molecular-weight lignin complexes may limit cell wall carbohydrate digestibility by restricting accessibility; however, bonding with the carbohydrates seems necessary for this effect.

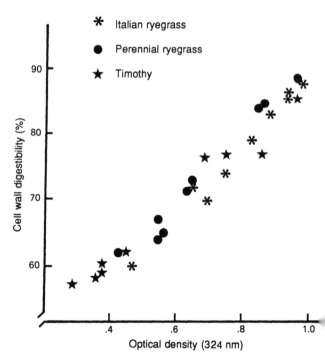

Figure 12.8. The relation of cellulase-releasable cinnamic acids (optical density at 324 nm) and digestibility of three grass species (Hartley et al., 1974).

The amount of enzymatically releasable ferulic acid declines as plants mature and is thus positively correlated with digestibility. A possible system for nutritive evaluation of grass is based on the UV absorbance of filtrates from cellulase-treated forage (Figure 12.8).

Types and concentrations of lignin vary from plant to plant; for example, the cellulase-soluble cinnamic acid fraction is much higher in the stalks of the low-lignin brown midrib mutant of corn compared with the normal isogenic cultivar (Hartley and Jones, 1978).

Lignin is most closely associated with hemicellulose, and one model of the polymerization and setting of the encrusting material in young plant cell walls involves the polymerization of low-molecular-weight lignin-hemicellulose conjugates. This might be the best explanation of the data plotted in Figure 12.9.

Linkages between lignin and cellulose or lignin and pectin have not been found, although linkages are known between phenolics and pectin (Fry, 1988). The galactouronide fraction is almost completely degradable (Gaillard, 1962), and cellulose indigestibility is closely associated with lignification. Perhaps there is a linkage that is destroyed in solubilizing the lignin-cellulose adduct.

12.3.1 Grass versus Nongrass Lignins

Grass lignins are considerably more soluble in alkali than are lignins from wood or nongrass forages (Figure 12.10). The solubility of grass lignin may be related to at least two factors: a high ester content and a lower content of methoxy groups. Lapierre et al. (1989) suggested that these factors are related to the digestibility responses observed when legume and grass straws are delignified. Alkali delignification in grasses is associated with an increase in digestibility of the remaining insoluble cell wall. Delignification in grasses results in solubilization of both lignin and hemicellulose. In legumes, cell wall digestibility is comparatively unresponsive, any increase in digestibility being due to solubilized carbohydrate components from the cell wall matrix. The mechanisms may differ; the response in grasses are probably due to cleavage of the ester links, while the response in dicots, in which lignin is presumably ether-linked, is due to cleavage of side groups from the carbohydrate complex and little or no cleavage of lignin-carbohydrate bonds.

Hartley (1983) was unable to demonstrate saponifiable linkages in grass and alfalfa that could be associated with digestibility in legumes. Alkali treatment of alfalfa cell wall resulted in only small releases of phenolic acids, and the digestibility increase was compa-

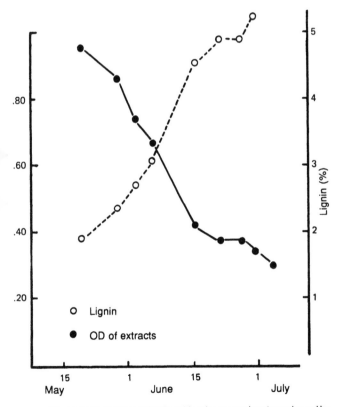

Figure 12.9. Relation of ferulic acid and p-coumaric esters released by cellulase as measured by UV absorption of extracts (declining curve) (Hartley et al., 1974). Positive curve shows lignin content changing with date of cutting of ryegrass.

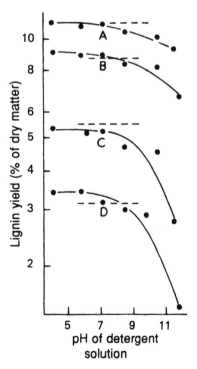

Figure 12.10. The pH of the detergent solution affects the recovery of Klason lignin in cell wall (from Van Soest and Wine, 1967). Logarithmic scale is used so that the curves indicate comparable proportional losses. Dashed line denotes lignin content of untreated samples: soybean hay (A), alfalfa (B), bermudagrass (C), and orchardgrass (D).

rably smaller. In other explorations of the angiosperms, comparing monocots (including grasses) and dicots, Harris and Hartley (1980; Hartley and Harris, 1981) came to the conclusion that ester groups are characteristic mostly of monocots, although a few dicots in the Carophyllaceae may have such linkages.

12.3.2 Lignin in Dicotyledons

Dicot plants include deciduous tree species such as oaks, maples, birch, and aspen, and also many vegetative species. Legumes are the main dicot forage plants; however, straw and stover can come from cotton, rape, and sunflower. Because lignins in these dicots are comparatively less soluble and are not attacked by alkali, no comparative studies similar to those on grasses exist, and it has been commonly assumed, perhaps without justification, that linkages and type of lignin in all dicots are similar to spruce lignin (a gymnosperm).

Harris and Hartley's (1980) survey of the dicots indicated that some of them, mainly in the Carophyllaceae, contained ferulic acid residues; but the majority of species surveyed did not. The ferulic acid moieties reported in alfalfa are probably a component of the phenolic extractives unassociated with the plant cell wall lignin. Legumes and most dicots contain smaller proportions of hemicellulose in their cell walls than monocots do, and those that form lignified walls tend to be more lignified than grasses.

12.4 The Nutritional Effects of Lignin

Lignin is generally regarded as the principal factor limiting digestibility, but it does not affect all feed components. The Lucas analysis shows that non–cell wall components are free from the effects of lignin, although lignin is often correlated with the digestibility of non–cell wall components. This occurs because of coassociation: protein and soluble carbohydrates decline as lignification proceeds. The best evidence that non–cell wall components are free from the effects of lignification is that up to 90% of the "plant cell wall" crude protein is removed from any lignified residue, even in the most mature straws, by a single pepsin digestion (except in tanniniferous or heat-damaged feeds). Further information comes from the treatment of graminaceous straws with alkali or other lignin-uncoupling reagents that cleave the bonds between cell wall carbohydrates and the phenolic component in grasses. The cell wall residue can be made completely biodegradable, while the freed phenolic component (soluble lignin) can remain in the feed but is indigestible in itself (Section 12.7).

According to an early theory, lignin encrusted, or covered, nutrients, rendering them inaccessible to degrading enzymes and thus indigestible. That concept is no longer accepted. On the other hand, encrustation could be a factor affecting availability of cell wall polysaccharides. There are also soluble indigestible lignin complexes. Rumen digestion releases soluble phenolics containing carbohydrate, some of which may remain resistant to digestion even in the soluble form (Neilson and Richards, 1978; Jung, 1988). Even phenolic acids esterified synthetically to cellulose reduce cellulose digestibility (Bohn and Fales, 1989).

The soluble phenolic-carbohydrate complexes isolated by Neilson and Richards may be similar to those found by Hartley. Soluble phenolic–protein interactions are not easily separated from the effects of tannin, which precipitates protein that tends to be recovered in crude lignin. This phenomenon occurs only in tannin-bearing plants but may have led to the errant conclusion that lignin retards protein digestion. In these instances digestion is limited by complexation, not by encrustation.

There have been reports of inhibition of cell wall digestion by soluble phenolics that occur naturally in forage (Jung, 1985). These substances seem to inhibit unadapted microbes and probably do not play a major role in limiting digestibility (Cherney et al., 1990; Dehority, 1993). Monomeric phenolic acids may be a source of phenylalanine and tyrosine for rumen bacteria.

Lignin is said to affect digestion rate and extent, but here again the literature is quite muddled; maturity effects are confounded with time and apparent increase in lignification. Thus it seems that digestion rate and extent are associated with the decline in quality of forages with maturity. If comparisons are made among forages of equal lignin content but of differing maturities, however, the only consistency is in extent of lignification; there is little association with rate of digestion. Grasses and legumes of equal digestibility are particularly dissimilar. The legume cell wall contains roughly twice the lignin and is much less digestible but ferments faster than grass at the same stage of growth. Perhaps the difference is the result of the grass's higher hemicellulose content, or perhaps the respective lignins are very different in their constitutions.

Kinetic studies on the rate and extent of fermentation associate the ultimate extent with relatively similar proportions of lignin in the final residue from a wide variety of lignified sources (Mertens, 1973; J. A. Chandler et al., 1980) at a content on the order of 35–40% lignin. The view that a constant ratio of lignin to carbohydrate sets a limit to digestion is not compatible with the surface-limiting concept suggested by Conrad et al. (1984). A constant lignin content of indegradable residues would result in a logarithmic power factor between undegradable residue and lignin content of cell wall of 1, whereas the Conrad model predicts a power slope of 0.66. The association of lignin and cell

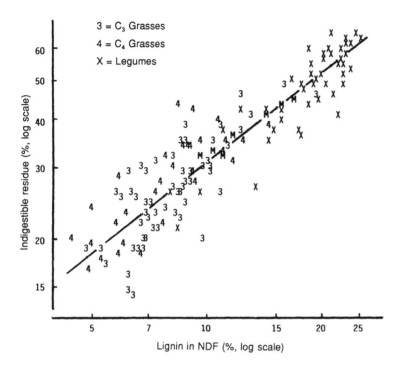

3 = C₃ Grasses
4 = C₄ Grasses
X = Legumes

Indigestible residue (%, log scale)

Lignin in NDF (%, log scale)

Figure 12.11. Relation between indegradable residues (after 96 h) in vitro and ratio of permanganate-determined lignin contents to NDF. (Logarithmic plot of data from Mertens, 1973.)

wall digestibility is not linear (Van Soest, 1967; Jung and Vogel, 1986). Figure 12.11 plots maximal extent of digestibility using data from Mertens (1973), and the overall slope is 0.76, which supports the surface limitation theory.

The figure does, however, support the view that lignification affects cell wall degradability in a similar way in all plants, and that differences in cell wall lignin content between dicot legumes and monocot grasses are not an important consideration. The data support using cell wall lignin content as a model for calculating digestibilities in summative systems (see Chapter 22).

The exponential association between lignin content and cell wall digestibility implies that a unit of lignin in immature low-lignin cell walls protects more carbohydrate than is protected in highly lignified mature cell walls. The lignin composition of immature tissues may be important. Jung and Deetz (1992) suggested that guaiacyl-to-syringyl ratios influence the lignin-digestibility interaction. Guaiacyl can cross-link, and syringyl groups cannot; thus guaiacyl would be expected to have a more detrimental effect on digestibility. This factor is probably less important in highly lignified tissue in which physical factors of surface are more limiting to the entrance of microorganisms (Akin, 1989; J. R. Wilson, 1993).

A mechanistic view arising from these observations is that steric hindrance or other inhibitory surface effects restrict digestion of carbohydrate existing near the lignin polymers. This is consistent with the difficulties experienced by lignin chemists in attempting to remove carbohydrate from lignin preparations by enzymatic means.

At the present time, the mechanism by which lignin protects associated carbohydrate from digestion is not entirely understood, and a number of questions remain:

1. Why are ester linkages in grass lignins resistant to esterases? Monophenolic esters appear to be fermented.
2. How are concepts of internal cell wall inaccessibility (encrustation or surface limitation) related to the ability of anaerobic fungi to penetrate lignified structures? (See Section 16.4.)
3. How is the digestibility of cellulose related to lignin if no bonding exists between the two?
4. How great is the ability of rumen organisms to adapt to and utilize low-molecular-weight phenolic substances? (Part of this is discussed in Section 12.6.)
5. What is the protein binding capacity of soluble lignins?

12.5 Lignin Assay Methods

Quantitative lignin assay methods can be classified into three basic categories (Giger, 1985): (1) gravimetric (these are all modifications of the original Klason procedure; see Section 12.5.1); (2) difference measurements after the removal of lignin (Section 12.5.3); and (3) measurements of absorbance, whether UV, NIRS, or NMR (Section 12.5.4; see Cyr et al., 1988). A nonquantitative assay class involves histological staining techniques and includes the phloroglucinol-HCl test for guaiacyl groups, chlorine-sulfite and coupling with diazonium compounds (Akin et al., 1990), and UV fluorescence (Hartley et al., 1990a).

Other assays include determination of carbon or energy contents of lignin isolates, since the aromatic nucleus is dense in both carbon and combustible energy relative to carbohydrate. Only fat and waxes have a higher caloric content than aromatic compounds. Fat is easily removed, in contrast to cutin and waxes, which tend to remain in Klason lignins.

Like fiber, lignin is not easily defined by a specific chemical characteristic. It is generally defined as the residue obtained by the Klason method or one of its many modifications. The possible nonlignin components of the Klason residue were described at the beginning of this chapter, and their presence has led to much criticism of the Klason method. For example, Englyst and Cummings (1988) did not consider lignin a part of dietary fiber. Alternative methodologies such as UV, NIRS, or NMR absorbencies depend on knowing which phenolics belong in lignin. These methods also require a calibration to convert absorbencies to organic dry weight, and the calculation usually falls back on some modification of the Klason method. The lack of an appropriate standard causes the various Klason modifications to give differing values on analysis of the same materials.

Lignin assays have been evaluated using the following criteria: (1) recovery in digestion balances, (2) degree of correlation of lignin with digestibility, (3) low nitrogen content of a lignin preparation, and (4) recovery of phenolic matter (Jung and Fahey, 1983). Although all these criteria do not define "true" lignin, they do impose an operational definition. Many attempts to improve lignin assay methods are based on reducing the number of procedural steps to reduce labor and increase laboratory economy and analytical precision. For a new lignin assay method to prove superior as a nutritional index of unavailability, it would have to improve on the association with cell wall indigestibility, as well as be competitive relative to time and effort.

12.5.1 Modifications of the Klason Procedure

The original Klason method is a treatment of solvent-extracted wood with 72% sulfuric acid followed by dilution with water and refluxing. Lignin is recovered as the final insoluble residue. The method's application to forages revealed a number of problems, of which protein interference is by far the most serious. Other problems involve the presence of low-molecular-weight phenolics and carbohydrate products. Many modifications have been devised over the years to overcome these problems. The excellent review by Sylvie Giger (1985) surveys most of these methods. Here I emphasize those currently in use. The older and more conservative methods prescribe predrying at 100°C followed by hot solvent extraction, usually with extensive refluxing. This is an unfortunate sequence because the heating tends to generate Maillard polymers that are difficult to remove or distinguish from lignin.

Many modifications of the Klason procedure involve the sequence of treatments. Most include solvent extractions followed by some kind of protease treatment, in turn followed by treatment with 72% sulfuric acid and then with dilution to 3% H_2SO_4 and prolonged refluxing. Alternatives to the protease treatment include extraction with acid detergent or correction of the nitrogen contamination by subtraction of N × 6.25. (The inadequacy of this calculation is discussed in Section 12.2.3.) The correction is sometimes applied even after protease treatment, since not all the nitrogen can be removed.

The final dilution of the sulfuric acid and refluxing causes a precipitation of matter which, in the opinion of Moon and Abou-Raja (1952), represents nonlignin phenolic matter. In general, the prolonged treatment of phenols with aldehydes in strongly acidic solutions promotes polymerization and precipitation. Phenolics can also arise from the heating of carbohydrate (Theander, 1980), and in this way lignin-like matter can be formed from a nonlignin precursor. The Maillard reaction also contributes to this lignin-like material, although this addition will contribute nitrogen to the residue.

The Klason residue also contains nonphenolic components such as cutin (Section 13.6). There is evidence for a cuticular fraction linked to lignin (Kolatukuddy, 1980). One procedure for cutin analysis consists of removing phenolic matter through oxidation and recording the residual nonphenolic matter in the Klason residue (Meara, 1955); however, tannin-protein complexes, Maillard products, and synthetic plastics interfere.

The suitability of a particular lignin assay method depends on the freedom from artifacts generated through the procedure. Thus, for example, if Maillard products are already present in a food sample through cooking or other heat treatment, the method that isolates them without generating artifacts best characterizes the material. If the method induces artifacts, the results may have no relevance to nutritive values.

12.5.2 Acid-Detergent Sulfuric Lignin

The application of acid detergent to a preparation of low-nitrogen fiber allows a fairly rapid determination of lignin in forages. The acid-detergent lignin method has therefore become the most popular procedure for lignin determination. The method is based on the systems of Sullivan (1959) and Moon and Abou-Raja (1952), but acid detergent replaces the pepsin digestion. Sulfuric lignin can be oxidized with permanganate in order to determine cutin and can be used with the ND-AD double sequential for the estimation of insoluble tannins (Chapter 13).

12.5.3 Difference Methods

In addition to direct gravimetric methods for assaying lignin there are methods based on other principles. Indirect methods measure lignin as loss in weight after its removal. Lignin can be estimated by subtracting the carbohydrate from the total cell wall fraction. Phenolics are much more easily oxidized than carbohydrates. Oxidizing agents such as permanganate (Van Soest and Wine, 1968) or chlorite (Collings et al., 1978) can be applied to cell wall preparations and the loss in weight recorded as lignin; alternatively, lignin can be dissolved by triethylene glycol–HCl (Edwards, 1973). Values obtained by the Edwards method and permanganate procedures are probably quite similar. Values obtained with chlorite may be higher and are based on NDF. Technically, these are methods for determining cellulosic carbohydrates.

Lignin's susceptibility to oxidation is increased by the presence of unsaturated double bonds and by the enolic nature of the phenolic hydroxyl, which is easily oxidized to a quinone. Further oxidation leads to vicinal diquinones that are cleaved to form aliphatic organic acids. The remaining double bonds are oxidized and cleaved. Permanganate oxidation visibly produces quite a bit of CO_2 from lignin. No lignin residue is left for further study, which may be an inconvenience.

Difference methods run the risk of removing nonlignin carbohydrates, in particular residual hemicelluloses and pectin contamination in ADF. The 72% sulfuric acid treatment causes some loss through solubilization, however, leading to lower values. The ratio of permanganate to acid-detergent sulfuric values is about 1.2:1 (Van Soest and Wine, 1968). The indirect methods do not measure cutin, and the Maillard polymers are incompletely removed from the cellulosic residue. The sequential treatment of ADF with permanganate oxidation and 72% sulfuric acid provides the basis for cutin determination. The permanganate and sulfuric acid steps can be reversed.

12.5.4 Spectrophotometric Methods

Spectrophotometric methods for measuring lignin content are based on UV absorbance. One procedure involves extraction of lignin by alkali and measuring UV absorbance in the filtrate. This procedure has several disadvantages. First, optical density is difficult to convert to absolute amounts of lignin in the absence of a standard, although regression with a gravimetric

method can be used. Second, high temperature and pressure are needed for quantitative extraction of lignin by alkali, particularly in the case of legumes. Alkaline solutions of lignin are, however, slowly oxidized unless they are protected by nitrogen gas, with consequent alteration of the optical density. Nonlignin components—proteins and other phenolics such as cinnamic acid esters, tyrosine, and tannins—are also dissolved in the process and provide interfering absorbance.

A much better system is the acetylbromide method of I. M. Morrison (1972), improved by Iiyama and Wallis (1990). Acetylbromide in acetic anhydride can dissolve most organic cell wall matter, and the acetylated phenolics can then be studied in these solutions. Solutions are read spectrophotometrically at 280 nm. Nonlignin material seems to cause little interference.

Bjorkman's ball-milling method (Lam et al., 1990) reduces carbohydrate by milling, and the residual lignin is dissolved in an appropriate solvent. Yields can be low relative to the net lignin in the material, however, and the product is never free of carbohydrates. This procedure has been used mainly for isolating naturally occurring lignins for structural chemical studies.

Because grasses are unique in their large number of ester groups bonding phenolic acids to other cell wall components, the net indigestibility of graminaceous straws can also be accounted for by the number of saponifiable groups, although it is not known whether all alkali-labile groups are of the ester type. Since all the grass phenolics are rendered alkali-soluble (virtually quantitatively), this offers a way of studying grass indigestibility. The system is not applicable to legumes (Hartley, 1983).

Figure 12.12 plots data from an experiment in which wheat and oat straws were treated with graded levels of alkali (NaOH). Alkaline treatment of graminaceous material saponifies lignin ester bonds, and the cleaved lignin becomes water-soluble, as shown by the solid line in the figure. Uncleaved lignin can be measured by saponification with excess NaOH. The data support the theory that alkali-sensitive linkages are quantitatively responsible for lignin's effect on digestibility, because the regression (dashed line) intercepts near 100% digestibility of NDF (Figure 12.12).

12.5.5 Fecal Recovery of Lignin

Nutritionists who want to use indigestible substances as markers or predictors have been prone to define lignin as indigestible and to seek a methodology that will produce fecal recovery or a marked relationship with digestibility. True lignin, however, particularly the lignin in immature grasses, while not utilizable in a nutritional sense, may not be easily recoverable in

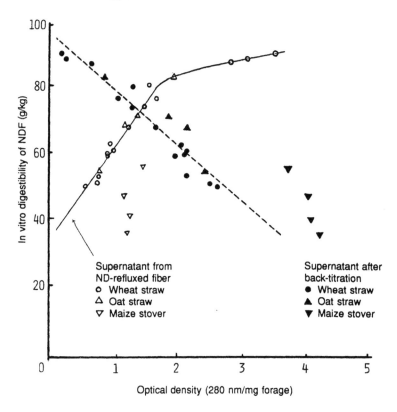

Figure 12.12. Ultraviolet absorbance (optical density per gram) of neutral-detergent extracts from alkali-treated NDF from wheat, oat straw, and maize stover (from Lau and Van Soest, 1981). Absorbencies of the sodium hydroxide extracts from the treated NDF (solid symbols, broken line) decline with in vitro digestibility. The open symbols (solid line) show soluble phenolics in the treated materials. The highest level of alkaline treatment increased optical density without a concomitant increase in digestibility, as shown by three open circles at upper right. The solid line is arbitrarily curved to indicate this discontinuity.

feces even though it is virtually indigestible. The other side of the problem is that some unmetabolized phenolic compounds may be lost in urine. Recovered lignin may include a variety of other materials as well. Nonlignin plant products, synthetic products produced in storage processing, such as phlobaphenes, and Maillard products could be present in feed or could even be generated synthetically in the lignin determination procedure. Crude lignin sometimes contains insoluble protein of animal origin such as skin, hair, and animal connective tissue when isolated from feces or feeds containing animal products. Also, most synthetic plastics added to the feed will be recovered in the Klason lignin fraction. These factors need to be considered in any situation in which animal, plant, and other substances are mixed in the diet.

Lignin has long been regarded as indigestible, and, indeed, there are no known anaerobic or mammalian enzymatic systems that degrade polymerized phenol. Some anaerobic fungi do possess lignase activity, however. There are methanogenic bacteria that can degrade simple phenols; however, such organisms do not live in the digestive tract, and gastrointestinal methanogenesis is confined to metabolism of methyl groups, formate, and CO_2. Other rumen organisms that metabolize simple phenolics probably use the phenyl groups to make aromatic amino acids.

Fecal recovery of lignin is a problem despite its indigestibility because lignins in very immature forages

appear to be somewhat digestible. Lignin recovery is generally better from animals fed grasses of higher lignin content and legumes and other nongraminaceous plants. As a practical matter, the use of lignin as an indigestible marker is limited to feeds having more than about 5% lignin in the dry matter.

There are several reasons for the apparent digestibility of lignin in some feeds, and they are not mutually exclusive. Prepared lignins from immature plants of low lignin content are apt to contain nonlignin components. The presence of a digestible contaminant would account for the higher apparent digestibility of lignin in low-lignin materials. Alternatively, lignins in immature grasses are polymerized to a lesser degree, and these low-molecular-weight fragments are absorbed and excreted in the urine, thus leading to apparent digestibility. Ruminant urine is high in phenols.

Methods that use much drying, heating, and boiling may generate Maillard products. Feeds that are high in true protein are more susceptible to the Maillard reaction than are the feces under similar conditions of treatment. The times for treatment with reagent in lignin recovery procedures are arbitrary. The permanganate oxidation, for example, may be a first-order reaction dependent on time. Therefore, higher amounts of lignin would need longer times for an equivalent degree of oxidation. Since the method is standardized to a fixed time of 90 min, this will led to a systematic apparent digestibility. One modification is to oxidize

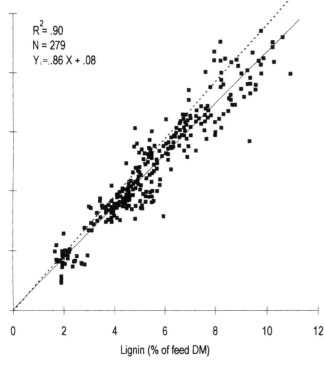

$R^2 = .90$

$N = 279$

$Y_i = .86 \, X + .08$

Lignin (% of feed DM)

Figure 12.13. Lignin data from Giger, 1985, recalculated in the form of a Lucas test. Recovery (indigestibility) of acid-detergent sulfuric lignin is 86% (solid regression line). The dashed line represents 100% recovery.

the fecal sample for 180 min, which is also arbitrary. Sulfuric lignin is a much better marker than is permanganate lignin.

The most insidious factor affecting recovery is lignin distribution relative to particle size. Lignin in feed tends to be associated with the coarser matter, while in feces it is associated with the finer material. Very fine lignin particles in feces may be lost during filtration. There is also a greater loss of fecal lignin in the Klason procedure because very fine lignin particles provide a larger surface area for chemical attack.

Finally, remember that the Lucas test for acid-detergent sulfuric lignin is zero digestibility (Chapter 22). Similarly, Giger (1985) could not find an association between lignin content and apparent digestibility, even though the individual variation included apparent values as high as 50% (Figure 12.13). These observations argue for a methodological explanation for the failure to recover lignin.

12.6 Chemical Treatments for Delignification

The removal of lignin from woody tissue is an old art developed in paper making, and most of the methods for improving low-quality materials for animal feed

Table 12.3. Chemical treatments of forages

Treatment	Action	
	Grasses	Legumes and dicots
Delignification		
NaOH	Sap	N
KOH	Sap	N
Ca(OH)$_2$	Sap	N
NH$_3$	Sap, A	N
Urea	Sap, A	N
Alkaline peroxide	Ox, Sap	Ox
Sulfite	Sulfon	Sulfon
Chlorite	Ox	Ox
Ozone	Ox	Ox
Hydrolysis		
Steam pressure	H	H
Acids	H	H

Note: Sap = saponifies lignin-carbohydrate ester bonds; N = no delignification, but cleaves hemicellulose side chains; A = ammonolysis of lignin esters with nitrogen entering the phenolic molecule; Ox = oxidation of lignin; Sulfon = sulfonation of lignin to soluble lignosulfonic acids; H = hydrolysis of structural carbohydrates to sugars.

derive from historical or current wood-pulping techniques. But forages, and particularly grasses, have characteristics not found in timber and wood. The difference between monocot lignins and dicot lignins is only beginning to be appreciated.

It was observed as early as 1880 that delignification increases the digestibility of cellulose. Since that time various methods have been applied to increase digestibility of wood, straw, and other highly lignified materials. Lignin can be removed by a variety of methods, producing variable yields and quality of products.

Most chemical treatments of low-quality forage and lignocellulose to make animal feed have been applied with little appreciation or understanding of the chemical mechanisms of action. Improving the availability of carbohydrates in lignocellulose and forage involves breaking the presumed lignin-carbohydrate bond. This can be done in several ways: (1) cleavage of the lignin-carbohydrate bond, releasing modified lignin; (2) oxidative destruction of phenolics, including lignin, resulting in the same effect; and (3) hydrolysis of the cell wall polysaccharide to sugars, leaving a lignified core. Table 12.3 lists chemicals used to treat forage; of these, the most popular are ammonia and urea treatments. If NaOH is used, there is a danger that the diet will have too much alkali, and problems in feed intake and water balance may ensue. Many of the potential feed materials are nitrogen deficient for rumen organisms, in which case the extra ammonia or urea may be beneficial. Economics is involved, of course; the cost of the chemicals may be expensive relative to animal responses to "improved" feed. Most chemical treatment systems are probably too expensive for people in developing countries to purchase.

12.6.1 Alkaline Treatments

The use of alkaline substances to increase digestibility is limited to monocots that contain a large number of phenolic bonds because dicots respond better to oxidizing systems (Barton et al., 1974; B. L. Miller et al., 1979; Soofi et al., 1982; Ben-Ghedalia et al., 1983; Mora et al., 1983; Gould, 1985; Chandra et al., 1985; Alexander et al., 1987; S. M. Lewis et al., 1987; and Mann et al., 1988). The more soluble alkalis, sodium and potassium, are more effective because of their solubility and higher base strengths, although calcium hydroxide becomes more effective when combined with ammonia or urea (Males, 1987).

Generally, the yield of digestible carbohydrate decreases with delignification of high hemicellulose-containing straws. A considerable loss of potentially digestible material occurs if the treated pulp is washed; that is, protein, sugars, hemicellulose, and soluble phenolics are lost in the wash. For this reason treated forage is usually unwashed, which leaves the chemicals of treatment and their products in the residual material. This mode of handling has generated its own set of problems.

Alkali treatments have yielded variable results, perhaps because the efficiency of the treatment depends on the proportion of the lignin-carbohydrate bonds that are broken (Lau and Van Soest, 1981) and the buffering capacity of the lignocellulosic residue (Dias–da Silva and Guedes, 1990). Most of the recommendations for upgrading products give fixed proportions of alkali to residue without considering buffering capacity. The variation in buffering capacities of straws is about 50%, and no delignification occurs unless a pH of 8 or greater is achieved. Fermentation is an acidifying process that can make alkali less effective.

Alkali treatment without washing has been presented as a feasible system, the advantage being that there are no waste residues. Some treatments neutralize the alkali with organic acid. Such preparations are accepted by the animal because the alkali serves as a rumen buffer. In these instances lignin content may not relate to digestibility because no lignin is removed (Rexen et al., 1975). The concentration of alkali (usually NaOH) is limited by the potential salt load of the diet.

The lignin in graminaceous straws becomes water-soluble after alkaline cleavage but does not become digestible. Furthermore, the dissolved phenolics dilute the neutral-detergent solubles and depress their digestibility (Figure 12.14). The phenolic content can be monitored by checking UV absorption at 280 and 314 nm. The 314 wavelength is mainly for cinnamic acid derivatives, and the 280 wavelength is for phenolics generally. J. D. Russell et al. (1988) studied the infrared spectra of treated forages. Spectra from treated

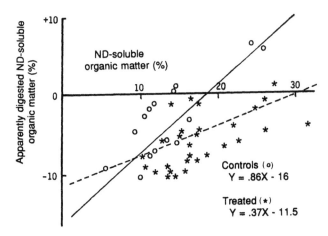

Figure 12.14. Relation between percentage of apparently digested neutral-detergent soluble organic matter and percentage of neutral-detergent soluble organic matter of alkali-treated straws and their untreated controls (from McBurney and Van Soest, 1984). The Lucas test indicates a drop in true digestibility from 86% in controls to 37% in treated straw.

forages differ from those of untreated forages such that special calibrations are required.

Alkali (OH^-) in the presence of oxygen, and in particular of peroxide, attacks phenolic groups. This is the basis of the alkaline peroxide treatment developed by the USDA. It is a very efficient system that works on all residues, but it is relatively expensive.

The greater susceptibility of most grasses to mild alkali treatment is probably due to their high proportion of ester linkages. Siliceous matter is important in rice and oat straw and is also readily attacked by alkali. In rice straw, silica is the main factor limiting digestibility (M. G. Jackson, 1977).

A side effect of alkali treatment is that carbohydrates may be chemically altered; while some hydrolysis may occur, crystallization (retrogradability) may cause the carbohydrates to become refractory to enzymes. The commonest result is an increase in the extent of digestion but a decrease in the digestion rate of available carbohydrate. This accounts for the often large increase in extent of in vitro digestibility of treated materials that may not be realized in vivo. Rate of passage from the rumen also appears to increase in animals feed treated straws because the lignin cross-bonding is removed and the fiber disintegrates through digestion into small particles that pass out of the rumen (Berger et al., 1979). It can be argued that this is evidence that ruminants require lignin for normal rumen function.

12.6.2 Ammonia and Urea Treatment

Ammonia and urea have been substituted for alkali as delignification agents both to reduce salt loads and to raise the nitrogen level of the resulting diet. Ammonia is a much weaker base than alkali, and so produces

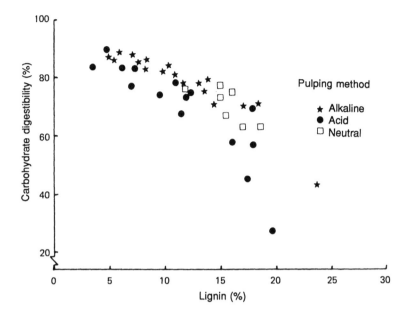

Figure 12.15. The relation between carbohydrate digestibility and lignin content in birchwood pulps obtained from digestion trials conducted in sheep. Alkaline treatments allow a higher digestibility in the face of increased residual lignin content. (Derived by Van Soest and Mertens, 1974, from the data of Saarinen et al., 1959.)

less delignification; however, the extra nitrogen it supplies can increase digestibility by overcoming nitrogen deficiency in the rumen. Thus, in order to evaluate the effects of ammonia and urea, isonitrogenous controls would have to be compared with treated products. Nitrogen supplementation often improves digestibility of low-nitrogen feeds (Chapter 14). Ammonia and urea (hydrolyzed) can promote saponification of phenolic ester linkages and thus enhance digestibility (Dias–da Silva and Guedes 1990), which is detectable by chemical procedures (Lau and Van Soest, 1981).

Urea hydrolyzes to ammonium carbonate, a weaker base than ammonia that yields ammonium hydroxide in water as the saponifying agent. Ammonium carbonates are more effective when used with calcium hydroxide, which removes carbonate and generates hydroxide (V. C. Mason, pers. comm., 1988). Attempts to improve efficiency of ammonia treatments by increasing the temperature are effective up to about 70°C. Higher temperatures result in Maillard reactions and ammoniolysis and create indigestible soluble nitrogen (Mason et al., 1989, 1990; Van Soest and Mason, 1991) and 4-methylimidazole, which is highly toxic to animals (Perdock and Leng, 1987). Production of 4-methylimidazole is enhanced by sucrose.

12.6.3 Other Treatments

Oxidative delignification, for example with ozone or chlorine dioxide, works on all plant lignins. Chlorine dioxide is generated from sodium chlorite and acid, and ozone requires electricity for its generation (Ben-Ghedalia et al., 1983). These agents generally oxidize and degrade lignin. Sulfite, on the other hand, creates soluble lignosulfonic acids and also increases

digestibility. Saarinen et al. (1959) compared various pulping systems relative to their efficiency in increasing digestibility (Figure 12.15). Chlorite delignification yielded the greatest amount of digestible pulp relative to original organic matter and involved the least damage to carbohydrate.

A number of research projects have examined the use of waste paper as feed. Commercial waste paper varies in digestibility depending on original treatment and composition (Table 12.4). Papers are very high in cell wall and often contain inert mineral additives. If heavy metals are present in the paper, health safety precludes using it for feed. Generally, newsprint is poorly delignified and the more digestible magazine papers are apt to contain heavy metals from colored inks. Pulping and drying may result in carbohydrate retrogradation, thus creating a slow rate of digestion, which may account for the observed poor acceptability by animals (Van Soest and Mertens, 1974).

Other methods that improve the nutritive value of lignified matter include milling, extrusion, exposure to

Table 12.4.   Composition (g/kg dry matter) and digestibility of various sources of paper

Paper	Ash	NDR	Cellulose	Lignin	Digestibility[a] (%)
Manchester Guardian	300	680	590	30	99
Solka floc (cellulose)	3	990	840	50	97
Whatman no. 41	2	980	890	50	91
Brown cardboard	17	940	720	120	77
Playboy	240	650	510	90	65
Christian Science Monitor	4	970	600	210	31
Washington Post	4	940	550	260	32

Source: Van Soest and McQueen, 1973.

[a]Digestibility of organic neutral-detergent residue (NDR) by incubation for 48 h with rumen organisms.

radiation, and acid hydrolysis (Fahey, 1992). These treatments attack the carbohydrate rather than the lignin. Very fine milling depolymerizes carbohydrate and offers a method for mechanical isolation of lignin. The fineness required makes the process impractical because the high rate of passage of fine particles usually decreases digestibility. Extrusion has been applied with the assumption that expansion would break encrusting lignin and thus increase digestibility. Digestibility is not increased unless lignin-carbohydrate bonds are broken. Structural carbohydrates are more susceptible to degradation by radiation than are aromatic substances. Exposure of lignified tissue to gamma radiation increases the solubility of carbohydrate and the digestibility of wood, but it is a very expensive process. Acid hydrolysis yields sugars and an insoluble lignin residue (Fahey, 1992).

12.7 Biological Decomposition of Lignin

That wood rots is common knowledge, but a comprehensive understanding of the biochemical process remains incomplete, despite recent progress (Kirk and Farrell, 1987). Understanding the biology of wood decomposition requires recognition of the physicochemical limitations of the respective processes as well as some ecological perspective. This is applied here in the context of anaerobic fermentation in the gut, and in particular in the rumen, but our perspective must also be broad enough to encompass the role of termites and other animals capable of destroying woody plant tissue (Chapter 5).

At issue is whether there is any real digestion of lignin in the gut, and whether that digestion yields any useful energy as ATP to the animal host, given that any digestion must involve symbioitic microorganisms inhabiting the gut. It is not necessary that biological delignification processes yield energy from lignin per se; the biological advantage could also come from the release of carbohydrate from the degraded cell walls.

The known biological decomposition paths of lignin are not rapid processes, and those that are anaerobic are undoubtedly slower than those that use available oxygen.[1] Animal digestive tracts, on the other hand, are limited by retention time, which is set by gut volume and energy requirements of daily food intake; they also tend to be anaerobic environments. Heat production in homeotherms, and thus in all large mammalian herbivores, is more critical than in nonhomeotherms, so it is interesting that the symbioitic association of

animal-microbe-fungus, the only well-attested animal system that digests lignin, is in nonhomeotherm arthropods, namely wood-destroying termites (Prins and Kreulen, 1991).

12.7.1 Aerobic Systems

While aerobic conditions do not occur in the rumen, aerobic delignification has been a subject of interest for the ways it might improve lignified cellulosic materials for feed. Aerobic destruction of lignin appears to involve peroxidative cleavage of phenolic rings after converting them to vicinal quinones. Once the ring is ruptured, the compounds can be metabolized as aliphatic fatty acids.

The primary destruction of lignin in nature involves the lignolytic capacity of wood-destroying fungi or bacteria (Kirk and Farrell, 1987). All known lignin-digesting organisms are aerobic. Wood-destroying fungi are classified into white rot and brown rot, and of these, the latter are better able to degrade phenolic matter (Zadrazil and Reiniger, 1988; Zadrazil et al., 1991). It is doubtful that any of these organisms derive much ATP from phenolic degradation and metabolism; more likely, delignification is but a means of access to carbohydrate. Thus the most efficient (so far) systems for biological delignification occur at the stage of development at which microbes have penetrated but little growth has occurred. No artificial system has been able to recover more than 50% of the carbohydrate, and the most efficient systems depend on pure culture. The system must be kept moist and aerated, and the incubation takes weeks, which involves expense.

Longer incubation results in the conversion of a major part of the carbohydrate to growth and metabolism by fungi.

12.7.2 Anaerobic Delignification

Anaerobic bacteria can metabolize phenols to some degree; for example, by cleaving the aromatic ring or, alternatively, using the phenyl ring for structural purposes. Phenolic cleavage can proceed anaerobically in simple phenols (Healy et al., 1980) but appears to be restricted in condensed polyphenols. Phenolic matter, including lignin, cannot be an energy source for anaerobic organisms because of its high carbon and hydrogen contents; instead simple phenolics may be more important as suppliers of phenylalanine or tyrosine.

The processes of anaerobic metabolism of phenols are slow; the induction time for ferulic acid, for example, is about 12 days, too long for any adaptation in a digestive tract. Larger-molecular-weight compounds would be expected to have even longer turnover times. Anaerobic bacteria might be expected to cope with

---

[1] Reported digestion rates for fungi, the most efficient delignifying organisms, are weeks and months (Zadrazil and Reiniger, 1988; Zadrazil et al., 1991), whereas animal retention times are hours or days.

some ester bonds, since no free oxygen is involved. Biphenyl linkages should be considerably more difficult to degrade.

The accumulation of polyphenolic matter in peat bogs has led to the theory that anaerobic degradation of phenolic compounds is limited to low-molecular-weight compounds, and this high-molecular-weight polymerized matter is not available to anaerobic metabolisms. Lignin appears to be biologically inert in anaerobic systems (Zeikus, 1980). This concept is safe as long as the definition is restricted to the unextractable polymer (Fukushima et al., 1991).

13 Plant Defensive Chemicals

Although lignin is the main factor limiting digestibility, other plant components involved in plant self-protection can also limit nutritive value. These components represent a very wide range of substances and can have diverse effects. Defensive substances act as inhibitors either by interfering in the animal's metabolism or by inhibiting rumen bacteria. There are a number of classes of such materials as well as various mechanisms for their action (Table 13.1). The phenylpropanoids include lignin, flavones, coumarins, isoflavones, and condensed tannins. Hydrolyzable tannins form another group, and others include a diverse class of nitrogen-containing alkaloids and the terpenoids.

The lipophilic terpenes include essential oils, saponins, steroids, latex, and rubber. Terpene-containing plants carry substances toxic to animals (saponins, steroids) and inhibitory to microorganisms (essential oils). Saponins have been blamed for bloat. Many familiar food flavors come from plants in this group.

Also in the lipid class are the cutins and suberins, the protective covering in all higher plants, which are composed of waxes and polymerized hydroxy fatty acids. Cutin is the indigestible part of the cuticle of leaf, stem, and root surfaces and the closely related suberin occurs with lignin in barks. Seed hulls, stone cells, and fruit pits often contain cutin as well as lignin. The cuticular surface acts as an impenetrable barrier to digesting microorganisms. The surface hairs and spines in many plants, particularly those adapted to aridity, are usually cuticular or siliceous or both.

Most nonlignin defensive compounds are of lower molecular weight than lignin and are metabolically more active. Not all compounds in these groups are toxic. For example, flower pigments (flavones), although relatively innocuous themselves are related to higher-molecular-weight tannins and polyphenols that may be inhibitors of digestion. The alkaloids are a diverse group with the common property of containing nitrogen. Some are phenylpropanoids or terpenoids, others are amino acid relatives; a few have important druglike or inhibitory actions in forage digestion (see Section 13.7). Probably more remain to be discovered in forage plants.

Protease inhibitors, including lectins, are hemagglutinins that occur widely in plants. Some are inactivated on heating. Since they are proteins, inactivation by rumen bacteria is quite likely, so these compounds have more serious effects in nonruminants than in animals with pregastric fermentation. The ability of proteins to reduce proteolysis and ammonia production in the rumen is relatively unexploited by ruminant nutritionists.

Silica in the form of biogenic opal forms part of the cell wall structure in grasses. Silica's inhibitory effect on digestion is discussed in Section 9.7.

13.1 The Ecology of Poisoning

Toxic substances defend plants against herbivores. The herbivores, in turn, use feeding strategies that minimize the effects of these secondary substances. Investment by plants in secondary protection systems

Table 13.1. Defensive compounds in forage plants

	Occurrence	Biological effect
Phenylpropanoids		
Lignin	All woody structures	Limits digestibility of cell wall
Proanthocyanidins (condensed tannins)	Many plants	Limit protein and carbohydrate digestion
Isoflavones	Legumes	Estrogenic, antifungal
Gallate esters		
Hydrolyzable tannins	Oak trees, many plants	Limit protein digestion? Toxic
Terpenoids		
Terpenes	Conifers	Toxic?
	Artemisia	Toxic (adaptable)
Rubber latex	Dandelion, etc.	Bitter taste
Saponins	Legumes, etc.	Toxic, bloat?
Waxes and wax polymers		
Cutin	Surface protection	Reduce digestion
Lectins		
Proteins	Widely distributed	Protease inhibitors
Phytohemagglutinins	Legume seeds	
Glucosinolates	Cabbage, etc.	Antithyroid
Silica		
Opal	Grasses, sedges	Inhibits digestion; sharp edges (decrease intake)

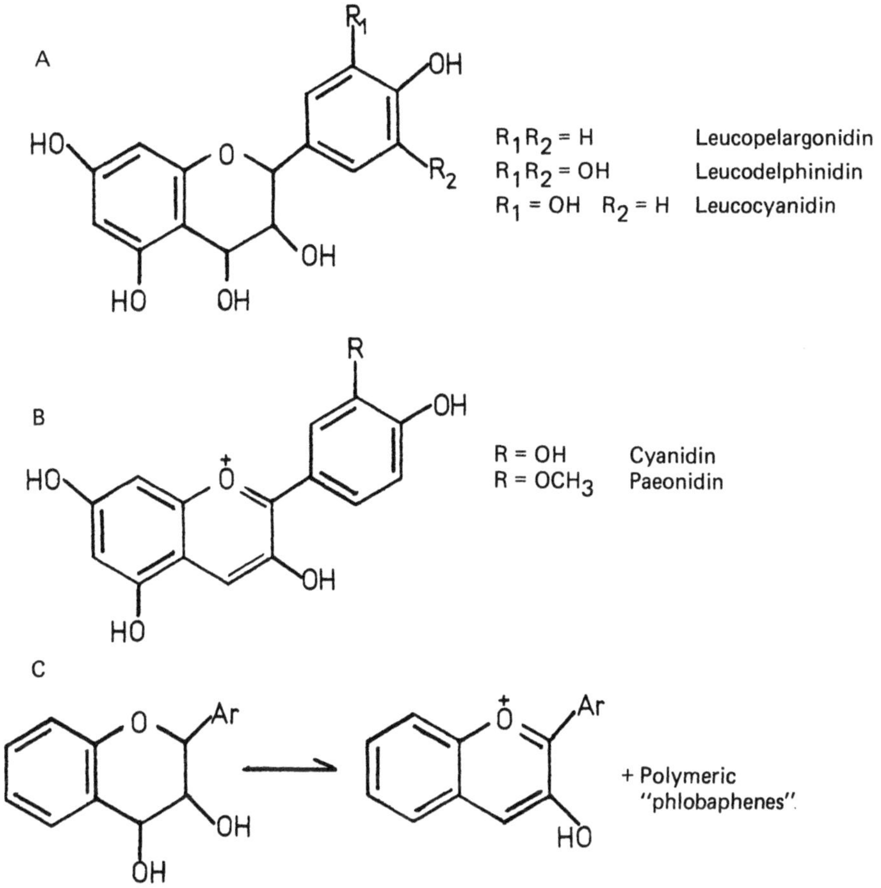

A

R₁R₂ = H   Leucopelargonidin
R₁R₂ = OH   Leucodelphinidin
R₁ = OH R₂ = H Leucocyanidin

B

R = OH   Cyanidin
R = OCH₃   Paeonidin

C

+ Polymeric "phlobaphenes"

Figure 13.1. Flavonoid substances in plants. (A) Leucoanthocyanidins. (B) Related flower pigments. Leucoanthocyanidins and condensed tannins having the flavan-3, 4-diol structure are converted into flavylium salts and are also polymerized by boiling with acid (C). These are monomeric relatives of the condensed proanthocyanidin tannins. (See Figure 13.3.)

can be energetically expensive, and such protection systems are often "turned on" only in response to overpredation or other stress-causing factors such as adverse weather and disease. Increased levels of tannins, estrogens, and cyanogens in forage plants are all associated with stress.

Animals in mixed grazing systems avoid ingesting too much toxic material (Rhoades and Cates, 1976), but if overgrazing removes most of the good forage, animals may be forced to consume plants containing toxic material. Ruminants often graze or browse lightly on potentially toxic plants, and adapted rumen organisms detoxify many but not all secondary substances (Allison, 1978; Carlson and Breeze, 1984; Krumholz et al., 1986; G. S. Smith, 1986; Cheeke, 1988). The cafeteria feeding style diversifies intake and minimizes risk (Kingsbury, 1978). The tendency toward single-species pastures (e.g., tall fescue) eliminates animal choice and thus emphasizes particular problems associated with given plant species.

13.2 Flavonoids

The simple flavonoid substances as a whole are relatively harmless, although they may contribute to a bit-

ter taste. Some of them are related to the condensed tannins, and these may have antinutritive characteristics. Many flavonoids occur as glycosides. Flavonoids arise from the shikimic acid pathway (Figure 12.1; see Stafford, 1989).

Figure 13.1 shows the structures of some plant flavonoids. The leucoanthocyanidins include the anthocyanidins (flower pigments) and the related flavones and flavonols. The catechins and epicatechins are flavan-3-ols, and leucoanthocyanidins are flavan-3,4-diols or flavan-4-ols. All these compounds are bases for the so-called condensed tannins, or proanthocyanidins, which tend to turn red when boiled with acid. The color varies depending on the degree of hydroxylation. Hydrolyzable tannins give no color. The type of tannin involved can be determined by chromatographic analysis of the colored components generated by the acid and their spectra. Tannins are often tightly complexed with proteins and polysaccharides and cannot be extracted into solution. They may be measured as "lignin" in conventional Klason-based procedures. The insolubility of the more reactive forms can cause color tests to give the wrong impression of the real tanniniferous activity in a forage plant because the less active components are more available for extraction.

13.3 Tannins

Tannins, a diverse group of polyphenolic substances of no single biogenic origin, are common factors affecting food taste and protein availability. The classic definition of tannins is that they convert hide into leather, and all the early work on tannins was devoted to this commercial application. The science of animal nutrition has brought other aspects of tannins to the fore. Tannins may bind to salivary proteins, thus producing the familiar astringent taste. Tannins may also be effective enzyme inhibitors. Their diversity may mean that some tannins react specifically with certain proteins and not with others, and this leads to the question of whether they are general precipitants and inactivators or specific inhibitors of enzymes. Tannins are thought to be among the plant protective factors elaborated to prevent predation by herbivores generally, although insects may be the major source of pressure (Swain, 1977, 1979). Many plants increase their tannin or anthocyanin content in response to stress or death of plant tissue. The autumn color change in leaves is evidence of condensed tannin formation. Many tannin products inhibit cellulolytic organisms (Mandels and Reese, 1963).

Haslam (1989) defined tannins as water-soluble polymeric phenolics that bind to proteins, but many related phenolics such as the anthocyanidin flower pigments do not fit this definition because they have little antinutritive effect. Many soluble phenolics analogous structurally and chemically to tannins do not precipitate proteins, although they may inhibit enzymes through complexation. There are also insoluble polyphenolic cell wall fractions that react as tannins (Reed et al., 1982). These phenolics are probably cell wall–linked. Most soluble polyphenolics have a bitter or astringent taste. Some are antimicrobial, and they may also bind to carbohydrates and other nonproteinaceous polymers. The insoluble tannins tend to appear in the lignin fraction.

13.3.1 Classification of Tannins

A tannin is any phenolic compound of sufficient molecular weight that contains enough phenolic hydroxyls to form strong complexes with protein and other macromolecules. Conventional classification of tannin recognizes two major groups: hydrolyzable tannins and condensed tannins (proanthocyanidins). This division is an oversimplification because some tannins contain functional properties characteristic of both groups, while other diverse polyphenolics that have tannin-like properties do not fit into either category— as, for example, the soluble lignins.

Hydrolyzable tannins split under mild acid or alkaline conditions into sugars and phenolic carboxylic acids, most of which are either gallic acid or its derivatives; condensed tannins do not hydrolyze under these conditions. Hydrolyzable tannins, which are also cleaved by hot water and tannases, consist of a carbohydrate core with phenolic carboxylic acids bound by ester linkages (Figure 13.2). There are three, possibly four, classes of hydrolyzable tannin: gallotannins, including common tannic acid (glucose and gallic acid); ellagitannins (ellagic acid and glucose); tara-gallotannins (gallic acid and quinic acid as the core); and, perhaps, caffetannins (caffeic acid and quinic acid).

Condensed tannin phenolics are biphenyl condensation products of phenols. The biphenyl linkage is resistant to cleavage by hydrolysis, hence the term *nonhydrolyzable*. It is obvious, however, from an examination of the structures of the ellagitannins and gallic acid that the former are condensed forms of the latter (Figure 13.2). Likewise, within the proanthocyanidins a variety of compounds exist through biphenyl condensation of the simple triple ring unit.

Proanthocyanidins, like leucoanthocyanins, turn red when treated with acid and form an oxonium group with depolymerization. The depolymerization may be incomplete, with more condensed products leading to insoluble red matter (Figure 13.3). This is the basis of the butanol-HCl test. Proanthocyanidins have been categorized on the basis of the flavonol monomer present in the tannin, although the polymerized forms also yield color in hot acid. All proanthocyanidins are composed of di- or tri-hydroxybenzyl nuclei bonded via carbon atoms or chains, or aliphatic heterocyclic structures. The basic unit is the flavonoid nucleus, which consists of a phenylpropanoid unit and another attached phenyl ring. The formation of the oxonium group is catalyzed by acid, and as a result, reflux with the acid detergent in the ADF procedure results in a pink or red color when colorless condensed tannins are present. This color is a presumptive test for proanthocyanidins.

Insolubility is a major problem because proanthocyanidin characterization depends on obtaining solutions of the respective polyphenols. The insoluble forms may have greater biological activity than the more easily extracted ones. Insoluble condensed proanthocyanidins occur in many tropical and temperate plants, including cassava (*Manihot esculenta*) and *Glyricidia* sp., and also in prepared feeds that originally contained soluble tannins, such as grape waste. If these insoluble tannins are present, the reaction with butanol-HCl may give only red-dyed fiber and limited amounts of color in solution that can be read spectrophotometrically.

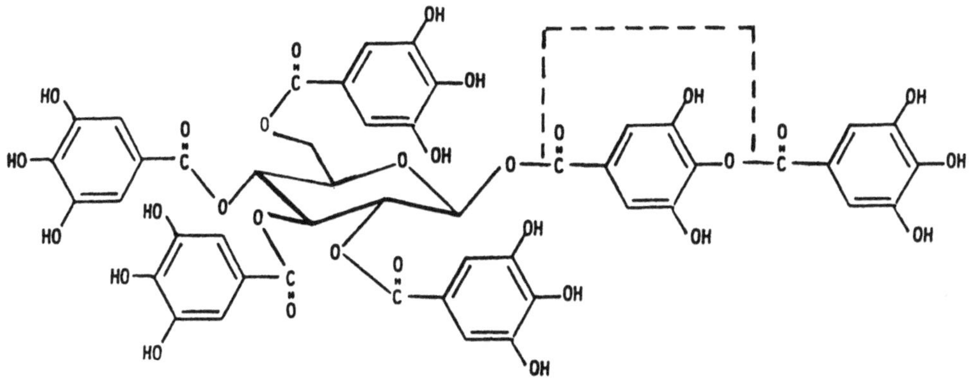

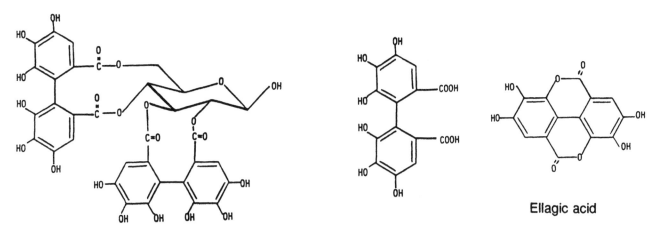

Gallic acid

Gallotannin

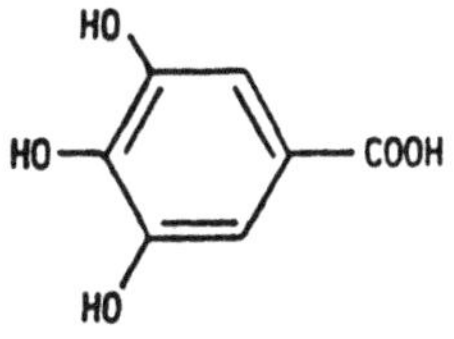

Ellagitannin

Hexahydroxydiphenic acid

Ellagic acid

Figure 13.2. Hydrolyzable tannins: gallotannin and ellagitannin and their hydrolysis products (see Hagerman et al., 1992). Dashed line outlines repeating gallic acid unit. Hexahydroxydiphenic acid spontaneously dehydrates to the lactone form as ellagic acid.

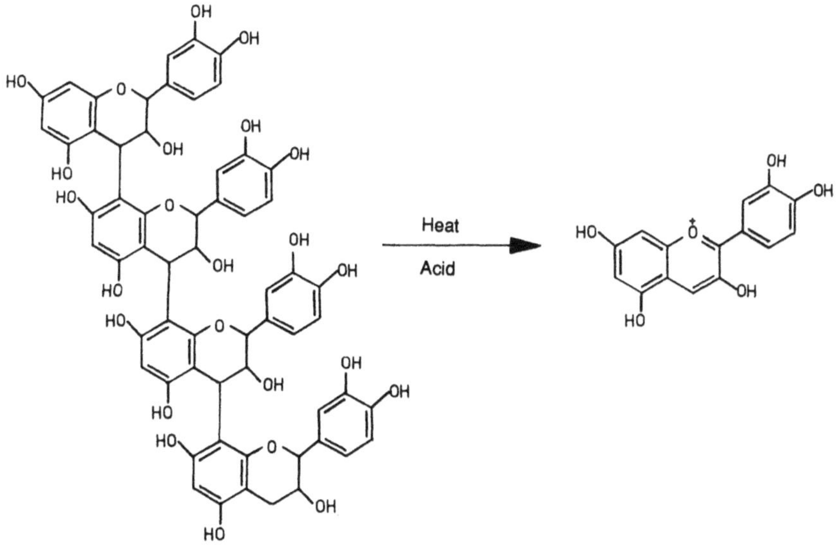

Heat

Acid

Figure 13.3. Structure of proanthocyanin tannins (see Mueller-Harvey and McAllan, 1989). Heating with strong acids causes some depolymerization and the formation of colored anthocyanin products.

13.3.2 Tannins in Plants

Not all plants contain tannins. Tannins have been selected against in common food plants such as hybrid maize, alfalfa, and most of the domestic grasses. Food plants that do contain tannins include red wheat (in the bran; the tannins cause the color), barley grain, and bird-resistant sorghum grain (Hewitt and Ford, 1982). Bird resistance is related to tannin content, and tannin has been selected for as a means of crop protection (Reed, 1987), but the tannins lower the protein quality for animal and human use (Hulse et al., 1980). Many warm-season legumes and browses contain tannins, which may be the most important antiquality factor in these plant species (Reed, 1986).

Many legume seeds, including fava beans (*Vicia faval*), red beans (*Phaseolus vulgaris*), and the pericarp (skin) of the peanut (*Arachis* sp.) contain tannins (Utley and Hellwig, 1985). Colored seed coats of grains are often an indication of the presence of tannin (Theander et al., 1977). Acorns are very high in hydrolyzable tannins.

The best-known characteristic of tannins in foods or feeds is the astringency caused by the precipitation of salivary mucoproteins. Tannins are partly responsible for the flavor in beer, wine, tea, and some fruit juices. They are formed in black tea, probably by enzymatic oxidation resulting in polymers of epigallocatechin (Haslam, 1989). Apples, cranberries, and grapes have high tannin levels. Tannins in red grapes are mainly in the skin and are the source of the tannins in red wine, which are mainly of the condensed type (proanthocyanins). They contribute to astringency, color, flavor, and storage stability. The amount of tannin varies considerably with the type of grapes used to make the wine (Haslam, 1989).

The astringency tends to decrease with ripening in bananas, persimmons, peaches, plums, apples, and stone fruits in general. The decrease in astringency is likely due to polymerization to less-soluble and reactive products that are recovered in crude lignin. Sucrose formation from starch along with the loss in astringency are responsible for the desirable flavor in ripening fruit. Oxidative polymerization of hydrolyzable tannins produces resistant condensed structures that are chemically and technically in a different class from regular condensed tannins.

13.3.3 Tannins and Lignin

Tannins are formally distinguished from lignin by their relative solubility and their ability to combine and form leather-like precipitates with proteins. Proteins bound in this form are probably resistant to attack by proteases, particularly so if they are combined with condensed tannins; hydrolyzable complexes may be digestible. It is possible that any phenolic compound in sufficient quantity can precipitate proteins out of solution. Thus the definition could also depend on solution conditions. The tannins that are not soluble but still react with proteins (e.g., those in *Glyricidia*) present another case, which is less well documented because the insoluble tannins often escape measurement. The choice of tannin assay may resolve some of this confusion (Hagerman and Butler, 1989). However, these methods work only if the tannin is extractable.

Lignin is covalently linked to cell wall carbohydrate and condensed with ether and biphenyl linkages into an insoluble complex. Some tannins are associated with the cell wall and do not extract with aqueous acetone or methyl alcohol solvents, but they are partly released when treated with acid, unlike lignin. In cassava leaves, for example, acid boiling removes much of the tannins that would otherwise be isolated as lignin. These tannins do not show up in normal tannin assay procedures. Solubilization of tannin appears to require acid hydrolysis and thus produces the difference between lignin content of neutral-detergent (ND) and acid-detergent (AD) prepared residues. If ND-extracted tissue is dried, oxidative polymerization of the polyphenolic material may result. Drying can therefore lead to a resistant product which acid boiling can no longer dissolve. The general susceptibility of polyphenols to oxidative condensation on drying leads to an indirect method of tannin analysis using sequential detergent extraction. This method is not very satisfactory, but it can be used when nothing else will work.

Lignin as defined by Klason isolation is extractive-free acid-insoluble matter (Section 12.1.2). The permanganate refinement of the Klason method that limits the definition to oxidizable phenolic matter does not distinguish tannins from lignin. Hence confusion arises. Tannins are supposed to be extractable but sufficiently polymerized to complex protein and form precipitates. These precipitated products then defy extraction and remain with crude lignin and are determined as such. Insoluble tannin complexes usually turn red when treated with butanol-HCl, and the crude lignin fraction from unheated feeds has an unusually high nitrogen content. Since both tannins and lignin are polyphenolic products, and isolated forage lignins always include some nitrogen and possibly condensed or Maillard products, the distinction between tannins and lignin becomes blurred.

All lignins contain the vanillin group either in ferulic acid or coniferyl alcohol. This gives the well-known red reaction with acid phloroglucinol (1,3,5-trihydroxybenzene; Figure 13.4). Proanthocyanidins that contain a 1,3,5-substituted ring turn red in acid vanillin. Vanillin does not stain lignin; thus a color distinction can be made to detect proanthocyanidin residues in cell wall preparations.

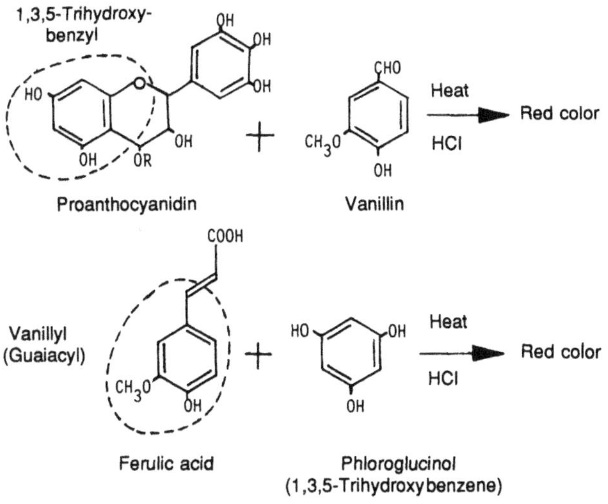

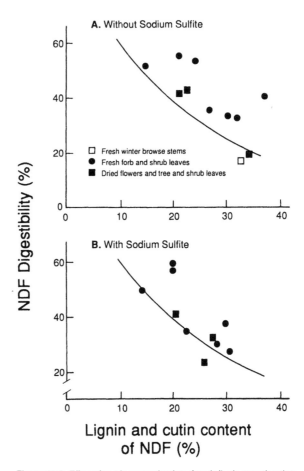

Figure 13.4. Comparison of the vanillin-HCl test for tannins (A) and the phloroglucinol test for guaiacyl (vanillin) residues in lignin (B). Each reaction is the inverse of the other. Proanthocyanidin tannins contain a 1,3,5-trihydroxybenzyl group that reacts with vanillin in acidic solution; ferulic acid and other guaiacyl components of lignin contain a vanillyl group.

Figure 13.5. Effect of tannin contamination of crude lignin on estimation of digestibility with the summation equation (from Hanley et al., 1992). The line in each figure is the expected digestibility for deer consuming nontanniniferous forages. Sodium sulfite removes some of the contaminating tannins and improves the relationship, as shown in the lower figure.

Lignins extracted by alkaline cleavage can form soluble products with tanning properties (McBurney, 1985). This may well be one of the limitations of delignification in which the lignin product remains in the treated forage residue. Tanning capacity is speculatively related to the size of the polyphenolic molecule. Less polymerization leads to weak complexes with protein, and overpolymerization leads to large complexes with limited or no solubility that will appear as lignin. The reactivity of large insoluble polymers is not well understood. Large size might reduce reactivity, and the occurrence of insoluble protein-tannin complexes may involve condensation after the formation of a protein-tannin complex involving an initially soluble tannin. On the other hand, insoluble reactive tannins are known. Any insoluble phenolic complex is recoverable as crude lignin. The contaminating tannin does not reduce carbohydrate digestibility as much as lignin does (Hanley et al., 1992), causing an overestimation of digestibility. The use of sulfite in preliminary extraction of neutral-detergent fiber (NDF) will reduce this error somewhat (Figure 13.5). Much of the tannin-protein complex contaminating the crude lignin seems to be dissolved by sulfite.

13.3.4 Biological Effects of Tannins

The tannins may include a group of cellulase inhibitors; these have been reported in many wild plants browsed by ruminants (Mandels and Reese, 1963; Robbins et al., 1975). Such inhibitors are generally absent from domesticated forages, although they are present in crown vetch, bird-resistant sorghums (Reed

et al., 1987), and *Sericea* (Smart et al., 1961). Extraction of the soluble tannins can improve digestibility of the cell wall. Adding the extract to a standard forage in an in vitro rumen culture reduces cell wall and protein digestibilities (Robbins et al., 1975; Van Hoven and Furstenburg, 1992). Whether the reduction in digestibility constitutes true inhibition (i.e., specific interference with enzymes) or an unspecific precipitation of proteins is not known.

The content of tannin-like substances varies among the browses and forages that contain them; tannin levels respond to disease, stress, and attack by fungi. A postulated model of this interaction is shown in Figure 13.6 (Mandels and Reese, 1963).

The fate of tannins following ingestion may vary depending on the type of tannin. Most tannins form complexes with protein in saliva or the rumen contents. Hydrolyzable tannins hydrolyze in gastric acidity beyond the rumen, releasing protein, amino acids, and small units of phenolics that likely pass to the urine

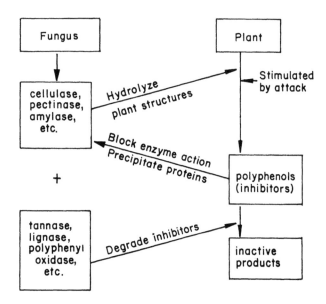

Figure 13.6. The interaction of plant and fungal pathogens. (Modified from Mandels and Reese, 1963.)

(A. K. Martin, 1982). The effects of condensed tannins on protein digestion usually are more negative than the effects of hydrolyzable ones. It should be understood, however, that this applies to preformed complexes such as those that might occur in dried or prepared feed. Drying promotes the combination of tannin and plant protein before ingestion, while the ingestion of fresh tanniniferous herbage may have a less serious effect. Proanthocyanidin tannins are kept in special organs in the leaves to prevent their interference with the plant's own metabolic apparatus, and this factor

opens the door to the tannin-tolerant animal browsers that have tannin-binding proteins in their saliva (Austin et al., 1989). In these instances the salivary factor binds the tannins and spares valuable forage protein.

Attempts to use tannins to promote rumen escape ("bypass") of feed protein (Chapter 18) have had limited or no success. This bypass may occur when fresh forage is ingested before tannin-protein complexes can oxidatively cross-link, as in the case of dry feeds. Post-ruminal digestion of proteins complexed with hydrolyzable tannins is more likely than those complexed with condensed (nonhydrolyzable) tannins. Some tannin complexes may dissociate because of gastric acidity (Mangan, 1988).

Another possible effect for tannins is related to the generally increased salivary flow in animals feeding on tanniniferous forage (Figure 13.7). Microbial protein synthesis provided with adequate salivary urea is a detoxification mechanism that provides extra protein to the animal. A number of studies indicate an increase in microbial protein after feeding moderate levels of tannin (Beever and Siddons, 1986). Paradoxically, nitrogen balance improves in animals that are fed tannins (Terrill et al., 1989), although digestibility of structural carbohydrates may be depressed. This effect is more serious in frozen material than in dried forage (Table 13.2). The increase in microbial protein may be due to the greater liquid turnover resulting from the increased salivary flow.

The increase in fecal nitrogen with tannin feeding can be interpreted as due in part to increased production of microbe-generated protein (Makkar et al., 1988a, 1988b). Whether this represents a net increase

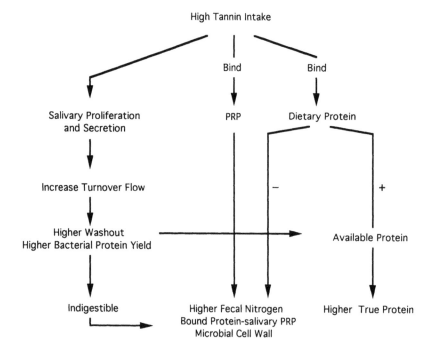

Figure 13.7. Proline-rich salivary proteins (PRP) interact with tannins, and rumen microorganisms respond by increasing the net protein flow and utilization, which is paradoxically associated with increased loss of fecal nitrogen.

Table 13.2. Apparent digestibility and nitrogen balance of high- and low-tannin field-dried (FD) and fresh-frozen (FF) *Sericea lespedeza*

Constituent	High tannin FD (%)	High tannin FF (%)	Low tannin FD (%)	Low tannin FF (%)
Dry matter	45	41	42	46
Neutral-detergent fiber	37	13	32	27
Hemicellulose	53	16	45	33
Cellulose	46	33	43	41
N intake	24	28	26	29
Fecal extraction	12	16	13	13
Total extraction	14	17	15	14
Absorbed N retained (g/kg)	87	91	86	90
N digestibility	53	43	49	56
NDIN[a] digestibility	44	24	32	27

Source: Modified from Terrill et al., 1989.
[a]NDIN = neutral-detergent insoluble nitrogen.

in available protein to the animal is not clear; the possible increase in microbial efficiency may be offset by a reduction in carbohydrate fermentation (Barry and Manley, 1986; Barry et al., 1986).

Polyethylene glycol forms complexes with tannins and has been fed to animals to reduce the inhibitory effects of tannins (Larwence et al., 1984).

13.3.5 Herbivores' Adaptations to Tannin

Wild ruminants such as deer and antelope preferentially consume plant species high in phenols. The fecal losses of metabolic nitrogen are higher in white-tailed deer than in sheep or cattle (Austin et al., 1989). The fecal nitrogen complex probably represents indigestible tannin-mucoprotein complexes.

Simply producing tannins does not guarantee protection against herbivory; many species of ruminants and nonruminants possess the salivary protection factor and may actually specialize on tanniniferous foods. Indeed, some items on the human food list (e.g., red wine) are chosen for their tannin content, for the added flavor. Wild selector ruminants seem to have specialized in tanniniferous browses and forage (Hofmann, 1989) and have detoxification mechanisms involving salivary proteins and urea recycling (Butler, 1989).

Long-term ingestion of tannins induces enlargement of the salivary glands, although not equally across animal species. Many ruminants and monogastrics possess proline-rich proteins that specifically bind tannins (Mehansho et al., 1987), but these binding factors seem to be absent or reduced in the salivas of sheep and cattle (Austin et al., 1989; Figure 13.8).

The tannin-binding factors in saliva that protect protein prevent tannin-protein complexes from forming and thus increase protein digestibility. The true digestibility of protein is probably even higher than it might first appear to be because the salivary binding

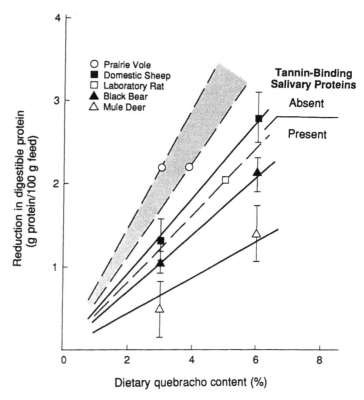

Figure 13.8. Quebracho tannin fed to various animal species reduces digestible protein. Evidence indicates that tannin-binding salivary proteins protect digestible plant protein. (Compiled by Robbins et al., 1991.)

sacrifices nonspecific nitrogen and nonessential amino acids in the form of glucosamine and proline. Fecal loss of nonspecific nitrogen sources bound to the tannins will be part of the difference between the true and apparent digestibility values.

13.3.6 Tannin Assays

Numerous methods have been devised to quantitatively or qualitatively determine the presence of tannins and related phenols in plant material. Tannin analyses may be based on precipitation with protein, adsorption of tannins on insoluble protein (i.e., the old hide powder method or polyvinylpyrrolidone), precipitation with heavy metal salts, formation of colored products by oxidation of the tannins, and UV measurement. Chromatographic separation (HPLC) mass spectrophotometry, fast atom bombardment mass spectrometry, and NMR (Haslam, 1989) are used for specific identification of molecular structure and are then followed by UV spectrophotometry of tannin compounds (Mueller-Harvey et al., 1987). Not all these methods are discussed here, but it is important to recognize the varied objectives of the respective methods. The difficulties of analyses for nutritional purposes result from the alteration of food or feed on drying, similar alterations in sample handling, and the difficulty of extraction. Tannins are fairly sensitive to pretreatments and storage. Usually the main objective is to assay the sample in a condition similar to the state in which it is consumed. The problems encountered and their solutions can thus best be discussed in terms of sample preparation (collection, storage, and extraction), separation into different classes, and analysis. Sample handling can be extremely important because of the reactivity of the polyphenols. The purpose of the investigation should determine the type of analysis that is used.

Protein Precipitation or Adsorption Methods

The fact that tannins precipitate many different compounds is the basis for one of the oldest methods of analysis. Precipitation with a standard protein (e.g., gelatin) is a practical approach that bases the assay on the presumed antinutritional properties associated with tannin (Hagerman and Butler, 1978). The precipitation or binding methods may correlate very well with other methods of analysis such as the vanillin-HCl method and the Folin-Ciocalteu method (Horvath, 1981). The hide powder (powdered animal hide) method, which has been used in the tanning industry for years, measures the loss of polyphenolics from solution. For example, permanganate (a general oxidant not specific to phenolic substances) has been used to oxidize the phenolics before and after precipitation, the tannin being considered equal to the loss. While it has been the standard for many years, this method gives very little information on the qualitative variation among the tannin compounds. The protein used is skin (useful for the leather industry), which may not relate well to the effect that a tannin compound has on digestive enzymes, salivary proteins, feed proteins, and intestinal mucosa.

An alternative is to use gelatin instead of hide powder. The gravimetric method consists of weighing the gelatin before and after it is added to the sample solution. J. Löwenthal substituted the titrimetric use of permanganate for weighing and added an indicator (indigo) to help determine the end point in the titration. This method is an old AOAC (Association of Official Agricultural Chemists) method for measuring tannins in tea. Although it measures oxidizable matter in general, the Lowenthal method does seem to correlate with astringent taste and fruit ripening. It gives no information on quantitative differences in tannin compounds.

Another protein precipitation method uses casein precipitation for measuring the tannins in beer. Another analysis of this type, developed by Bate-Smith (1973), is called hemanalysis. The technique uses he-moglobin (at 500 nm) because it is so easy to measure. A dilute solution of fresh blood (denatured blood does not work since soluble protein is required) is mixed with the sample solution, and the change in absorption after centrifugation is reported in terms of tannic acid equivalence (TAE). The hemanalysis was modified by W. T. Jones et al. (1976) who lowered the temperature and added a buffer. Their results indicate that hemanalysis may be useful in questions concerned with physiological effects. They found the TAE to be correlated with palatability and molecular weight. The amount of tannin precipitated by a standard protein can be measured directly using the method developed by Hagerman and Butler (1978) or indirectly with the radial diffusion method (Hagerman, 1987). The latter is simple, suitable for large numbers of samples, and not affected by solvents used for tannin extraction. Tannin extracts are placed in a well and then diffuse into agar gel, which contains protein. Tannin-protein complex is noted by the formation of a precipitated ring, the area of which is linearly related to the amount of tannin in the extract. Both condensed and hydrolyzable tannins can be determined with this method.

The protein-precipitating capacity (biological activity) of the tannin can be determined by the amount of protein precipitated by extracts from tannin-containing samples, although soluble tannins cannot be measured. Radiochemical (Hagerman and Butler, 1980) or colorimetric methods measure the protein precipitates (Bate-Smith, 1973; Schultz et al., 1981; J. S. Martin and Martin, 1983; Asquith and Butler, 1985; Makkar et al., 1987; and Marks et al., 1987). The radiochemical method is the more sensitive (Hagerman and Butler, 1980), but it requires special equipment and techniques. The blue bovine serum albumin method (Asquith and Butler, 1985) is probably the most convenient, but it is not very sensitive.

Adsorption on Synthetic Polymers

Most investigators use polyvinylpyrrolidone (PVP) in this assay (Andersen and Todd, 1968). Its structure is similar to the urea-formaldehyde complex, which has also been used. The larger PVP polymers are insoluble in water and can be used as adsorbing agents followed by elution of tannins from a PVP column. One of the problems with the PVP method is that the adsorbed tannins are not completely recovered and thus are lost for further study. One way to use PVP is to measure the gravimetric organic matter lost on passing a tanniniferous solution through a column.

Metal Salts

Many metal ions form colored complexes with phenols (e.g., trivalent iron in ink manufacture). Various iron salts, stannous chloride, lead acetate, potassium dichromate, potassium antimonate, copper acetate, mercuric oxide, aluminum hydroxide, zinc acetate, lime water, and rare earth acetates (Reed et al., 1985) have been used to precipitate tannins and other phenolic compounds. The color of the precipitates can be used to distinguish among tannins. Lime water precipitates, for example, are white, red, and blue, although these colors are unstable due to oxidation. Lead acetate has been used to separate orthodihydroxyphenols from other polyphenols (Andersen and Todd, 1968). Decomposition of the tannate precipitate by strong chelating agents such as EDTA allows further study of the isolated tannin. Metal ion precipitation may fail, however, because of the variability in the chelating power of the various tannins. Thus, while trivalent ytterbium precipitates polymerized tannins, lower-molecular-weight compounds, especially those without vicinal hydroxylation, do not participate (Lowry and Sumpter, 1990). It is probable that the phenolics not precipitated by ytterbium have little or no protein-binding capacity.

There is a significant correlation between the amount of ytterbium precipitate and the gelatin turbidity test for tannin. Solutions of ytterbium-precipitated phenolics from samples with a high content of proanthocyanidins precipitate more gelatin than those with a low content of proanthocyanidins at similar levels of ytterbium precipitate.

Methods Based on Phosphomolybdic Acid

Polyphenolic compounds generally possess reducing power, and this provides a means to measure tannins. The most convenient measurement of reducing power is spectrophotometric absorbance of phosphomolybdic acid, in which molybdenum is reduced from valence 6 to 5. The blue pentavalent molybdenum is proportional to the amount of reductant. This method, known as the Folin-Ciocalteu procedure, is an adaptation of the Folin-Dennis method in which the reagent has been modified to make it more sensitive for phenolics. Although it is the least sensitive tannin assay available, the Folin-Dennis method is the assay currently recommended by the AOAC for measuring tannins in tea, and it has been used by plant breeders to select low-tannin varieties of legumes and sorghum (Burns, 1971).

The molybdenum reagent has also been used for analyzing protein, sugars, vitamin C, divalent tin, and iron, all of which can be converted into the reducing equivalent of molybdenum. One problem in applying the molybdenum reduction to tannins is maintaining specificity; a second problem is that partially oxidized tannins have a lower reactivity relative to less polymerized phenols. Thus, Horvath (1981) was unable to

find a correlation between the capacity for blood coagulation and values obtained by the Folin-Ciocalteu procedure. The reaction is not stoichiometric; more likely it depends on the oxidation-reduction potential of the various phenolics (Goldstein and Swain, 1963). The relationship between absorptivity and concentration is almost linear; only the slopes vary with the different phenolics, the more oxidized having lower reactivity.

Protein precipitation is an important aspect of the ecological and nutritional effects of tannins. Precipitation of phenolics with ytterbium is a gravimetric measurement that can be related directly to protein precipitation capacity through the preparation of solutions of ytterbium-precipitated phenolics.

Indirect Measurement by Sequential Extraction

Active tannins are often insoluble and resistant to extraction, particularly in dried feeds and in certain tropical species such as cassava and *Glyricidia*. The failure of extraction techniques has resulted in overlooking the existence of certain tannins and their effects. Thus indirect methods may have to be applied. Although they also have limitations, these methods offer some appreciation of the fractions overlooked by extractive methods.

Some tannins are soluble in detergent solutions, but this varies according to whether acid detergent or neutal detergent or both is used, and with the dryness of the residues. If tannins dissolve in ND, sequential in vitro digestion will show an increase in cell wall digestibility relative to an unextracted control (Robbins et al., 1975). Any tannins remaining in the NDF become unextractable with sequential acid extraction as a result of oxidative polymerization as the NDF dries. Thus the residue obtained by sequential extraction first with ND and then with AD (ND → AD) differs in composition and yield from that obtained by the reverse order, AD followed by ND (AD → ND). The detergent procedures also influence the digestibility (Rebolé et al., 1988).

Figure 13.9 outlines the substances that should be present in each residue after extraction. It is important to realize that some tannin complexes may be insoluble in ND, AD, or both. The presence of tannin complexes is indicated by a large difference between ADF and AD → ND residues; the difference between the two residues should equal the insoluble tannin complex, although interfering factors complicate the result. Tannins soluble on an initial ND extraction will be lost, although the more reactive forms will be less soluble. Pectins, biogenic silica, and tannin complexes (soluble in ND and insoluble in AD) would also be removed by ND, making it difficult to use this as the sole measurement of tannins. One possible way to adjust for these

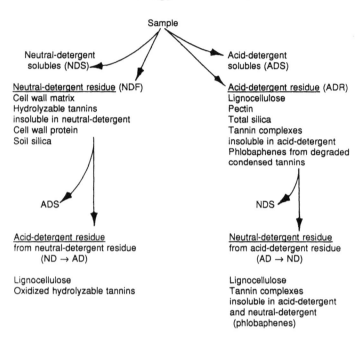

Figure 13.9. Double sequential analysis for tannins (from Van Soest et al., 1987).

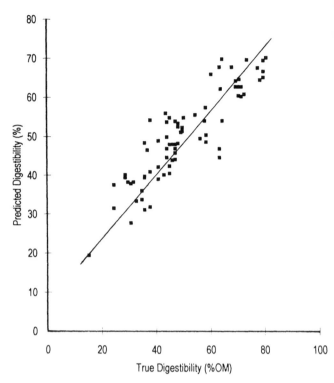

Figure 13.10. Regression of predicted digestibility of organic matter using a summative equation adjusted for (ND → AD) − (AD → ND) (from Conklin, 1987). Inclusion of this correction improved the correlation from 0.87 (unadjusted summative equation, see Section 25.5.3) to 0.90. Regression, Y = 0.83X + 6.9.

interfering substances is to compare the residues obtained by ND → AD and direct ADF.

The acidic conditions used in the AD extraction induce a reaction similar to that seen in the butanol-HCl test for condensed tannins, and AD extraction similarly results in red colors. Proanthocyanin tannins and hydrolyzable ones, too, tend to be dissolved in acid detergent. Neutral-detergent extraction should remove pectins, silica, and some condensed tannins, but the conditions of the ND extraction (high heat in the presence of oxygen, at pH 7) are such that polymerization tends to occur and acid-resistant complexes form. The detergent may break apart some of the weaker tannin-protein complexes. Hydrolyzable tannins that remain in the ND fiber will be hydrolyzed in the AD, and the condensed tannins will yield anthocyanins and phlobaphenes. The only components that would be present in the ND→ AD residue and not in the AD residue would be products polymerized by drying the ND residue and resistant to sequential AD, and components of a tannin complex (insoluble in ND) that are also insoluble in AD.

Sequential extraction gives the following results:

1. The direct AD residue minus the AD → ND residue is a measure of the amount of pectin, biogenic silica, plus tannin complexes soluble in ND after AD treatment.

2. The residual difference between NDR and ND→ AD is used to estimate the amount of hemicellulose (Chapter 10), but it will also measure the amount of cell wall protein and tannin complexes that are soluble in AD from the NDF.

3. The residual difference between NDR and AD→

ND may be a method of quantifying hemicellulose, cell wall protein, and tannin complexes that are insoluble in ND but soluble in AD; however, tannin complexes formed in AD and during drying of the prepared residues, and that are insoluble in ND, would reduce the ND − (AD → ND) value (e.g., phlobaphenes).

4. The ND → AD and AD residual difference is a measure of the amount of tannin complexes retained in ND on hot-drying that are subsequently insoluble in AD. Pectin and biogenic silica decrease the difference between NDR and ADR. Positive values were obtained from maple leaves (Horvath, 1981).

5. The residual difference between ND → AD and AD → ND is a value that results from the balance between tannin complexes formed in ND or during drying that are subsequently insoluble in AD, and tannin complexes resulting from AD conditions and insoluble in ND from the ADF after drying. Positive values therefore indicate that more complexes are produced in ND or on drying that are then insoluble in AD than complexes that are the result of AD extractions and drying that cannot be removed by ND.

Adding the (ND → AD) − ADR difference to the summative equation (Goering and Van Soest, 1970) improves the correlation with in vitro digestibility (Figure 13.10).

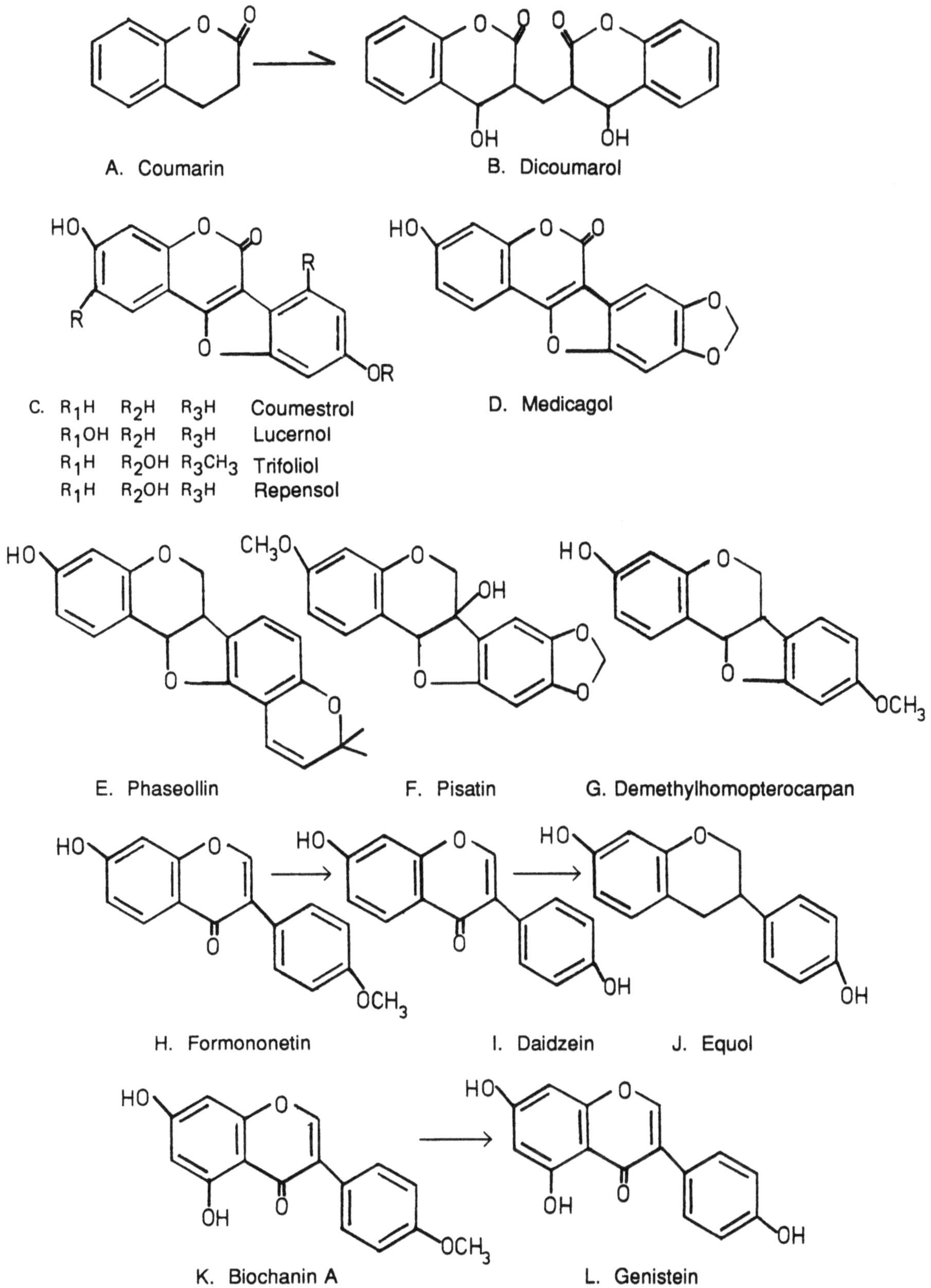

Figure 13.11. Coumarins (A, B) and isoflavonoid compounds (C–L) in forages. Coumarin (A) is a volatile component of clovers responsible for the odor of new-mown hay. It is metabolized to dicoumarol (B), an anti–vitamin K factor in molded or spoiled hays. The coumestan groups (C, D) causes the estrogenic activity of alfalfa and white clover. The pterocarpans—pisatin (E), phaseollin (F), and demethylhomopterocarpan (G)—are constituents of peas, alfalfa, and clovers with antifungal activity. They increase in diseased forage plants (Barnes and Gustine, 1973; Wong, 1973). Formononetin (H) occurs in subterranean clover and is metabolized to daidzein (I) and equol (J) in the rumen. Biochanin A (K), also in clovers, is metabolized to genistein (L). Equol and genistein are responsible for clover's observed estrogenic activity in sheep (Shutt and Braden, 1968).

The ND → AD sequence gives higher values for crude lignin when tannins are present. These tannins remain tightly bound to cell wall and protein and thus likely have major effects on digestibility. The relationship between the expected digestibility correction for these tannins is highly correlated with in vitro digestibility. These procedures for assaying insoluble tannins complement analyses for soluble tannins that can be extracted with 70% acetone or methanol, followed by precipitation with ytterbium acetate.

13.4 Isoflavones and Coumarins

Isoflavones are phenylpropanoid compounds in which substitution occurs at the alpha carbon (carbon 3 in the flavone ring) of the phenylpropanoid group, in contrast with anthocyanidins and proanthocyanidins, which are linked through carbon 3 in the flavone ring (see Figure 13.11). Many isoflavones are metabolically active substances, often with hormone activity (Setchell and Adlercreutz, 1988). The simplest member of this series is coumarin, which is widely distributed in legumes. A volatile substance with the aroma of new-mown hay, coumarin is common at least in the clovers (Cansunar et al., 1990). Sweet clover hay poisoning is caused by dicoumarol, a fermentation product of coumarin in moldy hay (Figure 13.11). The hemorrhagic action of dicoumarol and its derivatives is due to specific anti–vitamin K activity. Other relatives of coumarin have toxic effects in animals as well (Wong, 1973), but their importance in forages and effects on ruminants remain to be investigated. Many of these substances have antifungal properties.

Of particular importance is the estrogenic activity of several isoflavones present in legumes. Many isoflavones (Figure 13.11) have a low but variable level of estrogenic activity sufficient to cause problems in grazing animals on legume pastures. These substances can affect fertility in male sheep and cause sterility in females. The estrogenic activity is to some degree the result of rumen metabolism of isoflavones since the fermentation products can be more active than the parent compound. Equol is the dominant rumen product in sheep grazing on clover pastures in Australia and may be responsible for reproductive problems (Nilsson et al., 1967; Shutt and Braden, 1968). Estrogenic substances in legume hay may be responsible for more efficient use of feed in beef cattle and lactating dairy cows.

The concentration of isoflavonic estrogenic substances in alfalfa and clovers depends on the incidence of foliar disease or stress, either of which may increase levels by as much as an order of magnitude. The specific effect on foliar disease of the estrogenic substance is not understood. Other polyphenols and tannins may

respond to plant stress in a similar way; a general theory of pathogen-plant interaction has been postulated to account for this variation in forage quality (Figure 13.6).

13.5 Terpenoids and Essential Oils

The terpenoid class includes volatile terpenes, saponins, and steroids. Essential oils are a diverse group of organic substances in plants that have the common property of volatility and solubility in organic solvents. Esters, ethers, phenols, and members of the cinnamic acid family related to lignins are essential oils. Terpenes and essential oils are all low-molecular-weight compounds. Of interest here are substances that exhibit antimicrobial activity, and these include phenolics and probably some terpenoids. Some plants high in essential oils—for example, sagebrush (*Artemisia*)—are known to possess anticellulolytic principles, but browsing ruminants can adapt and detoxify these compounds. Adapted animals that consume sagebrush excrete detoxified products in the urine; the volatile terpenes in conifers are probably dealt with the same way (Oh et al., 1968).

Saponins can be toxic to nonruminant herbivores and have been suggested as factors causing bloat (Section 15.7.1) in ruminants. The complexity of bloat etiology and the likely existence of more than one kind of bloat have left the contributory role of legume saponins in doubt. Saponins are widely distributed in legumes (Oleszek, 1988) and can cause the formation of stable foams; they also promote hemolysis of red blood cells. Sarsaponin is a slobber factor that has been used to study salivary flow and passage (Goetsch and Owens, 1985). Saponins inhibit growth in many organisms, including the cellulolytic fungus *Trichoderma viride,* which is used to assay for saponin. It is not known whether saponins inhibit rumen bacteria. Rumen bacteria apparently have some ability to detoxify them.

13.6 Cutin

The major nonphenolic fraction of crude lignin is cutin, which is separated from crude lignin as the fraction resistant to oxidation (Meara, 1955). Cutin is not removed by most delignifying agents and is thus a component of crude cellulose preparations. There are two main fractions: the cuticular waxes, sparingly soluble but extractable with hot nonpolar solvents, and a polymerized fraction containing esters of hydroxy long-chain acids and alcohols (Table 13.3). The polymeric fraction is saponified by alcoholic alkali solu-

Table 13.3. Long-chain constituents of leaf waxes and cutin

Waxes
Even[a]
$CH_3(CH_2)_nCO_2H$
$CH_3(CH_2)_nOH$
$CH_3(CH_2)_nCHO$
$HO(CH_2)_nOH$
$HO(CH_2)_nCO_2H$
$HO_2C(CH_2)_nCO_2H$
Odd[a]
$CH_3(CH_2)_nCH_3$
$CH_3(CH_2)_nCHOH(CH_2)_mCH_3$
$CH_3(CH_2)_nCO(CH_2)_mCH_3$
$CH_3(CH_2)_nCX(CH_2)_4CX(CH_2)_mCH_3$
$CH_3(CH_2)_nCOCH_2CO(CH_2)_mCH_3$

Constituent fatty acids of cutin[b]
$HO(CH_2)_{15}CO_2H$
$HO(CH_2)_6CH(CH_2)_8CO_2H$
| OH

$HO(CH_2)_{17}CO_2H$
$HO(CH_2)_8CH(CH_2)_7CO_2H$
| OH

$HO(CH_2)_8CH\text{-}CH(CH_2)_7CO_2H$
| OH OH

$HO(CH_2)_8CH\text{—}CH(CH_2)_7CO_2H$
\ /
O

$HO(CH_2)_8CH\text{=}CH(CH_2)_7CO_2H$
cis

Source: Eglinton and Hamilton, 1967.

Note: In the natural unsaponified wax the alcohol and carboxylic acid functions are often present as esters.

[a]The *n* and *m* are alternatively odd or even, the chain length being in the range C_{20}–C_{37}, generally C_{29} or C_{30}. CX is C=O or CHOH.

[b]Bonding includes polyesters and linkages to polyphenols or lignin (Kolattukudy, 1980).

Table 13.4. Lignin and cutin content (percentage of cell wall) of some fibrous feeds

Feed	Cell wall (% DM)	Lignin Acid-detergent	Lignin KMnO_4	Cutin	In vitro holocellulose digestibility[a] (%)
Wheat straw	81	9	14	0.2	58
Timothy (late)	65	11	16	0.9	54
Alfalfa stems	71	16	20	0.4	49
Alfalfa (late)	50	17	19	2.4	50
Sunflower seed hulls	70	26	24	7.5	34
Cottonseed hulls	90	22	13	14	47
Peanut hulls	90	39	17	22	13
Castor seed hulls	80	79	7	71	37
Rice hulls	77	18	9	9	0[b]

Source: Van Soest, 1969b.
[a]Corrected for cutin content.
[b]Contains 22.9% silica.

that of lignin (Table 13.4). The cuticular layer appears to be a barrier to rumen organisms and is apparent on histological examination of leaf and stem sections.

13.6.1 Waxes

Free waxes, a significant component of the cuticular membrane of green forages, tend to be odd-chain-length hydrocarbons C_{21}–C_{37}. They are extractable with nonpolar solvents and separated by suitable chromatography (Spencer and Chapman, 1985). Waxes have become more interesting recently because they are fingerprints of the plant family and species, and at least the higher members ($>C_{33}$) are indigestible. As a result, they could be used as markers in nutrition studies on grazing animals (Mayes et al., 1986).

13.7 Alkaloids

Alkaloids include a wide class of heterocyclic nitrogenous compounds that can exhibit pharmacological activity as well as inhibit digestion. The only common factor among them is that their molecules contain nitrogen. The habit-forming drugs such as morphine and codeine and toxic substances in the potato (solanine) and other members of the Solanaceae (e.g., atropine in deadly nightshade) are examples of this class. Other alkaloid compounds such as caffeine and theobromine in coffee and tea are relatively nontoxic by comparison, although these also have pharmacologial activity. Not all alkaloids are toxic; for example, betanin, the red pigment of beets, is a phenylpropanoid nitrogenous compound providing mainly color and flavor.

The presence of toxic alkaloids in certain forages can cause problems for grazing and browsing ruminants. Toxic alkaloids in cultivated grasses include the indolalkylamine compounds in reed canarygrass, the

tions. Some of the waxes can be extracted with ether but are present in acid-detergent fiber and Klason (sulfuric acid) lignin. Cutin is not measured by the permanganate or other indirect lignin assay procedures, but it can be partitioned from crude lignin and measured by sequentially treating acid-detergent fiber with 72% sulfuric acid and permanganate (Van Soest and Wine, 1968).

The proportions of cutin and lignin vary in different plant species and among tissues of the same plant (Table 13.4). The cuticular surface of leaves offers a barrier to digestion and is the main indigestible fraction of leafy nonlignified vegetable tissues. Thus unlignified vegetable cell walls vary in digestibility according to cutin content; for example, curly thin leaves of lettuce are less digestible than the thicker fleshy tissue of cabbage. Cutin in seed hulls, pollen grains, and barks is often combined with lignin, causing low digestibility. Indigestibility of hulls and bark cannot be accounted for by true lignin alone (Enzmann et al., 1969), a factor favoring the use of acid-detergent lignin in the analysis of any feed material containing significant cutin content, particularly in concentrate feedstuffs. The relation of cutin to digestibility differs quantitatively from

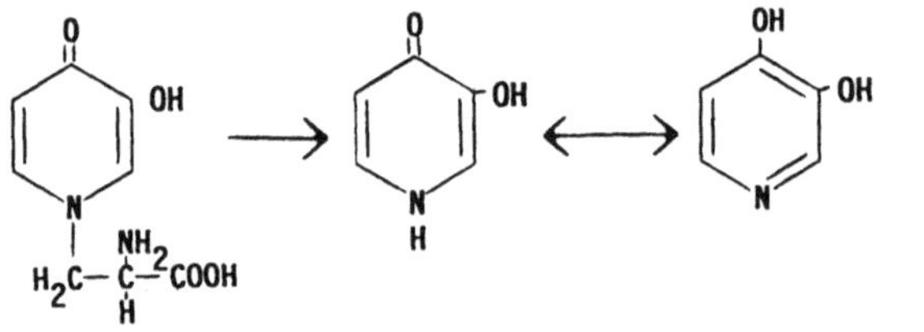

Figure 13.12. Mimosine and its metabolic products (see Tangendjaja et al., 1985).

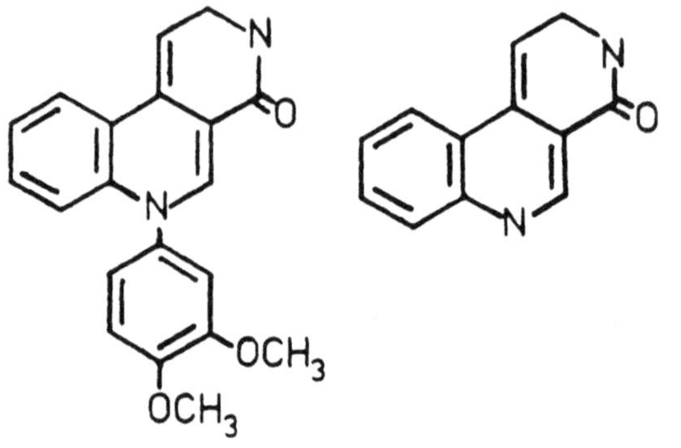

Figure 13.13. Perloline alkaloids in fescue (from Bush and Buckner, 1973).

perloline group in fescue, and the ergot-like substance that causes fescue foot. The most interesting alkaloid in tropical legumes is mimosine in *Leucaena*.

13.7.1 Mimosine

Mimosine is an alkaloid found in *Leucaena leucocephala,* a legume widely distributed in tropical areas as a fodder tree. *Leucaena* poisoning causes loss of hair and death in animals that have overgrazed it. The compound responsible is mimosine (Figure 13.12). For a time Australian agronomists selected for low-mimosine strains, but rumen bacteria capable of degrading and utilizing mimosine have been found (Jones and Megarrity, 1983; Tangendjaja et al., 1985), and more recent work on mimosine has shifted to inoculating animals with bacteria that can detoxify it.

Mimosine is metabolized by rumen bacteria to dihydroxypyridine (DHP), a goitrogen (Figure 13.12). This conversion explains the sometimes mixed symptoms of *Leucaena* poisoning. Adapted rumen organisms that can metabolize DHP to harmless products were first found in goats in Hawaii. They have also been found in Indonesia (Tangendjaja et al., 1985).

Mimosine toxicity is interesting because it is the first clear-cut case of nonadaptability of rumen bacteria. Rumen organisms can usually adapt to single compounds after exposure for a few weeks or more, but this is not the case with mimosine, the degradation of which depends on specific organisms.

13.7.2 Fescue Toxicity

Tall fescue has long been known to be an unpalatable grass. Fresh forage intake trials have shown consistently lower consumption of immature forage. In fact, tall fescue is one of the few plants in which voluntary intake increases with maturity. The reason is that perloline alkaloids are apt to be present in higher concentrations in less mature forage (Figure 13.13). Perlolines inhibit rumen organisms, reducing cellulose digestion in vitro (Bush et al., 1970), but they are not responsible for fescue foot (J. A. Jackson et al., 1984). The reduced intake of immature fescue may be the

result of the limiting effect on rate of digestion exerted by perloline compounds.

13.7.3 Fescue Foot and the Endophyte

In addition to its perloline alkaloids, tall fescue is also noted for causing "fescue foot," indicating the existence of an ergotamine type of alkaloid. Such alkaloids have not been found in fescue per se, but they are found in endophytic fungal infections associated with the grass (Bacon and Siegel, 1988). Animals grazing on tall fescue pastures may exhibit ergot-like symptoms with necrosis in the feet, tail, and ears. The incidence is low in proportion to the amount of fescue grazing, yet occurrences tend to cluster. The toxicity has been identified as coming from an endophytic fungus transmitted through seed, and endophyte-free fescue has been produced (Bacon, 1988). The endophyte-free plants are chemically less well defended, and the improved animal productivity is offset somewhat by reduced yield of forage (Bacon et al., 1986).

One way of managing endophyte-infected fescue is interseeding with legumes (Bowman, 1990), although the legume can be grazed out. Heavy fall grazing, adequate potassium, phosphate, and lime are compensatory factors, but not nitrogen fertilization.

13.7.4 *Phalaris* Staggers

Reed canarygrasses contain indolalkylamine alkaloids that are potentially toxic to herbivores. Their relation to staggers is, however, not entirely clear. *Phalaris* staggers is characterized by the degeneration of nerve tissue and is partly responsive to cobalt. The nervous aspects of the disorder distinguish it from a simple cobalt deficiency. The symptoms of *Phalaris* staggers can also be confused with those of grass tetany.

The structures of the gramine alkaloids found in reed

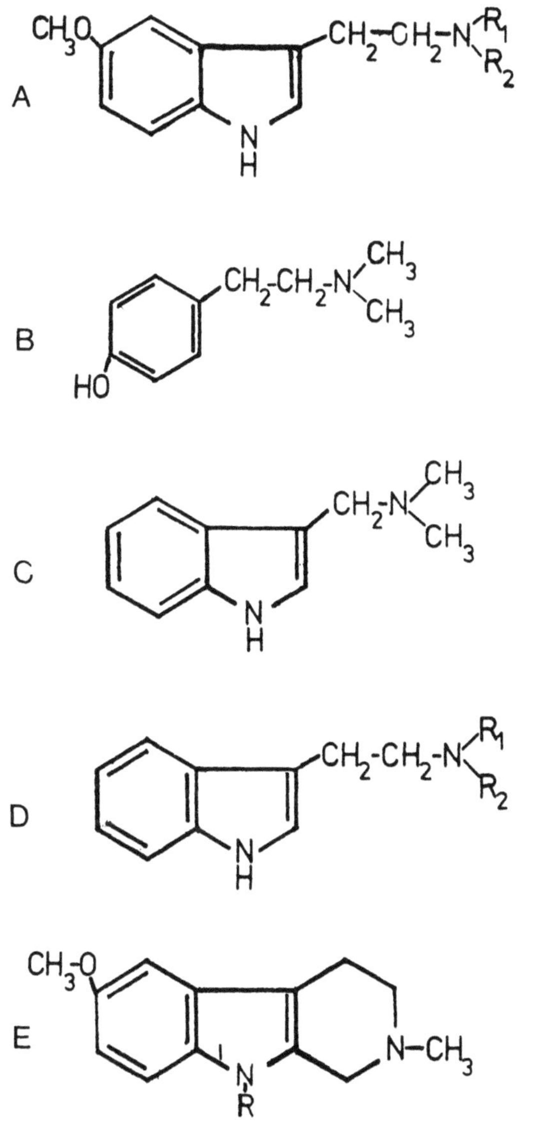

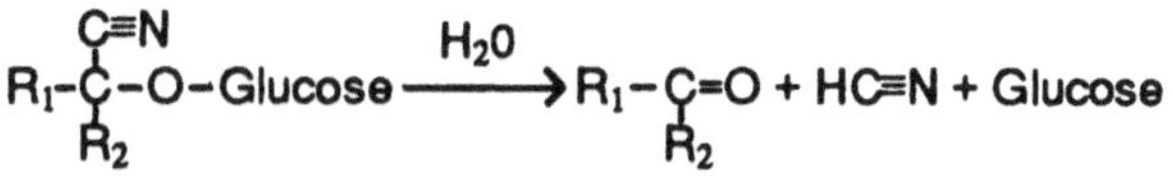

Figure 13.15. Structure of cyanogenic glucosides (from Barnes and Gustine, 1973). R_1 may be phenyl (amygdalin in *Prunus* species) or *p*-hydroxyphenyl (dhurrin in sorghum and sudangrass). R_2 is H in this example, but R_1 and R_2 are methyl and methyl, and methyl and ethyl, respectively, in linamarin and lotaustralin from white clover and lotus.

Strains of *Phalaris* species vary in alkaloid content (Marten et al., 1981), and it is possible to select for and against alkaloids. Alkaloids, isoflavones, and tannins may be involved in disease resistance in plants and could conceivably be selected for through programs concerned with disease resistance, although this association is hard to prove.

13.8 Cyanides

The cyanogenic glucosides are toxic to animals when hydrocyanic acid is generated (Conn, 1978). Although this acid is very toxic to animals, it has little effect on anaerobic organisms, including rumen bacteria, some of which can probably use it as a nitrogen source. Cyanogenic glucosides have a common general structure (Figure 13.15). Hydrogen cyanide is released through enzymatic action after injury to the plant or through microbial activity in the rumen (Majak and Cheng, 1987). Soluble carbohydrates reduce cyanide toxicity by enabling rumen bacteria to better metabolize cyanide. Cyanide is also detoxified by conversion to thiocyanate, which is aided by the addition of sulfur compounds (thiosulfate) into the rumen (Vennesland et al., 1982). Each plant species contains its own peculiar group of secondary substances, so animals can minimize toxicity by selecting a mixed intake. There is considerable variation in the toxicity of cyanogenic plants. The secondary compound is characteristic not of healthy plants but rather of stressed ones. Prevention of poisoning depends on recognizing the environmental factors that cause cyanide accumulation (R. L. Reid and James, 1985). Cyanide in cassava can be reduced by sun drying (Gómez et al., 1984).

13.9 Organic Acids

Organic acids are not ordinarily secondary compounds (although some can be), but they are undoubtedly important in the intermediate metabolism of plants, and they are often ignored in forage analysis. They are usually included with the soluble carbohydrates when these are estimated by the difference in organic matter not included in protein or cell wall. The content of organic acids is higher in immature forages

Figure 13.14. Alkaloids found in reed canarygrass (from Marten et al., 1981). (A) 5-Methoxy-N-methyltryptamine ($R_1 = CH_3$, $R_2 = H$); 5-MeO-N,N-dimethyltryptamine (R_1, $R_2 = CH_3$). (B) Hordenine. (C) Gramine. (D) N-methyltryptamine ($R_1 = CH_3$, $R_2 = H$); N,N-dimethyltryptamine (R_1, $R_2 = CH_3$). (E) 2-N-methyl-6-methoxytetrahydro-B-carboline (R = H), 2,9-N-N-dimethyl-6-methoxytetrahydro-B-carboline (R = CH_3).

canarygrass are shown in Figure 13.14. These substances are associated with poor intake of the grass. They are chemically related to lysergic acid (LSD) and hallucinogenic drugs sometimes used by South American Indians. Gramine alkaloids are also probably the cause of *Phalaris* staggers in sheep exhibiting extensive disorders of the central nervous system. Gramine alkaloids have been reported to limit digestion by rumen organisms in vitro, but the concentrations required are higher than those that ordinarily occur in forage. Cattle seem less adversely affected by the alkaloids than sheep.

Quinic acid

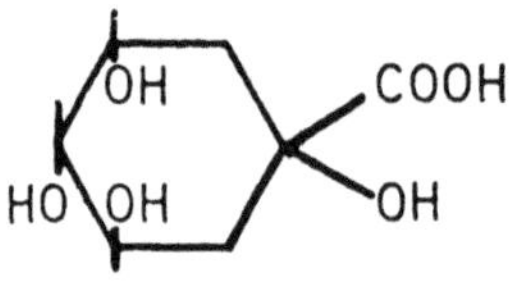

Shikimic acid

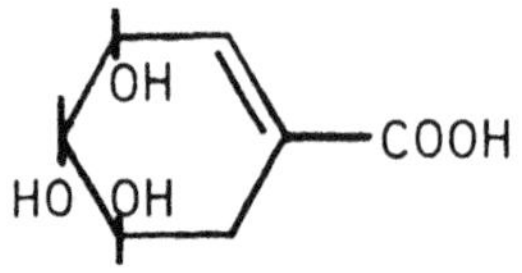

Citric acid $HOOC-CH_2-\overset{\overset{OH}{|}}{C}-CH_2-COOH$
$\underset{COOH}{|}$

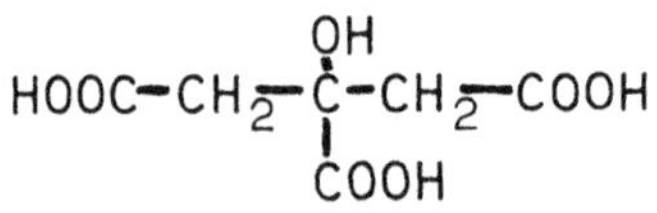

Aconitic acid (trans)

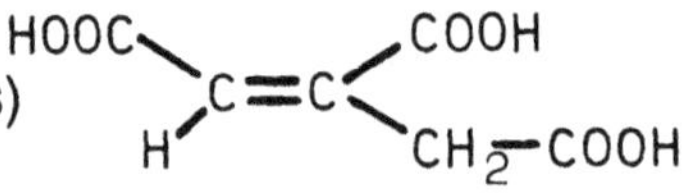

Succinic acid $HOOC-CH_2-CH_2-COOH$

Malic acid $HOOC-CH_2-\overset{\overset{OH}{|}}{C}H-COOH$

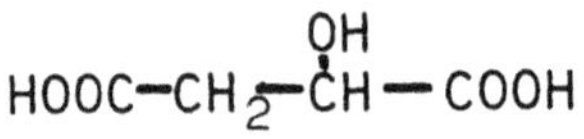

Fumaric acid

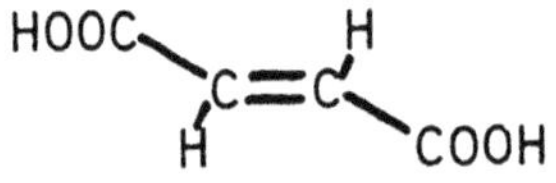

Oxalic acid $HOOC-COOH$

Ascorbic acid

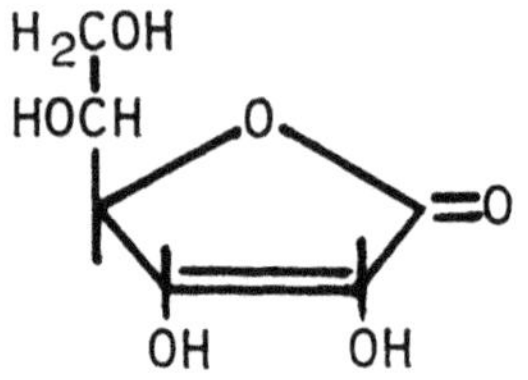

Figure 13.16. The more important organic acids in plants.

(Fauconneau, 1959). One reason organic acids are overlooked is that they are a sundry collection, not easily determinable as a group, and individual detection and measurement is apt to be tedious. Except for certain acids that can be toxic or cause nutritional problems, most natural plant acids are highly fermentable and yield energy to rumen microbes (Russell and Van Soest, 1984). They tend to disappear both in the rumen and in the silo (McDonald, 1981). Because many organic acids are fairly strong di- or tricarboxylic acids (Figure 13.16), they exist in the plant as salts of potassium or calcium, and thus have buffering value upon fermentation to acetate and bicarbonate, which are weaker acids (see Section 14.4.3). The presence of organic acids in forage has been associated with grass tetany (see Section 9.11.1).

13.9.1 Nitrate Toxicity

Heavy use of nitrogen fertilizer on grasses leads to luxurious growth, an increase in the nonprotein nitrogen fraction of the plant, and an accumulation of nitrate as the potassium salt. The accumulation of nitrate is promoted by low temperatures and lack of light (i.e., cloudy weather), which reduces plant metabolism. Nitrates per se are not very toxic. They are normally reduced to ammonia in the rumen and then metabolized in the usual fashion; however, this reaction sometimes fails due to a lack of reducing power or insufficient available carbohydrate in the diet, and nitrite is produced. Nitrite is toxic to rumen organisms and to the animal host. Cellulose digestion is inhibited in its presence. Nitrite absorbed across the rumen wall combines with hemoglobin to form methemoglobin, thus reducing the oxygen carrying capacity of the blood. In acute nitrite poisoning the animal dies from oxygen deprivation. Prevention or treatment depends on reducing nitrite by administering sufficient sugar or starch to increase microbial nitrogen requirements. Avoiding the overfertilization of cool-season pastures will also help. If dietary soluble carbohydrates are adequate, ruminant animals will tolerate quite large amounts of nitrate in the diet.

Nitrites are also associated with nitrosamines formed from the addition of nitrite to secondary amines. Nitrosamines are considered carcinogenic and can form when nitrites are used in the meat-curing process (National Research Council, 1972). Some forages also form toxic nitrocompounds; β-nitropropionic acid has been identified in vetch.

14 Forage Preservation

The conservation of forage for winter and dry-period feed is essential for efficient animal agriculture. The object of preservation is to conserve digestible nutrients as efficiently as possible. Traditionally this has been accomplished by making either hay or silage. The common methods of preparing hay and silage do not increase the nutritive value of a crop; they only attempt to conserve what was available in the growing plant when it was harvested. In the case of silage the price is the loss of some energy from sugar to form preserving lactic acid. There are efforts under way to combine ensiling with chemical treatments that improve quality, but it is not always clear whether a treatment is meant to improve fermentation (Section 14.7) or chemically alter the cell wall structure of the forage. Chemical delignification treatments designed to improve digestibility are discussed in Section 12.6.

Obtaining the highest forage quality with maximum nutrients depends on harvesting the crop at the optimum stage of growth. Nutrients usually reach their maximum at the critical point of heading or incipient

flowering (Figure 14.1), although this may vary with environment and the particular plant species. In perennial crops the yield of digestible nutrients generally peaks well before the maximum in dry matter. Annual crops such as corn tend to be grown to maximum maturity and yield. In this case, cob production offsets decline in stalk and leaf quality.

14.1 Hay versus Silage

Hay is traditionally defined as forage that has been air-dried to a sufficiently low moisture content that it becomes stable to ambient conditions. The moisture level is generally below 15% and in practical terms is in equilibrium with humidity. Preservation depends on preventing biological processes such as molding and fermentation through lack of water. Osmotic pressure is undoubtedly a factor in the inhibition, since salting is a common and effective method of controlling slightly moist hay.

Silage, or more correctly *ensilage,* means the composting of fresh forage in an anaerobic system and its preservation by means of an acidic fermentation of the sugars present in the forage. Preservation is thus based on a pickling process, which in turn depends on the production of lactic acid by bacteria. These bacteria derive only 2 ATPs per mole of glucose and thus are less efficient than other rumen microbes that produce volatile fatty acids (VFA) and derive 4 ATPs per mole of glucose (Chapter 16), the consequence being that more acid is produced per mole of sugar. Lactic acid is also a much stronger acid than other VFAs. Thus the art of ensiling promotes lactic acid fermentation at the expense of other energetically more costly fermentation.

Silages may be wilted, which reduces the risk of adverse fermentation and produces a higher-dry-matter silage or hay-crop silage (haylage) with 50–65% moisture content. (Figure 14.2). This, if well managed, often produces a more palatable feed than directly cut silage. In North America the practice of wilting has produced feeds of such widely varying moisture content that a continuum exists between hay and silage.

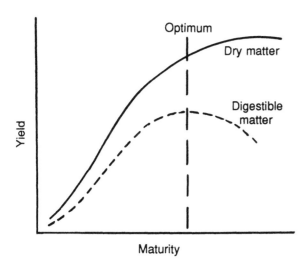

Figure 14.1. The relation between yield of dry matter and digestible matter and the stage of plant growth. The decline in digestibility with maturity causes digestible matter to peak well before dry matter production peaks. The yield of digestible matter is the product of dry matter yield and concurrent digestibility.

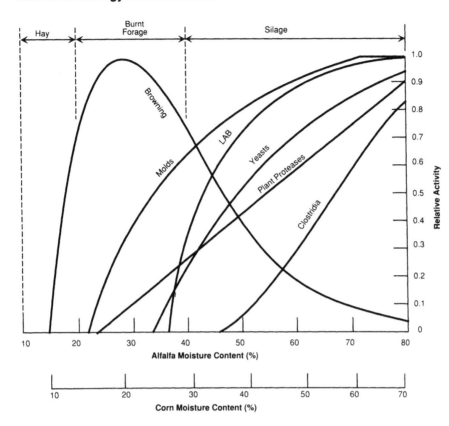

Figure 14.2. The relation between moisture content and biological processes (fermentation, molding, etc.) in cut forage (from Pitt, 1990). There is no clear-cut distinction between low-moisture silage and moderately wilted material. The higher-dry-matter silages are prone to molding and heating and do not develop much acidity (see Figure 14.8). *Lactobacillus* (LAB) are less inhibited by moderate wilting than are other bacteria. High-moisture silages are more dependent on fermentation for preservation, and higher water content allows expenditure of more feed energy in the fermentation process.

Material containing more than 15% water is unstable and tends to undergo fermentation and spontaneous heating, which can be controlled by excluding oxygen or using preservatives.

Traditionally, silage is made in regions or seasons of wet weather, and hay is made in dryer conditions. There are many regions where production of hay at the optimum stage of forage quality is inconvenient because of the weather, and it is still a common practice for hay to be made from mature forage when silage would have been a better choice. Wilting offers a choice: if dry weather holds, hay can be produced, otherwise hay-crop silage can be made if it rains.

14.1.1 Factors Affecting Nutrient Losses

During the cutting, wilting, and field-drying of hay, respiration, microbial activity, and mechanical action result in a selective loss of the most nutritious components (Figure 14.3). The extent to which these factors affect the nutrients varies with dry matter content and management. Hence, the largest harvest of forage nutrients is in direct-cut green herbage. The preparation of this material as silage, however, will produce losses through fermentation, oxidation, and effluent. Recourse may be made to the use of preservatives for silage or to artificial drying, but both options add economic expense against which benefits must be weighed.

Harvest losses tend to be larger with hay than silage. Leaf shatter is more important when plants are dry, respiration is more serious at higher moisture, and effluent loss becomes important at moisture contents above 70%. Effluent losses are promoted by finer chopping and the higher pressure found in storage towers. One solution to the problem of effluent in climates where wilting is ineffective is to collect and feed it to cattle (Steen, 1986).

14.1.2 Respiration

Whether hay, hay-crop silage, or silage is made, one of the major factors affecting quality is respiration in the dying plant. If any management is to be optimal, it is essential that postcutting plant respiration be minimized. Respiration in dying plant tissue involves the dissipation of sugar and the hydrolysis of proteins. These processes do not reduce digestibility, but they limit effective preservation and ultimately the efficiency of nutrient use in the rumen (R. Anderson, 1985). Respiration is more serious in silage, in which the sugar-to-protein ratio is important for good fermentation. Plant respiration is inhibited by rapid drying, crushing, or maceration. Biological activity beyond the death of plant tissue involves facultative and aerobic organisms, including mold, that further degrade plant nutrients.

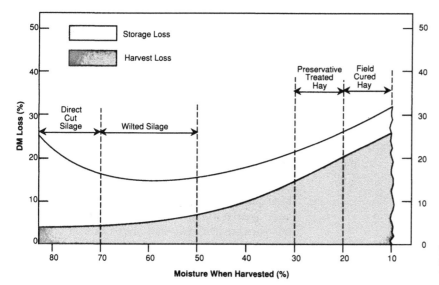

Figure 14.3. Loss of dry matter in relation to moisture content at harvest (from Pitt, 1990; based on Hoglund, 1964).

Respiration is inhibited by reduced moisture and wilting (Figure 14.2), although considerable activity can occur at dry matter contents of 50% or greater. Dry matter loss is promoted at higher temperatures and moisture contents (Figure 14.4).

Proteolysis can be very active at the low temperatures that tend to reduce moisture loss and drying (Licitra and Van Soest, 1991). The potential increase in nonprotein nitrogen (NPN) is larger than most published accounts indicate. Adverse drying conditions can also lead to development of high NPN values and the loss of sugar. Of all the factors affecting quality of conserved forage, postharvest respiration is the greatest. There are various ways of enhancing rate of drying, however.

14.2 Hay

Preservation of forage as hay is based on desiccation to halt biological processes and limit the action of microorganisms. Therefore, efficient preservation depends largely on rapid drying. Rainfall and humidity delay the drying and increase the amount of handling required. Both high relative humidity and, less important, low temperature retard drying. As Figure 14.5 indicates, making hay at relative humidities greater than 60% is difficult.

The effects of molding are to some extent unpredictable. Molding can lower feed value through loss of digestible nutrients and decrease palatability because of the musty odor it produces.

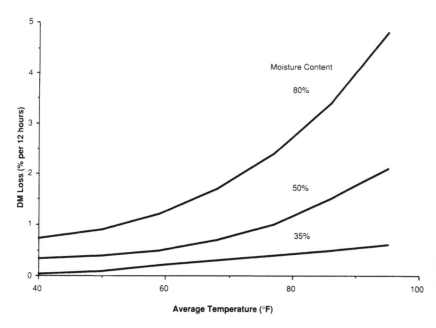

Figure 14.4. Rate of dry matter loss from plant respiration in the field based on forge moisture content and average temperature (from Pitt, 1990; based on Rotz et al., 1989, and Rücker and Knabe, 1986).

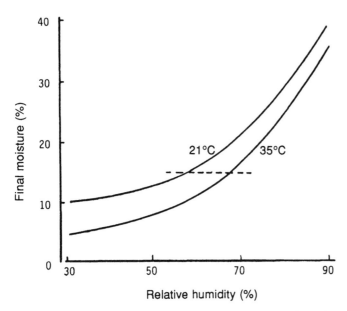

Figure 14.5. Final moisture content of hay and relative humidity in relation to ambient temperature (based on Pitt, 1990). Dashed line indicates 15% moisture, the maximum level allowable for safe preservation.

Heat is often associated with microbiological activity, particularly if hay is baled and stored somewhat damp, but the results of spontaneous heating can be variable. If the size of the bales and their ventilation is such that heat serves to drive out moisture, preservation is aided. This is more apt to occur in smaller bales that are piled loosely. Very large bales lose heat and moisture less efficiently and really should not be used in humid conditions (Baxter et al., 1986). On the other hand, large-baled silage properly sealed with a plastic cover can be an effective means of preservation.

Excessive heating, however, induces a nonenzymatic browning reaction (Section 11.7) with a resultant loss of digestible carbohydrate and protein. This Maillard reaction causes a caramel, tobacco-like odor and a darkening that reduce palatability. The potential for heat damage increases with increased bale size and forage mass.

Perhaps the most serious loss in weathered hay is weight loss through respiration in which available carbohydrates are consumed. Leaching is probably minor and is seen in bales that have been left in the field during rainfall. Heat damage may not occur even when mold is present because the greatest activity is in the outer layers. Generally NDF, lignin, and even nitrogen contents are elevated through the disproportionate loss of CO_2.

Other losses occur through chemical oxidation aided by sunlight (photo-oxidation). Prolonged exposure during curing bleaches pigments, resulting in a loss of carotenoids and vitamin A activity. At the same time

vitamin D activity is enhanced through ultraviolet radiation. The unsaturated leaf lipids oxidize and polymerize into resins, a process that continues during storage in the presence of oxygen (Couchman, 1959) and is reflected in a decline of ether extract content with storage.

14.2.1 Drying

The removal of moisture by heat or other means can help to dry hay or can be a means of direct preparation of dehydrated forage. The starting material must be of high quality to be worth the expenditures of energy and cost. Special hay driers are sometimes used, although this is expensive. Two basic types of heat-drying are practiced. One is the partial field-curing of hay, after which it is either baled while moist and dried by passing heated air through stacked bales or, especially in humid areas, dried loosely by means of a mechanical flue system. Use of this system has waned with the rising cost of energy. The other type is artificial dehydration, usually conducted in drum driers. This is the primary source of leaf meals for poultry and swine supplements.

Drying may be aided by crimping to break stems and other large structures to expose a greater surface area and aid evaporation. In humid weather, however, ensiling or the more expensive alternative of artificial dehydration may be necessary. Tedding and raking hasten drying but may promote leaf loss. Field losses under humid conditions are usually more severe in hay making than ensilage, although losses during storage may offset the latter's advantage of a more efficient harvest.

Other experimental systems for dewatering include the use of chemical wilting agents such as diquat, paraquat, formic acid, potassium carbonate, propionates, and ammonia that can be sprayed on the crop before harvest (T. R. Johnson et al., 1984; Pitt, 1990). This approach has been attempted mainly in humid areas where wilting before ensiling is limited by weather conditions. Desiccants may entail a lower energy cost than heat drying and can allow greater management flexibility in preparing either silage or hay, but most of these treatments have not gone beyond the experimental stage. The improper use of ammonia can promote the Maillard reaction (Mason et al., 1990).

14.3 Silage

The simplest and oldest method of making silage is by packing direct-cut herbage into pits, stacks, or tower silos. The ensiled material passes through a sequential fermentation (Figures 14.6 and 14.7), the first stages of which involve death of plant tissue and rapid

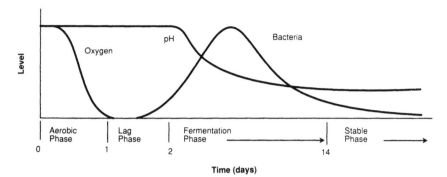

Figure 14.6. Short-term changes during ensiling of forage (from Pitt, 1990).

exhaustion of oxygen followed by proliferation of bacteria and the development of an acid fermentation (McDonald, 1981). Further fermentation depends on the composition of the forage, the pH attained, and the availability of oxygen and water. The best silages are those in which the original forage composition is least altered. Good preservation by fermentation depends on the production of lactic acid to stabilize the silage at a low pH, which in turn depends on an adequate supply of sugar to produce sufficient fermentation acids to overcome the potential buffering capacity of the forage.

The most important factor affecting silage quality is the availability of oxygen, which promotes respiration, molding, and, consequently, heating and Maillard reactions, even though the latter do not formally use oxygen. This heat can lead to spontaneous combustion. The major variables affecting oxygen availability are packing density and the permeability of the walls of the silo. For example, a bulk density of 20 lb/ft³ has 4 times the potential dry matter loss of a silage packed to 60 lb/ft³. The rate of oxygen diffusion is a function of the permeability of silo walls: more permeability, more dry matter loss and heating. Trench silos with concrete walls backed by packed earth probably transmit less oxygen than tower silos.

Heat production within a silo is also significantly affected by the thermodynamics of the structure, which involves solar radiation and color of the cover. Larger masses are more prone to heating because there is less surface for radiant losses. The problems of solar heating are much more serious in tropical climates, where solar radiation is known to have caused spontaneous fire in silages. Dark colors (e.g., black plastic) trap heat.

Losses in feed energy through the ensiling process occur in several ways: through initial plant respiration; through anaerobic fermentation; through aerobic decomposition, particularly at surfaces; and through effluent loss, especially in high-moisture direct-cut silages (Figure 14.3). Plant respiration and fermentative losses that produce CO_2 and heat are termed *insensible losses* and are often ignored. The total energy balance of silages is rarely measured, although it may be quan-

titatively important (Pitt, 1990). Insensible losses increase with the degree of fermentation and are much larger during aerobic respiration. Digestible fractions are preferentially lost compared with plant cell wall because cell wall is less fermentable. In low-nitrogen forages, nitrogen content may increase through selective retention by silage fermentation.

14.3.1 Storage

The tower silo is the traditional structure for silage preservation in the United States, although a variety of structures are used. The tower system aids in packing and preserves optimally if air can be excluded. Air exclusion becomes more difficult at higher dry matter content, and under these conditions exclusion of air entering through leaks and diffusion through walls becomes more important. Permeability problems are probably less severe in bunker silos provided the sealing on top is effective. Bunkers are more difficult to pack, however.

Good silage can be made in any structure if proper ensiling management is practiced. Trenches and above-ground bunkers are cheaper to build, but greater

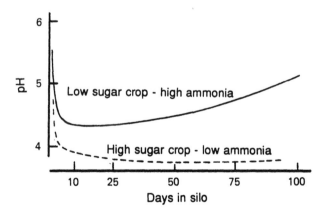

Figure 14.7. Longer-term changes during fermentation of direct-cut silages (from Wilkins, 1974). The sugar-protein ratio is an important factor influencing development of pH in the silage. Sugar promotes lactic acid production, while proteins degrade to ammonia and fatty acids of lesser strength than lactic acid. Since lactic acid is the stronger acid, a rise in pH is favored through production of fatty acids in combination with the neutralizing action of the ammonia.

care is needed in packing, and the top should be sealed with plastic or some other cover for protection from the elements and solar heating. The moisture content should be kept above 45% to limit heating and the Maillard reaction. No structure is foolproof relative to heating and the production of poor silage, however. Management and good procedure are the only means of avoiding these problems.

Effluent losses occur in forage with more than 30% moisture content and vary depending on the type of storage. Taller structures such as towers induce more pressure and therefore tend to suffer greater effluent loss with higher-moisture forage. The liquid fraction of silage juice contains sugar, organic acids, and other valuable substances important in silage fermentation. In general, the water-soluble components are lost in the effluent, causing a proportional increase in the less-fermentable water-insoluble fractions, particularly plant cell wall constituents. Because of the higher pressure in a tower silo as opposed to a trench structure, a higher dry matter optimum exists for the former compared with the latter.

14.3.2 Analysis of Silage

Too often an apparently digestible product is associated with poor intake and animal efficiency (C. Thomas and Thomas, 1985). This problem has much to do with the quality, state, and preservation of the protein and nonprotein nitrogen fractions in the silage. A second problem is that of assessing dry matter and digestible energy in situations in which considerable digestible organic matter is in the form of volatile substances.

The simplest criteria of silage quality to measure are pH and dry matter content (Figure 14.8). To be of the best quality, the pH of low-dry-matter silages should be less than 4.4. A higher pH in low-dry-matter silages indicates proteolytic fermentation and the development of amines and butyric acid. In higher-dry-matter silages (water content <65%), pH is less important. High-pH silages may be of good quality. Development of acidity is inhibited by lack of water and high osmotic pressure; thus, the pH of high-dry-matter silage is inversely related to water content. Lower-moisture silages (<40%) are more susceptible to molding and heating because oxygen removal is more difficult.

Estimating the water content of fermented products is one of the more difficult analytical problems. Conventional analysis of fresh forage usually commences with oven drying to fulfill the dual purpose of determining moisture as well as preparing a dry sample for grinding and subsampling for further chemical analyses. Volatile substances are lost in the drying process, resulting in overestimation of water content and underestimation of dry matter intake in feeding stud-

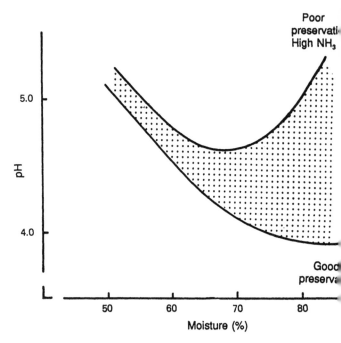

Figure 14.8. Observed pH range and moisture contents of silage. High-moisture silages are unstable with respect to pH, and a wide pH range is found, with good quality associated with low pH. In high-dry-matter silages pH is a less useful criterion of quality since deficiency of water restricts fermentation and acid production, causing an inverse relation between water content and pH. High pH at high water content is associated with proteolysis and very poor preservation. Low pH is associated with lactic acid production and better preservation.

ies. Since the volatile substances have a caloric value higher than that of the nonvolatile dry matter, artificially high efficiencies have been reported for silages. The direct determination of water and the estimation of volatile organic substances present can resolve these errors. The simplest and most often used method for direct water measurement is toluene distillation. Water is distilled at 117°C with toluene, collected, and measured in a receiver. Most volatiles, however, are distilled over in the process. Acetic and butyric acids and their ammonium salts are mostly distilled into the water phase, as are ethanol, ammonia, and other volatile amines. Some lactic acid is distilled. Sometimes the distillate is titrated for acetic acid to make an estimated correction, or, alternatively, the products are distilled into a dry ice trap. It is likely that all such corrections underestimate the true volatile content and therefore overestimate the actual water. Other methods of direct water determination include Karl Fischer titration, estimation of water by gas-liquid chromatography, and titration through a saponification reaction requiring water (Hood et al., 1971).

Volatile components can be directly determined on wet silage—volatile acids by gas-liquid or liquid column chromatography, and ammonia nitrogen by distillation with magnesium oxide. Ethanol, acetoin, and other volatile neutral products can be determined easily by gas chromatography. Bomb calorimeter determination of energy on wet silage gives accurate values for energy content, although the analytical error may be high. This procedure requires the use of added primer to provide sufficient energy for complete combustion of the silage.

For correct analysis of silage, most determinations involving volatile products must use wet matter. These assays include total nitrogen, ammonia nitrogen, nonprotein nitrogen not ammonia, volatile acids, and caloric value. Analyses of nonvolatile constituents (i.e., fiber, lignin, mineral components, etc.) can be performed on a dried residue provided the drying procedure does not create artifacts.

14.3.3 Tropical Silages

Most tropical grasses do not produce good silages. Many are low in sugar, tend to be coarse, do not pack well, and often have a high water content. In addition, the higher environmental temperatures in the tropics promote extensive fermentation. Legumes, while high in protein, generally have higher nonprotein cellular contents compared with grasses and may preserve better if the water content is not too high (Ashbell and Lisker, 1988). A further factor contributing to problems of preservation in warm climates is the thermodynamics of storage structures. Dark colors tend to absorb and trap heat under conditions of high solar intensity, and silo fires caused by solar heat are known. A roof or some other adequate cover may help to reduce the heating problems.

Tropical legumes are as high in protein as temperate legumes, but are lower in sugar. Most also contain tannins that inhibit proteolysis (Albrecht and Muck, 1991), leading to some compensation for the sugar loss; however, condensed tannin protein complexes are probably indigestible.

14.4 Microbiology and Chemistry of Silage Fermentation

Silage fermentation, like rumen fermentation, is anaerobic, and the fermenting organisms must derive their energy from the substrate. It is important, therefore, to consider the possible problems for rumen organisms subsisting on silage in which the most available substrates have already been metabolized. The silage organisms have had first choice, and this effectively decreases the amount of fermentable substrate available to the rumen organisms. Anaerobic organisms obtain oxygen for metabolism from the metabolites, leaving reduced compounds which must be regarded as waste products of anaerobic metabolism. Of

Table 14.1. Lactic acid–producing bacteria in silage

Homofermentative	Heterofermentative
Lactobacillus plantarum	Lactobacillus brevis
L. casei	L. buchneri
Pediococcus cerevisiae	L. fermentum
P. acidilactici	Leuconostoc mesenteroides
Streptococcus faecalis	
S. lactis	
S. faecium	

Source: Information from McDonald, 1981.

the fermentation products, lactic acid alone retains some fermentable energy for rumen organisms.

Protein and sugar may be spared if lactic acid–producing fermentation develops with rapid reduction in pH to exclude other fermentations. A rise in osmotic pressure will restrict fermentation, but lactic acid–producing organisms are more tolerant than nonlactics to osmotic effects and low pH. Starch, pectin, and hemicelluloses may survive a fermentation dominated by lactic acid–producing organisms but may be fermented to varying degrees if fermentative bacteria other than the lactic acid bacteria recover and promote a rise in pH. Cellulose and lignin alone are relatively stable to silo fermentations and are lost only in situations in which aerobic molding can occur. Differential metabolism leads to an increased concentration of refractory substances that are associated with poor feed value. This effect is significant in poorly preserved forages (J. W. Thomas, 1978).

14.4.1 Microbiology

While bacteria on fresh forage are mainly aerobic, they are normally quickly replaced by anaerobes as silage fermentation proceeds. This can occur only after good packing and elimination of air, however, because the survival of aerobic organisms depends on oxygen, whose presence promotes maximum degradation and poor preservation. Since the silo is a comparatively unbuffered system, single classes of organisms vie for control of the substrate (Leibensperger and Pitt, 1987); lists of the most important microorganisms in silage are given in Tables 14.1, 14.2, and 14.3 (Pitt, 1990). If lactic acid–producing organisms dominate, then pH is low and preservation is maximal. Nonlactic fermenters

Table 14.2. Clostridial species in silage

Species	Characteristics
Clostridium tyrobutyricum	Ferments sugars, lactic acid
C. sphenoides	Ferments sugars, lactic acid
C. bifermentans	Ferments amino acids
C. sporogenes	Ferments amino acids
C. perfringens	Ferments sugars, lactic acid, and amino acids; produces toxins

Source: Information from McDonald, 1981.

Table 14.3. Microbial genera associated with aerobic deterioration

Fungi		Aerobic bacteria
Yeasts	Molds	
Candida	Aspergillus	Acetobacter
Cryptococcus	Fusarium	Bacillus
Hansenula	Geotrichum	Streptomyces
Pichia	Monascus	
Saccharomyces	Mucor	
	Penicillium	
	Rhizopus	
	Trichoderma	

Source: Information from McDonald, 1981.

Table 14.4. Characteristics of silo and rumen fermentation

	Good-quality silage[a]	Normal forage-fed
pH	3.8	6–7
Microbial species	Few	Many
Cell synthesis	Limited, <5%	20–40%
Available cellulose digestibility[b]	0%	90%
End products	Lactic acid, less acetic acid, minimal butyric acid CO_2	Acetic, propionic, b and some lactic a CO_2 and CH_4

[a]Characteristics of a moderately low dry matter, high-quality silage.
[b]Potentially digestible cellulose. There is some fermentation of hemicellulos pectin in the silo.

(e.g., *Clostridia* and *Pseudomonas*) metabolize lactic acid but are inhibited by low pH and osmotic pressure (Langston et al., 1958). As a result, there is competition between lactic acid–producing and lactic acid–utilizing bacteria. The factors that favor one or the other are important in determining the course of fermentation and, consequently, good preservation (which may be defined as preservation of organic matter and original composition). Metabolic factors promoting losses disproportionately remove the more available components and are the principal means of altering silage composition. If oxygen is available, aerobic molds and bacteria develop.

The aerobic metabolism of acids by molds causes pH to rise with consequent deterioration of the silage. Aerobic metabolism allows the complete metabolism of lactic acid and other metabolizable compounds to CO_2 and water with greater heat production. The heat of this metabolism can induce Maillard reactions and is responsible for the sequence of events leading to heat damage and silo fires.

Fermentation also results in the production of neutral compounds such as ethanol, acetoin, and its reduced product, 2,3-butylene glycol (Fennessy and Barry, 1973). Ethanol is probably the most important neutral product quantitatively and is characteristic of low-protein silages (e.g., corn silages), often forming up to about 2% of the dry matter. The concentration of 2,3-butylene glycol ranged from trace amounts to 2 or 3% of dry matter in the limited number of grass silages that have been surveyed.

14.4.2 Silage Fermentation versus Rumen Fermentation

The ecology of silo fermentation differs from rumen fermentation in that single groups of organisms tend to develop and take over the substrate. Lactic acid–producing organisms accomplish this with a resultant rapid acid production and drop in pH. These organisms are less efficient than many other organisms in terms of cell yield from a given amount of substrate because much of the energy of the carbohydrate fermented is

retained in the lactic acid. Their need for control of the substrate is a consequence of their lower efficiency (cell growth), so a larger portion of the fermented substrate is necessarily produced in the form of acid that inhibits competing organisms. Organisms, including yeasts, that produce ethanol are similar in this regard except that osmotic pressure and inhibitory levels of ethanol are the means by which they control the substrate.

End product removal in the rumen has important effects on ecological balance. The salient effects are apparent when silage fermentation is compared with rumen fermentation (Table 14.4). Both systems have the same substrate available and tend to be anaerobic; however, the silo presents the characteristics of a poorly buffered batch culture. The dominating feature of its fermentation is the growth of organisms that compete for control of the substrate. Under ideal conditions in the silo, the lactic acid–producing bacteria will drive down pH so as to shut out other competitors for the substrate. No cellulose is fermented in a normal silage, whereas cellulose digestion is a dominant feature in a normal rumen. Silage fermentation is characterized by low efficiency and microbial yield. The higher microbial yield characteristic of the rumen is important in supplying the host animal with microbial protein. What is optimal for silage preservation is reversed in the case of the rumen. The important point here is that the more energy is expended in silo fermentation, the less there is for the rumen. Fermentation acids other than lactic acid contain no energy for rumen bacteria to use.

14.4.3 Organic Acids in Silage

The organic acids in silage are not the same as those found in forages (Figure 13.16). The principal acids in forage are citric, malic, and aconitic, with lesser amounts of quinic, succinic, and oxalic, of which shikimic and quinic are important in the synthesis of phenolic substances (Dijkshoorn, 1973). Most of these acids are di- or tricarboxylic and exist as mono- or dibasic salts of potassium and calcium in the plant.

ble 14.5. Some fermentation pathways of heterolactic bacteria

	Dry matter lost (% of nutrient)
rbohydrates	
glucose → lactic acid + ethanol + CO_2	24
3 fructose → lactic acid + acetic acid + 2 mannitol + CO_2	5
ganic acids	
malic acid → lactic acid + CO_2	33
2 malic acid → acetoin + 4 CO_2	67
malic acid → acetic acid + formic acid + CO_2	21
2 citric acid → 2 acetic acid + acetoin + 4 CO_2	46
citric acid → 2 acetic acid + formic acid + CO_2	14
2 citric acid → 3 acetic acid + 1 lactic acid + 3 CO_2	30
ino acids	
2 serine → acetoin + 2 CO_2 + 2 NH_3	42 (58)[a]
arginine → ornithine + CO_2 + 2 NH_3	5 (24)[a]

Source: Information from McDonald, 1981.
[a] If NH_3 is lost.

These acids are generally fermented to acetate and CO_2 and possibly butyrate (Table 14.5), all of which are weaker acids, so pH can rise as a result of the decrease in total acid groups relative to bases (Playne and McDonald, 1966). Thus, organic acids can influence the course of silo fermentation by moderating pH and may modify the importance of the sugar-protein ratio in pH balance. The fermentation products are mostly monocarboxylic fatty acids. In addition, malic and citric acids possess considerable fermentable energy for anaerobes that is expended and not present in the products. Organic acids are present in maximum concentrations in immature forages and decline with age and increasing maturity of forage. They are apt to be present in considerable amounts when forage is harvested for silage.

While fermentation acids such as acetic, propionic, and butyric represent considerable feed energy to the aerobic metabolism of the animal, they offer no available fermentative energy to rumen anaerobes, for which these acids are excretion products. They may be used only for synthetic reactions to form protein or fat, provided an external energy source is available. The conversion of soluble forage components to fermentation acids may provide an energy substrate problem for rumen organisms forced to subsist on a prefermented diet.

Buffering capacity is determined by the inorganic bases (potassium and calcium), protein, and the capacity for ammonia production. Acid salts of malic acid and citric acid ferment to the neutral salts of lactic and acetic acids. Thus, organic acids present as potassium salts also provide buffering capacity. Deamination and decarboxylation of amino acids produce ammonia and amines that neutralize acids. Consequently, a vital forage characteristic influencing the course of fermentation is the sugar-protein ratio. High sugar content fa-

vors acid production; nitrogen-fertilized forages, which tend to be lower in sugar content, may develop higher pH through ammonia production and are more difficult to preserve.

14.4.4 Proteolysis

Crude protein in fresh forages is composed of about 20–30% nonprotein nitrogen (NPN), 60–70% true available protein, and 4–15% unavailable nitrogen recovered in acid-detergent insoluble nitrogen (ADIN). Proteolysis is the hydrolysis of protein to amino acids with resulting increase in NPN at the expense of true protein. The principal protein in forages, the photosynthetic enzyme ribulose 1,5-diphosphate carboxylase, is highly susceptible to proteolysis (Fairbairn et al., 1988). The factors affecting proteolysis are numerous. Rapid wilting retards proteolysis due to the inactivation of plant proteases (Muck, 1987). On the other hand, rapid reduction in pH of wetter forage also retards proteolysis (McKersie, 1985). Proteases are active in cut forage, so prolonged wilting during unfavorable drying conditions promotes proteolysis. The NPN of fresh forage consists of nitrate and nonspecific amino acids. Nitrate tends to disappear during fermentation, and deamination of amino acids promotes a rise in ammonia and amines (McDonald, 1981). Good-quality silages are low in ammonia, and amino acids dominate the NPN fraction. Lower-quality silages range between two extremes: higher-moisture silages that develop increased levels of ammonia, amines, and butyric acid; and higher-dry-matter haylages in which fermentative activity is inhibited but is replaced by molding and heating, leading to high levels of ADIN. Heat reduces proteolysis (Charmley and Veira, 1990). Ammonia appears if there has been significant molding. All these low-quality silages exhibit degradation of the available true protein into NPN, ADIN fractions, or both. Well-preserved wilted silages generally have better-preserved protein fractions, although there may be considerable hydrolysis to amino acids and peptides (Papadopoulos and McKersie, 1983).

Some of the principal degradation reactions of fermentation are shown in Table 14.6. Fermentation results in the loss of CO_2 and the enrichment of hydrogen and carbon in the volatile products. The combustible energy per unit weight is higher in ethanol and volatile fatty acids than in carbohydrate. Consequently, the digestible energy loss in fermentation is less than the dry matter loss indicated in Table 14.6. Development of a nonlactic fermentation will convert lactic acid to butyric acid, and acetic acid may be used in the process. This reaction alone will cause a rise in pH.

Proteolytic degradation involves deamination of amino acids with the resultant formation of isoacids and diamines such as histamine, cadaverine, and pu-

Table 14.6.   Some products of clostridial fermentation

2 lactic acid → butyric acid + 2 CO_2 + 2 H_2
lactic + acetic acids → butyric acid + CO_2 + H_2
Amino acids
   Energy-yielding pathways
      3 alanine → 2 propionic acid + acetic acid + 3 NH_3 + CO_2
      alanine + 2 glycine → 3 acetic acid + 3 NH_3 + CO_2
   Decarboxylation
      histidine → histamine + CO_2
      lysine → cadaverine + CO_2
   Deamination
      valine → isobutyric acid + NH_3
      leucine → isovaleric acid + NH_3
      arginine → putrescine + CO_2 + NH_3

Source: Information from McDonald, 1981.

trescine. The latter are significant components of the nonprotein nitrogen fractions of silage.

The crude protein fraction of high-moisture silage is generally extensively fermented, with a resultant increase in water-soluble nonprotein nitrogen to about two-thirds of the total silage nitrogen. In good-quality silage, about half of this nitrogen occurs as amino acids, some bound in peptides. Ammonia and non-volatile amines (principally cadaverine and putrescine) make up much of the remainder (Table 14.7). There is often a significant unidentified soluble fraction in badly fermented or heated silages. The recovery of amino acids is decreased in higher-pH silage; in particular, methionine, cystine, and tyrosine tend to be lost due to heat damage in higher-dry-matter silages.

The degradation of the protein fractions results in the same relative loss in fermentable energy that was earlier described for the carbohydrates. Proteolytic silage organisms dissipate the energy and ATPs that would have been available to rumen bacteria. The depletion in value of the nitrogen fractions also includes the loss of available nitrogen to Maillard products and ADIN. There is also a rise in unidentifiable soluble nitrogenous fractions that may have some relationship to the ADIN (McDonald, 1981). Indigestible soluble nitrogen fractions have been identified in heated silages with high ammonia content (Van Soest and Mason, 1991).

Table 14.7.   Composition of nitrogen fractions in grass silage

Composition	Good	Spoiled		Heated
Dry matter (%)	17.3	18.9	23.8	19.2	24.4
pH	3.9	4.9	4.2	5.7	4.0
Total N (% DM)	0.51	0.51	1.07	0.80	0.69
NPN (% of total N)	66	45	83	59	33
Amino acids[a]	57	30	48	14	41
NH_3[a]	18	43	30	64	24
Nonvolatile amines[a]	14	13	8	6	Low
Unknown N[a]	11	14	14	16	35

Sources: Hughes, 1970, 1971.
[a]Percentage of nonprotein nitrogen (NPN).

Table 14.8.   Energy values of soluble components in forages and silages for rumen metabolism and metabolizable energy

Substrate	ATPs from rumen fermentation	ATP relative to glucose[a] (%)	ME (kcal/g)	ME conse from gluc (%)
Primary				
Starch	4	100	4.0	107
Glucose	4	100	3.7	100
Citric acid	2–3[b]	60	2.5	—
Malic acid	2–3[b]	60	2.4	—
Glutamine	2	50	3.5	—
Asparagine	2	50	2.4	—
Protein	1–2[b]	40	4.0	—
Nucleic acids	2.4	40	1.5	—
Nitrate	0	0	0	—
Ether extract	0.8	20	4.0	—
Fermentation products				
Lactic acid	2	50	3.6	97
Acetic acid	0	0	3.5	63
Butyric acid	0	0	6.0	78
Ethanol	0	0	7.1	97
2,3-Butylene glycol	0	0	0?	0?
Ammonia	0	0	0	0
Amines	0	0	0?	0?

[a]Based on theoretical yield of product from glucose in silo fermentation (McDo 1981).
[b]Variable value depending on disposal of metabolic hydrogen.

14.4.5 Energy Balance

The relative energy values of original forage components and their fermented counterparts in silage are shown in Table 14.8. Number of ATPs available to fermentation and net caloric value were calculated from biochemical pathways and *Handbook of Physics and Chemistry.* The fermentation value of protein and amino acids is taken at 40% of carbohydrate, based on the results obtained by J. B. Russell et al. (1983) using rumen bacteria. The value of ether extract is based on 50% galactolipid (Chapter 20). Only the galactose and glycerol fractions provide fermentable energy. The value of nucleotide bases is assumed to be zero, and the fermentable energy in nucleic acids is limited to that from ribose and deoxyribose (McAllan, 1982). Organic acid components of fresh alfalfa have somewhat lower ATP and energy values than do carbohydrates but still have considerable fermentation potential (J. B. Russell and Van Soest, 1984).

The sum of the ordinary forage components reported in the literature does not equal the total solubles because there is a miscellaneous group of minor components that collectively add up to a significant percentage of dry matter. In the case of fresh forage these components include lignans, flavones, isoflavones, glycosides, and terpenoids, including saponins (Chapter 13). These substances have little fermentative potential in the silo or in the rumen, except for the sugar portion of glycosides, which likely disappear in the silo. Silage contains an unaccounted-for fraction of

Table 14.9.   Estimated composition of cell contents
of fresh alfalfa

Component	% of dry matter	ATP fraction relative to glucose[a]	Caloric equivalent[a] (ME, kcal)
Starch	2	2	8
Sugars	7	7	26
Pectin	8	8	30
Organic acids	10	6	25
Amides + amino acids	4	2.4	10
Nitrate	0.5	0	0
Soluble protein	11	4.4	44
Peptides	0.5	0.2	2
Nucleic acids	3	1.2	45
Ether extract	3	0.6[b]	12
Ash	7	0	0
Miscellaneous	6	1	?
Total	59	32.2	162
ME[c] (Mcal/kg)			2.74

Sources: Sugar and starch values: McDonald, 1981; Muck, 1987;
Muck and Dickerson, 1988; organic acids: Dijkshoorn, 1973; amides
and amino acids: Brady, 1960; Hegarty and Peterson, 1973; Ohshima
and McDonald, 1978; Merchen and Satter, 1983.

[a]Calculated from data in Table 14.8.

[b]Fermentation potential from glycerol and the galactose portion of
galactolipid.

[c]In the neutral-detergent solubles.

Table 14.10.   Estimated composition of soluble compounds
in a highly fermented alfalfa silage at 30% moisture

	Dry matter (%)	ATP fraction relative to glucose[a] (%)	Caloric equivalent[a] (ME, kcal)
Starch	1	1	4
Sugars	3	3	11
Pectin	1	3	11
Organic acids	1	0.6	2.5
Amides	1	0.6	2.4
Protein	3	1.2	12
Peptides	3	1.2	12
Amino acids	4	1.6	16
Amines	1.4	0	0
Ammonia	0.6	0	0
Unaccounted[b]	9	0	0
Lipid	2	0	9
Ash	7	0	0
Acetate	3	0	10
Butyrate	3	0	18
Ethanol	1	0	7
2,3 Butylene glycol	2	0	0
Lactate	7	3.5	13
Total	53	14.7	127.9
ME (Mcal/kg)	—	—	2.41
ATP % of fresh alfalfa		46	
ME % of fresh alfalfa			86

Sources: Sugar and starch values: McDonald, 1981; Muck, 1987;
Muck and Dickerson, 1988; organic acids: Dijkshoorn, 1973; amides
and amino acids: Brady, 1960; Hegarty and Peterson, 1973; Ohshima
and McDonald, 1978; Merchen and Satter, 1983.

[a]Calculated from data in Table 14.8.

[b]Probably including the remnants of nucleic acids (McDonald,
1981).

products arising from fermentation (McDonald, 1981).

The fermentation products make considerable metabolizable energy available to the cow but are depleted fermentable sources of ATP for microbial growth. The volatile acids and ethanol concentrate energy but represent large losses in dry matter and recovery of energy (McDonald, 1981). Products such as 2,3-butylene glycol are of unknown value. There are no reports in the literature of its metabolism in animals.

Table 14.9 shows the ATP equivalent (glucose = 100) and the metabolizable energy (ME) value for estimated components in soluble compounds from fresh alfalfa. Values were calculated according to composition and summed to give net available ATP.

The ATP equivalent and metabolizable energy content of silage solubles are shown in Table 14.10. The values were calculated the same way as those in Table 14.9, using ATP equivalents and energy values from Table 14.8. These calculations are for a highly fermented alfalfa silage with fermentative losses of about 10% of soluble dry matter, estimated from McDonald (1981). The total available ME in soluble matter is 86% of that available in fresh alfalfa, while the ATP equivalent for fermentation is only 46%. The calculations in Table 14.10 represent an average highly fermented alfalfa silage; thus it can be expected that less-fermented silages that are well preserved will have ATP and ener-

gy recoveries intermediate between the values for unfermented alfalfas in Table 14.9 and the values in Table 14.10. Very badly preserved silages will have even lower recoveries of ATP and energy than indicated in Table 14.10.

Both the silo and the rumen are largely anaerobic systems, so the limitations of energy for microbial growth are similar and to a degree competitive. As I said earlier, the silage fermentation organisms may deplete the forage substrate of fermentable soluble carbohydrates and protein, leaving little for rumen organisms to use. While the VFAs do not supply energy for microbial growth and protein yield, they may supply growth factors to certain rumen bacteria. Microbial protein is the single largest source of amino acids for dairy cows, and it appears that fermented silages are not an efficient source of microbial protein. This means, in turn, that the utilization of the NPN components in the silage by rumen bacteria is also inefficient because of the lack of fermentable carbohydrates for the conversion of NPN to microbial protein. The load of extra ammonia and related amines is a further burden on the metabolism of the cow and may be one of

Table 14.11. Reported values for microbial yield
(g N/100 g true digestible organic matter)

Diet	Range	Mean	Reference
Unsupplemented silages	1.1–2.7	1.57	Beever et al., 1977; Kelly et al., 1978
Silage + supplement	2.6–3.1	2.85	Stern and Hoover, 1979
Unsupplemented forages[a]	1.6–4.9	3.03	Stern and Hoover, 1979
Forage + supplement[a]	1.7–4.7	2.78	Stern and Hoover, 1979
High concentrates	1.3–2.6	2.11	Stern and Hoover, 1979

[a]Excluding straw-based diets.

the reasons for poor feed intake and poor performance in animals fed poor-quality silages.

These results illustrate the problems encountered in formulating silages into complete rations. All feeding formulas assume soluble matter in feeds to have equivalent value, which leads to the overestimation of silage relative to unfermented feeds. The overestimation is far more serious for rumen fermentation than it is for supply of energy to the animal. The calculation may explain the low microbial yield reported for unsupplemented silages in the literature (Table 14.11).

Supplementation with concentrate probably adds sufficient fermentable carbohydrate and improves the microbial output, but this interaction is at the expense of feed energy in the supplement. Although the difference between supplemented silages and supplemented forages is not apparent in the means in Table 14.11, it must be recognized that the laws of energy conservation and thermodynamics prevail. The large range in microbial yield for supplemented forages in Table 14.11 probably represents the optimization of dietary balance for microbes. In this case supplemented forage appears superior to supplemented silage. The possibly less efficient use of concentrates by microbes may be considered a cost of a more convenient, less expensive feed storage system. Table 14.11 indicates that microbial efficiency may be lower for concentrates relative to forage. This may have to do with the higher maintenance cost of starch-digesting bacteria compared with cellulose digesters (J. B. Russell et al., 1992).

The loss of ME may seem trivial for high-quality silages (Table 14.8), but the greater the extent of fermentation and dissipation of residual energy in fermented carbohydrate and protein, the more difficult it may be to overcome the deficiency in fermentable carbohydrate.

14.5 Low-Moisture Silage

Wilting of grass or legume forage can improve silage quality, reduce or eliminate effluent loss, and restrict fermentation through increased osmotic pressure, provided the silage can be kept anaerobic. As a result, these silages develop less organic acid and sta-

bilize at a higher pH. If silage is well preserved, sugars and protein are less completely fermented and more likely to survive to the finished product—animal feed. Well-prepared silages (more than 30% dry matter) are generally more palatable and have better intakes.

Silage makers in North America have tended to prefer wilting to the use of preservatives; however, the unrestrained use of wilting has generated a new set of silage quality problems. Higher-dry-matter silages have a greater tendency to produce heat due to a lower specific heat of the mass, and to mold because of the greater difficulty of excluding oxygen. Heat capacity of forage is related to its water content such that an equivalent amount of fermentative heat causes a greater temperature rise in drier material, other factors being equal. Thus, heat damage is greater in higher-dry-matter materials (Figure 14.2). Molding probably follows a similar distribution. Water at less than 30% content in the forage increases the likelihood of heating; higher moisture content is associated with less heat damage. Water is necessary for aerobic respiration and for the Maillard reaction, and storage of silage at 30% moisture represents the greatest hazard for spontaneous heating and possible fire.

The development of high-dry-matter silage and its quality depend on two factors: thermodynamic effects and the supply of oxygen. In contrast to wetter silages, significant amounts of heat may be generated from nonfermentative processes such as solar radiation, which become more important because of the lower specific heat of silage matter, and from nonbiological reactions, including the Maillard reaction (Section 11.7).

All chemical reactions approximately double in rate with each 10°C rise in temperature. In temperate countries, where food is stored in autumn and winter, the cooler temperatures probably contribute to stability and preservation of silage. The Maillard reaction does not require oxygen per se, although oxygen will participate in the reaction by combining with unsaturated compounds, making more heat available and thus increasing the possibility of heat damage. The presence of oxygen greatly increases the potential heat. The elimination of oxygen goes far toward preventing heat damage and possible fire.

As temperature rises in a hay-crop silage, various stages are encountered. Up to about 50°C, normal fermentative processes operate. Above this temperature the Maillard reaction becomes much more important, although it may be detectable at temperatures as low as 30°C if these are sustained for long periods. Many organisms are killed as the temperature rises, but thermophilic organisms are viable up to 70°C. Thermophilic fermentation tends to produce proteolysis and little acid and may be responsible for poor recovery of true protein. At higher temperatures sterilization occurs and heat generation is confined to nonbiological processes.

Most hay and low-moisture silages do not burn but instead stabilize at some degree of heating. Factors influencing the extent of heating depend on the susceptibility of the forage crop to the Maillard reaction and on the thermal insulation of the stack or silo. Generally, larger structures expose less surface for heat loss and more heat is retained internally, thus inducing higher temperatures. The accessibility to oxygen, the moisture content, and the possibility of moisture and heat loss through steam produce a highly unpredictable state of affairs.

The most stable hay-crop silages have a moisture content in the range of 50–60%. Hay-crop silage in the range of 20–50% moisture will not necessarily suffer heat damage, but the likelihood of heating is greater. The degree of heat damage cannot be predicted solely from moisture content. The best indexes of heat damage in silage are analyses of the finished product for lignin and acid-detergent fiber nitrogen (Yu and Thomas, 1976).

Overheating causes losses in digestible nutrients through respiration and in digestible protein through binding in the Maillard reaction. The generally unobservable losses of vaporization are probably more serious, although the effect on protein is the most obvious. Since heat coagulates protein, animals fed a mildly heated feed may show increased protein utilization despite the reduction in protein digestibility. Extensive protein damage in feed is certainly detrimental, but it is difficult to determine the point of diminishing returns.

14.6 Silage Intake

The subject of nitrogen fractions always comes up in discussions of the problems of silage quality and feed consumption. Poor-quality silages are usually poorly consumed relative to hay or forage of comparable digestibility. In normal forages, the limiting factor to intake appears to be NDF (see Chapter 21). The intake of poor silage does not reach this limit; other factors must explain the low intake (Waldo and Jorgensen, 1981). In general, three hypotheses have been proposed to account for low intake of poorly preserved silage: (1) a toxic substance, perhaps an amine, is produced by the fermentation; (2) the high acid content of extensively fermented silages reduces palatability; and (3) the depletion of readily fermentable substances deprives rumen organisms of critical energy substrates needed for growth. Silage intake is inadequately understood, and other mechanisms than these could be involved. For example, it may be that alteration in physical structure renders silage more difficult to ruminate (Udén, pers. comm., 1988). This concept requires acceptance of the hypothesis that rumination time limits intake.

The first hypothesis is supported by experiments in which silage juice added to hay resulted in depressed intake (Moore et al., 1960). Other studies added suspected fermentation products such as ammonia or amines to the diet (Clancy et al., 1975) or silage extracts (Buchanan-Smith and Phillip, 1986). The hypothesis that silage acidity reduces palatability is supported by increased intake when buffers are fed to neutralize the acid. The first two hypotheses are in many ways similar in that they posit substances in silage inhibiting intake. Many silages that are low in dry matter and high in pH are not eaten. Excessively acid silages are usually made from high sugar–low protein forage at higher moisture contents. In particular, acidity appears to be a problem in higher-moisture corn silage, in which increased intake in response to limestone, ammonia, or urea addition is due to neutralization of acid. Addition of buffers in the ensiling process increases lactate but does not reduce free acid. It may be that the first and second hypotheses describe different populations of silages.

The third hypothesis is more inclusive. Since volatile fatty acids have little fermentative energy, the extensive conversion of sugar to these products in the silo deprives rumen organisms of ATP needed for growth. Microbes' use of ammonia and other nonprotein nitrogen compounds for synthesis depends on rapidly available energy from carbohydrate or lactate. The fermentation of cellulosic carbohydrate is too slow to support utilization of soluble nonprotein nitrogen. In the absence of energy for its conversion, ammonia is absorbed from the rumen and produces a burden on animal metabolism. Ammonia is potentially toxic, but it is converted to urea by the liver and excreted through the kidneys. This process costs energy, and the extra load decreases animal efficiency. At the same time, lack of carbohydrate limits protein synthesis from ammonia, and the feed source of protein is badly degraded, the result being that the animal is deprived of essential amino acids despite an abundance of nitrogen from silage (Hawkins et al., 1970). This situation usually responds to dietary supplementation with concentrates, which also tend to increase intake. Even protein supplements may increase intake. Carbohydrate promotes rumen conversion of NPN to microbial protein, and supplementation with insoluble protein supplements improves animal amino acid balance through rumen escape.

An alternative suggestion (Bergen et al., 1974) is that silage nonammonia NPN is available at a slower rate to rumen bacteria than urea or ammonia, thus limiting digestion rate. In this instance a lesser response to supplemental carbohydrate would be anticipated. Since concentrate supplements ordinarily contain both protein and carbohydrate, a clear-cut separation of the alternatives is not possible.

14.7 Silage Additives

The composition and nutritive quality of silages can be altered considerably by the addition of various materials at the time of ensiling. Additives have two main purposes in silage: to influence the course of fermentation so as to favor preservation, and to alter the composition to support a better nutritive value. The many kinds of substances that have been added to silage can be classified into several categories: stimulators of fermentation, inhibitors of fermentation, and substances whose principal purpose is to alter composition (Figure 14.9). Components of some silage additives cannot be classified into any of the above categories and have little value; These include flavor compounds and carriers (J. W. Thomas, 1978).

The available preservatives are quite varied and are based on different strategies, or on no apparent strategy at all in certain cases. Figure 14.9 includes desiccants used for hays because it is difficult to distinguish between inhibitors of fermentation and desiccants that

may operate on the common factor of osmotic pressure.

14.7.1 Fermentation Inhibitors

The class of additives that has received the largest and most sustained interest is the industrial acids. The original studies of mineral acid application were carried out in Finland by A. I. Virtanen (1933), who received the Nobel Prize in 1945 for his effort. The silage made when mineral acid is added is termed *AIV silage*. The application of sulfuric, hydrochloric, or phosphoric acid mixtures to bring the pH to 4 essentially blocks all fermentation, thus eliminating fermentative loss and effecting preservation of nutrients. The problem with using mineral acids is the load of acidic anion they present to animal metabolism. This can be partly ameliorated by adding lime at feeding or by using metabolizable organic acids as preservatives. The AIV process, while used extensively in Scandinavia, has been little used in the United States and in areas where there has been considerable interest in organic acid additives.

Organic acids have the advantage of being metabolizable and providing feed energy. They are, however,

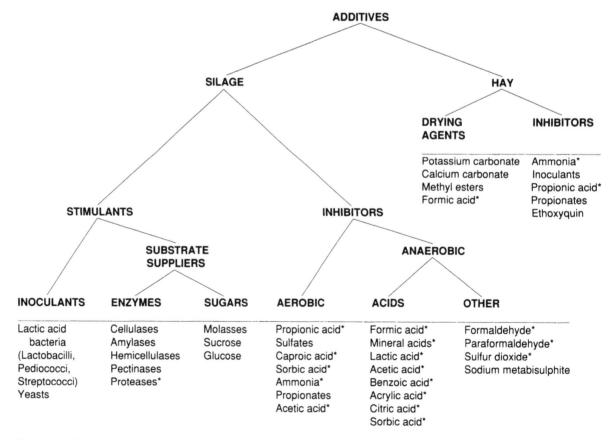

Figure 14.9. Silage and hay additives used to promote preservation (from Pitt, 1990). Not all the additives listed are effective. Asterisk indicates compounds dangerous to human health.

considerably weaker acids, and more acid is required to achieve an equivalent pH. As one proceeds up the fatty acid series, physical properties change. Formic acid is a considerably stronger acid than higher members of the series. As molecular weight increases, a greater quantity of the acid must be used to obtain an equivalent amount of acidity. Solubility becomes limiting when the chain length is above six carbons. Some organic acids possess considerable ability to limit fermentation, however, which must be ascribed to properties other than their acidity (Table 14.12). Propionate has the best overall protective effective against *Clostridia,* yeasts, and molds and is favored for higher-dry-matter silages. Butyrate is ineffective against the *Clostridia* that are the principal producers of butyrate and is also readily metabolized by mold. Odd-carbon acids seem to have special antifungal activity. Inhibition in most cases is more effective at lower pH, indicating that it is the free acid that acts as the inhibitor.

The practical choice of which compound to use in ensiling involves a number of factors. Commercial availability and cost essentially have limited large-scale usage to formic, acetic, and propionic acids. Acids of four to six carbons are precluded, probably by their unpleasant odor. Formic acid is most commonly used on high-moisture silages, while its better inhibition of mold and other aerobic organisms has made propionic acid the favorite for use on hay-crop silages. It is often blended with cheaper acetic acid to reduce the danger of low milk fat when fed to lactating dairy cattle (see Section 20.7).

Another class of additives includes various salts, but these are still experimental and are probably not used to any significant extent in silage production. Salts help preserve silage through increased osmotic pressure and the inhibitory properties of the anion. If the anion is metabolizable, as in ammonium lactate, there is a hazard of pH rise through the buffering of the ammonium ion. Metabisulfite salts, on the other hand, contribute considerable acidity as sulfurous acid. Bisulfite is also a very effective inhibitor of the Maillard reaction and related nonbiological heat-damage reactions involving carbonyl compounds complexed with bisulfite.

Antibiotics have also been studied as additives, but these are impractical on the basis of present food and drug regulations. Their success is severely limited by the difficulty of providing adequate inhibition in widely differing types of silages against many microbe types. Adaptation of organisms and inactivation of antibiotics are problems.

Among the additives intended to alter the amount of crude protein and its character are urea and ammonia, which are often added to corn silage. In wetter silages, urea tends to hydrolyze to ammonia, and the buffering creates the possibility of greater fermentation and, perhaps, inferior quality. The addition of ammonia narrows the margin of the sugar-protein ratio required for safe preservation. For a crop such as corn this is not a problem provided good management techniques have been applied. In a well-managed system ammonium lactate is the principal fermentation product and pH does not rise. If fermentation results in poor-quality silage, inadequate sugar or aeration is the cause.

Urea recovery and good quality are consistent in well-packed corn silages above 35% dry matter. Urea does not increase true protein, but it may retard proteolysis (Lopez et al., 1970). Hydrolyzed urea appears as ammonium lactate, which acts as a preservative by increasing the osmotic pressure. Ammonia and ammonium ion seem to be reasonable inhibitors of mold at higher dry matter contents. In diets high in corn grain and corn silage the supply of cationic buffers for rumen function can be critical, and these are sometimes supplied in the form of limestone added to silage at feeding.

Adding nonprotein nitrogen compounds to silage does not increase true protein. Any microbial synthesis of protein, which is negligible, is offset by hydrolysis of plant proteins. The production of ammonia is favored by the buffering needed to offset acid production, which becomes a limitation to fermentation. While pH is not altered, net buffering capacity in the rumen may be increased.

Overall, the most successful silage preservatives appear to be the ones that restrict fermentation rather than those that enhance it. This view is consistent with conditions promoting the maximum recovery of the original forage feed energy stored in the silo (J. W. Thomas, 1978).

14.7.2 Fermentation Stimulators

Fermentation stimulants restrict fermentation to lactic acid production, minimize loss of nutrients, and create a low pH in the final silage. Stimulators include carbohydrates such as molasses, enzymes, and lactic organisms. Carbohydrate increases the sugar-to-

Table 14.12. Minimum concentrations (mmol/l) at which organic acids inhibit microorganisms

Compound	Lactic acid–producing bacteria	Spore-forming bacteria		Yeasts	Molds
		Clostridia	Bacillus		
Lactate[a]	12	—	—	>250	>250
Formaldehyde[a]	1	—	—	6	8
Formate[b]	150	50	19	>200	200
Acetate[b]	280	47	<12	281	188
Propionate[b]	250	<8	<8	125	63
Butyrate[b]	200	>1000	125	47	12
Valerate[b]	110	31	8	39	16

Sources: Woolford, 1975a, 1975b.
[a]Tested at pH 4.
[b]Tested at pH 5.

protein ratio, rendering proteolytic fermentation less likely. Hydrophilic products such as beet pulp contribute carbohydrate and soak up water, reducing potential effluent losses. The simplest additions to silage are feedstuffs such as molasses, whey, cereal concentrates, and hay. Molasses, lactose, and starch provide a source of fermentable carbohydrate that aids lactic acid fermentation.

14.7.3 Inoculants

The addition of microorganisms to promote lactate fermentation has had a checkered history. Early efforts failed to appreciate the microbial ecology of silage, in particular the ecological competitions in the early stages of cutting, chopping, and ensiling. Native facultative organisms tend to proliferate on the tissues of the dying plant and can reach numbers on the order of 10^6 per gram. Most of the early inoculants failed to contain competitive quantities of lactate-producing organisms and thus failed to produce any consistent effect on silage quality (J. W. Thomas, 1978). Silage inoculants do work as long as the necessary conditions are met (Pitt and Sniffen, 1988).

The use of microbial inoculants presumes that the conditions for their successful growth exist in the cut forage. These conditions include an adequate sugar content and a stacking arrangement that excludes oxygen. Prolonged handling in the ensiling process increases the number of nonlactic bacteria and makes the action of the lactics more difficult. The amount of sugar required varies with the silage dry matter; more is needed in wetter silages that have greater fermentation potential. Also, more sugar is needed in higher-protein forages and those with a higher buffering capacity.

The favored inoculants involve strains of *Lactobacillus plantarum* and *Pediococcus* that are homofermentative; that is, they produce only lactic acid. The concentrations used should be 10^6 per gram or higher. This level will guarantee silage preservation in about 85% of cases.

14.7.4 Cellulases

Enzymes as potential additives to control silage quality received a bad reputation in the 1960s and 1970s due to ineffective products that did not work well in the silage fermentation (J. W. Thomas, 1978). Newer cell wall enzymes, cellulases, hemicellulases, and pectinases are now available that are stable in silage fermentation and can work over the range of pH encountered in silage. The problem with applying these new products is understanding their effects and the silage response.

The original hypothesis was that glucose released by cellulase would promote lactic acid fermentation and

aid preservation. The negative side was a worry that releasing carbohydrate from lignified cell wall would reduce the quality of the fiber. While it was known that cellulases release glucose, the effect of hemicellulases and pectinases—the release of the respective component sugars of pectin and hemicellulose—was more problematic. The destruction of pectin might be associated with cell wall collapse and consequent effluent in immature higher-moisture forages. This has indeed been seen in ensiled higher-moisture immature forages treated with enzymes.

The observed effects of cell wall–degrading enzymes go beyond what was expected, though. Cell wall content (NDF) is often but not always lowered, and the increased available glucose does seem to promote lactate fermentation. There is, however, an accumulation of soluble carbohydrates that may be hemicellulosic oligosaccharides unusable or poorly usable by silage bacteria but beneficial to the rumen. In addition, cell wall–degrading bacteria increase the initial rate (i.e., reduce lag), but not the extent, of fermentation (Harrison, 1989). The latter effects may be related to the amorphogenic action of cell wall–degrading enzymes (see Figure 11.8). Under these conditions the cell wall seems to "crack" so that it becomes susceptible to further enzymatic attack. The main effect is on initial rates and reduction of lag, and it occurs even when NDF content of forage has been reduced. The overall rumen fermentative efficiency can be expected to respond to available soluble carbohydrate and reduced lag, provided the released carbohydrates can survive the ensiling. The great problem of all systems that make carbohydrate available to fermentation is conserving that available energy for the rumen. Thus improved forage quality for silage demands conservation systems that will preserve it.

14.8 Other Methods of Forage Processing

There is continuing interest in adapting various systems of forage preservation and feed handling to multipurpose use. These applications involve dehydration, pelleting, and separation of plant parts. Because legumes and immature grasses contain much valuable protein concentrated in leaf tissue, the fibrous stem portion could be separated and fed to ruminants, leaving the higher-quality leaf to be fed to nonruminants.

14.8.1 Dehydration

Dehydration systems are often combined with pelleting or wafering processes. Using fuel to remove water is an expensive process, but it can be extremely efficient in conserving forage quality. Cubing and pelleting are frequently used to reduce volume and costs

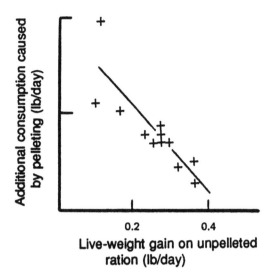

Figure 14.10. Pelleting rations of differing quality affects both rate of gain and intake in sheep (modified by L. A. Moore, 1964, from Minson, 1963). The best responses to pelleting occur in animals fed rations that promote limited gains. These are diets high in good-quality forage and sufficiently digestible to withstand the passage losses attendant on pelleting; that is, the increased intake more than offsets digestibility depression. Improvement may not occur in animals fed pelleted low-digestibility forages.

of storage and shipping. The consequent reduction in particle size generally promotes a higher voluntary intake, which increases animal efficiency, but decreases the effective fiber.

The increase in intake and reduced digestibility of pelleted diets depends largely on the plant cell wall, which is the slowest-digesting fraction and prone to competition between digestion and passage. Decreasing the particle size reduces volume and eliminates the work of rumination, but the finer particles in the feed are subject to a faster rate of passage. Feeding completely pelleted diets causes the elimination of the coarse layer in the rumen that may serve as a filter, the result being an increase in mean fecal particle size.

The increase in efficiency resulting from decreased feed particle size is largely due to the greater level of intake, but there are other changes, too: an increased rate of passage, depression in digestibility of fiber, decreased methane production, and higher ratios of

propionic acid to acetic acid in the rumen. These changes tend to increase efficiency of digested energy for fattening but depress milk fat in lactating cattle. In some cases the depression in digestibility due to the increased rate of passage may offset the increased efficiency. The increase in intake and consequent gross efficiency does vary (Figure 14.10). Generally, high-quality forages with low cell wall content show less animal response than medium-quality forages containing higher amounts of cell wall. The intake increase involves the reduction in density and the more rapid rate of passage of small feed particles.

The increased passage from the rumen of fine particles may include insoluble forage proteins, which may be an important factor in the efficiency of pelleted diets that increase rumen escape. Pelleted, heat-dried forages contain protein in an insoluble form that is more slowly fermented than unpelleted forages and may be even less available because of Maillard products. The effect on digestible protein is more variable because the balance of reduced fecal metabolic losses may be offset by the increased fecal loss of Maillard products.

Other factors resulting from pelleting are more difficult to control or describe. The heat of friction can denature proteins and gelatinize starches. These effects may be important in altering the intake and digestibility of concentrates and very high quality forages.

14.8.2 Dewatering and Leaf Protein Concentrates

The preparation of high-quality leaf proteins may involve dewatering systems that macerate and squeeze out plant juice. The dewatered residue, which can contain 20–40% true protein, is enriched in fiber and plant cell wall constituents (Collins, 1985a). This residue is a valuable ruminant feed because the efficiency of its utilization is aided by a moderate reduction in particle size through these treatments. The juice can be fed directly to swine or processed into leaf protein concentrate, possibly for human use but more likely for nonruminant supplementation.

15 Function of the Ruminant Forestomach

Understanding rumen function requires an understanding of the anatomy and physiology of the gastrointestinal tract, subjects so extensive as to merit books in their own right. The development of ruminant function in the young animal, the control of rumen fermentation, and the effect of diet involve an intimate integration of digestive physiology and nutrition. Students are referred to the extensive background literature on these subjects (Church, 1988; Tsuda et al., 1991). This chapter on rumen function must necessarily provide a synopsis.

Rumination is, literally, chewing of the cud, an activity limited to true ruminants and the tylopods. Foregut fermentation in nonruminants does not involve regurgitation or chewing of food. The discussion in this chapter is limited to the true ruminants; nonruminant herbivores are discussed in Chapter 5.

15.1 Development

The infant ruminant is functionally a monogastric animal with all the usual dietary requirements for vitamins and amino acids characteristic of nonruminants. Calves, and most likely the young of other ruminant species, lack sucrase and secrete limited amounts of amylase, and so are unable to use sucrose and starch. Thus milk replacers must be based upon glucose or lactose. The infant depends on its mother's milk until the development of ruminant fermentation allows digestion of carbohydrates other than glucose and lactose. The ability of the lower tract to utilize starch develops later.

The blood sugar concentration of infant ruminants is similar to that of nonruminants. As the animal grows, this level declines to the adult level and the rumen attains and develops fermentation. The two processes do not seem to be interdependent, as blood sugar declines even when the development of ruminant function is prevented.

At birth the rumen is essentially undeveloped and forms only a small proportion of the total stomach (Table 15.1). The increase in relative size of the reticulorumen rises from 25–35% at birth to 62–80% in the adult. At the same time the lower digestive tract is proportionally diminished (Section 17.2). Rumen development depends on access to a fibrous diet and inoculation by rumen bacteria. Specifically, rumen wall development depends on stimulation by volatile fatty acids (VFA), whose production requires the requisite bacteria and substrate.

If calves are maintained on a milk diet, rumen development can be greatly retarded. Leakage of milk into the rumen promotes a lactic acid type of fermentation with little VFA production. Suckled milk is bypassed to the abomasum via esophageal groove closure. This reflex is elicited by suckling and other stimuli. It has been possible to train lambs to retain this ability into adulthood (Ørskov et al., 1970). This function seems to be retained in adult selector ruminants (Hofmann, 1989).

In adults, the reticulorumen is the largest stomach compartment (Table 15.1). The omasum in sheep, goats, and many African antelope is relatively smaller than in cattle (Hofmann, 1973). The omasum seems more significant in reports of tissue weight because the amount of tissue per unit volume is much larger relative to that of the reticulorumen because of its internal structure.

R. G. Warner and W. P. Flatt (1965) elucidated the mechanisms that stimulate rumen development. Placing sponges or some other coarse material that simulates hay in the rumen of calves has no effect on development, which is characterized by the proliferation of the rumen wall and the development of papillae (Figure 15.1). Development *can* be elicited by placing dilute buffered solutions of VFAs in the organ. Butyrate is more effective in this regard than propionate, followed by acetate. This order is the same as that in which the acids are known to be metabolized by the rumen epithelium. The rate of VFA absorption is markedly influenced by rumen development. This difference is probably a function of both VFA concentrations and epithelial surface area.

230

15.1. Stomach compartment sizes in young and adult ruminants

	Body weight (kg)	Reticulorumen		Omasum		Abomasum	
		(g)	(%)[a]	(g)	(%)[a]	(g)	(%)[a]
e							
rth	24	95	35	40	14	140	51
ult	325	4540	62	1800	24	1030	14
rth	4	9	25	2	6	25	69
ult	67	1010	80	102	8	145	12
p							
th	6	19	32	5	8	36	60
ult	62	919	73	119	9	226	18

ource: Adapted from Lyford, 1988.
Percentage of the total stomach.

15.1.1 Inoculation

Young ruminants probably acquire rumen bacteria mainly through feed and interanimal contact. Anaerobic bacteria similar to those found in the rumen occur in nature, particularly in manure and soil. Despite the sensitivity of many rumen organisms to temperature and oxygen, they can be transferred via saliva and feed from one animal to another and escape down the digestive tract to inoculate the lower tract of ruminants and nonruminants alike. Inoculation probably depends on the survival of only a few cells. It is not possible to prevent the ultimate development of the rumen or intestinal fermentation merely by isolating an animal, although isolated calves may not develop the correct protozoa. Some years ago there was a flurry of studies on rumen inoculation as a method of hastening rumen development; however, no consistent advantage of inoculation was ever established.

The concept of manual inoculation emerged from the notion that rumen malfunction might be caused by a defect in the microbial population. Rumen populations tend to be similar in animals on a given diet, although many microbe species may occur in relatively small numbers, and any of these may respond to a dietary change. The presence or absence of a certain organism is determined by its access to favorable

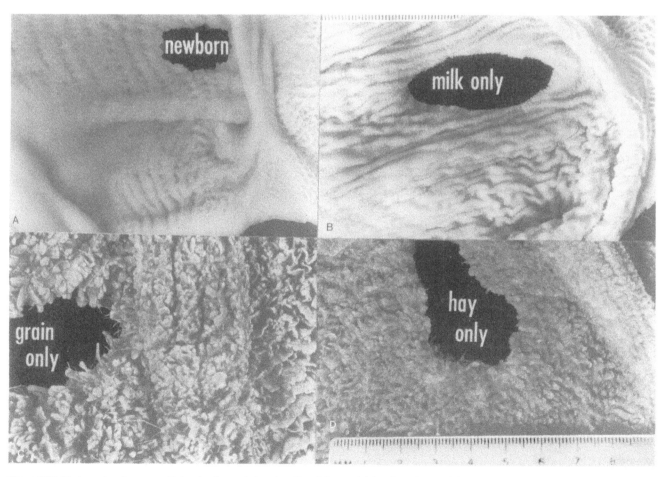

Figure 15.1. Photographs of rumen wall showing the ventralmost portion of the cranial dorsal sac just caudal to the reticuloruminal fold. The reticuloruminal fold and reticulum are visible in the righthand portion of each picture. (A) Newborn. (B) Milk-fed to 13 weeks. (C) Grain-fed to 13 weeks. (D) Hay-fed to 13 weeks. (Photos courtesy of R. G. Warner; see also R. G. Warner and Flatt, 1965.)

growth conditions (Hungate, 1966). Existing bacteria can adapt or mutate to accommodate a new substrate and changes in rumen conditions. The normal adaptation period is about one to two weeks. Evidence that inoculation can have a positive effect does exist. For example, in the case of abrupt dietary change from hay to concentrate, rumen adjustment is facilitated by inoculation with rumen contents from an animal already on the diet. Another example is the case of fasted ruminants in which the microbial population may be diminished. Inoculation at the time of refeeding may establish a normal rumen pattern more quickly than in uninoculated animals.

An additional question concerns the probability of differences among microbial populations in widely dispersed animals. Hungate (1966) summarized the evidence for the presence and absence of certain protozoa in New Zealand ruminants. Differences among flora in the digestive tracts of various herbivore species also occur, but many such differences may be accounted for on the basis of different habitats and dietary characteristics of the species (Chapter 16).

15.2 Anatomy

In popular parlance, ruminants are given credit for having four stomachs. In reality they have but one, which is divided into several compartments, the exact number depending on the species. The true ruminants, including sheep, cattle, goats, deer, and antelope, have four compartments. The tylopods—camels, llamas, and related species, often considered pseudoruminants —have a three-compartmented stomach, the omasum being absent.

Figure 15.2 shows the lateral and medial surfaces of the ruminant stomach. The four compartments, in the order of general discussion, are the rumen, reticulum, omasum, and abomasum. The rumen and reticulum, often considered a single organ (reticulorumen), are separated by the reticuloruminal fold. The separation is only partial; free exchange of contents is still possible. It is in these two sacs that the major portion of fermentative activity and absorption of nutrients occurs. The reticular fold is probably an important sorting device for heavier matter that has sunk to the bottom of the rumen. Whole corn and items of high specific gravity (mostly foreign objects) land predominantly in the cranial sac of the rumen. The cranial pillar prevents their passage into the rest of the rumen. At the time of the next cycle, cranial sac contraction dumps them into the reticulum, where they remain due to the reticuloruminal fold.

The total stomach occupies approximately three quarters of the abdominal cavity, from the seventh or eighth rib caudally to the pelvis. It occupies much of the left half of the cavity but extends over the median plane to the right. Its dorsal surface is suspended from the sublumbar muscles and by peritoneal and connective tissue.

Pillars divide the rumen into discrete sacs. During contraction (shortening) of these pillars, sacs become

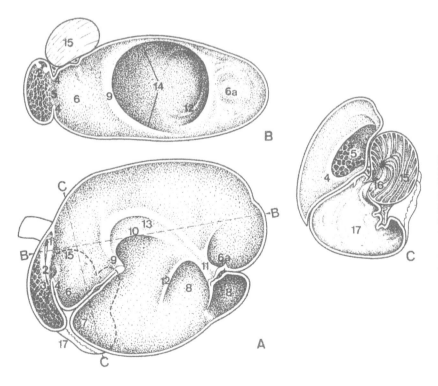

Figure 15.2. The ruminant stomach (from Hofmann, 1988). (A) Longitudinal section, left side. (B) Horizontal section along axis B–B in (A). (C) Transverse section along plane C–C in (A). Structures and compartments are as follows: (1) cardia, (2) reticulo-omasal groove, (2a) reticulum, (3) reticulo-omasal orifice, (4) cranial sac of rumen, (5) dorsal sac of rumen, (6) caudodorsal blind sac, (7) ventral sac of rumen, (8) caudoventral blind sac, (9) reticuloruminal fold, (10) cranial pillar, (11) right longitudinal pillar, (12) caudal pillar, (13) dorsal coronary pillar, (14) ventral coronary pillar, (15) omasum, (16) omasal canal, (17) abomasum. See Sisson and Grossman, 1953, and Sellers and Stevens, 1966, for more information.

smaller and the more liquid ingesta are circulated and generally forced upward through the floating mat of more solid ingesta. Although it is part of the fermentative mechanism, the reticulum is frequently the site for the accumulation of foreign objects such as hardware because of its proximity to the cardia and its generally ventral position. The esophagus terminates at the cardia, which is at the juncture between the reticulum and the rumen. The esophagus also serves as the cranial end of the reticular (esophageal) groove, which, in the adult bovine, extends for 17–20 cm ventrally to the reticulo-omasal orifice.

The forestomach compartments (rumen, reticulum, and omasum) all originate from the same embryological tissue. The organs are lined with nonglandular, non-mucus-producing, keratinized stratified squamous epithelial tissue. As such, the forestomach was long considered incapable of absorbing anything, but today it is recognized as a major site of nutrient absorption. The rumen is lined with finger-like projections called papillae which vary in shape and size (up to 1.5 cm in length). They are larger and denser in the ventral regions where nutrient concentration, and therefore absorption, is most pronounced. The surface of the reticulum contains elevations that provide compartments of four to six sides resembling a honeycomb.

15.2.1 Nerve and Blood Supply

The stomach is innervated by the vagus (both sensory and motor pathways; see Figure 15.3) and sympathetic nerves. The rate of reticuloruminal contractions is controlled by distension of the reticulorumen through reflexes, by the sight of food, and by rumination, all mediated via the vagal nucleus in the brain. Under physiological conditions, abomasal distention inhibits only omasal motility (Sellers and Stevens, 1966). Thus the rate of reticulorumen contractions can be increased by reticular distension. Conversely, abomasal distension suppresses reticuloruminal contractions and abomasal secretion of acid. Iggo and Leek (1970) identified tension receptors in the medial wall of both the reticulum (near the reticular groove) and the cranial-dorsal sac. There are also chemical receptors in those sites which respond to pH.

The blood supply to the forestomach originates from the abdominal aorta by way of the celiac-cranial mesenteric trunk (Figure 15.3). This artery enters the rumen between the esophagus and the mid-dorsal sac. Four branches then feed to the various parts of the rumen as follows: (a) the common hepatic artery supplies the cranial surface and also the pancreas, liver, and gall gladder; (b) the right ruminal artery supplies the right rumen surface plus the pancreas and omentum; (c) the left ruminal artery supplies the left rumen surface plus the reticulum and esophagus; and (d) the

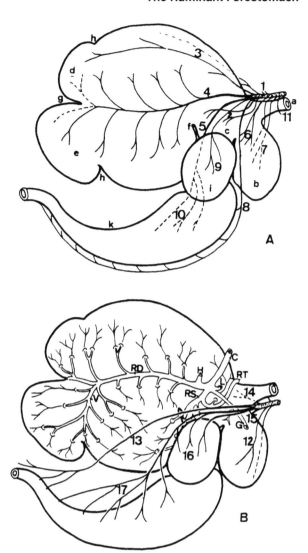

Figure 15.3. Distribution of nerves and blood vessels on the right side of the ruminant stomach (from Hofmann, 1988). (A) Dorsal vagus trunk. (B) Ventral vagus trunk and arteries. Rumen nerves, blood vessels, and lymph nodes are lodged in the grooves; those of the other compartments are in the mesenteries (omenta). Vagal branches: 1–11 in (A), 12–17 in (B). Arteries: C = celiac; G = left gastric; H = hepatic; L = lienal; RD = right ruminal; RS = left ruminal; RT = reticular (supplies atrium also); V = right ventral coronary.

left gastric artery supplies the omasum and abomasum. Venous drainage is by way of four veins: the right ruminal, left ruminal, omaso-abomasal, and reticular veins. All empty into the hepatic portal vein, which feeds directly into the liver.

15.2.2 Rumen Wall Structure and Musculature

The reticuloruminal wall (Figure 15.4) consists of a serous membrane, a muscular tunic, and the epithelium. The epithelium is the site of absorption, active transport of sodium and chloride, and passive transport of VFAs, water, and other substances such as urea.

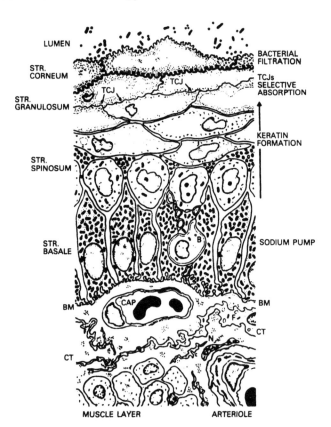

Figure 15.4. Cross section of the fully developed rumen wall depicting the types of cells and layers present (from D. H. Steven and Marshall, 1970). Details of the cell junctions are omitted. B = branching cell; BM = basement membrane; CAP = capillary; CT = connective tissue; F = fibroblast; N = nerve trunk; TCJ = tight cell junction.

The rumen musculature consists of two layers: the ruminal pillars are foldings of the oblique muscle layers, and the reticular groove consists of internal muscle fibers running along the length of the groove and forming a loop around the cardia (Hofmann, 1988).

The cells of the ruminal surface tend to be keratinized and have some of the character of ordinary skin, but the ruminal surface differs from skin in its electrical resistance and its permeability and transport qualities. The keratinization is normally offset by specialized cells that are important in absorption of VFAs. Specialized bacteria also adhere to the rumen wall and are involved in its metabolism.

15.2.3 Surface Structures

The luminal surface includes papillae in the rumen and reticular ridges in the reticulum. The papillae in-

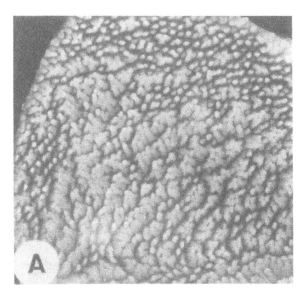

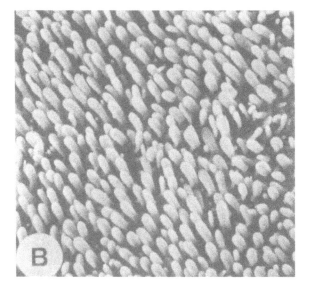

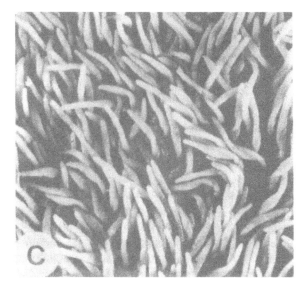

Figure 15.5. Filiform ruminal papillae (oval or circular cross section) are indicative of unstimulated ruminal blood flow and lack of butyric and propionic acids. (A) Atrium papillae of a one-day-old goat kid (intermediate feeder). (B) Dorsal wall papillae of a roe deer (concentrate selector; winter). (C) Atrium papillae of a Bohor reedbuck (grazer; drought); also called "hunger papillae." (From Hofmann, 1988.)

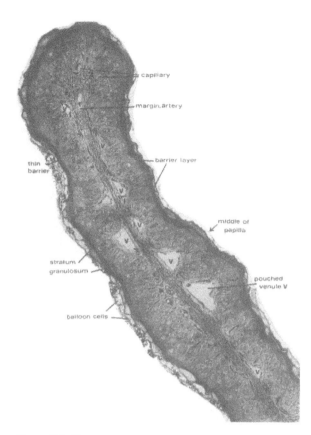

Figure 15.6. Microstructure of an absorptive ruminal papilla (semithin transverse section, ×97 (from Hofmann, 1988). Note the few epithelial cell layers with denser barrier and superficial balloon cells containing ruminal bacteria, and the extensive vascular system (mainly venules with fenestrated endothelium; i.e., absorptive type).

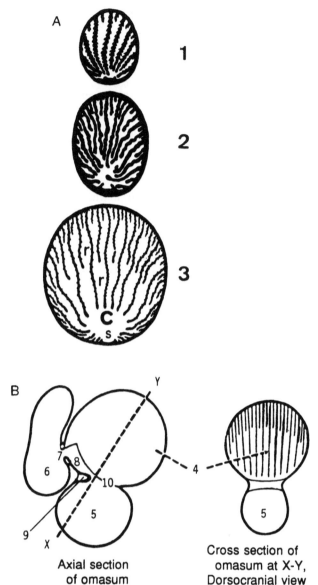

Figure 15.7. (A) Cross sectional of the omasum from a concentrate selector (1), an intermediate feeder (2), and a grazer (3). Note interlaminar recesses (r), omasal canal (C), and omasal groove (s). (From Hofmann, 1988.) (B) Axial and cross section of the omasum; X–Y indicates plane of reference. Structures visible include omasal body and lamina (4), abomasum (5), reticulum (6), reticulo-omasal orifice (7), omasal canal (8), omasal pillar (9), and omaso-abomasal orifice (10). (Modified from Sellers and Stevens, 1966.)

crease the absorptive surface of the rumen. The reticulated surface of the reticulum may be involved in the sorting and handling of particles that pass near the reticulo-omasal orifice.

The papillae differ among animals of the same species on different diets as well as among species with different feeding strategies. In cattle, the papillar structure is apt to degrade under conditions of high-concentrate feeding, leading to parakeratosis (Section 15.7.3). Starvation also reduces the papillae (Figure 15.5). The internal structure of a papilla is shown in Figure 15.6. Generally, concentrate selectors, which harbor higher VFA concentrations than grazers (Chapter 4), have more papillar surface (in a smaller rumen) that is more generally distributed. The state of the papillae and their surfaces are described by the surface enlargement factor, which is calculated as twice the papillary surface area divided by the basal surface area (Hofmann, 1988).

15.2.4 The Omasum

The omasum, the third compartment, is characterized by the presence of a large number of leaves

(Figure 15.7), which may absorb some water and nutrients and prevent passage of large particles of digesta. Much fluid can bypass the omasum via the omasal canal if the esophageal groove is closed. One of the omasum's major functions is to pump digesta from the reticulum into the abomasum (C. E. Stevens et al., 1960).

The omasum, which is peculiar to the true ruminants, is a small, compact, oval organ connecting the

Table 15.2. Omasum contents (dry weight as percentage of body weight) for five ruminant species in the wet and dry seasons

	Wet season		Dry season	
	Mean	SE	Mean	SE
Thomson's gazelle	0.007	0.008	0.028*	0.008
Grant's gazelle	0.006	0.008	0.032*	0.007
Kongoni	0.069	0.011	0.132*	0.011
Wildebeest	0.122	0.010	0.274*	0.011
Steer	0.203	0.011	0.535*	0.011

Source: Reed, 1983.
*P < .05.

reticulorumen to the abomasum. In true ruminants the arrangement of the organs is such that ingesta flow from the reticulo-omasal orifice through the omasal canal to the omasal-abomasal orifice and into the abomasum. The reticulo-omasal orifice is adjacent to the posterior end of the esophageal groove, so that when the groove is closed, ingesta can pass directly to the omasum and bypass the reticulorumen (Figure 15.2). The interior structure of the omasum consists of leaves attached to the distal wall relative to the neck of the organ. The omaso-abomasal orifice has no sphincter to limit backflow, although contraction may allow the folds of the omasum to act in this way.

The omasum has received much less attention than the reticulorumen, and as a result, its role is not entirely clear. It may serve as a filter pump to sort out the liquid and fine digesta for passage on to the abomasum. The filtration is conducted by leaves attached to the distal wall, which do not allow coarse fiber to enter the distal portion of the organ.

The relative size of the omasum varies among ruminant species. It is generally smaller and perhaps less functional in concentrate selectors (Hofmann, 1989) and also in smaller ruminants. The logarithmic association with body weight is considerably greater than 1 (Reed, 1983). Thus the sheep's omasum is both actually and relatively smaller than that of bovines (Table 15.1). The difference is also apparent in African antelope, which show seasonal changes in omasum size that depend on diet quality (Table 15.2).

In cattle, the omasum is probably an absorptive organ. As much as 30–60% of the water entering the organ is absorbed, along with a considerable amount of VFAs (40–69%), sodium, potassium, and other ions. The principal effect of this action is to reduce the net volume entering the abomasum and remove VFAs (lowering their concentration). The absorptive role may be less important in sheep, goats, deer, and antelope, in which the organ is relatively smaller (in proportion to body size). Perhaps omasal function is more important in grazing species that require this anatomical development to process the large quantities of fiber they consume.

Small concentrate selector ruminants are intolerant of high-fiber diets, which, if imposed, lead to impaction of the omasum (Hofmann, 1973). Impaction of the omasum has also been observed in cattle fed high levels of coarse rice hulls.

15.3 Eating and Ruminating

Ruminants tend to ingest feed rapidly and to ruminate it later. Feed is ingested, chewed, mixed with saliva, and rolled as needed to form a bolus, which is swallowed and ejected with some force into the anterior rumen. The time spent eating is affected by the nature of the feed or forage and the time required to prehend it and reduce it to a swallowable bolus. Concentrates or pellets, being denser and already in a relative fine state, are eaten more rapidly than coarse forage, which requires more chewing. The morphology of the forage also affects the prehension and selection of food for ingestion. Forage or browse that must be picked leaf by leaf limits the eating rate. Browsers are

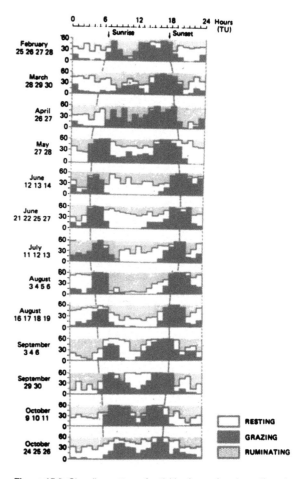

Figure 15.8. Circadian pattern of activities in grazing sheep (from Arnold and Dudzinski, 1978). White areas indicate resting; black areas, grazing; and stippled areas, ruminating.

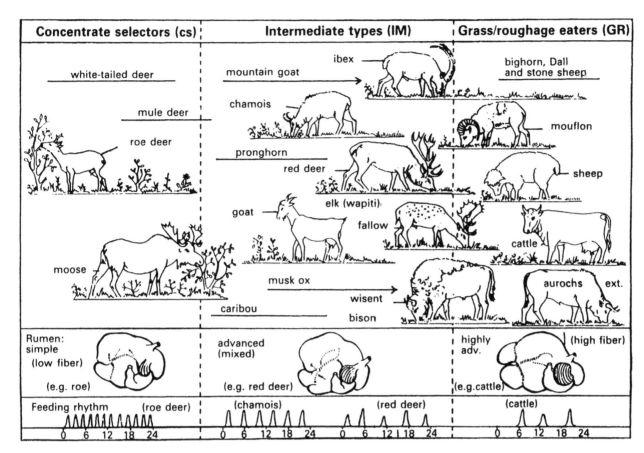

Concentrate selectors (cs)	Intermediate types (IM)	Grass/roughage eaters (GR)

Figure 15.9. The position of European and North American ruminant species along the continuum of morphophysiological feeding types (from Hofmann, 1988). The farther the baseline of a species extends to the right, the greater its ability to digest fiber in the rumen. Feeding on plant cell contents implies shorter intervals between meals since the simple rumen of concentrate selectors (CS) has fewer food passage delay structures than that of grazers (GR).

much more efficient at picking their food than are grazers.

The eating rate and the total time spent eating affect the way time is distributed among eating or grazing, subsequent rumination, and other activities in the day of a ruminant (Figure 15.8). An increase in time spent on eating and rumination necessarily decreases time spent on other activities. The diurnal pattern is also affected by day length and season.

If the ingestion rate is slow, fermentation is continuous and there are no peaks in acid production. Rapid eating allows more material to be fermented simultaneously. Rapid ingestion thus results in a more synchronized peaking of fermentation and an acid production that must be balanced by buffering mechanisms, the most important of which is ensalivation.

Eating and rumination frequency patterns vary among ruminant species according to feeding habit (Hofmann, 1989). Concentrate selectors tend to have simpler rumens and eat and ruminate frequently (Figure 15.9). In very small antelope such as the suni and dik-dik, eating and ruminating are distributed almost evenly over the entire daylight period (Hoppe, 1977).

The longer-spaced diurnal patterns of rumination are more characteristic of the larger grazers, while intermediate feeders fall between concentrate selectors and grazers in behavior (Figure 15.9).

A reason for this variation in behavior is that concentrate selectors eat a high-quality diet leading to much higher VFA production rates. Eating constantly allows the animal to avoid peaks of acid production that could be pathological. It may be that the function and purpose of rumination in small concentrate selectors is primarily to regulate the rate of release of nutrients for fermentation. These animals tend to consume whole seeds, fruits, and small plants, and the lack of chewing, which means that cell surfaces are not punctured, both limits and regulates the fermentation rate and VFA production.

15.3.1 Rumination

Rumination is the postprandial regurgitation of ingesta followed by mastication, reforming the bolus, and reswallowing. The process is cyclic and is closely integrated with reticuloruminal motility (or cycles; see

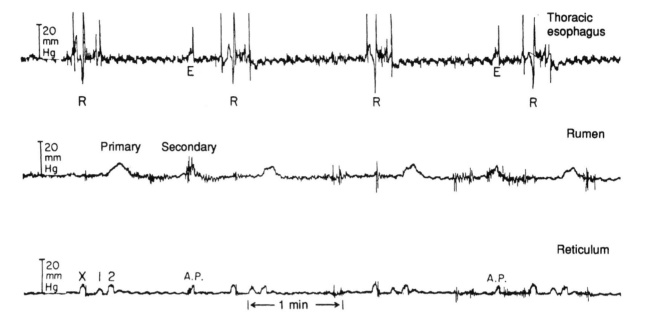

Figure 15.10. Pressure recordings of three complete cycles of reticuloruminal contractions during rumination (from C. E. Stevens and Sellers, 1968). The double reticular contraction (1 and 2) and primary and secondary rumen contractions are evident in the first cycle, as is the extrareticular contraction associated with regurgitation (X). The esophageal pressure changes during regurgitation (R) consist of a small positive pressure wave followed by a negative deflection due to inspiration against a closed glottis, and then, immediately, a large positive deflection due to antiperistaltic esophageal contraction. Regurgitation is usually soon followed by deglutition (not labeled) of the fluid expressed from the bolus at the beginning of mastication. A wave of deglutition, carrying the bolus to the rumen, is also visible on the esophageal trace at the end of each cycle of rumination and just before the next regurgitation. Note the close integration of the rumination and reticuloruminal cycles. Eructation (E) and the associated increase in pressure due to abdominal press can be seen on the reticular trace (A.P.) and also superimposed on the secondary wave of rumen contraction.

Figure 15.10). Rumination commences with an extra-reticular contraction, which concentrates digesta and fluid near the cardia. At the same time an increased inspiration of air against a closed glottis reduces the pressure in the thoracic esophagus (Sellers and Stevens, 1966). Ingesta are thus sucked into the esophagus and then moved by rapid antiperistaltic esophageal contractions to the mouth, where excess liquids are swallowed and mastication commences. Mastication reduces particle size, extracts soluble contents with saliva, and enriches the fiber content of the bolus. The chewed mass is remixed with saliva, reformed into a bolus, and swallowed (Schalk and Amadon, 1928; Church, 1975). The cycle is repeated but may be interrupted by other activities.

More time is normally spent chewing during rumination than during eating. Also, the rate of chewing is generally slower and more deliberate during rumination, with more time spent per unit of ingesta. Chewing patterns as related to eating, drinking, and rumination are shown in Figures 15.11 and 15.12. Chewing activity is more intense initially and diminishes as the bolus is masticated. The amount of time spent ruminating is influenced by the nature of the diet and appears to be proportional to the cell wall content in coarse forages. Feeding concentrates or finely ground or pelleted hay may greatly reduce rumination time (Figure 15.13), and feeding forages with high cell wall content tends to

increase the rumination time. Increasing the intake tends to reduce the time spent ruminating per gram of

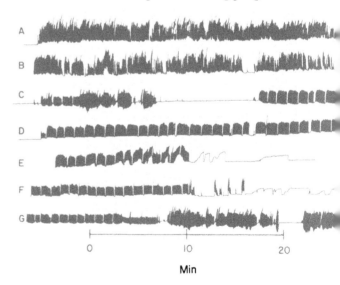

Figure 15.11. Patterns of eating and rumination in cattle (courtesy of J. G. Welch, University of Vermont). (A and B) Eating sequence; each oscillation represents a jaw movement. Eventually the animal begins to tire (B and C). (C) Mixed rumination and feeding is followed by a lapse and more rumination. (D) The onset of rumination in a different animal. (E and F) Termination of rumination followed by drinking. (G) Termination of rumination followed by a return to feeding. The average for each bolus is 62 chews in 56 sec. Individual rumination periods are highly variable and may last up to 2 h.

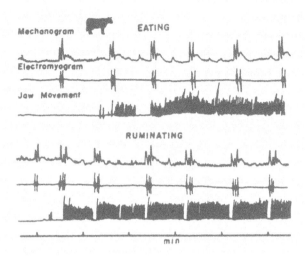

Figure 15.12. Pressure recordings, electrical activity, and jaw movements detected by pressure changes in a balloon fixed on the halter of a cow starting to eat hay (above) and ruminating (below) (from Ruckebush, 1988).

feed, a factor probably responsible for the increase in mean size of fecal particles at higher intakes. The total amount of time sheep and cattle ruminate seems to have an upper limit of about 10–11 h per day (Bae et al., 1979; Welch, 1982). This limit along with time spent eating are probably factors limiting consumption of coarse forages.

Rumination appears to be induced by sensors in the rumen wall, which is innervated principally by the dorsal trunk of the vagus nerve. Sectioning the nerve abolishes the response. Rumination can be stimulated by tactile means or by the pressure of coarse material; hence, the popular term *scratch factor* to describe the dietary characteristic probably responsible for induc-

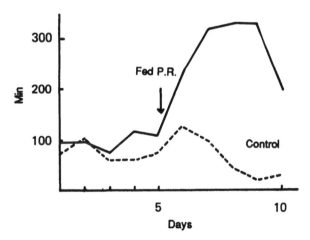

Figure 15.13. Rumination time in minutes of sheep fed 50 g/day of 5-cm-long polypropylene ribbon for 3 days compared with controls fed no polypropylene (from Welch and Smith, 1971). All animals received 400 g/day each of both alfalfa meal pellets and pelleted concentrate during the treatment period. Four rams were used in a single reversal design with 60 days between periods 1 and 2.

ing normal rumination. Lack of stimulation may be responsible for the low level of rumination in animals on concentrate and pelleted diets. The scratch factor is related to both particle size and cell wall content of the diet. Plastic or other devices added to concentrate feed or placed in the rumen have had variable success as rumination stimulators. The sites of maximal sensitivity are in the cranial sac of the rumen and regions of the reticulum.

15.3.2 Rumen Motility

Contraction and relaxation of the reticulorumen wall and pillars move and mix the ingesta. The motion can be divided into primary contractions, which affect the whole reticulorumen, and secondary movements, which affect only a part of the organ (see Figures 15.14, 15.15, 15.16, and 15.17). The motions require up to 50 sec to complete and occur in total cycles whose length depends on the activity of the animal: whether eating, ruminating, resting, and so on. Church (1975) described representative patterns for bison, cattle, deer, sheep, and goats.

Primary contractions associated with ruminal mixing begin with an initial sharp contraction of the reticulum and reticuloruminal fold followed by a second, more powerful contraction of the reticulum, with the wave of contractions passing over the rumen. This raises the cranial sac and causes the cranial pillars and the caudal and dorsal coronary pillars to contract, and also compresses the dorsal sac of the rumen (partly by contraction of the longitudinal pillars).

Rumen mixing and rumination together promote the turnover of indigestible residues, which, if allowed to accumulate, would clog the rumen. Material requiring further rumination is selectively regurgitated; finer material and liquid are allowed to flow from the rumen through the reticulo-omasal orifice. It has been commonly supposed that the size of the omasal orifice sets the limit on passage, but the selection of finer material probably occurs through other sorting mechanisms such as occlusion of coarse matter in the floating mat of fiber. A finely ground diet increases the rate of passage and results, paradoxically, in an increase in mean size of particles in the feces. The positive effect of adding a small amount of coarse fiber to a concentrate diet may be related to reestablishment of the mat and its consequent stimulation of the rumen wall and sorting of rumen particulate matter.

15.3.3 Particle Size Reduction

Most of the published observations regarding the effects of diet on the rate of passage have been determined with the stained particle technique as described by Balch (1950) and others. Although this method

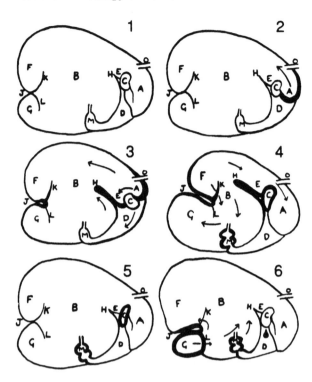

Figure 15.14. The movements of the stomach, shown from the right side (from Phillipson, 1939). (1) The stomach at rest. (2) The reticulum in the first stage of its contraction. (3) The reticulum in the second stage of its contraction. The anterior blind sac of the rumen is relaxed; the anterior and posterior longitudinal pillars are contracting; the omasum is moved downward and forward and the body of the abomasum is lifted up. (4) The reticulum is relaxed. The anterior blind sac of the rumen is contracted; the anterior and posterior longitudinal pillars of the rumen are fully contracted, as are the dorsal and ventral blind sacs of the rumen; the ventral and dorsal blind sacs are relaxed. The omasum is pear-shaped. Strong waves of peristalsis appear in the pyloric antrum of the abomasum. (5) The stomach at rest. The omasum is elongated. (6) The ventral blind sac of the rumen together with the longitudinal and ventral coronary pillars are contracted. A = reticulum; B = rumen; C = omasum; D = abomasum; E = anterior blind sac of the rumen; F = dorsal blind sac of the rumen; G = ventral blind sac of the rumen; H = anterior pillar; J = posterior longitudinal pillar; K = dorsal coronary pillar; L = ventral coronary pillar; M = pyloric antrum of the abomasum; O = esophagus.

yields only relative data and is incapable of quantitative marker recovery, it does provide valuable information.

Ration composition and form have important effects on passage. Generally, grinding forage increases the rate of passage. Concentrates, which usually have smaller particle sizes than forages, are associated with faster passage. Remember that pelleted forage and concentrate diets are often consumed in greater amounts, and the intake factor alone will be responsible for some of the increased passage. Particle size per se does tend to have its own effect on passage (Table 15.3), however, with smaller particles passing faster than larger ones. Larger particles are filtered by the rumen mat and disintegrated through rumination. Finely ground whole diets cause cessation of rumina-

Table 15.3. Effects of alfalfa hay particle size on retention time and fiber digestibility

Feed	Mean size (μm)	5% transit (h)	Retention 80 − 5 (h)[a]	Fiber digestibility (%)
Long hay	—	22	54	44
Coarse grind	434	16	39	34
Medium grind	393	16	44	31
Finely ground	280	13	27	22

Source: Rodrigue and Allen, 1960.
[a]Retention time is defined in Section 23.4.1.

tion and the relative elimination of the floating mat of fiber that separates the liquid and gas phases of the rumen.

The floating rumen mat is one of two sorting mechanisms for particles (the other is in the omasum), and its elimination allows the escape of intermediate-size particles that would otherwise be entrapped in the mat.

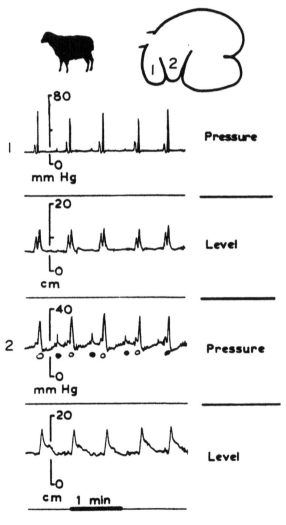

Figure 15.15. Comparison of simultaneous recordings of pressure and vertical displacement of the reticulum and the cranial sac of the sheep rumen during primary (1) and secondary (2) ruminal contractions (from C. S. W. Reid and Cornwall, 1959).

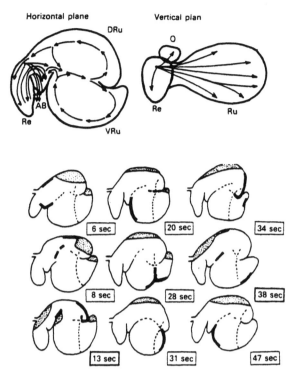

Figure 15.16. Movement of digesta in the reticulorumen as seen radiographically in the horizontal and vertical planes (from Wyburn, 1980). (Top) arrows indicate direction of movement. (Bottom) Main contraction sequences of the sheep's reticulorumen as indicated by X-radiography. Time (sec) indicates the interval after the reticular movement, and the contracting region of the reticulorumen wall is indicated as a heavy line. The gas bubble (stippled area) is brought over the cardiac orifice at 13 sec in the case of a primary contraction and during the secondary ruminal contraction at 38 sec. AB = abomasum; DRu = dorsal rumen; O = omasum; Re = reticulum; Ru = rumen; VRu = ventral rumen.

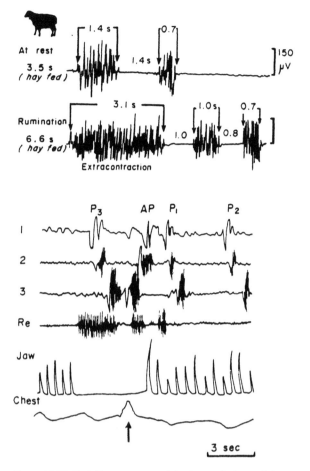

Figure 15.17. (Top) Electromyogram of the sheep reticulum wall showing the biphasic activity in the animal at rest and the superimosed extrareticular contraction during regurgitation. (Bottom) Events in the esophagus, reticulum, jaw, and chest associated with regurgitation (arrow). The esophageal electromyograms were recorded from electrodes placed at an equal distance on the esophagus, near the glottis (1), at the entry of the chest (2), close to the cardia (3), and on the reticulum (Re). The regurgitation of digesta (AP) is followed by swallowing just the excess liquid on two occasions (P_1 and P_2), and later, the bolus (P_3). (From Ruckebush, 1988.)

The operation of the mat differs from the filtration mechanisms of the omasum in that retention is based on occlusion and entrapment. Specific gravity and particle sizes are involved. Particles light enough to float are collected in the mat, which selectively retains them for further rumination (desBordes, 1981).

There is a direct relation between cell wall intake and rumination (Figure 15.18). Increased intake of cell wall promotes more rumination but decreases time spent ruminating per unit of cell wall, a factor that may be associated with the larger mean fecal particle size noted in animals on a pelleted diet and after increased intake of long forage. Heifers fed alfalfa pellets and chopped alfalfa hay showed mean fecal particle sizes of 0.36 and 0.30 mm, respectively (data from L. W. Smith, cited by Van Soest, 1966).

Increased intake increases rumen contents, or fill. The increased intake can be regarded as pressing both gastrointestinal volume and passage, these being the principal means of relief. Rate of digestion is less responsive since it is largely predetermined by ration composition. Balloons or plastic ribbon placed in the rumen also result in increased volume and passage rate (Balch and Campling, 1965; Welch, 1967). The need for food causes some compensatory expansion to allow for the greater ballast. Filling the rumen space tends to produce a smaller reduction in intake than would be expected, although intake and probably rumen volume require some time to adapt to new feeding conditions.

Plant cell wall represents the structural volume of the feed and as such is a major determinant of rumen volume. Removal of digestible and soluble contents from the interior of plant cells does not diminish their effective volume; the cells simply become filled with gas and water. Reduction in volume occurs only when the cell walls are destroyed by the processes of rumination and digestion. This is termed the *hotel effect* (Van Soest, 1975). Once the cell wall structure is destroyed,

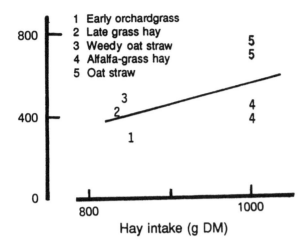

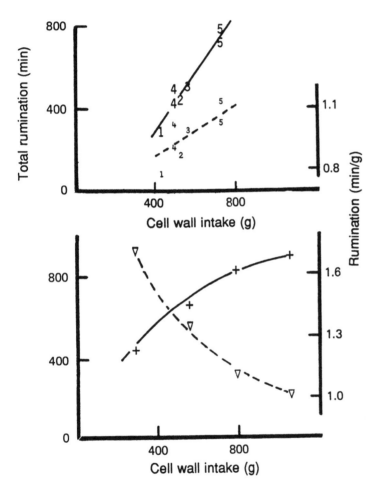

water held in the sponge of coarser cell wall material becomes available for absorption or passage.

15.3.4 Selective Retention

The selectively retained coarse particles become finer particles after rumination and digestion. These fine particles have a delayed passage, with their retention time being an inverse function of the rate of their production. The fine particles produced by rumination and digestion statistically outnumber those initially in the feed, such that they dominate the fine cell wall fraction of feces and contain an unusually high proportion of lignin (Figure 15.19). Since they have undergone a long period of digestion, fine fecal particles are higher in lignin content and take longer to appear in the feces than coarser material. (See also Figure 23.10.)

Although it is difficult to measure under in vivo conditions, the rate of particle breakdown is of major importance in the alleviation of rumen fill and, consequently, in feed intake. Evidence of its significance includes the very strong association of rumination time with cell wall intake (Figure 15.18), the strong association of feed cell wall content with voluntary intake, and the generally poor relation between rate of digestion and intake. The rate of particle breakdown or ease of rumination is a property of the feed composition, in particular its cell wall content, and the physical proper-

Figure 15.18. The effect of coarse forage on time spent ruminating in sheep. (Top) Rumination response to consumption of five forages at various intake levels (data from Welch and Smith, 1969). (Middle) The relation between rumination time and cell wall intake of the same forages (large numbers and solid line). Note the much closer relation with cell wall intake than with dry matter (upper figure). The time spent ruminating per gram of cell wall consumed (dashed line and small numbers) is greater when more mature forage is consumed (data from Welch and Smith, 1969). (Bottom) Increasing the intake of an individual forage (+) is associated with decreased rumination time per gram of cell wall consumed (▽) (data from Bae et al., 1979).

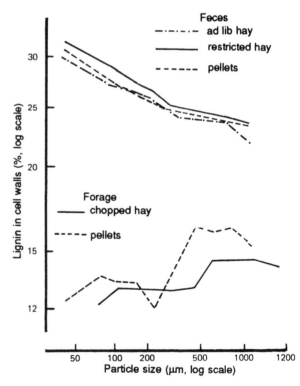

Figure 15.19. Relation between lignin content in the cell walls of alfalfa fed in different forms to heifers and fecal particle size (see Van Soest, 1975). Both axes of the graph use a logarithmic scale.

ties of the fiber that influence the ease or difficulty of comminuting fibrous particles into smaller ones.

When Welch (1967) placed polypropylene ribbons into the rumen, he found that ribbons longer than 9 cm were not ruminated and remained perpetually in the rumen. Longer ribbons that were ruminated passed into the feces as finer material than was the case with shorter ribbons. Since the fine ribbon particles in the feces can only have originated from larger pieces, the difference between appearance rates in feces of larger versus smaller particles is a relative measure of the comminution rate.

15.4 Structure of Ingesta

Rumen contents do not have a uniform composition. They are in the form of stratified layers showing ventral-to-dorsal differences as well as differences between anterior and posterior and between reticulum and rumen. Rumen contractions mix the contents, promoting turnover and accessibility of coarse floating matter for rumination. The mixing is inadequate to randomize distribution of particulate matter, although liquids may be mixed somewhat more efficiently.

The structure and composition of rumen contents are markedly influenced by diet. Coarse hay diets produce contents with a large, dense floating layer beneath the gas dome with relatively liquid contents and suspended fiber beneath (Figure 15.20). The floating mat is composed of the more recently ingested forage. As fermentation proceeds, digestion and rumination reduce particle size; fiber particles become waterlogged and tend to sink. The increase in apparent density is partly due to the loss of cellular gas space. Particles that settle to the floor of the rumen and have an optimum density are most likely to be selectively passed to the omasum. The optimum specific gravity for selective passage obtained with plastic particles appears to be about 1.2 (desBordes, 1981). Very dense objects (e.g., stones or pieces of metal) may be too large or heavy to escape. In animals fed higher-quality diets the floating mat is diminished, and it may be altogether eliminated in animals fed pelleted and concentrate diets. Rumen contents of animals fed concentrate are generally more viscous than those that receive only forage. Viscosity may affect mixing of liquid and the diffusion of VFAs toward the rumen wall.

Rumen fluid has the least amount of dry matter when coarse forage is the diet. Under most conditions of feeding, the fluid is relatively enriched by the cell wall components of the feed, which have a slower turnover rate. The rumen contents of sheep and cattle are similar in composition, while those of deer and other browsers may be somewhat higher in dry matter. Browsing species tend to have a proportionally smaller rumen content relative to body weight, which may be compen-

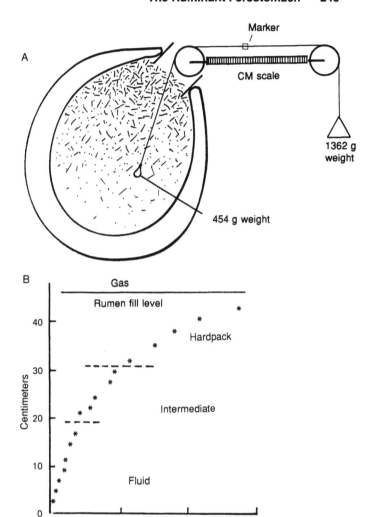

Figure 15.20. Measuring the consistency of rumen ingesta (from Welch, 1982). Figure A shows the device used to obtain the data shown in figure B. (A) Cross section of an open fistulated rumen with the measuring device in place. Support for the 3-lb weight is removed, and the 1-lb weight ascends through the rumen contents. The thicker and tighter the rumen ingesta pack, the longer the ascension time. (B) The ascension time curve obtained from the rumen of a cow fed corn silage. Rumen contents of animals fed coarse forages are stratified into a top layer (hardpack), an intermediate layer, and a fluid layer beneath. The thickness of the hardpack varies with the amount of coarse cell wall in the diet. Rumens of animals fed concentrate or pelleted forage are much more homogeneous, and the time for weight ascent is much shorter with these rations than with hay. Ranges of ascent times: grass hay 300–900 sec; corn silage, 90–200 sec; alfalfa pellets, 4–21 sec; and high concentrate, 60–130 sec. Higher feed intake also promotes a longer ascent time for any given ration. Ascent time decreases after feeding, rumination, and some rumen emptying.

sated for by a higher concentration of dry matter. The higher concentrations of solids and VFAs in rumens of concentrate-fed animals are often offset by a smaller rumen volume, so total content of dry matter or VFAs is not necessarily greater, and indeed may be less in these animals than in animals fed coarse forage (Balch et al., 1955).

15.4.1 Gases

The dome of gas in the upper part of the rumen is composed mainly of CO_2 and methane, the proportions of which depend on rumen ecology and fermentation balance. Ordinarily, the proportion of CO_2 is twice or three times that of methane. Small amounts of other gases may occur, including hydrogen and hydrogen sulfide. Nitrogen and oxygen are swallowed (as air) during feeding. Nitrogen may constitute as much as 10% of total gases during and after feeding. Most of the nitrogen is eructated with the fermentation gases (some is inhaled in the process). Some of the oxygen is absorbed or used by facultative organisms.

Although oxygen is toxic to obligate anaerobic bacteria, introduction of oxygen into the rumen through a fistula has little effect on fermentation. Accessibility of oxygen is limited to the surface and a few centimeters below it, and the hardpack layer of floating ingesta probably acts as a barrier and a metabolic sink for oxygen. Facultative organisms in this outer layer use oxygen rapidly and help maintain a low redox potential. Another possible source of oxygen is through diffusion from the blood across the rumen wall. The extent to which this occurs in not known. Oxygen in the rumen serves as a hydrogen acceptor and thus could be important in the fermentation balance.

Gases produced in the rumen are eliminated by eructation (a kind of silent belching) and, to a significant extent, also by absorption across the rumen wall and exhalation via the lungs. In the case of methane, the latter may account for 30% of the amount produced (Hoernicke et al., 1965). The fate of CO_2 is more complicated because of the pooling and recycling of animal metabolic carbon as urea and bicarbonates in saliva with that produced by the rumen organisms. Eructation is necessary to maintain rumen balance. Failure to eructate results in bloat, which can be fatal (Section 15.7.1). Eructation occurs through a slight variation of the normal reticular contractions whereby the area near the cardia is cleared of ingesta, the reticuloruminal fold and cranial pillar acting as a dam to hold back liquid. Relaxation of the cardiac sphincter, abdominal pressure, and contraction of the dorsal sac of the rumen combine to force gas into the esophagus, to be transferred to the mouth by antiperistaltic contractions (Sellers and Stevens, 1966). When the pharyngoesophageal sphincter is opened, the gas passes into the nasopharynx. Some gas passes into the respiratory passages. Gas pressure in the rumen stimulates eructation, and ingesta in the area of the cardia inhibit it.

15.5 Volatile Fatty Acids

The VFAs produced as end products of anaerobic microbial metabolism provide the ruminant with a major source of metabolizable energy. Removal of these acidic products is vital for the continued growth of cellulolytic organisms in the rumen.

The principal fatty acids, in descending order of usual abundance, are acetic, propionic, butyric, isobutyric, valeric, and isovaleric. The proportions of acetic, propionic, and butyric acids can be markedly influenced by diet and the status of the methanogen population in the rumen. Protozoa may also contribute significantly to the balance. Other organic acids may appear as products of microbial metabolism. Lactic acid is important when starch is a part of the diet, and is itself fermented to acetate, propionate, and butyrate. It appears only as a transient product 1–2 h after fermentation (Figure 15.21). Succinate and formate produced by some rumen species in pure culture do not normally appear as products in mixed cultures.

Rumen concentrations of VFAs are regulated by a balance between production and absorption whereby increased production rate induces higher VFA concentrations (Giesecke, 1970). Since production rates vary diurnally as a consequence of eating patterns, rumen concentrations and pH also vary. The pK's of the VFAs (4.8–4.9) are very much lower than normal rumen pH. The pattern following a meal shows a rise in VFAs and a drop in pH, followed by a slow recovery to the original conditions (Figure 15.22). Fermentation peaks about 4 h after feeding on a hay diet but occurs sooner if the diet contains much concentrate. The peak is largely a function of non–cell wall fermentation. The maximum quantity of cellulose digested from a meal occurs later, between 6 and 18 h after ingestion, depending on the rate of digestion.

15.5.1 Measuring VFA Production

The habit of expressing VFAs as molar proportions rather than normal concentrations has been responsible for much confusion. Molar proportions are valid only if presented along with total acid concentrations. The molar proportions of glucogenic propionate to nonglucogenic acetate or butyrate is of physiological significance to the animal; however, its value as a measurement is offset by the problem that a natural rise or drop in the amount of one acid requires a statistical change of opposite sign in the other acids. It was by this means that the erroneous theory that acetate deficiency caused milk fat depression in lactating ruminants arose (Section 20.7). The drop in molar proportions of acetate is usually caused by the dilutory effect of a large increase in propionate.

Rumen concentrations of VFAs depend on the amount of VFAs absorbed (Giesecke, 1970); however, the quantitative relationship depends on rumen pool size and turnover. Concentrate-containing diets may exhibit higher concentrations of acids relative to the amount absorbed than forage diets due to a smaller

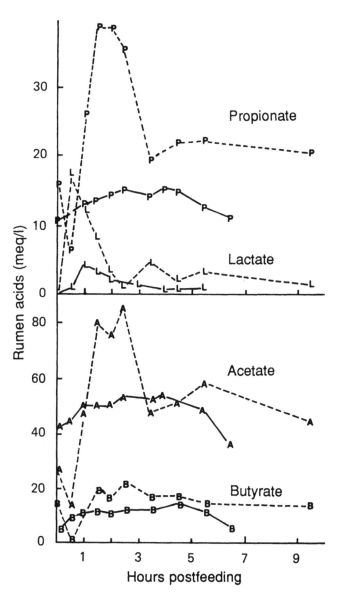

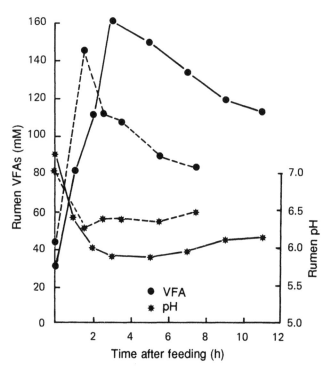

Figure 15.22. Rumen pH and VFA concentrations in two sheep fed chopped alfalfa (from Briggs et al., 1957). Note the different response in the two animals. Rumen pH is a negative mirror of acid production and disappearance.

Figure 15.21. Variation in concentrations of rumen acids with the time after feeding. Dashed lines denote high-concentrate rations; solid lines denote all forage. (Compiled from Van Soest, 1955, and Waldo and Schultz, 1956.)

rumen volume and a smaller pool that turns over more rapidly (Bauman et al., 1971). Studies utilizing ^{14}C-labeled VFAs and their rates of disappearance as measures of VFA production and absorption do not provide information on the conversion to other acids or microbial cellular products; this diversion is instead measured as absorption.

Absorption of VFAs

Acids are absorbed across the rumen wall largely in the free form, apparently without active transport. There may be considerable metabolism of the acids (particularly butyrate) in the wall, however, leading to a differential decline in concentration and more rapid

absorption. At normal rumen pH, only small amounts of VFAs are present in the free acid form. The removal of free acid is balanced by formation of more free acid through the reversal of the ionization equilibrium by mass action. The proportion of free acid is favored by lower pH and higher concentrations of VFA. The pH of the blood is ordinarily more alkaline than that of the rumen, favoring movement of acid toward the blood through the free energy of neutralization. This gradient similarly discourages flow of fatty acid anion. Thus rumen pH influences rates of VFA absorption (Bergman, 1990; Dijkstra et al., 1993). A high rumen pH narrows the rumen to blood gradient and increases anion absorption, which has been observed to be about one-half of the acetate absorbed at pH 7.06 (Ash and Dobson, 1963). Bicarbonate, sodium ions, and some urea flow in the reverse direction, toward the rumen (Figure 15.23). The mechanism of VFA absorption in the lower tracts of ruminants and nonruminants is similar to that in the rumen (Sellers and Stevens, 1966).

Cannulae in the portal vein show low values for butyrate and propionate relative to acetate because of the selective removal and metabolism of the former as they pass through the rumen wall. The flow of acids is not in order of their molecular weights, which suggest that diffusion is not a limiting factor.

The major factor affecting quantities of VFAs absorbed is their concentration (Figure 15.24). Therefore, it follows that the order of absorption will be

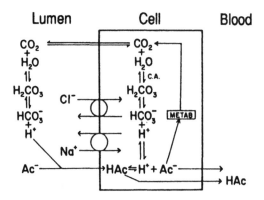

Lumen   Cell   Blood

Figure 15.23. Hypothesis for the mechanisms of VFA, Na, Cl, and HCO_3^- transport by rumen epithelium and colon mucosa (from C. E. Stevens, 1988). High levels of CO_2 produced by microbial fermentation in the lumen allow for its rapid hydration in the absence of carbonic anhydrase (C.A.). This provides H^+ for the nonionic diffusion of acetate or other VFAs into the cell and releases HCO_3^- into the lumen. Similar intracellular hydration of CO_2, derived from metabolism of VFAs and other substrates, is catalyzed by carbonic anhydrase, providing HCO_3^- and H^+ that can be exchanged for the Cl^- and Na^+ in the lumen. The relatively low levels of Cl^- normally present in the lumen could result in the more rapid secretion of H^+ than HCO_3^- into the lumen, which would aid in VFA absorption and favor release of cellular HCO_3^- into the blood. Acetate is transported to the blood by diffusion of both the dissociated and undissociated forms of the fatty acid. Transport of Cl^- and Na^+ to the blood (not depicted) is accomplished by diffusion of Cl^- down its electrochemical gradient and Na^+-K^+ ATPase transport of Na^+.

acetate, propionate, butyrate, provided that the absorption rates are calculated as quantities per unit time. If rates are compared in terms of absolute kinetics (i.e.,

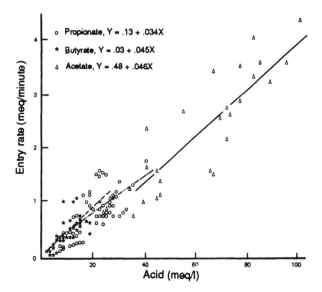

$\circ$ Propionate, Y = .13 + .034X

$*$ Butyrate, Y = .03 + .045X

$\triangle$ Acetate, Y = .48 + .046X

Entry rate (meq/minute)

Acid (meq/l)

Figure 15.24. Relation between measured entry rate and concentration in ruminal fluid for acetate, propionate, and butyrate. The lines indicate the relation between production and concentration for each VFA. The solid line indicates the relation after results were combined (summarized by Leng, 1970). The individual kinetic rates of absorption (slopes of the lines in the figure) are less important than concentration in determining net amounts absorbed. Entry rate and concentration are better correlated in animals and diets with similar rumen pool sizes and turnovers.

independent of unit concentration), the rates are more equal.

Other factors influencing absorption are rumen pH, surface, and volume (Dijkstra et al., 1993). Lower pH increases the proportion of unionized acid and promotes absorption. Papillar surface is maintained by VFA concentrations and ruminal wall metabolism of butyrate and propionate. High-concentrate diets tend to promote lower rumen volumes leading to less than expected amounts of VFA absorbed relative to ruminal concentrations (see Table 19.3).

15.6 Regulation of the Rumen Environment

Ecological conditions within the rumen must be kept within limits to maintain normal microbial growth and metabolism and thus the well-being of the host ruminant. Cellulolytic organisms grow optimally at pH 6.7, and deviations substantially higher or lower than this are inhibitory. The range for normal activity is about ± 0.5 pH units. In particular, pH below 6.2 inhibits the rate of digestion and increases lag (Richard Grant and Mertens, 1992). The osmolality of rumen contents is maintained within narrow margins by the large volume of isotonic saliva, rapid absorption of water from hypotonic solutions, and iso-osmotic absorption of water along with sodium, chloride, VFAs, and other substances. A low redox potential is maintained by the presence of fermenting digesta and is also required for the continued maintenance of the host.

Rumen pH is maintained to a major degree through the high buffering capacity of saliva and the removal of VFAs through absorption. There is also evidence that the rumen epithelium can secrete bicarbonate (A. Dobson, 1959; other factors involved in buffering are discussed in Section 15.6.2). Lowering the pH interferes with rumen fermentation and may lead to acidosis in the host. It may also allow facultative lactic acid–producing organisms to proliferate if there is too much starch in the feed. Osmotic pressure works against the flow of water across the rumen wall. If the osmotic pressure in the rumen exceeds that in the blood, water will flow toward the rumen. Ordinarily osmotic pressure is lower in the rumen, and water is lost to the blood. Osmotic pressure promotes the flow of liquid out of the rumen to the omasum; VFA absorption helps to keep osmotic pressure within the necessary limits. Resorption of sodium, chloride, phosphate, and other salivary ions is necessary for the maintenance of electrolyte, water, and acid-base balances in the host; it occurs in the rumen, omasum, and other sites farther down the digestive tract (A. Dobson, 1959). The entrance of these inorganic ions is enhanced by rumination and ensalivation, and they in turn promote liquid turnover and washout of finer particulate material.

Table 15.4. Characteristics of ruminant salivary glands

Gland	Calf		Sheep		Cell type	Factors governing volume	Approx. rate of flow (l/day)	Saliva type
	Weight (g)	% of total weight of salivary glands	Weight (g)	% of total weight of salivary glands				
th parotid	63.5	32.2	23.5	29.3	Serous	Continuous flow when denervated; respond to stimulation by mouth, esophagus, and rumen	3–8	Fluid and isotonic; strongly buffered with HCO_3^- and HPO_4^{2-}
th mandibular	64.0	31.6	18.2	22.6	Mixed	No flow when denervated; strongly stimulated by feeding; little or no response to stimulation by esophagus or reticulorumen	0.4–0.8	Variably mucus and hypotonic; buffered
th sublingual	11.3	5.6	1.3	1.6	Mixed	Continuous flow when not stimulated; little or no response to stimulation by esophagus or reticulorumen; other reflexes not studied	0.1 (?)	Very mucus and hypotonic; weakly buffered
bial	8.9	4.4	10.9	13.5	Mixed	Little or no flow when not stimulated; little or no response to stimulation by esophagus or reticulorumen; other reflexes not studied	?	Very mucus and hypotonic; weakly buffered
th ventral buccal	13.5	6.7	5.9	7.3	Serous	Continuous flow when denervated; responds to stimulation by mouth, esophagus, and reticulorumen	0.7–2.0	Fluid and isotonic or nearly so; strongly buffered with HCO_3^- and HPO_4^{2-}
th medial buccal	13.1	6.5	6.0	7.5	Mucus	Very slow continuous flow when not stimulated; responds to stimulation by mouth, esophagus, and reticulorumen	2–6	Very mucus and isotonic or nearly so; strongly buffered with HCO_3^- and HPO_4^{2-}

Source: Kay, 1960.

The rumen is more or less a continuous fermentation system, although its continuity is perturbed by meals, leading to cyclic patterns. Continuity requires that all substances entering the system via the diet, or saliva, and those produced by fermentation, be either absorbed or passed down the digestive tract. Net exit must balance net entry. Any imbalance leads to abnormal or pathological conditions. Gas production beyond the limits of the eructation capacity leads to bloat. Interference with the rumination process may lead to rumen impaction and an "off-feed" condition. Reduction in intake also can result from a dietary deficiency of the nutrients necessary for microbial growth and the maintenance of a normal rate of fermentative digestion. Generally, digesta are eliminated and kept in balance with intake through disappearance of dry matter via digestion and passage. The passage of indigestible residues is assisted by rumination and comminution to a particle size that will pass.

The cyclic pattern following eating shows a significant but normal diurnal variation. This cyclic variation may be important for some of the more fastidious rumen protozoa. Some tend to accumulate starch and burst after the host eats much concentrate. These populations (and methanogenic organisms as well) probably recover during the slower phases of rumen fermentation that precede the next meal. Survival of slower-growing species such as large protozoa require turnover times that are not in competition with generation time. Liquid ordinarily passes too quickly for these organisms to use it, and their maintenance thus depends on occlusion in the mass of fibrous matter with slower turnover. Pelleting and grinding of food disturbs this balance and often results in reduction or elimination of rumen protozoa.

15.6.1 Saliva

Saliva is produced by the parotid and other glands (Figure 15.25 and Tables 15.4 and 15.5). The parotid is rich in mineral ions—particularly sodium, potassium, phosphate, and bicarbonate—which provide buffering capacity. Ruminant saliva is also rich in mucins, giving it viscosity and, perhaps, resistance to the formation of foam in the rumen. It does not contain any amylolytic activity, although there is some lipase, which is important in newborns. Ruminants produce a large amount of saliva every day (sheep produce 15 l/day or more, and cattle produce 180 l/day or more), and animals depend on recycling the mineral bases it contains, particularly sodium. About 70% of the water entering the rumen comes from salivary secretion (Church, 1988).

Table 15.5. Relative composition of ruminant saliva

Gland	Saliva produced (l/day)	Na+ (meq/l)	K+ (meq/l)	HCO₃⁻ (meq/l)	HPO₄²⁻ (meq/l)	Cl⁻ (meq/l)
Sheep						
Parotid	3–8	147–185	5–31	91–125	25–71	9–16
Inferior molar	0.7–2	175	7–10	97–110	44–51	7–12
Palatine	2–6	179	4	109	25	25
Submaxillary	0.4–0.8	4–16	10–25	5–14	2–10	7–15
Sublingual	0.1	16–47	6–25	8–18	0.3–2	16–40
Labial	(small)	29–47	3–9	2–4	2–10	34
Calves						
Parotid	—	163–168	6–14	88–94	17–47	16–34
Submaxillary	—	11–24	24–41	5–8	0.4–4	6–15
Inferior molar	—	151–156	6–18	77–95	18–54	12–21

Source: Kay, 1960.

Urea in saliva varies in concentration, and small amounts of sulfate, calcium, and magnesium are usually present in saliva as well. Salivary composition is affected by many factors. The composition of secretions by the various glands that produce saliva varies depending on the rate of secretion. Increasing the rate of secretion by a factor of 5–10 causes a drop in potassium and phosphate and an increase in sodium and bicarbonate up to 10-fold (Kay, 1960). Generally, the sum of cations and anions tend to remain constant and equal to each other. Composition is also affected by sodium depletion and salt intake.

The flow of saliva is stimulated by eating and ruminating, although some flow continues constantly. The rate of eating is important in determining the buffering capacity of the feed-saliva mixture (Table 15.6, Section 15.6.2). The intake rate is faster for concentrate feeds, which also tend to ferment more rapidly. Faster eating rate in combination with maximum flow decreases the amount of saliva per gram of feed. Total salivary flow also is related to time spent eating and ruminating. Thus high-concentrate and pelleted forage

diets are characterized by less net flow (Bauman et al., 1971). The combination of these factors with salivary flow leads to lower rumen pH in animals fed concentrates. Changes in rumen flora may follow the reduction in rumen buffering capacity and the slower turnover and washout characteristic of high-concentrate diets.

Saliva composition reflects the ruminant's need for mineral balance and recycling as well as its symbiotic relations with the rumen bacteria and their requirements for optimal growth. Apart from their need for a fairly stable pH and osmotic pressure, bacteria are tolerant of wide ranges in ion concentrations. Ruminant saliva is relatively low in phosphate and high in carbonate, but rumen bacteria can tolerate a wide range in the ratio.

15.6.2 Buffering

The total buffering system in the rumen includes not only the saliva but also the feed. The nonprotein nitrogen fractions of forage are rich in glutamate, aspartate, glutamine, and asparagine. Feed proteins may also contribute to buffering capacity (Tables 15.7 and 15.8). Lactate and VFA buffering occurs at a pH too acidic for any practical affect in the rumen. The plant cell wall also has a cation exchange capacity that contributes to rumen buffering (McBurney et al., 1983). The net buffering capacity in the rumen varies with the

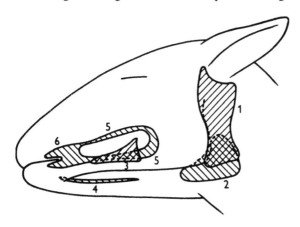

Figure 15.25. The main salivary glands of the sheep (from Kay, 1960). 1 = parotid; 2 = submaxillary; 3 = inferior molar; 4 = sublingual; 5 = buccal; 6 = labial. Glands are paired on both sides of the head and mouth.

Table 15.6. Effect of ration on saliva production and eating rate

Feed	Eating rate (g food/min)	Salivary production (ml/min)	Salivary production (ml/g food)
Pelleted ration	357	243	0.68
Fresh grass	283	266	0.94
Silage	248	280	1.13
Dried grass	83	270	3.25
Hay	70	254	3.63

Source: C. B. Bailey, 1958.

15.7. Factors contributing to rumen buffering

	Promoted by	Buffer source
...out (passage)	Osmotic pressure	Dilution
	Feed intake	
...rption	VFA concentration	Removal of free acid
...a	Coarse fiber and	Bicarbonate
	rumination	Phosphate
	Cation exchange	Neutralization
...ral salts of plant	Forage composition	Fermenting of plant acids
...ganic acids		to CO_2
...ein	NH_3 production	Neutralization
...bial efficiency	Microbial growth	Diversion of carbon to
		cells instead of acids

feed. Salivary buffering capacity also varies depending on the feed source (Figure 15.26). The pK's of various acids in feed and rumen contents are given in Table 15.8. Urea provides buffering through its conversion to ammonium bicarbonate and allows for nitrogen recycling, which is important in the economy of protein and nitrogen balance in the ruminant.

15.6.3 Rumen Volume and Liquid Flow

Rumen volume can be measured either directly by emptying the contents through a rumen fistula, by measurement at the time of slaughter, or by the dilution technique with a liquid marker. The latter procedure involves the same principles and assumptions as rate of passage measurements. Failure of the rumen to equilibrate and postprandial variation in rumen contents are the most important causes of measurement errors. The lack of equilibration is more serious for particulate matter than it is for liquid components because rumen volume must be assumed to be constant over the period of dilution measurement, and the marker (particularly polyethylene glycol) may not penetrate all the water space, particularly that within living microorganisms. These factors contribute to systematic errors.

Data collected by Colucci (1984) indicate that the

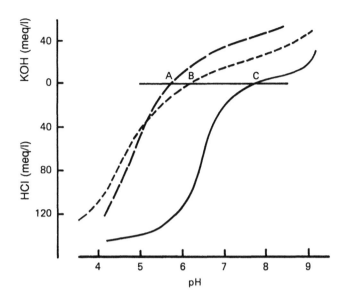

Figure 15.26. The buffering capacity of rumen liquor and ruminant saliva (modified from Turner and Hodgetts, 1955). (A) Rumen liquor from sheep on alfalfa and ryegrass pasture. (B) Rumen liquor from sheep fed wheat and oat chaff. (C) The buffering capacity of parotid saliva. Note that the higher-quality alfalfa-ryegrass diet produces a more acidic rumen compared with the wheat and oat chaff diet; however, the legume-containing diet has greater buffering (steeper curve) in the region of pH 5.

systematic error of marker measurement varies depending on the intake. At low intakes the marker method overestimates water space in the rumen, while at high intakes it underestimates. If the marker is unable to enter the cells of living rumen organisms, as much as 20% of the fluid space may not be measured. Loss of marker through absorption could add to the error. Higher osmotic pressure promotes marker absorption, leading to overestimation of kinetic rate and rumen volume (A. Dobson et al., 1976).

Rumen volume tends to increase with ad libitum feeding (Colucci, 1984). This stretch factor counterbalances the increased rate of passage. The net effect is that increases in passage rate with incremental feed intake may be less than anticipated.

Table 15.8. Approximate pH (pKa) of maximum buffering capacity of various metabolites and feed components

Reaction	pH (pKa)
Phosphoric acid, 2d hydrogen	7.1
Carbonic acid, 1st hydrogen	6.4
Acetic acid	4.8
Propionic acid	4.9
Butyric acid	4.8
Formic acid	4.0
Lactic acid	3.9
Feed components	
Glutamate (2d hydrogen)	5.6
Forage (mean pH, H_2O extract)	5.5
Citrate (3d hydrogen)	5.4
Aspartate (2d hydrogen)	5.2
Malate (2d hydrogen)	5.1
Alfalfa protein isoelectric point	4.5

15.7 Rumen Dysfunctions

Diseases of the rumen are most often related to diets that deviate from the diet to which the ruminant species is evolutionarily adapted. Problems thus arise from feeding high-concentrate diets to dairy cattle, beef cattle, and sheep, and from abnormal levels of nitrogen fractions (urea, ammonia, and nitrate) from inadequate feed mixing or overfertilization of crops. Pathologies such as bloat result from single-species pastures or too much of one kind of feed, and thus have an ecological basis.

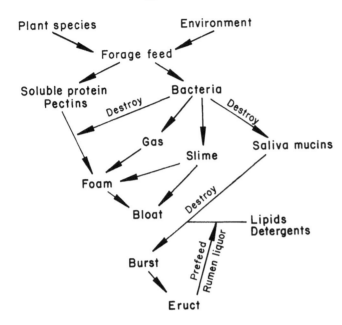

Figure 15.27. Hypotheses explaining bloat. The various models are not mutually exclusive, and their relative importance may vary with the animal's diet.

15.7.1 Bloat

Bloat is the distension resulting from the failure to eliminate gas produced in the rumen. Pressure levels in the normal animal are low, but they may rise to 100 mm mercury in cases of acute bloat.

The main recognized categories of bloat are legume bloat, frothy bloat, and grain bloat, although other forms have been described (Essig, 1988). Bloat may be acute, subacute, or chronic, reflecting the severity of the condition. Acute bloat is often fatal. The nomenclature considers both dietary and animal factors. For example, an animal described as a "chronic bloater" may reflect genetic factors predisposing a tendency toward bloat (C. S. Reid et al., 1975).

Bloat is relatively unknown in wild ruminants and in domestic animals on the range. It is likely to be the result of humans' mismanagement of the normal balance between diet and animal. Like some other nutritional problems, bloat is associated with grazing on certain plant species and with intentional or inadvertent feeding of high-starch diets.

Bloat is generated by a complex interaction among environment, animal/plant species, and rumen microbes. Factors leading to stable foam promote bloat; those that reduce foam by causing bubbles to burst reduce it. Foam stability is a function of surface tension. The more important bloat factors are outlined in Figure 15.27.

Rapid fermentative production of gas is required to make foam. Pectin, abundant in legumes, produces much gas and has been suggested as a cause of foam. It

should be pointed out that rapid fermentation could also destroy the foam-promoting factor, and gas production, although required, is insufficient in itself as a principal factor. Foam stabilization probably involves soluble proteins in fresh forages as well as rumen bacteria that can ferment the protein and also ferment protective salivary mucins. Bacteria are seen as a force both promoting and destroying conditions for bloat because they can ferment soluble protein as well as salivary mucus and may produce slime themselves. The slime increases the tendency for foam.

Legume or Frothy Bloat

Legume bloat occurs in animals grazing on legume pastures, usually white clover or alfalfa. Not all legumes produce bloat; it is unknown on pastures of tropical legumes or temperate pastures of trefoil, sanfoin, or vetch. These forages contain tannins that inhibit bloat through protein precipitation (W. T. Jones and Lyttleton, 1971).

The rumen of a bloated animal usually contains much foam, hence the description "frothy bloat." Only limited attention has been paid to the factors promoting formation and stability of the foam, which appears to inhibit eructation and elimination of gas. Substances in legumes that might contribute include proteins and pectins, since both increase viscosity and foam stability.

Foam stability involving protein as the cause of legume bloat has received more support, particularly since detergents and oils are effective foam suppressants and also decrease surface tension and form complexes with proteins (Laby, 1975). Legume hays do not produce severe bloat, perhaps because proteins are denatured to the insoluble form in the hay-curing process.

Legumes contain other components that have been suggested as contributing factors as well, including saponins and the amines produced from protein (e.g., histamine, tyramine, etc.), which might have toxic effects on the animal and on rumen motility (see Church, 1975, for discussion).

Grain Bloat

Feeds containing large amounts of concentrate or pelleted diets often cause bloat, usually of the chronic variety. This type of bloat seems to be different from that seen in animals on legume pasture (Bartley et al., 1975). In grain bloat the rumen contents are characterized by high viscosity and foam resulting from the production of extracellular slime by amylolytic bacteria. The mucin fraction of ruminant saliva, which may ordinarily protect the animal from bloat, might be inactivated by rumen organisms with mucinolytic activity. The smaller amounts of saliva per unit feed and low rates of rumen turnover characteristic of high-concentrate diets could be responsible. A large production of rumen acids, particularly lactic acid, may also reduce rumen motility.

Acute and Chronic Bloat

In some animals subacute bloat conditions occur continuously, a possible cause of discomfort but not of acute distress. In New Zealand, geneticists have shown the existence of sheep strains more tolerant to legume bloat conditions. Grain bloat is commonly chronic and less often acute compared with legume bloat.

An animal in a state of acute bloat is in distress and may become prostrated and die. High pressures of air or oxygen in the rumen introduced experimentally do not produce the distress of an equivalent pressure of CO_2. The final stages of bloat may involve pressure on the heart and cardiovascular collapse.

Prevention is the most efficient method of handling bloat. This may be accomplished by good feeding management or the administration of oil or detergents that reduce surface tension and foam stability in the rumen. The detergents also form complexes with the proteins involved in producing foam. Periodic drenching of the animals or spraying oil or detergents on legume pastures may prevent bloat. Polyoxaline is an approved detergent for treating bloat. Allowing animals to graze only when they can be observed helps managers detect the onset of bloat. In severe acute bloat, the last resort is to puncture the rumen and allow the gas and foam to escape (Essig, 1988).

15.7.2 Acid Indigestion

Most rumen disorders involve some disruption of the balance and control of the internal rumen environment. Imbalances may develop as a result of the sudden introduction of feed or substances to which the rumen flora are unaccustomed, leading to a rapid change in fermentation that cannot be controlled. Acute (acid) indigestion and urea and nitrate toxicities all result from imbalances.

Starch or cereal concentrates ingested in large amounts provide the substrate for rapid proliferation of facultative organisms that produce lactic acid and low cell yields (Allison et al., 1975). Lactic acid is a considerably stronger acid than volatile fatty acids (pK 3.9 vs. 4.8 for acetic acid) and is produced in both natural (D) and unnatural (L) forms by bacteria. In severe cases lactate may constitute 50–90% of total rumen acids. Succinate and formate may also occur in substantial quantities, although normally they appear only in trace amounts in the rumen. Rumen pH may drop to as low as 4, causing severe rumenitis. If large amounts of lactic acid are absorbed across the rumen wall into the

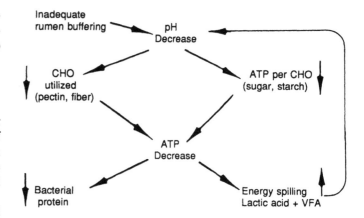

Figure 15.28. Metabolic activities in the rumen that might lead to decreased synthesis of microbial protein and increased acid production (modified from Strobel and Russell, 1986). The system is self-inductive in the face of inadequate rumen buffering.

blood, systemic acidosis can result. The hyperosmolality of the rumen contents and uncompensated acidosis cause systemic dehydration; death results from failure of hemoglobin to carry oxygen (Huber, 1976).

Similar disorders occur if animals break into a field of corn or if high-quality hay is given to hungry animals. Acute indigestion does not occur if the new diet is introduced gradually so that rumen flora are allowed to adapt. All-concentrate rations may produce a chronic acidosis leading to rumen parakeratosis. Treatments for acidosis include the administration of carbonate salts and evacuation of the rumen.

Lactic acid–producing organisms are typically more tolerant to lowered pH and thus do not reduce their utilization of substrate. The metabolic pathway yielding lactate provides only 2 ATPs per mole of sugar for microbial growth, however, whereas normal VFA production (acetate, propionate, and butyrate) affords 4 ATPs per mole of sugar. Thus fiber-digesting, VFA-producing bacteria theoretically have twice the capacity of lactic acid–producing organisms for production of microbial protein. The behavior of the important lactic acid producer *Streptococcus bovis* in response to pH shift is instructive. At normal rumen pH this organism shifts to VFA production so as to more effectively compete with the more efficient non-lactate producers, but it reverts to lactate production at lowered rumen pH. Since lactic acid is a much stronger (10 times) acid than VFAs, rumen acidity is greatly increased, and the feedback of this acidity on the induction of more acid production is likely a major factor in the development of rumen acidosis (Figure 15.28). This sequence can be prevented by providing adequate-quality fermentable fiber to maintain cellulolytic organisms and buffering capacity. Of the nonfibrous carbohydrates only starch and sucrose produce lactic acid. Pectin, hemi-

Table 15.9. Effect on fermentation of some carbohydrates of lowering rumen pH

Rumen factor	Sucrose or starch	Pectic or xylan (hemicellulose)
Lactic acid	Increase	None
ATP/unit digested	Decrease	Unchanged[a]
Energy spilling	Unchanged	Increase
Digestibility	Unchanged	Decreased

Source: Strobel and Russell, 1986.

[a]There will be a net reduction in ATP from pectin and xylan because of a reduction in digestion.

cellulose, and other more fiber-like polysaccharides produce acetate but no lactate (Table 15.9).

15.7.3 Parakeratosis

The rumen epithelium is responsive to fermentation acids, the production of normal volatile fatty acids being necessary for normal development of papillae. Excess production of lactic acid or a high concentration of acid in the diet, in combination with less saliva and buffering capacity per unit of feed, results in lower rumen pH. These conditions are unfavorable to the rumen lining and lead to a dark, abnormal appearance and atrophy of the papillae. In severe cases of rumenitis the lining may actually be sloughed away. These conditions of chronic acid production are also related to displaced abomasum syndrome.

Treatment or prevention consists of feeding carbonate buffers or enough coarse forage to induce rumination and neutralization of rumen contents. Other treatments have included plastic particles added to feed with the object of stimulating rumination, rumen motility, and mixing. Added plastic is apparently ineffective in alleviating milk fat depression in lactating cows.

15.7.4 Urea Toxicity

Urea toxicity occurs when large amounts of urea are ingested, followed by enzymatic hydrolysis to ammonia and CO_2. Ammonia is toxic to animal cells at quite low levels (Visek, 1978). Some evidence indicates that it is also carcinogenic. Like nitrate, urea itself is not toxic. It is the substance formed when ammonia derived from proteolysis is detoxified or from feeds high in nonprotein nitrogen and ammonia (e.g., high-moisture, high-nitrogen silages). Chronically high blood urea levels limit intake, and therefore an acute stage of toxicity is rarely reached.

The efficient utilization of urea as a nitrogen source depends on an adequate supply of fermentable carbohydrate to increase microbial needs and provide for conversion of urea nitrogen into microbial protein. There is no set level at which urea in the diet will cause toxicity. A more accurate indicator of status is blood ammonia. Toxicity symptoms occur at blood levels higher than 0.5 mg/100 ml, becoming more severe at higher levels. Blood ammonia levels of about 4 mg/100 ml are lethal. Such levels are reached when rates of rumen production and absorption of ammonia overwhelm the liver's capacity to form urea.

Treatment includes the administration of soluble carbohydrate or, in extreme cases, the evacuation of the rumen. Attempts at prevention have included the use of urease inhibitors to slow the rate of release of ammonia. Practical prevention includes admixture of the nitrogenous source with sufficient soluble carbohydrate or starch and ensuring that feed is always well mixed.

16 Microbes in the Gut

The microbes that inhabit the gastrointestinal tracts of herbivorous vertebrates are the main agents for the digestion of complex carbohydrates in ingested plant material. The biology and ecology of these complex microbe populations is remarkably similar in most of the animal species that carry them. This applies to both rumen and hindgut fermentations. Some animal species may contain specific protozoa or bacteria, but most of these represent substitutes for organisms found in other species that accomplish the equivalent biochemical work in their respective systems. Further, the specification and adaptations of these respective populations are determined more by diet and turnover time than by the species of the animal host, whether ruminant or nonruminant. The peculiar similarities in microbial nutritional requirements and ecology permit the application of rumen microbiology to the study of gut microbes in nonruminant animals. Despite the comparative uniformity of anaerobic environments in the digestive tracts of large herbivores, a remarkable variety of organisms manage to coexist, including eubacteria, archaebacteria (methanogens), fungi, and protozoa. For readers who want to learn more about rumen microbes, a classic text is Hungate, 1966. A more recent book covering rumen microorganisms is Hobson, 1988.

16.1 History

The history of rumen microbiology largely dates from the doctoral work of Kaars Sijpestijn (see Hungate, 1966) in wartime Holland and Robert Hungate's successful application of the anaerobic roll-tube technique in 1947. Until then, no one knew how cellulose was anaerobically fermented in the rumen. The mystery resulted from a lack of understanding of strictly anaerobic bacteria and their peculiar requirements. Successful investigation required a variety of new techniques that allowed in vitro culture. The roll-tube technique of Hungate (1950) was a breakthrough in studying anaerobic rumen bacteria.

Investigations of the variations in volatile fatty acid

(VFA) production by animals on different diets were greatly accelerated by the development of chromatographic techniques for separating the acids. The use of in vitro fermentation as a quantitative tool was made possible by the procedure developed by J. M. A. Tilley and R. A. Terry in 1963 and more recently by systems for continuous culture (Hoover et al., 1976; Czerkawski, 1984). Such systems depend on adequate anaerobic technique but use mixed whole cultures and do not rely on the more fastidious isolation techniques. Rumen microbiology has remained a rather specialized field because of the uniqueness of the problem and the rigorous training required for adequate control of anaerobiosis.

16.1.1 The Gut Environment

Microorganisms can live in most segments of the animal gut; however, certain constraints confine the digestion of complex carbohydrates to larger compartments and relatively neutral pH. In general no organism can survive in the system unless its generation time is less than the mean retention time of the organ. Some organisms can survive in situations of rapid transit or turnover through attachment to the gut wall or large particles of food.

The microbial population of the rumen is regulated by the peculiar ecological balance that tends to prevail there. The rumen and the hindgut contain a number of unique environmental features that cause them to differ from most other anaerobic systems. The system is essentially isothermal and is regulated by the homeothermic metabolism of the host animal. There is a relatively constant influx of water and ingesta, the fermentation of the latter giving rise to a considerable amount of acid. The pH (6–7), however, remains relatively constant because fermentation acids are removed by absorption across the rumen wall or neutralized by salivary buffers. In the lower tract fermentation acids are similarly absorbed, and buffers, including urea and bicarbonate, diffuse across the gut wall. The osmotic pressure is normally below isotonic levels. Ionic concentrations are regulated through dilution, absorption, and passage. Similarly, undigested substrate is ultimately passed down the digestive tract after comminution through rumination. End products and wastes are removed. All factors relative to the microbial environment, in fact, are regulated within narrow limits. The rumen has some aspects of a continuous culture system, but pulsative eating may generate some of the characteristics of batch cultures.

Rumen microorganisms are predominantly strict anaerobes. Some oxygen can be tolerated, however, as long as the fermentation is sufficiently active to facilitate the disposal of oxygen and the potential (Eh) of the medium remains within normal limits (-250 to -450 mv). Oxygen introduced through feed and water may diffuse across the rumen wall and affect organisms near the wall. These small quantities of oxygen are metabolized rapidly, and oxygen serves as an electron acceptor. Most rumen fermentations in healthy well-fed animals are protected by the symbiotic system, in which oxygen consumption by facultative anaerobes helps maintain the low Eh. When the fermentation is diluted (making substrate scarce) or facultative organisms are isolated, a much greater sensitivity to oxygen becomes apparent. Oxygen sets an unfavorable balance for anaerobic metabolism, including the synthesis of methane. Methanogenic bacteria are a normal component of rumen flora and have an important influence on fermentation balance.

The tolerance of the rumen system to oxygen and other electron acceptors generally is increased by availability of rapidly fermentable substrate. For example, a well-fed rumen is relatively unaffected by a leaky or open fistula, a property not shared by the average in vitro rumen culture, in which substrate concentrations are much more dilute, microbial concentrations are lower, and a greater sensitivity to oxygen poisoning prevails.

Maximum extraction of available energy requires maximum degradation of the cellulosic substrate. This goal tends to be accomplished with minimal acid yield. While some acid production is necessary for energy transduction and ATP production, maximal efficiency converts maximum ATP energy and maximum substrate carbon into microbial cells. Because acidic end products are buffered and removed, pH is regulated in the rumen. However, acid-producing organisms in some fermentations use acid production to dominate other organisms. Lactic acid fermenters are characteristic of this phenomenon. Acid production is accomplished at the expense of microbial efficiency and cell growth, since total products must balance the substrate metabolized. Lactic acid–producing organisms are not very important in a forage-fed rumen except when the diet contains a large amount of soluble carbohydrate or starch (Section 16.9.1).

Generally, the dominant rumen species are those that convert substrate energy into cells, resulting in less acid produced from a given amount of substrate. Natural selection has also favored maximum biochemical work (Hungate, 1966). The neutral conditions of the rumen allow many types of organisms to grow, and individual groups and species develop complementary and syntrophic relationships. The buffering of the rumen favors pH-sensitive groups that digest cellulosic carbohydrates.

16.1.2 Inoculation

The factors that govern the transfer of microorganisms from one animal to another are not well understood and have not been well investigated. Although rumen microorganisms are obligately anaerobic, they are sufficiently resistant so as to be transferred by feed, saliva, and perhaps air from one animal to another. Isolation of calves may prevent faunation by protozoa but the development of bacterial fermentation is not delayed. Anaerobic fungi are viable in feces for a considerable time, and fecal contamination or even coprophagy is another means of transfer. The benefits of rumen inoculation of young ruminants from adult donors are doubtful since young animals will become inoculated sooner or later anyway, a factor undermining controlled experiments. Organisms are sufficiently resistant to gastrointestinal action that the cecum and other sites in the lower tracts of ruminants and nonruminant herbivores also become inoculated.

16.1.3 Anaerobes in Other Environments

Anaerobic organisms with similar ecological adaptations and nutritional requirements to ruminant microbes exist in silos, silage pits, sewage, and other environments. Their microbiology has been explored because of the interest in methanogenesis. The most significant difference between a sewer and the gut is that there is usually longer turnover time in the former, which allows more complete extraction of substrate energy and the proliferation of methanogenic species that have longer generation times than the retention times of most animals. In particular, those that ferment acetic acid require about a 4-day retention to survive. The result is that most fermentation products like VFAs do not accumulate in a methane fermenter; they are degraded to methane and CO_2.

16.1.4 Cellulolysis and pH

A common generalization is that low pH (<6) inhibits cellulolysis under normal conditions, but there are also conditions in which substrate inhibition occurs at pH well above 6 (Section 15.7.2). Nevertheless, cellu-lolytic activity is certainly reduced at lower pH and must be considered separately.

The lack of anaerobic cellulolytic systems stable in acidic pH raises a question on the limitations of evolu-tion. Anaerobic organisms have existed for a very long time. Why have none evolved to take advantage of acidic conditions? Another observation: there are no anaerobic consortia that ferment cellulose to lactic acid or ethanol. A suggested reason for the formation of these products is that the process is energetically un-feasible. Low pH tolerance requires regulation of inter-nal cell pH in an acidic medium, whereas production of ethanol or lactate allows only 2 ATPs for growth. Or-ganisms living on slowly degrading substrates in a mixed competitive system cannot compete with more efficient organisms that produce acetate.

Rumen cellulolytic organisms tend to protect their food resources by attaching to the plant cell wall and storing their energy within cells. As little as possible is secreted into the medium where competitors could grab it, although spillover has been noted. The highest efficiency of substrate utilization is required for com-petitive growth. The overall competitive ecology in a normal rumen is parallel to that in the tropical rain forest, where many species of organisms compete and survive. Nutrients in these systems are generally stored in plant biomass. Existing organisms must die for their stored nutrients to be released back into the system. This situation prevails just as much in the rumen when the diet consists of poor-quality forages deficient in nitrogen. In such situations microbial growth and cel-lulose digestion are limited to the rate of turnover of the microbial mass. So it is in ruminants fed low-quality straws, but rumens in animals fed more and better diet may deviate from this ecology.

For some time the prevailing opinion has held that lactic acid production from fermentation of starch and sugar lowers rumen pH and inhibits digestion of cellu-lose. This inhibition becomes important at pH below 6.2 (Richard Grant and Mertens, 1992). On the other hand, a grain-fed dairy cow's rumen may show reduc-tion in cellulose digestion without any change in pH (Murphy, 1989). Total cellulolytic numbers may be as high in the grain-fed animal as in the high-forage-fed animal.

16.1.5 Is the Rumen Fermentation a Continuous System?

Rumen and lower tract fermentations are often thought of as continuous fermentation systems because they continue perpetually with apparent turnover. Many in vitro continuous systems are conducted on the basis of this assumption. The variable that makes it less of a continuous system is the feeding frequency of the animal. Every time the animal eats, pulses of potential substrate are sent down the gut. The influence of pulsa-tion may be more important in the rumen than in the lower tract, since there is some evidence that pulses become diffused after passing through one or more digestive compartments. Infrequent feeding does re-sult in expending most of the fermentable energy be-fore the next meal, however, more so if the feed has a rapid potential fermentation rate. Pulsative feeding could create cyclic perturbations of the microbial pop-ulation through shifts in the relative proportions of such populations resulting from death and lysis of those that exhaust their substrates. This constitutes the "feast or famine hypothesis" (McBurney et al., 1987) and poses an interesting problem. If an animal is con-tinuously fed, the famine part of the cycle is elimi-nated; however, the evened-out turnover may reduce microbial growth efficiency, so there has to be some balance between the two.

16.1.6 Genetic Engineering

Genetic manipulation of bacteria has become an im-portant research goal. The introduction of the proper genetic code into the appropriate organism in a pure culture system is the basis of the production of many proteins of industrial importance, including somatotro-pin. Introducing genes to produce specific hormones or enzymes taxes the energy metabolism of the cells and makes them less likely to survive in normal compet-itive environments. This further applies when the pro-tein manufactured is of no use to the cell. Thus the potential for genetic manipulation of rumen organisms to produce a species that would be more efficient in digesting and utilizing feed has been viewed with some skepticism. Hungate pointed out in a comment (at the Second Ruminant Physiology Meeting in Iowa in 1964) that rumen bacteria, with their ability to mutate and given that their numbers are on the order of 10^9/ml or more, can be expected to show optimum adaptation to any substrate in a week or two. This same argument can be applied to the introduction of a genetically manipu-lated organism into the mixed rumen environment. If the genetic alteration were of no advantage to its surviv-al, the respective organism would either die or would mutate within a couple of weeks. Some researchers have attempted and are attempting genetic manipulation of rumen bacteria (Hespell, 1987; J. B. Russell and Wil-son, 1988). These efforts may be very interesting in elucidating how rumen organisms operate and how they digest cellulose, but the likelihood of a breakthrough in digestive efficiency is dim.

Rumen organisms operate only in consortia; that is, in mixed systems. Their ecology is opposed to single-

Table 16.1.   Relative volumes and number of microbial organisms

Group	Number per ml	Mean cell volume (μ^3)	Net mass[a] (mg/100 ml)	Generation time	% of total rumen microbial mass
Small bacteria	1×10^{10}	1	1600	20 min	60–90
Selenomonads	1×10^8	30	300		
Oscillospira flagellates	1×10^6	250	25		
Ciliated protozoa					10–40
Entodinia	3×10^5	1×10^4	300	8 h	
Dasytricha + Diplodinia	3×10^4	1×10^5	300		
Isotricha + Epidinia	1×10^4	1×10^6	1100	36 h	
Fungi	1×10^4	1×10^5		24 h	5–10

[a]Estimated cell weight per 100 ml rumen fluid, assuming a density of 1.0.

cell, pure-culture systems. Therefore, the problems of interspecies competition and syntrophism must be considered in any genetic manipulation of rumen organisms.[1] Genetic transfer between strains may also be a factor. Evolutionarily, rumen organisms are more than 3 billion years old; they represent some of the most ancient taxa existing on earth, the methanogens being the oldest. Cellulose may have existed for a billion years, or at least since plants and animals diverged. Gut digestion by microbes may well have occurred in dinosaurs (100–200 million years ago). This being the case, why, after all this evolutionary time, might there still be advantages to manipulating rumen bacteria? Did they not achieve the most fruitful association long ago? There is abundant evidence that rumen fermentation systems are limited by the physicochemical nature of the substrate. Why has not evolution overcome these limits? No biological system known can degrade polymeric lignin anaerobically, although such systems have been diligently sought. My point here is that it would be far more profitable to genetically alter the plants to make them more amenable to digesting microorganisms rather than the other way around.

The only case so far in which introduction of new microorganisms has had a significant explainable effect is that of the mimosine-degrading organisms that detoxify *Leucaena*. These organisms are absent in many ruminants (Section 13.7.1).

16.2 Microbe Species

The constancy of the rumen environment and the regular influx of highly fermentable feed as substrate favor a highly active fermentation in which diverse species are involved in the common activity of degrading carbohydrates and protein. Groups with a common net feeding strategy are called *consortia*, and their common feeding strategy is *syntrophy*. The constant

conditions restrict the number of niches relative to some other anaerobic situations such as soil and sludge fermentations, but contrast with others like the silo or pickling that tend toward monoculture. The anomaly or pathology of acidosis is ecological drift of the rumen fermentation towards single or limited culture. Pulsative feeding may induce some of the cycles that in a lesser degree mimic the direction of such fermentations.

Rumen bacteria are predominantly strict anaerobes, although a few facultative anaerobes do exist (Stewart and Bryant, 1988). The requirement of anaerobiosis and the peculiar nutritional requirements have made the pure culture and study of rumen organisms a special field. The few facultative species are not important in terms of normal rumen function, although they may be easier to isolate. Facultative types sometimes become important in cases of rumen dysfunction.

Estimates of total numbers and types of microorganisms are usually based on direct counts. Table 16.1 lists microbial organisms known to be found in ruminants. Small bacteria account for about half of the total biomass in a normal rumen but are responsible for a very much greater share of the metabolic work. Metabolic activity is generally inversely related to organism size. Larger organisms, which constitute a smaller but still significant portion of the microbial mass, include the selenomonads, oscillospira, flagellates, protozoa, and fungi.

The small bacteria are quite diverse. They range from those that digest primarily carbohydrates (cellulose, hemicellulose, pectin, starch, and sugars) to those that may use cellulodextrins, pentoses, glucose, lactate, succinate, formate, or hydrogen as energy sources. Species may be considered in terms of the substrate used, products formed, or growth requirements. Many of the important species are very specialized and have numerous growth requirements that must be supplied by the general fermentation. A great number of these species and strains have been isolated (see Hungate, 1966, for a more complete description). Representative species that perform various functions are listed in Table 16.2.

---

[1] The ability of pure cultures to explain rumen digestion is limited. For example, only whole mixed cultures are useful in measuring the extent of degradability of lignocellulosic materials.

Table 16.2. Major species of primary rumen bacteria, and their substrates, products, and requirements

Species	Substrate									Products	Requirements
	C	Hm	Pectin	Starch	Sugars	Lipids	Protein	Acids	H₂		
Structural CHO fermenters											
Ruminococcus albus	H F C	F X								1,2,Et,H₂,CO₂	NH₃,CO₂,Br,V,2±
R. flavafaciens	H F C	F X								1,2,Su,H₂,CO₂	NH₃,CO₂,Br,Sta
Fibrobacter succinogenes	H F C	H Hm		F Dx						1,2,Su	NH₃,CO₂,Br,2,5,V,Sta
Butyrivibrio fibrisolvens	H F C	F X					F Pr			1,2,4,Et,La,H₂,CO₂	NH₃,CO₂,Br,V,Sta
Eubacterium cellosolvens	H F C	F X					F Pp			1,2,4,La,CO₂	
Pectinolytic species											
Succinivibrio dextrinosolvens		F Pn	F Pc							1,2,Su,La	Sta
Lachnospira multiparus	F Cb		F Pc							1,2,Et,La,CO₂,H₂	2,V,Sta
Nonstructural CHO fermenters											
Bacteroides ruminicola	F Cb	F X	F Pc	F S	F Hx		F Pr			1,2,3,Su	
B. amylophilus				F S			H Pr			1,2,Su	NH₃,CO₂
Selenomonas ruminantium	F Cb	F Pn		F S	F Hx	F Gl	F Pr			2,3,4,Su,La,H₂	2,CO₂±
Streptococcus bovis	F Cb		H Pc	F S	F Hx		F Pr			1,2,Et,La	
Succinomonas amylolytica				F S	F G					2,4,5,Su,H₂	
Eubacterium limosum	F Cb	F Pn	F Me		F G Fr			F La	U H₂	2,4	
Megasphaera elsdenii				F Ml	F Su	F Gl	F Pp	F La		2,3,4,5,6,H₂,CO₂	
Lipolytic species											
Anaerovibrio lipolytica					F Fr	F Tg	A	F La		2,3,Su,CO₂,H₂	A,V
Proteolytic species											
Peptostreptococci sp.					Fr		F Pr A			2,4,Br,NH₃,CO₂	
Clostridia sp.	F Cb	F X±	(F Pc)	F S	F Sc Fr		F Pr A			1,2,4,Br,Et,La,H₂,NH₃,CO₂	
Organic acid fermenters											
Megasphaera elsdenii				F S	F Ml	F Gl	F Pp	F La		2,3,4,5,6,H₂,CO₂	
Veillonella alcalescens								F La	U H₂	2,3,H₂,CO₂	
Hydrogen utilizers											
Methanobacterium ruminantium									U H₂	CH₄	2,CO₂,Br,He,NH₃,V
Vibrio succinogenes									U H₂	Et,CO₂	

Sources: J. B. Russell and Hespell, 1981; Stewart and Bryant, 1988.

A = amino acids	H = hydrolyzes substrate but does	S = starch
Br = branched-chain fatty acids	not use products	Sc = sucrose
C = cellulose	H₂ = hydrogen	Sta = stimulated by amino acids
Cb = cellobiose	He = heme	Su = succinate
Cf = cellulosic fragments	Hm = hemicellulose	Tg = triglyercides
CO₂ = carbon dioxide	Hx = hexose	U = utilizes
Dx = dextrins	La = lactate	V = vitamins
Et = ethanol	Ma = malate	X = xylan
F = ferments and utilizes substrate	Me = methanol	1 = formate
Fu = fumarate	Ml = maltose	2 = acetate
Fr = fructose	Pc = pectin	3 = propionate
G = glucose	Pn = pentose	4 = butyrate
Gl = glycerol	Pp = peptides	5 = valerate
	Pr = protein	6 = caproate
		± = only in some strains

Some strains are limited in their spectrum of energy sources; others are more versatile. There is considerable overlap of function such that disappearance of one species or group is not likely to have much effect on overall rumen function. Microbes that utilize products of others as an energy source may represent 60% or more of the total numbers and constitute a very important group in terms of net rumen output. These groups interact with primary carbohydrate-fermenting species to adjust the products, to provide growth factors for the primary groups, and to maximize the efficiency of the whole culture. These interactions are very complex and involve not only the utilization of products but also the stimulation of other strains to alter *their* products.

Rumen bacteria produce products in pure culture that are never seen in mixed culture because the syntrophic relation among species involve shared metabolism and cross-feeding. Such products include ethanol, succinate, and, particularly, formate. Ethanol is probably not produced in co-culture, while formate may be rapidly scavenged as an intermediate in cross-feeding. Other intermediates include hydrogen involved in propionate metabolism and in production of methane (see Section 16.9.1).

16.2.1 Syntrophy and Consortia

Let us recall the analogy of the tropical rain forest. In a nutrient-bound system, some individuals must die to provide nutrients for the growth of others. If all the nutrients remain bound in biomass, growth is impossible. The interdependence of the component species results in beneficial associations directed toward the use of the feed resources. This concept defines syntrophy, which differs from symbiosis in that more than two species are involved and the mutual benefit is directed to nutrient energy resources. Syntrophic groups that are devoted to fiber digestion, for example, can include cellulolytics, hemicellulolytics, and secondary fermenters, including methanogens. Some of the component species can be involved in cross-feeding and nutrient recycling of branched-chain acids, heme, CO_2, hydrogen, and other growth factors (Section 16.6).

Competition affects individuals of the respective species and leads to protective strategies, which include attachment to the substrate and the storage of energy within cells. Generally, any free substrate is prey for competing organisms. Competition can also be between consortia that use different feeding strategies.

16.2.2 Fiber Digestion

Cellulose digestion is commonly regarded as the rumen's principal function, and so it is in grazers (bulk and roughage) that receive a diet sufficiently high in cellulosic matter. Cellulolytic microbes tend to diminish in the presence of competitive substrates such as starch. A number of rumen species are facultative consumers of cellulose and starch. Because of this it is difficult to associate net numbers of potentially cellulolytic microbes with net cellulose digestion. The competing strains tend to compete for the more rapidly and easily degradable substrates. The cellulolytic, hemicellulolytic, and pectinolytic organisms are all inhibited by low pH (Richard Grant and Mertens, 1992). Not all inhibition of cellulose digestion in animals on high-grain diets is the result of low pH, however, and substrate competition is more important than is realized.

Many fiber-digesting rumen bacteria may require B vitamins, ammonia, CO_2, and acetic, valeric, isobutyric, isovaleric, and 2-methylbutyric acids, which are supplied in the general fermentation. About 30% of the strains require heme along with a spectrum of minerals (Section 9.2).

A number of *Ruminococcus* strains digest cellulose and hemicellulose; however, cellobiose or glucose remain their preferred energy sources, and the pentoses are excreted into the medium to be scavenged by other bacteria (Table 16.2). These bacteria require isovalerate, 2-methylbutyrate, and isobutyrate.

Many of the rumen species that attack cellulose and starch will not grow in pure culture in a medium where nitrogen is supplied in the form of free amino acids; however, they will utilize either ammonia or peptides (3 or 4 to 12 units or longer). If ammonia is supplied, branched-chain fatty acids are needed to satisfy re-

quirements or amino acid synthesis. *Fibrobacter* and several other genera also require CO_2 as a growth factor. Ammonia is a preferred substrate for protein synthesis by the cellulolytic, methanogenic, and some amylolytic bacteria, although many eubacteria exhibit extensive proteolytic activity and many species prefer amino acid or peptide nitrogen.

16.2.3 Proteolytic Organisms

The catabolism of protein and amino acids to produce ammonia in the rumen is of special interest in regard to the protein economy of the ruminant as well as the potential problem of excess ammonia (Chapter 18). Ammonia is also required by many primary rumen microbes that ferment carbohydrate, some of which also require or are stimulated by amino acids, peptides, and isoacids derived from valine, leucine, and isoleucine. Thus it is necessary that some protein be sacrificed and fermented in the rumen to meet the requirements of the net rumen fermentation. The animal receives microbial protein produced as a primary output; the escape, or "bypass," of dietary protein from the fermentation is a secondary matter.

Various rumen species have been considered responsible for ammonia production; protozoa are often cited (Section 16.2.3), but other proteolytic bacteria exist as well (Table 16.3). Of these, *Peptostreptococcus* and *Clostridium* have a capacity for ammonia production about an order of magnitude greater than most other rumen species. The overall contribution of these strains as well as of the protozoa needs further evaluation.

The metabolism of protein by mixed rumen organisms proceeds by hydrolysis through peptides to amino acids. Amino acids may be used in microbial cellular protein synthesis, or possibly for small peptides. Amino acids in excess of microbial needs are oxidatively deaminated to ammonia and carboxylic acids (Figure 16.1).

The availability of carbohydrate promotes the use of ammonia in amino acid synthesis and microbial

Table 16.3. Ammonia production by mixed pure cultures of rumen bacteria

Organism	Specific activity[a]
Clostridium sp. (strain R)	427 ± 31.0
Peptostreptococcus sp. (strain C)	346 ± 9.0
Mixed rumen bacteria	30 ± 5.7
Megasphaera elsdenii (B159)	19 ± 0.4
Selenomonas ruminantium (HD₄)	15 ± 0.9
Bacteroides ruminicola (B₁4)	11 ± 1.0
Streptococcus bovis (B1)	Not detectable

Source: J. B. Russell et al., 1988.
Note: Ammonia from 15 g of Trypticase per liter over 6 h.
[a]Measured as nmol of ammonia produced per mg of protein per minute.

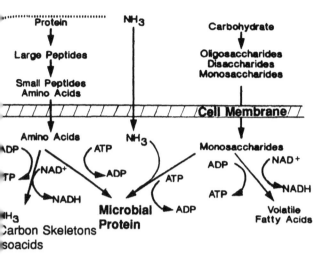

Figure 16.1. Utilization of protein by rumen bacteria in relation to supply of carbohydrate (modified from Nocek and Russell, 1988). Carbohydrate is the source of derived ATP needed for microbial utilization of protein, peptides, and ammonia.

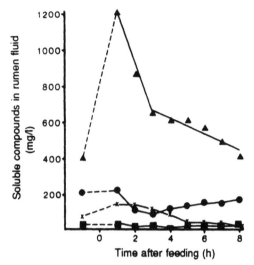

Figure 16.2. Protein (●), peptides (▲), ammonia (×), and RNA (■) concentrations in the rumen after feeding of a normal dairy ration (from Chen et al., 1987a).

growth. The optimal level of rumen ammonia is often said to be 10 mg/100 ml; however, this cannot be regarded as a fixed number, because the bacteria's capacity for protein synthesis and ammonia uptake depends on the rate of carbohydrate fermentation, and faster rates elicit greater efficiency and relatively higher ammonia tolerance.

The speed of conversion of peptides to ammonia is sufficiently slow to allow a significant pool of peptide nitrogen in the rumen, and peptide levels are commonly greater than the ammonia level (Figure 16.2). The increase in peptides may peak 2–4 h after feeding, depending on the type of protein fed, and can represent up to a third of the total rumen nitrogen at that maximum. Chen et al. (1987a) stumbled across this observation when they obtained discrepant results between trichloroacetic acid and tungstic acid precipitations. Trichloroacetic acid has a higher peptide cutoff (ca. 10 amino acids) than tungstic acid, whose cutoff is about 2–3 amino acid peptides (Greenberg and Shipe, 1979). These observations have exposed the probability of cross-feeding of peptides in bacterial nutrition; that is, one group produces peptides to be utilized by another. Peptides vary in their potential rate of hydrolysis. Hydrophobic peptides rich in glycine and the isoamino acids are hydrolyzed more slowly than hydrophilic peptides (Chen et al., 1987a).

These relationships are relevant to the conversion of isoamino acids to isoacids, which are essential factors for the growth of many cellulolytic rumen bacteria. The oxidative deamination of amino acids requires the transfer of hydrogen via NADH, and deamination is not favored in the anaerobic environment of the rumen because it is an oxidative process. The transfer of hydrogen is inhibited by ionophores that bind sodium and

potassium and is probably a reason for the inhibition of proteolytic bacteria by these agents. The saturated isoacids are the least favored amino acids for oxidative deamination, so the process of isoacid production can become limiting in the rumen. The general fermentation is usually responsive to isoacid supplementation (J. B. Russell and Martin, 1984; and J. B. Russell, 1987).

Response to isoacids supplement depends on a fermentation with a potential rate of carbohydrate fermentation sufficient to create a microbial need for the cofactors. The slow fermentations observed with low-quality forages and straws are likely to be comparatively unresponsive, since in these slower fermentations the fermenter can recycle the isocarbon skeletons sufficiently to meet the rumen requirements. Gorosito et al. (1985) found that digestion of alfalfa was more responsive than digestion of wheat straw in in vitro situations. Because of the limitation on recycling carbon skeletons, net protein will not resolve the isoacid requirement, and this situation can be made worse if too much protected protein or ionophores are in the feed.

16.2.4 Secondary Fermenters

Rumen fermentation involves primary fermentation products such as hydrogen as well as formate and succinate utilization. Lactate becomes an important product if animals are fed high-starch diets. Refermentation (i.e., cannibalism) of other organisms and fermentation of their cellular matter by protozoa is considered in Section 16.3.

Organisms that utilize lactate or hydrogen (hydrogen utilizers include methanogens) are important in hydrogen and carbon balance for microbial efficiency of carbohydrate conversion. Lactate production includes only 2 ATP per mole of glucose, and lactic acid has potential fermentable energy in the rumen supporting secondary fermenters.

Lactic acid production is characteristic of high-starch diets, and it is under these conditions that the lactate fermenters flourish (see Table 16.2). Two metabolic pathways exist, the succinate and the acrylate (Section 16.7). The succinate pathway favors more acetate production than does the acrylate. The acrylate pathway and the propionate it produces can become a dominant feature of rumen fermentation but is probably relatively minor under high-forage feeding conditions. Normal lactate fermentation follows the succinate pathway. Bacteria that ferment lactate via the acrylate pathway use some hydrogen to form propionate, but the methanogens are the major hydrogen utilizers under normal forage feeding.

16.2.5 Methanogens

The methanogens constitute a unique group of organisms that have been excluded from the true bacteria (Woese, 1987). Although their classification continues to be refined, two orders apparently exist in the rumen (Stewart and Bryant, 1988):

Methanobacteriales
    Methanobacteriaceae
        Methanobacterium formicicum
        Methanobrevibacter ruminantium
Methanomicrobiales
    Methanomicrobiaceae
        Methanomicrobium mobile
    Methanoplanaceae
        Methanosarcina bacteri
        Methanosarcina majei

Methanogens are members of the urkingdom Archaebacteria. The Eubacteria, or "true bacteria," are in another urkingdom, and plants, animals, fungi, and protozoa belong to the urkingdom Eukaryota. It may be that the Archaebacteria are more closely related to the Eukaryota than to the Eubacteria, with which they have long been confused (Woese, 1987). The methanogens are the strictest anaerobes, and their welfare can be impaired by factors affecting hydrogen supply in the rumen and by methanogen inhibitors. The ionophores are not true inhibitors. They restrict hydrogen production, thus starving the methanogens. True inhibitors include halogenated methane and methyl derivatives.

16.2.6 Methane Production for Energy

Anaerobic fermentations in sewers produce methane, a valuable fuel resource. In contrast to the rumen, sludge fermentations convert 90% of the substrate to

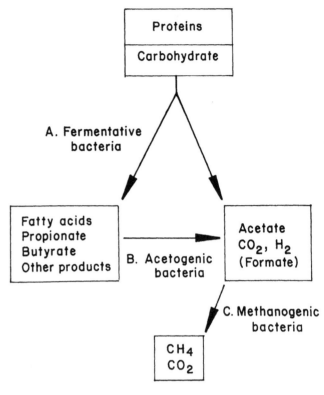

Figure 16.3. Metabolic sequence of methane production in sludges (modified from Bryant, 1979). Primary fermentative bacteria produce acetate, which can be converted by methanogens; however, hydrogen-producing acetogenic organisms are required to convert propionate and butyrate to forms utilizable by methanogens. Step B is not thermo-dynamically feasible unless step C is operable. Acetogens therefore depend on methanogens in a symbiotic relationship.

methane and CO_2, the free energy of the substrate being largely recovered in the methane (Bryant, 1979). In this fermentation process all fatty acid products and many other organic substances (W. C. Evans, 1977) are converted to methane and CO_2, the sole end products.

The net conversion is the result of a complex symbiotic system in which three microbial groups participate (Figure 16.3). Primary fermentations convert carbohydrate to VFAs, and a secondary fermentation converts propionate, butyrate, and longer-chain fatty acids to acetate and CO_2. Methanogenic organisms then convert VFAs to CO_2 and water. All methanogenic organisms can reduce CO_2 to methane, but not all can degrade acetate and other organic molecules.

If acetate were converted to methane in the rumen, this would represent a serious energy loss for the animal. Fortunately for the host, acetate-fermenting methanogens have a generation time of about 4 days, but rumen turnover time for the slowest fractions (fiber) does not ordinarily exceed 2 days, so these methanogens cannot survive.

A fourth group of syntrophic bacteria has a similar problem. These bacteria, closely related to the acetogens, degrade phenols and aromatic acids by hydrogenating the aromatic nucleus and splitting the cyclohexane products (benzoate and phenol to heptanoate and adipate, respectively), which are in turn converted to acetate. These reactions are responsible for degradation of monomeric phenols and the monomeric ferulic and *p*-coumaric esters. *Eubacterium oxo-reductans* degrades gallate, pyrogallol, phloroglucinol, and quercetin (Krumholz and Bryant, 1986). The time necessary for the species to adapt to dimers and trimers—up to 12 days—exceeds rumen retention times (Healy and Young, 1979). Since generation time exceeds retention, no degradation of condensed phenolics should occur in the rumen (W. C. Evans, 1977). True digestion of lignin appears to be very limited even in long-term (4 months) methane fermentations (J. A. Chandler et al., 1980). Lignin continues to protect associated cell wall carbohydrate in such long-term fermentations, as evidenced by the observation that the methane yield is highly and negatively correlated with lignin content.

16.2.7 Acetogens

A group of bacteria that seem to compete with methanogens metabolize CO_2 and hydrogen to acetate (H. G. Wood, 1991). These bacteria occur in the rumen (Genthner et al., 1981), in wood-digesting termites (Breznak and Switzer, 1986; Breznak and Kane, 1991), and in the hindgut of some humans. About 20% of people produce methane, most of the rest produce some hydrogen, which is in part converted to acetate (Lajoie et al., 1988). Since acetogens have been found in animals as diverse as humans and termites, it seems likely that they occur in the guts of other species. An intriguing aspect is whether acetogens can be manipulated to displace methanogens. The whole problem of energy efficiency and methane production could be solved without unbalancing the acetate-propionate ratio (H. G. Wood, 1991).

16.3 Protozoa

Protozoa are the most conspicuous organisms in the rumen, and also in the guts of some large monogastric herbivores, yet their role has been debated for decades and has recently become controversial. While they appear to form a large proportion of the rumen biomass (20–40% of net microbial nitrogen) their output may be minimal because of high retention and slow turnovers. The generation time of protozoa is relatively long, and their survival in the rumen depends on strategies that reduce their washout. Further, their predation

on rumen bacteria is blamed for apparent rumen inefficiency. The evaluation of protozoal contributions has been blurred by nutritionists' tendencies to promote oversimplistic and sweeping hypotheses that may be valid in some feeding situations but not in others.

A problem in assessing the role of protozoa arises from the difficulty in growing them apart from the bacteria on which they obviously depend. Defaunation (see below) tends to produce an adaptation in the bacteria in which the metabolic roles of protozoa are taken over by other species, so no changes may be evident.

The most influential aspect of protozoa is their predatory role. They will eat bacteria and almost any particle—starch, chloroplasts, lignified vesicular thickenings, plastic particles—of equivalent size. The difficulty in assessing their cellulolytic ability is connected with the difficulty in growing them free from bacteria, and it is still possible to wonder whether, for example, some of them are "miniruminants" harboring cellulolytic and other bacteria that continue to metabolize after being engulfed. Most scientists agree that some species of protozoa are indeed cellulolytic, but it may be their engulfment of starch and protein fragments, in addition to bacteria, that is the more important role in rumen metabolism.

16.3.1 Types of Rumen Protozoa

Rumen protozoa are divided into two general groups, holotrichs and entodiniomorphs, the latter having the greater number of species (Table 16.4). These protozoa, living in an anaerobic environment, produce volatile fatty acids and hydrogen to maintain the carbon balance (Section 16.7.1). The hydrogen is probably fed to symbiotic methanogens (Figure 16.4) that live attached to the protozoa's surfaces. Protozoa's role in hydrogen balance, and therefore VFA ratios, is a factor that cannot be overlooked (Section 16.3.4).

A significant number of protozoa are devoted to either starch or sugar digestion. Protozoa do not appear to produce lactic acid from starch or sugar, in contrast to rumen bacteria that specialize on these substrates, and thus the engulfment of starch grains by protozoa may have a major effect in moderating rumen acidosis. If protozoa engulf too much starch, lysis may result; this effect could be one of the reasons why high-grain diets are inimical to ciliated protozoa and lead to their elimination from the rumen. The loss of ciliated protozoa is probably a factor promoting milk fat depression in dairy cattle, as their loss (along with their attached methanogens) promotes propionate production by the bacterial population (Section 16.9.1). One curious aspect of this loss of ciliated protozoa is that it seems to be related to individual animals. High-grain diets do not always cause loss of protozoa or milk fat depression (Eadie and Mann, 1970). Lactating cows on high-

Table 16.4. Classification and characteristics of rumen protozoa

Genus	Probable main carbohydrate substrate	Cellulose digestion	Products[a]	Approximate generation time (h)
Holotrichs				
Isotricha	Starch and sugars[b]	0	2,3,La,H_2	48
Dasytricha	Starch and sugars	0	2,3,La,H_2	24
Entodiniomorphs				
Entodinia	Starch	0(+)	1,2,3,4,(La)	6–15
Epidinium	Starch, hemicellulose	0	2,3,H_2,(1,3,La)	
Ophryoscolex	Starch[b]	0	2,3,H_2,(3)	24–48
Diplodinium		+		
Eudiplodinium		+	H_2,fatty acids	
Polyplastron		+		48

Source: Hungate, 1966.
Note: Morphological features of the rumen protozoa are described in Hungate, 1966, and Church, 1975.
[a]Abbreviations are from Table 16.2; parentheses indicate a minor product.
[b]Pectinolytic, but do not seem to utilize products.

grain diets that maintain ciliated protozoa tend to be high in butyrate rather than propionate, a characteristic consistent with resistance to milk fat depression.

Entodinia species, which are more tolerant of rumen acidity and have a more rapid rate of growth than other genera, do persist on high-grain rations. Many rumen protozoa do not survive if in vitro cultures cycle (turn over) more rapidly than once each 24–48 h (Hungate, 1966). Since liquid turnover time in the rumen is on the order of 10–20 h, selective retention of protozoa in the fibrous mass may be important. The larger ciliated protozoa may contribute little protein because their washout rate is very slow (Weller and Pilgrim, 1974), although pelleted or high-concentrate diets, especially at high intakes, reduce the selective retention of fiber and may cause the elimination of protozoa through washout.

16.3.2 Metabolism

Protozoa are responsible for extensive ammonia production in the normal rumen (A. C. I. Warner, 1956) and could certainly supply at least some of the ammonia required by microbes; however, J. B. Russell et al. (1988) showed that there are also bacteria (*Peptostreptococcus*) that can produce ammonia. Protozoa are also known to metabolize bacterial and dietary nitrogen extensively and may contain 10–40% of the total rumen nitrogen (Weller et al., 1958; P. C. Thomas, 1973). Rumen microbiologists still do not know whether protozoa can sequester microbial capsular nitrogen available through their own cell synthesis.

Protein, either of plant or bacterial origin, is probably a major energy source for rumen protozoa. Other energy sources include sugar (holotrichs), starch, and probably cellulose and hemicellulose for various entodiniomorphs that are able to attack those substrates (Table 16.4). Starch is an important substrate for most protozoans, including *Epidinium* and *Ophryoscolex*.

Holotrichs use sugars and other soluble feed components, and entodiniomorphs depend on particulate food sources, which may be chloroplasts, fibrous particles, or bacteria.

16.3.3 Defaunation

Although it may be difficult to achieve, defaunation may be accomplished by withholding feed, administering various chemicals in the feed, sterilizing evacuated contents, or raising young ruminants in strict isolation (Jouany et al., 1988). Many of the chemical treatments result in a significant percentage of animal deaths.

Defaunation does not affect rumen digestion very much, but it may influence the balance of products. High-grain diets when protozoa are present in the rumen produce high proportions of butyrate, whereas defaunated rumens show high proportions of propionate (Eadie and Mann, 1970). These shifts in acid proportions are consistent with the known products of rumen protozoa. Bacteria are more numerous in defaunated rumens (Hungate, 1966), which indicates either competition for energy sources with protozoa, predation by the protozoa, or both.

16.3.4 Role of Protozoa in Overall Rumen Balance

The proposed beneficial effects of defaunation are questionable. Theoretically, the loss of the energy and protein in bacteria engulfed by protozoa should decrease rumen efficiency, decrease rumen microbial protein output, and increase products, unless the protozoa make some contribution to rumen fermentation. Indeed, some experiments do show lowered efficiency; however, the results were obtained with low-quality, high-forage diets (Bird and Leng, 1978).

Although protozoa may form a considerable part of the rumen biomass, their turnover is slow, leading to

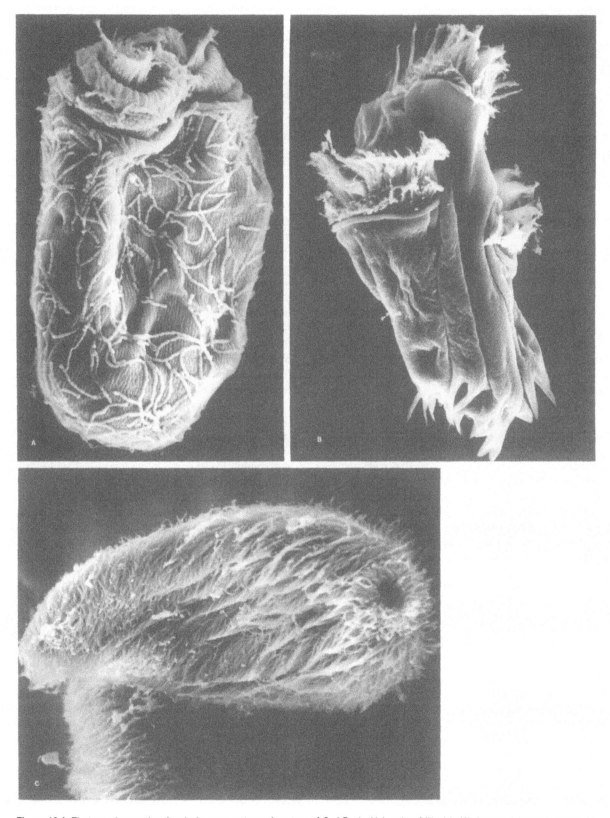

Figure 16.4. Electron micrographs of typical rumen protozoa (courtesy of Carl Davis, University of Illinois). (A) An entodiniomorph, genus *Diplodinium*. Note the bacteria adhering to the body of the protozoan. These are probably methanogens that live symbiotically with their host (Vogels et al., 1980). ×1065. (B) Entodiniomorph, genus *Ophyroscolex*. ×1065. (C) A holotrich, genus *Isotricha*. note the ciliated surface of this protozoan. ×440.

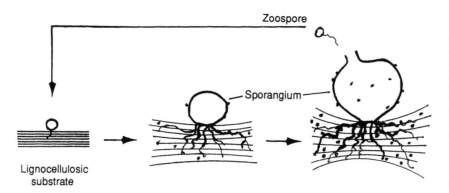

Figure 16.5. The life cycle of anaerobic rumen fungi that live on lignocellulosic fiber (courtesy of J. B. Russell). The fungi appear to be able to digest and utilize available cellulose and hemicellulose and may have an advantage over cellulolytic bacteria in being able to penetrate lignified cell walls and thus gain access to "encrusted" cellulosic carbohydrates. Their breaching of the cell wall may help other bacteria to gain access and promote the rate of digestion.

small outputs. The modifying role of protozoa is more subtle and complex, however, and no overall generalization of their role is adequate. Digestibility of the diet is not much altered in their absence, although it might be slightly less (Jouany et al., 1988), and their effect on overall nitrogen metabolism is variable. The methanogens they harbor are yet another factor (Figure 16.4A). As noted, the protozoa can have significant effects on VFA ratios, particularly in animals on high-grain diets, in which low acetate-propionate ratios have been associated with the absence of ciliates in the rumen (Eadie and Mann, 1970).

A more balanced view (Veira, 1986) is that the results of defaunation depend on rumen balance and type of diet. Defaunation seems to increase microbial yield and output of net protein. Whether this is beneficial will depend on the energy-protein balance and requirements of the animal, and whether protein is limiting. Many low-quality straw diets are limited in net true protein, and defaunation may be beneficial. With high-quality diets, however—particularly those that contain large amounts of starch—the protozoa may have another and potentially beneficial role in regard to acidosis. Protozoa engulf particles indiscriminately, and their engulfment of starch introduces a marked lag effect in starch fermentation.

16.4 Fungi

The fact that anaerobic fungi occur in the rumen has been known only since the 1970s (Orpin and Joblin, 1988; Theodorou et al., 1991). Before that the swimming body (zoospore) was mistaken for a flagellated protozoan. Because the zoospore is strongly attached to ingested plant fibers, most early preparations that filtered rumen contents unwittingly disposed of the fungi in the castaway filtered fibrous material. Because they are so "new" on the microbiological scene, there are yet more surprises to be found among this group of interesting organisms.

Like protozoa, fungi exhibit low rates of turnover in the rumen, and their output in terms of total mass is probably small, although it is not accurately known. Fungi are much more particle-associated and involved in fibrolytic digestion than are protozoa. There is no evidence that they can digest lignin, and it is improbable that any anaerobic organisms can degrade more than simple phenols. Still, their rhizoids can penetrate the innermost recesses of plant cell wall, and lignified surfaces seem to offer no barrier, so their potential to materially influence digestion of lignified structures may be important. Their major role may be in causing or facilitating cell wall disappearance.

There are a number of species already identified in at least four genera: *Neocallimastix*, *Caecomyces*, (formerly *Sphaeromona*), *Pyromyces* (formerly *Pyromonas*), and *Orpinomyces* (polycentric fungi) (Theodorou et al., 1992a). Unlike the protozoa, fungi are comparatively easy to grow. The life cycle involves a fruiting body (sporangium) that is established via free-swimming zoospores derived from ruptured mature sporangia. The zoospores attach to fiber and develop sporangia and rhizoid filaments, which penetrate the lignocellulosic matrix (Figure 16.5). Enzymes are secreted from the rhizoids, and nutrients are transported to the sporangium. Fungi secrete a more soluble cellulase complex than do rumen bacteria, and the mechanisms of enzyme-substrate interaction differ as well. Fungi also produce cysts, although the conditions promoting that are not well understood. No cysts have been found in the rumen, but they abound in the lower tract from the omasum onward and also occur in feces. The cysts are resistant to air and drying and are probably a means of transmission between animals. Cysts occur in ordinary air-dry ruminant feces in sufficient amounts to allow the substitution of fecal matter as inoculum for in vitro digestion of forages (Theodorou et al., 1988; A. Milne et al., 1989). The rate of digestion with fungi alone appears to be slower than that with cellulolytic bacteria, but the extent of digestion does not seem to be affected. The fungi attack coarser particles and ferment them more rapidly than bacteria, the advantage arising from the penetration of the cell wall structure by the hyphae. The fungi have less advantage with finer particles (M. K. Theodorou, pers. comm., 1989).

Anaerobic fungi have been found in many species of ruminants and other herbivores (A. Milne et al., 1989), but they may be restricted to species with sufficient retention time to allow the fungal life cycle to compete with turnover and that consume significant amounts of lignified coarse cell wall. Since the fungi are associated with the coarse lignified matter, fine-ground and high-concentrate feeds might be expected to be inimical. There is evidence that inoculating the rumen with fungi can improve fiber digestion and also intake and growth in young ruminants (D. E. Beever and M. K. Theodorou, pers. comm., 1989). The fungi produce VFAs, gas, and traces of ethanol and lactate (Orpin and Joblin, 1988). Their contribution to microbial mass may be small, however, since, like the protozoa, they stay in the slower-moving ingesta to avoid washout.

16.5 Culture Techniques

Some of the systems for in vitro evaluation of forages and feeds have already been discussed in Section 8.5. The following is a discussion of some basic problems in culturing rumen organisms.

Rumen organisms are very sensitive to their environment, and successful cultures must satisfy both environmental and nutritional requirements. Substrate concentrations are about 10% or more of the volume in normal rumen contents of sheep and cattle. At these substrate levels osmotic pressure and acids would be limiting if allowed to accumulate; however, acidic products of fermentation are readily absorbed across the rumen wall, and fresh buffering arrives regularly in the form of saliva. Rumen fermentation is normally limited by substrate, not by products.

In vitro cultures must be diluted by about an order of magnitude with respect to both feed and organisms to avoid hyperacidity and high osmotic pressure since no analogous system of absorption exists. In a batch culture, all products accumulate in the system, thus the maximum level of acidic products must remain below the limiting levels of osmotic pressure and buffering capacity. Since available substrate essentially determines products, substrate concentration should not exceed 1% of culture volume and can be less.

Dilution of the substrate greatly increases the susceptibility of the fermentation to traces of oxygen because substrate concentration is related to maintenance of redox potential. Special precautions must be taken to protect dilute concentrations of rumen organisms from oxygen contamination. The degree of caution necessary varies with the type of culture. Pure cultures are more sensitive than mixed cultures, and some strains (e.g., methanobacteria) are more fastidious than others.

In media preparation, most of the oxygen is removed by boiling followed by gassing out with CO_2. Any residual oxygen is removed with sodium sulfide or cysteine. If media are sterilized, reducing agents are often added afterward. The redox indicator resazurin is useful for detecting traces of oxygen. This indicator is colorless in the reduced state but turns pink in the presence of even a small amount of oxygen. Commercial sources of CO_2, nitrogen, and other gases usually contain traces of oxygen and may need to be purified. This is accomplished by passing the gas through a tube of heated copper turnings, but this precaution may not be necessary if the gas is used only to prepare inoculum and media for whole culture systems (Bryant, 1972). Several methods can be used to separate organisms from ingesta. Direct filtration on cheesecloth fails to catch many organisms, especially cellulolytics. This preparation, however, is commonly used for in vitro fermentations designed to measure digestibility. A more complete recovery of bacteria can be obtained by blending rumen contents for several minutes in a food blender gassed with CO_2. Treatment with 10% methanol in water promotes detachment. The blending method causes many fragments of chloroplasts from plant tissue to enter the filtrate and destroys many of the protozoa; the fungi remain largely attached to the fiber. Thus, other techniques must be used to recover protozoa and fungi. Rumen fluid must be filtered free of plant fiber if it is to be used in any assay for plant cell wall carbohydrate (e.g., the Tilley-Terry procedure). The blended preparation is suitable for quantitative enumeration and culture of bacteria.

Diluted rumen organisms can be kept in media consisting only of buffer, but the incubation system must supply their nutritional requirements. Adequate buffering and mineral requirements are supplied by mineral solutions. Branched-chain acids, vitamins, and other factors may be added directly, or the culture organisms can take them from filtered rumen fluid (about 20% of the medium volume). Filtered rumen fluid has the disadvantage of oversaturating various substances, but it is important if maximal substrate degradation is the object. Mixed cultures tend to supply their own growth requirements once off to a good start and given an adequate substrate. Filtered sterile rumen fluid is a source of medium in studies meant to isolate and identify specific types of organisms, and it can be added to media designed to enrich specific and individual nutritional requirements. Other media that attempt to support the growth of diverse species in mixed cultures simulate as closely as possible the normal composition of rumen fluid.

16.5.1 Batch Cultures

In vitro rumen digestions conducted with mixed cultures (Tilley and Terry, 1963) have been used to measure digestibility (Section 8.5) and rates of digestion. Anaerobic technique is critical in measurement at short fermentation times (6–24 h) because of the lag that can be induced by careless handling of inoculum. Such effects are usually overcome by 48 h, the usual time period for measurement of digestibility. A 96-h digestion period allows the use of a very small amount of inoculum, the deficiencies of which are overcome during the long fermentation period (den Braver and Eriksson, 1967). At such long times of fermentation, however, the extent of substrate degradation often exceeds that observed in vivo. Individual microbial strains or combinations of strains can be used to study microbial interactions in the digestion process.

16.5.2 Continuous Fermentation

Batch cultures' similarity to the rumen environment ends as available substrate is used up, and end products and organisms that cannot be removed may be recycled into the fermentation. In the normal rumen, turnover is an important variable. Continuous fermentors have the objective of simulating rumen conditions. Continuous fermentation is less satisfactory for measuring digestibility of substrates but has been important in studying conditions for the maintenance of methanogenic organisms and protozoa and measuring microbial synthesis.

Continuous cultures deviate from the rumen in several aspects. In the rumen, removal of excess acid is via absorption across the rumen wall, a method impossible to employ in a glass vessel where removal is entirely dependent on washout. The rumen is also very selective in its retention of fiber and the relative ease of liquid or fine particle washout. This selectivity is very difficult to reproduce in vitro. There are several kinds of continuous fermentation systems, including simple chemostatic systems that depend entirely on turnover of liquid and systems that attempt to accommodate solid material.

Chemostatic systems, in which there is complete control of pH, flow, and turnover of liquid, work well with soluble media and soluble substrates. The turnover required to prevent end product buildup is, on the whole, high for cellulosic carbohydrates. Older systems, such as the type described by Slyter and Putnam (1967), suffer a considerable loss of potentially digestible cellulosic carbohydrate. No presently available device can simulate the absorption of VFAs, rumination, and the complex differential passage that occurs in the living animal.

Two types of continuous fermentors are currently in use. In the Rusitec system, developed by Czerkawski (1984, 1986), feed in synthetic cloth bags is placed in the fermentor with inoculated flowing buffer solution. Liquid turnover can be regulated, and turnover of solids depends on the placement and removal of the bags. Bags must not be overfilled lest microenvironments develop within them.

In the other continuous fermentor, described by Hoover et al. (1976), both liquid and solids exit via overflow and the differential rates of solids and liquids depend on settling and control of the mixing rate. Liquid rates are precisely regulated, and solid turnover is measured.

Because there is no absorption of end products, all continuous systems must have sufficiently rapid turnover to remove VFAs in the outflow. Nevertheless, continuous systems are the only ones that allow study of the influence of feed quality and digestive turnover on net microbial efficiency. The alternative is duodenal fistulation in live animals.

16.6 Ecology and Consortive Associations

The concentrations and relative proportions of the various classes of rumen organisms vary according to the quality and composition of the diet. From the viewpoint of microbial nutrition there are basically two groups: those that ferment the primary nutrients in feed and those that ferment the (secondary) products of the first group. This second population performs a vital function by removing end products and cycling essential factors back to the members of the primary group; for example, proteolytic species feed isoacids to cellulose digesters, which require them (Table 16.5).

Interactions of this sort have further effects on the distribution of fermentation products. An example is the interaction between *Fibrobacter succinogenes* and *Selenomonas ruminantium* (Figure 16.6). Fermentation by the cellulolytic organism produces succinate, acetate, and formate. When the two species are grown together, however, the end products include propionate and CO_2, succinate being the intermediate in the formation of propionate. Interspecies transfer of hydrogen is another important interaction. In this connection the methanogens are most important. The utilization of hydrogen reduces hydrogen concentration. Since it is a mobile equilibrium, the reaction under certain conditions will favor accumulation of hydrogen on the right:

$$NADH + H^+ = NAD^+ + H_2 \qquad (16.1)$$

The influence of hydrogen concentration on the free energy of this reaction is shown in Figure 16.7. The

16.5. Interaction between the cellulolytic *Ruminococcus albus* (strain 7) and ⌐arbohydrate-fermenting, protein-catabolizing species *Bacteroides ruminicola* ⌐n 118B) grown in mixed culture in a defined medium with cellulose as the ⌐y source and casein as the nitrogen source

	Growth (absorbance at 600 nm)		
Deletions	Strain 7	Strain 118B	7 + 118B
	0.74	0.00	0.70
+ isobutyrate + methylbutyrate	0.01	0.01	0.65
⌐tyrate + methylbutyrate	0.00	0.05	0.65
	0.01	0.00	0.70
⌐se replacing cellulose			
⌐⁺₄ + isobutyrate + methylbutyrate	0.00	0.60	0.68
⌐ne	0.85	0.80	0.81

⌐urce: Bryant and Wolin, 1975.
⌐te: The complete medium contained 0.1% ground cellulose (filter paper), N-free ⌐als, a mixture of B vitamins, 0.2% casein, 0.05% cysteine, sodium bicarbonate ⌐, CO_2 gas, 4.0 mmol $(NH_4)_2$, CO_4, 1.0 mmol DL-2-methylbutryate, and isobuty- ⌐Absorbance was determined after shaking and allowing time for cellulose parti- ⌐o settle in tubes.

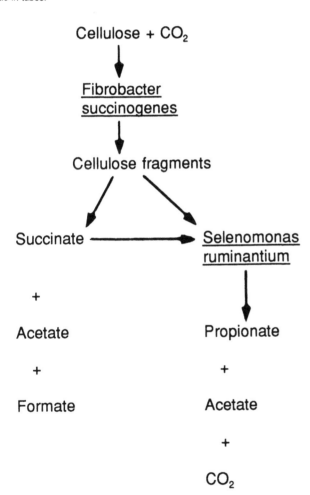

Figure 16.6. A two-series microbial interaction involving propionate formation from cellulose via the extracellular intermediate succinate (after Bryant and Wolin, 1975). Succinate and formate appear as products only when *Fibrobacter* are grown in pure culture. *Selenomonas* depend on carbohydrate fragments and succinate supplied by *Fibrobacter*. In combined culture, propionate, acetate, and CO_2 are the only products. *Selenomonas* also supply the CO_2 required by *Fibrobacter*.

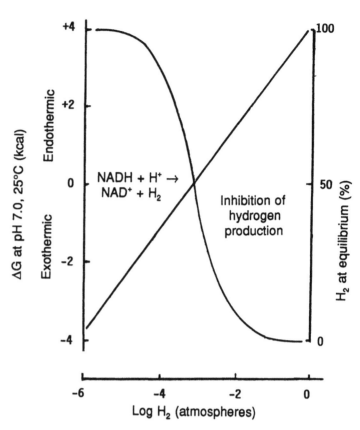

Figure 16.7. Effect of the partial pressure of H_2 on the free energy of oxidation of NADH to NAD and H_2 (straight line) (modified from Wolin, 1975). Hydrogen genesis is favored by low hydrogen concentrations and negative (exothermic) ΔG. Accumulated hydrogen markedly retards its own production (curved line).

normal rumen level is about 10^{-4} atmospheres and favors a net production of hydrogen. Thus a two-species culture of *Ruminococcus* and *Methanobacterium*, respectively hydrogen-producing and -utilizing organisms, stimulates a fourfold or larger increase in hydrogen production. Hydrogen appears as a product of the binary fermentation, which also shows a much increased acetate production, the reasons for which lie in the fermentation balance (Section 16.7.1).

The production and utilization of hydrogen causes a drastic reduction or disappearance of other products such as ethanol, lactic acid, succinate, propionate, and formate, and an increase in acetate, hydrogen, and CO_2 production by the fermentative bacteria. Ethanol is produced only by pure strains of certain rumen organisms and is not likely to be a significant intermediate in normal rumen fermentation. The same is true of lactate, with the exception that it often appears as a product in diets high in starch or soluble carbohydrate. In such situations lactate utilization becomes important. Formate is an important substrate for methanogenic bacteria and never appears as an end product in the mixed rumen population.

16.6.1 Attachment

Attachment is a microbial strategy whereby organisms sequester their substrate and avoid being washed out in flowing systems. Attachment is important in a wide variety of situations, among them the intestinal lining, gonorrheal infection, tooth decay, and organic matter decay in streams and rivers. Many types of rumen bacteria attach to the surfaces of feed particles; some attach to the gut wall. Attachment is an important factor in competition that allows microbes to achieve dominance over substrate and environment.

Adhesion can involve various mechanisms, including fimbria, proteinaceous surface structures (important in protein-degrading bacteria), the glycocalyx (characteristic of many cellulolytic bacteria), and lectins, which serve as intermediates. The adhesion process begins by ionic or hydrophobic association involving van der Waals forces. Cation exchange is important because both microbial cell wall and feed particles are negatively charged (Stotzky, 1980). Covalent ions such as magnesium or calcium, or the double ion layer effect of potassium can neutralize the repellent ionic charge. There are also specific adhesion recognition systems that allow an organism to seek out its favored substrate (Latham et al., 1978).

The glycocalyx, which consists of external filaments that can conform to the substrate surface, may serve as a reserve for captured substrate (Pell and Schofield, 1993a). It must be considered a part of the cellular mass, and not a product. Attachment is at broken edges of plant surfaces, less frequently on lignified and cutinized surfaces, which offer less available substrate or are inhibitory in some way. Cell wall–digesting organisms etch pits into available surfaces (Figure 16.8). Some digest hemicellulose and excrete pentosan fragments in order to get at the cellulose that is their main substrate. Other secondary organisms metabolize the pentosans and other products of the primary fermentation. No free cellulase occurs in the rumen, although significant amounts of soluble hemicellulase and amylase are present (Chesson and Forsberg, 1988). Although there are rumen bacteria (*Ruminococcus*) that cause clear zones on cellulose-agar plates (Leatherwood, 1969), most cellulolytic activity on crystalline substrates is associated with attachment,

and factors influencing detachment may inhibit cellulolytic capacity.

Cellulolytic organisms frequently attach to their substrate, and those that degrade starch and protein probably do as well. Cellulolytic attack also means hemicellulolytic attack because of the intimate association of these carbohydrates in the cell wall, and cellulolytic and hemicellulolytic organisms coexist symbiotically. Generally, attachment is more common on slower-digesting substrates. Organisms that digest unlignified amorphous cell walls such as those in vegetables may secrete extracellular enzymes because the environment is less restrictive; however, E. J. Morris and Cole (1987) indicate that adhesion is mandatory for crystalline cellulose digestion and is beneficial in the case of amorphous cellulose. *Ruminococcus* and *Fibrobacter* play this role in the rumen by attachment and the etching of pits in the substrate (Figure 16.8).

Cellulose digestion involves membrane-attached endoglucanases (cellulosomes) that "crack" crystalline cellulose, create amorphogenic zones, and break the cellulose into insoluble eight-sugar pieces (Coughlan and Ljungdahl, 1988). Exoenzymes (cellobiosidase) remove cellobiose units from the nonreducing end of the chain (see Figure 11.8).

Competition for surface attachment sites may result in replacement of older cells by younger ones, so the surface is the site of more active metabolism and cell synthesis. Easily degradable substrates mean a rapidly disappearing surface and more fermentable substrate fragments appearing in solution, thus favoring extracellular secretion of enzymes and more unattached bacteria that can ferment fiber. Attachment to a completely degradable substrate may eventually result in stranded and unattached bacteria. Also, slowly digesting portions of the substrate that escape the rumen cause the microbial population to depend more on the rapidly degradable portions. The addition of finely divided cellulose with an inherently slow digestion rate may be inhibitory in itself because attachment promotes removal of the requisite organisms through washout before the substrate can be efficiently attacked.

The maximum rate of digestion is determined by the substrate's characteristics and physicochemical properties. Organisms can grow only at the rate allowed by

Figure 16.8. Electron micrographs showing microbial attacks on structural matter of forages (courtesy of D. E. Akin). (A) Orchardgrass; 12-h digestion; epidermis and mesophyll tissues. Note the digestion of mesophyll and inner part of epidermis cell wall by free enzymes apparently secreted by unattached bacteria. ×2,770. (B) Coastal bermudagrass; 12-h digestion; parenchyma bundle sheath cell walls under attack by many types of rumen bacteria. ×1,385. (C) Coastcross-1 bermudagrass; 12-h digestion; parenchyma bundle sheath cell walls under attack by rumen bacteria. Capsules of diplococci are visible. ×1,924. (D) Coastcross-1 bermudagrass; 6-h digestion; parenchyma bundle sheath cell wall and encapsulated cocci. ×7,273. (E) Coastal bermudagrass; 12-h digestion; epidermis; attack by encapsulated coccus and splitting off of the cuticle. Other bacteria are adhering as well. ×7,811. (F) Coastal bermudagrass; 12-h digestion; attack on parenchymal bundle sheath by encapsulated cocci. Note the lack of degradation of the lignified inner bundle sheath. ×6,095. (G) Coastal bermudagrass; 24-h digestion; digestion of easily degraded phloem and mesophyll, initial digestion of the more slowly degraded parenchyma bundle sheath and epidermis, and resistance to digestion of the lignified vascular tissue, sclerenchyma, and cuticle. ×210. (H) Orchardgrass; 24-h digestion; residue consisting of lignified vascular tissue (xylem and inner bundle sheath), sclerenchyma, and cuticle. ×184.

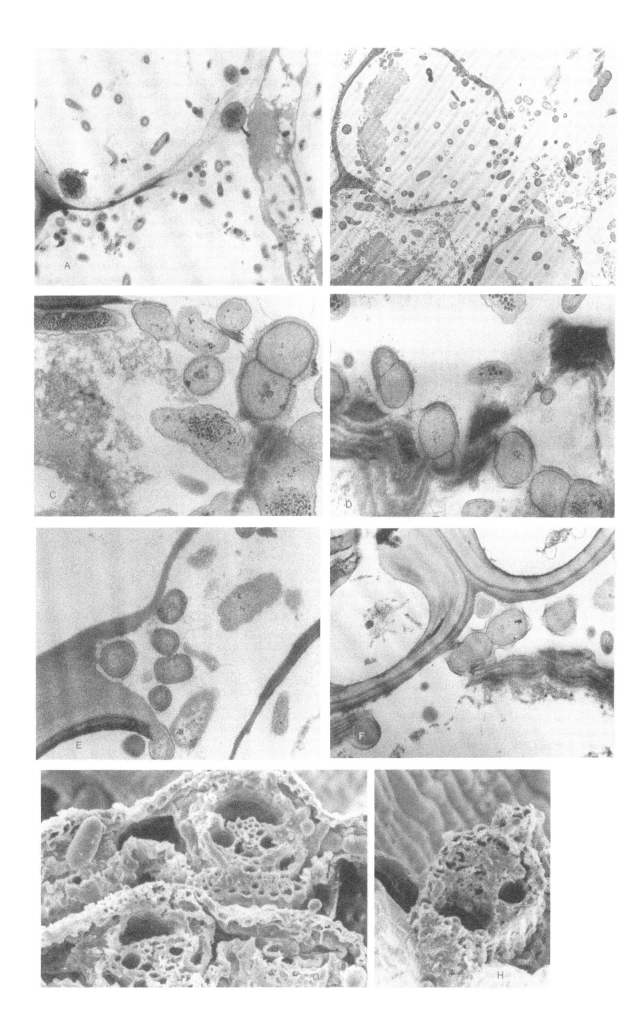

the substrate. The ratio of digestion rate to passage rate is therefore a fundamental factor determining balance of the microbial population and its efficiency. Free unattached bacteria tend to wash out at the liquid rate of passage, which is usually faster than the particulate passage (which is, in turn, inversely related to particle size). These variables explain the adverse effects of a finely ground diet.

16.6.2 Compartmentalization

A number of ecological zones exist in the rumen, and microbe populations differ relative to the various sites (Czerkawski, 1986; Czerkawski and Cheng, 1988). The zones include the rumen wall, where urea and some oxygen diffuse across and ureolytic bacteria attach. Oxygen in this zone is rapidly soaked up and does not affect the internal space. Because of diffusion and flow, the parts of the rumen farthest from the absorptive and secretory surface are apt to be higher in VFAs and lower in pH, particularly in the viscous contents of grain-fed animals.

Rumen stratification also results in compartmentalization. Recently ingested feed sorts into soluble and insoluble fractions. The soluble matter diffuses and flows with the liquid as pushed by the ruminal contractions; it contains a pool of largely noncellulolytic bacteria that degrade the water-soluble components. Some fine particulate matter will be in this pool, which is too fine to be captured by the filtration mechanisms in the floating mat.

The recently ingested fibrous insoluble matter has a low functional specific gravity and is too large for passage, so it floats on the surface of the ventral rumen and forms the mat. Most of the rumen protozoa and fungi probably live in this zone, whose fate is colonization by these microorganisms and rumination into smaller particles. Waterlogging, digestion, and particle comminution lead to high functional density of this matter, which ceases to float and sinks into the lower, more watery part of the rumen. This matter still has considerable fermentable fiber and is perhaps the richest source of viable cellulolytic organisms. Insoluble matter with a functional density of about 1.2 and particle size less than 4 mm in cattle and 2 mm in sheep is prone to passage. This fiber material is relatively low in available carbohydrates and high in lignin.

16.7 Metabolic Pathways

The main routes of rumen bacterial metabolism are shown collectively in Figure 16.9. Individual strains may be able to metabolize carbon only through certain pathways, the intermediary products being picked up

and utilized by other strains. Rumen bacteria do not have a complete citric acid cycle because they cannot oxidize acetate to CO_2 and water. In fact, portions of the cycle—namely, malate to succinate—tend to operate in reverse compared with the path in aerobic organisms. Formation of oxaloacetate is an important step and may be accomplished by phosphoenolpyruvate (PEP) fixation of CO_2 and to a lesser degree (depending on availability of methylmalonate) by transcarboxylation. Energy conserved by entrapment in ATP can be used for cellular growth above maintenance requirements. Two ATPs are derived from the formation of lactate, which contains still more fermentable energy. The overall conversion to acetate involves about 4 ATPs, assuming that hydrogen transport is not limiting. The pathways shown in Figure 16.9 do not indicate all cases of energy conservation. For example, in the hydrolysis of cellulose by *Ruminococcus*, energy in the glycosyl bond is conserved in the formation of glucose-1-phosphate. The export of free un-ionized acids may be promoted by lipophilic membranes, and thus may provide free export of protons.

The metabolism of oxaloacetate to succinate is the main route used by rumen organisms to synthesize propionate. The direct synthesis of propionate from pyruvate via the acrylate pathway is favored in the rumens of animals fed high-concentrate diets. *Megasphaera elsdinii* is probably the main organism responsible for propionate synthesis through acrylate. Under these conditions net propionate production is greatly increased relative to production in forage-fed animals (Davis, 1967; Bauman et al., 1971).

Ethanol, if it is formed at all, is formed at the expense of acetate, both being derived from acetyl-CoA. The formation of acetate from pyruvate produces at least 1 mole of ATP, whereas the formation of ethanol from pyruvate produces none. This competition favors acetate production over that of ethanol. Butyric acid is produced via the condensation of two molecules of acetyl-CoA, and n-valerate from propionyl-CoA and acetyl-CoA. Isobutyrate, isovalerate, and 2-methylbutyrate arise from the degradation of valine, leucine, and isoleucine.

$$\text{valine} + 2H_2O \rightarrow \text{isobutyrate} + NH_3 + CO_2 \tag{16.2}$$

$$\text{leucine} + 2H_2O \rightarrow \text{isovalerate} + NH_3 + CO_2 \tag{16.3}$$

$$\text{isoleucine} + 2H_2O \rightarrow \text{2-methylbutyrate} + NH_3 + CO_2 \tag{16.4}$$

The production of these acids is extremely important. They are growth factors for many cellulolytic organisms, and other species use them for long-chain fatty acid synthesis and, sometimes, for amino acid

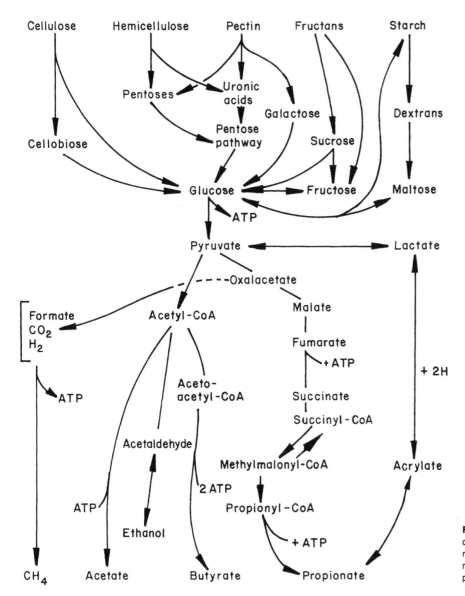

Figure 16.9. Pathways of carbohydrate metabolism in the rumen. Some methanol and acetate may be derived from the hydrolysis of pectins and hemicellulose.

synthesis through essentially the reverse of equations 16.2, 16.3, and 16.4. CO_2, ammonia, and the requisite carbon skeletons are required and fixed in the formation of amino acids. Because of their saturated side chains, the isoacids and their amino acid relatives are among the more hydrophobic substances among the VFAs and amino acids, respectively (Chen et al., 1987b). Hydrophobic peptides containing these amino acids are metabolized more slowly. The reduced rates of metabolism, along with the fact that amino acid catabolism that proceeds via oxidative deamination is an uphill reaction in anaerobic environments, causes these amino acids and their products to become limiting to cellulolytic fermentation. The problem is more severe in rapid fermentations and when ionophores are present in feed. Ionophores block transport of hydrogen ions, increase the intracellular NADH-NAD ratio, and (Section 16.9.2) render the recycling of NAD to

NADH more difficult. This limits oxidative deamination and is the likely reason why proteolysis is inhibited and rumen escape of protein is promoted by ionophores (J. B. Russell, 1987).

Anaerobic rumen organisms cannot derive energy from long-chain fatty acids, but they may synthesize them for metabolic needs or scavenge them from dietary sources. Long-chain fatty acids in bacterial lipids are unique in their content of odd-carbon and methyl-substituted fatty acids. Longer-chain fatty acids are synthesized through the condensation of acetate units with any of the linear or branched acids. Thus, pentadecanoate is formed from five acetate units plus n-valerate (required by some bacteria), and methyl-substituted acids from acetate units plus the isoacids and 2-methylbutyrate. Some methylated acids (e.g., phytyl alcohol) may arise from metabolism of plant terpenoids.

Table 16.6.   Oxidation equivalents of organic substances

Compound	Formula	Net oxidation equivalent per mole	Oxidation units per carbon
Glucose	$C_6H_{12}O_6$	0	0
Acetic acid	$C_2H_4O_2$	0	0
Propionic acid	$C_3H_6O_2$	−1	−0.33
Butyric acid	$C_4H_8O_2$	−2	−0.5
Carbon dioxide	CO_2	+2	+2
Methane	CH_4	−2	−2
Stearic acid	$C_{18}H_{36}O_2$	−16	−0.81
Nitrate	$(H)NO_3$	+4[a]	—
Sulfate	$(H_2)SO_4$	+4[a]	—

Sources: Wolin, 1960, 1975.

Note: The net consumption of hydrogen in conversion of CO_2 to CH_4, NHO_3 to NH_3, and H_2SO_4 to H_2S is uniformly 8 H.

[a]Assuming the oxidation values of NH_3 and H_2S to be zero.

16.7.1 Fermentation Balance

Stoichiometric laws of chemical balance apply to anaerobic fermentative systems since all oxygen as well as microbial substance and products must be derived from the substrate. This may be stated in equation form:

$$C_6H_{12}O_6 + NH_3 \rightarrow \text{microbes} + CH_4 + CO_2 + \text{VFAs} \qquad (16.5)$$

It is assumed that the substrate can be reduced to the equivalent of carbohydrate (glucose) and ammonia and that reduced products such as alcohols are unimportant in the rumen. From the equation it is clear that a change in the amount of any one product will alter the balance of the other products. The basic problem in anaerobic metabolism is the shortage of oxygen and the excess of reduced cofactors in need of sinks for disposing of hydrogen.

Organic compounds can be rated according to their oxidation state, which is calculated as the number of oxygen atoms they contain minus half the number of hydrogen atoms. Table 16.6 lists oxidation equivalents for glucose and its fermentation products. On a per carbon basis, CO_2 and methane represent the highest and lowest possible oxidation states, respectively. Other organic substances lie somewhere between these limits. Fermentation products may be considered relative to glucose. Those with negative oxidation values can serve as hydrogen sinks, while the production of substances with positive values represents a cost of hydrogen disposal. Values for organic substances containing nitrogen or sulfur may be calculated with an amine (NH_2) or sulfhydryl group (SH) equivalent to a hydroxyl group (OH), the value of which is 0. The oxidation value of sulfate and nitrate is +4 relative to the 0 value of hydrogen sulfide or ammonia, to which these ions are respectively reduced in the rumen. Sul-

fate and nitrate can be important sources of sulfur and nitrogen in water and forage, respectively.

The advantage or disadvantage of synthesizing individual products may be compared on the basis of the net hydrogen balance of the route of synthesis. For example, the formation of acetic acid requires the production of CO_2:

$$C_6H_{12}O_6 \rightarrow 2C_2H_4O_2 + 2CO_2 + 8H \qquad (16.6)$$

The net balance thus requires the disposal of eight hydrogen atoms. The synthesis of butyric acid has a more favorable hydrogen balance because this acid is formed from two acetyl-CoA molecules. Hydrogen is used in the reduction of acetoacetate and crotonate. The net equation is:

$$C_6H_{12}O_6 \rightarrow C_4H_8O_2 + 2CO_2 + 4H \qquad (16.7)$$

In the case of propionic acid, two pathways are available—synthesis via succinate (eq. 16.8) and synthesis by the acrylate pathway (eq. 16.9).

$$\tfrac{1}{2}\text{glucose} \rightarrow \text{pyruvate} + (CO_2) \rightarrow \text{fumarate} + (2H) \rightarrow \text{propionate} + CO_2 \qquad (16.8)$$

$$\tfrac{1}{2}\text{glucose} \rightarrow \text{lactate} - H_2O \rightarrow \text{acrylate} + (2H) \rightarrow \text{propionate} \qquad (16.9)$$

$$\text{Net: } C_6H_{12}O_6 \rightarrow 2C_3H_6O_2 + 2[O] \qquad (16.10)$$

$$2H_2O + C_6H_{12}O_6 \rightarrow 2C_3H_6O_2 + 4H \qquad (16.11)$$

In the succinate route, the formation of oxaloacetate requires CO_2, which is again lost on decarboxylation of methylmalonate. Thus the net carbon balance (eq. 16.10) is the same, yielding a net excess of two oxygens, which when traded through two molecules of water are equivalent to a net disposal of four hydrogens (eq. 16.11). The efficiency of hydrogen conservation is related to the lack of net CO_2 production.

Methane formed from formic acid (eq. 16.12) or CO_2 (eq. 16.13) is a major hydrogen sink:

$$HCOOH + 6H \rightarrow CH_4 + 2H_2O \qquad (16.12)$$

$$CO_2 + 8H \rightarrow CH_4 + 2H_2O \qquad (16.13)$$

With the stoichiometric balance one can calculate the equilibrated amounts of the products: acetate, propionate, butyrate, CO_2, and methane (Wolin, 1960). If the fatty acid ratios are known, the ratio of methane to CO_2 is fixed. Conversely, if the ratios of the gases are known, the proportions of fatty acids are fixed with one degree of freedom. The equations for balancing this system are algebraic with two unknowns (Y and Z) and are set so that the balance of net oxidation equivalents is zero.

$$Y + Z + Ma + Mp + Mb = 0 \qquad (16.14)$$

where Y, Z, Ma, Mp, and Mb are moles of CO_2, methane, acetate, propionate, and butyrate, respectively.

Substituting oxidation values per mole from Table 16.7 for each product into equation 16.14 gives:

$$+2Y + -2Z + OMa + -1Mp \\ + -2Mb = 0 \qquad (16.15)$$

Acetate drops from the equation because its value is zero. Simplified, the equation becomes:

$$2Y - 2Z - Mp - 2Mb = 0 \qquad (16.16)$$

Solving for Z:

$$Z = Y - \frac{Mp}{2} - Mb \qquad (16.17)$$

All methane must arise at the expense of CO_2, which is a product of the formation of acetate and butyrate. Butyrate in turn is formed from two moles of acetate. Therefore:

$$Y + Z = Ma + 2Mb \qquad (16.18)$$

Elimination of Z by algebraic substitution from equation 16.17 gives:

$$Y + \left(Y \frac{- Mp}{2} - 2Mb \right) = Ma + 2Mb \quad (16.19)$$

And solving for Y gives:

$$Y = \frac{Ma}{2} + \frac{Mp}{4} + \frac{3Mb}{2} \qquad (16.20)$$

For example, we wish to calculate molar proportions of CO_2 (Y) and methane (Z) in equilibrium with the rumen molar proportions of 65% acetic acid, 20% propionic acid, and 15% butyric acid. Substituting these figures in equation 16.20 gives:

$$Y = \frac{0.65}{2} + \frac{0.20}{4} + \frac{(3)(0.15)}{2}$$

$$Y = 0.60 \text{ moles } CO_2$$

Substituting in equation 16.18 gives:

$$0.60 + Z = 0.65 + 0.30;$$

$$Z = 0.35 \text{ moles methane}$$

It is possible to calculate the molar proportion of glucose from which these products arise by summing the moles of carbon in the product:

Acetate	$2 \times 0.65 = 1.30$
Propionate	$3 \times 0.20 = 0.60$
Butyrate	$4 \times 0.15 = 0.60$
CO_2	$1 \times 0.60 = 0.60$
Methane	$1 \times 0.35 = 0.35$
Net moles carbon	3.45
Moles glucose ($\div 6$)	0.575

A simplified example (Table 16.7) illustrates the principles involved. The production of propionate and butyrate together balance oxidation equivalents, but acetate production requires an output of one more mole of methane per mole of acetate to balance the oxygen output in CO_2. An increase in acetate output (Case 2) subsequently increases methane production at the expense of CO_2.

Four assumptions were made in developing the stoichiometric balance (Wolin, 1960): (1) all excess hydrogen appears as methane, alternative sinks described below, not being considered; (2) microbial yield is not considered; (3) synthesis of microbial cells would provide a sink for hydrogen in protein and lipids that are more reduced than carbohydrate; (4) oxidation products in feed that may serve as electron acceptors, such as nitrates, sulfates, or unsaturated compounds, are ignored. Oxyanions are generally reduced to the hydride form: sulfate to H_2S and nitrate to NH_3. The final result is that the net balance of methane production will likely be less than the theoretical calculated value. Suppression of methanogenesis will make alternative sinks more important.

The Wolin calculation ignores the production of cells and the fermentation of noncarbohydrate substrates. These products and substrates can be considered in larger, more complex models with more degrees of freedom. These models generally require

Table 16.7.   Theoretical stoichiometric carbon-hydrogen balance equations describing conversion of glucose in the rumen

Case 1
glucose → 2 acetate + 2 CO_2 + 8 H
glucose → butyrate + 2 CO_2 + 4 H
glucose + 4 H → 2 propionate
CO_2 + 8 H → CH_4 + 2 H_2O

Net 3 glucose → 2 acetate + butyrate + 2 propionate + 3CO_2 + CH_4 + 2H_2O

Case 2: Acetate production increases threefold and propionate and butyrate are unchanged:
3 glucose → 6 acetate + 6 CO_2 + 24 H
glucose → butyrate + 2 CO_2 + 4 H
glucose + 4 H → 2 propionate
3 CO_2 + 24 H → 3 CH_4 + 6 H_2O

Net 5 glucose → 6 acetate + butyrate + 2 propionate + 5 CO_2 + 3 CH_4 + 6 H_2O

Source: Baldwin, 1970.
Note: In Case 1 the acetate-to-propionate ratio is 1:1 and the methane-to-glucose ratio is 1:3; in Case 2, acetate-to-propionate is 3:1, and methane-to-glucose is 3:5.

computers and can provide interesting perspectives and insights into the problem of rumen balance (Rice et al., 1974).

Production of hydrogen gas can occur if the hydrogen concentration rises about 10^{-3} atmospheres (Figure 16.7). This sometimes happens when rumen production of methane is suppressed. Hydrogen can also be disposed of by deposition in unsaturated lipids, although this does not represent a very important sink because the amount of unsaturated fat in the diet is seldom great enough to offer sufficient capacity, and also excess unsaturated acids inhibit methanogenic fermentation and induce a rise in propionate (Section 16.9).

The general application of carbon balance for accounting for 2C–3C ratios and methane outputs is subject to some limitations. Any channeling of available hydrogen into products other than propionate (or odd-carbon fatty acids) will cause the system to underpredict propionate or overpredict methane outputs. These other products might include lipid or protein synthesis in microbial growth or other reduced products such as hydrogen, H_2S, and nitrates reduced to ammonia.

Microbial species and their respective metabolisms have a great deal to do with fermentation balance. Thus it is likely, for example, that high production of propionate is related to metabolism of lactate and its rate and amount of production. Propionate is produced by two pathways, the usual one via succinate and from lactate via acrylate, the latter being much enhanced under conditions of high-concentrate feeding. Under these circumstances *Megasphaera* and allied species that possess the acrylate pathway become important utilizers of lactate. As already noted, ciliated protozoa engulf starch particles and thus are another moderating factor.

Fermentation rate determines the potential rate of hydrogen supply. It is significant that most dietary problems resulting from high propionate levels are associated with high-concentrate diets, which cause fast rates of fermentation. On the other hand, one would expect that it might be more difficult to alter VFA ratios on poor-quality, slow-fermenting diets by antimethanogenic substances, since in these cases it is easier to dissipate the slower production of reducing units (NADH) into the various competing pathways.

Fast-fermenting carbohydrates do not always promote lactic acid and propionate production. The main carbohydrate source for lactate is starch (and possibly sucrose, which can be associated with butyrate production). Antimethanogenic compounds can promote both propionate and butyrate production at the expense of acetate (Armentano and Young, 1983). Pectin is usually fermented faster than starch, but the rumen organisms that use pectin yield no lactic acid and give high yields of acetate (Strobel and Russell, 1986). It is for this reason that high-pectin (and also hemicellulosic sources) feeds such as sugar beet pulp are valuable for reducing rumen acidosis (see Chapter 15) while still maintaining the fast and efficient rumen digestion associated with concentrate feedstuffs. The net increase in overall rumen efficiency if acetate and methane are diverted to propionate and butyrate is about 6% (Armentano and Young, 1983).

16.8 Microbial Efficiency of Growth

Anaerobic fermentation converts substrate into microbial cells and chemical products. Heat loss (5–7%) is probably not an important variable. Microbial efficiency, which is defined as the proportion of substrate energy fixed into cells, is inversely related to volatile fatty acid production. This efficiency does not necessarily relate to the efficiency of the animal host, since the fermentation acids are available to the animal while methane is not.

Microbial yield determines the microbial protein available to the animal. The net metabolizable protein received by the animal is the sum of the true digestible microbial protein and the feed protein that escapes the rumen. Chapter 18 discusses ruminant nitrogen metabolism in more detail.

Microbial yield is calculated as grams of cells per mole of glucose or per 100 g of fermented feed. The expression "per unit feed digested" is less standardized and not without problems (see below). The yield per mole of glucose is expressed as $Y_{glucose}$. Fermentable feed may be converted by calculation into estimated glucose units (Wolin, 1960). The microbial yield per unit of glucose is divided into two components: the ATP yield and Y_{ATP}. The ATP yield is moles of ATP formed from the fermentation of one mole of glucose, and Y_{ATP} is grams of microbial cells formed from one mole of ATP.

The yield and source of ATP for mixed rumen organisms is not known, although it may be fairly well defined for individual species grown in pure culture. The synthesis of one mole of acetate is thought to produce two ATPs, and one mole of propionate synthesized via the succinate route produces three ATPs; the acrylate route may only produce one ATP, although this is not certain. Butyrate formation may produce three ATPs per mole, and the formation of a mole of methane produces one ATP. In a normal mixed culture of rumen organisms producing methane, the yield from one mole of glucose may be four ATPs depending on the products formed and the balance of rumen species (Bergen and Yokoyama, 1977).

16.8.1 Maintenance

Rumen microbes live in a homeothermic environment provided by the host, and as de facto homeotherms their energy metabolism involves maintenance costs of metabolism and efficiencies of cellular growth above maintenance. Like that of the host, microbial efficiency is affected by maintenance requirements and substrate supply above maintenance. Obligate rumen anaerobes tend to have lower maintenance requirements and higher Y_{ATP} efficiencies for growth compared with facultative organisms in the intestinal tract, perhaps because they use preformed peptides and spend less energy on regulating intracellular pH. The net efficiency of cell growth for mixed rumen cultures is about 19–20 g of cells per mole of ATP (Hespell and Bryant, 1979). This value is substantially higher than the often quoted Elsden constant of about 11–12 g. Because maintenance cost varies, actual rumen yields of $Y_{glucose}$ fluctuate. Maintenance cost varies with species and environment. Maintenance cost of *Streptococcus bovis* is six times that of *Fibrobacter succinogenes*. Rumen pH is a major environmental factor affecting yield.

16.8.2 Measuring Yield

The output of microbial cells relative to the utilization of a substrate has been measured in vitro through controlled fermentation studies in chemostats, and in vivo by means of marker and labeling systems. The in vitro measurements allow a much greater degree of control and precision than in vivo ones. In vitro measurements cannot replicate the ecological conditions of the rumen, fermentor outputs thus are not identical with in vivo outputs.

Estimating microbial yield in vivo presents further problems of measurement and calculation. Microbial protein must be distinguished from dietary protein, and the amount of diet fermented in the rumen must also be known. In the context of rumen dynamics, estimates of rumen turnover or washout are also needed. Ingesta sampled at a point below the rumen and above the sites of absorption (usually with a duodenal fistula) provide information on microbial components and undigested feed.

Microbial protein can be separated from dietary protein by means of various procedures. These include feeding the animal a protein-free diet, using internal markers that occur only in bacteria, such as diaminopimelic acid (DAPA) or D-alanine, and measuring nucleic acids that occur in much higher concentration in bacteria than in the diet (Ling and Buttery, 1978). A more sophisticated (and complicated) separation method is the use of ^{35}S (Beever et al., 1974) or ^{15}N isotopes in inorganic compounds that are metabo-lized by bacteria into their amino acids. The isotopic composition of pure microbial protein must be known to calculate the ratio of bacterial to dietary protein in the outflow. This requires mass spectrophotometry or radioactive isotopes.

The measurement of microbial output relative to feed digested requires representative sampling of the rumen outflow. The principal error arises from the potential separation of liquid and particles through filtration effects. Often, only one marker is used, either particulate or liquid, while bacteria are distributed between particles and liquid, leading to variation in the ratio of bacterial mass to markers. Measuring both particulate and liquid passage could lead to more control of this variable as well as providing information regarding rumen dynamics, since bacteria attached to particles are more affected by particulate flow and those suspended in liquid are more subject to fluid washout.

Yield is often calculated as microbial crude protein per unit of apparently digested matter. Expression as crude protein is not always accurate because nonprotein nitrogen (NPN) in bacteria (i.e., nucleic acids and cell wall components [peptidoglycans] such as muramic acid and DAPA) make up 20–50% of the microbial nitrogen. The correction factor, 6.25, underestimates microbial mass because bacteria are about 10% nitrogen. Yield expressed on the basis of apparent digestibility means that the microbial yield is included in the undigested residue, and therefore the digestible matter that supported growth is increasingly underestimated as efficiency of microbial yield increases. This causes microbial yield to appear larger than it really is. The problem is resolved if yield is expressed per unit of truly digested matter. The literature on microbial yield presents a mixed situation, where yield is sometimes calculated per unit of apparent digested matter and, in other cases, per unit of truly digested matter. Comparison of literature values requires a conversion to a common base.

16.8.3 Variation in Yield

Historically, animal nutritionists have regarded microbial yield values as constant for a given organism or particular fermentation. The prejudice arose because of reliance on too many batch culture experiments. Elsden and Walker (1942) found that lactic acid–producing organisms obtained two ATPs from one mole of glucose (grown under glucose limitation) and produced about 12% cells from substrate, corresponding to a Y_{ATP} of about 7 g. Many values for rumen organisms are biased by the expectation that these limits apply to the rumen fermentation. Since 1970, how-

Table 16.8. Effect of dilution rate on fermentation products in continuous fermentor containing a mixed rumen population and using glucose as an energy source

Product	Dilution (turnover) rate (h)		
	0.02	0.06	0.12
Acetate[a]	1.18	1.11	1.13
Propionate[a]	0.16	0.22	0.26
Butyrate[a]	0.23	0.18	0.15
Methane[a]	1.67	1.34	1.04
ATP[a]	5.59	5.19	5.03
Y_{ATP}[b]	7.5	11.6	16.7
$Y_{glucose}$[c]	42.4	60.2	83.9
Nitrogen in cells (%)	9.9	—	12.0
g crude protein[d]/100 g glucose	14.5	—	34.0

Source: Isaacson et al., 1975.

[a]Moles produced per mole glucose fermented.

[b]g cells/mole ATP.

[c]g cells/mole glucose.

[d]Total cell N times 6.25. The data do not consider the proportion of true protein in cells.

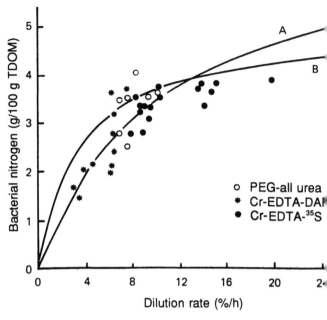

Figure 16.10. Relation between in vivo microbial yield and liquid turnover rate as summarized from the literature (courtesy of P. H. Robinson). Yield values have been recalculated uniformly to g N × 6.25 per 100 g true digestible organic matter (TDOM). Values obtained using Cr-EDTA and diaminopimelic acid (DAPA) are denoted by an asterisk, and those obtained with polyethylene glycol and all-urea diets are open circles. Line A is the best fit to Michaelis-Menten kinetics: $1/y = 0.14 + 0.015(1/x)$, $R^2 = 0.76$. Line B is the regression obtained by van Nevel and Demeyer (1979) using a chemostat. Note that overestimation of rumen turnover by the liquid markers could lead to substantial error at low dilution rates but negligible error at high dilution rates. Maximal yield (y_g) and maintenance (M) cost can be calculated from the above regression according to the equation $1/y = (M/K) + (1/y_g)$ (Hespell and Bryant, 1979), where K is the dilution rate. Maintenance (slope of regression) is 0.015 g N/100 g carbohydrate fermented; and maximum yield is 7.11 g N/100 g carbohydrate fermented.

ever, yields up to three times greater than that found by Elsden have been obtained.

These more recently obtained values are sufficiently variable that they challenge the concept of a constant microbial yield. Hespell and Bryant (1979) designed a model and a basis for expecting variation in microbial yield. Major factors influencing microbial yield appear to be rate of fermentation and washout or passage rate. These factors impinge on the maintenance cost and the efficiency of microbial growth above maintenance. Cell yield efficiency increases as the dilution (washout) rate is increased. The fundamental principle is that the mean age of the microbial population is decreased at higher washout rates, and the younger population is subject to lower death and predation rates. Dying cells are more likely to be retained in the fermentation and their nutrients recycled or, alternatively, to be washed out and counted as part of the yield. Recycling in the fermentation is diminished by increased turnover of the fermentation up to at least 12% dilution rate. Theoretically, very high rates may cause plateauing and inefficiency through loss of unfermented substrate and competition of washout with cell generation time. High rates of passage induce escape, a loss that reduces digestibility.

Microbial yield increases as dilution rate is increased (Table 16.8, Figure 16.10). In vivo measurements generally agree with results from continuous fermentors. Observed values are in the range of 20–45 g cells/100 g carbohydrate fermented. The yield and productivity of rumen bacteria can be related to the concept of maintenance. Mature vegetative cells producing products but no longer growing may be likened to a mature animal fed at a maintenance level. The turnover rate required for optimal microbial output is analogous to the increased feed intake required for above-maintenance production. The composition of

bacteria varies with increased turnover in the fermentation (Table 16.8) as the proportions of products vary.

While published values for microbial yield do reflect problems in measurement, there can be no doubt of the real variation in yield caused by diet and feeding conditions. Averages of microbial yields for different classes of diets show marked variation. Silages and high-concentrate diets show lower yields than do forages or mixed diets. Purified diets are highly variable, probably indicating the wide variation in quality that falls under this description. Properly supplemented forage diets may give higher yields than either concentrate supplement or unsupplemented forage (Chamberlain and Thomas, 1979). Increased intakes increase both yield and escape through the passage effect, thus confounding yield and rumen escape effects. The lower values associated with concentrate diets may reflect slower rates of passage relative to digestion rate than are probably characteristic of high-grain diets. Since cell wall intake increases passage, addition of forage to high-concentrate diets is likely to increase microbial efficiency and offers the possibility that efficiency of microbial protein output might be increased considerably if the rumen turnover and nutrient supply could be regulated properly.

In the context of the interactions among yield, escape, and balanced nutrition of rumen organisms, these reported effects are open to an alternative interpretation; namely, that the increase in metabolizable protein is due in part to an increase in microbial yield. Such variation might also be considered in light of attempts to define a microbial constant and to view responses to added high-quality protein as evidence of rumen escape (Burroughs et al., 1975; Satter and Roffler, 1975). For example, the addition of casein to a purified diet increases microbial yield (Hume et al., 1970). A dynamic modeling of the response of metabolizable protein to feed intake in dairy cattle predicts microbial yield to contribute more to protein than escape does (J. B. Russell et al., 1992); this has also been observed experimentally (P. H. Robinson, 1983).

16.8.4 Composition of Rumen Organisms

Bacteria and protozoa that pass from the rumen to the lower tract represent a major portion of the host's diet and the largest part of the protein nitrogen supplied to the animal. The composition of this microbial matter is thus an important variable in the animal's nutrition. The indigestible portion of microbes is the cell wall membrane. In bacteria this is composed of peptidoglycan. Protozoa, being larger organisms with less surface area to total mass, are more digestible than bacteria. Microbial cells also contain about 10% nucleic acids (Table 16.9), which leaves roughly one-half to two-thirds of the microbial nitrogen in the form of true protein. (Values for amino acids in bacteria are given in Table 18.7.)

The habit of presenting total microbial nitrogen as crude protein (N × 6.25) is not satisfactory for estimating cell yields or for evaluating the nitrogen as protein available to the animal host. The distribution of nitrogen in protein, RNA, DNA, and cell wall membrane (peptidoglycan) varies among rumen species and within individual animals depending on conditions of growth. About one-third of microbial nitrogen exists in the indigestible cell wall, which has a nitrogen content of only 7% (requiring a correction factor of 14). The nitrogen content of true microbial protein is about 15%. Thus, using 6.25 as a correction factor underestimates net organic matter. Since determination of microbial cell composition yields so much more meaningful data and the analytical techniques are rather well established, a more complete description of microbial nitrogen ought to become standard (Hespell and Bryant, 1979). Unfor-

Table 16.9. Composition of microbes (on a dry matter basis unless otherwise indicated)

Constituent	Bacteria Probable[a]	Bacteria Range	Protozoa Range
Total nitrogen	10[b]	5.0[c]–12.4[d]	3.8–7.9[d]
True protein	47.5[e]	38–55	—
RNA	24.2[e]	—	—
DNA	3.4[e]	—	—
Lipid	7.0[e]	4[f]–25[e]	—
Polysaccharide	11.5[e]	6–23[e]	—
Peptidoglycan	2	—	0
Nitrogen digestibility	71[g]	44–86[g]	76–85[h]

[a]Many discordant values have been recorded, possibly reflecting contamination or inclusion of plant material.
[b]Isaacson et al., 1975.
[c]R. H. Smith and McAllan, 1973.
[d]Weller, 1957.
[e]Summarized by Hespell and Bryant, 1979.
[f]Abdo et al., 1964; also reported 6% crude fiber.
[g]Bergen et al., 1968; values as percentage of total N.
[h]Bergen et al., 1967.

tunately, total N × 6.25 is still a commonplace calculation in the rumen literature. These problems should be kept in mind in evaluating published microbial yield and composition data.

Rumen organisms contain significant quantities of lipid, storage carbohydrate, and minerals. Like the cellular contents of forage, microbial cell contents are entirely available to the animal and represent an important transfer of food energy from the rumen. The microbial starch may be a significant source of carbohydrate for the host and may reduce the requirement for glucose derived from gluconeogenesis. Microbial lipids are also significant; many unusual long-chain fatty acids in ruminant lipids are derived from rumen microbes.

Two important variables affecting the determination of microbial composition are stored carbohydrates and the problem of isolating rumen organisms free of plant particles. Storage of carbohydrates will lower nitrogen content of organisms and is probably the major factor influencing variation in protozoal nitrogen (Weller et al., 1958). Feed contamination is probably a more serious problem, particularly when centrifugal techniques have been used. Rumen ingesta contain many feed particles in the range of bacterial size (Van Soest, 1975; Pichard, 1977), including highly lignified cell wall fragments and fragments of chloroplasts. One must treat data reporting a crude fiber content of rumen bacteria (Abdo et al., 1964) or high ash values (R. H. Smith and McAllan, 1973) cautiously, since pure bacteria are essentially soluble in neutral or acid detergent, while fine plant particles from ingesta may have a high fiber and ash content.

Isaacson et al. (1975) conducted experiments with purified substrate in a chemostat and found that nitrogen in microbial matter increased by more than 20% as turnover increased from 2 to 12%. Faster turnover of fermentation contents decreases maintenance and recycling of microbial matter in the fermentation because washout of microbes increases at the expense of predation. Microbial cell wall (of lower nitrogen content than the cell contents) from cultures with low turnover may dilute the microbial matter and be partly responsible for the lower nitrogen content of microbial matter. These walls are refermented by other bacteria in the ecosystem; however, refermentation does not greatly alter the metabolic output.

Pittman et al. (1967) suggested that stored material may include peptides in addition to starch. Microbial strains (e.g., *Bacteroides*) that utilize peptides without hydrolysis might enhance their apparent yield by such storage. Peptides appear to be catabolized rapidly to deaminated products, ammonia, and CO_2, and bacterial amino acids and peptides are resynthesized from the carbon skeletons, CO_2, and ammonia.

16.9 Factors Influencing Rumen Fermentation

The amount and composition of the diet are external variables that affect rate of digestion, rate of passage, and therefore turnover of the rumen contents. The intake of the animal is set by its needs and the composition and availability of feed. Dietary composition generally determines the distribution of the microbial population that digests feed nutrients in the rumen. Thus, high-protein diets favor proteolytic organisms, and high-starch diets that are low in fiber are associated with a large population of starch utilizers. Whether cellulolytic organisms occur in reduced numbers in animals fed a high-concentrate diet will depend on fiber particle size and passage rate. With a small amount of coarse forage, passage of fiber may be slow and cellulolytic organisms comparatively numerous through cumulative retention of the substrate (Bryant and Burkey, 1953). Diets low in fiber tend to have high rates of digestion and acid production, which place a greater load on the rumen buffering system. Such conditions favor species capable of rapid metabolism and tolerant to some downward drift in rumen pH. Generally, cellulolytic and methanogenic organisms are less tolerant to such changes and may decrease in number (Slyter, 1976).

Rapid rates of fermentation depend on sufficient nutrition for rumen microorganisms. Adequate external sources of nitrogen, sulfur, and essential minerals are required for optimum carbohydrate utilization. Sodium, potassium, phosphate, and bicarbonate contribute to the buffering system and are recycled through saliva. Urea is hydrolyzed to ammonium bicarbonate, which also contributes to buffering. If any essential nutrient is lacking or the rumen environment is not optimal, the rate of digestion will slow and either intake or digestibility will decrease. An important consideration here is whether limiting nutrients influence the animal or its bacteria first; that is, whether the requirement of the host or the bacteria is most limiting. The same principle applies to the influence of fermentation inhibitors on intake and digestibility. The most important factors influencing rate of digestion are the intrinsic properties of the feed carbohydrates and protein.

The overall rate of passage is affected by rumination of plant cell wall, the passage rate of generated fine particles and rumen organisms, and the washout of liquid as affected by osmotic pressure (Chapter 15). The washout rate from the rumen may be considered to be in competition with the generation time of rumen organisms. Slow-growing species may be washed out in a high-intake, high-rate-of-passage situation. They may survive, however, by attaching themselves to slower-moving fibrous matter or in pockets of ingesta that have a longer turnover time than the main rumen mass. High rates of passage promote faster digestion rates in the rumen through selective removal of the more slowly available substrates (Chapter 23). Removal of these substrates is likely to affect microbial species selectively. It is usually cellulolytic digestion that is affected.

Fine grinding and pelleting of feed increases dietary density and intake and promotes more rapid passage of insoluble matter. If the entire diet is ground, the floating mat normally found in forage-fed rumens is destroyed, allowing the easier passage of coarser particles. Selective retention of fiber (e.g., coarse lignified particles for further digestion and rumination into finer matter) is a normal ruminant function that tends to be obliterated in animals fed ground or high-concentrate diets.

16.9.1 Diet, Rumen Acids, and Methane

The methanogenic bacteria are the most sensitive to changes in rumen environment and are affected by many dietary factors. Since the methanogens are the principal utilizers of hydrogen, their welfare affects the entire rumen metabolism and the carbon balance. Reduction of methanogenesis tends to promote hydrogen production and shifts the carbon balance toward propionate, which requires less transport of hydrogen (Section 16.7.1). A considerable list of dietary variables is associated with a rise in propionate (Table 16.10). There is an increase in energy efficiency when the rumen propionate-to-acetate ratio is increased. Part of this effect is due to the decrease in methane and the fact that most diets producing this shift tend to be consumed at high intakes. The problem of acetate-to-

16.10. Factors influencing the proportion of rumen propionate
...so probably depress methanogenesis

Dietary variable	Animal response	Factor affecting methane-producing bacteria
...sed intake	—	Increased passage
...d pelleted feed	Depress rumination, increase intake	Increased passage
...tarch or grain	Increase intake	Lower rumen pH, rapid rate of fermentation
...d or gelatinized feeds	Increase intake	Rapid rate of digestion
...urated oil	Reduce intake	Toxic?
...d pearl millet	Reduce intake	Toxic?
...ounds with haloge-...d methyl groups	Possible drug action	Inhibition
...nores[a]	Increased efficiency	Decrease in hydrogen producers that supply methanogens

...rces: Van Soest, 1963a; Davis and Brown, 1970; Czerkawski, 1972.
...onensin, Lasolacid.

propionate ratio and metabolic efficiency is discussed in Section 24.6.6.

The factors influencing methanogenic bacteria include those that create a less favorable rumen environment for methanogenesis through increased rate of passage and rate of digestion, depression of rumination, and depression of rumen pH. These conditions favor propionate-producing bacteria over acetate producers, making less hydrogen available to methanogens. Increased rumen turnover may also restrict methanogens through competition with generation time. Ionophores also limit hydrogen transport and availability to methanogens. True antimethanogenic drugs inhibit methane production per se and consist primarily of halogenated methanes or halogenated methyl derivatives.

In general, animal response to methanogenic inhibition depends on channeling carbon and hydrogen, which ordinarily appear as methane, into propionate, thereby increasing the metabolizable energy available to the host. The influence of methanogenic inhibition on microbial yield is complicated and in need of more study. It appears, however, that agents with antimethanogenic ability are general microbial inhibitors and may reduce both cellulose digestion and microbial yield (van Nevel and Demeyer, 1979). Inhibition of cellulose digestion would reduce metabolizable energy and offset any advantages of increased propionate. On the other hand, general inhibition would promote rumen escape of protein and available carbohydrate. The advantages of antimethanogenic agents are likely optimal in high-quality diets and minimal on diets high in digestible cell wall and NPN.

The total metabolizable microbial products are distributed between cell yield (protein) and VFAs. Suppression of methanogenesis will cause the loss of some ATP that may be compensated for by ATP derived in propionate production. Nevertheless, there may be a resultant decrease in microbial protein because high-grain diets tend to have lower microbial yields than do high-forage diets (Table 14.11); however, this may not be caused by a lesser methanogenesis, only associated with it.

Another problem with suppressing methanogenic bacteria is that alternative hydrogen sinks will compete with propionate. Such sinks may include cellular synthesis (particularly of lipid), hydrogen production, and multiplication of other rumen organisms that utilize hydrogen. Increased hydrogen production could still entail a loss of energy to the animal. The amount of hydrogen produced is usually less than the reduction in methane output, the difference being accounted for in the movement of hydrogen into other products such as propionate or lipids (Bryant and Wolin, 1975).

Shifts in the balance of rumen acids may be associated with the protozoal population as well as with methane-producing bacteria. In earlier studies of the effects on rumen propionate of feeding high-grain and pelleted rations, animals did not always respond in the expected manner. Instead of high propionate levels, these animals showed high butyrate levels (Jorgenson and Schultz, 1965). Eadie and Mann (1970) showed that ciliated protozoa are associated with butyrate production (Section 16.3.4). Perhaps the protozoa provide the fermentation with additional hydrogen sinks through lipid synthesis. Adding acetate to high-grain diets often results in increased butyrate production. This is yet another example of the general principle that the appearance of a new product elicits a fermentation that will make use of it.

The decline in methane-producing bacteria in situations of high substrate turnover may seem to associate greater efficiency with lower methanogenesis; however, it is the turnover that reduces the relative demands of microbial maintenance and increases Y_{ATP}, although ATP yield from glucose may decrease. Therefore, the effects of methanogenic inhibitors have to be considered in the context of rumen turnover.

Dietary Changes

The abruptness of a dietary shift is a major factor in determining the degree of perturbation of the rumen fermentation and potential digestive upset. Abrupt changes in diet cause the organisms present to shift their fermentation balance; this is followed by adjustments of the microbial species to the new situation. In the case of abrupt shift to a higher-quality diet, the imbalance of microbial species may open the door to adventitious facultative organisms that seek to dominate the fermentation through acid production and reduction in rumen pH, leading to rumen upset. The new diet usually imposes new rates of fermentation and rumen turnover, which are major factors in the adjustment. These events usually take about a week (Grubb and Dehority, 1975). Longer-term adjustments can induce microbial mutations that take several weeks to achieve a balance (Hungate, 1966). It usually requires several weeks on a high-concentrate diet for a lactating cow to develop milk fat depression.

16.9.2 Antimethanogens: Ionophores

Agents that inhibit methanogenesis may produce results by affecting methanogens directly or by affecting the syntrophic groups on which they depend. Thus, halogenated methyl compounds (chloral hydrate dibromomethane, etc.) may inhibit methanogens, while other agents, such as chlorate, may disturb the oxygen balance and be generally inhibitory. Many of these agents also inhibit cellulose digestion (Prins and Seekles, 1968).

Ionophoric agents such as Monensin or Rumensin and Lasolacid probably work by inhibiting hydrogen producers such as *Ruminococcus* and *Butyrivibrio* and thereby favor propionate producers. Ionophores are chelators of sodium, potassium, and other cations that inhibit hydrogen production by limiting counterflow of sodium and potassium through microbial membranes (M. W. Smith, 1990). The suppression of hydrogen transport diverts carbon from acetate production toward propionate production and increases irreversible acetate loss in the synthesis of butyrate (Armentano and Young, 1983). The reduction in available hydrogen causes the decrease in methanogens (Chen and Wolin, 1979). Monensin probably acts similarly to Lasolacid in that it appears to inhibit lactate producers, thereby reducing acidosis in animals on high-grain diets (B. Scott et al., 1979). Rumensin also retards proteolysis, leading to increased escape of dietary protein and probably decreased microbial yield. Anabolic agents would thus seem to be most successful in inhibiting methanogenesis in animals on high-quality diets. Methane inhibition should have a beneficial effect in situations in which energy is limiting and the animal benefits from the diversion of feed energy from methane to propionate. Nutritional situations in which protein is limiting or dietary efficiency depends on efficient use of NPN would respond negatively or not at all to ionophoric agents (van Nevel and Demeyer, 1979).

The energetic efficiency of methanogenic inhibition depends on the diversion of feed energy from methane to propionate and escape of available carbohydrate and protein from the rumen. If metabolizable protein is unaltered, the nutritive ratio is widened and any animal response is dependent on increased metabolizable energy. Growing beef cattle do not have high protein requirements. Except for very young animals their requirements are probably less than the microbial yield (see Figure 18.8), and they are more limited by energy intake than by protein. It therefore fits that this class of animals is probably the most responsive to methanogenic inhibition. Dairy cattle are not as responsive because of the antilipogenic properties of propionate.

Finally, it must be noted that most current methods of manipulating the rumen environment, whether by pelleting or by feeding grain, protected ingredients, or anabolic agents, act through reducing fermentive efficiency such that the ruminant is more directly dependent on and responsive to dietary quality and, in particular, to proteins. A consequence of this management is that it brings ruminants into competition with people and nonruminants for dietary resources. The future of ruminants as utilizers of noncompetitive food resources depends on maximizing rumen efficiency and output. Even if fermentation inhibitors allow the escape of potentially digestible matter from the rumen, this effect may go unnoticed or even appear to be beneficial because the escaped nutrients may still be digested efficiently in the lower tract. The effect on forage diets is uncertain, however, because the beneficial effects of rumen escape are limited to protein, and passage of available cellulosic carbohydrate will be a net loss.

17 The Lower Gastrointestinal Tract

The lower part of the ruminant gastrointestinal tract receives much less attention than the reticulorumen. Judging from the emphasis, one would think ruminants had no lower tract. Comparative studies of ruminant and nonruminant herbivores (Argenzio and Stevens, 1984; C. E. Stevens, 1988) do provide some understanding of the functions of the ruminant lower tract and the ruminant-like fermentations of the cecum and large bowel. The general aspects of lower tract digestive physiology are covered extensively elsewhere (see Swenson and Reece 1993; L. R. Johnson, 1987). Studies of the lower digestive tract have been dominated by physiologists with limited appreciation for nutrition, and the available literature deals more with nervous control and gut motility relative to flow than with any quantitative assessment of the contribution of digestive capacity and absorption in the respective compartments. This chapter emphasizes the peculiarities of the ruminant lower tract relative to other species, although I have resorted to the literature on nonruminants where necessary.

The deficiency of specific information about the ruminant lower tract leaves the door open for speculation. Some of my speculations could be wrong, and the reader is cautioned to distinguish between direct observations and speculation intended to stimulate more work in a neglected area.

17.1 Overview

Lower tract digestion in ruminants is comparable in many respects to that of nonruminants and performs the same functions of digestion and absorption of nutrients. A number of exceptional conditions apply in ruminants, however, arising from the consequences of the pregastric fermentation in the reticulorumen. Certain food constituents, such as carbohydrates, are normally in very low supply in the lower tract due to their removal by fermentation in the rumen, and only very small amounts reach the abomasum. The digesta are also in a relatively dilute medium. Other components, such as fat and protein, have been greatly altered by

fermentation. Fats have been hydrolyzed and glycerol fermented, and the partially hydrogenated fatty acids are present as soaps, particularly as calcium and magnesium salts (Chapter 20). Microbial protein is ordinarily the dominant nitrogenous constituent of the ruminal outflow (Chapter 18). The removal of fermentable carbohydrates from dietary fiber by rumen fermentation decreases the role played by the cecum and colon relative to this function in hindgut fermenters.

The fluidity of the digesta may make increased capacities for absorbing water and small molecular species necessary. The flow problem may be ameliorated by the larger reservoir of the reticulorumen, which tends to diminish the pulsative effect of meals such that there is a more or less continuous flow of digesta into the lower tract unlike the pulsative flow seen in monogastrics (K. J. Hill, 1961). This contrast is more pronounced in the abomasum and duodenum and becomes less important as ingesta proceed down the digestive tract. Digestive enzymes, secreted more continuously in ruminants, may not reach the concentrations observed in monogastric species. Likewise, tissue concentrations of enzymes tend to be lower in ruminants than in nonruminants. In contrast with nonruminants, ruminants show no rise in blood glucose after eating, and gluconeogenesis increases.

The small intestine of a ruminant grazer on its normal forage diet receives very little available carbohydrate in the form of sugars and starch, and ruminants appear to have less carbohydrase capacity than monogastric species. Most of the fermentable matter has already been extracted in the rumen, so the grazing ruminant's small intestine receives only the more slowly fermentable matter that has escaped digestion. Ruminant concentrate selectors may not be so extreme in this respect (Hofmann, 1988).

Ruminants can be compared both with other ruminants and with nonruminants, including foregut and hindgut fermenters (Chapter 5). Ruminant and nonruminant anatomies are compared in Figure 5.3.

Ruminant species that evolved in arid or semiarid environments have a longer colon, which aids in the resorption of water. Concentrate selectors have a larger

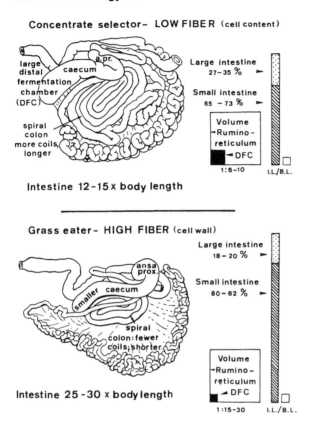

Figure 17.1. Anatomical differences between midguts and hindguts of two feeding types (from Hofmann, 1988).

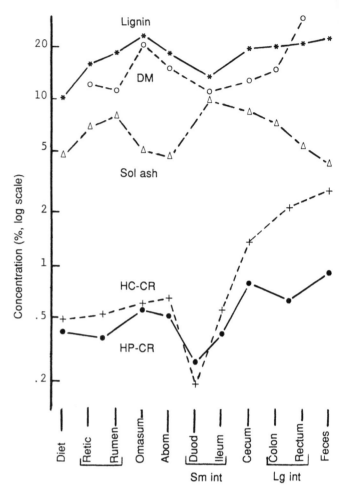

Figure 17.2. Composition of the ingesta in the compartments of the sheep digestive tract. Lower portion of the figure shows the relative concentrations of chromic oxide marker in two diets, pelleted ground alfalfa (HP-CR) and high concentrate HD-CR (Vidal et al., 1969). Upper portion shows values for lignin, soluble ash, and dry matter (DM) for sheep fed coarse Rhodes grass (*Chloris gayana;* Rogerson, 1958). Changes in lignin and chromium oxide are comparable since the vertical scale is logarithmic. Increases in marker concentration indicate digestion and absorption; decreases indicate dilution by secretion. The latter effect is particularly evident at the duodenum-ileum. Note the high lignin contents in the rumen and omasum of sheep fed coarse forage. This probably reflects selective retention of the coarser, more lignified fiber.

lower tract and a larger small intestine (Figure 17.1) than grazer species (e.g., buffalo). This observation can be connected with a number of problems of temperate cattle, which, like the buffalo, are basically grazers. Intensively fed cattle (beef cattle in feedlots and high-producing dairy cattle) are fed a great deal of concentrate, which taxes their ability to utilize the more resistant starches. Hoffman (1988) classified concentrate selectors as intermediate between nonruminants with pregastric digestion and the more typical grazers. This range of capability also exists for the pregastric fermenters, depending on the degree to which they rely on pregastric digestion.

17.1.1 Passage and Mixing

The volume of reticuloruminal contents is determined by the quantity of feed and water ingested plus salivary dilution and is diminished by absorption. The flow of fluid and selected finer particulate matter past the omasal orifice leads to further secretions, peptic digestion, and absorption. Secretions dilute the ingesta, and digestive absorption and resorption concentrate undigested fiber and lignin. These details have been studied in sectioned tracts of slaughtered animals using inert markers such as chromium and lignin (Figure 17.2). A drop in concentration of marker is indica-

tive of secretions. The ratios of nitrogen, dry matter, and minerals give some idea of the composition of the secretions. Absorption generally results in an increase in the concentration of markers. The small intestine appears to be very efficient in resorption of secretions, a process continued in the colon.

The flow of ingesta beyond the duodenum tends to be plug flow, which minimizes pooling of contents and is driven by the peristaltic contractions that push ingesta down the tract;[1] however, the propulsive forces may not act equally on liquid and particles of differing

---

[1] Plug flow assumes minimal pooling of ingesta, so sequentially administered labeled particles should appear in the administered order at the end of the flow. In kinetic flow ingesta are pooled and

Table 17.1.   Rates of passage of ^{14}C-labeled wheat plant fed to sheep with respect to particulate fractions

Source of cell wall	Percent of particles > stated size of screen (μm)								
	840	595	420	297	210	149	105	74	Pan (<74)
Differential rate of change/day (dpm ^{14}C/g CW)									
Ruminal	0.82	0.65	0.50	0.51	0.48	0.46	0.37	0.41	0.39
Duodenal	0.57	0.52	0.50	0.40	0.39	0.41	0.40	0.40	0.42
Fecal	0.54	0.38	0.43	0.38	0.34	0.35	0.43	0.28	0.30
Relative rates of disappearance									
Rumen to duodenum[a]	0.1	0.3	0.5	0.9	1.2	1.5	1.7	1.8	3.5
Duodenum to feces[b]	1.0	1.0	1.7	1.1	1.0	0.8	0.9	0.8	0.8

Source: L. W. Smith, 1968.

[a]Ratio of proportional fraction size in the duodenum to that in the rumen. The rumen retains coarser particles and contains proportionally more than postruminal sites.

[b]Ratio of proportional fraction size in feces to that in the duodenum. There is relatively little sorting of particles in the lower tract of ruminants.

sizes. The push on the boli may involve filtration effects such that liquid can gush backward while coarser matter is held by filtration effects and is pushed forward. This phenomenon, called *streaming* in human nutrition, causes liquid markers to have broader excretion curves relative to insoluble particulate matter. Finer particles may tend to follow the liquid. The data shown in Table 17.1 suggest somewhat more rapid passage of coarser matter than of fine particles. This could be a composite of the physical effects of selective filtration and rumination in the rumen; that is, fine lignified particles represent the principal extracted fraction and the coarser particles are a minority that escaped the process. These larger particles thus present more potentially fermentable matter to the colon. The sorting of matter in the lower tract is minor compared with the output from the rumen, where the selective retention of coarser matter induces great disparity in passage rates (Table 17.1).

Continuous recording in hay-fed sheep showed diurnal variation in passage; 50% more flow occurred at night than during the day. This effect is related to ruminating patterns, which are more dominant in nighttime (Ruckebusch, 1988).

Flow through the tubular small intestine is essentially plug flow, which allows more efficient absorption per unit of retention time. Passage through the lower tract generally involves less mixing and sorting of ingesta compared with the rumen, although some mixing occurs in the abomasum, cecum, and colon. The cecum and colon in monogastric herbivores (e.g., horse and rabbit) do show kinetic pooling. Very little information is available on the cecum and colon of sheep and cattle, although Ruckebusch (1988) suggested that colonic flow in the sheep is sequential relative to the order of administered labeled particles, and thus is possibly plug flow.

17.1.2 Regulation of Lower Tract Functions

The rumen, reticulum, omasum, and abomasum are specialized parts of a single primordial organ, so regulation of the abomasum is related to that of the reticulorumen. The reticulorumen and omasum and probably the abomasum depend mostly on neural control (Grovum, 1984); hormones become more important farther down the tract, from the duodenum onward.

The regulation of the digestive tract involves blood supply, innervation, and a complex of hormone secretions that affect gastric and intestinal secretions and interact with the nervous system to regulate muscle contractions. The blood supply is from the celiac artery, which usually originates in a common trunk with the cranial mesenteric artery from the aorta. There are six major branches of the celiac artery in ruminants: a small phrenic artery; the hepatic artery serving the liver and abomasum; the splenic artery, which detaches from the right ruminal artery (the main one for the rumen); the left ruminal artery serving the reticulum; the left gastric artery serving the abomasum, the omentum, and some parts of the reticulum and omasum; and several small pancreatic branches.

Beyond the abomasum, the ruminant circulatory system is more similar to that of nonruminant artiodactyls such as the pig. The initial part of the duodenum, the liver, and the pancreas are supplied by the celiac artery, and the rest of the duodenum, the small intestine, and part of the colon are supplied by the cranial mesenteric artery. The descending colon and most of the rectum are supplied by the caudal mesenteric artery; part of the rectum is served by the internal iliac artery. Most of the digestive tract is drained by the hepatic portal vein, which makes the liver the general metabolic center for absorbed nutrients (Hofmann, 1988).

The lower digestive tract is enveloped by outer longitudinal and inner circular layers of smooth muscle. Between these two layers lies the myenteric plexis, which, with the submucosal plexus constitutes the en-

passage depends on a random statistical escape from the pool that is related to concentration (see Section 23.5.3).

Table 17.2. Stimulatory effects of dietary components

Stimulus	Gastrin	CCK	Secretin	GIP
Peptides and amino acids	+ +	+ +	0	+
Fatty acids	0	+ +	+	+
Carbohydrates	0	0	0	+ +
Acid	I	+	+ +	0
Distension	+	0	0	0
Nerve (vagus)	+ +	0	0	0

Source: T. R. Houpt, Cornell University.
Note: CCK = cholecystokinin; GIP = gastrin-inhibiting peptide; I = inhibition.

teric nervous system (ENS). Mechanical stretch (i.e., distension) stimulates receptors in the submucosal plexus, causing increased muscle contractions that can be reinforced by bulky ingesta.

The extrinsic nervous system probably serves a regulatory function. Contraction patterns persist after section of the splanchnic and vagus nerves (Ruckebusch, 1988). The contractions and digesta flow are regulated by interactions of the entire nervous system and hormone secretions that involve feedback loops. These mechanisms interact with dietary components (Tables 17.2, 17.3, and 17.4). Table 17.4 is based on human medical data, although many of the factors are derived from comparative animal studies. The relevance of the total list to ruminants is relatively unexplored.

Tactile and pressure stimuli are probably of major importance in stimulating flow distally along the lower digestive tract and may inhibit flow proximally from that region. Cholecystokinin appears to regulate food intake. Fill pressure may involve both opiates (positive for intake) and gastrin (Tables 17.2 and 17.3), which are integrated in the overall interaction between the nervous system and hormonal regulation.

The passage of digesta down the digestive tract from one compartment to another is promoted when the more distal segment is empty and prepared to receive them. This has led to the suggestion that this mechanism regulates feed intake, but it is likely that this effect is but one of the factors contributing to net intake (Chapter 21). The interaction of the regulatory hormones and neuropeptides involved in propelling the digesta implies that food intake is in part controlled by

Table 17.3. Physiological action of secreted hormones

Physiological actions	Gastrin	CCK	Secretin	GIP
Acid secretion	+ +	I	I	0
Pancreatic HCO_3^- secretion	0	+	+ +	0
Pancreatic enzyme secretion	+	+ +	+	0
Bile, HCO_3^- secretion	0	+	+ +	0
Gallbladder contraction	0	+ +	0	0
Gastric emptying	+	I	0	0
Insulin release	0	0	0	+
Mucosal growth	+ +	+	0	0
Pancreatic growth	+ +	+	+	0

Source: T. R. Houpt, Cornell University.
Note: CCK = cholecystokinin; GIP = gastrin-inhibiting peptide; I = inhibition.

Table 17.4. Hormones of the digestive tract

Hormone	Secretory location	Action
Cholecystokinin	Duodenum, jejunum, ileum	Stimulates pancreas hibits food intake
Endorphins Enkephalins Histamine Bombesin Serotonin	Antrum, duodenum, jejunum, ileum, pancreas, colon	Opiate activity; hista is an antagonist[a]
Gastric-inhibiting peptide	Duodenum, jejunum, ileum	Inhibits absorption o ter and electrolyte
Gastrin	Antrum, duodenum	
Glucagon	Pancreas, ileum, colon	Inhibits gastric and creatic secretions erates motility
Insulin	Pancreas	Glucose utilization
Motilin	Duodenum, jejunum	Stimulates contracti fundus and antrur creases gastric e ing
Pancreatic polypeptide	Pancreas	Stimulates bicarbon cretion by pancre
Secretin	Duodenum, jejunum	Stimulated by prote
Somatostatin	Body + antrum, duodenum, jejunum, ileum, pancreas, colon	Inhibits acid secreti motor activity, pa ic secretion
Vasoactive intestinal peptide	Ileum, pancreas, nerve plexuses	Relaxation of esoph and pyloric sphin

Sources: Ganong, 1987; L. R. Johnson, 1987.
[a]Occupies receptor sites and reduces action.

feedback signals from the transpyloric flow that involve particular nutrients at these sites.

17.2 Development

The digestive tract beyond the omasum forms the major part of the total digestive tract of infant ruminants, which are functionally nonruminants in their digestion characteristics. Rumen development is stimulated by fermentation acids, particularly butyrate and propionate (Chapter 15). The relative sizes of the abomasum, small intestine, cecum, and colon also change as the animal develops (Figure 17.3). The cecum and colon grow at the fastest rate, followed by the small intestine and the abomasum.

Neonatal ruminants can close the esophageal grooves, thereby bypassing ingested milk to the abomasum. This ability can be maintained in adult animals through training. Hofmann (1988) said that concentrate selector ruminants may maintain this ability into adulthood. In these instances the need for selective passage of seeds and other concentrate-bearing matter is involved.

17.2.1 The Effect of Fiber in the Lower Digestive Tract

Ruminants begin to nibble at forage at an early age. While most of the attention paid to this activity has

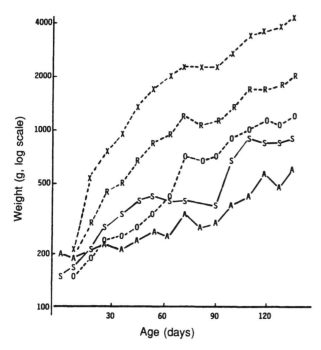

Figure 17.3. Digestive organ growth in lambs from birth to 10 weeks of age (modified from Wardrop and Combe, 1960). A = abomasum; S = small intestine, colon, and cecum; O = omasum; R = reticulum; and X = rumen. Data are plotted on a semilogarithmic scale to reflect proportional growth.

focused on development of the rumen, fiber apparently acts on the lower tract as well.

Feeding alfalfa to young growing pigs elicits the development of the colon, cecum, and small intestine; abomasal growth seems to be unaffected (Table 17.5). The effect of fiber on growth of the colon and cecum are generally attributed to VFAs, particularly butyrate and propionate, which are major energy sources for these organs and stimulate their growth. Insoluble coarse particles might stimulate the gastrointestinal wall (the so-called scratch factor), and insoluble undigested fiber may be responsible for the sweep and flow through the digestive tract. Human dietary fiber studies show that undigested fiber is responsible for relieving constipation. Only undigested matter promotes passage (Van Soest, 1988). On the other hand, the fermentable fiber promotes microbial fermentation and growth, which is in turn affected by the digestive flow and turnover.

Wanderley et al. (1985) studied steers fed high-grain and forage diets and found that indigestible marker flow was markedly affected in the high-forage diet (Table 17.6). The effects of fiber on the small intestine are not entirely known. The fact that fiber has a significant effect should be stimulus for further work.

17.3 The Abomasum

The ruminant abomasum is unusual because it features relatively large spiral folds whose function is not

Table 17.5. Effect of level of dietary alfalfa on gastrointestinal tract weight of young pigs

Level of alfalfa meal (% of diet)	Organ weight (% of body weight)				
	Stomach	Small intestine	Cecum	Colon	Total
0	0.75	1.98[a]	0.18[a]	1.40[a]	4.30[a]
20	0.79	2.09[a,b]	0.17[a]	1.60[a,b]	4.65[a,b]
40	0.76	2.33[a,b]	0.20[a,b]	1.79[b,c]	5.08[b,c]
60	0.77	3.57[b]	0.22[b]	2.02[c]	5.53[c]

Source: Kass et al., 1980.

Note: Each value is a mean of 16 animals, 8 slaughtered at 48 kg, 8 slaughtered at 89 kg.

[a,b,c]Means within each category in the same column not bearing common superscripts are significantly different (P < .01).

entirely clear (Hofmann, 1988). Ingesta flow in ruminants requires a more or less continuous secretion of gastric juice, in contrast with nonruminant species, in which entrance of ingesta to the abomasum is intermittent. The abomasum is divided into the cardiac region near the omasal entrance, the fundus (the main body of the organ), and the antrum near the pylorus. Generally the acid-secreting area is in the antrum, the upper part being non–acid secreting.

In contrast to that by nonruminants, secretion of gastric juice by ruminants does not respond to cephalic stimuli; however, the presence of VFAs and lactic acid stimulates gastric secretions and contraction in the organ. The evolution of this response is undoubtedly related to the lack of association between dietary apprehension and the appearance of ingesta in the abomasum; typical of simple-stomached animals.

Peptic digestion in ruminants includes the digestion of microbial cells that arrive from the rumen in a virtually living state. Acid and pepsin are accompanied by lysozyme secretions that lyse the bacteria and thus speed digestion of microbial protein. Ruminant lysozyme differs from that of most other species in having an amino acid sequence that provides resistance to pepsin digestion (D. E. Dobson et al., 1984). This enables digestion of microbial protein at several times the rate that would occur in a monogastric species.

Table 17.6. Average daily flow rates and marker recoveries in four steers

	Grain diet	Forage diet
Marker recovery (%)		
Cr_2O_3	92.6	91.6
Lignin	86.9	90.7
Digesta flow (l/day)		
Cr_2O_3	46.4	76.1
Lignin	49.8	72.3
Dry matter flow (g/day)		
Cr_2O_3	2299	2583
Lignin	2566	2509

Source: Wanderley et al., 1985.

17.3.1 Gastric Emptying

Ingesta probably remain in the abomasum only a short time (1–2 h) relative to volume of flow, but the time could be significantly affected by the fiber level in the diet. The passage of digested food past the pylorus and into the duodenum is influenced by a number of factors. Gastric emptying is inhibited by the presence of digested products and a high osmotic pressure in the duodenum. Gastric motility and emptying are also inhibited by acid or fat in the duodenum. These effects probably depend on a vagal inhibition reflex involving afferent and efferent fibers of the vagal nerves (Ruckebusch, 1988).

Dietary fiber delays gastric emptying in humans (Roe et al., 1978). Fiber particles are retained by virtue of their water-holding capacity and may entrap other soluble matter in the mass. Gels formed from pectin or starch may have an even greater effect (D. J. A. Jenkins et al., 1987; P. J. Wood, 1990). In monogastric species the delay in emptying is associated with delayed absorption of glucose and other water-soluble substances (J. W. Anderson, 1985). In humans coarse dietary fiber appears to delay gastric emptying through the formation of gel or by a "logjam" effect with coarser fibers at the pylorus. Thus in ruminants fiber could have an effect on abomasal retention and the efficiency of digestive action.

17.3.2 Displaced Abomasum

High acidity in the digesta escaping the rumen of animals on very high concentrate diets may contribute to the displaced abomasum syndrome observed in dairy cattle that have been overfed on concentrates. In this condition the abomasum becomes enlarged with fluid and gas and displaced dorsally, either left or right, within the abdomen. It may accumulate fluid and gas after displacement. Pressure from the rumen may sustain the displacement. Left displacement tends to be a chronic condition in which digesta passage is not blocked, while right displacement with torsion may result in an acute blockage of digesta flow. There are several theories that explain the cause of displaced abomasum, but the causal factors are certainly related to large intakes of starchy concentrates and insufficient intake of coarse fiber.

17.3.3 Pregastric Fermentation

Pregastric fermentation is not limited to ruminants (see Chapter 5). The mammalian stomach is lined with cardiac mucosa which secretes bicarbonate and mucus, proper gastric mucosa (which secretes HCl and pepsinogen and mucus), and pyloric mucosa (which secretes some bicarbonate and mucus). Some carnivorous species (monotremes) and some omnivores (pig) and monogastric herbivores (horse), along with polygastric herbivores, have an additional region of stratified squamous epithelium (C. E. Stevens, 1988). In the hamster (Ehle and Warner, 1978) and the vole, the stomach is organized into two compartmental areas that tend to separate peptic digestion from a type of fermentation. Fermentation apparently can occur anywhere in the digestive tract where favorable retention and pH (> 5) allow microorganisms to proliferate and regenerate.

17.4 Postpyloric Digestion and Absorption

After passing from the abomasum, the acidic digesta enter the duodenum and are mixed with biliary and pancreatic secretions that neutralize the gastric acid and provide bile for homogenization of fat, and pancreatic amylolytic, tryptic, and lipolytic enzymes that hydrolyze starch, protein, and triglycerides. Most soluble carbohydrates and lipids (triglycerides and galactolipids) are destroyed in the rumen and thus do not reach this stage of digestion, a factor that may account for the lower quantities of lipolytic and amylolytic enzymes secreted by the ruminant pancreas (K. J. Hill, 1961). Whether the capacity for induction exists is unknown. Fatty acids, although they may be hydrogenated, are not digested in the rumen. The products of triglyceride fermentation are primarily saturated, or *trans*, calcium soaps (Chapter 20). The more saturated acids may form very insoluble soaps that reach the feces. Ruminants are able to absorb triglycerides (Wrenn et al., 1978).

Ruminants do not secrete invertase (sucrase). Thus, if sucrose reaches the lower tract, fermentation in the cecum and colon is its only possible fate. Schingoethe et al. (1988) speculated that the capacity for amylolytic digestion of starch is limited in cattle and other grazing species.

About 70–90% of starch is digested in the rumen, and about half of the remainder is digested in the small intestine. Although increasing the level of starch promotes a decline in starch digestibility, the amounts digested in the small intestine, fermented in the colon, and voided in the feces increase with the level of starch intake (Carpenter, 1976; J. R. Russell et al., 1981). Starch digestion in the intestines is low at least partly because the rumen has first chance at the more available fractions, and only the less digestible starch is passed on down the tract.

17.4.1 Duodenum

The flow and composition of digesta passing through the duodenum are of considerable interest, particularly in regard to passage studies and the problem of measuring microbial protein and the escape of unfermented residues from the rumen. Duodenal fistulae and external cannulae with pumping and sampling devices designed to determine net flow and composition have provided important data on the net output of protein and amino acids from the rumen before absorption in the small intestine (P. H. Robinson, 1983). The results give more information about rumen output than about duodenal function.

The ingesta arriving at the duodenum are in a finer state of division than rumen ingesta. There is little sorting of particulate matter, and liquid and particles tend to flow together. This factor probably affects the entire lower tract. L. W. Smith (1968) found that larger particles may flow through the lower tract slightly faster than smaller particles and liquid (Table 17.1). This, as mentioned, may be caused by the backflow of liquid during peristalsis. Coarser particulate matter may be held back by filtration effects.

17.4.2 Peristalsis

Peristalsis is a recurrent series of propulsive contractions by a ring of muscle pushing ingesta before it. Distension is probably the initiating stimulus. Stimulation induces contractions in both directions along the intestine, although contractions that pass toward the mouth tend to die out more rapidly than those in the caudal direction. Contractions tend to segment the contents into boli, and also to mix chyme with the digesting feed and bring the digesting solution to the mucosal surface to facilitate digestion (L. R. Johnson, 1987). Peristalsis is under neural control.

17.4.3 Absorption in the Small Intestine

As the digesting food passes down the intestine, the end products of digestion are brought into contact with the intestinal villi. Glucose and amino acids are absorbed by active transport. In ruminants, only small amounts of sugars, fatty acids, and glucose from starch escape the rumen, and the main activity of the ruminant small intestine is to absorb amino acids. The efficiency of amino acid absorption in ruminants is likely very high and may be underestimated, partly because of the problems with measuring microbial composition. The habit of expressing amino acids as a proportion of total microbial nitrogen is unfortunate because it is confounded with nonprotein nitrogen (NPN) and indigestible capsular fractions from bacteria (Chapter 18).

As the digesta flows down the tract, absorption concentrates undigested matter, so the composition and rate of the flow, which is generally plug flow, differs greatly between the upper and lower parts. There may be a small degree of microbial fermentation in the ter-

minal ileum. Because of the removal of digested contents, rate of flow decreases toward the end of the small intestine, most severely in animals fed high-concentrate diets. Indigestible fiber promotes flow toward the terminal ileum and in the colon. Radio-opaque marker in sheep moves from the duodenum to the proximal ileum in 20–40 min but requires 2 h or more to pass the terminal ileum. Contractions tend to phase out after traversing 60% of the small intestine (Ruckebusch, 1988).

17.4.4 Fiber and Starch

Undigested fibrous residue has a major role in pushing digesta down the intestine. This is important in both the small and the large intestines. In the small intestine the sweep of fiber pushes some starch into the cecum and colon. This effect, seen in human subjects with ileostomies, indicates that the sweep of escaping starch may be proportional to the fiber load (McBurney et al., 1988).

Ruminants normally eat large amounts of lignified fiber, which could be seen as a mechanism limiting starch digestion. High-starch–low-fiber diets may thus benefit starch digestion, although there appears to be a limitation to amylolytic capacity. High-producing dairy cattle are probably most sensitive to limitations on starch utilization because they also consume considerable starch in the form of concentrate.

17.5 Large Intestine

The ruminant colon is spiral. Motility and regulation are probably similar to that seen in the pig, about which quite a bit is known. Fermentation in the rumen removes most of the available carbohydrate, so fermentation in the colon and cecum is normally limited to reabsorption of electrolytes and water and breaking down slowly digesting residues that escape from the reticulorumen. This may not apply to concentrate-fed grazers such as sheep and cattle, which may be in an abnormal situation relative to their ancestral dietary adaptations.

17.5.1 Fermentation in the Colon and Cecum

The ruminant lower tract has been overlooked because of the assumption that the rumen efficiently removes the fermentable matter. It is evident from Figure 17.2, however, that considerable matter disappears in the lower tract. Water, minerals, and nitrogen are absorbed, as, probably, are VFAs. Moreover, there appears to be a significant and variable escape of potentially fermentable carbohydrate from the rumen, including structural carbohydrate and starch. The es-

Table 17.7.   Sites of disappearance of digestible cellulose and hemicellulose

Site	Species	Diet	Cellulose	Hemicellulose	Reference
Rumen-abomasum	Sheep	Restricted to grass	93	86	Beever et al., 1972
Small intestine			2	0	
Cecum + colon			5	14	
Rumen-abomasum	Sheep	Ad lib hay	93	83	Beever et al., 1972
Small intestine			2	8	
Cecum + colon			5	9	
Rumen-abomasum	Sheep	Pelleted	80	71	Beever et al., 1972
Small intestine			2	5	
Cecum + colon			18	24	
Postrumen	Cattle	Restricted alfalfa	9	34	Waldo, 1970
Postrumen		Ad lib alfalfa	10	31	
Postrumen	Sheep	Alfalfa	10	20[a]	Hogan and Weston,
Postrumen		Ground alfalfa	10	33[a]	1967
Postrumen		Wheaten hay	10	0[a]	
Postrumen		Ground wheaten hay	31	41[a]	

Note: Values are percentages of total disappearance.
[a]Calculated from original cell wall values.

capes of cellulose and hemicellulose are not equal; the latter tends to be greater and leads to more hemicellulolytic than cellulolytic digestion in the lower tract (Table 17.7). This relatively greater digestion of hemicellulose in the lower tract of the ruminant is parallel to the generally higher hemicellulose digestion seen in pigs and rodents (Chapter 5). This observation leads to the question of hemicellulose alteration during peptic digestion. Furanosidic linkages in hemicellulose may also be sensitive to very weak acid, and this might explain the greater fermentation of hemicellulose in the large bowel. If gastric acid cleaves bonds that rendered hemicellulose unavailable in the rumen, these fractions would become available to lower tract fermentation.

The capacity of the ruminant lower tract can be tested by postruminal feeding through fistulae or other devices designed to bypass the rumen. Calves fed postruminally, through an abomasal cannula, showed an extensive capacity for fiber digestion (29% of alfalfa and 38% of solka floc digested), although this capacity was less than in oral feeding (49% and 52%, respectively). Postruminal feeding of starch or sucrose in sheep stimulates lower-tract fermentation and promotes increased fecal nitrogen excretion via microbial products (Table 17.8).

Fecal nitrogen elevated by postruminal fermentation of carbohydrate bears on the question of ruminant efficiency and the desirability of protecting feed nutrients from rumen fermentation to allow their escape to the lower tract. In respect to the utilization of true protein and perhaps lipids, there seems to be no doubt that escape from the rumen to the lower gastrointestinal tract results in more efficient use. The same factors cannot apply to carbohydrate, however. The ruminant's digestive secretions are of distinctly lower capacity than the nonruminant's, and the escape of digestible carbohydrate to the large bowel and cecum will result in fermentation, production of microbial protein, and nitrogen lost in feces. Of energy-containing carbohydrates, probably only the VFAs will be absorbed. The nitrogen loss in feces may not be serious if the diet has adequate nitrogen since the fecal nitrogen's source is mainly urea diffusing from the blood toward the bowel. This urea would compete for nitrogen recycled to the rumen through saliva or coprophagy. Methods for protecting protein from rumen degradation by means of heat or formaldehyde may inadvertently "protect" some carbohydrates as well.

17.5.2 Specializations

Ruminant species that are adapted to dry environments excrete pelleted feces with high dry-matter content. This feature is also seen in desert-adapted rodents. Resorption of water inhibits colonic fermentation, and this water-conserving adaptation may compete with other adaptations to trap calories through hindgut fermentation.

Data from monogastric species such as pigs and horses indicate that butyrate and other VFAs stimulate mucosal growth. The same nutritional relationships

Table 17.8.   Influence of rumen and postrumen digestion of sucrose on rumen and cecal fermentation in sheep

Treatment	pH		Viable count (10[7]/ml)		Fecal N excretion (g/day)
	Rumen	Cecum	Rumen	Cecum	
Grass pellets	6.6	6.9	59	7	8.9
Plus sucrose in feed	6.4	6.6	430	5	9.9
Sucrose postruminal[a]	6.6	5.6	130	180	10.9

Source: Ørskov et al., 1972.
[a]250 g sucrose in solution fed to lambs conditioned to close esophageal groove.

are therefore relevant to ruminants. Ruckebusch (1988) noted that high levels of butyrate inhibit motility and may be responsible for cecal dilation. This observation has several qualifications. One is that the levels of butyrate imposed were unphysiological, particularly in view of the normal limitation of the ruminant colon to slow-digesting substrates. On the other hand, fermentation of starches and sucrose yields high concentrations of butyrate in nonruminants. Thus if overfeeding or misfeeding leads to significant flow of these carbohydrates to the lower tract of ruminants, these overloads might be an explanation of the cecal dilation. There have been attempts (controversial) to relate cecal pH to starch passage through lactic acid production.

18 Nitrogen Metabolism

The nitrogenous components of the diet support the protein metabolism of the rumen organisms and their host, but the interactions of diet, microbes, and animal host that determine the net supply of protein to the host are complex. Section 16.8 described the factors affecting rumen microbes' efficiency from the standpoint of their energy requirements. This chapter characterizes the nitrogenous constituents of the diet and their use.

The diet that the ruminant actually receives differs from that which is eaten, particularly with regard to carbohydrate and protein. The rumen fermentation may largely destroy available carbohydrates, requiring gluconeogenesis on the part of the animal; in the case of protein, the dietary source may be more or less destroyed but compensated for by microbial protein synthesis. Excessive protein in high-protein diets is turned into ammonia, which is absorbed and wasted as urea in urine. On the other hand, low-protein diets can be supplemented by microbial synthesis using endogenous recycled urea, leading to more protein presented to the intestines than was fed to the animal.

Thus there are two counteractive processes: the degradation of dietary protein and the synthesis of microbial protein, which may use either degraded dietary protein or nonprotein sources of nitrogen. To overcome the problem imposed by the first process, manipulation of ruminant diets has tended to emphasize the feeding of degraded protein that will largely escape fermentation and ensure the passage of dietary protein and amino acids to the small intestine. This approach emphasizes the nonruminant aspects of nutrition and in a sense seeks to convert the cow into a pig. The unbalanced emphasis on undegraded protein arose because of concepts of rumen nitrogen metabolism that regard microbial output as a function of rumen fermentable matter at a relatively constant efficiency, the logic of which promotes an emphasis on so-called bypass (actually protection from rumen fermentation and consequent escape) of less soluble proteins. Actually, protein solubility is also important for microbial efficiency, leading to confusion as to which process protein solubility might favor—lower tract digestion or microbial output.

Technically, bypass is the passage of ingested diet down a closed esophageal groove into the abomasum, a phenomenon first noted in infant ruminants. Ørskov et al. (1970) trained sheep to retain this reflex into adulthood. Unfortunately, the term has been erroneously applied to the technique of feeding slow-digesting proteins that may pass undegraded from the rumen. In this case, the proper term is *escape,* and it is so used throughout this book.

Bypass, which is important in immature ruminants, has also been noted in some adult concentrate selectors, including some antelope, deer, and possibly goats (Chapter 4). Since true bypass has been demonstrated in some adult ruminants as well, the distinction between *bypass* and *escape* becomes all the more important.

The technology of protecting feed from rumen degradation by processing as well as monitoring solubility is relevant only to the escape process. Much of the research on the ruminant digestive process has focused on grazing species, particularly cattle and sheep, which tend to maximize fermentative digestion. Models from these species may not be applicable to less-studied species, many of which do less grazing and more selecting. This subject is dealt with in Chapters 3 and 4.

18.1 Forms of Feed Nitrogen

The forms of nitrogen ingested by ruminants are largely the forms elaborated by plants. True protein makes up about 60–80% of the total plant nitrogen, with soluble nonprotein nitrogen (NPN) and a small amount of lignified nitrogen making up most of the remainder. Plants are comparatively low in nucleic acids; however, fermented by-products (ca. 4–5% of total nitrogen) that may be enriched in microbial matter may contain more. Heat damage may decrease the availability of true protein to both the microbes and their host.

Plant proteins can be classified into two groups: proteins of leaves and stems and storage proteins of the

Table 18.1. Nitrogen content of proteins

Source	Nitrogen (% of protein matter)	Factor[a]
Wool	17.8	5.61
Vicia faba (broad bean)	16.8	5.95
Spinacea oleracea (spinach)	16.3	6.13
Medicago sativa (alfalfa)	15.8	6.33
True microbial protein	15.0	6.67
Brassica oleracea (cabbage)	14.7	6.80
Zea mays (corn leaves)	14.4	6.94
Microbial cell wall	7.1	14.00

Sources: Chibnall and Glover, 1926; Salton, 1960.
[a]Value required to convert amount of nitrogen to equivalent weight of protein.

Table 18.2. Composition of soluble nonprotein nitrogen in fresh forage

	Grasses (% DM)	Alfalfa (% DM)
Non–amino acid bases	1–25	—
Basic amino acids	1–15	3–4
Aminobutyric acid	5–28	12–19
Glutamine	10–25	1–2
Glutamic + aspartic acids	5–20	12–15
Asparagine	1–4	24–38
Other amino acids	7–25	7–12
Nitrate	10–25	2–4
Total NPN[a]	14–34	25–38

Sources: Brady, 1960; Hegarty and Peterson, 1973; Ohshima and McDonald, 1978.
[a]NPN = nonprotein nitrogen, as percentage of total plant nitrogen.

seed. The former represent the actively metabolizing matter of the living plant; the latter are reserves. This classification has given rise to a misconception over the quality of plant proteins. Leaf proteins are, in fact, of the highest quality. Essential amino acids mean little to the plant, which is autotrophic and capable of synthesis of all organic essentials. Seeds thus have evolved with aberrant ratios of amino acids that often emphasize insolubility and intractability to predators.

Leaf proteins may appear to be of lower biological value when crude protein is expressed as total N × 6.25, which belies the variety of the forms of nitrogen in the plant. Even when the expression is restricted to true protein, the proper factor varies according to the protein source (Table 18.1). Plant and bacterial protein factors are likely to be more variable than factors for proteins of animal origin.

The soluble NPN fraction of fresh forage is composed of peptides, nitrate, and nonessential amino acids, with glutamine, asparagine, and aminobutyric acid often predominating (Table 18.2). The NPN fraction is often increased at the expense of protein, especially in fermented crops such as silage.

Nitrogen fertilization also may alter the nitrogen distribution in plants. Low environmental temperatures and increased pools of nitrate and amino acids cause the proportion of NPN to increase in grass fertilized with nitrogen. These tendencies, along with the decline in soluble carbohydrate associated with nitrogen fertilization, can result in nutritional problems for the ruminant such as inefficient nitrogen utilization and possibly magnesium tetany.

Fermented feeds have a completely different composition of NPN (Section 14.4.4). Plant organic acids tend to ferment away and are replaced by lactic acid and VFAs plus ammonium salts. Amines also appear in the less well preserved products.

18.1.1 Protein Classification

In biochemistry it was once customary to classify proteins according to their physical solubilities. This classification is still relevant in the context of industrial extraction of proteins from plant sources and also in appreciating the contribution of proteins to dietary characteristics. It should be understood, however, that the old classification system is not useful for distinguishing pure proteins in the modern sense, since the respective fractions obtained from any one plant usually are mixtures (Altschul, 1958).

Seed proteins, the main plant proteins described in this context, include *albumins,* which are water-soluble but alcohol-insoluble; *globulins,* insoluble in water and alcohol but soluble in salt solutions of medium strength; *prolamines,* soluble in alcohol but insoluble in water and salt solution; and *glutelins,* soluble only in dilute alkali. Most seed proteins contain some of these fractions. Graminaceous cereals tend to contain large amounts of prolamines and glutelins, which characterize the insoluble and hydrophobic character of corn and wheat proteins and possess slow rates of hydrolysis in the rumen, although oats are high in globulins. Dicotyledonous sources, particularly legumes, tend to contain potentially more soluble globulins and albumins, but these are sensitive to heat denaturation, which will render them water-insoluble. The physical effect of solubility in the rumen is relevant mainly to albumins, since ionic concentrations are too low for solution of globulins and pH is too neutral to affect other proteins. The solubility characteristics of the four protein classes are shown in Table 18.3.

Table 18.3. Solubility of proteins in various media

	Alcohol	Water	Salt	Alkali
Globulins[a]	I	I	S	S
Albumins[a,b]	I	S	I	S
Prolamines[c]	S	I	I	S
Glutelins[c]	I	I	I	S

Source: Information from Frear, 1950.
Note: I = insoluble; S = soluble.
[a]Characteristic of legume seeds.
[b]Leaf proteins.
[c]Characteristic of cereal seeds.

Note the contrast between seed proteins from plants in the grass and legume families.

18.1.2 Seed Meals

The principal protein supplements of plant origin fed to animals are derived from the cake or meals left after the oil has been extracted from oil-bearing seeds. All are dicotyledons. Many (particularly the legumes) tend to contain inhibitory substances that can be destroyed by heat or removed by extraction; thus nutritive quality is improved by some heat treatment. This general improvement of quality in heated soybean and similar meals may not be for the same reason in ruminants and nonruminants. Nonruminants are more sensitive to antiquality factors such as the antitrypsin factor in soybean meal or gossypol in cottonseed, which have relatively little effect on ruminants because of the detoxifying ability of the pregastric fermentation. Heating may improve protein quality for ruminants because of protein denaturation and reduction in solubility and degradation rate, which favors more rumen escape of intact undegraded proteins and better utilization of released peptides, isoacids, and ammonia for microbial synthesis.

Most seed-derived proteins have a lower amino acid quality than leaf proteins. Seed proteins are compounds stored for the infant plant, which will be capable of synthesizing all its organic essentials. Unfavorable amino acid ratios and poor solubility as well as antiproteolytic factors in seeds make them less desirable to animal predators.

18.1.3 Leaf Proteins

Leaf proteins are of higher quality than the storage proteins found in plant seeds, although they are diluted with variable amounts of NPN. Up to 90% of the true protein in alfalfa leaves may be a single complex of photosynthetic enzymes. Leaf proteins can be divided into cytoplasmic and chloroplastic proteins, the nucleoproteins of the nucleus, and the extensin proteins of the plant cell wall. The latter two are usually present in much lower concentrations. Cytoplasmic and chloroplastic proteins are essentially soluble in the plant cell contents but are permanently precipitated by mild drying or heat. Hay proteins, for example, are almost all water-insoluble. These proteins also exhibit isoelectric points and minimum solubilities at a pH of about 4.

18.1.4 Cell Wall Protein

The cell wall–associated proteins, sometimes called *extensins* because of their probable role in fiber cross-linking, are much less soluble and are recovered in neutral-detergent fiber (NDF), as also are some of the heat-denatured cytoplasmic and chloroplasmic proteins. Extensins are probably covalently linked to polysaccharides associated with the plant cell wall, which may account for their insolubility. The nitrogen content of NDF in feeds is greatly increased by heating, which promotes denaturation of albumins, but not necessarily in acid-detergent fiber (ADF), which requires the Maillard reaction to render protein recoverable in ADF. The ADF nitrogen is relatively indigestible and is poorly used, if at all. Protein that is insoluble in neutral detergent but soluble in acid detergent appears to have high digestibility, however, although it usually digests at slower rates than the fractions that are soluble in neutral detergent. Usually, a substantial portion of feed nitrogen is water-insoluble but still soluble in neutral detergent (Pichard and Van Soest, 1977).

18.2 Protein Solubility

The custom of dividing feed nitrogen into the two categories of nitrogen-soluble and nitrogen-insoluble in rumen buffer is utterly inadequate because such a simplistic division does not distinguish NPN from true protein in the soluble fraction, nor does it account for unavailable nitrogen in the insoluble fraction. Also it is assumed that insolubility confers slow-degrading characteristics. Alfalfa leaf protein is denatured in hay making but remains rapidly degradable, while blood sources of soluble serum albumin are slowly degradable. For these reasons solubility must be interpreted carefully and its application should be restricted to a respective true protein source.

The need to distinguish soluble NPN from true protein is exemplified by the postulated effects of ionophores that inhibit proteolysis. Feeding ionophores should be beneficial for animals fed on highly soluble true proteins, but ionophores are ineffective for most silages, in which NPN constitutes 99% of the soluble nitrogen (Pichard and Van Soest, 1977). The general view that soluble proteins are rapidly degraded to ammonia in the rumen cannot be supported because an important peptide pool forms for some hours after proteins are ingested.

Tables 18.4 and 18.5 give a more relevant classification of feed nitrogen. Slow-degrading proteins may be more accurately assayed by enzymes in vitro or by the judicious use of nylon bags (Section 8.5.2; measurement of the respective fractions is also discussed in Chapter 8). A simplistic concept of nitrogen solubility portrays rumen degradability as a tabular number, but rumen degradability is actually a variable whose value depends on the competition between rates of digestion and passage (Chapter 23). The classification in Table 18.5 may require sequential analysis (Chapter 10) for

Table 18.4. Partition of nitrogen and protein fractions in feedstuffs

Fraction	Abbreviation	Estimation or definition	Enzymatic degradation	Classification[a]
Nonprotein nitrogen + peptides	NPN	Not precipitable	Not applicable	A
True soluble protein	BSP	Buffer soluble and precipitable	Fast	B_1
Neutral-detergent-soluble protein	IP[b] − NDIP	Difference between IP and protein insoluble in neutral detergent	Variable	B_2
ND-insoluble protein	NDIP − ADIP	Protein insoluble in ND but soluble in acid detergent	Slow	B_3
AD-insoluble protein	ADIP or ADIN	Includes heat-damaged protein and nitrogen associated with lignin	Indigestible	C

Sources: Pichard and Van Soest, 1977; Chalupa et al., 1991.

[a]Soluble fraction A contains NPN, peptides, and some soluble peptides. Fractions B_1, B_2, and B_3 represent decreasing available fractions of true protein. C represents the unavailable lignified or heat-damaged protein fraction.

[b]IP = CP − NPN − BSP.

certain feedstuffs, such as corn gluten meal, which tends to have a higher acid-detergent insoluble nitrogen (ADIN) fraction than neutral-detergent insoluble nitrogen (NDIN) as a result of detergent precipitation of an acidic complex from the feed. Preextraction with neutral detergent will remove this interference.

Insoluble nitrogen-containing compounds may have a special significance for the ruminant because solubility generally (but not necessarily) renders the nitrogen more available for microbial metabolism. The competition between passage and rumen digestion for potentially digestible substrate determines the proportion of unfermented feed passing to the omasum and abomasum. This passage is important in regard to potentially digestible true protein since it will determine the amount of unaltered dietary protein and amino acids that reaches peptic digestion. This aspect is not necessarily positive since the less-digested insoluble fractions may have unbalanced amino acid ratios.

Drying any plant material causes a general reduction

in the solubility of cytoplasmic proteins through denaturation. The degree to which solubility is reduced depends on the circumstances—in particular, the amount of heat applied. Two fundamental types of chemical reaction are involved: physicochemical reduction of solubility (denaturation) and formation of links with other substances. Synthetic means of reducing protein solubility include heat and acid coagulation; for example, the use of formic acid in ensiling or the use of formaldehyde and tannins to form leatherprotein complexes. The latter can induce indigestibility. Generally, condensed tannins have negative effects on nutrition (see Chapter 13).

18.2.1 Measuring Protein Solubility

Laboratory methods for measuring protein solubility and degradability lack standardization. Techniques that measure nitrogen solubility in various aqueous solutions must be classified according to

Table 18.5. Average distribution of protein and nitrogen fractions in some feedstuffs

Feedstuff	Crude protein	NPN (% CP)	BSP (% CP)	IP − NDIP (% CP)	NDIP − ADIP (% CP)	ADIP (% CP)
Distillers' dried grains (DDG)	27	2	1	45[a]	22	20
DDG + solubles	29	17	2	18	41	21
Brewers' dried grains	29	4	2	54	27	13
Corn gluten meal	66	3	1	94[a]	0	2
Corn grain	9	10	5	70[a]	10	5
Beet pulp	8	3	0	45	41	11
Corn gluten feed	22	55	0	37[a]	5	2
Corn silage	9	44	4	36[a]	8	8
Alfalfa hay	20	27	3	50	10	10
Soybean meal	52	13	12	71	3	2

Sources: Pichard and Van Soest, 1977; Krishnamoorthy et al., 1983a.

Note: Abbreviations are defined in Table 18.4.

[a]Corn products contain zein, a slow-degrading prolamine protein that is soluble in neutral-detergent.

whether they separate soluble NPN from protein and whether a real solution of true protein actually occurs. Buffers tend to dissolve mainly NPN in forages and some protein in certain concentrates, and confusion of NPN with soluble protein is a hazard of the buffer solubility methods. The detergent reagents are much more powerful extractors of protein. Protein precipitation methods—for example, trichloracetic acid (TCA) and tungstic acid procedures—separate NPN and peptides from true protein. The cutoff for tungstic acid is about 2–3 amino acid units; for TCA, about 10.

Hot-water extraction of forage produces results similar to other buffers since leaf proteins are coagulated by heat and only NPN and peptides dissolve. In contrast, some protein from concentrate feedstuffs is soluble in cold buffers, and even more may be soluble in hot solutions. Thus the cold extraction with rumen-type buffers dissolves very little true protein from dry forages and silages (Pichard and Van Soest, 1977), although cold-water extraction of fresh forage may dissolve as much as 50% of the cytoplasmic protein (high in albumins). Solubility is promoted by agitation and maceration and in some of these proteins is pH-sensitive, higher pH giving greater solubility.

The simplistic division of feed nitrogen into soluble and insoluble categories overlooks NPN, peptides, true protein, and unavailable nitrogen. A proper analysis would distinguish TCA- and tungstic acid–precipitated matter in the solubles from other nonproteinaceous matter. Kinetic analysis applies only to the true protein pool. Since simple methods exist to assay all the entities deemed to be of nutritional significance, there seems little excuse to exclude these nitrogen fractions from biochemically rational analyses.

18.2.2 Rate of Protein Digestion

A more sophisticated approach for measuring protein degradability is to measure the rate of digestion of true protein. The main approaches have been with nylon bags in situ and enzymatic digestion in vitro. Neither system is without some problems. In situ cloth bags may be invaded by rumen microbes, which attach themselves to the particulate matter inside the bags. As a result, the undigested fraction is overestimated and the true digestion rate is underestimated. The effect is more serious with higher-fiber feeds.

Enzymatic procedures require the use of commercial enzymes, which differ from microbial enzymes in their pH optima. The most commonly used enzymes are pepsin, which requires a low pH (Lovern, 1965); ficin (Poos-Floyd et al., 1985); and a protease from *Streptomyces griseus* that requires a basic (~8) pH (Krishnamoorthy et al., 1983b). The maximum rate obtained with saturated enzyme on substrate is faster than rates actually seen in the rumen. Reducing the

enzyme level to limit the rate results in a nonideal assay system; that is, a small variation in enzyme activity or quality results in large analytical variations. It appears that there is a considerable lag in the rumen relating to initiation of proteolysis, and introducing an appropriate lag function into the in vitro system (see Section 22.4.2) might allow for a more satisfactory enzymatic system. Nevertheless, Krishnamoorthy et al. (1983b) were able to show a significant correlation between the in vivo measurements and predicted rumen escape of undegraded protein.

18.2.3 Effects of Heating

The production of insoluble protein through heat or by complexing decreases the rate of proteolytic hydrolysis, not only through reduced accessibility of the substrate but also through the formation of linkages resistant to enzyme attack. Producing resistant linkages risks permanently reducing the availability of protein, which then becomes a part of the ultimately indigestible residue (fraction C in Table 18.4). At least two feed treatment systems (formaldehyde treatment and tannin complexes) operate in the hope that linkages resistant to rumen degradation will be broken by gastric acidity and enzymatic digestion, making the protected protein available for lower tract digestion (Tamminga, 1979). In the case of heat coagulation or denaturation, solubility and accessibility are merely reduced. The Maillard reaction involves condensation of amino groups with carbonyls and dehydroreductones derived from carbohydrate followed by polymerization into a lignin-like matrix (Section 11.7). The condensation in formaldehyde treatment is identical with the first step of the Maillard reaction and may be somewhat reversible. In the case of the overall Maillard reaction, however, polymerization results in permanently bound and indigestible nitrogen. The ADIN appears to have some digestibility (Weiss et al., 1986; Van Soest and Mason, 1991). Some of the unavailable Maillard nitrogen can appear in the soluble fraction, but it seems to be relatively unutilizable (Nakamura et al., 1991). Unmetabolizable Maillard products may even appear in urine. Overall, the unmetabolizable forms of absorbed Maillard products probably balance the lack of fecal recovery of ADIN, and the ADIN remains the best available assay for unavailable nitrogen. The ADIN is highly correlated with indigestible nitrogen (J. W. Thomas et al., 1982).

The advantage of using heat to protect protein lies in the lower energy input required for denaturation compared with the amount necessary to cause the Maillard reaction. The limiting step in the Maillard reaction is the formation of active intermediates from carbohydrate that can condense with available amino groups. Water content and pH could also be limiting. Denatura-

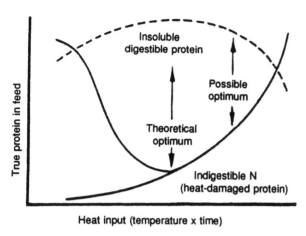

Figure 18.1. The theoretical relation between protein solubility and heat (from Van Soest and Sniffen, 1984). Maximum utilization of insoluble denatured protein should occur at the intersection with the heat damage curve, allowing maximum digestion of insoluble protein; however, further heating may improve animal response even though protein digestibility is significantly reduced.

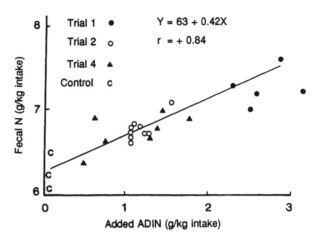

Figure 18.2. The relation between total nitrogen and added dietary acid-detergent insoluble nitrogen (ADIN) (from Van Soest and Mason, 1991). The regression includes pooled data from Nebraska trials 1, 2, and 4, with three controls and 21 diets containing distillers' grains.

tion has a sharp inflection at the threshold temperature (50–60°C), while the Maillard reaction shows a slow exponential increase with temperature (Figure 18.1). The result is a level of heat input that has a potential for maximum heat coagulation and minimal (but probably measurable) damage through the Maillard reaction. The main difficulty in manipulating this system is that the potential susceptibilities of forages and feeds to the Maillard reaction vary by an order of magnitude (Goering et al., 1972, 1973). Any lack of control of heat input can cause heat damage and loss of quality. The optimum balance between heat denaturation and damage has not been established, but theoretically it is the point of maximum insoluble but still digestible protein.

18.2.4 ADIN and Digestibility

About 70% of low dry matter silages and dehydrated forages show measurable heat damage (Goering, 1976; Yu and Thomas, 1976), but in many instances the damage is at a level low enough that the beneficial effects of low protein solubility through denaturation may override the loss of digestible nitrogen. Beever et al. (1976) reported improved growth of lambs fed dehydrated ryegrass in which protein digestibility had been reduced 10 units through heat damage. Figure 18.1 indicates diminishing returns with added heat input in processing. For this reason Goering (1976) considered only materials containing more than 0.3% of feed nitrogen in acid-detergent fiber as possessing a detrimental level of heat damage. This cutoff is arbitrary and is substantially higher than the normal range for undamaged feedstuffs (0.05–0.2%).

There have been suggestions that ADIN may not apply to concentrate feeds. Britton et al. (1986) and Klopfenstein and Britton (1987), who fed distillers' grains to sheep in isonitrogenous diets, could not find an association between protein digestibility and ADIN intake; however, a reanalysis (Figure 18.2) of their data using Lucas models shows a significant relationship (Van Soest, 1989).

While the negative association between ADIN and nitrogen digestibility has been substantiated, ADIN has been shown to be about 60% digestible in distillers' grains. The regression slope for ADIN represented in Figure 18.2 (Van Soest, 1989) is an estimate of the recovery of ADIN contributions, and is thus an estimate of the true indigestibility of ADIN. The slope is about 40% and is significantly less than 1, which would be required for theoretically total indigestibility. Several interpretations are possible: (1) There is a true partial digestibility of ADIN (see Sections 22.2.1 and 22.2.2). (2) There is a compensatory reduction in metabolic fecal nitrogen with increments of ADIN as a result of reduced microbial yield. This needs to be explored with samples that have been adequately analyzed. (3) The ADIN in distillers' grains may contain products that are absorbed from the small intestine but not metabolized in tissues and lost in urine.

The behavior of ADIN in distillers' grains seems different from that in silages as well as in treated ammoniated forages. The latter products demonstrably contain indigestible soluble nonammonia nitrogen, and ADIN underestimates unavailability in ammoniated forages (Van Soest and Mason, 1991).

All feeds and forages, whether heated or not, appear to contain an unavailable fraction of nitrogen (Figure 18.3). The ADIN in most feeds appears to be essentially indigestible (Chaudry and Webster, 1993). Study by the Nebraska group (Nakamura et al., 1991) suggests that the apparently digestible ADIN is poorly

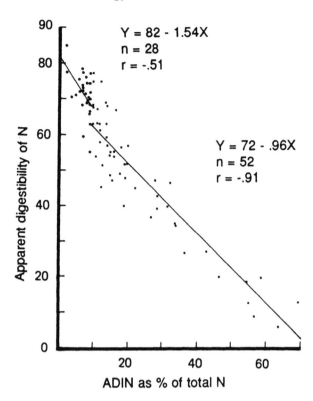

Figure 18.3. The relation between apparent digestibility of nitrogen and the nitrogen bound in acid-detergent insoluble fiber (ADIN) (courtesy of J. W. Thomas, Michigan State University; see Yu and Thomas, 1976). Values, obtained from 80 digestion trials with sheep or cattle, are expressed as a percentage of total feed nitrogen. The hays and silages without heat damage (n = 28) give a different regression line from forages (n = 52) that were heated.

used by the animal. The quantitative subtraction of ADIN from crude protein to estimate available nitrogen seems justified (Waters et al., 1992). In normal feeds, the unavailable nitrogen ranges from 3 to 15% of the total nitrogen. Normal distillers' grains are at the upper level of this range, at a level that may promote benefit from cation exchange and microbial efficiency without being high enough for much negative effect. Moderate levels of damage may increase microbial output and nitrogen efficiency, while yet higher levels become damaging (Chaudhry and Webster, 1993). Thus it is possible that the practical optimum may be a little to the right of the theoretical optimum relative to heat input.

18.3 Microbial Nitrogen Requirements

Microbes are the sole means of converting NPN to high-quality protein, but they are also responsible for degrading high-quality dietary protein. The possibility of manipulation and maximization of microbial yield has been overshadowed in animal nutrition in favor of protein protection and rumen escape. This latter strate-

gy, however, makes the ruminant more dependent on dietary quality and brings it into competition with non-ruminants. Increasing the microbial efficiency should make the ruminant more independent of food resource competition and allow more efficient use of forage.

Optimization of microbial yield requires optimal utilization of nitrogen, achieved through manipulation and control of rumen outflow using protein sources with the right rates of degradation. Since yield increases with intake, it is associated with increased escape because of increased passage and rumen turnover. The efficiency responses of ruminants at high intake levels are complex. High intake promotes both escape and increased protein synthesis on a diminishing portion of the dietary carbohydrate that is fermented. Feeds are not equal with respect to rumen escape and the degree to which they depress digestibility, and these factors significantly affect the productive energy of practical ruminant diets (Chalupa, 1977).

The net result of this pattern of digestion is that diets containing protein and NPN amounts within a certain range can result in more or less the same amount of metabolizable protein being available to the animal. Factors influencing this quantity are apt to be internal. The catabolism of dietary protein and the microbial yield determine the ability of the microbes to compensate for low-protein diets and reduce high-protein ones, while only the escape of unfermented feed protein will tend to override the leveling effect. The limit of microbial compensation in ruminants fed low-nitrogen diets is set by the efficiency of recycling of urea, and on high-nitrogen diets, by microbial catabolism of protein. This range of crude protein intake extends from about 8 to 16% of the diet and can result in roughly the same net protein output by the animal (Jacobson et al., 1970). Increases in dietary protein or nitrogen intake are generally balanced by increased urinary loss. This leveling effect is characteristic of many ordinary ruminant diets but can be modified by a number of factors.

The absolute amounts of protein needed for ruminants' physiological functions of maintenance, growth, reproduction, and lactation are not known. This uncertainty is accounted for by the failure of dietary intake to correspond to the amount of protein the animal actually receives.

The efficiency of the rumen fermentation in yielding microbial protein is driven by two major forces: (1) the rate of fermentation, which sets the amount of feed per unit time and is the functional plane of nutrition for microbes; and (2) the rate of passage, which favors rumen loss of slower-fermenting substrates and removes the more mature organisms, reducing the median age of the microbe population. It also reduces predation on bacteria by protozoa, leading to greater growth potential on a given amount of substrate.

The interaction between passage time and diet quality can make microbial efficiencies highly variable. The microbial requirement for nitrogen may exceed that of the protein equivalent for the animal host, or, alternatively, in poorly balanced rumens, the host animal's requirements may exceed the rumen requirements. Microbial species requirements must also be considered. For example, digesters of nonstructural carbohydrates prefer peptides, while cellulolytic organisms depend more on ammonia and isoacids. Availability of peptides probably improves microbial efficiency.

18.3.1 Bacterial versus Host Nitrogen Requirements

The most important factor determining the amount of metabolizable protein available to the animal is the yield of the rumen fermentation. The nitrogen bound in microbial cells represents the microbial requirement that can be satisfied by an exogenous feed source and endogenous recycling of nitrogen. This requirement is not constant and is affected by available substrates and turnover. If the microbial nitrogen requirement differs from that of the animal, one of six theoretical nutritional situations may arise (Table 18.6), depending on which is the more limiting situation: microbes or ruminant host. Each situation responds to addition of dietary NPN, protected protein, or fermentable carbohydrate. Underfeeding will not limit the rumen fermentation as long as recycling of urea through saliva can satisfy requirements. There is, however, a level (approximately 6–8% crude protein) below which recycling does not satisfy requirements, and either intake or digestibility falls off. Adding starch or sugar is detrimental because it increases microbial requirements at a time when nitrogen supply is limited (see Figures 18.4 and 18.5).

Any benefit gained from overfeeding dietary nitrogen is related to the extent to which extra energy is needed by the bacteria and the animal. Protein in excess of the microbial requirement is fermented to ammonia and VFAs (Figure 18.6). Adding starch or some other digestible carbohydrate should elicit a comparable effect. Excess dietary NPN or protein that is fermented is converted to ammonia, which is absorbed, converted to urea by the liver, and excreted in urine. Protein supplied to the animal in excess of its requirements is similarly catabolized by the animal for energy and the excess nitrogen excreted in the urine. Adding fermentable carbohydrate will increase the microbial requirement and promote utilization of excess ammonia.

The intermediate situation, in which either the animal or the microbial requirements may be limiting, is the more common situation. If the microbial requirement is greater, added NPN supports increased micro-

Table 18.6. Comparison of microbial and host animal requirements

	Animal requirement less than microbial requirement	Animal requirement greater than microbial requirement
Too little nitrogen, inadequate for microbes and animal	Exogenous NPN beneficial; protected protein (for escape) beneficial; added starch or sugar probably detrimental	Exogenous NPN beneficial but less satisfactory than protected protein; added starch or sugar very detrimental
Intermediate level of nitrogen, inadequate for upper requirement	Exogenous NPN beneficial; addition protected proteins (for escape) no effect except as added energy; added digestible carbohydrate improves animal performance via supplied energy	Exogenous NPN has no effect or is detrimental; protected protein gives positive animal response; starch beneficial through greater microbial use of recycled urea
Nitrogen in excess of microbial and animal needs	Exogenous NPN has no effect or is detrimental; adding protected protein (for escape) has no effect; added digestible carbohydrate improves animal performance via supplied energy	Exogenous NPN detrimental; protected protein gives positive animal response; added digestible carbohydrate beneficial for greater microbial synthesis and energy supply to animal

bial synthesis, and beneficial effects are limited to stimulation of rumen fermentation to yield more energy and perhaps increase feed intake. Excess protein can be used by the animal as energy, and the effect may be confused with the effects of supplemental concentrate. It is important to note that all cases of animal response to added dietary NPN require that the nitrogen available in the rumen be below the microbial requirement.

18.3.2 Rate of Fermentation

There are important correlations between rates of fermentation of the respective carbohydrates and mi-

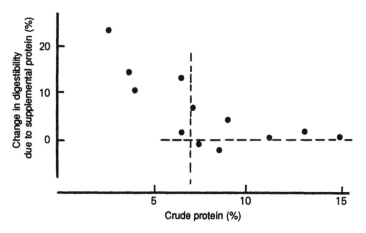

Figure 18.4. Effect of protein supplementation on forage digestibility (from Ellis and Lippke, 1976). Protein supplementation increases digestibility only when forage contains less than 7% crude protein (vertical dashed line). Inadequate dietary nitrogen limits both digestibility and intake (Figure 18.5). The limitations are determined by microbial requirement and efficiency of use of recycled nitrogen.

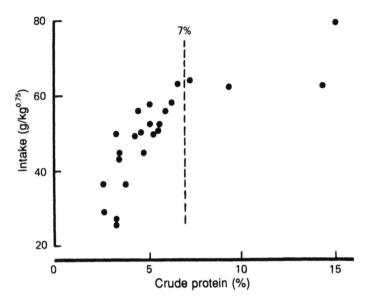

Figure 18.5. Relation of animal intake to crude protein content of tropical forages. Cell wall content of tropical grasses is more constant and therefore is a less important variable influencing intake (Milford and Minson, 1965). At concentrations above 7%, crude protein is not well related to intake; however, a precipitous decline in intake occurs when forages contain less than 7% crude protein.

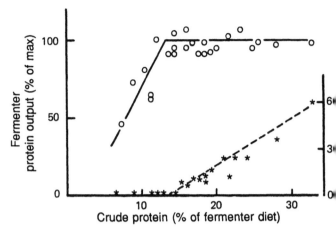

Figure 18.6. The response of a continuous culture of rumen organisms to added nitrogen in the feed (from Satter and Slyter, 1974). Upper line shows the microbial yield, and the lower line the concentration of ammonia in the medium. This figure illustrates the capacity of rumen organisms to utilize and respond to crude protein up to their required level, which is set by the energy input and the turnover. Substrate and turnover are constant in the continuous culture, leading to a fixed microbial requirement and yield. Ammonia is a detectable product only when the crude protein input exceeds microbial needs. These data have been misused to postulate a fixed microbial yield of 13–14% crude protein from the digestible feed and a lack of response to added NPN above this level, but the reader should be aware of the factors influencing variability in microbial yield. The above figure was obtained at a single constant flow. Variation in turnover and rate of fermentation causes microbial yield to vary.

crobial efficiencies (i.e., production of microbial protein per unit of feed digested in the rumen). The rate of fermentation sets the amount of feed energy per unit time for rumen bacteria; faster digestion rates provide more food. The effect is similar to that of level of nutrition for animals, whereby the extra feed dilutes maintenance requirements, leaving more for growth and production (Sniffen et al., 1983).

The digestibility of various carbohydrates versus the associated rumen microbial yield is shown in Figure 18.7. These relationships, which follow the Michaelis-Menten equation, were determined for the starch digester *Streptococcus bovis,* which requires six times as much energy for maintenance as cellulolytic bacteria need. Pectin is invariably the most rapidly degraded complex carbohydrate, while starches and celluloses are quite variable according to source; hence their quality becomes important. Complementation with different carbohydrate sources may be beneficial provided that competition between substrates is not severe. Sugars and rapidly degrading starches appear to inhibit cellulose digestion, but pectins may not act in this way. Pectin is present in large amounts in citrus, beet pulp, and alfalfa; there is virtually none in most grasses and corn silage.

18.4 Animal Protein Requirements

It is important to consider nutritional situations in which the animal host's requirements for protein ex-

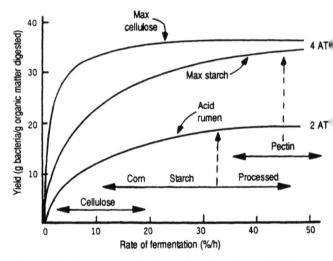

Figure 18.7. The amount of microbial protein produced per unit of feed fermented in the rumen affects the rate of fermentation (from Van Soest et al., 1991). Cellulolytic bacteria are more efficient because their maintenance cost is lower. The rate of digestion dilutes the maintenance cost for all rumen bacteria, and faster-fermenting carbohydrates improve rumen efficiency (Sniffen et al., 1983). When large amounts of starch are added to the diet, rumen fluid digestion rates increase and starch-digesting organisms such as *Streptococcus bovis* switch from acetate production, from which they derive about 4 ATPs per unit of glucose fermented, to lactate production, from which they get only 2 ATPs per unit of glucose. In this case the microorganisms sacrifice the efficiency of ATP production for the sake of increasing lactic acid, which makes the environment more favorable for their exclusive growth.

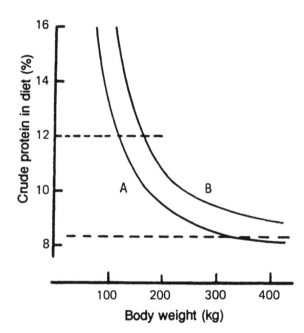

Figure 18.8. The relation between crude protein requirements (expressed as a percentage of NRC-recommended dry matter intakes) and body weight. (A) Steers gaining 0.75 kg/day. (B) Large-breed dairy heifers. Dashed lines represent microbial yields. The lower line, for poorer-quality diets, assumes an average microbial yield of 1.9 g nitrogen per 100 g true digestible organic matter (Table 14.11). Upper dashed line represents the microbial yield for a supplemented diet. Forage-based diets often exceed this level. Microbial yield will exceed protein requirements in animals that weigh more than 120 kg provided the diet is of reasonable quality. Poor-quality diets (low in energy) may be inadequate for any level of body weight and animal growth potential.

ceed those of the microbes. Comparing animal protein requirements with expected microbial yields from fermented ingested feed allows an estimate of the situation. With high-quality diets, only very young animals (Figure 18.8) and lactating females are likely to require protein in excess of expected microbial yields.

About one-third of the dry matter in milk is protein, and it is easy to understand how lactation is associated with an increased dietary requirement of protein. Whether energy or protein is the limiting factor at high levels of milk production is unknown. Microbial yield increases with intake and passage and is therefore correlated in the incremental response to feed intake leading to escape of dietary protein. It is probable that the animal body has a limited reserve of protein available for secretion as milk protein, and amino acid reserves may support gluconeogenesis for lactose production.

In growing animals beyond the stage of initial growth demand, microbial output exceeds animal requirement. It is in this sense that beef cattle's lack of response to urea supplementation on rations above 10–12% crude protein should be understood. Responses to added protein above the animal requirement are possible, however, since satisfaction of the microbial requirement can improve rumen efficiency. In this case

the consequent animal response is an answer to increased metabolizable energy.

18.5 Recycling and Use of NPN

Recycled nitrogen occurs in the form of urea, which is present in saliva and can diffuse across the rumen wall, and (much overlooked) the mucins of saliva, which contain some protein and much glucosamine polysaccharide. Mucins contain significant fermentable energy, and most normal gut microorganisms can facultatively digest them. The mucins may be a reason why starvation does not cause the loss of microbial species, although net numbers are greatly depressed. Mucins have been called "animal fiber" since they are secreted by the digestive tract in all mammals and seem to be fermentable mainly by gut microorganisms in both ruminants and nonruminants.

Most nitrogen utilized by rumen organisms is in the form of ammonia, and it is into this common substance that much nitrogenous matter in feed is converted. Bacteria are extraordinarily efficient at scavenging ammonia if it is in short supply but will allow it to accumulate in the liquid if the supply exceeds their needs (Figure 18.6). Nitrogen promotes microbial growth up to the limit of the microbial nitrogen requirement, which is set by the available fermentable carbohydrate, the ATP yield, and the efficiency of conversion to microbial cells.

Nonprotein nitrogen compounds such as urea and amides are converted in the rumen to ammonia, which is either utilized there or absorbed across the rumen wall. Blood ammonia levels normally remain low because the liver rapidly converts ammonia back to urea (a detoxified form), a conversion that costs the animal about 12 kcal/g of nitrogen (Tyrrell et al., 1970). Excess dietary NPN causes ammonia production beyond the conversion capacity of the liver, resulting in a rise in the blood ammonia concentration. Excessive levels may result in a rise in pH and impair the capacity of blood to expel CO_2. Since the level of ammonia in the blood tends to be lower than that in the rumen, and the level of urea lower in the rumen than in the blood, the potential for a perpetual exchange cycle exists, particularly under conditions of overfeeding of dietary nitrogen. This argument assumes that these relatively small molecules will passively diffuse across membranes (T. R. Houpt and Houpt, 1968).

Figure 18.9 gives an overall perspective of the shift in the nitrogen cycle that occurs when dietary nitrogen is increased. At low levels of nitrogen intake, a large portion of nitrogen metabolized within the animal is recycled, probably largely through the rumen, and very little appears in the urine. The net flow of NPN shifts from the rumen toward the urine as dietary nitro-

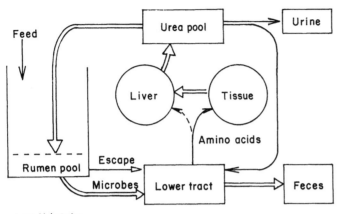

A. Low N intake

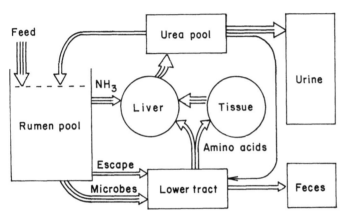

B. High N intake

Figure 18.9. Comparison between nitrogen metabolisms at low nitrogen intake (A) and at high nitrogen intake (B), assuming equal dietary energy. An increase in dietary nitrogen causes increases primarily in the rumen pool and urine output. Nitrogen recycling and the size of the urea pool are comparatively inelastic (see Figure 18.11). Excess amino acids absorbed from the lower digestive tract are metabolized for energy.

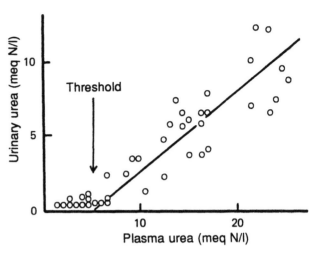

Figure 18.10. The relation between plasma urea concentration and urine concentration (based on Cocimano and Leng, 1967). The urinary threshold, the intercept of regression of blood level on urinary concentration, is about 5 meq/1 of blood. The model assumes an increase in urea pool size in the body at levels above 5 meq.

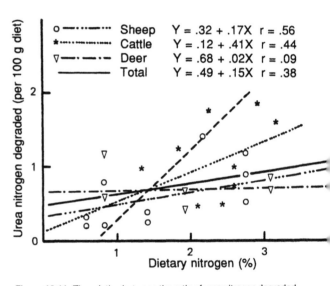

o ----·---- Sheep	Y = .32 + .17X	r = .56
* ----·-·---- Cattle	Y = .12 + .41X	r = .44
▽ --·--·--·- Deer	Y = .68 + .02X	r = .09
———— Total	Y = .49 + .15X	r = .38

Figure 18.11. The relation between the ratio of urea nitrogen degraded daily and the nitrogen content of the diet, showing comparative unresponsiveness of urea recycling to nitrogen intake. Data are based on experiments with deer (Robbins, 1973), cattle (Mugerwa and Conrad, 1971), and sheep (Cocimano and Leng, 1967). The positive intercept is an estimate of recycling at zero nitrogen intake, and the slope is the increment in nitrogen recycling due to the dietary nitrogen. The dashed line is the Lucas test regression line for dietary nitrogen (Y = 0.93 − 0.58X). Values above and to the left of this line indicate a greater degradation than the dietary intake of digestible nitrogen. Cattle were fed corn silage and urea at higher dietary nitrogen levels, which probably influenced the regression to a steeper slope here. Sheep were fed forage-concentrate diets, and deer were fed purified soy diets, which were probably lower in NPN and contained no urea. The disparity in the diets does not allow accurate comparison of the deer, sheep, and cattle data; however, the deer do have the highest intercept (0.68 at zero nitrogen intake), and it is possible that some wild ruminants may be better at nitrogen recycling than domestic ruminants.

gen is increased. An increase in dietary nitrogen intake is associated with a larger urea output from the liver, while an above-threshold blood plasma level is associated with greater urinary excretion rate (Figure 18.10). As dietary nitrogen increases, the proportion of total urea that is degraded declines, with the balance being lost in the urine. At low dietary intakes of nitrogen, the recycling can be very efficient, with most of the urea coming from the endogenous metabolism of tissue and absorbed amino acids. The escape of feed protein from the rumen fermentation removes the possibility of adding nitrogen to the urea pool for cycling.

The amount of urea recycled is relatively independent of dietary nitrogen (Figure 18.11). Since the size of the urea pool in the body is under homeostatic physiological control, it tends to be constant. This nitrogen is present in amounts less than the total microbial requirement, and the remainder of the microbial requirement must be derived from feed nitrogen. The efficiency of urea reutilization falls off with increasing dietary nitrogen intake.

Some urea hydrolysis probably occurs in the lower tract rather than in the rumen (J. V. Nolan, 1975; Visek, 1978), and the ureolytic activity there may vary with the type of diet. The lower tract's ability to capture nitrogen depends on the presence of fermentable carbohydrate. Lower tract fermentation in monogastric animals tends to reduce urinary nitrogen and increase fecal nitrogen loss. Further, the escape of fermentable carbohydrate to the lower tract will determine the microbial population there and its capacity to metabolize nitrogen compounds. This effect is probably more pronounced at higher intakes. The ureolytic activity of the lower tract is the main nitrogen source for cecal and large bowel fermentations in nonruminant herbivores. Since the types of bacteria in the lower tract are similar to those in the rumen, their requirements, responses, and production are probably also very similar.

18.5.1 NPN Supplementation and Synchronizing Carbohydrate and Protein Use

Diets low in soluble carbohydrates and high in mature plant cell wall carbohydrates, such as straw, limit NPN use because of the low energy content and the slow rate of digestion of the carbohydrate that is available. In general, added urea is poorly utilized in such diets because ammonia is produced well before it is needed by the fermentation. Ammonia production peaks at fermentation times long before the maximum fermentation of the low-quality carbohydrates is achieved. The early peaking of ammonia promotes its absorption and loss from the rumen into the blood (Figure 18.12), followed by its conversion to urea by the liver and excretion in the urine. Recycled nitrogen is limited by the size of the metabolic pool and is of limited value for production.

This general problem of urea use has prompted a search for utilizable forms of NPN that have reduced rates of ammonia release. Among the substances that have been tested are biuret and less soluble forms of urea or ammonia. The preparation of products such as Starea and ammoniated rice hulls depends in part on binding nitrogen as an insoluble form through the use of heat, but the process risks the formation of unavailable products.

The availability of high-quality carbohydrate supplements has limited the use of this approach. The addition of rapidly degradable carbohydrate (viz., molasses) promotes utilization of urea, the microbial requirement being about 16% crude protein. In this instance carbohydrate and nitrogen become synchronized, with greatly improved microbial efficiency, pro-

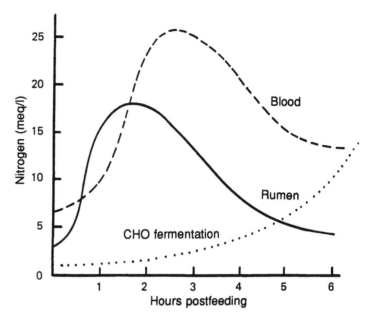

Figure 18.12. The influence of nitrogen supply on rumen ammonia and blood urea levels in the hours after feeding an unbalanced diet. Note that the blood urea concentration peaks about an hour after the rumen ammonia peaks. Carbohydrate digestion in poor-quality diets does not peak until after 4–6 h postfeeding, a time when most of the dietary urea will have been lost from the rumen and the blood.

vided that other requirements (e.g., for sulfur, isoacids, minerals, etc.) are also satisfied.

The problem of matching carbohydrate and protein sources applies to high-production systems as well, particularly high-producing dairy cattle. High levels of intake and production dilute recycled nitrogen to a point at which it becomes unimportant and the rumen becomes dependent on exogenous sources of soluble nitrogen and protein to supply microbial needs. This creates a need for matching sources for both nonstructural and fiber fermenters (Figure 18.13).

18.5.2 Peptides

Peptides supply the more rapidly fermenting organisms and spare the cost of amino acid synthesis by supplying preformed amino acids. Amino acid uptake by many bacteria is more efficient when the acids are in peptide form. Peptide levels can reach 30% of the total rumen nitrogen some hours after feeding (Figure 16.2). Hydrophilic peptides (those with functional groups that attract water, such as free amino or hydroxyl groups) are more rapidly degraded than hydrophobic ones. The more hydrophobic amino acids are those with saturated side chains, such as leucine, isoleucine, and valine. Because the hydrophobic peptides are digested more slowly, they are apt to persist somewhat longer in the rumen and contribute to cross-feeding by other microbes and also to wash out to the lower digestive tract.

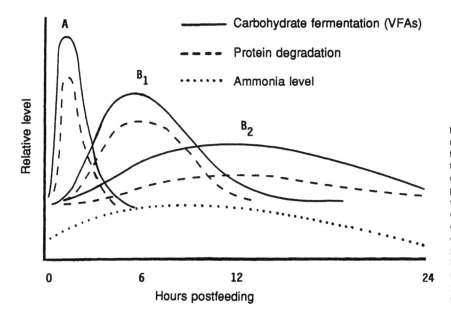

Figure 18.13. Model relation in which release of nitrogen fractions parallels fermentation curves for respective fast- and slow-degrading carbohydrates and optimizes microbial protein output. Peptide production, which is important for the fast pool, depends on the presence of degradable true protein. Ammonia can come from the NPN pool and from peptide degradation. The ammonia level does not rise much because it is competitively recycled into microbial protein. (See Table 18.4 for definitions of A, B_1, and B_2.)

18.6 Dietary Intake and Protein Escape

The protein that reaches the lower digestive tract of the ruminant is the sum of the protein that escaped the rumen fermentation plus the microbial yield. Factors affecting microbial yield involve microbial maintenance, dilution, and the Michaelis-Menten kinetics of growth. Escape often involves indiscriminate loss from the rumen, the consequences of which are variable, depending on the components and the circumstances. Escaped protein may be beneficial, since it will be utilized efficiently in the postruminal digestion as long as it contains essential amino acids. Escaped carbohydrate represents a loss in net energy to the microbes. On the other hand, increased rumen washout enhances microbial yield when expressed as a percentage of feed fermented.

Dietary nitrogen can escape rumen fermentation and pass to the lower tract in quantities sufficient to significantly modify the ruminant's efficiency. Escape can be altered by manipulating digestion or passage rates. It is likely that rumen escape is variable and depends on the type of protein and its rate of degradation, level of intake, rate of passage, and other factors, although there has been a regrettable tendency to regard rumen escape as a constant for a given diet (Chalupa et al., 1991). The National Research Council (NRC) treats escape as a constant, which is not realistic considering the increase in rumen passage with increasing feed intake.

The physical nature of dietary protein—that is, whether it is soluble and moves with liquid or is insoluble and moves with particulate matter—is critical. Particle size is also important. Fresh forages, which contain as much as half of their true protein in a water-

soluble form, ferment rapidly. Probably most of this protein is degraded in the rumen. The high-moisture silages present a similar situation, although most of their water-soluble nitrogen is NPN. In these instances escape of feed nitrogen is on the order of 10–30% of the total amount. Remember that about 5–15% of forage nitrogen is lignin-bound, totally indigestible, and an inevitable part of the dietary nitrogen passed down the tract. Consequently, the available escaped protein is the difference between the two and may be only 0–25% of the nitrogen in fresh forages and silages. Fresh forages with a larger cell wall–bound protein fraction, or which contain moderate levels of tannins, may promote more efficient nitrogen use.

High-fiber protein sources such as distillers' or brewers' grains present another case. If not too heat-damaged they are excellent sources of insoluble protein, prime for escape. The protein is precipitated on the surface of the fiber particles, leading to the movement of protein with undigested fiber to the lower tract. The loss in digestible fiber is less serious if the coarse fiber requirement of the rumen is satisfied.

The concentration of protein in the diet is also a factor influencing escape, since the passage of any ingredient from the rumen depends on its concentration in the rumen. The concentration of undegraded protein in the rumen will depend on the dietary content and level of the intake; thus the character and amount of the protein present will affect the quantity that can be degraded by rumen organisms. For these reasons it might be expected that good-quality feeds with high protein content will elicit the escape of larger amounts of dietary protein than feeds low in protein, and that less easily degradable proteins will also show greater escape. Also, increased feed consumption means fas-

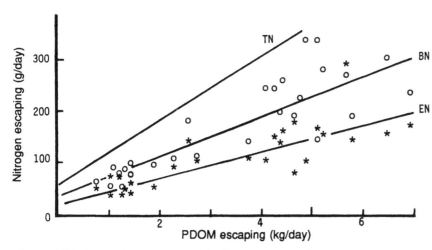

Figure 18.14. The relation between the net microbial nitrogen (BN) and feed nitrogen (EN) leaving the rumen and potentially digestible organic matter (PDOM) escaping the rumen of cannulated dairy cattle (using data from 27 balance experiments performed by P. H. Robinson, 1983). Rumen escape is influenced both by level of feeding and by feeding frequency. The true digestibility of microbial protein is about 70%. Digestibility of escaping feed protein may be similar since it contains the unavailable ADIN fraction. The total nitrogen (TN) leaving the rumen (upper line) is essentially the sum of bacterial and feed nitrogen escaping. (Data points for TN were omitted for clarity.) Open circles indicate BN values, and EN values are designated by asterisks. There is no tendency for bacterial protein to taper off at high levels of PDOM escape, even though at the higher intake (3–4 times maintenance) more than 60% of the available PDOM is leaving the rumen. Regression lines: TN = 61.5(PDOM) + 57, r = 0.93; BN = 38.5(PDOM) + 33, r = 0.87; and EN = 25.3(PDOM) + 20, r = 0.74

ter passage, and therefore greater escape. The same applies to ground and pelleted forages. Many dehydrated forages are reduced in particle size relative to their original form, and the reduced particle size and lowered protein solubility make protein escape more likely.

Treating forages with heat or formaldehyde may reduce protein solubility and increase the quantity of amino acid nitrogen digested in the lower tract (Barry et al., 1973; J. V. Nolan, 1975). In those cases it is difficult to say whether response or lack of response is due to escape, as microbial efficiency and synthesis may be affected by the reduced protein solubility. A moderate reduction in degradation rate should be beneficial, however, because this would provide a steadier supply of ammonia for the fermentation of the slower-digesting carbohydrates. On the other hand, if over-protection forces rumen organisms to become dependent on recycled urea, which is inadequate as a sole support of fermentation, their growth is generally slowed and intake or digestibility will suffer. Complete protection of dietary protein should lead to rumen nitrogen deficiency and responsiveness to NPN at high protein intakes.

18.6.1 Microbial Interaction

Improved animal responses when protected high-quality protein or amino acids are introduced into the lower tract result from a combination of two effects: the greater quantity of essential amino acids and the higher true digestibility of quality proteins compared with the crude protein in rumen organisms. Only about 60–70% of the nitrogen in rumen bacteria is in the form of true protein (which is, however, of high quality). The remainder is composed of nucleic acids and

indigestible cell wall peptidoglycans (Chalupa, 1972). Thus, if high-quality protein is fermented, a portion is converted to unavailable or unusable products (i.e., bacterial structure). The consequence is that promotion of protein escape can actually increase the true digestible protein available to the animal at isonitrogenous intake.

An alternative but not mutually exclusive interpretation is that slow-degrading protein may feed the fiber digesters—the most efficient rumen microbes—thus increasing microbial yield in addition to enhancing escape. Increases in microbial growth through passage and washout are offset by the concomitant loss of potentially digestible feed organic matter, largely carbohydrate, to which microbes are attached, such that the particle flow becomes a harvesting of bacteria. Experimental data from Cornell University (Sniffen and Robinson, 1984) indicate that this judgment is far too conservative. The range of microbial efficiencies (1.7–4.3 g microbial nitrogen per 100 g truly digested organic matter in the rumen) exceeds the effect of passage of potentially digestible organic matter (PDOM) from the rumen (Figure 18.14).

Microbial nitrogen exceeds the washout of feed nitrogen by an average factor of 1.5–1.6 and is not diminished at high feed intakes. These data indicate that the role of rumen bacteria in producing protein is grossly underestimated and explain why feeds containing significant amounts of low-level heat-damaged protein can elicit positive effects on protein nutrition.

18.7 Metabolizable and Net Protein

Traditionally, the quality of protein diets fed to ruminants has been evaluated by its digestible protein content and less often by its nitrogen balance. Neither is satisfactory for the purpose intended. Digestible protein considers only the balance of dietary intake and fecal loss. Feeding NPN in an improperly balanced diet will result in the unused NPN being lost in the urine, but it will be counted as digestible crude protein if only the fecal losses are measured. Nitrogen balance measurements cannot distinguish unused NPN from metabolized nitrogen. Studies that measure productive responses through substitution of protein sources in the diet may be better provided they do not involve an adjustment in net energy. For example, urea contains no metabolizable energy as compared with protein. Since energy is needed by rumen organisms to synthesize protein and the animal expends energy in recycling and excreting excess urea in urine, the problem in ruminant feeding studies of separating energy responses from dietary protein responses has no simple solution.

The concept of biological value, so important in nonruminant nutrition, has remained largely unapplied to ruminant studies because of the complicating factor of the rumen fermentation and the fact that the actual nature of the protein digested by the animal is unknown. Modification of the concept, however, has led to efforts to describe the real protein digested and utilized by the ruminant animal.

18.7.1 Metabolizable Protein

Metabolizable protein (MP) has been defined as the net quantity of true protein or amino acids absorbed in digestion. In ruminants, metabolizable protein represents the net quantity of feed and microbial true protein that is truly digested in the small intestine. While metabolizable energy and nitrogen balance are calculated by difference, values of MP are generally obtained by summing estimates of escaped protein plus microbial protein output from the rumen. Because microbes can use NPN and recycled nitrogen from saliva and rumen wall diffusion, animals at or near maintenance can obtain more MP than is in the diet. As intake and content of crude protein in the diet increase, the recycled nitrogen becomes diluted and is used less efficiently. Figure 18.15 applies a Lucas model to illustrate the effect—that more nitrogen is absorbed than can be accounted for by digestible nitrogen intake.

The intercept at zero nitrogen intake gives a positive output of amino acid, which must reflect the microbial use of recycled nitrogen. Total amino acid nitrogen (TAAN) absorbed exceeds the dietary intake of nitrogen below about 2% nitrogen in the diet, while nonammonia nitrogen (NAN) absorption exceeds the dietary intake of nitrogen up to about 3% dietary nitrogen. The quantity of nitrogen absorbed in excess of that in the diet must be derived from the microbial yield and recycled nitrogen. This may amount to as much as 1.5 g

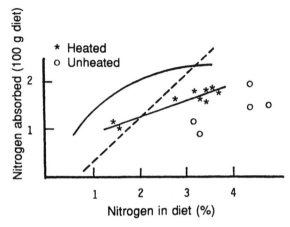

Figure 18.15. The amount of amino acid nitrogen absorbed in the intestines in relation to the nitrogen content of the diet (based on the summary by Armstrong, 1973). This graph is essentially a Lucas test (see Sections 22.1–4) showing that the amount of amino acid nitrogen exceeds the dietary intake (dashed line, same as in Figure 18.11) below a dietary nitrogen content of about 2%. Curved solid line is the approximate relation for the amount of nonammonia nitrogen (NAN) absorbed (Hogan and Weston, 1970), indicating greater amounts of NAN absorbed at dietary intakes below about 3% nitrogen in the diet. Straight solid line is the maximum nitrogen absorption for heated forages. Points below this line indicate the lower efficiency of unheated forages. Slope of the straight solid line indicates about 40% efficiency relative to dietary nitrogen, which is the sum of the responses due to microbial synthesis and rumen escape. If the average true digestibility of rumen microbes is assumed to be 70%, and of dietary protein, 90%, the efficiency range of digestible nitrogen will be 0.4/0.9–0.4/0.7 = 0.44–0.57, depending on the proportions of microbial and escaped feed nitrogen reaching the lower tract.

NAN and about 0.9 g of TAAN per 100 g of organic matter intake at dietary levels of 2–3% nitrogen. This emphasizes the large size of the microbial yield even under conditions of extensive protein escape from the rumen and the importance of recycling urea.

The negative intercept of digestible nitrogen reflects mainly microbial losses (Figure 18.15). These metabolic losses as a proportion of intake do not diminish at high intakes, when rumen bypass or escape is substantial. This is consistent with the expected high microbial yield at high feed intake. Since it is energy availability that sets the microbial requirement, the joint effect of the dietary energy and protein interaction is emphasized. Increasing dietary energy always reduces net metabolism of dietary nitrogen on the part of the animal, whatever the form, and this effect must be primarily from the microbial fermentation. The fermentation of carbohydrate causes more microbial nitrogen to be formed, the true digestibility of which is lower than the dietary source from which it was derived. Optimal nitrogen utilization may involve preferential retention of digestible carbohydrate in the rumen for fermentation and selective escape of true protein, the microbial requirements being met by dietary NPN and recycling of urea.

The feed (protein, carbohydrate, and NPN) converted to rumen microbial protein is the product of the amino acid content of the microbes (40–50%) and the efficiency with which feed nitrogen is converted into microbial matter. This efficiency may be very high if dietary nitrogen is limiting but is less efficient at higher nitrogen levels, probably varying relative to the microbial requirement. The escape of large amounts of feed protein may make the rumen more dependent on feed NPN and recycling. Thus, urea added to rations of 14–20% crude protein has elicited responses in high-producing dairy cattle (Kertz and Everett, 1975).

18.7.2 Net Protein

Net protein (NP), another estimate of protein value, is analogous to net energy since it represents protein in products (e.g., weight gain or milk; see Fox et al., 1990). Net protein has the advantage of being quantified as an output, although the efficiency of NP and amino acid absorption and the maintenance costs deducted from MP are not exactly known. Some information can be gained by comparing dietary protein input and output responses in terms of NP. These kinds of net protein data are open to modeling. Theoretically, NP use can be divided into the efficiency of absorption, which appears to be high, and tissue use of absorbed amino acids. If passage and digestion rates and the size of the turnover pool are known, some models of the digestive metabolic processes involving protein might be fairly accurate.

Unfortunately, the necessary information on amino acids is lacking. The net efficiencies of individual amino acids are very likely different (Clark et al., 1978). These values are confused by lack of specific information on the individual digestibilities of amino acids in the metabolizable protein presented to the intestines. This situation persists because of the continued use of the correction factor N $\times$ 6.25 for expressing bacterial nitrogen as crude protein. The true protein in microorganisms is confined to their cellular contents and constitutes the metabolic machinery of the cells. This protein is diluted by nucleic acids, which, although they are available to fermentation, are functionally an NPN source. The remaining nitrogen in the microorganism is indigestible and likely consists of nonspecific glucosamines, nonessential amino acids, and other nonspecific nitrogenous substances. Thus the division of essential amino acids by the total N $\times$ 6.25 has led to confusion. Despite the early work of Bergen et al. (1967), no modern study has followed up on the distribution of essential amino acids relative to the protease-degradable and undegradable parts of the microbial cellular mass.

Another problem involves the value of recycled nitrogen from mobilized tissue. The NRC assumes that the energy values of mobilized tissue and mobilized

protein (nitrogen) occur in a constant ratio. This assumption is physiologically impossible because of the disparity of energy contents in fat and protein and disagrees with the urea pool and recycling data shown in Figure 18.11. Ordinarily, the animal has a circulatory metabolic pool of urea and amino acids required for its maintenance functions. As I have pointed out, the metabolic pool is limited by the metabolic size of the animal (i.e., its food and water space), and expanding this pool in proportion to dietary intake at high levels of milk production is not possible. This means that the value of recycled nitrogen of all kinds declines with feed intake and level of production and that the intercept of the regression of available recycled nitrogen, including amino acid, upon mobilized energy must have a positive intercept. Although quantitative information is lacking, this hypothesis would go a long way toward explaining the utter failure of the mobilization value estimated by the NRC when applied to tropical lactating animals by Nicholson (1991), who found that the NRC mobilization formula grossly underestimated the quantity of endogenous nitrogen for these animals, which are more limited for protein than for energy.

About 65–70% of MP can be secreted as milk protein in early lactation. The mammary gland absorbs more of some essential amino acids (arginine, valine, isoleucine, and leucine) than are secreted in milk proteins, while other essential amino acids (phenylalanine, methionine, lysine, histidine, threonine, and probably tryptophan) appear to be passed on more or less quantitatively to the milk. Nonessential amino acids (glycine, glutamate, and aspartate) appear in milk protein in greater amounts than those absorbed, indicating net synthesis by the gland or the tissues of the animal.

18.7.3 Tissue Mobilization

Domestic ruminants probably do not have a protein "bank" analogous to fat stores. Tyrrell et al. (1970) noted that cattle in early stages of lactation mobilized up to 30% of the daily milk energy from fat depots but could obtain no more than 15% of their milk protein output from endogenous sources. When these values were employed in the Cornell carbohydrate-protein model (Fox et al., 1990, 1992) to low-producing lactating cows in Venezuela, producing 5–10 kg milk per day, considerable underestimation of the endogenous contribution to milk protein occurred (Nicholson, 1991). The problem involves amino acids and their metabolism.

Mobilizable amino acids are probably obtained from the metabolic pool of small nitrogenous compounds in the blood. Even at maintenance levels this pool, which includes amino acids and urea, has to exist to maintain

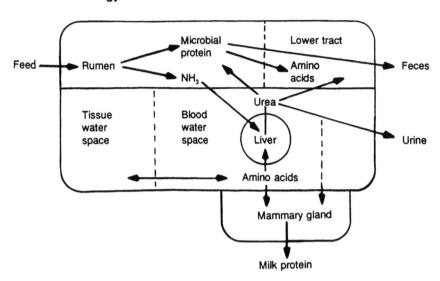

Figure 18.16. This model of the nonprotein nitrogen (NPN) pool in a lactating cow supports recycling of nitrogen to the digestive tract and supply of amino acids to tissues and organs. The pool's size is homeostatically regulated based on body size in relation to the digestive tract. (Areas in the figure are not indicative of the actual relative sizes of the respective water spaces or their nitrogen subpool in the animal.)

vital homeostatic functions such as gluconeogenesis, which depends on glucogenic amino acids. Some of the amino acids can come from microbial protein synthesized from the recycled NPN. Since the metabolic pool is limited by the host's metabolic size (muscle mass?) plus the capacity of the gastrointestinal tract to contain nitrogenous substances (in turn regulated by body size), it can be expected that the value of mobilized nitrogenous tissue is greater at low intakes and production than at higher ones (see Section 25.6.7). This expectation agrees with the argument in Section 18.7.2 and would follow the turnover law discussed in Chapter 23 and shown in Figure 18.16.

The net mass of the NPN pool, which includes all low-molecular-weight nitrogen compounds, is physiologically limited by the fluid mass of the body plus that of the digestive tract. This volume is essentially controlled by body size and by homeostatically controlled concentration limits. The logical metabolic consequence of this homeostatic regulation is dilution of its supply value at high intakes and production.

The NRC and other nutrition requirement systems treat mobilized tissue as having a relatively constant composition with a zero intercept at maintenance or zero tissue balance. In the context of the foregoing discussion this appears unreasonable and inconsistent with biological principles.

18.8 Amino Acid Requirements

Ruminants probably have amino acid requirements similar to those of nonruminants. This can be inferred from the specific requirements of the young animal with a nonfunctioning rumen compared with the amino acids found in milk. A comparison of the respective amino acid compositions of tissue and milk with that of rumen microbes and feed ingredients (Table 18.7) indicates that methionine, histidine, tryptophan, and, pos-

sibly, leucine may be limiting for milk production, and methionine for wool production. Microbial and animal tissue compositions also vary, however. Direct comparison of such values assumes complete efficiency in the utilization of amino acids, an assumption that is not totally realistic.

An important question concerns the quality of protein intended for escape in relation to microbial protein, the common denominator to which much feed nitrogen is converted. Certain amino acids are relatively abundant in rumen microbes; these include lysine, serine, threonine, phenylalanine, isoleucine, and valine (Table 18.7). A comparison of the amino acids in rumen organisms and those in egg protein (the nutritional standard for the biological value of protein) shows that threonine, phenylalanine, and lysine are higher in rumen bacterial protein, but valine and leucine are lower (Bergen et al., 1967). The high level of lysine in bacteria may be important in heat-treated feeds, in which lysine availability is important, and high-grain diets, especially those based on maize which are low in lysine.

18.8.1 Limiting Amino Acids

Experiments have been conducted in the interest of determining limiting amino acids. One way to determine limiting amino acids is to show that nitrogen retention is responsive to expected limiting acids administered through a duodenal fistula. The problem with this approach is that results may reflect inadequacy of the escaping dietary source, which is often from plant seeds that are unbalanced sources. Thus limitation becomes a relative thing. Many experiments have focused on lysine and methionine. The experience of C. J. Sniffen (pers. comm., 1990) is that diets can be designed in which protected lysine and methionine appear to be limiting; however, optimization of rumen fermentation and microbial protein synthesis

Table 18.7. Essential amino acids in protein

Protein	ARG	HIS	ILE	LEU	LYS	MET	PHE	THR	TRP	VAL
Cattle tissue[a]	6.3	2.4	2.7	6.4	6.2	2.2	3.3	4.0	0.7	3.8
Muscle[a]	7.7	3.3	6.0	8.0	10.0	3.2	5.0	5.0	1.4	5.5
Keratin[a]	3.8	1.0	5.0	10.0	3.2	1.0	3.7	7.2	1.4	6.0
Sheep wool[b]	—	1.3	2.4	5.9	3.6	0.5	2.9	5.8	0.7	4.5
Cow's milk[c]	3.7	2.7	6.0	9.8	8.2	2.6	5.1	4.6	1.4	6.7
Rumen microbes[d]	—	2.2	7.3	9.4	11.3	2.6	6.8	6.4	0.6	7.2
Bacteria[e]	9.1	2.3	6.4	7.3	9.3	2.6	5.1	5.5	—	6.6
Protozoa[e]	9.0	2.0	7.0	8.2	9.9	2.1	6.1	4.9	—	5.3
Insoluble proteins[f]										
Alfalfa	2.4	0.6	3.1	6.4	3.2	1.2	4.2	3.3	1.3	—
Corn grain	1.8	2.1	2.7	10.7	1.6	1.1	3.6	2.8	1.0	3.8
Corn gluten	3.2	2.4	4.3	16.2	1.2	2.1	6.5	2.9	—	—
Corn silage	1.9	1.1	2.4	6.4	2.1	0.8	2.9	2.1	—	3.2
Dried brewers' grain	2.6	1.5	3.5	8.5	2.1	1.3	4.8	2.8	—	3.9
Soybean meal	8.4	2.4	4.2	6.7	5.7	0.8	4.4	3.3	1.3	3.8

Note: Values are percentages of true protein.
[a]Average values from the literature summarized by Ainslie (1991).
[b]Wool also contains 7–13% cystine and 3.5% sulfur (Ryder and Stephenson, 1968).
[c]Average from Jacobson et al., 1970, and J. D. O'Connor, pers. comm., 1991.
[d]Burroughs et al., 1975.
[e]Purser and Buechler, 1966.
[f]Muscato et al., 1983.

usually abolishes such limitations. For example, the problem is expressed on high corn silage diets with insoluble seed proteins, but it is not apparent on diets with high-quality forage, such as alfalfa (C. J. Sniffen, pers. comm., 1990).

Virtanen (1966) designed experiments in which protein-free diets containing urea were fed to lactating animals. Blood levels of arginine, lysine, histidine, phenylalanine, leucine, isoleucine, and valine were found to be substantially lower in urea-fed animals than in animals fed a normal diet. Histidine, in particular, was much lower. The high protein content of milk and the large energy demands of lactation may have placed the amino acid requirements of lactating animals well above the yield from microbial fermentation. On the other hand, the microbial yield may have been reduced by the purified diet, which did not contain adequate isoacids or sources of phenyl groups for aromatic amino acids.

18.9 Dietary Additives

The beneficial effects of adding protected amino acids for lower tract digestion have already been mentioned, but there are other additives or supplements as well. These include isoacids, ionophores, tannins, and yeasts, and all have effects on rumen fermentation. The various additives operate in different ways, often not in the way inexperienced wisdom would have predicted.

18.9.1 Isoacids

Isoacids were discussed in Chapter 16 as essential factors for the growth of many cellulolytic bacteria and

as limiting rumen fermentation because they are not preferred for oxidative deamination. Isoacids are relatively ineffective in low-quality diets even when protein may be marginal, because recycling of the respective carbon structures is sufficient in slower fermentations. Attempts to market isoacids as a feed additive have not been successful, although, when added to well-constructed diets designed to optimize output of microbial protein, isoacids have positive effects on milk yield. Two problems have been failure in herds fed suboptimal forage and the handling of the noxious smell.

18.9.2 Ionophores and Protein Responses

Ionophores were discussed in Chapters 9 and 16 relative to their basic biological effects, which include chelation of sodium and potassium and interference with hydrogen transport in cells. Ionophores interfere with protein and amino acid degradation. Since ionophores inhibit proteolysis, it can be predicted that they may be valuable for treating rumen proteolysis that produces high concentrations of ammonia. Ionophores have also been suggested for treatment of acidosis. The problem here depends on the nature of soluble nitrogen, which is not necessarily all protein. The soluble nitrogen of hays and silages is virtually all NPN, although other supplemental feeds may contain appreciable true protein. Thus it can be expected that ionophores will be ineffective in conjunction with silages and more valuable when used with rapidly degradable concentrate sources.

Ionophores generally reduce microbial efficiency because they inhibit hydrogen recycling and thus add to the maintenance cost of the organisms. One problem

Table 18.8.   Fractionation of fecal nitrogen showing that indigestible and microbial nitrogen comprise the bulk of the total

Sheep no.	Ration[a]	Total fecal N	ADIN	NDIN	Probe-resistant N	Whole feces	BED	IB	BED-IB ratio
76	A	46	7.9	8.6	8.0	0.56	0.88	0.84	1.05
79	A	47	8.3	8.5	8.9	0.44	0.70	0.69	1.02
80	B	44	4.8	4.2	5.1	0.48	0.82	0.95	0.85
81	B	45	4.6	5.1	5.1	0.45	0.73	0.86	0.84
41	C	17	1.2	1.3	1.6	0.49	0.61	0.56	1.09
99	C	23	1.3	1.2	1.6	0.51	0.68	0.78	0.87
09	D	—	—	—	—	1.08	1.56	1.13	1.38
10	D	—	—	—	—	1.53	2.18	1.75	1.25
49	E	—	—	—	—	0.63	1.48	2.08	0.71
47	E	—	—	—	—	0.29	0.77	1.19	0.64

Source: Mason, 1969.

Note: ADIN = acid-detergent insoluble nitrogen; NDIN = neutral-detergent insoluble nitrogen; BED = bacterial and endogenous debris; IB = isolated bacteria. Values of whole feces, BED, and IB are diaminopimelic acid nitrogen as a percentage of total nitrogen. Values of total fecal nitrogen, ADIN, NDIN, and probe-resistant nitrogen are expressed as percentages of total nitrogen.

[a]A = clover-ryegrass hay; B = dried ryegrass; C = high concentrate; D = paper pulp; E = high starch, low cellulose.

with adding biologically active agents to the rumen is that most of them have a negative effect on microbial efficiency.

18.9.3 Yeasts and Enzymes

Yeasts, microbial products, and enzymes are said to improve cellulose digestion, feed intake, and animal performance. Live yeast and other species of bacterial cells have been added to feed on the assumption that their introduction into the rumen would add to the rumen fermentation. But yeasts, *Lactobacillus,* and other microbes do not survive in the rumen, although a small population can probably be maintained by continuous feeding. Since none of the added live organisms can digest structural carbohydrates, their role must involve some cross-feeding of essential factors that might be limiting to the main rumen population. The possibilities are quite broad and include amino acids, peptides, vitamins, and enzymes.

Enzymes have been added to the rumen to improve enzymatic efficiency. Since such additions are open to proteolytic degradation by *Streptococcus* and protozoa, the enzymes added would have to be resistant to degradation. Lectins, which are related to the antitrypsin factors in plant feeds, might have value in reducing proteolytic activity.

18.10 Composition of Nitrogenous Fecal Matter

The ultimate limit of adaptability to low-nitrogen diets is set by the obligatory nitrogen losses in the feces and urine. Fecal loss is less flexible than urinary loss, since the fecal nitrogen is largely indigestible micro-

bial matter, which tends to be in proportion to the intake of dry matter. Net fecal loss is a balance of factors: the net microbial yield, the digestibility of microbes, and the efficiency and maintenance cost of bacteria. The passage of potentially digestible carbohydrate to the cecum and colon can increase fecal losses through fermentation of resultant microbial matter (Ørskov et al., 1972).

Fecal nitrogen losses in ruminants average about 0.6% of the dietary dry-matter intake. This value is equivalent to about 3–4% of dietary protein and is an absolute minimum for dietary adaptation on the part of the animal, assuming no other losses through urine, shedding of hair, and so on. The real limit is somewhat higher than this, in the range of 6–8% crude protein in the diet. Below this level intake and efficiency of rumen digestion decline through lack of nitrogen for rumen microbial functions. Fecal nitrogen loss and negative nitrogen balance can be exacerbated by addition of highly digestible carbohydrate to low-nitrogen diets.

A variety of metabolic materials are excreted in the feces; these include microbial debris and endogenous substances (calcium and magnesium salts of fatty acids, bile salts, some sloughed-off animal cells, mucus, and keratinized tissue). Microbial debris, mainly bacterial cell walls and some bacteria, consists of soluble and insoluble nitrogenous matter. The soluble nitrogenous fraction includes considerable ammonia produced by microbial degradation. The insoluble nitrogenous matter in ruminant feces is mostly resistant to peptic digestion, indicating that it contains little true protein. There is no evidence of potentially digestible feed protein in normal feces. Any feed protein that arrives in the feces is quantitatively insoluble and either is keratin or a Maillard product or is bound to lignin and therefore resistant to peptic digestion. Table

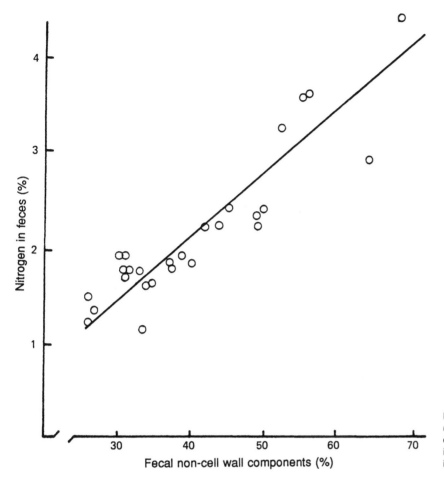

Figure 18.17 The relation of fecal non–cell wall matter to fecal nitrogen content. Slope of the line is 7%; the intercept at zero nitrogen (not shown) is negative (−0.62).

18.8 provides a breakdown of the fecal nitrogen fraction. The ratio of diaminopimelic acid in the metabolic matter to that in isolated bacteria indicates that microbial matter accounts for most organic metabolic matter in feces.

The largest sources of microbial matter in ruminant feces are the indigestible cell walls from rumen bacteria plus cells from fermentation in the lower tract. Whole bacteria are probably more significant in nonruminant feces. Endogenous excretions have been fermented to an undetermined extent such that the separation of endogenous (nonmicrobial) and microbial matter in feces is not of much physiological significance. Endogenous matter forms only 10–15% of the metabolic fraction. The nitrogen content of the organic metabolic matter of ruminant feces averages about 7% (Figure 18.17), a value in close agreement with the composition of microbial cell wall (Salton, 1960). The nitrogen content of metabolic organic matter in nonruminant feces is higher, about 9%, and probably contains a larger proportion of living microbial cells. All microbial matter, whether in the form of cell membranes or intact cells, must, along with some endogenous matter, be contained within the organic metabolic

output. Therefore, microbial matter is limited to an amount less than the metabolic constant (M_i).

The nitrogen concentration in the feces is increased through loss of fermentable matter from the rumen to the lower tract (Ørskov et al., 1972). This fraction can be directly accounted for by subtracting the nitrogen content in the neutral-detergent residue or acid-detergent residue (insoluble fecal nitrogen) from the nitrogen content of fecal matter. The increment due to lower tract fermentation is probably mainly microbial cells. Neutral-detergent fiber will contain keratins (skin proteins) in addition to insoluble feed proteins unless it is treated with sodium sulfite to dissolve insoluble keratin products of animal origin. These proteins are of endogenous origin if the diet is solely from plant sources, but if rations are supplemented with animal protein, the fecal keratins could be derived from both sources.

Microbial cell walls are composed of substituted glucosamine (muramic acid) polymers with attached peptides (Figure 18.18). Among the amino acids present is diaminopimelic acid, which is unique to bacteria. Aminoethylphosphonic acid is an unusual amino acid found in holotrich protozoa but not in bacteria

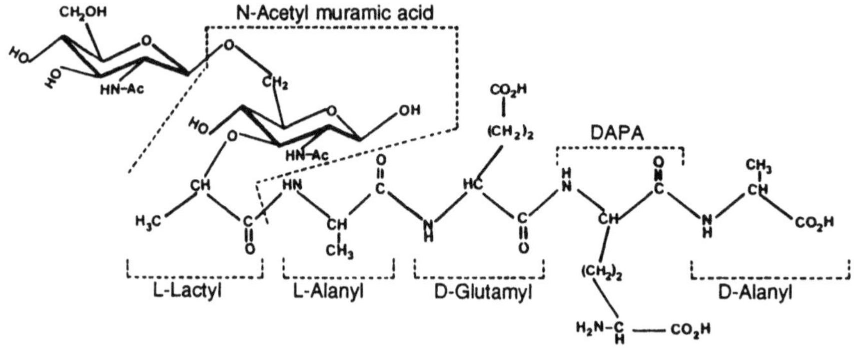

Figure 18.18. Structure of a typical bacterial cell wall component (muropeptide C6 from *E. coli*) (see Salton, 1960). Note the lactyl ether on the third position of glucosamine. Sugar rings are shown in spatial configuration.

(Abou Akkada et al., 1968). Another component group found in microbial cell matter in feces is the teichoic acids, which are polymers of ribitol and glycerol phosphates with alanine side chains. These are cell wall substances with a strong capacity for binding magnesium and other metal ions. The cation exchange capacity of microbial cell wall in rumen feces is about 20 meq/g.

18.10.1 Metabolic Fecal Nitrogen

Early nitrogen-balance studies showed that fecal nitrogen tended to be a relatively constant proportion of the dietary intake. Nutritionists were unaware of the microbial contribution to feces and fecal nitrogen and regarded the fecal nitrogenous matter as protein from dietary and endogenous origin. This view still persists in many general nutrition textbooks, but it must be regarded as fundamentally incorrect with regard to ruminant nutrition and probably with regard to nonruminant nutrition as well. The error arises from a failure to recognize that metabolic fecal nitrogen is mostly composed of microbial matter (Mason, 1979). True protein, if present in feces, either is confined to keratinized tissues, living microbial cells, or Maillard products or is incapsulated in seeds. In ruminants, the content of fecal true protein is very low.

A further error considers the large amounts of nitrogen in enzymes and other gastrointestinal secretions of most animals as part of fecal protein. Almost everything that is degradable is absorbed, however, and only the most resistant matter survives to the feces to become a part of the true endogenous fecal nitrogen. It is perhaps significant, then, that microbial cell wall residues constitute the major part of fecal nitrogen. The microbial cell walls have been suggested as possible substrates for refermentation in the lower tract. My own opinion is that microbial cell wall is not an important lower tract substrate. It is far more likely that the large proportion it forms of fecal nitrogen in ruminant feces indicates its resistance to degradation.

Much of the fecal nitrogen may originally have been

endogenously secreted, but by the time of excretion it has been largely converted to microbial matter through fermentation (Mason, 1984). Endogenous secretions may indeed provide the major substrate for lower tract fermentation. The total fecal organic matter fraction is a composite of microbial cell wall from the rumen and microbial matter arising from both endogenous secretions and dietary carbohydrate that has escaped the upper digestive tract.

The conclusion that true protein does not form a large part of fecal nitrogen does not conflict with the presence of a wide amino acid spectrum in fecal extracts. Amino acids are contained in the keratinized tissues, microbial cellular debris, and heat-damaged (Maillard) proteins of dietary origin. Much of the microbial amino acid may be in the D form. Thus the insoluble fecal nitrogen fractions may have a lower-than-expected biological availability for refeeding or fermentation (Mason, 1984).

Fecal nitrogen is often expressed as crude protein (N $\times$ 6.25), an expression that grossly underestimates the actual organic matter associated with metabolic fecal nitrogen. Since metabolic organic matter contains 7% nitrogen (Figure 18.17), it is apparent that the correct factor is about 14, more than twice 6.25. This underestimation accounts for about half of the apparent nitrogen-free extract fraction in feces, the remainder being composed of lignin and hemicellulose dissolved by alkali in the crude fiber determination.

Metabolic fecal nitrogen has been estimated by (a) measuring fecal nitrogen from animals on a nitrogen-free diet, (b) extrapolating digestible nitrogen to a zero intake of nitrogen (basically a Lucas test), (c) measuring fecal nitrogen using rations in which the true digestibility of protein is assumed or known to be 100%, (d) internally labeling protein, and (e) measuring the increment in fecal nitrogen obtained when dietary nitrogen is increased. Metabolic fecal nitrogen may be determined directly in any fecal sample as the difference between total fecal nitrogen and residual nitrogen in the neutral-detergent residue (Mason and Frederiksen, 1979). Microbial nitrogen occurs in much greater amounts in feces than does residual endogenous matter. Virtanen (1966) thought that the proportion was about 9:1, in agreement with Mason (1979). Although a great deal of metabolic nitrogen is secreted into the gut, most of this is fermented and resorbed, so the fecal composition of nitrogen consists of indigestible intractable structures in microbial cell wall, keratins, Maillard products, and other resistant debris.

19 Intermediary Metabolism

To comprehend and manipulate whole animal functions, one must understand the intermediate processes in animal metabolism. Biochemical pathways and metabolic regulation are extensively discussed elsewhere; this chapter gives an overview of the subject that integrates unique aspects of ruminants.

19.1 Ruminant Metabolism

The true metabolic diet of the ruminant is not what is eaten, but rather the combined fermentation products and unfermented feed that escape from the rumen. The net changes in the ingested feed include conversion of dietary protein and nitrogen into microbial protein and metabolism of carbohydrate into a variety of noncarbohydrate products. The problem posed for the ruminant is how to meet its requirements for nutrients that are destroyed by rumen fermentation. These may include glucose and other factors such as polyunsaturated acids. Some glucose can be supplied either through rumen escape or bypass, but the balance must be obtained by gluconeogenesis from other absorbed products. Low-quality forages probably cause the greatest metabolic strain because they contain little available carbohydrate or protein.

The normal ruminant diet is very low in available carbohydrate, and the animal carbohydrases are practically limited to those that split sucrose, lactose, and starch. These substances are largely fermented in the rumen to volatile fatty acids (VFA), however, and grazing ruminants seem to have adapted to diets perpetually low in available carbohydrate. Ruminants have efficient gluconeogenic capabilities, and the lower digestive tract has adapted to the lack of sugar and starch and may even have only a limited capacity to handle these compounds. If there is a shortage of available carbohydrate, the animal is unable to metabolize fatty acids beyond the ketone body stage, resulting in ketosis. No animal can synthesize glucose from fat, whereas in germinating plants, triglycerides stored in seeds are readily converted back to carbohydrate through the glyoxalate shunt pathway.

All animals require glucose as a common circulatory pool for energy. Entry rates of glucose are related to metabolic body size, and ruminants seem to differ little from nonruminants in this relationship (Figure 19.1). The replacement of glucose by xylose is limited (Table 19.1), but the glycerol portion of triglycerides is a very important gluconeogenic source, as are glucogenic amino acids. Elimination of all carbohydrate sources results in a rise in ketone bodies in the blood (Renner, 1971). If unused, these intermediate products of fat catabolism are excreted in the urine and are a prime characteristic of ketosis (Section 19.8).

Diets low or high in available carbohydrate result in metabolisms that represent the animal's adjustments to different feed mixtures.[1] Thus, high-starch diets lead to fat production in both ruminants and nonruminants, but the production of acetate units for fat synthesis is accomplished largely in the rumen in the former, as opposed to in the liver or adipose tissues in nonruminants. Absorption of carbohydrate in excess of energy requirements requires its dissimilation into noncarbohydrate products (principally fat), a process that normal nonruminant metabolism has no difficulty in accomplishing. The reverse process—net conversion of even-carbon fatty acids back to carbohydrate—is not possible, hence a shortage of available glucose causes problems for ruminants.

Newborn calves possess the enzymes to hydrolyze lactose but not sucrose or starch, and the capacity to hydrolyze starches may be limited even in adult ruminants. Excess starch in the diet escapes from the rumen and the small intestine and is fermented in the lower tract. Some of this starch will be lost in the feces. Much of the starch that escapes is of the resistant type and does not ferment at fast rates, so the likelihood of lower tract acidosis is small. Grazing ruminants do not depend on starch digestion or glucose absorption in the lower tract. Glucose is absorbed from the ruminant small intestine by a transport system that is less active than is characteristic of nonruminants.

[1] Available carbohydrate is defined here as glucose-yielding dietary sources actually absorbed by the animal.

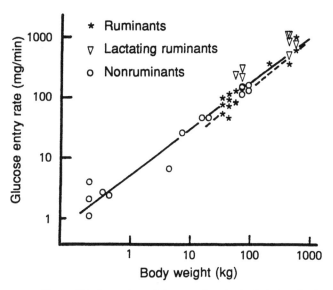

Figure 19.1. Comparison of glucose entry rates for fasted ruminants and nonruminants (data from Ballard et al., 1969; see also R. W. Russell et al., 1986). Solid line represents the regression for nonruminants. Dashed line represents the regression for nonlactating cattle and sheep, which appear to have glucose entry rates nearly as great as nonruminants. Slopes of the lines are roughly 0.75 power.

19.1.1 Development of Ruminant Metabolism

Infant ruminants are essentially nonruminants; they lack a fermentation and require dietary vitamins and amino acids. Their blood glucose level is high, like most nonruminants. As the rumen develops, the metabolism adjusts toward the adult condition with a decline in blood glucose—which does not seem to depend on rumen development, because its prevention does not limit this decline (Figure 19.2). The decrease in blood glucose concentrations seems to be an evolutionary adaptation of the adult ruminant to the perpetual necessity of carbohydrate conservation.

Adult ruminants maximize glucose conservation, and they have a number of metabolic alterations not seen in nonruminants that help in the process (Table 19.2). Among the metabolic changes associated with glucose conservation is the virtual elimination of citrate lyase and NAD-dependent malate dehydrogenase which prevent the direct conversion of glucose to acetate and fatty acids in ruminant tissues, thus conserving glucose for its most essential functions, particularly supplying energy for brain tissue. Ruminant tissue is able to take up glucose in the face of hypoglycemia that would produce coma in nonruminants.

Another difference between mature ruminants and nonruminants is the response of blood glucose to fasting. In nonruminants gluconeogenic mechanisms maintain blood glucose concentrations throughout a fast; in ruminants, however, concentrations decline within a day or two. The relative tolerance of ruminant tissues to low blood glucose may be possible because

Table 19.1. Response of chicks to glucose, lactose, and xylose when fed "carbohydrate-free" diets containing soybean fatty acids (SFAs)

Energy source	Supplemental CHO Type	Level (g/g SFA)	Average wt. at 4 wk (g)	Kcal consumed/g gain	Blood ketone bodies[a] (mg/dl)
	Treatment				
SFA	None	0	379	5.06	33
SFA	Glucose	0.035	422	4.80	12
SFA	Glucose	0.105	470	4.90	6
SFA	Lactose	0.034	410	4.91	28
SFA	Lactose	0.102	418	4.92	14
SFA	D-Xylose	0.030	420	4.96	29
SFA	D-Xylose	0.090	454	4.82	10
Soybean oil	None	0	519	4.84	21

Source: Renner, 1971.
[a]Total ketone bodies as acetone.

tissues can utilize other energy substrates. There is no great rise in the level of blood glucose after a meal in ruminants, as is the case in nonruminants, nor does the administration of large amounts of propionate cause any great rise in the blood glucose level. After a meal gluconeogenesis is generally suppressed in nonruminants, while it is increased in ruminants because of increased substrate availability.

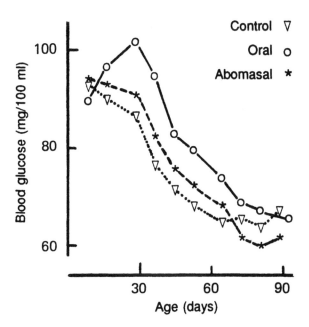

Figure 19.2. Blood sugar levels in calves fed orally and through abomasal fistulae with concentrate ration (from Nicolai and Stewart, 1965). Control animals were offered alfalfa hay. Abomasal-fed calves did not develop a rumen fermentation. Preventing rumen development did not prevent or change the decline in blood sugar with age. Most of the lactose of the high-milk diet would have been fermented to lactic acid (Lengemann and Allen, 1955), leading to the need for gluconeogenesis from lactate. A more definitive experiment might be done with germ-free calves or by administration of glucose via abomasal cannulae.

Table 19.2. Metabolic differences between ruminants and nonruminants

	Ruminants	Nonruminants
Blood glucose	40–60 mg/100 ml; falls on fasting; unchanged after feeding	80–100 mg/100 ml; unchanged on fasting; rises after feeding
Recycling glucose	Little recycling	Considerable recycling
Lipogenesis	Main source is acetate; enzymes lacking for synthesizing fatty acids from glucose	Glucose is a major source
Liver hexokinase	Virtually absent	Present
Response of brain to hypoglycemia	Uptake of glucose largely unchanged	Uptake of glucose decreases
Ketone bodies	Brain does not use; used by peripheral tissues	Can be used by brain

19.2 Metabolism of Volatile Fatty Acids

Intermediary metabolism in ruminants is largely the same as in other mammals. It differs principally in the quantities of carbon passed along certain pathways due to the very low net absorption of glucose and high absorptions of acetic, propionic, and butyric acids from the gastrointestinal tract. Glucose conservation and gluconeogenesis are special features of the general metabolic adaptations of the ruminant. The general metabolism of the VFAs is shown in Figure 19.3. In general, acetate and butyrate are the major energy sources (for oxidation), and propionate is reserved for gluconeogenesis. Acetate is the most important lipogenic precursor.

Acetic acid is the largest single dietary component absorbed (to the extent of at least 90%) from the reticulorumen. Propionic and butyric acids are also absorbed in large amounts. Volatile fatty acids are absorbed in the free form and, apart from rumen wall metabolism, pass into the hepatic portal blood to circulate as neutralized anions at blood pH. Rumen epithelium metabolizes considerable amounts of VFAs, the quantities being the largest for butyrate and the least for acetate. There may be a species difference with regard to use of propionate. Sheep rumen epithelium appears to metabolize larger quantities than that of cattle, in which the maximum amount metabolized may be only 3–5%. Portal blood circulates through the liver, which takes up most of the remaining propionate and butyrate such that acetate represents 90% or more of the VFAs in the peripheral circulation. Acetate is taken up and used for energy or for lipogenesis by various body tissues.

Both acetate and butyrate must be metabolized to acetyl-CoA in order to be used for energy by oxidation in the citric acid cycle (Figure 19.3). Administration of acetate does not give rise to extra ketone bodies as does butyrate, because it does not have to pass through that stage for metabolism. Acetate may enter fatty acid synthesis as acetyl-CoA or through carboxylation to

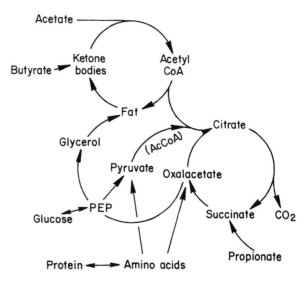

Figure 19.3. The main pathways for metabolism of volatile fatty acids. Butyrate and propionate are largely metabolized by rumen wall and liver, respectively, and acetate is largely utilized in the peripheral tissues. Ketone bodies and glucose are among the principal substances transported by the blood in shuttling energy from the liver to the tissues. Arrows indicate the main direction of energy flow. See Figure 19.4 for more on propionate metabolism and regulation of gluconeogenesis.

malonyl-CoA; alternatively, it may enter the citric acid cycle through condensation with oxaloacetate.

Butyrate is partly metabolized by rumen epithelium to ketone bodies—principally acetoacetate and (D-)β-hydroxybutyrate, which are interconverted in the liver. When energy stores need to be mobilized from body fat, as may occur in undernutrition or a situation of high energy demand, mobilized fatty acids are metabolized by the liver to (L-)β-hydroxybutyrate, which enters the peripheral circulation. Excess levels of ketone bodies that occur in animals under severe stress or under conditions of impaired ability to metabolize them result in the pathological condition known as ketosis (Section 19.8); however, the mobilization process is a normal one. Ketone bodies occur in substantial concentrations in the peripheral circulation and, like acetate, are utilized by the body tissues.

19.2.1 Propionate Metabolism

The major site of propionate metabolism is the liver, which takes up about 80% of the blood propionate load in one pass. Propionate is the only major VFA that is glucogenic. The metabolic pathway from propionate to glucose involves conversion to propionyl-CoA and carboxylation to methylmalonyl-CoA, which is followed by rearrangement of the carbon skeleton to form succinyl-CoA (Figure 19.3). This last step requires vitamin B_{12} as the coenzyme (Elliot, 1980).

Propionate carbon enters the citric acid cycle in the form of succinate and in a few steps is converted to the important intermediate oxalacetate. Several routes of metabolism are open from oxalacetate. It may be directly converted to glucose via a reversal of the Embden-Meyerhof pathway; it may be condensed with acetyl-CoA to form citrate; or the respective keto acids may be reductively aminated to form the glycogenic nonspecific amino acids aspartate, glutamate, and alanine (Bergman and Pell, 1985). Studies with labeled propionate indicate that about 50% is converted to glucose, with a portion appearing as lactate (Annison and Armstrong, 1970), but this is probably an underestimate because considerable crossover of carbon can occur. Steinhour and Bauman (1986) estimated the true value to be nearer 90%. Moreover, propionate may function in glucose conservation without being converted to glucose at all. Propionate will eliminate ketosis and spare insulin in diabetics (J. W. Anderson, 1985).

Comparison of the net daily glucose turnover with propionate production (Elliot, 1980) suggests a regression slope near the theoretical 1 mole of glucose per 2 moles of propionate. Amino acids probably are second in importance, but their contribution may be overestimated. If amino acids were to supply 20% of the glucose requirement of a cow producing 40 kg of milk, one-third of the digestible protein required for this level of production would have to be diverted to gluconeogenesis. Such a level would seem to be in competition with the protein needs of the lactating cow.

The metabolism of other odd-carbon fatty acids (e.g., valerate) parallels that of propionate and acetate, into which they are split. Generally, the metabolism of any long-chain, odd-carbon fatty acid involves production of at least one propionyl group on degradation. Branched-chain compounds and isoacids (e.g., isobutyrate, isovalerate, etc.) also yield a glucogenic moiety on metabolism.

Feeding or injecting propionate markedly decreases blood levels of ketone bodies. The mechanism by which this occurs is not well understood. In general, interpreting blood level responses caused by diet change is difficult. For example, both propionate and butyrate stimulate gluconeogenesis, but only propionate is involved in a net conversion to glucose. The effect of butyrate, which is ketogenic, is to stimulate production of oxalacetate, which is required for normal metabolism of ketone bodies in the citric acid cycle.

19.3 Entry Rates

Entry rate is defined as the net quantity of a given metabolite received by the body via blood per unit time. In the case of glucose this means the sum of glucose synthesized by the liver and kidney, plus any

absorbed from the digestive tract. In the case of the VFAs, it more often has been interpreted as the amount derived from rumen fermentation and absorption, but this is an oversimplification because acetate is also produced endogenously (Section 19.3.3). Measurements of the concentrations and amounts of VFAs are an important part of ruminant nutrition because these acids make such a large contribution to the dietary energy of the animal. Their measurement has been studied for more than 40 years and has presented pitfalls and misinterpretations.

The earliest methods for VFA determination were by fractional distillation, a procedure that was not always quantitative, which led to the expression of the respective acids in molar percentages. The use of molar percentage to predict net production of VFAs led to misinterpretations. Molar proportions are the sum of the expressed VFA values adding to 100. The hazard is that a degree of freedom is lost in the expression, eliminating the variation in rumen dilution. If there is an increase in real concentration of one acid, the proportions of the others must decline statistically even though their real amounts may have remained the same. This problem applies particularly to the association of VFA proportions with milk fat depression in dairy cattle (Chapter 20.7). When a large amount of grain and little forage are fed to lactating cows, the molar percentage of acetate declines, largely because of increased production of propionate, although the actual amounts of acetate produced may not have changed.

Rumen VFA levels are best expressed as concentrations (e.g., meq/liter dry matter). If molar proportions are absolutely necessary, total acid content should also be given. Still, proportions are important in understanding carbon balance and methane production in microbial metabolism (Section 16.7.1).

19.3.1 Measuring Net VFA Production

Rumen volume and turnover of each VFA must be known to determine the total daily output of VFAs. The pool size is the product of the concentration of VFAs and the rumen volume. The dilution rate of a labeled dose of a VFA follows first-order kinetics and is assumed to represent the production rate. Rumen volume can be estimated by extrapolation of the dilution rate to zero time (dosing time) concentration if one assumes instantaneous mixing of the tracer with rumen contents, although this may generate a small error because there is usually some lag and imperfect mixing (Bauman et al., 1971) as well as metabolic diversion in the rumen. It is better to estimate volume by direct rumen emptying (Colucci, 1984).

Examples of VFA production in the rumen are given in Table 19.3, which shows that VFA concentrations

Table 19.3.   Estimates of acetate and propionate production in lactating cows

	[14]C acetate		[14]C propionate	
	Normal	High grain	Normal	High grain
Milk yield (kg)	20.3	20.3	17.9	21.5
Milk fat (%)	3.2	1.6	4.0	1.9
Rumen fluid				
Volume (l)	34	37	40	42
Concentration (meq/l)	70	55	21	57
Molar percentage	67	53	21	47
Acetate-propionate ratio	3.3	1.3	3.3	1.0
Pool (moles)	2.39	2.06	0.83	2.42
Turnover time (min)	118	105	93	113
Entry rate[a] (moles/day)	29.3	28.1	13.3	31.0

Sources: Summarized from Davis, 1967, and Bauman et al., 1971.
Note: The acetate experiment (Davis, 1967) and the propionate experiment (Bauman et al., 1971) were done separately and have individual controls.
[a]Entry rate from the rumen; endogenous production is ignored.

and molar proportions may not be closely related to the actual amounts that are produced. The acetate pool in the rumen tends to be relatively smaller in grain-fed animals and turns over more rapidly, resulting in an underestimate of net daily yield of acetate relative to that of forage-fed animals. This effect is partly responsible for the confused theories regarding low milk fat syndrome. An incorrect assumption was made that the lower molar proportion of acetate indicated a deficiency of acetate for milk fat synthesis. Actually, the rumen production of acetate in grain-fed cows is about the same as acetate production in forage-fed animals, but propionate production is greatly increased by grain feeding.

Under many conditions rumen concentrations are indicative of net amounts absorbed (Figure 15.24). This relationship is most reliable when animals are fed similar diets and amounts and rumen volumes are not too different. Entry rates of VFAs are often closely related to rates of production, although Table 19.3 shows a case for acetate in which this was not true.

19.3.2 VFA Turnover

VFA production measurement errors result when the metabolism of the respective VFA by rumen organisms is not taken into account. Butyrate in particular is formed largely from acetate, but there is also some reverse flow to acetate and smaller interconversion with propionate in the rumen. The error, involving recycling of the ^{14}C label, is not severe at reasonable times after dosing of the marker but does increase with time after dosing. Turnover values for acetate and butyrate are on the order of 90–120 min in lactating cows (Davis, 1967).

Overall, interpreting turnover rates to predict daily production rates of VFAs probably results in some overestimate of rumen output. The use of labeled acids as tracers for the turnover of the respective rumen acids

may measure the net turnover of rumen VFAs, but such turnovers reflect the sum of all possible utilizations of the respective compound. Thus turnover data need to be interpreted carefully relative to the assumptions involved when one extrapolates the results to net rumen output. Apparent acetate turnover includes the conversion of ruminal acetate to butyrate and the utilization of acetate by many rumen organisms for synthetic purposes. In addition to ruminal interconversions, rumen wall metabolism of VFAs—in particular butyrate and propionate—further reduces the net VFAs received by the liver and peripheral tissues.

19.3.3 Endogenous Production of Acetate and Acetate Metabolism

Substantial endogenous acetate is produced from the metabolism of other substances, particularly long-chain fatty acids and amino acids. Endogenous acetate is maximized under conditions of fasting, undernutrition, and tissue mobilization. It forms about 30% of the acetate metabolized by sheep fed ad libitum (Bergman, 1975, 1990). Therefore, the net entry rate for acetate must consider two sources: absorption and endogenous metabolism.

The amounts of VFAs absorbed and utilized by animals may be viewed in terms of the proportions of the caloric intake formed by VFAs and the variation in net quantities produced in given dietary situations. The amounts and proportions of the VFAs in various diets are not constant and are influenced by microbial fermentation balance, the escape of nutrients from the rumen, and endogenous production (particularly acetate) by body tissues. Since most VFAs are produced by rumen fermentation of the diet, the contribution of rumen VFAs is lower as a proportion of net energy in a fasted animal than in one fed to satiety (Table 19.4). The endogenous production of acetate causes the net acetate utilization in the two dietary states to be fairly similar, as opposed to the case of butyrate or propionate. While butyrate is an important precursor of β-hydroxybutyrate (BHBA) in well-fed animals, in the fasting-state BHBA is derived from endogenous fatty acids that are mobilized and metabolized to BHBA in the liver. The recirculated BHBA is oxidized by peripheral tissues or excreted in the urine.

In ruminants on dietary intakes adequate for their physiological state and function, VFAs may be oxidized for energy, deposited in tissues, or used for milk synthesis. Therefore, the proportion of CO_2 supplied by metabolism of acetate, for example, is generally less in states of growth, body tissue gain, or lactation than in a purely maintenance state.

Ruminants have higher blood levels of acetate than nonruminants do; however, VFA production by cecal and lower tract fermentation in nonruminants gener-

Table 19.4. Rates of entry and oxidation of acetate, determined by isotope dilution

Animal	Nutritional or physiological status	Blood acetate (meq/l)	Acetate entry (mg · min^{-1} · kg$^{-0.75}$)	Contribution of acetate to CO_2 (%)
Ruminants				
Steer	Normal	1.57	10.6	—
Cows	Nonlactating	1.08	9.6	—
	Lactating	1.07	9.8	—
Sheep	Fed	1.00	10.8	32
	Fasted (24 h)	0.35	5.8	22
Sheep	Control	0.80	7.2	17
	Early pregnancy	0.68	6.1	18
	Late pregnancy, well fed	0.82	7.0	17
	Late pregnancy, underfed	0.53	3.0	9
Sheep	Normal pregnancy	1.14	11.7	13
	Pregnant, ketotic	0.70	6.1	9
Goat	Lactating, fed	1.60	15.8	27
	Lactating, fasted (24 h)	0.33	3.8	10
Steer	Fasted (24 h)	0.50	25.0	50
Jersey cow	Lactating, fed	1.71	36.2	—
	Lactating, fed	1.27	28.1	34
Nonruminants				
Pig	Fed	0.42	11.8	10
	Fasted (24 h)	0.37	8.7	19
Chicken	Fed	0.51	7.0	15
	Fasted (24 h)	0.42	5.2	10

Source: Annison and Armstrong, 1970.

ally contributes a smaller (but significant) fraction of the dietary energy.

19.3.4 Acetate-Propionate Ratios and Animal Efficiency

The acetate-propionate ratio and heat production have long been blamed as causes of caloric inefficiency. This concept was revived by MacRae and Lobley (1986), who suggested that using propionate in gluconeogenesis spares glucogenic amino acids. When acetate ratios are high, propionate may be insufficient for gluconeogenesis and glucogenic amino acids may be required to fill the gap. Use of glucogenic amino acids for glucose synthesis will detract from their use in proteins and increase the maintenance cost of metabolizable protein and the output of ammonia. Metabolism of ammonia might be responsible for added heat increment and lower efficiency of energy. Adding propionate has a positive effect on weight gain, even when metabolizable energy is held constant (T. C. Jenkins and Thonney, 1988). The efficiency of acetate may depend on the supply of NADH, which in turn depends on gluconeogenic sources (Black et al., 1987).

19.4 Gluconeogenesis

Ruminants' metabolic requirement for glucose is no different from nonruminants' requirement (Figure 19.1), but net absorption of glucose from the gastroin-

testinal tract is low in ruminants. Thus the ability to synthesize glucose from nonglucose precursors is mandatory. In addition, ruminants have evolved modes of metabolism geared toward glucose conservation. Glucose carbon can be supplied by propionate or any odd-carbon compounds or by glucogenic amino acids (Bergman and Pell, 1985). Carbon sources other than propionate include lactate, isoacids, valerate, glycerol from mobilized lipid, and nonspecific glucogenic amino acids (Figure 19.4). In animals on high-grain diets, another possible source is the lower tract digestion of starch. Of all these, propionate is the single most important carbon source. The liver and the kidney are the most important organs in glucose synthesis; however, lactating dairy cattle can produce significant amounts in the gut at high feed intakes (Wieghart et al., 1986).

Nerve and brain tissue must have glucose for energy. Glucogenic resources are also needed in the form of oxalacetate for normal oxidative metabolism of acetyl-CoA in the citric acid cycle, otherwise intermediate products of fatty acid metabolism (ketone bodies) will accumulate. In addition, glycogen reserves must be maintained in liver and muscle, and gluconeogenesis is involved in maintaining normal blood sugar. The utilization of glucose increases in pregnant females.

The requirement for glucose is maximum in pregnant or lactating females, and both conditions make the animal prone to ketosis. Pregnancy has little to do with ketosis in cattle, but it plays a major role in sheep. Ketosis in cattle occurs mainly in high-producing dairy cows. Glucose is the main source for the lactose syn-

Table 19.5.   The regulation of gluconeogenesis

Reaction	Designation in Figure 19.4	Regulation	
		Depressed by	Increased by
Propionate → succinate	—	?	?
Pyruvate → oxalacetate	(a)	Insulin	Glucagon
Oxalacetate → PEP	(c)	Insulin	Glucocorticoids[a]
Fructose-1,6-(PO₄)₂ → glucose-6-PO₄	(d)	Insulin	—
Glucose → glucose-6-PO₄	(e)	Insulin, epinephrine[b]	Glucagon

Source: Courtesy of E. N. Bergman.

[a]Secretion of somatotropin and ACTH by the anterior pituitary promotes glucocorticoid secretion by the adrenal glands.

[b]Catecholamines, including epinephrine and norepinephrine as well as thyroid hormone and somatotropin, promote lipolysis, releasing glycerol and free fatty acids.

thesized in milk. Approximately 1.5 units of glucose are needed to synthesize 1 unit of milk lactose, and the net daily requirement may be as much as 2.8–3 kg of glucose in the case of high milk production.

19.4.1 Regulation of Gluconeogenesis

Gluconeogenesis is under the control of the hormones insulin, glucagon, epinephrine, and the glu-

cocorticoids (Table 19.5). Insulin and glucagon function as in nonruminants. When the ruminant liver produces excess glucose, insulin promotes its utilization by other tissues. Glucagon and epinephrine are involved in the mobilization of glycogen, the former in the regulation of blood sugar level and the latter in quick energy release. Hepatic output of glucose appears to be stimulated by glucagon, which may interact with the cyclic AMP system to stimulate gluconeogenesis in states of starvation and energy mobilization. Glucocorticoids are longer acting and are involved in mobilizing glucogenic amino acids.

The regulation of gluconeogenesis in ruminants is not as well understood as it is in nonruminants. The ruminant system is better equipped to cope with a shortfall in gluconeogenic precursors than to deal with an excess, particularly of propionate. Because little glucose is ordinarily absorbed, the glucose entry rate must largely reflect gluconeogenesis. Differences between ruminants and nonruminants are listed in Table 19.2. The entry rate of glucose is closely related to digestible energy intake (Figure 19.5). Free fatty acids, amino nitrogen, and somatotropin levels (not shown) decline after feeding, and glucose and insulin increase.

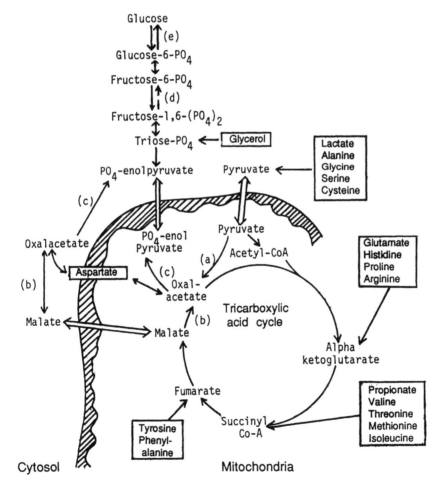

Figure 19.4. Metabolic pathways showing the sources for gluconeogenesis (courtesy of D. E. Bauman, Cornell University). Key for enzyme identification: (a) pyruvate carboxylase; (b) NAD-dependent malate dehydrogenase; (c) phosphoenolpyruvate carboxykinase; (d) fructose-1,6-diphosphatase; and (e) glucose-6-phosphatase. Mitochondrial boundary separates reactions occurring in the cytosol from those in the mitochondria. Metabolites that can cross this boundary are indicated by large arrows.

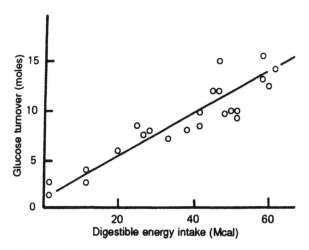

Figure 19.5. The relation between glucose turnover (entry rate) and intake of digestible energy (DE). The glucose turnover in ruminants reflects gluconeogenesis and is also related to rumen propionate production (Elliot, 1980).

Table 19.6. Comparison of fatty acid synthesis in animals

Species	Principal tissue site	Physiological carbon source	Major pathways of NADPH production[a]
Chick	Liver	Glucose	MTC
Pigeon	Liver	Glucose	MTC
Human	Liver	Glucose	MTC + PP
Rat, nibbler	Adipose + liver	Glucose	MTC + PP
Rat, meal-fed	Adipose	Glucose	MTC + PP
Mouse	Adipose	Glucose	MTC + PP
Pig	Adipose	Glucose	MTC + PP
Rabbit	Adipose + liver	Acetate	PP + IP
Guinea pig	Adipose	Acetate	PP + IP
Sheep	Adipose	Acetate	PP + IP
Cattle	Adipose	Acetate	PP + IP

Source: Courtesy of D. E. Bauman.
[a]MTC = malate transhydrogenation cycle; PP = pentose pathway; IP = isocitrate pathway.

Somatotropin appears to be low in well-fed animals as well. Injected propionate and butyrate, but not acetate, stimulate insulin release. The hyperglycemic effects of butyrate and propionate depend on glucagon secretion.

19.5 Lipogenesis

While absorbed VFAs represent the major source of dietary energy, dietary fats are comparatively unimportant sources because most ruminant feeds are quite low in lipids. Metabolism of lipids from tissue reserves is important in all animals, however, and triglycerides are the main form of stored energy that can be mobilized to counteract a deficiency in dietary energy. Lipogenesis is a major metabolic function in most herbivores because their dietary intake of lipids is so low.

The main site of lipogenesis in ruminants is the adipose tissues, where about 90% of fat synthesis occurs. The liver, which is the major site in some nonruminant species (Table 19.6), accounts for only 5% of fat synthesis in ruminants. Adipose tissue is the site of lipogenesis in herbivores of the orders Artiodactyla and Rodentia as well.

Acetate and, to a much lesser extent, available dietary fatty acids are the major sources of carbon for fatty acid synthesis in ruminants and some other herbivores. A pulse dose of labeled acetate in growing or fattening lambs becomes distributed such that 92% is in adipose tissues and only 5–6% is in the liver. Acetyl-CoA carboxylase activity in adipose tissue increases in proportion to the rate of fatty acid synthesis (Bauman and Davis, 1975).

The importance of acetate in ruminant fatty acid synthesis is related to the relative fermentation capacity. ATP production is a mechanism for regulating fatty

acid synthesis because ATP in excess of the cell's energy needs inhibits the production. In ruminants the isocitrate cycle yields an excess of ATP, while the metabolism of glucose through the pentose pathway results in a deficit of ATP. A shunt mechanism that can generate reducing equivalents without ATP is that between isocitrate and oxaloacetate within the mitochondrial citric acid cycle (Figure 19.6). The problem in fatty acid synthesis is to balance the ATP yield against the production of reducing equivalents required for hydrogenating the fatty acid chain. Fatty acid synthesis requires NADPH to supply reducing equivalents. In nonruminants, reducing equivalents are derived from the pentose shunt and from malate dehydrogenase, which is lacking in ruminants (Figure 19.6). Nonruminants depend on the conversion of glucose into fatty acids and on the operation of a complete pyruvate-to-citrate-to-oxaloacetate-to-pyruvate cycle. In functioning ruminants this cycle is inoperative and glucose is not converted to acetyl-CoA and fatty acids.

Thus, the synthesis of fatty acids in ruminant adipose tissue requires other sources of reducing equivalents. These reducing equivalents may be obtained from the pentose cycle, the oxidation of isocitrate in the cytosol, and the isocitrate-oxaloacetate shunt in the mitochondria. Since the pentose cycle involves the oxidation of nonglucose sugars to CO_2, the operation of the isocitrate cycle and the mitochondrial isocitrate-oxalacetate shunt will spare glucose.

19.6 Mobilization of Lipids

Lipids are mobilized to counter energy deficiencies and, in lactation, to meet energy needs. As such, lipid mobilization is a mechanism of homeostasis and homeorhesis. While underfeeding results in a reversal of the process of lipogenesis to meet the energy deficit, the mechanisms of fat lipolysis and mobilization in the ruminant are not entirely understood. Regulation of

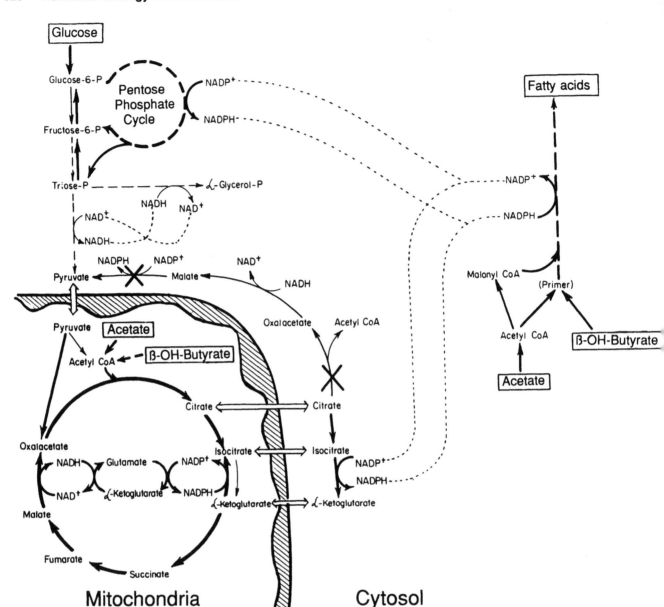

Figure 19.6. Pathways of fatty acid synthesis in ruminant adipose tissue (modified from Bauman and Davis, 1975). Negligible activity of ATP citrate lipase and NADP-dependent malate dehydrogenase in the cytosol is denoted by X. Ruminant tissue is not efficient at converting glucose into acetyl-CoA for fatty acid synthesis in the cytosol. Utilization of glucose for fatty acid synthesis that occurs in nonruminants and infant ruminants requires a translocation of mitochondrial acetyl-CoA to the cytosol as citrate. Citrate crosses the mitochondrial membrane and is cleaved to acetyl-CoA and oxalacetate in the cytosol. NADPH for fatty acid synthesis must also be generated in the cytosol, and the sources for adult ruminants are the pentose phosphate cycle and isocitrate. The impairment of glucose carbon utilization for fatty acid synthesis and the generation of a portion of the reducing equivalents from acetate via isocitrate represent a means for ruminants to conserve glucose.

lipolysis involves the cyclic AMP and second messenger systems (Figure 19.7). The first step in lipolysis is hormonal action at the cell membrane activating adenyl cyclase to form cyclic AMP, which induces hormone-sensitive lipase. Exactly which hormone activates the system in ruminants is not clear. There is considerable species variation with respect to fat mobilization responses to various hormones. Norepinephrine, epinephrine, and theophylline cause

about a threefold or fourfold increase in lipolysis in ruminants, while glucagon is marginal or insignificant in its effects. Insulin, propionate, and glucose appear to repress fat mobilization in ruminants. Epinephrine promotes lipolysis in rats, dogs, and humans but is less effective in pigs, rabbits, birds, and guinea pigs. Somatotropin has also been suggested as a fat mobilization trigger in ruminants.

Homeostasis is understood to be the metabolic regu-

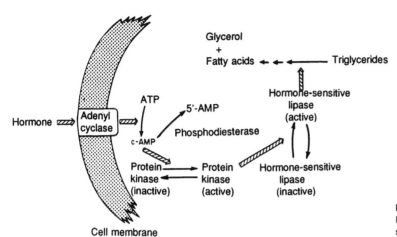

Figure 19.7. Suggested scheme for regulation of lipolysis (fat mobilization) in ruminant adipose tissue (from Bauman and Davis, 1975).

lation by which an animal maintains the status quo of energy balance. Homeorhesis adds the dimensions of increased energy expenditure required for growth, reproduction, and lactation (Bauman and Currie, 1980; Bauman, 1987). Homeostasis therefore must involve mechanisms of intake regulation, the full details of which are still unknown (Chapter 21).

The metabolic behavior of lactating ruminants strongly suggests control mechanisms for fat mobilization and its suppression. In particular, milk fat response to diet indicates the complexity that needs to be taken into account. For example, the lipid mobilization response is relatively rapid and is sufficient to cause a temporary increase in milk fat output in underfed lactating cows.

19.7 Somatotropin

Scientists' understanding of the role of peptide hormones has increased greatly in recent years. It was once thought that somatotropin (then called growth hormone) was involved in regulating growth and mobilizing fat from storage depots in the body. In fact, somatotropin is involved in the total homeostatic and homeorhetic control of energy metabolism (Bauman and McCutcheon, 1986; Gluckman et al., 1987).

The injection or implantation of somatotropin in cows elicits an increase in milk production (up to 20%), which increases the need for energy. Since energy efficiency is not altered (Peel and Bauman, 1987), this added demand is first supported by energy mobilization, but after some weeks mobilization is usually replaced with increased appetite and food intake (Table 19.7). Eventually energy intake and milk output reach a new balance. It is clear that somatotropin is involved with the adjustment of the "set point" of feed intake (Baile and Buonomo, 1987).

An account of this effect is of interest in discussions of food intake and its regulation (see Section 21.3). A

similar period of adaptation was observed by J. G. Welch when he permanently displaced part of the rumen capacity of sheep with plastic (see Figure 21.8). After some weeks the sheep had recovered much of their former feed intake. It may be that the added intake requirement could be only incompletely handled at the beginning of the somatotropin treatment, and that further adaptation by the rumen (a stretching factor?) eventually allowed the required intake. Unfortunately there are no data on rumen volumes in somatotropin-treated cows. Theoretically the increased intake should speed up passage and cause a possible decline in digestibility.

The higher feed intake required to cover lactation costs will narrow the margins of dietary quality that will allow maximum milk production, so successful

Table 19.7. Comparison between cows treated with somatotropin and genetically superior cows

Variable	Somatotropin-treated cows	Genetically superior cows
Feed intake	Increases over a several-week period to match increased milk production	Higher; increases to a peak over a several-week period following parturition
Digestibility of feed	Differences minor	Differences minor
Body reserves	Increased mobilization of nutrient reserves to support increased milk yields in first weeks of somatotropin administration	Greater use of body reserves in early lactation
Maintenance	No difference	Differences minor
Partial efficiency of milk synthesis	No difference	Differences minor
Mammary glands	Increased number of secretory cells and/or increased synthetic rate per cell is postulated	Larger quantities of secretory tissue; activity per secretory cell not known
Reproduction	Unknown; reproduction normal in well-managed herd	Improved management needed to optimize reproductive performance
Efficiency	Increased because maintenance represents a smaller proportion of consumed nutrients	Increased because maintenance represents a smaller proportion of consumed nutrients

Source: Peel and Bauman, 1987.

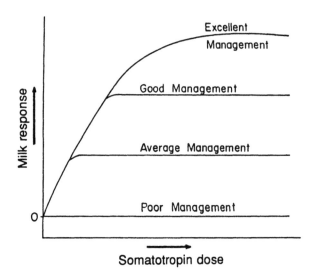

Figure 19.8. Proposed effects of management on the dose-response curve in somatotropin-treated dairy cows (from Bauman, 1987).

use of somatotropin will require adequate nutritional management. Underfeeding will induce a readjustment, perhaps stress and failure to respond (Figure 19.8). These concepts are consistent with the observed effects of limiting feed intake in lactating dairy cattle whose genetic potential is limited by inadequate environments (Figure 7.15).

19.8 Ketosis

Ketosis can occur in any animal species and arises from diverse causes. Ruminants may be especially prone to it because of their heavy dependence on gluconeogenesis. Nevertheless, ketosis is usually seen only in stressed animals under heavy demand for energy output, particularly pregnant sheep and lactating dairy cows (Foster, 1988). Not all ketotic conditions are pathological. For example, fat mobilization and the rise in ketone body level in blood is a common response to the stress of energy deficiency through starvation or increased energy demand. Fasting or starvation ketosis is regarded as a response of normal metabolism, while pathological ketosis, also referred to as spontaneous ketosis, suggests an impairment in the animal's metabolism (Baird, 1982; S. E. Mills et al., 1986b).

Fatty acid carbon must travel through butyryl-CoA and acetyl-CoA and must condense with available oxalacetate in order to enter the citric acid cycle. The oxalacetate must be derived from carbohydrate precursors (propionate or glucogenic amino acids) that are in short supply in underfed animals. When fatty acids are mobilized from adipose tissue, it leads to an increased pool of the ketone bodies: β-hydroxybutyrate, acetoacetate, and acetone. Ketogenesis is

normally regulated by the rate at which long-chain fatty acids are transported into the mitochondria (Aiello et al., 1984). During normal fasting ketosis the circulatory metabolite is BHBA, while in pathological ketosis the proportion of acetoacetate rises. Acetoacetate, the oxidized form of BHBA, is unstable, decomposing spontaneously into acetone and CO_2 at about 5% per hour. Acetoacetate and acetone are toxic, and elevated blood levels could be responsible for some of the pathology of ketosis. Accumulation in blood exceeding the urinary threshold results in a loss of ketone bodies in urine and milk, and thus constitutes a metabolic loss that tends to exacerbate the energy deficiency and the ketosis. Acetone may be recycled to the rumen, where it is reduced to the even more toxic isopropyl alcohol.

Because ketosis is a general response to any kind of stress that imposes an energy deficiency on an animal, it may often constitute a secondary response to other pathological conditions that tend to anorexia. Thus, the normal ketosis that arises in conditions of underfeeding or starvation may become pathological if ketone body levels in blood reach toxic levels.

The etiology of ruminant ketosis is often confused in practice. Inanition, or "off-feed," is a common response to illness or poor feed, which in turn stimulates mobilization of fat and a consequent rise in blood ketones. Ketosis arising from these effects is termed *secondary*. Thus a primary spontaneous ketosis is one in which another condition promoting the ketones is not apparent. Primary spontaneous ketosis is believed to be a genuine condition, but sometimes what is called spontaneous ketosis may actually be secondary, caused by some unrecognized primary factor.

The common types of spontaneous ketosis observed in ruminants occur in pregnant or lactating females. Examples are pregnancy toxemia, found most often in the late stages of pregnancy in sheep, and ketosis in high-producing dairy cattle, which occurs in early lactation. In goats, ketosis may appear during pregnancy or lactation. The syndrome always exhibits anorexia (reduced feed intake) and hypoglycemia, which could be a factor promoting the anorexia.

19.8.1 Hypoglycemia

Low blood glucose is a characteristic of fasted or underfed ruminants. Blood glucose levels are regulated by insulin and glucagon and liver glycogen, which becomes depleted after a day or so of fasting. The gluconeogenic sources such as propionate are shut off when no feed enters the system, and glycerol and amino acids are apparently inadequate sources even in physiologically normal animals.

The ruminant brain has no trouble coping with this situation. In fact, the administration of insulin dosages that would induce coma in nonruminants have no such effect in ruminants. The nerve tissue compensates for reduced blood glucose levels with a more efficient uptake of glucose. But extreme decreases of blood glucose to the point where nerve tissue cannot compensate will produce the familiar symptoms of hypoglycemia, including trembling, twitching, and coma. These conditions probably occur in the severe pregnancy toxemia of sheep, which is reported to cause blindness and brain damage.

19.8.2 Toxemia in Pregnant Sheep

Toxemia, the form of ketosis that occurs in pregnant sheep, is quite often fatal. The best remedy is prevention rather than treatment. Toxemia is more likely to occur in ewes carrying twins than in those carrying a single fetus because such ewes are generally under greater nutritional stress. The pressure of the fetuses, and perhaps of obesity, on the rumen may limit the fill, restrict intake, and help induce the condition. Ketosis depends on a reservoir of body fat, the required source for ketone bodies. Therefore, practical prevention is a matter of avoiding an overfat condition as well as poor health conditions that can be regulated by appropriate feeding. Treatments are similar to those applied to dairy cattle and usually include administration of glucogenic substances.

19.8.3 Bovine Ketosis

Ketosis in dairy cattle occurs largely during the first stage of lactation and is seldom fatal. Most cases of bovine ketosis regress spontaneously. The animal may return to eating after body stores of fat have been dissipated. Overly fat cows that are potentially high milk producers are most prone to the syndrome. Because bovine ketosis is a spontaneous condition, evaluation of treatments is difficult. Also, it may be confused with lack of appetite or any factor that reduces feed intake under the stress of high energy mobilization.

A wide variety of treatments have been applied with varying success. Oral administration of glucose is remarkably unsuccessful because the glucose is rapidly fermented, with a substantial yield of ketogenic butyrate. Intravenous glucose is more effective but often fails to sustain the animal, and there is a high degree of relapse. Feeding animals glycogenic materials that will not be metabolized in the rumen is more successful; these substances include propionate and propylene glycol. The latter is not fermentable and is converted to pyruvate after absorption.

Feeding methanogenic inhibitors is also effective. Chloral hydrate, introduced originally to treat nervous symptoms in ketosis (perhaps confused with grass tetany or magnesium deficiency), is far more effective

when administered intraruminally than by intravenous injection. A rapid increase in rumen propionate follows administration of any methanogenic inhibitor because imbalance of hydrogen transfer created by loss of methanogens in the general microbial metabolism forces carbon into propionate (Section 16.9.1).

19.8.4 Theories about Ketosis

A comparison of reviews published in 1956 and 1990 (Shaw, 1956; J. W. Young et al., 1990) shows very little progress in elucidating the etiology of ketosis. Various hypotheses have existed virtually unchanged over the years. Some of these hypotheses are presented here.

Not all animals are equally susceptible to ketosis, which suggests genetic differences in metabolism. Overfat animals and those with a genetic capacity for high fat mobilization may be more prone to the condition. Ruminant ketosis is characterized by low blood glucose and therefore presumably indicates a deficient supply of carbohydrate for the formation of oxaloacetate and the subsequent metabolism of ketone bodies in the citric acid cycle. Ketotic cows are unable to metabolize ketone bodies at the rate of normal fasted animals, indicating some deficiency relative to normal ruminant metabolism. There is also a definite limit to the amounts of acetoacetate and β-hydroxybutyrate that the animal can handle (Kronfeld, 1970). In ketosis, acetoacetate is produced by the bovine mammary gland, causing an elevation in the acetoacetate-β-hydroxybutyrate ratio. Since β-hydroxybutyrate is a normal metabolite of fat mobilization and oxidation and is less toxic than acetoacetate and acetone, its level in the blood is less indicative of a pathological condition than is acetoacetate.

Another aspect of ketosis is its association with fat mobilization and large amounts of circulating free fatty acids, which are presumably the source of at least some ketone bodies and are highly correlated with them in the blood. Fat mobilization is one response to stress, and one theory of ketosis is based on the concept of failure to adapt to fat mobilization and the resultant glucocorticoid insufficiency. Adrenal damage has been observed in cows that have undergone repeated bouts of ketosis (Shaw, 1956); however, liver impairment has also been observed (S. E. Mills et al., 1986b). Cortisone is one treatment, but it is no more effective than feeding glucogenic compounds.

An alternative theory is that ketosis is the result of a nutritional deficiency rather than a physiological one. This stems from recurrent suggestions regarding sulfur, cobalt, or vitamin B_{12} deficiency, or excess loading of the animal metabolism resulting from butyrate-producing or -containing rations. Such a hypothesis depends on the lack of specific coenzymes to conduct

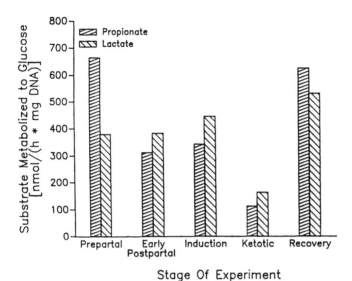

Figure 19.9. Capacity of liver slices from pregnant and lactating cows to convert propionate and lactate into glucose (from J. W. Young et al., 1990). Stages of experiment were prepartal overfeeding, ad libitum feeding for two weeks postpartum, induction of ketosis by dietary restriction and supplementation with 1,3-butanediol. Ketotic cows developed clinical ketosis, then recovered after treatment with dexamethasone and propylene glycol and return to ad libitum feed intake.

gluconeogenesis and provide a supply of oxalacetate, which in turn is affected by both the supply of glucogenic materials and ketogenic metabolites. High butyrate levels may come from high-sucrose diets or badly fermented high-moisture silage. These separate dietary effects have not been evaluated; in reality, it may be that a number of dietary imbalances contribute to the syndrome, all involving an excess of ketogenic substrates relative to available glucogenic substances. The hypothesis regarding vitamin B_{12} is of some interest because animals fed high-grain diets produce less true B_{12} and more inactive B_{12} analogues (possible inhibitors), along with more propionate (Elliot, 1980). Inadequate B_{12} would limit gluconeogenesis from propionate. Ruminants seem to have higher cobalt requirements than equivalent nonruminant B_{12} requirements.

Jerry Young and his students at Iowa State University have investigated the etiology of ketosis. (S. E. Mills et al., 1986a). They have developed a protocol that induces apparent pathological ketosis and involves moderate restriction of feed intake (ca. 20% below ad libitum), coupled with the feeding of 1 kg/day of 1,3-butanediol. This compound is not metabolized by the rumen and is an excellent precursor of ketone bodies in the liver.

The induction of clinical ketosis results in a reduction in gluconeogenesis that is not seen in normal fasted animals (Figure 19.9). This inadequate gluconeogenesis seems to be associated with a depletion of liver glycogen and a high level of mobilized triglycerides deposited in the liver. This limitation can be rapidly reversed with treatment, which usually involves administration of glucogens. The ultimate reasons for the depletion of glycogen and the gluconeogenic insufficiency remain uncertain.

20 Lipids

Fatty acids are the major components of lipids and of any dietary fraction, and they form the most important energy storage reservoir in most living organisms. This chapter is of necessity a summary of a voluminous literature. The reader who wishes to indulge in greater depth is referred to the books edited by W. W. Christie (1981) and J. Wiseman (1984).

20.1 Overview

Herbivores' diets are normally quite low in lipid because of the small quantity (1–4%) contained in most plant food sources. These dietary characteristics have required both metabolic adaptations and methods for conserving essential fatty acids. Plant lipids are altered extensively by the rumen fermentation, and the lipid actually received and absorbed by the animal differs from that ingested. The rumen is intolerant to high levels of fat, which may upset the fermentation. This situation in the functioning ruminant contrasts with that in the newborn ruminant, which ingests milk at about 30% or more fat in the dry matter, representing 50% or more of its caloric intake.

Other adaptations in regard to lipids probably have involved divergence in ruminant groups. For example, it appears that African antelopes have much higher vitamin E requirements than domestic ruminants (Dierenfeld, 1989). Such specializations may have much to do with rumen bypass or escape of lipid material.

In most metabolic systems fatty acids are derived from glucose. Dietarily derived glucose is scarce in ruminant metabolism, however and ruminants have evolved mechanisms for its conservation, the most important of which is the lack of pathways for converting glucose into fatty acids (Chapter 19).

About 90% of fat synthesis in ruminants occurs in the adipose tissue. The liver, which is the major lipogenesis site in many nonruminant species (viz., pig, human, rat, mouse, chicken, etc.), accounts for only 5% in ruminants (Bauman and Currie, 1980).

20.1.1 Plant Lipids

From a feed-oriented and quantitative point of view, lipids can be grouped into storage compounds in seeds (chiefly triglycerides), leaf lipids (galactolipids and phospholipids), and a miscellaneous assortment of waxes, carotenoids, chlorophyll, essential oils, and other ether-soluble substances. Some of these substances, particularly essential oils, which can have inhibitory activity, were discussed in Section 13.5. Triglycerides are generally confined to the seed portions of plants. As a consequence, they are the main lipid component of concentrate feedstuffs but are only negligible in forages. Leaf lipids are mainly galactolipids involving glycerol, galactose, and unsaturated fatty acids (Figure 20.1), with sulfonate groups also occurring. The leaf lipids are generally more polar than triglycerides and have a lower energy value than would be estimated by the 2.25 factor used to calculate total digestible nutrients (TDN). The galactolipid and triglyceride content is a good estimate of the total feed lipid utilizable by the animal.

The fatty acids associated with galactolipids and many of the triglycerides of seed organs are relatively unsaturated and contain high amounts of linoleic and linolenic acids. The galactolipids, a metabolically important fraction in the plant, are probably less variable in composition than the seed triglycerides, which are mainly storage compounds. Galactolipids are charac-

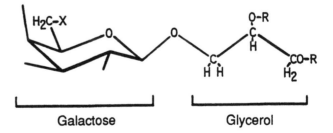

Figure 20.1. General formula for galactolipids in leaf tissue. Linolenic acid is the chief fatty acid in the R group. The X group may be either hydroxyl, α 1–6-linked galactose, or a sulfonic acid (Bonner and Varner, 1965).

Table 20.1. Content and composition of ether extract from forage leaves

	Percentage of the dry matter	Percentage of the ether extract
Ether extract	5.3	100
Fatty acids	2.3	43
Non–fatty acid		
Wax	0.9	17
Chlorophyll	0.23	4
Galactose	0.41	8
Other unsaponifiable fat	1.0	19

Source: Palmquist, 1988.

teristic of actively metabolizing leaves. Their concentrations decline with the age of the plant and vary with the proportion of leaves to stems and other metabolically inactive plant tissue (Table 20.1). The fatty acid composition of some of the more common plant lipids is shown in Table 20.2, and some fat feed supplements are listed in Table 20.3.

20.1.2 Storage and Deterioration

There is some alteration of the lipids in dead plant tissues (e.g., in hays and silages). Ether-soluble substances and carotenoids in forage decline with storage (Couchman, 1959). The decline is a result of the slow oxidation and polymerization of the unsaturated oils to form resins. This process is the same as the process that occurs in oils used in paints (e.g., linseed oil). The process proceeds via the oxygenation of double bonds,

the formation of free radicals, and polymerization by interchain cross-linking. The polymerized products are indigestible, generally insoluble in fat solvents, and in the analytical scheme become associated with the cutin fraction of crude lignin. These chemical processes occur in stored hay and in standing dry forages. They are characteristic of many grazing situations in arid climates.

Lipids must be sufficiently unsaturated in order to form resins (iodine no. > 140). Less unsaturated fatty acids polymerize to form gels and less completely altered products. Such changes are associated with rancidity and the development of peroxides, which may be quite toxic for nonruminants. Such alterations are less serious for adult ruminants because of the reducing conditions of the rumen. Rancidity also involves lipolysis and the formation of free fatty acids. Some of these changes can be associated with low palatability.

The unsaturated lipid fractions of forages can participate in the polymerization of carbonyl compounds and amino acids in Maillard reactions leading to the formation of artifact lignins. Such changes are characteristic of wet hay and high dry matter silages exposed to molding, oxygen, and high temperatures. Similar polymerizations can occur during dehydration and other processes that involve heating the feed.

Forage lipids are altered by anaerobic fermentations that occur during ensiling or in the rumen. Generally, these changes involve hydrolysis and hydrogenation of unsaturated fatty acid bonds, although other reactions

Table 20.2. The fatty acid composition of some typical feedstuffs

	Fatty acids (%)	Fatty acid content (g/100g FA)									
		14:0	16:0	16:1	18:0	18:1	18:2	18:3	20:0	20:1	22:1
Cereals											
Barley	1.6	—	27.6	0.9	1.5	20.5	43.3	4.3	—	—	—
Maize	3.2	T[a]	16.3	—	2.6	30.9	47.8	2.3	—	—	—
Milo	2.3	—	20.0	5.2	1.0	31.6	40.2	2.0	—	—	—
Oats	3.2	T	22.1	1.0	1.3	38.1	34.9	2.1	—	—	—
Wheat	1.0	T	20.0	0.7	1.3	17.5	55.8	4.5	—	—	—
By-products											
Gluten meal	1.3	—	17.2	0.9	0.8	26.7	53.0	1.4	—	—	—
Distillers' grains	10.5	—	15.6	—	2.7	24.2	54.5	1.8	—	—	—
Forages											
Citrus pulp	2.7	—	32.1	3.9	1.8	23.8	36.7	1.8	—	—	—
Dehydrated alfalfa meal (17%)	1.4	0.7	28.5	2.4	3.8	6.5	18.4	39.0	—	—	—
Cocksfoot	—	1.4	11.2	6.4	2.6	—	76.5	—	—	—	—
Perennial ryegrass	—	T	11.9	1.7	2.2	14.6	68.2	—	—	—	—
Pasture grass	—	1.1	16.0	2.5	2.0	3.4	13.2	61.3	T	—	—
Red clover	—	1.5	14.2	—	3.7	—	5.6	72.3	—	—	—
White clover	—	1.1	6.5	2.5	0.5	6.6	18.5	60.7	2.0	—	—
Oil seeds											
Cottonseed	18.6	0.8	25.3	—	2.8	17.1	53.2	T	T	—	—
Rape (Tower)	38.0	—	4.3	T	1.7	59.1	22.8	8.2	1.0	—	0.9
Rape (Target)	38.0	—	3.3	T	1.5	21.4	14.2	7.0	0.7	12.3	38.9
Soybean	18.0	T	10.7	T	3.9	22.8	50.8	6.8	T	—	—
Sunflower	34.7	T	5.5	—	3.6	21.7	68.5	T	T	—	—

Source: Modified from Palmquist, 1988.
[a]Value less than 0.5%.

Table 20.3. The fatty acid composition of some typical fat supplements

Supplement	Fatty acid content (g/100 g FA)													Iodine value[a]
	8:0	10:0	12:0	14:0	16:0	16:1	18:0	18:1	18:2	18:3	20:4	20:5	22:6	
Animal-vegetable blend	—	1.3	3.5	2.2	23.8	1.6	16.8	31.7	14.9	1.7	—	—	—	60
Beef tallow	—	—	—	3.0	25.8	6.1	18.8	39.7	4.5	1.0	—	—	—	50
Coconut oil	10.0	9.8	39.8	15.0	12.3	—	3.4	8.5	1.2	—	—	—	—	10
Peanut oil	—	—	—	—	11.5	—	3.0	53.0	26.0	—	—	—	—	90
Lard	—	—	—	0.9	24.4	6.5	10.6	38.4	19.3	—	—	—	—	65
Menhaden oil	—	—	—	11.9	23.2	16.4	5.6	15.3	2.7	1.9	—	—	—	160
Olive oil	—	—	—	—	13.0	1.0	2.5	74.0	9.0	1.5	1.6	11.5	7.6	85
Palm oil	—	—	—	1.5	42.0	—	4.0	43.0	9.5	—	—	—	—	50
Palm kernel oil	2.5	5.0	48.5	17.5	10.5	—	1.5	12.5	2.0	—	—	—	—	17
Safflower oil	—	—	—	—	8.0	—	3.0	13.5	75.0	0.5	—	—	—	140

Source: Palmquist, 1988.

[a] Values are a standard measure of unsaturation.

of considerable complexity do occur in the rumen. While saturation and reduction occur in fatty acids of ensiled forages, ensiling generally preserves the carotenoid fraction, maintaining the vitamin A activity of the forage better than in the case of hay, because of elimination of oxygen from the environment. This is probably because the conjugated double bond system of carotenoids is somewhat more resistant to oxidation and hydrogenation than are unconjugated unsaturates. This factor probably also operates in the rumen to allow escape of provitamin A activity to the lower tract.

20.2 Miscellaneous Components

Roughly half the ether-soluble matter in forages is composed of galactolipids and phospholipids; the remaining portion includes pigments, waxes, and essential oils. The pigments are chlorophyll, carotenoids, the related xanthophylls, and saponins, and the waxes are related to the plant cuticle. Essential oils include anything that is steam-volatile, including esters, terpenes, aldehydes, and ketones. Some of these substances were discussed in Chapter 13. Their value as an energy source is restricted to their contribution of aliphatic carbon chains entering the fatty acid pool in rumen and animal metabolism (Table 20.4).

The cuticular component of plants is composed of two main fractions: a low-molecular-weight fraction, consisting of alkanes, alcohols, ketones, and esters of long-chain alcohols, and a fraction of polymerized cutin compounds probably polyester in nature and closely associated with lignin. The lower-molecular-weight fraction is sparingly soluble in fat solvents and will be measured as ether extract.

Cuticular waxes are long-chain compounds (18–37 carbons). Generally, the alkanes are odd-carbon aliphatics, suggesting their formation through decarboxylation of the corresponding acid of one higher carbon number. Alcohols and aldehydes are generally

even-numbered, as are the acids. Alcohols are partly free and partly esterified. Lower-molecular-weight waxes have some solubility in ether, but higher-weight matter is mostly insoluble and is included with cutin and the crude lignin fraction. Radioactive tracers have shown that some of the cutin can be degraded by rats and also that alkanes up to about 30 carbons are absorbed by ruminants (Mayes et al., 1986). It appears that animals can at least partially degrade the smaller, less polymerized portions (Kolattukudy, 1980). Generally, the more insoluble waxes, and the alkanes in particular, are assumed to be essentially inert and lost in the feces. Many of the essential oils and lower-molecular-weight phenols do not offer any metabolizable energy; those that are absorbed are mostly excreted in the urine. Chlorophyll is indigestible, although the magnesium ion is removed and the porphyrin ring may be ruptured. The alcohols and acids are more likely to be metabolized. It is evident that the phytyl group, which is ester-linked in chlorophylls and vitamin K, is split off in rumen metabolism, hydrogenated, and converted to phytanic acid through oxidation of the terminal alcohol group (Keeney, 1970). It is then incorporated into ruminant fats as a 20-carbon branched-chain acid.

Table 20.4. Nutritive availability of miscellaneous lipids

	Anaerobic bacteria	Higher animals
Triglycerides	Hydrolyze and ferment glycerol; FA hydrogenated	Utilized
Galactolipids	Hydrolyze and ferment glycerol + galactose; FA hydrogenated	Utilized
Waxes	Very little fermentation	Probably low utilization
Chlorophyll	Very little fermentation	Some degradation; degraded products in feces
Carotenoids	Destructive hydrogenation	Utilization very low, 0.1% digestibility
Ca & Mg soaps	Produced	Partial digestibility
Free saturated fatty acids	Produced	Utilized

Phytanic acid is technically a terpenoid from the point of view of plant biochemistry.

20.3 Rumen Lipid Metabolism

Two interwoven aspects of rumen lipid metabolism have received limited attention: the microbial metabolism of feed lipids and the de novo synthesis by microbes of their own lipids. The utilization of fatty acids for metabolism by anaerobic bacteria is restricted. Rumen organisms are limited in their ability to use highly reduced substances for energy, and fatty acid use is limited to cell incorporation and synthetic purposes.

Microbial metabolism of galactolipids and triglycerides commences by splitting off the galactose moieties, then hydrolyzing the glycerol fatty acids, the glycerol and galactose portions being readily fermented to volatile fatty acids. Hydrolysis in the rumen proceeds rapidly after ingestion (Table 20.5). The accumulation of phospholipid is indicative of microbial synthesis. Liberated fatty acids are neutralized at rumen pH, probably mainly as the calcium salts, which have a low solubility and adhere to the surfaces of bacteria and feed particles (McAllan et al., 1983). Unsaturated fatty acids are hydrogenated and metabolized by certain rumen bacterial strains (Dawson and Kemp, 1970). The position of the remaining double bonds is altered, and generally the acids are converted to the more stable *trans* form. The double bonds in unsaturated rumen fatty acids become distributed in many positions along the chains; hydroxylation and a spectrum of ketoacids also occur. Since the *trans* acids are more difficult to hydrogenate, there is an accumulation of *trans* forms relative to *cis* forms (Table 20.6). *Trans* unsaturated acids have higher melting points than their *cis* relatives and are transported and absorbed by the animal in that form, contributing to the generally higher melting point of ruminant fats (Garton, 1964).

20.3.1 Microbial Lipids

The biosynthetic modification of lipids by rumen bacteria involves the formation of many odd-carbon

Table 20.6. Double bond positions of octadecenoic acids (18:1) in rumen lipids

Position	Rumen digesta cis	Rumen digesta trans	Dialyzable cis	Dialyzable trans	Nondialyzable cis	Nondialyzable trans
5–7	1	1	5	3	9	5
8	2	2	4	4	2	3
9	80	4	57	11	36	7
10	2	7	2	12	1	10
11	13	75	18	34	41	50
12	2	2	8	6	11	6
13–16	—	8	5	29	Trace	19

Source: Katz and Keeney, 1966.
Note: Values are weight percentage of total acids.

and branched-chain acids, probably through the incorporation of propionyl, 2-methylbutyryl, and 3-methylbutyryl moieties into the carbon skeletons. A number of rumen bacteria require *n*-valeric acid for biosynthesis of odd-carbon fatty acids (e.g., *Fibrobacter succinogenes*, *Treponema*, and *Selenomonas*). Propionate also can be incorporated into fats through the host animal's metabolism and may occur under conditions of grain feeding where propionate is produced in very large amounts (Garton et al., 1972; Duncan et al., 1974a, 1974b). The 15-carbon linear and branched acids are major components of microbial lipids. The relative concentration of odd-carbon and branched-chain acids is shown in Table 20.7. Branched chains contribute to the lower melting point of membrane lipids in the absence of significant quantities of polyunsaturated fatty acids.

The fatty acid composition of microbes is intimately bound up with the general rumen metabolism of feed lipid (Bauchart et al., 1990). Microbial lipids are the result of modification of dietary lipids and some synthesis, although it is difficult to tell how much each process contributes to the final product. Triglycerides are hydrolyzed to fatty acids, which may be precipitated as calcium salts. These insoluble forms become associated with the microbes and fine-particle fractions of the rumen. The fatty acid compositions of rumen bacteria and protozoa are shown in Tables 20.7 and 20.8.

While ordinary amounts of unsaturated lipids as they occur in forages are not important in rumen fermentation, excess unsaturated fatty acids and triglycerides can cause profound alteration in the fermentation balance through the suppression of methanogenic and cellulolytic bacteria; generally, all gram-negative bacteria are inhibited. Excess saturated fats have no comparable effect, and the mode of administration of unsaturated lipid has a considerable influence on the direction and magnitude of the change. Unsaturated oil dosed in frequent small quantities is much less apt to cause methane suppression than if administered in a single large dose. There may be a threshold of tolerance by

Table 20.5. Hydrolysis of lipids in the rumen

	Total lipid (%)	Glycerides Tri	Glycerides Di	Glycerides Mono	Phospholipid	Free fatty acids
Diet[a]	6.3	72.4	13.7	1.7	1.2	11
Rumen digesta, fresh	5.1	0.1	—	0.1	15.2	85
Rumen digesta, 1 h	6.2	30.4	1.7	—	7.1	61
Rumen digesta, 5 h	6.4	11.1	—	—	12.4	76

Source: Bath and Hill, 1969; see also Dawson and Kemp, 1970.
[a]Diet consisted of 1 kg chopped hay + 50 g palm oil.

Table 20.7. Fatty acid composition (percentage by weight) of lipids of mixed rumen bacteria

Fatty acid	Total lipid[a]	Total lipid[a]	Total lipid[b]	Unesterified fatty acid	Phosphatidyl-ethanolamine[c]	Phosphatidyl-serine[c]
11:0	T	—	—	—	—	—
12:0	T	5	1	2	1	T
12:0 br	1	—	—	—	—	—
13:0	T	—	1	1	1	T
13:0 br	1	—	1	1	T	T
14:0	2	4	4	4	4	2
14:0 br	2	—	1	1	2	1
15:0	4	—	8	7	11	6
15:0 br	10	—	13	10	18	14
16:0	35	25	31	30	31	29
16:0 br	1	—	1	T	2	1
16:1	—	T	4	2	5	3
17:0	2	—	—	1	2	2
17:0 br	2	—	—	—	—	—
18:0	32	21	15	23	7	13
18:0 br	—	—	T	T	T	1
18:1	4	20	6	6	7	11
18:2	3	6	3	2	3	4
18:3	—	T	1	1	1	1
20:0	—	—	T	—	—	T
Other	—	20	9	9	6	13

Source: Harfoot, 1981.
Note: T = trace; — = not detected.
[a]Bacteria from rumen of cattle (two separate studies).
[b]Bacteria from rumen of sheep.
[c]In the phospholipids, 13:0 br, 14:0 br, 15:0 br, and 16:0 br include 12:1, 13:1, 14:1, and 15:1 straight-chain acids, respectively.

the fermentation, and the administration of the oil via distributed feeding may allow time for metabolism and hydrogenation of the unsaturated fatty acids before inhibition can occur.

Unsaturated fatty acids do not compete with methane production for available hydrogen. The administration of relatively large amounts of oil traps only an insignificant portion of the available hydrogen pool. Upsetting the methane fermentation produces an ex-

Table 20.8. Fatty acid composition of lipids from holotrich and entodiniomorph protozoa

Fatty acid	Entodiniomorph protozoa total lipid	Holotrich protozoa total lipid
12:0	T	T
14:0	1	3
14:0 br	—	—
15:0	—	—
15:0 br	—	—
16:0	48	37
16:0 br	—	—
16:1	—	—
17:0	—	—
17:0 br	—	—
18:0	10	9
18:0 br	—	—
18:1	21	18
18:2	10	11
18:3	1	4
Other	9	18

Source: Harfoot, 1981.
Note: Values are percentages of total fatty acids present. T = trace; — = no data given.

cess of hydrogen and a resultant alteration in rumen fermentation balance toward production of a high proportion of propionate to maintain the fermentation balance. The excess propionate is associated with metabolic changes in the host's lipid metabolism (as seen in the low milk fat syndrome). Suggestions that unsaturated fatty acids themselves inhibit synthesis of milk fat by the mammary gland may be questioned since unsaturated oil protected to escape the rumen causes an increase in milk fat, as does the administration of saturated fat or graded doses of unsaturated oil below levels that will upset the rumen fermentation.

Ionophores added to feed interfere with hydrogen transport and probably inhibit hydrolysis of lipid in the rumen (Kemp and Lander, 1978). Increasing propionate can cause increased synthesis of long-chain methylated acids (Wahle and Livesay, 1985).

Unsaturated fatty acids are among the substances that have been suggested as additives to eliminate methane fermentation to reduce fermentation loss and increase animal efficiency. While they are among the least objectionable feed additives, their instability and lack of reliability have prevented their exploitation for this purpose.

20.3.2 Protected Lipids

Cheap fat by-products from industry allow the economical inclusion of fat in animal feeds as a rich source of energy. Giving fats to high-producing dairy cattle in

Table 20.9. Fatty acid composition of tissues from animals fed conventional diets or protected sunflower seed supplement

| Tissue | Weight percentage of major fatty acids | | | | | | | | | | | | | |
| | 14:0 | | 16:0 | | 16:1 | | 18:0 | | 18:1 | | 18:2 | | 18:3 | |
	C	P	C	P	C	P	C	P	C	P	C	P	C	P
Rump														
S	4	3	24	20	9	4	10	15	41	37	2	17	2	1
IM	5	4	29	22	4	3	18	21	33	31	2	14	2	1
Rib														
S	5	3	25	20	10	4	8	16	38	31	1	18	2	1
IM	5	2	27	18	6	3	13	22	40	34	1	20	2	1
Perirenal adipose	4	2	26	17	3	2	22	27	32	27	1	20	2	1

Source: Palmquist, 1988.

Note: C = pasture-fed animals; P = protected sunflower oil–fed animals; S = subcutaneous adipose tissue; IM = intramuscular adipose tissue.

early lactation, when they would otherwise need to mobilize body stores, is an attractive notion; however, feeding fat to ruminants, particularly lactating cows, has problems and limitations.

Rumen bacteria are not very tolerant to lipid. It provides them with little energy, and the more unsaturated fatty acids are decidedly toxic (McAllan et al., 1983), which leads to the need to protect the fatty acids so they pass through the rumen largely unmetabolized. Overfeeding protected fat leads to another limitation: the energy from fat dilutes that from carbohydrate, which is the chief energy source for rumen microorganisms, which in turn provide the largest single source of daily protein to the host. Thus overfeeding protected fat leads to a need for protected protein as well. Generally, it is less expensive to rely on the energy (VFAs) and microbial protein that are provided by the rumen when rumen fermentation and feed intake are optimized.

Protection can involve the use of insoluble calcium salts of the fatty acids. The salts of the more saturated fatty acids are so crystalline and insoluble that they may escape digestion altogether (Palmquist et al., 1986). The specific gravity and size of particles are important in determining escape from the rumen. Densities below 1 tend to float and not pass, and many lipid compounds have such low densities. Thus some failures have occurred. Unsaturated acids made available to rumen metabolism can cause lowered milk fat. If fatty acids pass through the rumen without affecting fermentation, however, they are used by the host animal in the usual manner and can be used to increase milk fat. Table 20.9 shows how the addition of protected sunflower seed oil to the diet affects the fatty acid composition of tissues.

Protection that would substantially alter ruminant fats toward unsaturation was suggested in about 1970 as a means of overcoming the objectionable hardness, saturation, and high *trans* acid content of ruminant fats (Keeney, 1970). This, it was thought, would help human heart disease and atherosclerosis. Wrenn et al. (1978) showed that protected unsaturated fat was passed directly to the adipose tissues and to milk. Practically, however, their effort was a failure because the unsaturated oil created unstable milk and off flavors.

More recently the interest in protected fat has returned, although for different reasons. Protection prevents fats from interfering in the rumen fermentation, and thus fat can be fed to dairy cattle at levels practical for energy supplementation. The degree of unsaturation needs to be controlled to avoid problems of milk quality. Substituting fat into the diet dilutes fermentable carbohydrate, however, and overfeeding fat will decrease rumen output and affect microbial protein.

Adding protected unsaturated oil to the diet causes modification of both tissue fats (Table 20.9) and milk fat. Ruminant fats are low in unsaturated fatty acids relative to plant lipid, and protection causes the composition of long-chain fatty acids in the milk to approach that in the diet. There is an elevation of triglycerides in the blood as well as in milk fat. Blood cholesterol is increased, but there is an increased fecal excretion of steroids.

Since the dietary intake of fatty acids is quite low, much of the circulating lipid may be endogenous or the product of lipoprotein metabolism. The elevation of blood triglyceride and its accumulation in depots when protected fats are fed indicate that the normal transport system for triglycerides and fatty acids characteristic of nonruminants is also operative in ruminants.

Increased levels of blood lipids are associated with increased milk fat. The fact that blood lipids decline in conditions of milk fat depression supports the hypothesis that the supply of lipid to the mammary gland is involved in the depression. It appears that ruminant lipid metabolism is similar to that of nonruminants apart from the modifying action of the rumen.

20.3.3 Essential Fatty Acids

How do ruminants derive their essential fatty acids? While plant lipids are rich in linoleic and linolenic acids, hydrogenation tends to destroy these compounds under normal rumen conditions. Ruminants, like other higher animals, are unable to synthesize essential fatty acids (EFAs), as evidenced by deficiencies that can be induced in calves.

Young ruminants are born with very low reserves of polyunsaturated acids compared with nonruminants (J. H. Moore and Noble, 1975). Under normal feeding the levels rise soon after parturition. Ruminants appear to have an EFA requirement about an order of magnitude lower than nonruminants (Mattos and Palmquist, 1977). In ruminants these essential acids are used sparingly and are selectively incorporated into cholesterol esters and phospholipids. The efficient utilization of essential fatty acids apparently evolved from the need to counterbalance rumen hydrogenation, thus representing yet another metabolic adaptation of ruminant tissues. An interesting point is that sufficient escape of unsaturated acids occurs under normal rumen conditions to satisfy the animal's requirement (Bickerstaffe et al., 1972; Murphy et al., 1987; Wu and Palmquist, 1991).

20.4 Animal Fats

Ruminant fats are characteristically hard (i.e., saturated) compared with those of nonruminants. It has long been known that nonruminant fats are much easier to alter through the feeding of the appropriate dietary lipid. This is not surprising in view of the considerable capacity of the rumen to alter and hydrogenate unsaturated fatty acids. Tables 20.10 and 20.11 compare the relative compositions of some animal fats.

Another ruminant peculiarity is the contrasting composition of milk fat compared with depot fats. Short-chain components are present in butter fat but largely absent in other ruminant lipids and in nonruminant milk fats. The quality and composition of ruminant milk fat is associated with the propionate and carbohydrate metabolism of the animal.

Ruminant lipids are unusual in their content of odd-carbon and branched acids, which reflects the absorption and incorporation of microbial lipids and some modified plant compounds (e.g., phytanic acid) into ruminant fat. Ruminant tissues have the ability to desaturate and alter the length of fatty acid chains. Some oleic acid is apparently derived from palmitic acid through this process.

The dietary lipid of all herbivores is very highly unsaturated, and one would expect the depot fats of nonruminant herbivores (without pregastric fermentation) to reflect this unsaturation. Dietary lipids consumed by omnivores and carnivores may be more variable, depending on the dietary source, but ruminant fats are the most difficult to alter via dietary means because of the saturating effect of rumen fermentation. Ruminant depot fats are no less alterable than those of

Table 20.10. Fatty acid composition of animal adipose tissues

Fatty acid	Cattle	Sheep	Red deer	Camel	Horse	Rabbit
14:0	4	1	10	6	5	2
14:1	1	T	1	2	—	—
15:0	1	1	1	2	—	—
15:0 br	—	1	2	—	—	—
16:0	24	23	36	30	26	22
16:0 br	T	T	—	—	—	—
16:1	6	2	6	4	—	—
14–16 uns	—	—	—	—	4	2
17:0	2	2	1	2	—	—
17:0 br	—	2	1	—	—	—
17:1	2	1	—	1	—	—
18:0	14	26	22	25	5	6
18:1 *cis*	—	34	18	—	34	13
18:1 mixed	44	—	—	27	—	—
18:1 *trans*	—	4	1	—	—	—
18:2	3	2	1	2	5	8
18:3	2	—	—	1	16	42

Sources: Gupta and Hilditch, 1951; Shorland, 1953; Christie, 1981.
Note: T = trace.

nonruminant species, however, if the rumen fermentation is bypassed.

20.4.1 Odd-Carbon and Methyl-branched Fatty Acids

Ruminant adipose tissue triglycerides are unusual in that they contain about 1–2% of odd-carbon and methyl-branched-chain fatty acids (methyl-BCFAs), which are derived from propionate and rumen bacterial lipids (Garton, 1965), respectively. A similar distribution occurs in milk (Keeney, 1970). The methyl-BCFAs found in ruminants that are fed diets adequate in roughage are mainly iso- and anteisoacids in 13–18-carbon chains with traces of other monomethyl-BCFAs (Garton, 1967). The enhanced availability of

Table 20.11. Triglycerides and cholesteryl esters in milks of various mammals

Fatty acid	Triglycerides					Cholesteryl esters	
	Cow	Sheep	Goat	Blackbuck antelope	Camel	Cow	Goat
4:0	3	4	3	4	—	—	—
6:0	2	3	3	3	—	—	—
8:0	1	3	3	2	1	—	—
10:0	3	9	8	5	1	3	8
12:0	3	5	3	3	1	4	4
13:1	—	—	—	—	—	11	2
14:0	10	12	10	16	11	7	9
14:1	1	1	1	—	—	1	1
15:0	1	2	—	—	3	2	6
16:0	26	25	25	37	26	27	26
16:1	2	3	2	2	11	12	4
17:0	1	2	1	—	T	T	4
18:0	15	9	12	7	12	7	13
18:1	30	20	29	19	30	14	14
18:2	2	2	2	2	4	10	3
18:3	1	1	—	—	2	—	—

Source: Christie, 1981.
Note: Values are weight percentage of total fatty acids. T = trace.

propionate in animals on high-grain or pelleted diets is associated with increased proportions, in adipose tissue of sheep and goats, of both odd-chain and methyl-BCFAs in addition to the iso- and anteisoacids normally found. Methyl-BCFAs may attain concentrations of about 15% of total fatty acids and are responsible for the soft back fat in sheep fed high-barley diets (Duncan et al., 1974a, 1974b). These methyl-BCFAs consist largely of 10–17-carbon chains, usually with a mono-methyl substitution on an even carbon in the chain ranging from C2 to C14; dimethyl and trimethyl substitutions have also been identified. Odd-chain acids arise when the propionyl group is terminal to the chain. Isoacids and acids methylated on odd carbons appear to be largely derived from microbial lipids and micro-bially altered plant lipids (e.g., phytanic acid).

The association between propionate availability and methyl-BCFA synthesis was clarified when it was shown that methylmalonyl-CoA, a metabolite of pro-pionate, in the presence of malonyl-CoA could be uti-lized by the fatty acid synthetase prepared from tissues of ruminant and nonruminant species to yield methyl-BCFAs. Methylmalonyl-CoA was also found to be a competitive inhibitor (with respect to malonyl-CoA) of fatty acid synthetase from several species (Scaife et al., 1978; Wahle and Paterson, 1979), and it has been sug-gested that an accumulation of methylmalonyl-CoA could cause low milk fat syndrome in dairy cows by inhibiting de novo fatty acid synthesis in the mammary gland (Frobish and Davis, 1977); however, this has not been verified and the etiology of the syndrome requires further investigation.

Although restricted-roughage high-grain diets result in enhanced propionate production in all ruminant spe-cies, the fact that cattle and red deer (unlike sheep and goats) do not accumulate methyl-BCFAs indicates there are possible differences in the mode or extent of propionate metabolism among ruminant species. The observation that fatty acid synthetase prepared from adipose tissue of cattle can utilize methylmalonyl-CoA (in the presence of malonyl-CoA) for methyl-BCFA synthesis is interesting because cattle do not accumu-late these fatty acids. It would appear that these differ-ences among ruminant species in their ability to syn-thesize methyl-BCFAs are related in some way to the mode of propionate metabolism and the availability to the fatty acid synthetase of methylmalonyl-CoA (Wahle et al., 1979).

20.5 Postruminal Fat Digestion

The digestion and absorption of fatty acids by rumi-nants occurs in the rumen, while in nonruminants li-polysis occurs farther down the tract, primarily in the small intestine near the site of absorption. In both groups the long-chain fatty acids are absorbed into the lymphatic system (Leat and Harrison, 1975).

Fatty acids are neutralized at rumen pH and pass down the system as soaps. Potassium soaps are readily absorbed from the small intestine of ruminants. There is evidence that calcium soaps (much less soluble) may escape absorption and appear in the feces in well-fed animals. Saturated fatty acids are absorbed more slow-ly than unsaturated acids, the ease of absorption de-creasing with increasing chain length. Still, ruminants ordinarily absorb most fatty acids, and true di-gestibility approaches 100%.

While lipids are largely hydrolyzed in the rumen, ruminants readily hydrolyze and absorb triglycerides that escape the rumen fermentation. Pancreatic juice supplies lisolecithin, which serves as a stabilizer for the lipid particles (micelle) in the lumen of the small intestine (Garton, 1965). Diverting the secretions re-sults in drastic reduction of chylomicrons in the lymph.

20.6 Synthesis of Milk Fat

The mammary gland in lactating ruminants is a ma-jor site of triglyceride synthesis. Fat accounts for about 50% of the calories in milk. Ruminant milk fat is pecu-liar in that its fatty acid composition deviates from that of the adipose tissue and from nonruminant milk fats. The main peculiarity is the content of short-chain vol-atile fatty acids (VFAs), which are absent from most other fats. The synthesis of the shorter chains is the main lipogenic activity of the mammary gland. The carbon source for the fatty acids in milk fat differs depending on the chain length (Figure 20.2). The 4-carbon primer accounts for butyrate, while longer chains (caproate, etc.) involve stepwise elongation by 2-carbon units. The contribution of acetate and β-hydroxybutyrate (BHBA) to long chains is very lim-ited. The mammary gland depends on circulating blood lipid as a source of long-chain acids, as is evident by the fact that feeding protected unsaturated lipid will both increase and alter milk lipid (Table 20.12). The malonyl-CoA pathway seems to account for fatty acids 10–14 carbons long, while 16–18-carbon chains origi-nate from circulating blood lipids.

The mammary gland utilizes BHBA in addition to acetate to supply the carbon for the short-chain acids in milk fat. The mammary gland fatty acid synthetase prefers butyryl-CoA as a primer, in contrast to the adipose tissues, which use acetyl-CoA for this pur-pose. The contribution of BHBA to net fatty acids is not a large one, however (Palmquist et al., 1969).

Milk fat synthesis is especially relevant to under-standing the fat metabolism of ruminants, particularly in regard to the milk fat depression phenomenon, which is not well understood. A number of apparently

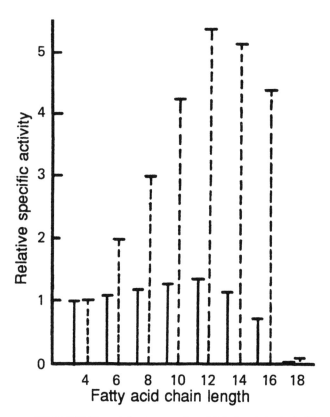

Figure 20.2. The relative incorporation of β-hydroxybutyrate (solid lines) and acetate (dashed lines) into milk fatty acids. β-Hydroxybutyrate serves as a primer for chain lengths of up to 16 carbons, and acetate accounts for elongation. Very little activity occurs in C_{18} acids, which, along with some C_{16} acids, are derived from blood lipids. β-Hydroxybutyrate is not a large source of carbon for milk fatty acids (3–8%), casting doubt on theories that its deficiency is a cause of depressed milk fat synthesis (Palmquist et al., 1969).

disparate changes in feeding management can elicit a reduction in the number of short-chain volatile acids and an increase in unsaturation of milk fat. These dietary changes include high-grain feed and underfeeding or malnutrition.

20.7 Low Milk Fat

Low milk fat is not a pathological situation, but it has received a good bit of attention for economic rea-

Table 20.12. Effect of feeding protected polyunsaturated lipid supplement (protected or unprotected safflower oil)

Animals	Linoleic acid (%)	
	Unprotected	Protected
Steers		
Inramuscular fat	5	23
Kidney fat	5	30
Subcutaneous fat	4	20
Lactating cows		
Milk fat	5	20–30

Source: Bauman, 1974.

sons. Practical concern is connected with the association of fat content with milk quality and the requirements of cheese making. The objective here is to discuss milk fat depression as an aspect of ruminant metabolism, offering insights into the remaining problems of ruminant carbohydrate and lipid metabolism that lack an adequate explanation and understanding.

When lactating ruminants are fed diets deficient in coarse roughage, a decline occurs in the concentration of fat in the milk. This may be accomplished by grinding the forage (Richard Grant et al., 1990) or by feeding high- or all-concentrate diets. In addition, several other conditions are associated with low milk fat: feeding unsaturated lipids (Shaw and Ensor, 1959), drugs that suppress rumen methane synthesis, certain forages (pearl millet) irrespective of physical form, and hot weather (Van Soest, 1963).

The alteration in composition affects mainly the fat, although there is some compensatory rise in lactose and protein, a change that deviates from the normal positive association of milk fat content with nonfat solids assumed in the 4% fat-corrected-milk (FCM) equation. As a result, the application of the FCM equation to estimates of energy in milk under conditions of high grain feeding may underestimate energy output. Tyrrell and Reid (1965) developed a special equation to describe the FCM value of low-fat milk. Dietary conditions associated with milk fat depression have been generally associated with greater efficiency in fattening. While the practical efficiency of lactation is depressed, the total animal efficiency is less altered because of the diversion of milk energy toward tissue lipogenesis. Substituting grain for forage suppresses fat mobilization and secretion, although the total fat produced (tissue plus milk) increases (Table 20.13).

There is considerable variation among individual animals as to the degree of milk fat depression on a given diet. Generally, the low-milk-fat breeds of cattle (Holstein) exhibit greater proportional depression than high-milk-fat breeds (Jersey or Guernsey); some individuals show no depression at all. Animals at the peak of lactation are more apt to exhibit depression than are animals at a lower level.

It is not known whether all forms of milk fat depression have the same etiology. The low milk fat induced by heat stress seems different in view of the low intakes and high proportions of rumen acetate evident in this situation (Moody et al., 1971), but the great majority of cases do involve the alteration of rumen fermentation toward high propionate-acetate ratios. It is with this form that the discussion of ruminant metabolism is mainly concerned. The common effect of diverse diets—including those involving ground forage, high concentrate, unsaturated oil (Shaw and Ensor, 1959), antimethanogenic substances, and pearl millet pastures (R. W. Miller et al., 1965)—is the reduction of

Table 20.13. Energy balance and output in cows fed rations containing 60, 40, or 20% alfalfa

	Alfalfa content of total ration		
	60%	40%	20%
Total energy consumed (kcal)	12,250	12,634	11,966
Milk energy (kcal/24 h)	13,525	13,209	10,643
Body tissue gain or loss (kcal/24 h)	−1275	−575	+1323
Milk yield (g/24 h)			
Fat	714[a]	627	489
Protein	632[b]	669	561
Lactose	853[c]	924	750
Net change (g/24 h)			
In body fat	−138[d]	−62	+144
In fat yield	576[d]	565	633

Source: Annison and Armstrong, 1970.

[a]Yield of milk × fat content.

[b]Yield of milk × protein content (orange dye method).

[c]Calculated from the milk energy by deducting energy in milk fat and in protein and dividing remainder by caloric value of lactose. Caloric values used were 9.21, 5.66, and 3.95 kcal/g for fat, protein (casein), and lactose, respectively.

[d]Assuming all body tissue energy to be fat with a caloric value of 9.21 kcal/g.

methane and the consequent elevation of propionate production necessitated by the disturbed carbon balance in the rumen.

20.7.1 Changes Associated with Low Milk Fat

Perhaps the most remarkable aspect of the milk fat depression syndrome is the associated increase of lipogenesis in the adipose tissue. Acetyl-CoA carboxylase levels are elevated in adipose tissue but not in the mammary gland. Adipose tissue is insulin-responsive, while the mammary gland is not. Acetate probably serves as the main source for this lipogenesis. The high energy efficiencies associated with high-concentrate diets in fattening animals probably reflect the same metabolic adjustment as that seen in lactating animals. Efficient acetate utilization appears to be related to availability of NADPH (Black et al., 1987).

The principal changes noted in milk fat depression are in the amount and content of milk fat and the volume of milk, with concentrations and net quantities of lactose and protein being much less affected. There is a small counterbalancing effect in which nonfat solids increase slightly in response to the fat depression. The proportion of shorter-chain fatty acids in milk triglycerides is diminished. There is also some depression in the concentrations of the blood lipids, ketone bodies, and acetate, although not all studies found these effects. Some authors reported elevation in blood glucose (Van Soest, 1963) and insulin (Richard Grant et al., 1990). The amounts of acetate and BHBA taken up by the mammary gland appear to be reduced (Van Soest, 1963; Davis and Brown, 1970).

The rumen production of VFAs generally shows increased total propionate and little change in acetate.

The lowered molar ratio of rumen acetate does not reflect a lesser yield of acetate, a rationalization often used to explain the lower blood acetate and milk fat depression. It is possible that there is either a decrease in endogenous acetate production—also questionable (Palmquist, 1972)—or a more competitive use of available acetate in response to the general metabolic suppression of fat mobilization associated with low milk fat.

20.7.2 Causes of Low Milk Fat

Several theories have been proposed to account for the low milk fat phenomenon. None has been demonstrated to describe the true mechanism of the effect, although some elements of an adequate hypothesis are apparent. The ultimate explanation must take into account lipid synthesis and mobilization. It is for this reason that the low milk fat problem remains of fundamental interest.

An adequate theory must account for the fact that the primary causative factor is produced in the rumen, and that it both promotes tissue lipogenesis, either directly or indirectly, and at the same time negates lipogenesis in the mammary gland. The theory must account for the depression in mammary output of both long- and short-chain acids. About half of milk fatty acids are formed de novo (the shorter ones) and the other half are taken preformed from the blood, yet both fractions are reduced in low milk fat.

Dietary Fat Deficiency

The oldest theory regarding milk fat was proposed by Boussingault (1845), who fed a low-fat and (inadvertently) low-fiber diet of beets and obtained milk fat depression. He concluded that the low milk fat was due to a dietary deficiency of fat. This, of course, is unreasonable in light of modern knowledge of metabolism. All animals in one way or another synthesize body fat from dietary carbohydrates. Still, the hypothesis is not unreasonable in the context that the mammary gland under conditions of milk fat depression may not receive sufficient fatty acids for lipogenesis. While ruminants are unable to directly convert glucose to fatty acids, acetate and butyrate are extensively used by the mammary gland for short-chain fatty acid synthesis. The source of long-chain acids appears to be blood lipids supplied either by dietary lipid or by mobilization from adipose tissue. Dietary lipids probably increase milk fat through this mechanism.

Acetate Deficiency

The second theory, historically, is the acetate deficiency theory, which was proposed on the incorrect assumption that net supply of acetate from the rumen was diminished and limiting. Molar proportions of acetate are indeed reduced in milk fat depression; however, this effect is actually produced by the large increase in net propionate production (see Table 19.3) and by the smaller and more rapid turnover in the rumen pool. Whatever factor accounts for the low milk fat, it must be caused by or associated with the increased production of propionate.

The acetate deficiency theory may still be viable in terms of the net supply to the mammary gland. Rumen yields of acetate do not seem to be altered, but endogenous acetate is likely reduced; in addition, increased uptake by adipose tissue may compete with the udder for acetate. Feeding acetate or butyrate, or infusing BHBA or 1,3-butylene glycol into milk fat–depressed cows has mixed results. Furthermore, the contribution of two- or four-carbon units to milk fat is limited. Sodium acetate will act as a rumen buffer, and generally feeding buffer salts increases the proportion of rumen acetate, decreases propionate, and increases milk fat. Whether this effect on the rumen is due to buffering, osmotic pressure, or liquid turnover is unclear, but it probably improves conditions for methanogenic organisms. Therefore, it is difficult to say that the increase in milk fat is caused by fed acetate per se.

The BHBA Theory

Alternatives to the acetate deficiency theory have involved looking for other metabolites that may be limiting in mammary synthesis of fatty acids. BHBA was early shown to be a primary factor in the synthesis of short-chain fatty acids in milk fat (Shaw et al., 1960). The difficulty with this hypothesis is that it does not account for the suppression of long-chain fatty acids that do not involve BHBA, and it does not consider Bauman and Davis's (1975) assertion that intravenous infusion of large amounts of BHBA has no effect on low milk fat. The amount of BHBA utilized for fatty acid synthesis in the mammary gland is quite small (3–8%) and is not particularly altered in milk fat depression (Palmquist et al., 1969).

The Vitamin B_{12} Theory

Another possibility is that a metabolic problem is induced by high propionate. Vitamin B_{12} is required for the metabolism of propionate and has a possible connection with ketosis. Rumen production of true vitamin B_{12} is low in animals fed high-grain diets, and under these conditions propionate yield is maximum. This theory argues that the excess propionate indicates that vitamin B_{12} is a limiting factor. Its deficiency would limit the conversion of methylmalonyl-CoA to succinate. Perhaps methylmalonate is lost in urine, or, alternatively, perhaps propionyl-CoA competitively

interacts with acetyl-CoA in fat synthesis. Duodenal true B_{12} is negatively correlated with milk fat concentrations (Elliot, 1980).

Insulin and Suppression of Fat Mobilization

Historically, the acetate deficiency theory was demolished by McClymont (cited by Van Soest, 1963), who found that intravenous glucose infusions produced low milk fat. Similarly, the feeding or intravenous infusion of propionate produced low milk fat. McClymont theorized that insulin production was stimulated either through glucogenesis from propionate or from glucose via lower tract digestion of starch, and indeed some studies indicate elevated insulin in milk fat depression (Jenny and Polan, 1975). Insulin has little effect on milk fat, however, and it tends to suppress lactation. Nevertheless, many nutritionists feel that any theory that accounts for low milk fat will have to consider the excess supply of propionate. Frobish and Davis (1977) infused glucose and propionate individually into the abomasum and failed to produce milk fat depression. They argued that McClymont's and other intravenous infusions were unphysiological; however, their observations are not in accord with other feeding studies with propionate in which depression was observed (Davis and Brown, 1970).

This leads back to the general concept of control of fat mobilization, for which there is presently no adequate explanation. The depression of both long- and short-chain acids in milk fat depression might be accounted for if mammary synthesis of milk fat is regarded as a function of mobilization. In this hypothesis the mammary gland is considered dependent on blood lipids and acetate, which are reduced by inhibition of fat mobilization and promotion of synthesis in adipose tissue. A difficulty with this hypothesis is that in milk fat depression lipogenesis is repressed in the mammary gland.

That fat mobilization and its regulation are involved is supported by the experiments of Vandersall et al. (1964), who fed grain with and without thyroprotein and showed that the thyroid factor abolished milk fat depression even though rumen propionate production remained high. Animals fed thyroprotein had increased metabolic rates and mobilized large amounts of adipose tissue, as evidenced by body weight loss. The prevention of milk fat depression lasted until body fat reserves were depleted, after which the depression returned.

Of the proposed theories, the hormonal inhibition of fat mobilization remains the most viable, although the precise interaction of hormones is still not well understood. Leek (1983) and Sutton et al. (1988) suggested that low milk fat is caused by increased insulin, which stimulates lipase activity in adipose tissue and reduces availability of fatty acids to the mammary gland. The ratio of insulin to growth hormone seems involved in this effect since somatotropin antagonizes insulin (G. H. McDowell, 1983). The combined hypothesis accounts for the promotion of fattening under high rumen propionate production as well as the suppression of fatty acid mobilization essential for the mammary synthesis. There are no reports that somatotropin will reverse milk fat depression, although such an effect would be expected. Thyroprotein will certainly accomplish this.

20.8 Analytical Problems

The standard AOAC method for measuring fats in feeds is to extract an oven-dried sample with dry ethyl ether. The anhydrous condition is intended to minimize the extraction of nonlipid matter, particularly sugars. The fact that forages contain galactolipids points to the lack of success of this method. Lipid can be determined as the evaporated residue or by differential weight loss of the sample on ether extraction. The results of these two ways of measuring do not always agree, as plants often contain significant amounts of volatile essential oils and other substances.

Another problem is the difficulty in extracting the lipids, which can resist solution in ether, leading to low values. Incomplete extraction occurs in baked products and in feces. In baked products, fats become associated with carbohydrate or protein through what may be the initial stages of the Maillard reaction. This degree of heating may not impair digestibility, but it may increase the probability of rumen escape. The ether-insoluble complex is destroyed by acid. In feces most of the undigested fatty acids are in the form of calcium or magnesium soaps that are also insoluble in ether. These soaps are generally formed in the hindgut (Grace and Body, 1979). Acid pretreatment releases these fatty acids to solvent extraction. Ether extraction thus may tend to overestimate the digestibility of fats by failing to measure this fecal fraction, causing a further error in the TDN value. When fats or oils are added to animal diets, the saturated acids may precipitate calcium, causing a lowering of its utilization.

For these reasons the AOAC ether extraction is not a good method for evaluating diets with added fat. It is also relatively useless applied to most forages. Only in the case of concentrate feeds in which the fraction represents mainly triglycerides are its values relevant. Since the use of ether extract has been a preliminary step in the outmoded crude fiber analysis, both could be discontinued in the analysis of most ruminant feeds. There are now more sophisticated procedures that can be applied for accurate lipid analysis (Sukhija and Palmquist, 1988).

21 Intake

The consumption of feed is fundamental to nutrition: it determines the level of nutrients ingested and, therefore, the animal's response and function. Digestibility and utilization of nutrients are in a sense only qualitative descriptions of the net food intake. Intake of feed is itself regulated and limited by the requirements of the animal's physiology and metabolism.

The two broadest aspects of the animal's response to feed are the quantity of indigestible residues, which push ingesta through the digestive tract, and the absorption of digestible and metabolizable nutrients, which then enter the animal's metabolic system. These two aspects have led to a branching in intake studies: one branch emphasizes the metabolic regulatory mechanisms of the animal, and the other emphasizes functions of the digestive system. Two publications that cover intake phenomena in detail are *The Voluntary Food Intake of Farm Animals* by J. M. Forbes (1986) and *Predicting Feed Intake of Food-Producing Animals* by the National Research Council (1987).

21.1 Terminology

The developing field of animal intake studies has led to the introduction of a somewhat varied terminology. Older publications tend to describe intake in terms of *plane of nutrition,* which refers to the level of digestible nutrient intake in relation to maintenance requirements. *Feed level* can be imposed on the animal by external circumstances such as availability of feed. If it is controlled in feeding trials, intake is said to be *restricted.* The amount of feed may vary from zero intake (termed *fasting*) upward to the maximum intake accepted by the animal. This maximum intake is termed *ad libitum* if feed is available to the animal at all times. If feed is limited to the amount that the animal will consume, some degree of restriction is implied.

Hunger and *appetite* are terms used to describe an animal's urge to eat. *Hunger* tends to denote a short-term effect, as might occur between or before meals and the onset of eating, while *appetite* has implications

concerning the amount eaten and the physiological factors contributing to a cessation of eating. This plateau of consumption leads to a plane of nutrition when it is acquired over repeated meals and has been described as a set point, or satiety, relative to satisfaction of the animal's urge to eat. *Satiety* is the theoretical level needed to balance energy losses and achieve optimal growth, produce milk, and perform work under conditions of a balanced diet. *Set point* and *plane of nutrition* refer to the functional level the animal achieves under the combination of restricting factors involving feed and environment.

Restricted and ad libitum intake are relevant only to stall feeding, where control can be exercised. In grazing conditions other physical factors that limit intake come into play. Here the terms *availability* and *accessibility* are often used. The availability or accessibility of forage is determined by stocking rate of animals and plant density and morphology. Other factors such as distance to water supply and terrain may limit the animal's feeding effort, and therefore its intake. Grazing always offers some degree of selection, even under conditions of limited intake; the degree depends on the density and morphological differentiation of the forage species. Thus accessibility and selection are interrelated. Plant diversity and animal selection are usually greater under range conditions and are at a maximum in browsing situations. Accessibility may also be related to managed feeding situations. Accessibility to feed may involve factors such as "bunk" (feeding) space, amount of time that feed is accessible each day, and crowding of animals. Competition among animals reduces accessibility.

Offering excess feed under any management gives the animal an opportunity to select palatable portions and leave the less desirable parts unconsumed. Selection is limited by the animal's ability to manipulate the feed and by the form of the feed. Smaller animals are generally better able to select feed; that is, browsers are better than grazers and sheep are better than cattle. Chopping and pelleting the feed render sorting more difficult, and selection is thus reduced.

There is a limit to the amount of feed or forage that

an animal will consume. The factors that cause cessation of eating are physiological and include appetite (the drive to eat), the metabolic requirement of the animal, and the quality of the feed. In very high quality feeds (e.g., concentrates), the metabolic requirement (i.e., set point) tends to be the limiting factor. In the case of most forage diets, the set point level of intake is not reached because limiting factors of feed quality intervene to impose a lower level of intake, and a lower plane of nutrition is reached. The detailed mechanistic hypotheses that effect this limit are several and include fill, time, metabolic factors, and homeostatic limits of the animal.

The unifying concept in intake is the plane of nutrition, which is set by the circumstances of the animal in its environment, whether artificial or natural. An animal will achieve the most advantageous food intake to satisfy its desires within the limits of its environment. This, however, does not assume that the animal will choose the optimum diet if given several options.

The plane of nutrition is achieved by a sequence of meals. Thus meal size and daily intake may not be equivalent. Meal size and number of meals are short-term phenomena contributing to daily intakes, which can in turn vary over time. These shorter-term effects have been comparatively less well studied than the longer-term effects.

21.2 Measuring Intake

The weight of feed an animal consumes in a given time period is its intake. It is a necessary measurement in digestion trials that use the total collection method. In the case of stall feeding, intake is simply the daily feed weight less the orts. Accurate measurement is much more difficult in grazing situations, and indirect methods have to be applied (Chapter 8).

Other aspects of food intake less often studied are meal size, number of meals per day, and rate of eating (Black and Kenney, 1984). Such parameters contribute to net intake but may be more reflective of choice and palatability.

Ad libitum intake is regarded as a parameter of feed quality; unlike digestibility, however, values of intake for respective feedstuffs are not generally available because there is too much variation in animals' appetites. Intake trials for evaluation of forages require standardization to remove as much animal variability as possible. Most forage trials have followed the standardization designed by Swift and Bratzler (1959). This procedure uses young male castrated sheep (wethers) of about 45 kg body weight placed on excess test forage. After a suitable period (3–4 weeks) the offered diet is controlled at a level such that orts are no more than 15% of feed offered. This situation is maintained

for at least a week, and the average feed consumed is recorded for that period. This system of ad libitum intake measurement may not be applicable to animals that consume tropical forages and browses because selective feeding is more difficult to avoid (see Chapters 6 and 7).

21.2.1 Palatability Trials

The trials mentioned above are designed for single, unsupplemented feeds. If choice is a factor, two or more individual feeds or forages are offered in a compartmented manger, and the separate intakes of the respective feeds are then measured. Similar experiments can be conducted under grazing conditions when randomized strips of several pure forage species are grazed by test animals. Such experiments allow the selective ingestion of different feeds, forages, or supplements. These are often called palatability trails, but factors of animal physiology other than taste and preference could play a role in the choice of feed. Ruminants do select feeds on the basis of flavor and color (Munkenbeck, 1988).

Some forages are consumed if they are provided as a sole choice but may be rejected if offered with alternative feeds. Palatability also affects the ranking of forages by animals with differing appetites. Animals with voracious appetites may discriminate less than finicky eaters with lower demands. A possible result is that intake of given sets of forages can be ranked differently by different groups of animals; for example, growing cattle versus lactating cows, or sheep versus cattle. This can be expected from differences in animal size and ruminating efficiencies.

21.2.2 Feed Refusal

Control of the quantity of orts is necessary to restrict selection. If a larger feed refusal is permitted, animals can choose a more digestible diet. Generally, stemmy portions of forage are rejected, a factor that can be reduced by chopping the forage. Trials standardized by a controlled refusal may be suitable for temperate forages and mixed diets. In the case of tropical forages, however, in which leaf and stem portions may be much more differentiated in respect to quality, orts may be much greater (up to 60%), and it may not be possible to restrict intake to a point at which no feed is refused (Olubajo et al., 1974). Thus the procedure is not suitable for selector feeders. Also, if mature tropical forages are offered (ground or finely chopped), values of digestibility and intake may not correspond to those obtained under practical feeding conditions. Since the rejected parts are usually of very low quality, undernutrition can result from feeding management that attempts to attain a more complete ingestion of the tropical forage (Owen et al., 1989).

Selecting uniform test animals does not entirely remove animal differences and preferences. Moreover, the application of intake data in a practical context requires an adjustment according to the animal's requirements, as they determine appetite. Practical application requires estimation of the animal's set point as well as its metabolic size and gastrointestinal capacity. The ultimate objective of these calculations—predicting animal intake in a given physiological situation—is still unresolved.

21.2.3 Body Weight

Two mathematical expressions are used to adjust for individual differences in body weight: intake of feed per unit metabolic body size (body weight $kg^{0.75}$) and percentage of body weight. The use of the former is based on the assumption that metabolic requirements are related to metabolic size. The second, more direct expression is easier to use and is favored by those who see little advantage in relating intake behavior to metabolic body size.

Gastrointestinal capacity and rumination are related to power 1.0 of body weight, while metabolic requirements may be related to power 0.75 or less, depending on whether the comparison is across or within animal species (Thonney et al., 1976). If intake depends on gastrointestinal fill and metabolic requirements, the best power fit could be variable depending on the character of animals and feeds. Studies for best fit of intake data with body weight have revealed powers from 0.5 to 0.8 (Colburn and Evans, 1968).

21.2.4 Relative Intake

Relative intake is a calculation that attempts to adjust for appetite differences among animals (Crampton et al., 1960). The animals are evaluated on a standard forage for appetite, and values of test forages are expressed per 100 units of the standard forage. Crampton chose alfalfa as the standard forage, with an intake of 80 $g/kg^{0.75}$. Unfortunately, alfalfa is a variable material, and individual harvests always differ in both digestibility and intake. A necessary requirement for a standard feed is that it can be readily reproduced anywhere.

Many nutritionists who use relative intake merely divide intake values by the value of Crampton's alfalfa standard. This is, in effect, multiplying them all by a constant, and it results in no correction for animal differences at all. A truly standard reference will remove large amounts of animal variation inherent in standard intake trials (Osbourn et al., 1974).

21.2.5 Eating Rate

The rate of eating (i.e., how rapidly the diet is ingested) can be expressed in grams per minute. The rapid ingestion of feed causes a pulsation in the rumen fermentation curve that may alter rumen ecology and fermentation balance. Regulated consumption tends to produce a more even and efficient rumen fermentation. Factors affecting eating rate include the density and the morphological nature of the diet (Black and Kenney, 1984). Palatability and appetite are also factors affecting the rate of ingestion.

21.3 Intake Regulation

Ruminants present special problems for the physiologist wishing to understand the mechanisms of response to dietary intake. One theory of dietary regulation in nonruminants is based on a concept of satiety: the animal eats until its metabolic requirements are filled, and excess circulating nutrients trigger the cessation of eating. The difficulty with this theory lies in the identity of the circulating triggering substance, which might be either an absorbed nutrient or a released hormone. In nonruminants, glucose absorption, subsequent glycemic rise, and stimulation of insulin and other secretory responses are believed to cause cessation. Functioning ruminants, however, do not show a postfeeding rise in blood sugar, regardless of the digestible carbohydrate intake. This is not surprising since most of the sugar and starch are fermented in the rumen to volatile fatty acids (VFAs), and the metabolic glucose requirement must be supplied through gluconeogenesis from other metabolites. A modified version of this theory, for which there is some evidence in ruminants, is that acetic acid, propionic acid, or possibly some other metabolite substitutes as the triggering substance. These metabolites in turn may stimulate hormonal peptides such as cholecystokinin (CCK), particularly from the hypothalamus. The presence of these metabolites generally leads to a cessation of eating, and some of them, particularly the fermentation acids, may be of short duration, so the factors affecting individual meals may vary during the eating cycle (Forbes, 1986). An alternative theory involves response to thermogenic effects. Heat increment may limit intake under conditions of heat stress (Section 21.3.1).

Another problem with determining intake regulation in ruminants is the mechanism by which they respond to low-caloric-density and poor-quality diets. Under the usual conditions of feeding management, ruminants rarely ingest sufficient energy to reach their true production potential, so the aspect of poor diet is of the greatest importance in understanding ruminant feeding

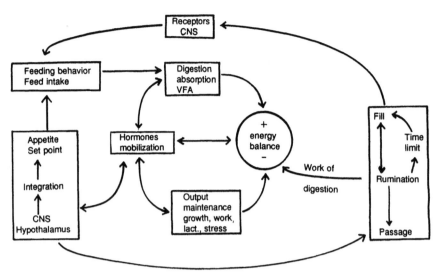

Figure 21.1. The various mechanisms postulated to regulate voluntary intake. Energy balance is governed by the input of digestible nutrients, the output, and the cost of work of digestion and metabolism. Hormonal factors in blood are a composite of responses to absorbed nutrients, components that regulate mobilization of energy from depots in case of negative energy balance, and the factors secreted by the anterior pituitary that regulates eating. The central nervous system (CNS) and hormones regulate gastrointestinal motility and probably passage, causing some alleviation of fill by passing coarser material at higher intakes and set points.

behavior. Several mutually nonexclusive factors are thought to prevent optimum intake, including the limitation of eating time and the concept that fill limits intake through gastrointestinal or rumen distension. The latter produces its effect either through discomfort or through some other physiological effect. The high acetic acid production resulting from high-fiber diets may be connected with the hypothesis of intake limitation by acetic acid or other VFAs. Another possibility is that low-quality diets may be deficient in nitrogen or some other nutrient. This might limit intake either through retarded rumen digestion (bacterial growth requirements are not met) or through a more direct effect on the animal's metabolism. The nutrient deficiency concept may not be a truly separate alternative because retardation of digestion will affect fill.

The sum of this discussion is that the control of feed intake is the result of various physiological interactions that are somehow integrated in the overall eating response. The tendency of various studies to focus on a single mechanism tends to obscure this integrative relationship, and diets can be designed to favor the operation of a given mechanism.

A difficulty in comparing supposedly alternative hypotheses is that although experimental diets may be designed so that any one of the respective factors is limiting, the demonstration of the operation of one mechanism does not exclude others that may operate under a different set of conditions or in combination under practical feeding conditions. It is in this context that efforts to demonstrate satiety (mainly limited to high-energy diets), fill, or too little time available for eating as factors that cause eating cessation must be understood. Individual authors may erroneously interpret these results as support for an exclusive hypothesis.

All of these effects on voluntary intake are integrated in the living animal (Figure 21.1), apparently under the

control of the central nervous system (CNS; NRC, 1987). For example, fill and time available for eating are offset by time spent ruminating, which reduces fill and allows more gastrointestinal space for feed consumption, but at the expense of eating time. The fill undoubtedly interacts with tension receptors that feed into the CNS to restrict or turn off feed intake (Forbes, 1986). The regulation of energy balance and the tissue mobilization processes are capable of counterbalancing any energy deficit for considerable periods of time. In the long term, energy expended may be reduced to balance the energy budget. This often involves a reordering of the set point.

The existence of a satiety control center in the hypothalamus has been demonstrated through implantation of electrodes in that organ. Initiation and cessation of eating can be produced by appropriate signals (Baile and Della-Fera, 1984). The question, then, is: What signals to the satiety center are responsible for controlling eating? These might be nervous stimulation, hormonal factors, or humoral levels of metabolites. There is also evidence that a hunger center exists in the brain (Forbes, 1986).

21.3.1 Humoral Factors for Intake Regulation

Several kinds of experiments have demonstrated the operation of hormonal factors in the regulation of intake. The administration of gold thioglucose to mice produces lesions in the anterior pituitary and results in uncontrolled intake. Ruminants and some other animal species are unresponsive to this treatment, indicating perhaps a different mode of intake regulation.

The hypothesis that some humoral factor regulates intake in ruminants has been demonstrated in several ways. Volatile fatty acids can limit intake and offer an attractive hypothesis since a major part of dietary energy passes through these intermediary fermentation products. Acetate and propionate are more effective than butyrate in eliciting cessation of feeding. The results with acetate, butyrate, and propionate do not rule out the possibility of some other metabolite being the regulatory agent or (even more likely) that these acids stimulate a hormonal agent. Evidence of humoral factors has come from experiments involving transcirculation of blood from fed and fasted sheep (Seoane et al., 1972). The intake of fasted sheep can be inhibited by this means, and the intake of animals fed ad libitum is increased (Figure 21.2). These observations are consistent with the demonstration that cholecystokinin octapeptide secreted by the central nervous system is the satiety-regulating factor in sheep (Baile and Della-Fera, 1984; NRC, 1987). Perhaps the secretion of cholecystokinin octapeptide is interrelated with or stimulated by the entry rates of VFAs.

A second group of hormonal substances that affect intake, the opioid peptides, cause full animals to eat more, perhaps by reducing the pain of gastrointestinal pressure (NRC, 1987). In addition, the liver appears to play a role in feed intake (Forbes, 1986).

The concept and demonstration of satiety limits have generally been restricted to ruminants fed high-concentrate diets and are probably of secondary importance in forage-fed animals. The satiety control is not always a fixed limit. For example, feeding a poor diet to a lactating cow will prevent her, probably because of fill imitation, from consuming enough feed to meet the requirements set by the level of production, which then falls as body reserves are depleted until the animal adjusts to the feeding situation.

Another example is the thermogenic increment after feeding. Animals eat less when in a state of heat stress, and it has been suggested that metabolic heat regulates satiety. Applying heat to the hypothalamus or heating the rumen through a fistula has variable effects on food consumption (Gengler et al., 1970), however. Internal heat is probably relatively unimportant under temperate feeding conditions, but it can be of major importance in warm environments.

The aim of most studies involved in metabolic control of intake is to discover some unifying hypothesis or underlying principle that accounts for variation in intake of feeds. An example of an extreme approach is that suggested by Ketelaars and Tolkamp (1992; Tolkamp and Ketelaars, 1992). They propose that the efficiency of oxygen utilization is a controlling factor. This efficiency varies quadratically with feed intake: as caloric density increases (from poor forage to high-energy feeds), the optima also increase. To promote this theory, these authors discard the concepts of fill, indigestible matter, and satiety, along with the evidence for them, discussed in this chapter. Available eating time is not even considered.

In summary, the regulation of intake is likely an

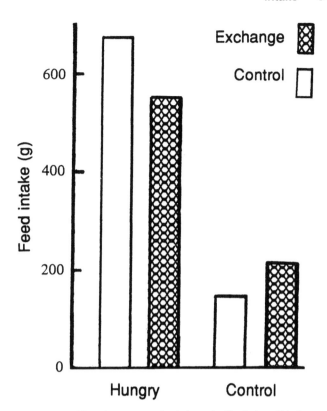

Figure 21.2. Effect of blood cross-circulation on feed intake by satiated and hungry sheep (from Seoane et al., 1972). Each bar represents the average 1-h feed intake following blood exchange with a satiated or fasted donor. Control values were obtained from self-circulation tests. The intakes of six hungry sheep were significantly depressed when they received blood from satiated donors. The intakes of satiated sheep increased significantly after receiving blood from hungry donors.

integration of various factors in the metabolic system of the animal. Specific situations can promote one factor to become dominant: for example, fill in pregnant or obese animals, time when forage is poor and sparsely distributed, or metabolic factors affected when feed is ground and rumination time reduced. None of the respective models are mutually exclusive, and the net limitation of intake could be a combination of these factors, the quantitative contribution of which can vary with diet and situation (Forbes, 1986). Any approach to establish a particular mechanism is hazardous because conditions can be established to make any one factor appear to be the major regulator of intake.

21.3.2 Caloric Density

The expected relationship between digestibility and intake may be positive or negative, depending on the logical argument. If one assumes that animals eat to satiety, then more of a less digestible diet would have to be consumed to achieve the required level of digestible calories obtained in smaller amounts of a more digestible diet. This postulates a negative relation between

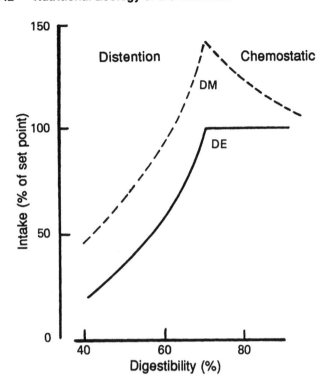

Figure 21.3. Relation between intake of dry matter (DM) and the digestible energy (DE) in the feed ingested. Data fitting this model are obtained when a concentrate diet is diluted with a bulky filler or coarse forage (Conrad, 1966; Baumgardt, 1970).

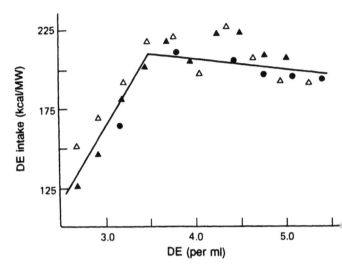

Figure 21.4. Relation between concentration of digestible energy (DE) and intake in sheep (from Dinius and Baumgardt, 1970). The different symbols represent different experiments. MW = metabolic weight, or body weight[0.75].

the amount of freely chosen feed and the digestibility of the diet. On the other hand, the assumption that poor-quality feeds contain factors that limit intake, such as bulk or dietary deficiency, presumes a positive relationship between intake and digestibility.

Both of these contrasting concepts have been shown to regulate food intake. The limiting factor of satiety is more important in intake of rations with high caloric density than in diets with low caloric density (Figures 21.3 and 21.4). Distension is the dominant factor affecting intake of lower-quality diets. This dual regulation has been demonstrated in sheep, rats (Baumgardt, 1970), and cattle (Conrad, 1966) and is most readily demonstrated by the dilution of a poor-quality forage with concentrate. Single-fed forages with differing quality do not fit this model (Section 21.4). The relation between digestibility and intake among coarse forages of varying quality is less clear, although the concept may apply to pelleted forages. The concept of eating to satiety assumes a constant intake of digestible energy, the net quantity eaten being influenced by dilution with indigestible bulk (Figure 21.3). This constancy is not often observed in high-energy diets, and digestible energy intake may decline with increasing energy content in the feed (Figure 21.4). A practical reason for this decline is the deficiency of effective fiber in concentrates, lack of rumination and conse-

quent potential rumen acidosis, and imbalance of metabolites (Grovum, 1987).

21.3.3 Intake Level and Animal Production

A higher energy demand requires greater rumen fill or faster passage such that fill becomes limiting at higher densities of dietary energy (Figure 21.5). The point of maximum dry matter intake has been the subject of several investigations. Conrad (1966) suggested that it occurred at about 67% apparent digestibility when concentrate-alfalfa combinations were fed to lactating dairy cattle. Other authors have suggested that this point is not fixed, but depends on the density of the diet, adequate fiber, and the energy demand (set point) of the animal. The optimum point of intake can also vary depending on ration and forage quality, as Figure 21.5 shows.

21.3.4 Intake and Diet Composition

It is often assumed that intake and digestibility of forages are directly related. This concept is implicit in the nutritive value index (NVI) proposed by Crampton et al. (1960). The NVI is the product of relative intake and digestibility. Although they are somewhat interdependent, intake and digestibility are separate parameters of forage quality. Intake depends on the structural volume, and therefore the cell wall content, and digestibility depends on both cell wall and its availability to digestion as determined by lignification and other factors. The independence of intake from digestibility is perhaps best demonstrated by examining the association between lignin content and intake (Figure 21.6),

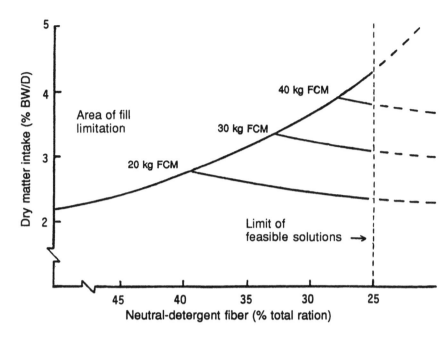

Figure 21.5. Theoretical solutions relating dry matter intake (as percent of body weight per cay [BW/D]), neutral-detergent fiber (NDF) content of diet, and milk production for a 600-kg cow producing varying amounts of 4% fat-corrected milk (FCM) (modified from Mertens, 1985). As the production level increases, the range of diets that can be formulated to achieve that level narrows. To obtain 75% of the fiber from forage the NDF content of the forage must be below 40% to formulate rations for production in excess of 40 kg FCM. Minimum allowable NDF content of the total ration is assumed to be 25%. Intersection of the curves occurs at points where physical and metabolic limits on intake balance and forage utilization is maximized. The decline in intake at high intakes represents a limitation due to fill. Declining intakes at low NDF beyond the respective optima are the result of metabolic limitations.

which is a net zero relationship (r = 0.08) in a sufficient population of mixed forages. Several features of the scattergram are pertinent. Legumes are generally high in lignin but are consumed at a high intake, and grasses are usually lower than legumes in lignin content but higher in cell wall. There are two groups of grasses. In one (timothy, brome, orchardgrass, etc.), intake is related negatively to lignin content and pos-

itively to digestibility, following the expectations of the NVI concept. The other, an unusual group including fescue, bluegrass, and probably reed canarygrass (not shown), tends toward a positive relation between intake and lignin content.

Fescue and reed canarygrass are known to contain alkaloids that inhibit rumen digestion or interfere with the physiological balance of the animal. The alkaloid concentrations decline with age, so immature forage is rejected but later stages of growth are accepted. These plants are characterized by a low mean lignin content. The net result is a highly significant and positive interspecies regression between intake and lignin content (Figure 21.6).

The significance of the positive interspecies association of lignin with voluntary intake reflects the tendency of lignin to be associated with forage density because maturity and age effects are removed as factors in the comparison by holding digestibility constant. Forages with high lignin content (chiefly legumes) tend to have low cell wall contents and to be consumed at higher intakes. The positive association between lignin and intake is largely a result of the legume-grass comparison.

The generalization derived from Figure 21.6 also applies to tropical forages. Lignin content is closely associated with digestibility, so it is not surprising that the relation of plant age to intake (Figure 21.7) is similar to the relation of lignin to intake. The contrast among tropical legumes and grasses involves the same interaction as in temperate forages.

The relation of various forage constituents to animal intake ultimately depends on their association with plant structure. Thus, one finds cellulose more closely

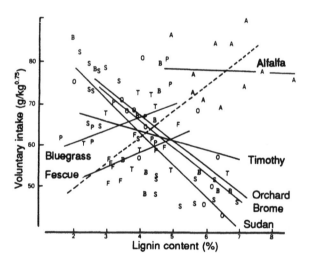

Figure 21.6. The relation between lignin content and voluntary intake of 83 temperate forages from West Virginia (from Van Soest, 1965a). The overall relation is not significant. Dashed line represents the regression for all species, with the within-species variation removed (r = 0.90, P < .01). The positive slope (dashed line) results from the interaction of highly lignified alfalfa (legume) and high intake and the low lignification of fescue, which is poorly eaten. Intakes of fescue and bluegrass are associated positively with lignin content and negatively with forage maturity. A = alfalfa; B = brome; F = fescue; O = orchardgrass; P = bluegrass; S = Sudan; T = timothy.

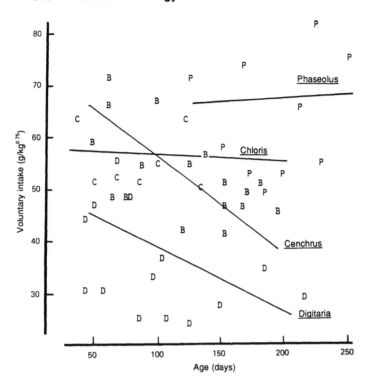

Figure 21.7. Age of tropical forages and voluntary intake (from Milford and Minson, 1965). Species shown: *Digitaria decumbens*, or Pangola grass (D); *Cenchrus ciliaris*, or buffelgrass (B); *Chloris gayana*, or Rhodes grass (C); and *Phaseolus atropurpureus*, or pigeon pea (P). Compare these distributions with those in Figure 21.6. A very close relation exists between age and lignin content in tropical forages.

associated with intake than with digestibility. Conversely, lignin is more closely associated with digestibility than with intake. The total structural matter—the plant cell wall—is the most consistent fraction related to intake because the cell wall contains the entire structural substance of the plant within which all other components are contained. Several hypotheses have been advanced to explain how cell wall affects intake. These are discussed in Sections 21.4 and 21.6.

The purported relation of water content of tropical forages to intake, for example, may be considered a function of structural volume if it is considered that the plant water is contained within the cell wall structure. The addition of water per se to the rumen has little effect on intake, as it is largely absorbed and removed; however, water retained in the coarse structural components of ingested forage can inhibit intake (Balch and Campling, 1965). It is generally assumed that voluminous feed reduces intake by means of fill, but bite size and the time needed to achieve sufficient ingestion of feed may also limit intake (Stobbs, 1973a, 1973b). Because bites and chews are associated with discrete times and amounts per bite depending on feed NDF content and density (Welch and Smith, 1969), available time rather than fill may be the intake limiter.

21.3.5 Dietary Deficiencies and Intake

Intake is most often discussed in connection with rations of varying caloric value and density. Nevertheless, the dietary concentration and intake of adequate levels of minerals and nitrogen also influence intake and animal function. The satiety hypothesis suggests that animals will seek to obtain the nutrients required to satisfy their needs. If the quality of the available diet is so poor that specific requirements cannot be met, the influence of the respective deficiency exerts its effect on the animal. The effect on intake is often adverse, since the ability of the deficient animal to cope with the diet is impaired. For example, low-nitrogen diets are associated with reduced intake. This depression in intake is associated with crude protein concentrations below 7% (Chapter 18). This level is below the nitrogen requirement of rumen bacteria, even supplemented by recycled urea, and the result is a depression in digestibility. The lack of a discernible relation between nitrogen content and intake or digestibility above this level suggests that the inflection point represents the critical level of nitrogen, below which compensation is not possible. The intake depression could be regulated either by nitrogen deficiency in the host or by retarded rumen fermentation and reduction in the ability to clear undigested bulk from the rumen. The fact that amino acid supplementation in the lower tract increases intake does not provide clear-cut evidence for distinguishing the role of bacterial fermentation from that of the animal's amino acid requirement, since such supplementation could add to the supply of recyclable nitrogen. The administration of digestible carbohydrate increases the microbial nitrogen requirement and thus aggravates the loss of fecal nitrogen. Also, the animal's restriction of intake is consistent with the conservation of nitrogen. Such factors may well be important in intake regulation in grazing and browsing herbivores.

21.3.6 Silage Intake

Silage intake often tends to be less than that expected for a hay of similar NDF content and digestibility. This lower intake is probably the result of metabolic imbalances induced by fermentation losses. These losses are discussed in detail in Section 14.4.5; the purpose here is to place silages in perspective relative to the intake of other forages. Experiments have shown that the pressed juice from silage contains factors that can reduce intake. These compounds seem to be ammonia and amines. The occurrence of such inhibitory factors is consistent with the concept of metabolic imbalance, because poorly fermented silages have lost much of their fermentable carbohydrate and protein, which is the source of ammonia and amines. Excess rumen ammonia can inhibit feed intake, but the situation can be ameliorated by proper supplementation with carbohydrate (England and Gill, 1985). On the other hand, infusion of amino acids via duodenal fistula increases silage intake. These observations are consistent with the hypothesis that silages are depleted in fermentable carbohydrate, leading to low microbial yield and poor utilization of the large amounts of nonprotein nitrogen (NPN) they can contain. Feeding carbohydrate promotes yield of microbial protein and use of the NPN. Making protein and amino acids available to the lower tract overcomes the animal's nitrogen deficiency.

21.4 Fill

Considerable evidence exists that dietary bulk and consequent distension of the digestive tract limit intake. This effect is demonstrated by feeding low-density diets or by displacing gastrointestinal space with inert material such as balloons, sponges, or plastic ribbon. Figure 21.8 shows the effect on intake of permanent displacement of a portion of rumen volume. Plastic ribbon longer than 9 cm is not ruminated or passed through the digestive tract. When a sufficient amount is added to the rumen through a fistula, intake is initially reduced, although it tends to recover somewhat with time, presumably through stretching of the rumen wall into the abdominal cavity. The addition of a sufficient amount of ballast in excess of that to which the rumen can adapt results in permanent lowering of intake. Evidence that fill limits intake is also supplied by the increase in intake obtained by feeding ground or pelleted forage diets. Grinding and pelleting increase feed density and rate of passage.

If gastrointestinal fill limits intake, there must be some equilibrium between the degree of fill and stretch of the gastrointestinal organs and the rate at which the contents can be disposed of through digestion or passage. Greater tolerance to a high-bulk diet would be provided by increased rumination or chewing or by increased stretch of the organs to allow a larger digestive pool. It is likely that in the case of increased intake the volume of digesta expands to a new equilibrium, which is balanced against some increase in passage rate (Balch and Campling, 1965; Colucci et al., 1982).

Figures 21.9 and 21.10 show the effect of a permanent displacement in the abdomen in fat animals and in pregnant animals, respectively. The problem is more severe in multiparous females and promotes toxemia and ketosis in pregnant ewes. The exact physiological mechanism that acts to limit intake in response to fill is not known. Intake may respond to discomfort or the humoral intake-regulating factor. Fill changes over time because rumen volume does not remain at a constant maximum even in nutritionally limited animals (Gill et al., 1988).

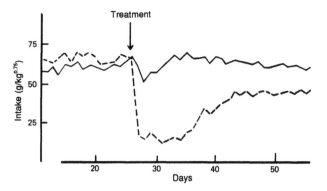

Figure 21.8. Hay consumption by wethers that have had 150 g of 30-cm-long polypropylene fibers placed into their rumens via fistula (dashed line) compared with untreated controls (solid line) (from Welch, 1967). Plastic fiber of this length is not ruminated and remains perpetually in the rumen. The eventual increase in treated animals' intake indicates adjustment of the rumen volume to meet appetite demand.

21.4.1 Plant Cell Wall

Forage intake is usually regarded as being less than the optimal amount, especially with lower-quality forage, and fibrous bulk is considered the limiting factor. Whether this suboptimal intake is actually due to the volume of plant cell wall is open to doubt. Possibly the effect attributed to cell wall is due to an interaction among fill, rumen stretch, time available for eating,

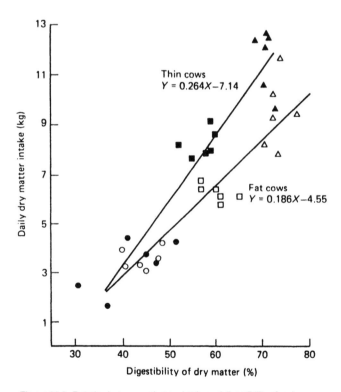

Figure 21.9. Relation between voluntary intake and digestibility of various diets fed to fat cows (open symbols) and thin cows (solid symbols) (from Bines et al., 1969). Symbols: ○, ●, straw; □, ■, hay; △, ▲, concentrates.

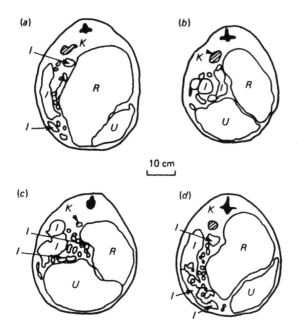

Figure 21.10. The anterior aspect of sections of ewes in the region of the third lumbar vertebra: (a) 88 days pregnant with a single fetus; (b) 111 days with one fetus; (c) 143 days with one fetus; (d) 95 days with twins. I = intestines; K = kidney; R = rumen; U = uterus. (From Forbes, 1986.)

and energy density, which can be independent from cell wall content. These problems are discussed below.

The concept of bulk volume limiting intake is founded primarily on dilution studies in which caloric density was reduced by the addition of bulky material. If concentrate is diluted with coarse forage, density and cell wall content vary inversely with one another, their respective effects being associated and inseparable.

The importance of plant cell wall as a restrictive determinant of forage intake was demonstrated experimentally by Osbourn et al. (1974) and Mertens (1973). In vitro rumen rate-of-digestion trials support its importance, since the maximum correlation of in vitro digestion with intake is at a fermentation time of 6 h, while the maximum for digestibility is in excess of 36 h. After 6 h, very little cell wall matter has been fermented but the cell contents have disappeared, so 6-h digestion corresponds more or less with NDF, which consistently gives the highest correlation with intake (see Figure 8.4).

Osbourn et al. (1974) fed sheep a standard forage to directly determine relative intake. Figure 21.11 compares the relations they found between cell wall and directly measured or relative voluntary intakes. A comparison of the scattergrams shows the large reduction in variation among high-quality forages when intake is expressed as a ratio to a standard forage against which all the experimental animals have been calibrated. Therefore, the variation in direct intake mea-

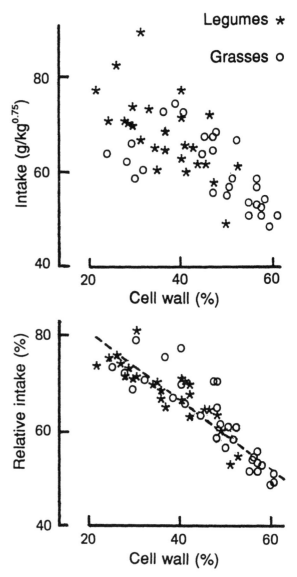

Figure 21.11. Relation between the observed organic matter intake and the cell wall content of 56 forages fed to sheep (from Osbourn et al., 1974). Lower figure shows the effect of correcting intake relative to a common standard forage fed to all animals. Relative intake of organic matter was calculated according to Crampton et al. (1960), and a standard reference forage was fed to all the experimental sheep to remove interanimal differences. The wider scatter at low cell wall content in the upper figure indicates the greater range in appetite of animals fed higher-quality forages.

surements appears to be largely due to variation in the ability of individual animals to consume forage. Intake of coarse forages, however high in digestibility (85%), is still limited by their cell wall content (Figure 21.11). This contrasts with the relation presented in Figure 21.4 in that no inflection point is seen. The experimental conditions described by the two figures differ in that the inflection in Figure 21.4 is at the point where bulky dietary matter is diluted with concentrate, whereas im-

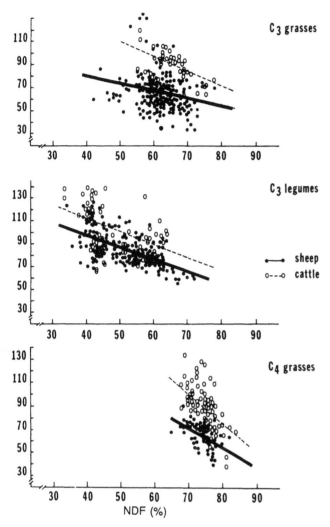

Figure 21.12. Daily dry matter intake (DMI) in sheep and cattle related to concentration of neutral-detergent fiber (NDF) in different classes of forage (from R. L. Reid et al., 1988). Note the narrow range for C_4 grasses.

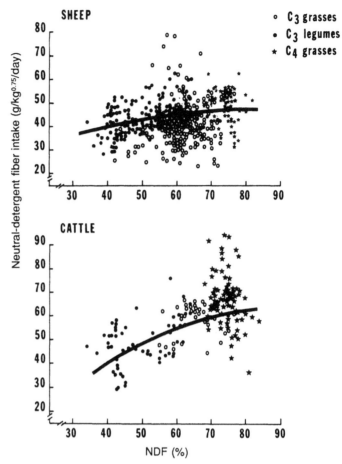

Figure 21.13. Daily neutral-detergent fiber intake (NDFI) by sheep and cattle versus concentrations of NDF in forages (from R. L. Reid et al., 1988). Note the greater ability of cattle to compensate for NDF level.

mature forage, although low in cell wall, may be still limiting to intake because of the bulkiness of its large, thin-walled cells.

The variation in intake is better expressed among animals fed high-quality forages (Osbourn et al., 1974; R. L. Reid et al., 1988; Figures 21.11 and 21.12). Poor-quality forage restricts individual expression, and hence variability in intake is largest for high-quality forages. The differences in animals' ability to eat are likely the result of a combination of factors. Test animals are not equal in growth potential and appetite, and rumination capacity and rumen size are variable. The differences between cattle and sheep are particularly interesting.

None of the data on pure forage diets reveal any tendency for a satiety plateau as exemplified in Figures

21.3 and 21.4. The effect of satiety in restricting ruminant intake to a plateau appears to be limited to concentrate feeds. While satiety does affect the intake of forages, it seems to affect the slope of the cell wall intake relationship such that expression of animals' desire for feed is wider at low cell wall content than at higher levels.

21.4.2 Cell Wall and Feed Volume

A broad spectrum of data relating NDF and voluntary intake was summarized by R. L. Reid et al. (1988), who gathered data on sheep and cattle and on C_3 legumes and grasses and C_4 grasses (Figure 21.13). The NDF intake of sheep is more restricted than that of cattle as cell wall increases. Cattle seem to be better able to compensate for poor forage quality, particularly in the case of C_4 grasses. This relatively higher intake of high-cell-wall C_4 grasses has important implications for the tropics. Possibly cattle, with their relatively large rumens, can better accommodate fibrous forages.

Cell wall density is related to lignification (see Figure 10.6). This leads to a potential contradiction of the bulk theory of fill limitation since (1) plant cell walls are not of uniform density, and (2) mature lignified walls are much denser than immature ones. Tropical C_4 grasses tend to have high NDF contents that vary considerably in lignification and thus presumably in cell wall density. Though unstudied, this could be an important factor contributing to the variation shown in Figure 21.13. Both forage density and plant cell wall density have lower correlations with voluntary intake (r = 0.3–0.4) than does cell wall content (0.76; van der Aar and Van Soest, unpublished 1978 data).

The reason for the failure of density to account for cell wall intake is not known, although some obvious points can be made. Immature, voluminous, thin-walled cells are not only more digestible, they are also more likely to collapse during rumination or pelleting than thicker-walled, denser, more lignified cells. Thus, the bulkiness may be offset by higher digestibility and volume decrease after grinding.

21.4.3 Fine Grinding

The intake of forages and other fibrous feeds can be increased substantially by grinding or pelleting them before feeding. The reduction in particle size and the collapsing of the cell wall structure increase the density of the feed. There is also a reduction in rumination time with concomitantly more time available for eating. The greater density will allow faster rates of ingestion and less rumen volume. The degree of improvement in intake will vary depending on the nature of the feed. The greatest improvement in intake after pelleting is seen in bulky forages of good digestibility (e.g., grass hays).

Some of the effects of grinding have already been discussed in Section 14.8. The response in tropical and more mature forages may be limited, as lignified walls collapse less on pelleting, and the intake of digestible nutrients may remain low. This may explain the effects described in Figure 14.10. Finer particles induce less rumination and have faster rates of passage, thus the penalty on digestibility that results from the passage and loss of potentially digestible fiber may offset the advantage of increased intakes of some high-cell-wall forages.

21.5 Intake Models

Thus far I have presented two major concepts regarding limitations of feed intake—a hormonal and physiological explanation and the fill limitation theory.

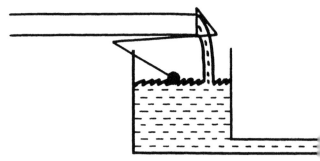

Model A: Restricted liquid outflow

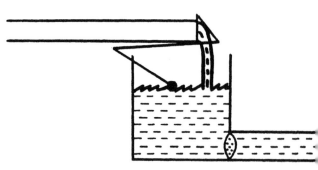

Model B: Restricted particle flow

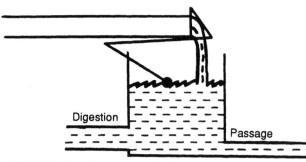

Model C: Digestion rate regulation

Figure 21.14. Three hydraulic models that explain how fill limits intake (from Mertens, 1973). (A) Fecal output regulates fill. (B) Rumen function and outflow of reduced particles sufficiently small to pass the filter regulate fill. (C) Rate of digestion limits fill in competition with and added to rate of passage.

These ideas are not mutually exclusive, and each is dominant under differing feeding conditions, although both likely operate to some degree under most conditions. The fill limitation principle can be explained by several models. Also, the fill models are vulnerable to alternative hypotheses, particularly ones involving time.

21.5.1 Physical Models of Fill

Fill limitation models include the following: (1) rumen fill is limiting, (2) lower tract fill and fecal output are limiting, (3) slow rate of digestion at any site limits intake, and (4) rumination rate reduces volume and

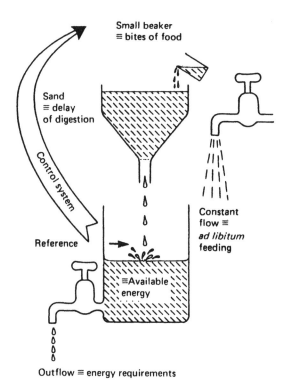

Figure 21.15. Hydraulic model of the control of food intake (see text for explanation) (from Forbes, 1986).

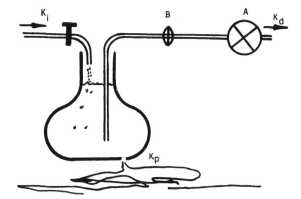

Figure 21.16. The leaky carafe model of digestion (from Van Soest et al., 1983a). Passage (K_p) is represented by a leak in the bottom of the carafe, and digestion (K_d) by a pump (A) insulated from pressure by a transducer (B). Increasing the intake (K_i) will fill the jug to a higher level, placing greater pressure on the leak and increasing passage flow out of proportion to the filled volume. Thus intake increases passage but not digestion and causes potentially digestible matter to be lost. The digestion rate can be altered only by changing the quality of the feed.

limits intake (Figures 21.14, 21.15, and 21.16). Demarquilly et al. (1965) suggested that the volume of indigestible matter is limiting in the lower digestive tract, and Mertens (1973) pointed to the limiting effect of rumen fill volume. S. S. Gill et al. (1969) proposed rate of digestion in the rumen as a limiting factor. This last hypothesis regarding rates of digestion is not supported by the data of Mertens on 187 forages in which rates of digestion showed a correlation of only 0.4 with intake compared with -0.76 for cell wall (Van Soest et al., 1978a).

Limitation by fecal output (Demarquilly et al., 1965) presupposes a limited flow capacity in the lower tract (Figure 21.14). This model requires a constant maximum fecal output set by the limit of lower tract flow. Mertens's data, however, yield an r^2 of 0.33, a significant positive association with fecal output and intake. This is in contrast with the more constant behavior of cell wall intake. It is possible that fecal output could become limiting in high-fiber pelleted diets. Rate-of-passage studies do not indicate a great variation in lower tract passage, which exhibits the slowest lignin rates of passage compared with the rumen rates. An alternative model has a filter that limits output from the rumen (Figure 21.14). This model fits both situations in which rumination time is limiting and those in which fill is limiting. If rumen fill is limiting, rumen volume is a maximal constant with distension to the limit of tolerance in balance with the set point. Mertens

(1973) assumed that forage volume is set by cell wall content and that the restriction of rumen volume will limit the volume of feed consumed. This general concept presents difficulties in that plant cell wall volume decreases with plant age and maturity while the intake of cell wall is relatively constant.

Figure 21.14C assumes that digestion rate sets passage. This would describe many nonruminants on concentrate diets in which the largest disappearance is due to digestion. This model is not compatible with the hotel theory of cell wall volume (Section 10.5.1).

Figures 21.15 and 21.16 describe modifying characteristics of the pregastric digestion. Delay in rumen output leads to pooling and more constant flow (Figure 21.15). The regulation of rumen volume is set by the rate of inflow over that of outflow. If the inflow increases, the depth of the filled container increases and pressure rises to increase outflow. This principle is carried a step further in Figure 21.16, in which a small rise in the neck of the container causes a relatively large increase in pressure. This might be compared to the rumen pushing against the ribs and other organs.

21.5.2 Indigestible Bulk

Closely related to the fill concept is that of limitation by indigestible residue. This is the oldest of the current hypotheses of intake of forages, originally suggested by Lehmann (1941). This concept is closely related to the limitation by fecal output hypothesis (Demarquilly et al., 1965), shown in Figure 21.14, model A. The difficulty with this concept is that output of net indigestible matter decreases with higher-quality forages (Ketelaars and Tolkamp, 1992) and is less consistent than NDF content (Mertens, 1973). The indigestibility of NDF is not well related to its intake (see Section 22.3.1). The output of indigestible matter is increased by fine grinding of the feed.

21.5.3 Time, Eating, and Rumination

Animals require time to eat and time to ruminate, both of which are related to net NDF consumption (Welch, 1982). Digestion and passage are the means by which rumen fill is alleviated. These represent true competitive routes of flow from the digesting organ. While rate of digestion has been directly related to intake, the mathematical logic of modeling requires integration of the rate to estimate the net decrease in amount of rumen ingesta. Most rumen digestion rate measurements are on a gravimetric rather than a volumetric basis. The decrease in volume of indigestible particulate matter likely depends on chewing and rumination.

The hypothesis of fill as an intake limiter is supported by the intake depression seen when a fetus or extra body fat presses on the rumen (Figures 21.9 and 21.10). Water-filled balloons and plastic pieces too large to ruminate or pass also depress intake (Figure 21.8). Limiting animals' access to feed can also reduce intake, however, and M. Gill et al. (1988) found that the rumen does not have a constant maximum fill between meals. A time limitation model can be constructed using Welch's rumination data (see Figure 4.11) in which individual maximum rumination rate (g NDF/min) depends on animal size.

The NDF content of forages is the most consistent feed component associated with intake. The negative association has usually been interpreted as a fill effect. Cell wall volume, however, is less well related to intake than is cell wall (NDF) contents (van der Aar and Van Soest, unpublished 1978 data). Immature bulky cell walls are consumed to about the same extent as dense mature lignified ones (Mertens, 1973; see also Figure 10.6). Rumination (g NDF/min) has an upper limit related to body size (see Figure 4.11) so that time spent ruminating high-NDF forages becomes a time constraint because net ruminating time competes with eating time.

Time is more apt to become an intake limiter in low-quality or sparse pastures, and bite size and plant morphology are factors here.

21.6 The Influence of Intake on Digestibility

The physical necessity of ingesta flow to offset the filling effect of feed ingestion presupposes the escape of potentially digestible matter in feces. The sequence of ruminant digestion suggests that substances escaping both rumen and lower tract digestion will be those

Table 21.1. Values reported for digestibility depression in sheep and cattle

Feed	Cell wall (NDF)	Depression[a]		Species
		DM	Per unit NDF	
Molasses	0	+0.8	—	Cattle[b]
Clover, white	36	−1.4	−3.5	Sheep[c]
Alfalfa	52	−2.9	−5.6	Cattle[b]
Alfalfa	50	−2.1	−5.2	Sheep[c]
Alfalfa	(55)	−2.5	−4.6	Sheep[d]
Ryegrass	58	−3.2	−4.2	Sheep[c]
Bermudagrass (coastal)	74	−3.8	−6.4	Sheep[c]
Corn silage	(45)	−3.7	(−8.2)	Cattle[b]
Oat forage	66	−6.4	−11.7	Sheep[c]
Sorghum	62	−6.5	−12.1	Sheep[c]
Wheat bran	(45)	−3.7	(−8.2)	Cattle[b]
Oat grain	(31)	−4.1	(−9.0)	Cattle[c]
Barley grain	27	−3.6	(−13.3)	Cattle[b]
Corn grain	13	−3.3	(−27)	Cattle[b]
Corn grain	(13)	−2.7	−21.0	Sheep[d]
Soybean meal	(14)	−4.1	(−29)	Cattle[b]

Note: Values in parentheses are estimates or interpolations.

[a]Depression in digestibility per unit of maintenance calculated from data in Tyrrell and Moe, 1975a; J. B. Robertson and Van Soest, 1975; and Riewe and Lippke, 1970.

[b]Tyrrell and Moe, 1975a.

[c]Riewe and Lippke, 1970.

[d]J. B. Robertson and Van Soest, 1975.

exhibiting the slowest digestion rates. Escape from any digestion compartment is a function of the competitive rates of digestion and passage.

Despite the expectation that digestibility will decrease as passage rate increases, measurements have not substantiated this. Factorial experiments with two or three levels of intake have often been inconclusive because the depression (often not larger than 2–4 units of digestibility) was difficult to detect. A more sensitive measurement method is to feed at multiple levels and solve the problem by regression or covariance. Such studies, however, rarely conform to the statisticians' requirements for the randomization of the time variable through replications. Most digestion trials are treated as absolute determinations, with the assumption that they are sufficiently repeatable to be presented as averages in feeding tables.

Riewe and Lippke (1970) studied the effect of intake restriction on digestibility of forages, and J. B. Robertson and Van Soest (1975) and Tyrrell and Moe (1975a; references cited by Van Soest, 1975) studied intake restriction in other feeds. These studies are singled out above others because the components responsible for the depression were determined. Data from these studies support the contention that non–cell wall constituents (cell contents) contribute minimally, and the major part of the dry matter depression is accounted for by cell wall (Table 21.1, Figures 21.17 and 21.18). Starch in high concentrations may become a significant factor at high levels of intake in dairy cattle, but

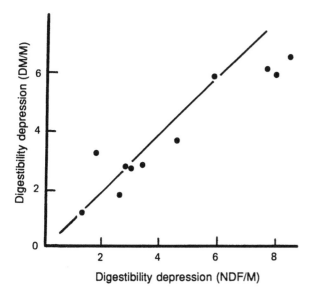

Figure 21.17. The relation between depression in digestible dry matter (DM) per unit of intake (in multiples of maintenance [M]) and the depression in the change of digestible amount of cell wall (NDF) per unit of intake (from Van Soest, 1973a). Solid line denotes 100% equivalence. The change in digestible NDF accounts for most of the decreased indigestibility with increasing level of intake. The data points include both forage and forage-concentrate diets.

even here the contribution of the cell wall remains a large factor.

While it is generally accepted that digestibility is depressed as intake increases, the reasons for the varia-

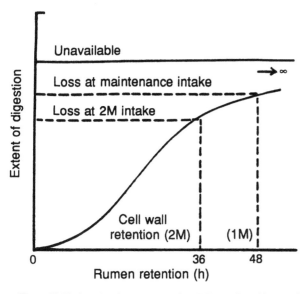

Figure 21.18. A method for estimating digestibility decline with level of intake from rate-of-digestion curves (modified from Mertens, 1973). Rumen turnover time can be estimated from the lignin rate of passage or by relating intake to rumen fill. Digestion rate is the natural logarithm of the slope of the cumulative digestion curve. Dashed lines represent integrations at maintenance intake (M) and at twice-maintenance intake (2M). This model is not mathematically precise; see Chapters 22 and 23 relative to pooling and plug flow.

tion in the degree of depression are not well understood. Some authors have treated digestibility depression as a constant for all feeds; others (Wagner and Loosli, 1967) have said that depression is negligible, while still others have argued that adding concentrate supplement depresses forage fiber digestion (Kane et al., 1959). Such views ignore the substantial cell wall content of concentrates, which is experimentally more depressible than that of many forages (J. B. Robertson and Van Soest, 1975; Colucci et al., 1982).

Rate of fermentation depends on intrinsic properties of the cell wall carbohydrates rather than lignification. The rate of passage of the average forage diet indicates a rumen retention time of plant cell wall of about 40–50 h. Doubling the intake will decrease this time to about 30–35 h in sheep. The cumulative integrated digestion curve in Figure 21.18 indicates that the cell walls most susceptible to digestibility depression are those exhibiting substantial digestibility increases between 30 and 48 h of fermentation. This would account for the generally greater depression observed in grass-fed animals compared with those fed legumes.

Digestibility depression is a function of the competition between digestion and passage, and it has the greatest effect on the slowest-digesting fractions in the plant cell wall. Digestibility depression is inversely related to lignification and to rate of digestion. The relation of lignification to digestibility depression depends on cell wall digestibility. In order to be depressible, cell walls must be digestible; the more digestible the cell wall, the greater the potential for digestibility depression through the effect of intake level, physical form, passage, or concentrate addition. Digestibility depression is proportional to digestible cell wall and rate of passage and inversely related to rate of digestion. Therefore, cell wall characteristics are of prime importance.

Table 21.2 ranks fibrous feedstuffs and their digestible cell wall carbohydrates. Feeds ranking high in the list would be expected to have large digestibility depressions. There are large differences among feeds, and forages do not always have the highest amounts of digestible cell wall. This quality is exhibited by the by-product feeds, particularly the unlignified hulls, brans, and pulp products. These feedstuffs are often more finely divided than coarse forages, and this factor—in addition to their high digestibility—contributes to the high digestibility depressions reported for them. Within forages, legumes have lower digestibility depressions than most grass forages, and those forages with the highest cell wall content do not necessarily have the greatest amount of digestible cell wall. As forages mature, the increase in cell wall is offset by a decline in digestibility through associated lignification (Figure 21.19).

Another variable that affects digestibility depression

Table 21.2.   Fibrous feeds ranked according to their digestible cell wall content

Feedstuff	Cell wall (% DM)	Crude lignin (% DM)	Apparent digestibility of DM (%)	Digestible cellulosic CHO[a]
Soybean hulls	63	2.0	80	55
Soybean flakes	57	1.7	81	51
Dried beet pulp	60	2.7	78	51
Corn bran	51	1.7	82	47
Corncob meal	85	4.5	47	45
Cottonseed hulls	90	23.0	31	34
Malt sprouts	45	1.1	72	31
Alfalfa stems	71	12.0	45	29
Timothy (mature)	65	7.0	47	25
Wheat bran	43	3.8	70	25
Wheat straw	81	7.4	47	22
Alfalfa, late cut	50	8.7	54	20
Citrus pulp	23	2.6	84	19
Alfalfa, early cut	36	5.3	68	17
Oat hulls	77	6.0	44	17
Sunflower seed hulls	71	19.0	31	15
Sesame seed meal	17	2.1	83	13
Rice bran	24	4.3	66	8
Peanut hulls	91	31.4	4	7
Rice hulls	77	14.3	8	0

Source: Van Soest, 1969b.

[a]Units of TDN arising from cellulose and hemicellulose per 100 units DM.

is the rate of fermentation, which changes when rumen pH changes or a starch substrate is present to compete. The substrate competition effect is particularly significant since in vitro digestion with inoculum from concentrate-fed animals exhibits inhibited cell wall digestion at reduced pH (Richard Grant and Mertens, 1992). The degree of digestibility depression varies depending on the substrate. The cell wall sources most adversely affected are unlignified cell walls of the concentrate type such as soybean hulls and corn bran (Jeraci et al., 1980). The depression of forage cell wall digestibility, although significant, is smaller. This observation suggests that concentrate fiber contributes more to digestibility depression than does forage when concentrate is added to forage diets. This difference may be related to the faster passage of the finer concentrate fiber.

21.7 Modeling and Prediction of Intake

The difficulty of predicting and accounting for feed intake makes it hard to predict animal output. This field as yet has no adequate mechanistic models, and existing intake prediction systems are mainly of a stochastic nature. Two mechanistic models have been formulated mathematically, one based on integrated fill (Mertens and Ely, 1982) and the other on the satiety model (Section 21.3), but neither is adequate because each assumes a single cause and ignores interactions with other contributing effects.

Another model is the additivity hypothesis of Forbes

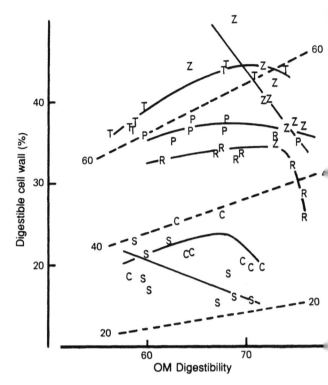

Figure 21.19. The relation between amount of digestible cell wall and organic matter (OM) digestibility in various forages (courtesy of D. J. Thomson). Forages with high digestibilities contain more total digestible fiber. Digestible cell wall is a function of cell wall content and lignification. Both increase with plant maturity, leading to a tendency toward quadratic relationships between digestible cell wall and dry matter digestibility. Note the lower digestible cell wall content of legumes. Forages with a high digestible cell wall content should elicit a greater change in digestibility with level of intake. Dashed lines represent the approximate isoclines for digestibility of cell wall at 20, 40, and 60%. C = red clover; P = perennial ryegrass; R = Italian ryegrass; S = sainfoin; T = timothy; Z = corn plant.

(1986), whereby the intake response is a summative integration of causal impulses. This idea is valuable in accounting for effects arising from more than one cause, but unfortunately it has not been formulated mathematically or used much by people involved in the prediction of intake. This is an area badly in need of more work.

21.7.1 Prediction

Feed intake prediction is a complicated subject, and here I do no more than provide a critique on the general problem. The reader is referred to major review publications by the National Research Council (1987) and the Agricultural Food and Research Council's Technical Committee on Response to Nutrients (1991).

Predicting feed intake under practical conditions poses quite different problems from those encountered in experiments designed to understand what factors actually affect feed intake. Standardized intake trials observe the effects of feed quality and compensation, but animals must adjust their requirements, hunger, and feed intake to account for environmental conditions. Satiety is driven by need, so lactating females need more than growing young animals, which need more than healthy adults. The most successful approach has been to recognize environmental factors affecting food intake (e.g., heat stress reduces intake, and cold may promote it) and integrate them into a system that also considers the physiological state of the animal—whether lactating, growing, or under environmental stresses.

Applied feeding management attempts to optimize diets so that other environmental constraints and the animal itself become the limiting factors of feed intake. As a consequence, feed quality, such as NDF level, is often overlooked as a contributing factor. This has been overcome to a limited extent by classifying types of diets, for example, forage, hay, silage, and concentrate. Prediction of intake by growing animals is based on body weight and expected growth poten-

tial, which takes into account age, sex, breed, pregnancy, lactation, feed additives, growth stimulants, season of the year, and environmental temperature. Such equations and adjustments work only for high-quality diets, since ration composition is not considered. When diet quality does become limiting, NDF is the most important dietary component (NRC, 1987).

Dairy cattle achieve very high levels of milk output, and this leads to intakes that are much higher than those seen in nonlactating animals. Because the pattern of the lactation curve leads to changing energy demand offset in part by tissue mobilization, the level of milk and the stage of lactation become the dominating features affecting feed intake. Failure to optimize dietary composition will ultimately lower milk output and reduce the demand for feed intake, which will adjust itself to an equilibrium with the lower output of milk. Thus arguments regarding satiety and animal physiology over dietary factors such as NDF become circular.

22 Mathematical Applications: Digestibility

Digestion balances have been a common means of diet evaluation, to the extent that digestibility values are now as much attributes of a feed or diet as compositional values are. Yet such a static view overlooks the dynamic aspects of gut flow and the equilibrium between digestion and passage rates that the digestion coefficients truly reflect. This chapter gives a mathematical analysis of the digestion process. The more dynamic aspects of passage pushed by feed intake and modified by gut volume are treated in Chapter 23.

22.1 Digestion Balances

The quantification of fecal output is the basic measurement of any digestion trial. Usually it is necessary to know something about fecal composition to characterize a particular feed and its individual ingredients. Feces are composed of undigested diet plus metabolic excretions. Metabolic excretions are mostly microbial matter derived from the fermentation of feed ingredients plus a small amount of endogenous secretions that are the residual unfermented substances derived from sloughed-off cells and the animal's waste products.

This metabolic matter constitutes the difference between true and apparent digestibility. Since the fecal quantity of true indigestible residue is necessarily less than the total quantity of feces, it follows that true digestibility is a larger number than apparent digestibility. It is the habit of nutritionists to speak in terms of digestibility (that which has disappeared) rather than in terms of indigestibility (fecal recovery), the consequence being that metabolic and endogenous quantities appear as negative numbers. Actually the mathematics of digestion balances requires the use of indigestibility. The result of the inversion to digestibility is that the concepts of variation in digestion trials are very often distorted. Variations in biological and chemical measurements are usually related to the size or quantity of whatever is measured, this being the statistical concept of coefficient of variation. Thus it

follows that variation in digestion balances is related to the amount fed and the amount of feces collected, so that in terms of digestibility, a highly digestible diet (a large amount has disappeared) will be less variable than a less digestible one (Figure 22.1); that is, variation is inversely related to digestibility. The minimum variability associated with carefully controlled digestion trials conducted under restricted intake is about 2 units of total digestible nutrients (TDN); variability usually increases with poor-quality feed, greater intake, and diversity among animals. The difference of about 2 units of digestibility can be taken as the lower limit of biological significance in discussing the digestibility of feeds.

Another aspect of digestion coefficients is their mathematical limits. A true digestibility value cannot be less than 0 or greater than 100%. Metabolic quantities are negative numbers relative to digestibility measurements and can never be meaningfully positive. A positive number signifies that the animal is excreting less than nothing, clearly an impossibility. On the other hand, apparent digestibility could be negative if a small intake, low digestibility, or both were associated with a relatively large metabolic loss of the constituent. Often this occurs in the case of apparent nitrogen digestibility of poor-quality or damaged feeds; however, apparent digestibility cannot be meaningfully greater than 100%.

Estimates of partial digestibility (digestion coefficients for the dietary components), true digestibility, and metabolic amounts occasionally exceed the above-stated limits of biological meaning. These occurrences usually arise from interactions between feed ingredients (associative effects) or the association of metabolic output with ration quality and composition.

An important use of true digestibility (as obtained from the Lucas test) is to ascertain the nutritive ideal of feed constituents in order to provide a biological basis for classifying feed composition. Metabolic fecal nitrogen is important as an internal indicator that can serve as an indirect means of estimating digestibility and intake when total collections are inconvenient.

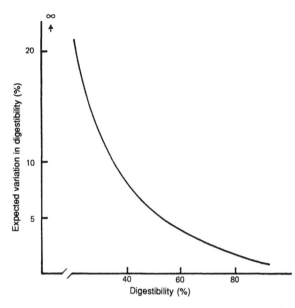

Figure 22.1. Influence of a constant variability in fecal output on the apparent coefficient of variation (CV) of digestibility. A variation of 5% on the true indigestible residue and 7% and a constant M_i (Section 22.1.1) of 15 is assumed. Coefficient of variation is obtained by dividing the estimated error (standard deviation) by the digestion coefficient. Digestion coefficients are more difficult to reproduce when less digestible feeds are used.

22.1.1 Digestion Trials

Digestion trials are conducted in order to evaluate diets with regard to nutrient availability. In practice, trials are conducted with many animals and a number of rations. The feed to be tested is mixed and randomly bagged before the feeding trial to eliminate variation in dietary composition as much as possible. A preliminary period of at least two weeks—so that feed residues from the previous diet are eliminated from the digestive tract—is used to establish the level of intake. The diet is then fed at a level at which orts are absent or their quantity is controlled. The feed may be chopped to eliminate selection, and the orts from the previous meal mixed with the next. The collection should continue for 5–10 days to ensure a constant average fecal excretion to minimize the effect of diurnal variations. Feces must be collected daily and mixed and sampled to represent an average for the whole collection period for each animal. Samples must be dried at temperatures below 65°C to avoid formation of artifacts, and if nitrogen balances are to be accurately measured, a frozen sample is required to retain any volatile nitrogen, which would be lost on drying. Alternatively, feces may be acidified for this purpose before drying.

Variation in digestibility in carefully controlled trials will be about 2% of the intake. Digestion coefficients are often viewed as numerical constants in the same sense as chemical determinations of feed compo-

sition. This view is not entirely justified. Digestion coefficients may be relatively constant for a given set of environmental conditions, but a feed may be more or less digestible if fed at a different level of intake or in a finely divided form.

The digestibility of organic nutrients in feedstuffs is commonly expressed as the proportion (usually a percentage) of matter disappearing in the balance between feed and feces. For a mathematical treatment, however, coefficients are recommended because they are much simpler; they are used exclusively in this treatment.

22.1.2 Definitions of Algebraic Terms

Definitions of the algebraic terms used in this chapter and in Chapter 23 are as follows:

A	Available potentially degradable substrate
a	Subscript denoting apparent (i.e., inclusive of the metabolic fractions)
b	Subscript denoting microbial matter (included within M_i)
C	Concentration
C_{fr}	Concentration of undigested feed in feces
C_{fv}	Concentration of feed in the rumen
C_{mr}	Concentration of metabolic and endogenous matter in feces
C_{mv}	Concentration of net metabolic and endogenous matter in the rumen
C_{mxr}	Concentration of the metabolic fraction of component x in feces
C_{xi}	Concentration of component x in the intake
C_{xp}	Concentration of component x in the outflow from the rumen
C_{xr}	Concentration of x in feces
C_{xv}	Concentration of component x in the rumen
C_{zi}	Concentration of the basal diet in a ration containing a supplement
D_a	Apparent digestibility coefficient
D_f	True digestibility coefficient
D_{xa}	Apparent digestibility of feed component x
D_{xf}	True digestibility of feed component x
e	Subscript denoting endogenous origin (included in M_i)
F_e	Endogenous urea and salivary secretions into the rumen
F_i	Feed intake (g/day)
F_{xi}	Intake of dietary component x (g/day)
f	Subscript denoting the true feed component
i	Subscript denoting the intake
k	Rate constant, reciprocal of turnover
k_d	Fractional rate of digestion
k_f	Rate of disappearance, sum of digestion and passage
k_p	Rate of passage

M   Metabolic amount

M_i Ratio of fecal metabolic matter output to the feed intake

M_{xi} Ratio of fecal metabolic matter due to x to net feed intake

m   Subscript denoting metabolic component

n   Subscript denoting nitrogen

o   Subscript denoting the initial amount at time zero in a kinetic system

P_{fr} Passage of undigested feed from the rumen or to the feces, as specified (g/day)

P_{mr} Passage of net microbial and endogenous matter to feces (g/day)

P_{mxr} Passage of metabolic and endogenous component x (g/day)

P_r Fecal output (g/day)

P_v Rumen passage output (g/day)

P_{xfr} Passage of undigested component x (g/day)

P_{xr} Passage of undigested component x to feces (g/day)

P_{xv} Passage of undigested component x from the rumen

Q_{fv} Quantity (g dry matter) of undigested feed in the rumen

Q_{mv} Quantity (g dry matter) of metabolic matter in the rumen

Q_{mb} Net microbial matter in the rumen (g/day)

Q_{me} Net endogenous matter secreted into the rumen (g/day)

Q_v Net quantity (g dry matter) in the rumen

Q_x Rumen content of component x

R_a Apparent indigestibility coefficient

R_f True indigestibility coefficient

R_{sf} The true indigestibility of cell contents, or solubility

R_{xa} Apparent indigestibility of component x in a feed

R_{xf} True indigestibility coefficient

R_{za} Apparent indigestibility of a basal diet

S   Substrate

s   Subscript denoting cell contents

T_a Apparent turnover in days; equals Q_f/F_i; in hours, 24 Q_f/F_i

T_f True turnover of feed in the rumen

T_{mp} Clearance turnover (outflow) for metabolic and endogenous matter

T_p Clearance turnover (outflow) in days; equals Q_v/P_v

T_x Rumen turnover for dietary component x

T_{xp} Clearance turnover (outflow) of component x

t   Discrete lag; transit time; an interval of time

U   Undegradable fraction or marker

u   Subscript denoting lignin or any other obligately indigestible matter

v   Subscript denoting quantity present in the rumen

w   Subscript denoting cell wall or NDF

x   Subscript denoting component

z   Basal diet

Apparent digestibility of the dry matter (D_a) is then algebraically expressed as:

$$D_a = \frac{F_i - P_r}{F_i} \qquad (22.1)$$

where F_i is the average daily dry matter intake and P_r is the average quantity of undigested dry matter voided daily. Mathematical manipulation requires the use of indigestibility coefficients for the interconversion of partial digestion coefficients. The apparent indigestibility coefficient (R_a) is:

$$R_a = \frac{P_r}{F_i} = 1 - D_a \qquad (22.2)$$

22.2 Partial Digestion Coefficients

It is often necessary to know the digestibility of a ration ingredient that cannot, for logistical or practical reasons, be fed in an undiluted form. Such a situation is represented by the digestibility of compositional components such as protein and fiber or by the addition of a supplement to a basal ration where the partial digestibility of the supplement is to be calculated.

If the proportion (content) of ingredient x in the feed dry matter is $C_{xi} = F_{xi}/F_i$, and fecal content of undigested x is $C_{xr} = P_{xr}/P_r$, division gives:

$$\frac{C_{xr}}{C_{xi}} = \frac{P_{xr}/P_r}{F_{xi}/F_i} = \frac{P_{xr}F_i}{F_{xi}P_r} \qquad (22.3)$$

Since the partial indigestibility of x is $R_{xa} = P_{xr}/F_{xi}$, substituting from equation 22.2, it follows that:

$$R_a C_{xr} = R_{xa} C_{xi} \qquad (22.4)$$

From this relationship the partial digestibility of any ingredient can be obtained provided the feed and fecal compositions are known in addition to the digestion balance.

If the digestibility of a supplement (x) is to be found, the fecal content of the supplement (C_{xr}) is unknown and must be found by difference. The basal unsupplemented diet (z) is fed in a separate trial to find the indigestibility of the basal diet (R_{za}). In the supplemented trial, $R_a C_{xr} = R_{xa} C_{xi}$ and $R_a C_{zr} = R_{za} C_{zi}$. The sum of these two forms of equation 22.4 reduce to:

$$R_a = R_{xa} C_{xi} + R_{za} C_{zi} \qquad (22.5)$$

which shows that the ration indigestibility is the sum of the indigestible amounts of its components.

When the value of the basal component is determined in a separate trial, it is assumed that there is no interaction as a result of the supplementation and that the digestibility of the basal component is unaltered by the addition of the supplement. Interactions among

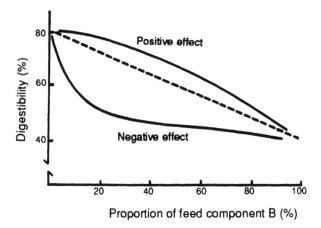

Digestibility (%)

Positive effect

Negative effect

Proportion of feed component B (%)

Figure 22.2. Associative effects that occur when poor-quality feed (B) is substituted into high-quality feed (A). Dashed line represents expected digestibility if no associative effects occur. Positive and negative effects are denoted by the upper and lower curved lines, respectively. When straw is supplemented with protein, a positive association may be observed since the addition of an increment of nitrogen allows better utilization of the straw. On the other hand, negative associative effects are often observed when high-grain diets are diluted by pelleted or finely chopped forage.

feeds (called associative effects) are very common and often lead to under- or overvaluation of supplements. For example, the addition of a small amount of coarse low-quality forage such as straw to an all-grain diet may give a response greater than the digestible energy value of the straw. Examples of positive and negative effects are shown in Figure 22.2. Interactions can be tested by feeding the supplement at differing ratios to the basal diet in a series of trials. The digestible amount of a constituent is its feed content times its partial digestibility and is the basis for TDN calculations.

22.2.1 Fecal Composition and True Digestibility

Feces contains not only undigested food residues but also microbial and endogenous matter arising from the process of digestion. The calculation of true digestibility depends on proper fecal analysis and the application of the partial digestibility equation in a special sense. At the time of excretion most of the nondietary matter in the feces is microbial debris. It is difficult to distinguish endogenous matter from matter contributed by the digestion and fermentation processes since much endogenous matter is likely to be fermented; for example, the mucins and urea that flow into the rumen. Consequently, the fecal composition may be divided into two fractions that constitute the whole:

$$P_{fr} + P_{mr} = P_r \text{ or } C_{fr} + C_{mr} = 1 \qquad (22.6)$$

where C_{fr} is the fraction of undigested food residue and C_{mr} the concentration of metabolic microbial and en-

dogenous matter in the feces. This division distinguishes the true indigestible residue (P_{fr}) from the portion of feces not of dietary origin (P_{mr}). The apparent indigestible matter is the entire feces.

If equation 22.6 is multiplied by apparent indigestibility (R_a), then,

$$R_a C_{fr} + R_a C_{mr} = R_f - M_i = R_a \qquad (22.7)$$

where R_f is the true indigestibility coefficient and M_i is metabolic matter as a proportion of the intake.

The individual terms in equation 22.7 are equal to their respective dietary equivalent or concentrations so that:

$$R_a C_{mr} = \frac{P_r P_{mr}}{F_i P_r} = M_i \qquad (22.8)$$

$$R_a C_{fr} = \frac{P_r P_{fr}}{F_i P_r} = R_f \qquad (22.9)$$

Similarly, the multiplication of any fecal component concentration by the apparent indigestibility coefficient converts that quantity into units of the dietary intake.

22.2.2 True Partial Digestion Coefficients

When this principle is applied to partial digestion coefficients with the purpose of distinguishing between apparent and true partial digestibilities, a problem of denominator arises, since all partial digestibilities are in terms of the ingredient rather than in units of the total diet and are as a result nonadditive. Various fecal components—particularly nitrogen, ether extract, and ash—have endogenous or metabolic contributions that are theoretically additive components of the net M_i of the feces when expressed either as amounts or as fecal concentrations.

Partial digestibility for any component or ingredient x in a diet is defined in equation 22.10:

$$R_{ax} = \frac{P_{xr}}{F_{xi}} = \frac{P_{xfr}}{F_{xi}} + \frac{P_{mxr}}{F_{xi}} = R_{xf} + \frac{P_{mxr}}{F_{xi}} \qquad (22.10)$$

The metabolic term reduces to M_{xi}, which is the partial metabolic component of x expressed in units of feed intake, in which form it can be added with other indigestible dietary components:

$$\frac{P_{mxr}}{F_{xi}} = \frac{P_r C_{mxr}}{F_i C_{xi}} = \frac{R_a C_{mxr}}{C_{xi}} = \frac{M_{xi}}{C_{xi}} \qquad (22.11)$$

The resultant forms of equation 22.11 may be expressed in terms of either indigestibility or digestibility. Most nutritionists think in terms of digestibility (what disappeared) rather than feces (what was collected and measured), such that the inverted form of digestibility causes the metabolic term to be a negative number.

$$R_{xa} = R_{xf} + \frac{M_{xi}}{C_{xi}} \qquad (22.12)$$

$$D_{xa} = D_{xf} - \frac{M_{xi}}{C_{xi}} \qquad (22.13)$$

When the metabolic amount (M_x) is expressed as a proportion of feed intake (f), total fecal metabolic matter (M_i) is the sum of these respective M_x's for individual fecal ingredients—as, for example, metabolic fecal nitrogen, ash, and lipid.

Since the metabolic fraction of feces represents a nondietary quantity, true indigestibility is always a smaller number than apparent indigestibility, while true digestibility is always greater than apparent digestibility by the metabolic amount.

22.2.3 Nature of Partial Digestion Coefficients

The expression of partial digestion coefficients in terms of units of x rather than f causes the main inconvenience of partial digestion coefficients. That is, the magnitude of the difference between D_{xa} and D_{xf} is influenced by dietary concentration of x.

The most important case for this is nitrogen or dietary crude protein. Fecal metabolic losses of nitrogen tend to be relatively constant (although there is a true variance), so apparent digestibility of nitrogen (D_{na}) is largely determined by the ratio of metabolic fecal nitrogen and dietary nitrogen concentration. This is shown in Figure 22.3 where the apparent digestibility drops to minus infinity at zero nitrogen intake. Zero digestibility balances occur near 4 units of crude protein, which is about the average fecal nitrogen multiplied by 6.25. At the same levels of dietary protein the true digestibility, D_f, remains comparatively constant.

Historically this state of affairs has led to considerable misunderstanding of protein availability in low-quality forages such as straws. In the case of wheat straw, for example, the true digestibility is on the order of 85–90%, whereas the apparent digestibilities are often negative. Thus the nutritional shortcoming of such feeds is that they are deficient in nitrogen, not that their nitrogen is unavailable. An exception is the case of heat-damaged or tanniniferous feeds, in which the availability of protein is indeed lowered. Such feeds will show large deviations from expected values in the analysis for ideal feed fractions.

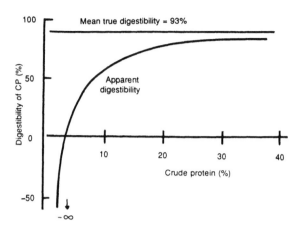

Figure 22.3. Relation between apparent digestibility of crude protein (CP), true digestibility of crude protein and crude protein concentration in the diet. Curves are based on the equation: Y = 0.93X − 3.6, where Y = apparent digestible protein and X = crude protein.

22.2.4 Partial Digestibilities of Subfractions in Supplements

Feed supplements cannot be used alone as a diet, and so their evaluation is rather difficult. Generally, supplement analysis requires two separate sets of digestion trials: one on the unsubstituted basal diet and another on a diet including the supplement. A more fundamental problem exists in the case of evaluating, for example, a protein supplement because one must deal with the protein component of the supplement, which is in turn a component of a total diet. Determining the digestibility of a supplement or the protein in a supplement involves assumptions about the constancy of its partial digestibility in a complete diet. Usually diets are made as isonitrogenous as possible, and urea or a reference protein is added to the control diets and replaced on an equivalent nitrogen basis with the protein supplement in the substituted experimental diets. The assumption can be made that urea is completely available, and thus any change in the digestibility of nitrogen in the substituted diets reflects the availability of nitrogen in the supplement. Equation 22.14 is generally used for components in supplements when digestibility of the supplemental protein equals 100 minus the difference between the digestibility of the treatment groups and the nitrogen digestibility of the supplemented diet divided by the nitrogen percentage of the protein source in the supplement times the nitrogen supplied by the supplement. In terms of the notation listed below, where $S_x = C_{pxs}/C_{pb}$:

$$D_{fxp} = 100 - \frac{D_{ap(b+x)x} - D_{apb}}{S_x} \qquad (22.14)$$

This equation contains additional tacit assumptions that became apparent in its derivation (Van Soest, 1989). The definitions for the additional terms are as follows:

C_{pb} — Concentration of protein in basal diet

$C_{p(b+x)s}$ — Net concentration of protein in mixed substituted diet(s)

D_{apb} — Apparent digestibility of protein in the basal diet, fed singly

$D_{ap(b+x)s}$ — Apparent digestibility of protein in the substituted diet(s)

$C_{p(b+x)s}$ — Concentration of net protein in the mixed substituted ration

C_{pbs} — Contribution of protein from the basal source in the substituted diet(s)

$C_{pxs} = C_{px}S_x$ — Contribution of protein from the experimental supplement in the mixed substituted rations (these contributing concentrations are equal to the respective protein content of the component times its level of substitution)

$D_{fp(b+x)s}$ — True digestibility of substituted protein in the mixed substituted diets

D_{fbp} — True digestibility of the protein in the basal diet

D_{fxp} — True digestibility of the experimental protein

M_{pb} — Metabolic fecal nitrogen in the basal diet

$M_{p(b+x)s}$ — Metabolic fecal N from the mixed substituted diets

$$D_{apb}C_{pb} = D_{fpb}C_{pb} - M_{pb} \quad (22.15)$$

$$D_{ap(+x)s}C_{p(b+x)s} = D_{fp(b+x)s}C_{p(b+x)s} - M_{p(b+z)s} \quad (22.16)$$

Subtracting equation 22.16 from equation 22.15 gives equation 22.17:

$$D_{apb}C_{pb} - D_{ap(b+x)s}C_{p(b+x)s} + (D_{fbp}C_{pb}) - (D_{fp(b+x)s}) - M_{pb} + (M_{p(b+x)s}) \quad (22.17)$$

The term $D_{fp(b+x)s}C_{p(b+x)s}$ contains the experimental component and can be partitioned into additive subcomponents as given in equations 22.18 and 22.19.

$$D_{fp(b+x)s}C_{p(b+x)s} = D_{fxp}C_{pxs} + D_{fbp}C_{pbs} \quad (22.18)$$

$$C_{p(b+x)s} = C_{pxs} + C_{pbs} \quad (22.19)$$

Substituting equation 22.18 into 22.17 and rearranging to obtain the value of $D_{fxp}C_{pxs}$ yields equation 22.20.

$$D_{fxp}C_{pxs} = D_{ap(b+x)s}C_{p(b+x)s} - D_{apb}C_{pb} + D_{fbp}C_{pb} - D_{fbp}C_{pbs} - M_{pb} + M_{p(b+x)s} \quad (22.20)$$

Division by C_{pxs} and factoring gives:

$$D_{fxp} = \underbrace{D_{ap(b+x)s}\left[\frac{C_{p(b+x)s}}{C_{pxs}}\right]}_{\text{Term A}} - \underbrace{D_{apb}\left(\frac{C_{pb}}{C_{pxs}}\right)}_{\text{Term B}} +$$

$$\underbrace{D_{fbp}\left[\frac{(C_{pb} - C_{pbs})}{C_{pxs}}\right]}_{\text{Term C}} + \underbrace{\frac{(M_{p(b+x)s} - M_{pr})}{C_{pxs}}}_{\text{Term D}} \quad (22.21)$$

The equation can be divided into terms involving various interactions. Terms A and B describe the differences between basal and substituted diets that can be attributed to differences in the protein sources. Term C in equation 22.21 describes the differences in protein level and digestibility between the basal and experimental diets and is the basis for the adjustment between groups for differences among controls. Term D describes similar differences associated with metabolic fecal nitrogen. Although methods now exist for the assay of true digestibility and metabolic quantities by fecal analysis (Mason, 1979), conventional application has not taken advantage of such methods, but rather has involved further assumptions about isonitrogenous diets. These assumptions are that perfect isonitrogenicity allows $C_{p(b+x)s} = C_{pb}$; $C_{pxs} = C_{pb} - C_{pbs}$; and $M_{p(b+x)s} = M_{pb}$. Introduction of these equivalences reduces equation 22.21 to give equation 22.22. Terms C and D are canceled to unity and zero respectively.

$$D_{fxp} = \frac{C_{pb}}{C_{pxs}}(D_{ap(b+x)s} - D_{apb}) + 1 \quad (22.22)$$

Equation 22.22 is similar to equation 22.14, and this derivation exposes the following assumptions:

1. The variation in ration mixing that affects nitrogen content (i.e., the coefficient of variation of the isonitrogenous diet) is insufficient to affect apparent nitrogen digestibilities.
2. The balance of the substituted supplements does not alter fecal metabolic nitrogen.
3. The proportional substitution of the supplement has no important variance. Note that in this case there is no obvious method of quantitative analysis of a mixed feed to determine its subcomponents relative to sources of nitrogen.
4. The digestibility of the basal diet is not altered when it is fed along with the supplement. This is not analyzable except as experimental error.

5. The digestibility of the unsupplemented control is treated as a constant; its variation is ignored, and the relative control is set at 100.

6. The error of the partial digestion coefficient is increased by the reciprocals of the coefficients of substitution for the supplement and its proportion of contributed protein.

22.3 The Test for Uniform Feed Fractions

H. L. Lucas was the first to perceive that a rational system of feed analysis could be developed if feed fractions could be found for which the indigestible or digestible amounts could be predicted from their composition (Lucas, 1964; Van Soest, 1967). This type of analysis has been useful in distinguishing feed fractions that behave similarly or uniformly across diets and therefore become descriptions of choice for feed quality or lack of quality. This concept is also applicable to rates of turnover and digestion of components in the rumen or other compartments of the gut. This statistical analysis of feed fractions incorporates the concepts of true digestibilities and metabolic fractions and the additivity of digestible amounts. A reliable prediction of the digestible amount will depend on the uniform availability of the nutrient fraction in question. Therefore, the purpose of the statistical analysis is to discover feed entities that have uniform true digestibilities in all feeds. This information is obtained by performing a regression of digestible amount on intake of the entity, a procedure long used to estimate metabolic fecal nitrogen. In Lucas's application its interpretation is reoriented.

Let equation 22.13 be multiplied by C_{xi}:

$$C_{xi}D_{xa} = C_{xi}D_{xf} - M_{xi} \qquad (22.23)$$

and followed by differentiation of the digestible amount ($C_{xi}D_{xa}$) with respect to C_{xi}:

$$\frac{d(C_{xi}D_{xa})}{dC_{xi}} = D_{xf} - \frac{dM_{xi}}{dC_{xi}} \qquad (22.24)$$

If the metabolic amount (M_{xi}) is a constant, dM_{xi}/dC_{xi} will be zero; plotting digestible amount versus intake of the component will yield a regression slope that estimates the true digestibility. The negative regression intercept will be an estimate of the metabolic amount (integration constant).

The purpose of the analysis is to discern feed fractions that show constant true digestibilities over a wide range of feedstuffs. Feed fractions showing variable true digestibilities are regarded as nonideal. Criteria required of an ideal fraction are a low standard deviation of the regression slope and a zero or negative

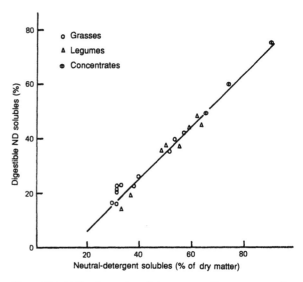

Figure 22.4. Uniformity test for cellular contents of forage measured by the dry matter dissolved in neutral detergent (from Van Soest, 1967). True digestibility (regression slope) is 98 ± 2%; metabolic loss is 12.9% of dry matter intake (negative intercept). The amount of digestible neutral-detergent solubles is the product of the partial apparent digestibility and the content of neutral-detergent solubles in the respective diets.

intercept. Regression slopes greater than unity and positive intercepts have no biological meaning. Variation in metabolic amount is assumed to be random and not correlated with the feed content of the fraction being tested; however, a significant regression of M_{xi} with true digestibility ($dM_{xi}/dC_{xi} \neq 0$) will bias the true digestibility estimate and intercept value of M_{xi}, while random variability in M_{xi} will increase the standard deviation from the regression and the error of the intercept at zero content.

22.3.1 Application

Figures 22.4, 22.5, 22.6, and 22.7 show applications of the Lucas method. The regression for digestible cellular contents (Figure 22.4) shows this fraction to be an ideal one, as indicated by a low standard deviation of the regression slope and the fitting of a

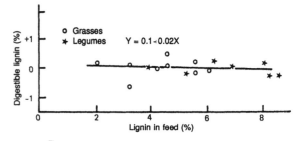

Figure 22.5. Uniformity test for acid-detergent sulfuric lignin, showing nutritional uniformity (based on data from Van Soest, 1967). Although the correlation is not significant, the standard deviation of the regression slope is 2%, indicating uniformity. Permanganate lignin is less uniform and shows a positive slope.

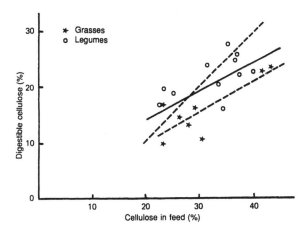

Figure 22.6. Test for cellulose showing nutritional nonuniformity. True digestibility estimated to be 50% for the combined groups, with a standard error of ± 13.5.

diverse group of forages, with no tendency for legumes and grasses to form separate regression lines. The negative intercept (12.9) indicates a large and significant metabolic fraction. The true digestibility of all cell contents is not significantly less than complete, which is further evidence of the freedom of non–cell wall matter from the effects of lignification. Since the cell wall fraction of the feed can have no metabolic quantity (i.e., there is no endogenous fiber), the value of M_s for neutral-detergent solubles must estimate the net metabolic loss of the total diet (M_i) and therefore represent the difference between apparent (D_a) and true (D_f) digestibility of the diet.

The acid-detergent sulfuric lignin test (Figure 22.5) shows that lignin is an ideal fraction because its true

digestibility does not significantly differ from zero and there is no metabolic fraction. The slightly negative slope is an indication of an analytical problem in lignin determinations; lignin recovery is incomplete in feces derived from immature forages low in lignin, and apparent digestibility can often be as much as 40%; but negative balances are also seen (Giger, 1985). A contaminating fraction of about 1–2% of the dry matter might produce a positive effect on apparent digestibility of lignin. The fact that lignin behaves as an ideal fraction does not remove the problems associated with its use as an internal marker.

The principal nonideal components are the structural carbohydrates. The cellulose test (Figure 22.6) shows a scattering of points and different regressions for grasses and legumes. A better result is obtained from legumes with more constant lignification.

Hemicellulose also exhibits fundamentally nonideal behavior (Figure 22.7). The overall correlation is 0.94, which might be taken as an indicator of ideality, but as the figure shows, this number is an artifact that arises out of an interaction between legumes with low hemicellulose content and grasses with high hemicellulose content. Within each group the error and scattering are no less than in the case of cellulose. The decline of hemicellulose digestibility with plant age causes mature plants to contain less digestible hemicellulose relative to content, a factor biasing the true digestibility to a low estimate and causing a positive intercept. A positive value for M_i has no meaning since this would indicate a metabolic excretion of less than nothing. Correlation is really of no value in evaluating a Lucas regression. For example, in the case of lignin, uniform behavior is associated with an insignificant correlation value of 0.13 (Figure 22.5), while the high correlation (0.94) in the case of hemicellulose is associated with nonuniformity. Generally speaking, most fractions of positive digestibility will show significant correlations whether they are ideal or not.

Chemical analyses of feces and microbial products do not show any metabolic structural carbohydrate in feces; therefore, the apparent digestibilities of cellulose and hemicellulose must be taken as true digestibilities, a fact that applies as well to lignin, and thus to the whole plant cell wall.

A summary of Lucas tests for animal species and various feed fractions culled from the literature is given in Table 22.1. Protein and cellular contents show important metabolic fecal fractions and high true digestibilities. Metabolic losses as evidenced by the intercept values are larger for cattle than for sheep or goats. The variation extends to slopes greater than unity, however; that is, to values that can have no biological meaning. Values of true digestibility greater than unity would mean amounts of indigestible matter of less than zero in the feces. This anomaly arises from

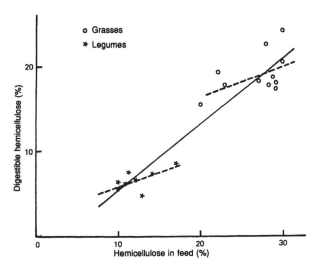

Figure 22.7. Test for hemicellulose, showing nonuniformity (from Van Soest, 1967). The spurious high correlation results primarily from the statistical interaction of the values for grasses and legumes. Standard error of the combined true digestibility (0.79) is ±6.7 units of digestibility.

Table 22.1.   Values of true digestibility and endogenous excretion for different fractions (Lucas test)

Animal	Fraction	Slope[a]	Intercept[a]	Reference
Cattle and sheep	NDS[b]	98	−12.9	Van Soest, 1967
Sheep	NDS	83	−10.7	Combellas et al., 1971[c]
Sheep	NDS	116	−18.2	Aerts et al., 1978
Horses	NDS	100	−12.7	Fonnesbeck, 1968[c]
Cattle	NDS	—	−15.2	Van Soest, 1966
Cattle	NDS	109	−18.0	Aerts et al., 1978
Goats	NDS	100	−9.9	McCammon-Feldman, 1980
Steers	NDS	109	−14.8	Arroyo-Aguilu and Evans, 1972
Cattle	Protein	92	−3.87	Lucas and Smart, 1959[c]
Sheep and goats	Protein	95	−3.72	Lucas, 1959[c]
Sheep	Protein	80	−1.60	Combellas et al., 1971[c]
Horses	Protein	80	−3.30	Slade and Robinson, 1970[c]
Horses	Protein	90	−2.61	Lucas and Smart, 1959[c]
Horses	Protein	82	−3.20	Fonnesbeck, 1968[c]
Capybaras	Protein	87	−2.50	Gonzalez and Parra, unpubl. results, 1972[c]
Guinea pigs	Protein	94	−5.50	Slade and Robinson, 1970[c]
Rabbits	Protein	92	−5.00	Slade and Robinson, 1970[c]
Swine	Protein	96	−4.08	Lucas and Smart, 1959[c]
Deer	Protein		−4.72	Robbins, 1973
Cattle	Ether extract	100	−1.19	Lucas and Smart, 1959[c]
Sheep and goats	Ether extract	95	−0.93	Lucas and Smart, 1959[c]
Horses	Ether extract	70	−0.68	Lucas and Smart, 1959[c]
Horses	Ether extract	75	−1.2	Fonnesbeck, 1968[c]
Swine	Ether extract	99	−1.10	Lucas and Smart, 1959[c]
Horses	Soluble CHO	100	−5.1	Slade and Robinson, 1970[c]

[a]From equation Y = a + bx, where a = endogenous fraction in g/100 g intake, b = true digestibility in %.
[b]NDS = neutral-detergent solubles.
[c]As cited by Parra (1978).

the nonuniform distribution of M_i, which tends to be larger in poor-quality feeds with high cell wall content and smaller in more concentrated feeds.

Feed nitrogen is not completely available, and this leads to estimates on the order of 0.9 for true digestibility. Unheated forages contain insoluble indigestible nitrogen (5–15% of total nitrogen) associated with the lignin fraction. The proportion of unavailable nitrogen can be much increased through heating or improper feed preparation.

Evaluation of lignin is subject to methodological problems. The Klason sulfuric procedure affords better recovery than the permanganate method, which yields low values for samples with high lignin content. This technique shows an apparent endogenous excretion which is not real.

The Lucas test initiated a new way of looking at the classification of feed components. The inconsistency of M_i renders the test less useful in deriving true digestibilities and metabolic amounts than it would otherwise be; however, it has been a means of validating direct analyses of feces for metabolic and feed components. The Lucas test has added to the evidence that suggests the complete availability of cell contents, allowing the use of fecal non–cell wall content as an estimate of total metabolic matter. Direct chemical analysis shows the general absence of cell contents in feces; the only exceptions are starch and heat-damaged

proteins, both of which can be measured. Direct analysis is always preferable to a statistical method.

An important generalization is that no structural carbohydrate exhibits nutritional ideality (pectin may be an exception, but it is not recovered in the NDF fraction). All ideal fractions are either completely digestible or wholly unavailable, and all fractions with significant but incomplete true digestibilities are nonideal and are contained within the plant cell wall. Since the plant cell wall does not possess a significant metabolic fraction, it follows that the entire metabolic fraction in feces is dissolved by neutral-detergent reagent. The Lucas method thus offers a new means of studying metabolic fecal fractions and true digestibility.

22.3.2 Summative Systems

The neutral-detergent reagent divides feed organic matter into two basic fractions: that which is completely available and that which is not—namely, insoluble plant cell wall. The neutral-detergent reagent also dissolves metabolic material in the feces and isolates undigested feed residue so that true digestibility and metabolic fecal output can be directly obtained by fecal analysis. Starch is the only feed fraction likely to appear in fecal neutral-detergent solubles, but it is generally absent from forage diets and otherwise can be accounted for by means of direct chemical analysis.

The results from the general Lucas analysis allows the simplification of the summative rule of additive components for three components: cell wall, solubles, and a metabolic fecal loss.

$$D_a = D_w C_{wi} + D_s C_{si} - M_i \qquad (22.25)$$

The value of M_i can be treated as a constant 10–14 units of digestibility, and the value of D_s is not significantly different from unity, thus leaving the digestibility of cell wall as the only unknown variable. This observation reiterates the fact that it is the insoluble cell wall that is responsible for most of the variation in digestibility, and it is the cell wall that becomes the object of interest in most in vitro systems that attempt to estimate animal digestibilities.

Equation 22.25 offers a rational means of predicting digestibility because digestibility of cell wall is related to lignification. Two systems for this purpose have been developed (Goering and Van Soest 1970; Conrad et al., 1984; see Section 25.5.3). The Conrad system is of particular interest since it has been applied to the problem of net energy (NE) estimation (Conrad et al., 1984; Weiss et al., 1992).

22.4 Rates of Digestion

Rate of digestion refers to the quantity of feed that can be digested per unit of time; it is essentially a physicochemical function of the diet. Mathematical analyses of kinetics and rates of digestion and fermentation have received considerable attention (Ørskov and McDonald, 1979; Bu'Lock, 1987; van Milgen et al., 1991). The composition of the diet and its quality, deficiencies, excesses, and availability of nutrients determine the speed of digestion. Generally, soluble components such as sugars are fermented very rapidly, and less soluble substrates are attacked more slowly. Insoluble proteins, for example, tend to be attacked more slowly than soluble protein and nonprotein nitrogen (NPN). Structural carbohydrates such as cellulose are fermented more slowly than insoluble storage carbohydrates (e.g., starch). The fermentation of complex carbohydrates depends on adequate microbial nutrition (i.e., a nitrogen supply and cofactors).

The consumption of a meal begins a fermentation curve that rises to a maximum at a time after feeding that depends on the diet and the speed of eating. This gives rise to pulsation in fermentation rates between meals, the amplitude of which is influenced by the eating rate and meal frequency. Thus, the eating rate influences the rumen environment and probably affects the efficiency of rumen fermentation. Numbers of organisms and microbial mass (see Figure 8.2) are likely to lag behind the respective peaks of volatile fatty acid (VFA) production.

The disappearance of feed during digestion is a composite of the digestion rates plus passage. Soluble matter ferments rapidly, leaving more slowly available insoluble matter to dominate the later phases of digestion. Fermented matter gives rise to bacteria such that the net disappearance of dry matter or organic matter underestimates the true digestion. Because the bacteria are a part of the insoluble phase, they are incorporated in dry matter measurements of substrate utilization. Unless corrected, this inclusion confounds efficiency estimates of microbial utilization of substrate since products and unused substrate are combined.

22.4.1 Measuring Digestion and Fermentation Rates

Fermentation rates can be measured by batch cultures, incubation in cloth bags in the rumen or cecum, enzyme solubility, and the Hungate zero time method. Digestion or fermentation must be measured at different times in order to estimate rate of change. End point measurements in various procedures include residual substrate, gas production, VFA production, and residual dry matter. Rate estimates obtained by measuring insoluble matter, gas, or VFAs may be less than accurate because these products are produced at the expense of microbial yields and efficiency. For this reason an end point measurement that does not confound substrate with products is preferred. The general problems with the measurement and culture techniques are described in Sections 8.5.1–8.5.3.

Hungate Zero Time

The object of the Hungate zero time procedure is to assay the fermentation rate occurring in the rumen at the time of sampling. It is a useful measurement when associated with direct assay of rumen composition. Combining these kinds of information permits calculations of VFA turnover. A representative sample of rumen contents is incubated with measurements made at specified times up to about 6 h after sampling. The rate of fermentation declines as the rapidly fermentable matter is used up and as acidic end products accumulate to raise osmotic pressure and decrease pH to inhibitory levels. The zero time value is obtained by graphical extrapolation to obtain the maximum initial rate (Figure 22.8). Rates may be measured as CO_2 production, VFA production, or as the disappearance of substrate.

The Hungate technique tends to measure the more rapidly available substrate and is not useful for estimating the digestion of cellulosic carbohydrates, which tend to show peak fermentation 6–18 h after feeding. It

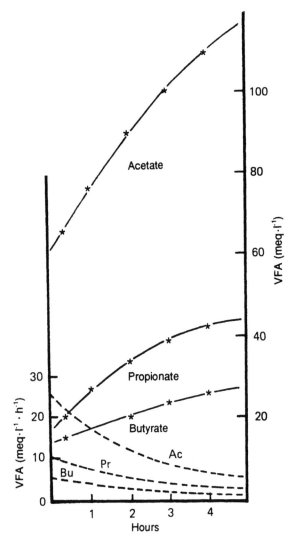

Figure 22.8. Rate of VFA production estimated by time extrapolation (based on Hungate, 1966). Solid lines show cumulative concentrations as measured at various incubation times. Amounts of VFAs produced per hour (dashed lines) decline as the substrate is exhausted and end products accumulate. Extrapolating the amounts produced per hour to zero time (the time the ingesta was removed from the rumen) estimates rates of VFA production in the rumen.

is intended to measure rates as they vary with time after feeding and to reflect the changes in available substrates. The different rates reported in the literature tend to reflect ruminants' feeding behavior and selectivity rather than the peculiar efficiencies of each species. The faster rate of rumen fermentation observed in animals on pelleted feeds and concentrate diets reflects the higher concentrations of available substrate in rumen contents and, perhaps, the removal of the more slowly digestible residues through passage.

Calculation of Rates

The digestion rate of a single pure feed ingredient depends on two main factors: (1) its amount and (2) its

intrinsic properties, which determine the magnitude of the rate constant. If the bacteria are in equilibrium with the substrate, and if the enzyme supply is limiting, no lag will occur. Equation 22.26 also assumes constant volume or pool size. The kinetics are first order if these conditions are satisfied, and the rate is directly proportional to the fractional rate constant (k_s) and the concentration of substrate (S) (= cell contents):

$$\frac{dS}{dt} = -k_d S \qquad (22.26)$$

Integrating equation 22.26 leads to the expectation that concentration will decline logarithmically with time (t). These kinetics apply generally to uniform substrates. Nonuniform fractions such as cell wall require special treatment (Section 22.5).

The turnover time, or the average time feed remains in the digestive compartment, is equal to the reciprocal of the disappearance rate constant under equilibrium conditions assuming first-order kinetics. It is the time in which 63% of the substrate will be digested or transferred. The half-life, or the time for 50% turnover of substrate ($t^{1/2}$), is also a measure of the substrate survival:

$$t^{1/2} = \frac{\ln 2}{-k} = 0.693T \qquad (22.27)$$

where k is the rate constant and T is the turnover time.

Nylon Bag Systems

Measurements of digestion rate using cloth bags with controlled pore size are popular. Usually dry matter disappearance is the end point, but this is less than ideal because the bag contents are contaminated with fine particles of cell wall and attached bacteria (see Figure 8.1). Alternatively, bags are opened and boiled in neutral detergent. Criteria for standardization of nylon bags and their management have been suggested by Nocek and English (1986).

Analyses of nylon bag data often use the "abc" system devised by Ørskov et al. (1988), in which disappearance with time equals $a + b(1 - e^{-ct})$, where a has immediate solubility (underestimated if there is lag) and b is the available fraction of the insoluble remaining matter (c). This problem can be negated if a zero measurement in distilled water is made. An indigestible component (U) must be subtracted, as in the Waldo model (Section 21.5.3), otherwise true linearity cannot be achieved. Practical values of U can be obtained by suitable iteration procedures. This system has the inconvenience that lag affects can be confused with extent of digestion.

The nylon bag system has been applied to protein degradability, supported by arguments that commer-

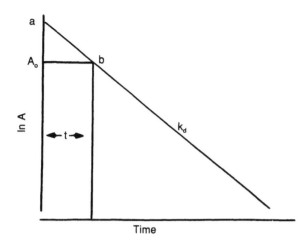

Figure 22.9. Method of estimating discrete lag. Intersection point b represents the time after feeding or inoculation when the regression slope k_d interacts with the initial amount of substrate (A_o) offered. Projection allows the lag delay to be expressed in hours, which can be added to turnover time ($1/k_d$).

cial enzymes used in vitro, even with rumen fluid, cannot exactly replicate the rumen environment and microbes. There have been attempts to correct for microbial matter on the basis of DAPA (Varvikko, 1986).

22.4.2 Lag

If the supply of substrate in pure systems exceeds the capacity of the microbial population and its enzyme supply, the rate of digestion tends to be of zero order—that is, directly proportional to time and slower than the intrinsic limit set by the substrate—and the amount digested per unit time is less than it would be if enzymes were adequate. In a microbial system, population growth will cause enzymatic production to increase and ultimately to saturate the substrate, but the net effect is a lag.

Various models have been proposed to describe lag. These models fall into two categories: discrete lag, where lag is a finite unit of time; and kinetic lag, which allows a lag pool to decrease exponentially with time (Van Milgen et al., 1991). Some models have been based on empirical experience in applying a regression equation involving logarithmic disappearance with time. Others are more rigidly logical but lack experimental evidence to support them. This section presents some current lag models, but the subject is undeveloped and the models are not in all cases mathematically or experimentally validated.

The inadequacy of the rumen fermentation to handle substrate at the maximum possible rate occurs particularly directly after feeding and is likely to account for part of the usually observed initial lag in the fermentation rate. Variation in the fermentation rate during and after feeding may be influenced by the ingestion of air (oxygen), by temperature (i.e., cold water or feed),

and by changes in osmotic pressure. Lag may be due to factors other than microbial capacity, including physical factors such as rate of hydration of the substrate, rate of solution, the attachment of cellulolytic species to the fibrous components, and nutrient limitations (Pell and Schofield, 1993a). Some substrates, such as purified celluloses, have intrinsically long lag times. In vitro systems may have artificially induced lag because of poor anaerobic technique or lack of cofactors in the medium.

Discrete Lag

The simplest and most often applied model is discrete lag. It is not a true mechanistic model because it involves no questions regarding cause. Discrete lag results from the projection of the logarithmic rate (y-axis) on time (x-axis), as shown in Figure 22.9. This model assumes that no ingesta leave the rumen during the lag phase. Since the regression is $\ln A = a - k_d t$, where ln is Naperian logarithm, the value of lag (t) is:

$$t = \frac{\ln (a - A_0)}{k_d} \qquad (22.28)$$

where $a - A_0$ is from the y-axis and t is in hours.

Kinetic Lag Using Compartmental Digestion

A nondiscrete lag model (van Milgen et al., 1991) recognizes that the size of the lag pool (L) decreases with time. The substrate early in digestion is the sum of a lag pool (L), a pool available to digestion (D), and an indegradable fraction (U). The size of L is assumed to decline according to first-order kinetics; the decline probably involves hydration, microbial attachment, and enzyme induction:

$$\frac{\partial L}{\partial t} = -k_1 L \qquad (22.29)$$

and

$$L = F_d e^{-k_1 t} \qquad (22.30)$$

The pool available to digestion (D) increases at the expense of the decline in L.

$$\frac{dD}{dt} = k_1 L - k_1 D + U \qquad (22.31)$$

On integration this gives:

$$D = F_d \left[\frac{k_1 (e^{-k_1 t} - e^{-k_d t})}{k_d - d_1} \right] \qquad (22.32)$$

The residue (R) is the sum of the lag compartment, the digestion compartment, and the indegradable fraction.

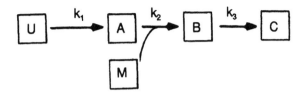

Figure 22.10. Kinetic model of lag involving hydration of feed particles and induction of microbial attachment and digestion (from Allen et al., 1981). Unavailable substrate (U) becomes available (A) and interacts with bacteria (M) in a second-order system (B) to yield digested matter (C). Rates of conversion are represented by k_1, k_2, and k_3; k_1 is limited by hydration, k_2 by microbial attachment, and k_3 by digestion.

Substituting the functions from equations 22.31 and 22.32 yields:

$$R = F_d e^{k_1 t} + F_d \left[\frac{k_1(e^{-k_1 t} - e^{-k_d t})}{k_d - k_1} \right] + U \quad (22.33)$$

which, rearranged, gives:

$$R = F_d \left(\frac{k_d e^{-k_1 t} - k_1 e^{-k_d t}}{k_d - k_1} \right) + U \quad (22.34)$$

This equation yields a sigmoid curve with an asymptote at U. It appears to be more successful in describing the type of disappearance curves seen in nylon bag measurements than the discrete systems described above (van Milgen et al., 1991).

Mechanistic Model of Lag

Most practical systems that model lag involve discrete estimates. Neither discrete lag nor its saturation form are true turnovers, however; that is, their reciprocals do not represent true first-order rates. Discrete lag has been used because it is easy to estimate by regression procedures. A more mechanistic model proposed by Allen et al. (1981) involves kinetic rates of hydration and microbial growth and attachment to feed particles, as shown in Figure 22.10.

The difficulty with this model is its lack of information on the rates of hydration and microbial growth and induction of enzymes, plus the inconvenience of second-order kinetics. Such concepts are important in pointing the way for future work, however, and perhaps with sufficient information less exact relationships can be replaced by more exact ones.

22.5 Rates of Digestion of Cell Wall

Rates of digestion are often presented as cumulative curves (Figure 22.11). These are not rates at all, of course, and the representation as curves has masked the mathematical problem of analysis of the kinetic rate of digestion, which is a function of the slope of

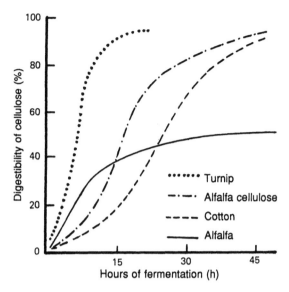

Figure 22.11. Cumulative digestibility curves for various types of cellulose in the rumen (from Van Soest, 1975). Note that the initial fermentation of native lignified cellulose (alfalfa) proceeds more rapidly than fermentation of the delignified alfalfa cellulose and the unlignified crystalline cellulose in cotton. Mature vegetable celluloses are both uncrystalline and unlignified and ferment very rapidly, as exemplified by turnip.

these curves. They are useful to demonstrate the diversity in plant cell walls, however. Unlignified cell walls in turnip and cotton show asymptotic maximum digestibility at widely different times. Delignified alfalfa cellulose has a slower fermentation rate (but a greater extent) than intact cell wall. Mature lignified cell walls have lower extents of digestion than younger plants of the same species, but the more lignified legume cell wall often ferments faster than the less lignified grass cell wall at a similar growth stage. Generally, the more lignified cell walls exhibit curvilinear behavior. This is a case of second-order kinetics (see Section 22.5.3). Digestion curves for naturally occurring hemicellulose and cellulose are very similar and essentially identical with the curves of the cell wall in which they are contained. Isolation causes hemicellulose to become partly soluble and very rapidly fermentable, however, while cellulose may undergo crystallization and ferment more slowly.

The fermentation of nonuniform plant cell wall offers special problems. The cumulative curves (Figure 22.11) reveal the occurrence of lag and incomplete digestion in most substrates, features that cannot be resolved by chemical fractionation of the cell wall in the hopes of obtaining uniform fractions. (Lag is treated in Section 22.4.2)

22.5.1 Ultimate Extent of Digestibility

Ultimate extent can be regarded as a separate entity isolatable only by biological means. As a result, there have been different approaches to its measurement.

The existence of an unavailable portion of cell wall causes first-order rates to become apparently curvilinear, a great inconvenience because all modeling systems depend on linear functions, or at least functions that can be linearized in one way or another.

The estimation of ultimate extent has been approached in several ways. One is by long-term digestion either with nylon bags or with Tilley-Terry-type batch systems. Fermentation times may need to approach two weeks depending on the substrate, and the residue may need to be extracted with neutral detergent to remove microbial contamination. Such values are useful in using the indigestible residue as an estimate of ultimate extent (U) or as an internal marker or in turnover calculations (Section 23.3).

The amount of absolutely indigestible matter may be estimated from a regression of the reciprocal of residual substrate on the reciprocal of time (Figure 22.12). Such plots are not truly linear but may be sufficiently so at long fermentation times to permit extrapolation of maximum digestibility at infinite time.

Long-term fermentations (i.e., 40 days in methane fermentors) have given ratios of about 2.4 for residual and undigested cell wall to Klason lignin (J. A. Chandler et al., 1980). Extrapolations to ultimate extent using Mertens's method yield values greater than 2.5–3.0. As a result, some researchers have calculated the indigestible residue as a function of lignin using one of these factors.

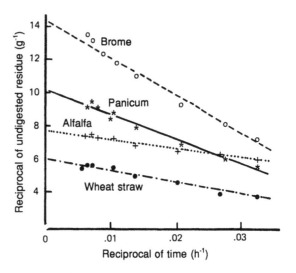

Figure 22.12. Expanded double reciprocal plot of undigested residue versus time (from Mertens, 1973). Values for times earlier than 24 h (1/t > 0.04) are nonlinear and have been excluded. Attempts to use this model with *Trichoderma* cellulase have been unsuccessful because the values did not give linear plots (McQueen, 1974). In some instances, cellulases (particularly grass cellulases) did not approach the maximum extent obtained with mixed cultures of rumen bacteria.

22.5.2 First-Order Systems with an Indigestible Pool

Equation 22.35 represents an attempt to correct the curvilinearity obtained when the log of residual digestible cell wall concentration is plotted against time.

$$\frac{d(W - U)}{W - U} = k_d dt \qquad (22.35)$$

where W is cell wall or NDF and U is the ultimately indegradable residue. An example of this method of correcting data is shown in Figure 22.13. Less lignified cell walls do not necessarily produce faster rates of digestion (Figures 22.13 and 22.14).

The value for U used in the correction is often the in vitro cell wall remaining undigested at 48 or 72 h. The residue of substrate at these times of fermentation may not be completely depleted, but using a U value based on longer digestion leads to undercorrection; that is, the residual corrected values still exhibit second-order behavior (Figure 22.15). It is likely that the cell wall contains more than one pool of available carbohydrate; however, the existence of a slow-digesting fraction beyond 72 h has received less attention because it is beyond the mean retention time of fiber in most ruminant diets. Using the value of an insufficiently digested

residue will result in overcorrection and convex curves, while undercorrection by means of a high digestion value will produce concave curves. The values giving the straightest lines are obtained from residues still containing significant digestible matter.

22.5.3 Alternative Models of Cell Wall Digestibility

The problem of residual curvilinearity in first-order plots of cell wall digestion data is applicable to other models as well. The three systems presented here are not mutually exclusive.

Multiple Pool Models

The multiple pool model proposes that the curvilinearity describing cell wall digestion rates results from different pools of available cell wall carbohydrate of inherently different and varying quality, as between leaves and stems and between vascular tissue and parenchyma. This view is consistent with agronomic observation showing varying cell wall quality between leaves and stems and of plant parts of different ages within the aerial plant structure.

Mertens attempted a three-pool system involving two first-order pools with a fast and a slow rate, respectively, plus an indigestible residue estimated by the extrapolation method. This procedure, if combined with a lag function, is difficult to handle because of the number of degrees of freedom, so estimates of component rates have considerable variability.

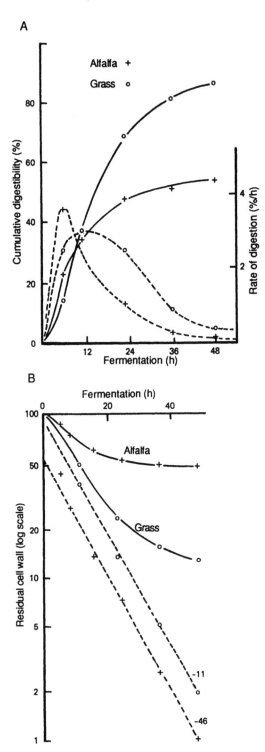

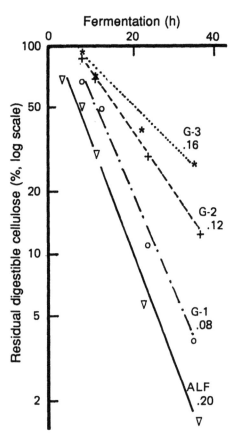

Figure 22.14. Digestion rates of cell wall for three grasses of different ages compared with alfalfa (from Van Soest, 1973b). Digestion rates are faster for immature grasses and alfalfa, despite its high lignification. Lignin-to-cellulose ratios are given in the figure.

Figure 22.13. (A) In vitro fermentation times for alfalfa and or-chardgrass (from Van Soest, 1973b). Solid lines represent cumulative curve of digestion (extent). Dashed lines represent differential slopes of the solid lines and are the true rates of digestion. Note the greater lag, slower rate, and greater extent of digestion of grass compared with alfalfa. (B) The semilogarithmic plot of residual cell wall digestion is derived from data presented in (A). The indigestible amounts at 72 h (11% and 46%, respectively) are used as the basis for correction in the Waldo model (equation 22.35). Corrected values are obtained by sub-tracting 46 and 11, respectively, from each point on the solid-line curves for alfalfa and grass.

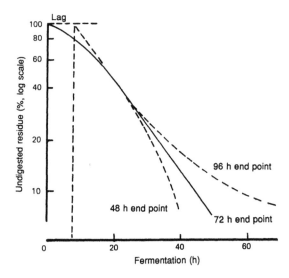

Figure 22.15. The effect of digestion end point correction (see Figure 22.13) on Waldo's estimate of kinetics of digestion (Waldo et al., 1972). Overcorrection results when a short time value (48 h) is used, and undercorrection results from the use of a long time (96 h) (Mertens, 1973). The vertical dashed line indicates discrete lag.

Second-Order Systems

A problem with the multiple pool system is that it takes no account of lignification. There have always been suggestions that lignification may affect rate of digestion, the concept being that as digestion proceeds, residual lignin concentration rises in the remaining undigested residue and proportionally inhibits the rate of digestion. The alternative is to apply second-order models. The inclusion of unavailable lignified residue and the substrate itself leads to a second-order model that accounts for the curvilinearity seen in the application of first-order models and also is consistent with the observation that more lignified substrates tend to be more curvilinear. The equation is obtained by considering available cell wall (A_w), which is equal to $(W - U)$:

$$\frac{dA_w}{dt} = kA_w \left(\frac{A_w}{A_w + U} \right) = -k \frac{A_w^2}{A_w + U} \qquad (23.36)$$

The rate of digestion (dA_w) is related to the amount of A_w and its concentration in the lignified substrate. Note that in an unlignified substrate $U = 0$, and the equation collapses into a single first-order system. Integrating and solving for the integration constant that turns out to be U yields:

$$\ln A_w + \frac{U}{A_w} = U - kt \qquad (22.37)$$

Unfortunately, the result is a transcendental equation that is not easy to analyze. It has been solved by iteration procedures and appears to explain quite well the later stages of digestion (Fadel, 1984). It requires a lag term to fit earlier times of digestion.

Models Based on Available Surface

Surface area has long been suggested as a factor limiting digestion, but good evidence to support this has been lacking. Significant differences in surface due to particle size do not appear to exist over the range of normal feed particle size, and the available rate information seems to indicate that particle size of forage plant cell walls is associated with lag instead of kinetic rate. The problem may be that available surface cannot be distinguished from lignified surface; further, while available surface may be not as limiting as other factors such as crystallinity or hydratability at early times of fermentation, it may become so at later times when available substrate and surface become exhausted. Supporting evidence comes from the observation that extent of digestion (indigestibility, R) is highly correlated with the two-thirds power of the ratio of lignin to cell wall content (Conrad et al., 1984; Figure 12.11 adds further corroboration).

The ratio of surface to volume is theoretically the two-thirds power, but it may lie nearer to three quarters (Figure 12.11), a relation that is also the basis for the three-quarter-power association between heat production and body weight in homeothermic animals. A surface-limited model for digestibility (Weiss et al., 1992) is presented in Table 25.7.

Fadel's (1984) surface model assumes that the rate of digestion is limited by available surface. Note that the Conrad-Weiss system does not treat rate, but rather says that the extent of digestibility is limited by surface. If the rate of digestion is limited by the available substrate (dA_w), expressed in units of mass or weight, the relationship is, according to Fadel (1984),

$$\frac{dA}{dt} = -kA^{0.66} \qquad (22.38)$$

the solution of which is:

$$W = (A_0^{0.33} - kt)^3 + U \qquad (22.39)$$

where W is the amount of cell wall remaining and A_0 is the initial amount of available substrate.

22.6 Particle Size and the Rate of Digestion

An expectation arising from the surface model is that grinding feed should increase the rate of digestion because it increases surface area. This is a somewhat simplistic view, considering the complexity of feed particles and the lack of any good physical evidence for increased surface or faster rates in smaller particles. The argument that grinding increases surface and digestion rate is beset with several problems:

1. Less lignified material is more easily broken, so grinding is selective for more easily digestible fractions.
2. The cell walls of forages are porous and the major effect of grinding may be to expose internal space rather than to increase net surface; this is the hotel theory.
3. Practical feeding of finely ground cell wall is associated with a reduction in digestibility through increased passage. If there is an increased rate of digestion, it must be smaller than the increase in rate of passage.
4. Very fine grinding involves the cleavage of chemical bonds. Ball milling of cellulose can produce reducing sugar.
5. Plant cell walls and seeds have different physical properties. The best positive evidence for association of particle size with rate of digestion may be from the breaking of seed coats or the digestion of starch granules.
6. The above effects are not necessarily operative over the same range of particle sizes.

The relationship between particle size and rate of digestion probably varies with type of materials, such as plant seeds and vascular forage tissue. Seed coats are barriers to invading organisms and enzymes and are a part of the plant's protective system. Vascular plant tissue in leaf and stem, on the other hand, represents the transport and metabolic part of the plant, and movement of nutrients in and out of cells is a prime function. These entrances and exits are the route of entry for invading organisms and enzymes. If this were not so, it would not be possible to prepare plant cell walls by enzymatic digestion or with nonhydrolytic reagents such as neutral detergent. The complete availability of cell contents is also consistent with the above view. M. I. McBurney (pers. comm., 1983), using argon adsorption and polyethylene glycol assays, did not find an increase in surface in milled forage cell walls. The evidence for loss of internal space is perhaps best represented by the loss in water-holding capacity after milling (i.e., loss of the sponge). The increase in effective density resulting from the decrease in water holding and the smaller particle size promote passage out of the rumen. If the rate of digestion increased to the same extent as passage, digestibility would be unchanged, a prediction of the escape equation (Section 23.5).

Ultrafine grinding is known to disrupt glycosidic bonds but does not break bonds in aromatic molecules such as lignin. This effect is significant at the submicrometer level, which is far smaller than that involved in any ordinary milling of feed. Attempts to increase digestibility of lignified materials by grinding, extrusion, and exploding have generally been unsuccessful. On the other hand, heating or grinding starch grains causes a dramatic increase in rate of digestion. Processed starches may have the largest potential for lactic acid production in the rumen. Sequeira (1981) milled alfalfa and wheat straw cell walls to a range of six sizes, from 6 mm to 0.25 mm, taking care that there was no sorting of particles and that all milled products had the same theoretical composition. The effects of this milling were seen mainly at early times of digestion. The interpretation is that grinding over the range of sizes studied affected lag but not rate of digestion.

The overall conclusion to this discussion of the effects of particle size on digestion is that milling forage to sizes of 100–10,000 μm has no significant effect on rate. Very much finer milling probably does increase rate, but that effect is due more to bond cleavage than to increase in surface per se. The milling of grains and starches is a different affair, but even in this case chemical effects are not easily separated from the postulated effects of surface.

23 Digestive Flow

The dynamic aspects of digestion include the flow and pooling of ingesta in the digestive tract. These processes involve rates of passage of undigested matter and turnover, which in turn involve the rate of digestion and net capture of nutrients (digestibility). It is essential that the animal maintain a nutrient intake that balances the costs of living, growth, and reproduction. The processes of turnover and washout have profound effects on gut microbes in the rumen and beyond and on their net production. Because the interactions between digestion and passage are complex, mathematical modeling of these dynamic processes is essential for understanding overall gut function and digestive efficiency. The principles discussed here may be applied to any animal digestive system; however, the focus is on the ruminant system.

23.1 Modeling the Digestive Process

Models are conceptual and mathematical representations of systems, and by making them modelers hope to gain a mechanistic understanding of the system in question. The ultimate goal in modeling is to design something that can be put to practical use. There are no less than four steps to making a model: understanding, mathematical expression, validation, and application. All this applies to modeling in general. When discussing ruminant digestion, however, a peculiar set of problems needs to be considered. From this point of view understanding of the digestive process is benefited by mechanical (physical) and chemical functions. This is in contrast to many areas of agricultural science that have depended on stochastic systems that are statistically efficient but not always revealing about causative mechanistic principles.

Rumen studies and animal nutrition in general have tended to remain empirical, a feature that has likely retarded progress. Earlier applications of statistics were mostly borrowed from animal breeders. The transitional application to nutrition became unqualifiedly empirical, probably because the statistical consultants, though good mathematicians, had little technical understanding of the problems at hand.

23.1.1 The Requirements of Mechanistic Models

A mechanistic model requires the input of known causative factors and provides an account, often on a balance basis, of their effects. Since digestive tracts are input-output flow systems, the laws of conservation of matter and energy apply. The type of matter—soluble, insoluble, digestible, or indigestible—needs chemical description, and the products and residues also need to be characterized. Knowing the value of the composite parts allows generalizations about feedstuffs of similar character. An account of input and output means distinguishing ingesta from microbial products and undigested ingesta in feces. Recycled endogenous matter is another distinguishable entity. The only means of discriminating among these components is proper chemical analysis and characterization.

There are nutritionists who opine that chemical analyses are unnecessary if they cannot be highly correlated with the objective result. For example, nylon bag rate and extent measurements on a simple dry matter basis suffice in their minds for practical feed evaluation. The first problem with this approach is that there is no description of the respective feed composition to be associated with the biological measurement. But there is another, more fundamental problem that involves the measurement of the fraction to which the rate and extent measurements apply, since few if any feedstuffs are chemically or physically uniform. The undigested residue will also be confounded between undigested matter and microbial production.

The fundamental scientific heresy is that understanding is not required for application. Many nutritionists have confounded high correlation with cause-effect. The ultimate weakness of this approach is that a fatal deficiency in understanding prevents further creative development. This has led to some conflict between efficient statistics and mechanistic models, which in turn become harder to evaluate as they become more complex. At the same time, any mechanistically oriented scientist must be wary of the stochastically oriented statistician and mathematician who can provide a higher r^2 with his stochastic model that does not explain cause.

23.2 Turnover and Flow

Turnover is the mixing effect that occurs when a flow enters a pool that has one or more exits to balance the input. When the concept of turnover and the mathematics that describes it are applied to the ingesta in the larger digestive organs such as the rumen or parts of the lower tract, which usually contain some residues from previous meals, a cumulative relationship forms a net meal. The disappearance of undigestible residues tends to obey the first-order law that output is proportional to residual concentration, and so it declines asymptotically (Figure 23.1).

Turnover can be estimated in two ways: either by dividing the gut content mass by the daily feed intake or by calculating the reciprocal of a rate of disappearance obtained by means of a marker dilution technique. In a perfect system these values would be identical; however, liquid in the gut often flows faster than solids, and rates of inflow and outflow vary diurnally with meals, resulting in pulsative flow. Nevertheless, the principle of turnover will still apply since as the intake expands during eating, outflow may not respond, and rumen volume will expand to regulate the balance. Information on diurnal variation is limited, and most passage studies tend to deal with the averaged daily intake, fill, and/or marker decline over a period of days to obtain an average rate.

23.2.1 Compartmental Sequences

The ruminant digestive process is a dynamic system in which feed flows into the rumen and liquid, bacteria, and undigested feed residues flow out of the rumen, through the omasum, and into the lower tract. Various digestive products are absorbed along the course: for example, volatile fatty acids (VFAs) from the rumen and amino acids and long-chain fatty acids from the small intestine. The animal's enzymes digest feeds at successive sites along the lower tract (abomasum, duodenum, and small intestine). This is followed by fermentation in the cecum and large bowel of matter that is resistant to or has escaped from previous digestion. This digestive sequence can be compartmentalized, with the first and largest compartment being the reticulorumen. There is a tendency to view the ruminant tract mathematically as a two-compartment system—the reticulorumen and the lower tract—no doubt because the turnover of markers, as observed by fecal output, does not allow the characterization of more than two compartments. Functionally, any of the anatomical compartments could provide an individual compartmentalizing effect, and information could be obtained on smaller compartments by slaughter and tying off the respective organs, use of content markers, ingesta volume, and chemical composition (Paloheimo and Mäkela, 1959; Vidal et al., 1969).

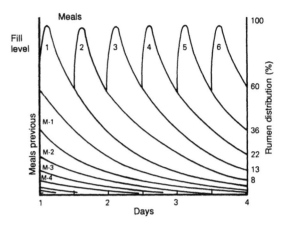

Figure 23.1. The influence of previous meals on cumulative fill, which is the sum of the residues from previous meals (M) remaining in the digestive tract or the rumen. The disappearance of ingesta through digestion and passage causes diurnal variation in cumulative fill. Residues from earlier meals are denoted M-1, M-2, and so on. Rumen distribution is the theoretical rumen composition over time, constrained by twice-daily feeding, 4-h lag to maximum intake, and mean retention time of 24 h for total dry matter. Figure 23.4 plots the mathematical treatment of an individual meal curve. The assumption is made in passage measurements that an individual pulse is representative of the other meal pulses.

Ingested feed and water disappear from a compartment either through digestion and absorption or through passage. Only escaping undigested matter passes down the tract to the next compartment. In this context the rate of disappearance within the compartment will be the sum of the rates of digestion and passage, which equals the rate of appearance in the outflows (absorbed products and passed materials). In a strict mathematical sense, rate of passage refers to the passage of undigested matter. On the other hand, rumen turnover (obtained by dividing rumen volume by intake) is a function of both digestion and passage rates. Passage and digestion offer competitive means for removal of digesta. The fractional rate of digestion compared with the rate of passage will determine the extent of digestion within a compartment and, generally speaking, in the entire digestive tract. These relative rates also determine the amount of potentially digestible matter escaping digestion and arriving in the feces. This concept is fundamental to understanding the response of digestibility to the level of feed intake and the factors influencing rumen escape of potentially digestible matter.

The rate of microbial growth is essentially limited (after the initial lag) by the digestion rate, which in turn is limited by the composition and structure of the feed. The distribution of the rumen flora will be determined by feed composition and the relative amounts of available substrates and their turnovers. Organisms with the enzymatic capability to cope with a substrate will tend to dominate; for example, starch-digesting bacteria will be the dominant microbes in animals fed a grain diet, and cellulolytic bacteria in those on a high-forage diet. The survival and persistence of specific microbe populations depends on their growth rate and generation time relative to the washout of substrate and passage flow. The influx of endogenous saliva and urea plus the growth of microorganisms cause the net rumen turnover to exceed that of ingested feed.

Finally, the escape of potentially digestible substances from the rumen allows the possibility that the non–cell wall matter (microbial and feed) will be digested by animal enzyme systems in the lower tract, and cell walls (plant and microbial) may undergo fermentation in the cecum and large bowel. The amounts of fermentable matter arriving at the lower fermentation sites will involve the same competitive factors of digestion and passage that determine escape from higher compartments. The potential for fermentation in the lower tract is limited by the rates of digestion of the already extensively digested matter arriving there. Turnover and passage rates determine the amount escaping in the feces.

23.3 Mathematics of Turnover

Turnover in the rumen (T) and other digestive compartments is often calculated as the contents (Q) of the compartment divided by the inflow per unit time (F) (Van Soest et al., 1992a). (For definitions of symbols, see Section 22.1.2).

$$T_a = \frac{Q_v}{F_i} = \frac{Q_{fv} + Q_{mv}}{F_i} \qquad (23.1)$$

The digestive pool (Q_v) is composed of matter introduced from the feed (Q_f) plus endogenous inflow from saliva and other nonfeed sources. The growth of rumen microorganisms utilizes both feed and endogenous components to produce a net metabolic fraction (Q_{mv}) composed of organisms (Q_{mb}) and residual endogenous matter (Q_{me}). Production values of net flow per unit time out of the rumen are denoted by P_v.

The net intake (F_i) into the rumen might logically be considered the sum of feed (F_i) plus endogenous flow (F_e); however, this would counter the historical tradition in nutrition to regard true digestibility in terms of undigestible dietary components residual in the digestive tract or the feces. It is noteworthy to mention that the digestive models applied by H. L. Lucas isolate the true digestibility factor from the endogenous. Moreover, it has become apparent that what was once regarded as entirely endogenous (H. H. Mitchell, 1962) is actually mainly composed, in ruminants at least, of microbial cells and their cell walls (Mason, 1984). These microorganisms arise from fermentation of both feed and endogenous secretions, the latter likely con-

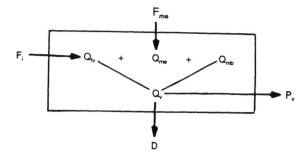

Figure 23.2. The relation of the feed input (F_i) to disappearance through outflow (P_v) and digestion (D). Inflowing saliva and urea (F_{me}) become confounded with feed in the production of microbial matter (Q_{mb}), which with the residual endogenous matter (Q_{me}) forms a net metabolic component (Q_{mv}). Q_{fv} is the net feed in the rumen, and Q_v is the total rumen contents. All quantities in the rumen pool are subject to variable digestion and passage.

tributing more to the balance at lower feed intakes than at high ones (Section 18.10).

Although rumen organisms are supported to a considerable degree by the host's intake, they cannot be considered a part of the feed fraction; rate and extent of digestion of substrate would be obscured by microbial efficiency and cellular growth if microbes were considered a part of the diet. Greater microbial efficiency will diminish apparent digestion of dry matter, causing it to appear smaller than the true extent to which feed was actually digested. The same effect will apply to the turnover, as indicated in equation 23.1, causing the apparent turnover time to exceed that of the actual turnover for the feed. Disappearance rates are the reciprocals of the corresponding turnovers, so rates of feed disappearance can be underestimated and thus fail to agree with estimated rates of passage and digestion obtained by other procedures.

The quantitative relationships governing concentration of feed input relative to the pool concentration are developed like those for digestion balances. Mathematical symbols are used as described in Section 22.1.2. Units of F_i and P_v are quantities per unit of time and are therefore differential rates, although at constant feeding levels values of F and P are constant; the value of Q, the mass of the pool, is timeless. The conceptual model is shown in Figure 23.2.

The balance between inflow and pool content is given by equation 23.2:

$$\frac{Q_v Q_x}{F_i Q_v} = \frac{Q_x F_x}{F_x F_i} \qquad (23.2)$$

The proof of this relationship is self-evident since it reduces to identity via cancellation. Substituting the algebraic equivalents from Section 22.1.2 gives equation 23.3:

$$T_a C_{xv} = T_{xa} C_{xi} \qquad (23.3)$$

which states the relationship between the partial turnover of ingredient x to the total. Note that this equation is parallel to that for digestion balance (equation 22.4):

$$R_a C_{xp} = R_{xa} C_{xi} \qquad (23.4)$$

and that the function of turnover (T_a) replaces indigestibility (R_a). If equations 23.3 and 23.4 are solved for C_{xi}, the relationship to outflow concentration can be obtained in the hybrid equation 23.5:

$$C_{xi} = \frac{T_a C_{xv}}{T_{xa}} = \frac{R_a C_{xp}}{R_{xa}} \qquad (23.5)$$

Rearrangement of equation 23.5 indicates the relationship to clearance turnover (T_p), which is the ratio of Q_v to the outflow (P_v) per unit time:

$$T_p = \frac{Q_v}{P_v} = \frac{T_a}{R_a} \quad \text{and} \quad T_{xp} = \frac{T_{xa}}{R_{xa}} \qquad (23.6)$$

Substitution into equation 23.5 yields:

$$\frac{T_a C_{xv}}{R_a} = \frac{T_{xa} C_{xp}}{R_{xa}} = T_p C_{xv} = T_{xp} C_{xp} \qquad (23.7)$$

Clearance turnover differs from net turnover by the factor of indigestibility. Thus, in the case of indigestible components turnovers are equal since indigestibility is unity. Clearance turnover is the reciprocal of the passage rate. For components with incomplete digestibility, P_v represents the passing or escaping fraction of these partially digestible components per unit time from the rumen.

23.3.1 Metabolic Components

The treatment of microbial and endogenous components is a special case of the additivity principle as stated in equations 23.8 and 23.9. As noted in equation 23.1, the additional metabolic matter in the rumen causes the apparent turnover time (T_a) to exceed turnover for the true feed component (T_f). The quantitative difference $(T_a - T_f)$ depends on the net concentration of metabolic matter in the pool (C_{mv}); its derivation follows. Substituting the true feed component (f) for x in equation 23.3: $T_a C_{fv} = T_f C_{fi}$; then $T_f = T_a C_{fv}$, since C_{fi} is unity (that is, the total of true feed in the dietary intake). Net rumen concentration (unity) is partitioned into true (C_{fv}) and metabolic (C_{mv}) factors: $C_{fv} + C_{mv} = 1$, and $C_{fv} = 1 - C_{mv}$. Substituting for C_{fv} in $T_f = T_a C_{fv}$ yields $T_f = T_a (1 - C_{mv})$. Removing the parentheses and rearranging gives equation 23.8, which indicates that the incremental effect of a metabolic component on net turnover is the product of pool concentration times the net turnover:

$$T_a = T_f + T_a C_{mv} \qquad (23.8)$$

The metabolic matter cannot have a turnover in the same sense as a dietary component because it is not part of the diet; however, its clearance turnover $(T_{mp} = Q_m/P_m)$ is an expression of its mean life in the ruminal pool and is the turnover that should have the most theoretical meaning relative to efficiency of microbial growth and general chemostatic kinetics. The relation of this turnover to the mathematical relations developed here is obtained from equation 23.7 as a special case for metabolic matter:

$$T_{mp} C_{mp} = \frac{T_a C_{mv}}{R_a}$$

$$C_{mp} = \frac{T_a C_{mv}}{T_{mp}} \qquad (23.9)$$

This equation contains all the components needed for microbial efficiency calculations. The efficiency (Y) (microbial matter per unit of true digestible organic matter, D_f) is shown in equation 23.10, where $M_i = C_{mp} R_a$:

$$Y = \frac{M_b}{D_f} = \frac{C_{mp} R_a}{1 - R_a + C_{mp}}$$

$$= \frac{T_a M_b}{R_a T_p (1 - R_a + M_b)} \qquad (23.10)$$

The most difficult of these terms to assay is T_{mp}, which requires an estimate of Q_m output. This generally has meant assay of abomasal or duodenal contents obtained by cannula. If the time could be estimated from general rumen turnover kinetics, the experimental problem could be reduced to an analysis of rumen contents.

23.3.2 Additive Components

Since the relations between the input and the pool components parallels that of the digestible components to one another in the context of the analysis according to H. L. Lucas et al. (1961), it follows that the additivity of subcomponents will also apply. The components of a diet comprise the sum of the products of the dietary components times their partial coefficients of indigestibility (Section 22.2). For any dietary component $(R_x C_{xi})$ a turnover component $(T_x C_{xi})$ also exists in the pool, so that equation 22.4 can be written relative to turnover components:

$$T_a = T_{x1} C_{x1i} + T_{x2} C_{x2i} + \\ T_{x3} C_{x3i} \ldots T_{xj} C_{xji} \qquad (23.11)$$

where C_{x1i}, C_{x2i}, and so on are the respective concentrations of X_1, X_2, and so on in the intake, and T_{x1}, T_{x2}, and so on are their respective turnovers. If all individual ingredient concentrations (C_x) are mutually exclusive and add to unity, the degree of freedom law applies and the last component is fixed if all others are known.

Lucas tests on digestible components indicate the fundamental separation of the cell wall from the non–cell wall and the attendant metabolic component associated with the non–cell wall, which is equal to the difference between total diet apparent and true digestibilities (Section 22.2.1), as summarized in equation 23.12, where T_{fs} is the true turnover of the solubles and T_{sa} the apparent turnover of the solubles (which is longer than T_{fs} because of microbial recycling). Net turnover T_a is then

$$T_a = T_w C_{wi} + T_{fs} C_{si} + T_{sa} C_{mv} \qquad (23.12)$$

The term C_{mv} includes both endogenously secreted matter such as saliva and microorganisms that have grown in the rumen, although these could be represented and described by their own respective turnovers.

The cell wall component of ingesta drives rumination, intake limitation, and solids turnover. The effect of other components within the non–cell wall or the cell wall on turnover can be expected to depend at least in part on the overall behavior of the associated fraction of which they are a part. This particularly applies to nitrogen and protein in the non–cell wall and lignin and cellulosic fractions in the cell wall.

Comparing equations 23.3 and 23.5 makes it apparent that ruminal concentrations of components with faster turnover than the total will be proportionately depleted and less than the dietary content, while components with slower than average turnover will become proportionately enriched in the rumen pool. This aspect emphasizes the significance of the ratio of ruminal or pool concentrations of subcomponents to the respective subcomponent concentration in the diet and is indicative of the relative turnover of any component with regard to the total.

23.4 The Rate of Passage

Passage, or transit, refers to the flow of undigested residues through the digestive tract. Outflow from the rumen includes bacteria and some potentially digestible feed residues in addition to unavailable lignified fiber. The true passage is that of the respective undigested fraction; liquids and solids can have individual passages. At subsequent stages more digestion of bacteria and feed matter occurs. The final fecal residue

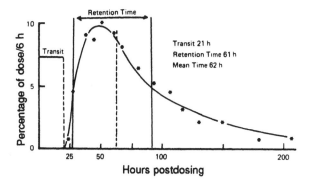

Figure 23.3. Comparison of methods of expressing passage data (see Van Soest et al., 1983b). Transit time is the time from administration of the marker to its appearance in the feces. Retention time is the time from appearance of 5% of the dosed marker to the appearance of 80% of the marker. Mean time is the integrated average of the distribution. Ascending curve estimates lower tract passage. Declining slope represents residence in the rumen. (Based on a ^{14}C-labeled cell wall experiment by L. W. Smith [1968]). Similar excretion curves can be obtained with nonruminants. In these species the ascending curve may signify passage through the upper tract, and the declining phase retention in the larger compartments represented by the cecum and large bowel.

is composed mainly of bacterial and plant cell walls and some endogenous matter. The microbial and endogenous components arise during the course of digestion and passage and, to a limited extent, counterbalance the disappearance of matter through digestion.

Two general methods exist to measure passage. The simplest is the measurement or estimation of rumen volume followed by division of that quantity by the intake to obtain a calculated turnover time (Paloheimo and Mäkela, 1959). The estimate is relevant to passage provided that an indigestible, recoverable reference substance forms the basis of the determinations. The other measurement system is the administration of a pulse dose of marker followed by frequent collections over a period of days. Dosage may be by mouth or fistula, and collection may follow at fistulated sites, in feces, or both. Appearance of the marker is followed by a rise to a maximum concentration followed by an asymptotic decline (Figure 23.3). The results of such measurements may be variable, and their interpretation is complicated. Section 8.4 discusses markers and their various usages; here it is important to state the requirements and limitations of the mathematical models applied to passage.

From turnover mathematics it is obvious that failure to recover the input (F) in the output (P) represents disappearance through digestion. As a result, it is imperative that markers be recoverable, else the rate of passage will be overestimated and confounded with rate of disappearance. For these reasons, nonquantifiable materials such as dyed particles are unsatisfactory recovery markers. A further requirement is that the respective markers must be associated with the fractions they are intended to measure.

Table 23.1. Evaluation of markers and their modes of expression

Marker	Expression	(R^2)[a]	Subjects (df = 11)	Periods (df = 3)	Diets (df = 3)
		Probability of a larger F			
Brillian blue	First appearance	0.72	.0001	.002	.46
Pellets	First appearance	0.62	.0001	.02	.07
	Mean time	0.72	.0001	.03	.002
	Peak time	0.55	.0001	.58	.01
	Turnover $(1/k_1)$[b]	0.37	.02	.32	.75
	Peak + turnover[c]	0.60	.0001	.03	.30
PEG	First appearance	0.61	.0001	.03	.16
	Mean time	0.73	.0001	.01	.0008
	Peak time	0.57	.0001	.06	.12
	Turnover $(1/k_1)$	0.43	.03	.57	.27
	Peak + turnover	0.75	.0001	.0005	.005
Chromium-mordanted	First appearance	0.63	.0001	.001	.16
wheat bran	Mean time	0.79	.0001	.002	.0002
	Peak time	0.67	.0001	.002	.04
	Turnover $(1/k_1)$	0.52	.0001	.64	.004
	Peak + turnover	0.73	.0001	.005	.001

Source: Van Soest et al., 1983b.

Note: Analysis of variance performed on data from a human dietary fiber study. Plastic pellets, brilliant blue, polyethylene glycol (PEG), and chromium mordant were administered twice in each feeding period for a total of 96 passage measurements. Only first appearance could be measured for brilliant blue, and it gave no significant data. The experiment was factorial, involving 12 subjects, 4 periods, and 4 diets.

[a] Proportion of total variance accounted for by subjects, periods, and diets.

[b] Turnover is the reciprocal of the k_1 value obtained by plotting the natural logarithm of marker concentration in the declining phase versus time.

[c] The sum of peak time plus turnover time.

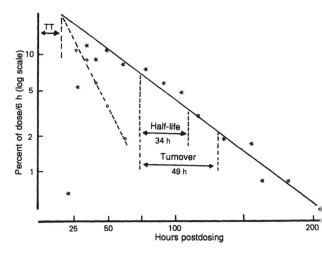

Figure 23.4. The differential excretion curve from Figure 23.3 plotted on a semilogarithmic scale (based on Grovum and Williams, 1973). Note that the descending phase is linear, allowing expression of rates as proportional constants. The slope of the declining phase (solid line) is 2% per hour. Turnover time is the reciprocal of the natural logarithmic slope. The ascending phase is made linear by plotting deviations (o) separately. These points yield the dashed line, with a 10% slope per hour and turnover time of 10 h. This turnover is probably a composite of rumen mixing and lower tract turnover. The intersection of the two lines occurs (theoretically) at the transit time (21 h), the first appearance of markers.

23.4.1 Presentation of Passage Data

Various conventions have been used to calculate passage, although not all describe the same thing. The mean time of residues in the digestive tract will be the integrated average of the curve in Figure 23.3. The retention time, or mean retention time as it has sometimes erroneously been called, is arbitrarily the time in hours between appearance of 5% and 80% of marker.

In studies using human subjects, and in some cases other nonruminants, a value termed *transit time* is measured. This is the time elapsed from feeding to the appearance of the marker in the feces. Such measurements make the unwarranted assumption that passage is a simple pulse through the tract and there is no pool intermixing to cause kinetic flow. Still, transit time is an important component of the calculation of true mean time. According to Faichney (1975) and Parks (1973), mean time is calculated as:

$$\sum_{j=1}^{n} \frac{t_j P_{uj}}{U_i} \qquad (23.13)$$

where t_j is the time between marker administration and the jth observation of defecation, P_{uj} is the amount of marker in the jth defecation, U_i is the total marker administered, and n is the number of defecations. The mean time thus is a composite of transit time, the lag effect (k_2), and the decay turnover (k_1). This value may be adequate in many simple passage studies, and it has

been widely used, particularly in studies of monogastric animals.

Table 23.1 compares the efficiency of techniques. Balch and Campling (1965) estimated retention time on the basis of stained particles but described their results only as retention time and did not analyze them mathematically. This expression has the great disadvantage of not being amenable to rate flow calculations, and alternative mathematical systems have been developed as a result (France et al., 1988).

23.4.2 Kinetic Analysis

Grovum and Williams (1973) applied a model for calculating turnover in which the ascending and descending phases are separated into k_1 and k_2 rates by means of curve peeling (Figure 23.4). This analysis involves resolving the output into two first-order pools according to equation 23.14:

$$C_{ux} = C_{uo} \frac{k_1}{k_2 - k_1} (e^{-k_1(T-t)} - e^{-k_2(T-t)}) \qquad (23.14)$$

where C_{ux} is marker concentration, C_{uo} is initial marker concentration in feces at time zero, k_1 and k_2 are the respective rate constants, and t is transit time. Half-life, the time it takes for half the marker to disappear, is 0.693 (ln 2) times the turnover time.

While mathematically satisfactory according to criteria of mechanistic modeling, and frequently used,

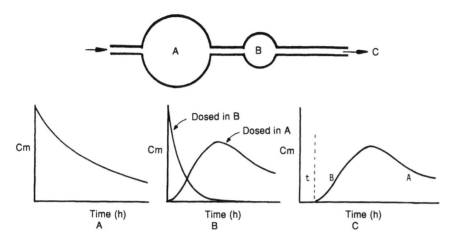

Figure 23.5. Pulse-dose marker flow and concentration (Cm) in compartments of unequal size. As in ruminants, the first compartment (A) is larger than the second (B). C represents fecal output, which shows a lag time (t). Dual curves in B indicate behavior varying according to whether marker was added to compartment A or B.

this equation does have some limitations. The main turnover time is an average over days, and there is some argument over which parts of the curve refer to the respective digesting pools in which k_1 and k_2 are operative. The system cannot resolve more than two compartments. There is also controversy as to whether k_2, supposed to represent the lower tract, includes a component from the rumen (W. C. Ellis et al., 1979).

A coarse fiber diet induces a pulsative curve in which part of the delay is related to selective retention of coarser particles (Section 23.7), so the case for particulate matter is that the second pool is in part a composite of a slower-moving retained pool in the rumen as well as in other organs of the lower tract.

Some statisticians solve equation 23.14 with the help of gamma functions, which normalize the logarithmic distributions, thus reducing error. The logical difficulty of this application is that variance is reduced by empirical means, while the intention of the equation is to provide a mechanistic explanation of passage phe-

nomena. The objectives of statistical efficiency and mechanistic understanding are sometimes incompatible.

23.4.3 Compartmental Sequences

Pulse-dose passage data analyzed according to equation 23.14 are open to alternative interpretations. The mechanics of sequences are related to the size of the pool and the relative rate of turnover. Thus, slower-digesting, longer-retained pools tend to dominate the passage kinetics and appear as k_1 in Figures 23.5 and 23.6. The consequence of having differing pool sizes is that the larger pool dominates the slower-moving phases, as k_1, regardless of its sequence in the system. If the pools are the same size, the data become analyzable according to equation 23.14. Alternatives would be to conduct direct measurements of rumen fill, either by fistula or slaughter.

W. C. Ellis et al. (1979) obtained a similar separa-

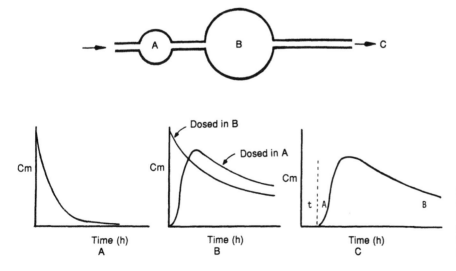

Figure 23.6. Pulse-dose marker concentration when compartment B is larger than A, as is the case in most simple-gutted animals with hindgut fermentation. Note that a transit time (t) appears as a result of delay.

tion of rates using a time-dependent model, although they did not interpret k_1 and k_2 as Grovum and Williams (1977) did. The latter assumed that k_1 represents rumen outflow and k_2 is lower tract turnover in the cecum and large intestine. Ellis's view is that k_1 represents rumination and consequent rumen escape, and k_2 represents rumen mixing. Grovum and Williams (1977) presented evidence that the k_2 rate is highly correlated with lower tract turnover but is an overestimate compared with measurements of the respective organ contents from slaughtered animals. Probably, k_2 is a composite of several turnovers in addition to rumen mixing. Grovum and William's data are based on liquid markers, while those of Ellis were obtained with particulate markers that mixed less rapidly.

23.4.4 Results of Passage Studies

Curve analysis and passage through the rumen and lower tract measured by means of duodenal fistulae indicate that the rumen contents are relatively slower moving compared with the lower tract contents. Particle size influences passage in a complex way, mostly according to the particle breakdown rate through rumination and microbial action. For this reason, rate-of-passage measurements using single markers often measure only the fraction to which the marker is most closely associated in terms of particle size and density.

Liquids and solids in the digestive tract represent separate pools, each influenced by a particular set of factors and their interactions. Water may be absorbed through the rumen wall, favored by lower osmotic pressure in the rumen relative to that in blood, and the flow of rumen water is generally in favor of absorption. The high concentrations of low-molecular-weight rumen metabolites present just after the consumption of a rapidly fermentable meal may violate this generalization. Because high osmotic pressure in the rumen retards the absorption of water, it follows that increased osmotic pressure promotes rumen washout, as demonstrated by the addition of soluble matter to increase liquid passage (D. G. Harrison et al., 1973). Washout affects the balance of the microbial population through its relation to generation time. Bacterial maintenance is decreased and microbial protein production is enhanced by increased washout. At still higher rates of washout, rumen efficiency might be decreased through loss of digestible substrate. It is not known how liquid influences particulate flow, although it may be important. Generally, increased intake will push passage of liquid and particles. Fine cell wall particles are more likely to wash out than coarser material. Finely divided matter is more apt to behave in a fluid manner when slurried with water, while coarser material is apt to collect, much like a logjam, at a sphincter or constriction and act as a filter.

23.5 Digestive Escape

Escape from digestion arises from competition between passage and digestion rates and can occur in any compartment in the digestive sequence. Thus the principles are applicable to both ruminants and nonruminants.

Mathematically, the treatment of escape is identical to that of indigestibility and markers. *Escape*, however, unlike indigestibility, can apply only to potentially available fractions and must be distinguished from obligately unavailable fractions. The equations below apply in this sense to the degree that the analyses applied to diet and rumen contents are consistent with the biological definition.

An equation estimating the digestion balance in the rumen can be derived by dividing the equation between intake and turnover (equation 23.3) by the analogous equation (23.7) for clearance:

$$\frac{T_x}{T_{xp}} = \frac{Q_{xv}P_{xv}}{F_xQ_{xv}} = \frac{P_{xv}}{F_x} = R_x \qquad (23.15)$$

The estimation of an escaping fraction thus requires the knowledge of the rumen digestion balance. It is better to estimate rumen escape on the basis of intake and rumen contents, which is possible by applying the internal marker principle. An internal indigestible marker (u), such as lignin or some other comparably recoverable fraction, that has constant composition in the diet is substituted for x in equation 23.15, and its indigestibility (R_u) will be unity:

$$R_u = \frac{T_u}{T_{up}} = 1 \qquad (23.16)$$

It is obvious that $T_u = T_{up}$; that is, net turnover of the indigestible matter equals its clearance rate. Therefore the clearance rate of undigested matter is estimated from the marker turnover, which is assumed to equal passage turnover (T_p). The digesting fractions must be in equilibrium with their requisite markers for accuracy and for the marker turnover to represent the clearance rate of associated digesting fractions. Substituting T_u for T_{xp} in equation 23.15 yields an equation from which rumen balance can be obtained if the intake, its composition, rumen volume and its composition, and the marker contents are known:

$$R_x = \frac{T_x}{T_u} = \frac{Q_{xv}F_{ui}}{Q_{uv}F_{xi}} = \frac{C_{xv}C_{ui}}{C_{xi}C_{uv}} \qquad (23.17)$$

This equation is consistent with the escape equation in common use (Broderick, 1978):

$$R = \frac{k_p}{k_p + k_d} \qquad (23.18)$$

where $k_p + k_d$ represents disappearance, the reciprocal of which is the functional turnover of the respective fraction (T_{fx}), and the estimated passage is T_u. Equation 23.18 has been used a great deal, but the assumptions involved have not always been apparent. These are that fraction x is available and in equilibrium with marker and that the marker turnover estimates clearance (equation 23.15). Furthermore, x must not be contaminated by metabolic interactions if accurate estimates of true turnover to be obtained.

Alternative forms of equation 23.18 can be used to calculate digestion

$$D_f = \frac{k_d}{k_p + k_d} = \frac{T_p}{T_p + T_d} \qquad (23.19)$$

and the expression for functional turnover (i.e., the turnover resulting from the equilibrium competition between passage and digestion) is the sum of the respective reciprocals:

$$\frac{1}{T_f} = \frac{1}{T_p} + \frac{1}{T_d} \qquad (23.20)$$

and

$$T_f = \frac{T_p T_d}{T_p + T_d} \qquad (23.21)$$

Turnover estimated by recoverable marker in feces is a passage measurement provided that the marker moves at the same rate as the ingesta. The ratio of rumen volume to intake reflects disappearance that incorporates the effects of digestion and passage unless the ratio is based on an indigestible and recoverable fraction. Since the rate of disappearance equals the rate of digestion plus passage, it is evident that total rumen turnover time is always a shorter interval than any of its component turnover pools.

Applying equation 23.17 to apparent turnover will allow an estimate of metabolic and microbial output from the rumen. Metabolic output as a proportion of intake (M_i) is formally the difference between true and apparent digestibility. Applying this relationship to equation 23.17 by substituting for apparent turnovers and true functional turnover of feed (T_f) provides an equation from which metabolic fractions, and therefore microbial output and efficiency, can be estimated:

$$M_i = \frac{T_a}{T_u} - \frac{T_f}{T_u} \qquad (23.22)$$

Another value needed for the calculation of microbial efficiency is the clearance rate for metabolic matter (T_{mp}). This is the ratio of metabolic matter in the rumen (Q_m) to that in the outflow (P_m). The rumen

contents must be measured and the output is calculated from M_i by multiplying it by the amount of feed intake:

$$Q_{mp} = F_i M_i \qquad (23.23)$$

Applying equation 23.22 to microbial output and efficiency requires analytical partitioning of the metabolic fraction in the rumen into microbial and nonmicrobial components. The same requirement will pertain to nitrogen fractions.

23.5.1 Application to in Situ Systems

Many applications of rumen disappearance involve pulse rate measurements over time with markers added to the rumen to assay passage or in nylon bags or in vitro to assay digestion. The particular problem here is applying this kind of data to the estimation of extent of digestion. If samples are placed in bags, the relevance of the data is in reference to the fraction dominating the disappearance. If first-order kinetics are applied, the assumption is made that the fraction is uniform and ideal. Most feeds show a more or less instantaneously soluble component that is wrongly assumed to have instantaneous digestion.[1] The insoluble fraction also contains an indegradable fraction that offers the same mathematical problems discussed in Chapter 22. We also assume here that the residue remaining is uncontaminated with microbial products.

For a first-order system with a discrete functional lag (t) the disappearance of an insoluble substrate corrected for U and the amount of available initial substrate is F_o. The equation assumes that passage occurs during the lag phase (t) of digestion so that the amount of feed remaining at lag time t is:

$$F_o e^{-k_d t} \qquad (23.24)$$

This quantity is subject to competition as described in equations 23.25 and 23.26 below.

The lag function in the equation assumes that passage can occur while there is a lag in digestion rate. The value of t can be a discrete lag or a kinetic one (Section 22.4.2). It could apply to proteins, starches, and fine fiber but not to the coarse fiber pool, which is too large to pass when ingested.

The integration to functional turnover time T_f is defined in equation 23.21. Note that functional turnover

[1] Soluble components are not instantaneously digested, although they seem to be when gravimetric analyses for insoluble matter are applied. These components can be treated in equations 23.26–23.28 in the same manner as insoluble ones provided that liquid passage is used for k_p and that a proper assay for the soluble component is applied. The net disappearance of whole contents will follow the summative rule (equations 23.11, 23.12). The indigestible pool (U) k_p must also be included.

time is always a shorter interval than its component turnovers of passage and digestion. The instantaneous rate of digestion in amounts per unit time is k_dF, and cumulative digestion offered to time T_f (with a discrete lag t) is:[2]

$$D_f = \int_t^T k_dFdt = F_0[1 - e^{-k_d(T-t)}] \qquad (23.25)$$

$$F_0D_f = F_pe^{-k_pt}\frac{k_d}{k_p + k_d}(1 - e^{(k_d + k_p)(T-t)}) \qquad (23.26)$$

The net digestibility at T_f for the respective feed fraction is:

$$D_f = \frac{k_d}{k_p + k_d}e^{-k_pt} \qquad (23.27)$$

Note that this equation is fundamentally the same as equation 23.18 but is arrived at by an alternative derivation. These equations apply to uniform pools of feed ingredients—for example, digestible cell wall and soluble components individually—and have been applied to describe the disappearance of soluble and insoluble components in the ABC classification of carbohydrates and proteins in the Cornell net carbohydrate and protein model (Fox et al., 1992). In this model the net feed function is the cumulative sum of the respective fractions, which have been individually evaluated with equation 23.27.

For the in situ bag or an in vitro fermentation the equation reduces to that described in equation 22.35. Here the soluble matter disappears instantaneously and must be corrected for. It has been the practice to plot cumulative digestion against time of fermentation of in vitro or in situ systems. It is important to note that extent of digestion at time t in the closed system without passage does not correspond to the expected extent of digestion in the rumen, where the solution is exposed to the competition of digestion and passage. The basic problem is that T on the x-axis in Figure 23.4 is not a turnover relative to a mixed-pool system; however, it may be applicable to plug flow (Section 23.5.3).

23.5.2 The Competition between Digestion and Passage

Ingested feed disappears from the digestive tract through two routes, digestion and passage. Consequently, these two processes compete for the same ma-

---

[2] I am indebted to J. France for clarifying equations 23.25–23.27.

terial, with some likelihood that potentially digestible matter will escape digestion and pass into the feces. The degree to which this happens depends in part on the relative rates of digestion and passage. The various feed components have inherently different digestion rates. Cell wall components digest slowly but also have slow passage rates (Figures 23.7, 23.8). Equation 23.25 would lead one to expect that digestible plant cell wall components will be the major fractions subject to digestive escape since their rates of digestion are in a competitive range with passage. Any slowly digesting fraction can be a likely contributor to digestibility depression.

The relation between the proportion of total available digestible feed, including the available digestible cell wall fraction (S_w), and the true digestion coefficient (D_f) of the total diet is given by equation 23.28, based on the digestible amount principle:

$$D_f = S_w\left(\frac{k_d}{k_p + k_d}\right) \qquad (23.28)$$

where the available portion of the feed fraction (S) is multiplied by the available proportion digesting to give its amount. The equation assumes that all digestion occurs in the rumen and is applicable to plant cell wall. The slower-digesting cell wall is the major factor in depressing digestibility. If it is assumed that there is no depression in S, equation 23.29 may be rearranged to show an important relationship:

$$S_w - D_f = D_f\left(\frac{k_p}{k_d}\right) \qquad (23.29)$$

which states that the escaping fraction $S_w - D_f$ is directly proportional to passage rate and digestibility and inversely proportional to the rate of digestion. Passage is a consequential function of intake because the consumption of more feed will pressure the flow of undigested residues. The only alternative relief is the limited expansion of digesta volume (pool size). The relationship applies to the escape of residues from the rumen—as, for example, insoluble protein—as well as to the escape of other potentially digestible organic matter into the feces. Overall, digestibility depression is limited to the slow-digesting fractions such as cell wall and some starch fractions. In order to exhibit a depression, a fraction must be potentially digestible. Relative escape is probably most closely related to feed content of potential digestible cell wall and the level of intake.

Highly lignified or poorly digestible feeds will exhibit less escape of available fractions because their D_f is low and this promotes less depression in digestibility

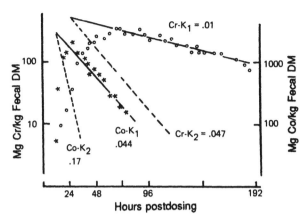

Figure 23.7. Differential rates of passage of Co-EDTA and chromium-mordanted cell wall in the cow fed timothy hay (from Van Soest et al., 1986). Liquid marker (Co-EDTA) passes more rapidly than particulate marker (Cr) in both the rumen (K_1) and the lower tract (K_2).

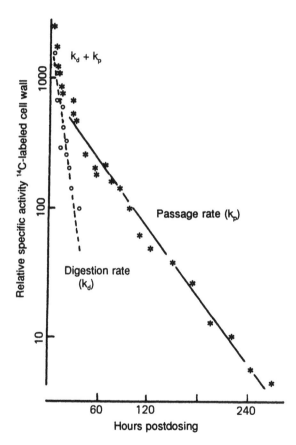

Figure 23.8. The disappearance of marker (uniformly ^{14}C-labeled wheat plant cell wall) dosed into the rumen of a sheep fed orchardgrass pellets (from L. W. Smith, 1968). The total disappearance can be separated into two exponential curves representing digestion and passage. The rate of digestion (k_d) is approximately 9.2%/h. At long times after dosing, the disappearance essentially represents passage since digestion rate has decayed. Second-order effects in the digestion rate cannot be discerned. The rate of passage (k_p) is 2.1%/h, predicting an escape of potentially digestible cell wall from the rumen of 18%.

with increasing intake. Depression is increased in animals fed forages of high digestible cell wall content. The loss in feces will be of unlignified carbohydrates through competitive passage, and it can be assayed by in vitro digestion of feces (Van Hellen and Ellis, 1973). Thus depression in digestibility is inversely related to lignin content and directly related to digestible cell wall (Table 23.2). In forage-fed animals, depression in digestible cell wall accounts for the depression in digestibility. In high-grain diets, starch contributes significantly to the depression as well.

Depression in digestibility as a function of increasing content of a dietary ingredient or with level of intake is an important phenomenon that needs to be taken into account when applying net energy values at above-maintenance levels of feeding.

23.5.3 Plug Flow

Plug flow is flow through a pipe without mixing. In this type of flow the passage time becomes equal to the retention time, and the substrate is not subject to competitive passage. Under these conditions the extent of digestion in relation to passage time is more efficient (Udén, 1989). Less undigested available substrate is lost at the end of the tube because there is no mixing and the ingesta are in contact with the digestive system for the full time of passage. The extent of digestion in a perfect system is:

$$D_f = \int_0^{T_p} \frac{-dF}{F}\, dt = 1 - e^{-k_d t} \qquad (23.30)$$

where F is feed and T_p is plug flow retention. Comparison of the efficiency of the mixed-pool model can be calculated from:

$$T_f = \frac{\ln E}{k_d} \qquad (23.31)$$

where E is the value of escape in the mixed-pool system. Theoretical results are compared in Figure 23.9.

Plug flow is probably dominant in the abomasum, small intestine, and colon of species that do not have a large volume in the latter organ. This could be important in ruminants, in whom retention can be relatively short in certain parts of the digestive tract, particularly the abomasum and small intestine. The extent to which the rumen deviates from a perfect mixing system (via selective retention) causes values of digestion extent to exceed those predicted by the turnover model, but they are still substantially less than values estimated by plug flow.

Table 23.2.   Digestibility versus intake of individual forages fed to sheep

Forage	Number of trials	r	Slope[a]		Lignin/ADF (%)	Cell wall (%)	Digestible cell wall[b] (%)
			DDM	DCW			
Alfalfa	15	−0.82	−2.1	−2.6	22	48	24
White clover	11	−0.60	−1.4	−1.5	16	37	24
Ryegrass	9	−0.85	−1.4	−3.0	11	56	45
Coastal bermuda	13	−0.94	−3.8	−4.7	13	75	53
Sorghum	14	−0.82	−6.5	−7.9	9	57	54
Oats	10	−0.94	−6.4	−7.7	7	65	55

Source: Modified from Riewe and Lippke, 1970.

Note: Intakes ranged from 0.5 to 3.0% of body weight.

[a]Digestible amount of dry matter (DDM) and digestible amount of cell wall (DCW) expressed as units decline per percentage of DM intake.

[b]Estimated from the summative equation (Goering and Van Soest, 1970).

23.6 Rumen Volume and Fill Factor

A valuable term relative to rumen fill and modeling of intake is the rumen fill factor (RFF), based on Waldo et al., 1972.

$$RFF = \frac{C_D}{k_p + k_d} + \frac{C_U}{k_p} \qquad (23.32)$$

C_D and C_U are concentrations of available and unavailable matter of the diet (F), and k_p and k_d are partial passage and digestion rates descriptive of the respective D and U fractions (which add to unity). In this form they conform to Lucas models.

The units of rumen fill factor will be in hours of turnover since:

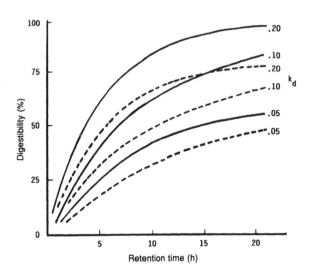

Figure 23.9. The extent of digestion in a mixed-pool system (dashed lines) compared with a plug flow system (solid lines) according to retention (passage) time (based on Udén, 1989). Curves for 0.05, 0.10, and 0.20 digestion rates (k_d) are shown.

$$RFF = F(C_D T_d + C_U T_{pU}) \qquad (23.33)$$

This form could be modified by some function of feed intake such as kg/day or by an expression of bulk hydrated volume.

Thus there is no volume component that could be multiplicatively added in the form of a coefficient relating volume to unit weight. Introducing a lag factor (t) into the full model has the following results. The equation adjusted for discrete lag (t) is:

$$\frac{D}{k_d + k_p} + t + \frac{U}{k_p} = RFF \qquad (23.34)$$

Because ln (a − D) (from the y-axis)[3] = tk_d, then:

$$t = \frac{\ln (a - D)}{k_d} \qquad (23.35)$$

Introduced into the Waldo equation for fill factor:

$$RFF = \frac{D}{k_d + k_p} + \frac{\ln(a - D)}{k_d} + \frac{U}{k_p} \qquad (23.36)$$

$$\frac{D}{k_d + k_p} + \frac{e^{a-D}}{k_d} + \frac{U}{k_d} = RFF \qquad (23.37)$$

This representation of the fill factor is in the form of weight of undigestible matter multiplied by units of time. Since the assumption is made that fill after eating is coarse unprocessed cell wall that is not likely to have an important passage until the particles are converted to a size that can pass, it seems appropriate to use the form expressed in equation 23.36. For soluble matter,

[3] See equation 22.28 and Figure 22.9 for definition of a and derivation of discrete lag.

however, and for escaping protein, starch, and so on, this form may be inappropriate. The general passage equation for mixing systems given in Section 23.4.2 includes this component.

23.7 Modeling the Particle Disintegration Process

The general results of rumination and particulate passage studies support the view that rumen fill limits voluntary intake of forage and that processes promoting disintegration of particulate matter may be specifically time-limiting (Welch, 1982). This hypothesis poses a viable alternative to the currently held theories of intake regulation in ruminants and suggests limits to the advantages of rumination.

The disintegration of ingested particulate matter in the reticulorumen has been the object of particle disintegration models. While it is likely that a continuous function exists so that as particle size decreases, the probability of further comminution decreases and rumen loss (passage) increases, no practical application of this concept has been made. Instead, compartments containing large, medium, and small particles have been conceived. The hotel effect may be restricted to lignified (i.e., cross-linked) structures that cannot be disintegrated by microbial digestion.

The fate of any ingested particle and of any particles produced by digestion or the rumination processes is threefold: disappearance (or loss in mass and size) through digestion, comminution into smaller fragments, and passage. This may be stated in equation form:

$$\frac{dC}{dT} = C(k_d + k_r + k_p) \qquad (23.38)$$

where k_d, k_r, and k_p are differential rates of digestion, disintegration, and passage, respectively, and C is the size of a particulate pool of defined particle size. All these rates are specific for a particle size and type, and a change in size or origin of particles involves a new set of rates.

Figure 23.10 illustrates a model for particle disintegration in the rumen. The smaller particle pools (below 300 μm) have larger passage rates than disintegration rates. Disintegration rates generally decrease on passing down from the larger particles to the smaller ones, while passage rates increase. Digestion rates are fastest for the smallest dietary particles; however, these are greatly diluted by lignified particles resulting from the rumination and digestion of larger particles.

The fine fast-digesting particles entering from the feed may not actually equilibrate with older fine lig-

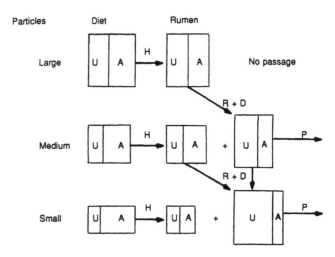

Figure 23.10. Model of the conceptual degradation and fate of ingested particulate matter (from Van Soest et al., 1986). Unavailable (U) and available (A) portions are shown in relative proportions. Finer fractions of the diet are less lignified and contain less U. All fractions are subject to hydration and microbial attachment (H). Hydration capacity is greater for larger particles, but the process is slower. Large particles are above the size limit for passage. The rumen retains larger particles, and the smaller ones are lost in passage in inverse proportion to their size. Large particles can disappear only through rumination and digestion (R and D), whereas medium and small particles disappear by means of rumination, digestion, and passage, with the latter becoming more important as particle size decreases.

nified particles produced by rumination, since the former will have a lower density and thus will float for a time. With their low lignification and high digestibility, however, they are relatively ephemeral in the end. For the larger particles, the rate of digestion (k_d) depends on particle size, which in turn depends on rumination rate and resistance to fracture (and lignification, etc.).

23.7.1 Integrated Model of Fiber Disappearance

Mertens and Ely (1979) and Allen and Mertens (1988) attempted to model fiber disappearance processes. The latters' version is shown in Figure 23.11. This model considers six pools of particles that are, respectively, (1) unavailable for attachment or escape but potentially digestible (UND), and (2) unavailable for attachment or escape and indigestible (UNI). These hydrate, and attachment is allowed in pools (3) available, nonescapable, but digestible (AND) and (4) attachable, nonescapable, indigestible (ANI). Rumination rate k_r produces pools (5) available, escapable, and digestible (AED) and (6) attachable, escapable, and indigestible (AEI). The lag function (k_a) is consistent with a discrete lag. This model requires knowledge of the factorial rates of hydration involving lag (k_a), rumination rate (k_r), and escape (k_e) for respective digestible and indigestible pools.

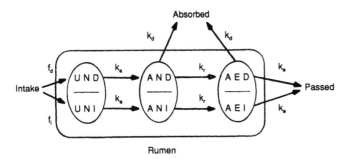

Figure 23.11. Model of fiber disappearance incorporating a lag phase with particles unavailable (U) and available (A) for attachment and passage (from Allen and Mertens, 1988). Nonescapable (N) and escapable (E) as well as potentially digestible (D) and indigestible (I) fiber fractions are included. Fiber fractions and rates are as follows: f_i = digestible fiber as a fraction of intake, k_a = fractional rate of availability, k_d = fractional rate of digestion, k_e = fractional rate of escape, and k_r = fractional rate of release from the nonescapable fraction to the escapable fraction. Note that the terms of this model are inconsistent with the general mathematical nomenclature (Section 22.1.2).

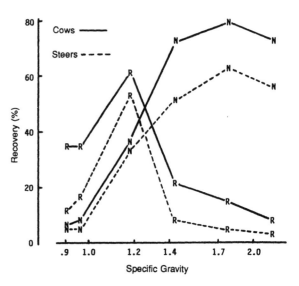

Figure 23.12. Ten-day fecal recoveries of plastic pieces with controlled specific gravity from Jersey steers and lactating Jersey cows (based on data in desBordes and Welch, 1984). N = nonruminated particles, and R = ruminated particles.

23.7.2 Density and Escape

The buoyancy of feed particles with a density below unity is a major factor limiting their escape from the rumen. Very dense particles are few, and there is a functional specific gravity optimum of about 1.2 (desBordes and Welch, 1984). Figure 23.12 shows recoveries of plastic pieces of varying specific gravities. Particles with density less than unity have very low recoveries and tend to remain in the rumen. Peak recovery for nonruminated particles was at about 1.77, and ruminated particle recovery peaked at about 1.17.

Very heavy material such as stones (specific gravity 2.6–3.0) or pieces of metal may tend to remain perpetually on the ventral rumen floor. Lighter matter may be heaved over the reticular pillar into the reticulum and thus have some probability of passage.

When a bolus is brought up, chewed, and swallowed, the finer particles thus generated are deposited near the omasal orifice, and the heavier ones are favored for passage down the canal. Particle density is inversely related to particle size, following the physical loss of interior cellular space that can hold gas and water. The absolute densities of lignin and cellulose are about 1.6, so particles favored for passage will still have some hydration and internal space.

Mathematical modeling of the associated processes of digestive flow is limited by lack of knowledge of the respective rates of particle size reduction and the coassociated rise in density. The above concepts may not apply to concentrate feeds, which are solider and are intrinsically heavier as ingested. Protected fat and protein are frequently added to feed without any attention to how they affect particle density. Fine, light-coated particles of lipid would tend to remain in the rumen.

24 Energy Balance

Energy balance studies have served many purposes, among them the assay of feed values and the determination of animal requirements and efficiencies of feed conversion. The efficiencies characteristic of digestion and metabolism involve the animal's metabolism, the feed composition, and the interactions between the two. Many standardized systems for calculating animal requirements pay only lip service to feed composition, the conflict in theory being whether the animal and its physiological requirements determine intake of metabolic nutrients or the feed itself imposes limitations. Although these problems have already been discussed in Chapter 21, here I should point out the short-sightedness of some animal nutritional physiologists who overlook the role of feed quality and its effect on intake of digestible energy. Thus the Agricultural Research Council (ARC), for example, has published a book on animal requirements and energy use that hardly mentions the physical factors such as neutral-detergent fiber (NDF) and bulk in dietary composition affecting intake and energy use (ARC, 1980). The National Research Council (NRC) has begun to recognize the importance of feed composition but has a long way to go to accommodate mechanistic concepts.

None of the official energy systems now in use give much recognition to mechanistic processes such as cause-effect relationships. Instead, they depend largely on empirical considerations. This chapter gives a brief history of the field along with a description of the principal methods for calculating energy balance and their drawbacks. The application of any energy balance system is inextricably involved with feed evaluation.

In general, the focus of studies of ruminant energy metabolism has been to establish constants for energy conversion and to understand the complexities of energy metabolism in the animal, and the composition of feedstuffs has been largely ignored. Any feeding system must take into account the availability of energy in the feed and its productive value. Sophisticated methods for calculating energy balances have been wedded to archaic and inefficient feed analyses such as the Weende system of proximate analysis.

Because energy balances are expensive and time-consuming to determine, it follows that the evaluation of animal requirements must fall back on prediction systems, and the value of the feeds on their composition. This problem is reserved for the last chapter. This chapter is concerned with the problems of energy balances and measurement and the utilization of this information in the various energy systems in use today.

24.1 Historical Background

The development of a true net energy system whereby animal requirements and feed evaluation are based on energy available for production was not possible before the twentieth century. For a long time the concept of net energy was regarded as an unreachable ideal.

Energy balances were first calculated in the eighteenth century by Antoine Lavoisier, who conducted the earliest respiration measurements. Their application to large animals was long delayed by the intensive labor requirement, lack of precision equipment, and sheer expense. The sequence of advances enabling the development of the net energy concept is diagrammed in Figure 24.1.

Early work in Germany demonstrated that the law of conservation of energy applies to energy balance studies with animals. The ability to account for all energy losses in large herbivores required the discovery of methane in the respiratory gases (Reiset, 1863; Popoff, 1875; as cited by Hungate, 1966), an understanding of the factors influencing body heat loss, and an understanding of the phenomenon known as heat increment. Heat loss was found to vary in proportion to body surface (metabolic size), and body heat production was found to be metabolically regulated to replace heat losses. *Specific dynamic action* describes the phenomenon of increased heat production by the body after ingestion of food and *heat increment* is defined as the difference between metabolizable energy (ME) and net energy (NE). O. Kellner (1912) showed that the heat increment varies with the amount and type of diet,

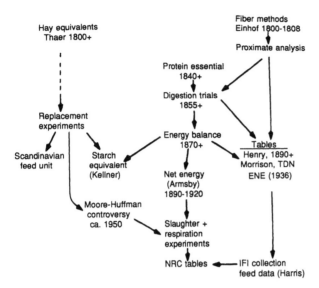

Figure 24.1. Historical sequence of interactions among systems for measuring animal nutrient requirements.

particularly with regard to its fiber content. The heat increment is unusually large in ruminants and is the prime reason for the use of a feed evaluation system based on net energy.

The insistence that feed evaluation systems should reflect the capacity of the feed to support a productive response in the animal was the primary motivation of Henry A. Armsby in developing his system of net energy at the turn of the century. The intent of NE is to express feed value as the energy that could be realized in animal product or work. Armsby's work was limited by the number of feeds that he was able to measure. Nevertheless, he demonstrated the associative effect among feeds and provided some of the values that enabled Frank Morrison to formulate tables of estimated net energy (ENE) for both feeds and animal requirements.

The net energy system received only limited use before 1950, largely because of the limited availability of feed values. A controversy between C. F. Huffman and his former student, L. A. Moore, was instrumental in changing the situation as well as generating a whole new attack on the net energy question. The controversy began when Huffman and C. W. Duncan (1949) published results of an experiment comparing alfalfa and corn in lactating dairy cattle. When an equivalent amount of corn (calculated on a total digestible nutrients [TDN] basis) replaced a portion of an all-alfalfa diet, an increased yield of milk was obtained. This caused Huffman to question the nutritional adequacy of the alfalfa. He postulated the existence of an unidentified lactation factor in corn. This view is historically interesting because of the emphasis on vitamin re-

search during that period. Moore et al. (1953) recalculated the energy values for corn and alfalfa on an ENE basis (using Morrison's ENE system) and showed that a pound of TDN in corn had a greater net energy value than a pound in alfalfa and thus was able to account for the increased milk yield. The greater fiber content of the alfalfa diet relative to the corn diet had caused a greater heat increment.

The inadequacy of the TDN system to account for production led a number of laboratories to invest in net energy research. An important result was the construction of energy balance chambers for dairy cattle by the U.S. Department of Agriculture (USDA) laboratory in Beltsville, Maryland, and the organization of the World Congress on Energy Metabolism. Laboratories in Europe, America, and elsewhere were also active, with the result that since 1960 information on energy balance in ruminants has multiplied.

From this work came the observation that the efficiency of energy use varies with the animal function using it. Hence, the net energy value of an individual feed varies depending on whether it is used for maintenance, lactation, growth, or fattening, with efficiency declining in that order. As a result, the ENE system has been replaced by systems that express food value in terms of the energy content of the specific product. Different approaches to this problem have led to divergent practices, such that the California system for beef cattle and sheep gives separate values for maintenance and gain while the Beltsville system expresses maintenance in units of milk energy, resulting in a single set of values for dairy cattle. The continuing development of these net energy systems is leading to efforts to reconcile and unify them.

In Europe, contemporary feed evaluation systems employ metabolizable energy rather than net energy. Since European feeding practices emphasize forages with only limited use of concentrate feedstuffs, metabolizable energy has appeared to be an adequate system for their needs.

24.2 The Concept of Net Energy

The laws of thermodynamics and conservation of energy inspired animal physiologists to study energy balance in animals in terms of the mechanical principles governing the efficiency of machines. Armsby in particular was trained in Germany in these principles. He spent years constructing an energy metabolism chamber[1] and provided the first values for feed values and animal energy costs in the United States.

---

[1] The chamber is now a museum piece at Pennsylvania State University.

The principles applied by Armsby and his associates assumed a constancy of efficiency above the maintenance cost of the machine and also regarded digestion coefficients with the same attitude of constancy. *Net energy* was defined as the energy value realized in the form of product, usually animal tissue or milk output. Work output (e.g., horsepower) received comparatively little emphasis then and now.

Defining net energy in the form of organic product requires the conservation of feed energy in the form of ATP in metabolism and the use of this ATP for synthesis of cellular matter. Biochemistry will allow only free energy to be used for that purpose, and any oxidation of food matter to yield energy involves a loss due to activation of the system, which ultimately appears as wasted heat, plus energy stored in the product. No chemical process can be spontaneous unless heat is produced, and all biological systems live on the free energy derived from spontaneous systems. The utilization of ATP involves the same activation and conservation in synthesized matter as it does in every synthetic reaction in living systems. The net energy balance will be the sum of all these processes.

24.2.1 Energy Costs of Maintenance

The greatest problem with using a net energy system to account for energy balance has to do with maintenance, because there is no product as such that can be evaluated other than heat loss. Therefore there is no logic in speaking of a net energy for maintenance, because obligatory heat losses from the activation and inefficiency of biochemical reactions cannot be distinguished from use of ATP and other stored energy for maintenance functions. Thus a net energy for maintenance is a contradiction in terms, even though it is a part of the U.S. beef cattle system.

In the classical systems, the energy cost of calculated or measured maintenance is deducted from the total energy balance to estimate NE. Practically, this means determining energy balance at two levels of intake: one at a fasting or near maintenance level, and another at a higher, ad libitum, level to determine the differential efficiency (slope of line BC in Figure 24.2). Technically, the efficiency of NE/ME is represented by this linear slope.

The measurement of energy balance at two levels of intake has practical limitations both in fattening animals and in the case of lower-quality feeds in which an intake differential sufficient for accuracy cannot be obtained because of the low ceiling on voluntary intake. A consequence is that in animal energy systems with maximum intakes not very much above maintenance, fasting is popular as the lower level of measurement, and the values obtained are less successful in accounting for maintenance at higher intakes in dairy cattle.

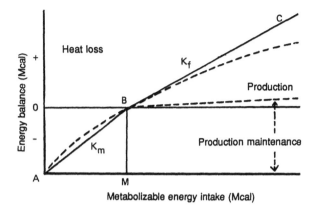

Figure 24.2. The relation between energy balance, metabolizable energy (ME) intake, fasting metabolism, maintenance, and gain (modified from Moe et al., 1972, and Webster, 1978b). The efficiency of energy use below maintenance level (K_m) is measured by the slope of line AB, and the efficiency of gain (K_f) by the slope of line BC. The intercept A is the fasting metabolism. This treatment of energy balance by the linear model may be somewhat arbitrary. Dashed line shows the curvilinearity probable for meat animals, although milk production is linear (see Figure 24.7). Various factors contribute to the diminishing returns. The area above line ABC will represent heat production provided ME balance (digestion balance) has been determined at each level of intake. Many net energy measurements are based on a single digestion balance (often at maintenance), and any depression in digestibility with level of intake will affect the slope of line BC. Further, energy deficiencies are greatest in animals exhibiting tissue mobilization and decrease in order from lactation and growth to fattening. The highest intakes are more likely to induce fattening rather than growth or milk production.

Heat production under fasting conditions involves energy conservation because of undernutrition, so the maintenance estimates are apt to be low. They may also be underestimated because such animals are confined and sedentary and are not spending any energy on feeding activity.

Feeding animals less than the amount necessary for maintenances causes them to mobilize fat to supply energy needs. This function occurs with a different efficiency. Moreover, underfeeding tends to elicit metabolic economies that depend on the supply of reserves. For example, a fat fasting sheep produces a higher energy output per unit of metabolic body size than a thin one of equal weight (Marston, 1951). Arguments for the existence of a true metabolic body mass different from body weight will not suffice, since by this logic the thin animal (with the smallest energy stores) should have the larger relative body mass, and therefore the higher metabolism. It appears that maintenance level is partly set by the reserves available to the animal.

The economy of underfeeding or feeding near maintenance is probably one reason for the higher value of NE of maintenance as compared with production in beef cattle and sheep, and is thus a reason to use different sets of values for maintenance and for gain. This dualism in the case of lactating animals is avoided by the NRC dairy system, which expresses maintenance cost in units of energy in milk, resulting in only one set of NE values.

In dairy cattle with a high metabolic set point, a sufficient number of observations of energy balance above maintenance can be obtained to establish a reliable regression of milk energy output on digestible energy intake. Maintenance can then be estimated from the intercept at zero milk output. Such measurements indicate higher maintenance costs during lactation than during the dry period.

Extrapolation does not truly solve the problem of calculating maintenance energy. One means of estimating the value more directly is to measure the amount of feed required to maintain mature animals at constant weight. This method cannot be applied to growing animals, however, for the process will result in undernutrition when the animal's stature increases at the expense of its own body. It also cannot be applied to dairy cattle, which adjust milk production and mobilization of body energy stores relative to their feeding level (see Section 24.7.1).

24.2.2 Efficiency

Animal efficiency may be measured as the balance between feed input and the output of work or products. Such efficiencies are gross efficiencies that increase with level of intake and result from the dilution of maintenance costs. Efforts to account for maintenance are usually arbitrary and often based on the animal's metabolic size. A complex set of factors determines the energy expenditures necessary to collect, ingest, and digest feed, however, and to utilize the nutrients for maintenance, work, growth, and reproduction—the functions necessary for the survival of the individual and the species—leading to variations in maintenance costs that are not strictly related to metabolic size. Meaningful NE values are possible only if the environment is controlled or predictable. Environmental cold and heat, work, and grazing are all factors that substantially increase maintenance costs. As a result, tabular values of maintenance costs are practically limited to feedlot or stall-fed animals.

Net energy systems, particularly the European ones, have tended to separate efficiency below maintenance (K_m in Figure 24.2) from efficiency above maintenance (K_f in Figure 24.2), reflecting the fact that efficiencies of production functions vary, with maintenance and lactation being more efficient than gain. This has led to the divergent U.S. systems for dairy and beef cattle and also to the ARC system.

These different systems have caused continuing conflict among nutritionists concerning the practical expression of feed values and efficiency. The resurgent interest in NE arises from the desire to account for animal response. Historically, it was the Huffman-Moore controversy that led to renewed interest in NE and contributed to the development of more accurate systems to estimate the efficiencies of animals in different physiological states. The metabolizability of a feed can be expressed in terms of energy (ME) and is comparatively easier to measure than NE. One reason many nutritionists favor ME is that all losses can be described in a material sense; that is, as feces, urine, and methane, although not many ME values have been so measured. Methane is difficult to measure without a respiration chamber; the alternative is to estimate methane by calculation. This is usually done by performing a regression on intake of digestible energy, which ignores volatile fatty acid (VFA) ratios and carbon balance (Chapter 16). As a result, such derived ME values may be no better than the TDN or DE values from which they were derived. The use of DE or ME is in conflict with the desire for a scale of feed values that directly relate to animal performance or output. In general, ME systems work well for nonruminants, in which the heat increment is smaller and less variable than in ruminants.

The supporters of NE base their case on the need for the feed evaluation unit to reflect the comparative productive value of feeds. Only the NE systems will do this. The energy required for work or growth in any productive animal is derived from the chemical energy stored in the organic compounds of the animal's feed. This energy, technically termed *free energy*, is liberated by oxidative processes. Only that portion captured in ATP can be used for other functions; the remaining feed energy is liberated as heat. Some of the free energy of the feed is used by the rumen bacteria for their growth and metabolism (ca. 5%). Not all the free energy in the chemical bonds of feed or microbial products is digestible or metabolizable, and the nonutilizable portion is lost in feces or as methane. Of the remaining metabolized energy, not all is available for productive purposes because a portion is used for maintenance, heat production, and animal activity.

Attempts to model animal efficiency through integration of the ATP stored from metabolism are also imperfect scientifically because it is impossible to determine the function for which the energy will be used. As I have mentioned, efficiency varies with type of productive output, the animal's state, and environmental conditions. It is also not certain to what extent the heat increment can replace the maintenance cost of feed energy in keeping the animal warm. The differing efficiencies of ME use can be analyzed with modern software and mechanistic computer programs, and metabolic functions of maintenance, heat increment, growth, pregnancy, and lactation can be dealt with as well as such environmental factors as cold, heat, and wind (Fox et al., 1990). Thus the observations of Moore et al. (1953) and Huffman could be reconciled with a system based on ME.

24.3 Energy Requirements

The problem of quantifying the amounts of feed nutrients required for specified animal functions has occupied nutritionists since the time of Albrecht von Thaer, and a notion has persisted that refined feeding studies might determine absolute requirements. As a matter of practical reality, it is obvious that the only unit of energy that might satisfy any concept of universal constancy will be NE, because ME is used with differing efficiencies for various functions. Lactation is more efficient than growth or fattening. Any quantification of requirements in terms of dry matter, TDN, or even ME will require some definition of the type of ration consistent with the tabulated figures. Generally, energy requirements inherit all the problems attendant on NE evaluation of feeds, regardless of the form in which they are expressed.

There is another fundamental problem related to the apparent efficiency of energy use at levels above maintenance. Underestimating the maintenance requirement will cause the apparent efficiency of a feed to be low, which has the same effect as overestimating the NE value of the feed. Unfortunately, in most cases of this sort, definitive information is lacking as to which is the error. Practically, this means that either one adjusts feeding standards to account for production or one adjusts feed values to accord with animal response relative to fixed standards. Respiration experiments could settle the issue; however, with estimated NE values and varying conditions of feeding, the urge to adjust feed values or standards to fit experience is not easily resisted. The result in either case is a set of relative requirements and feed values that may not represent the real quantities of energy involved. This problem exists in the current NRC beef cattle system.

The state of the animal is important. American systems do not distinguish between growth and fattening. Despite much work on gain in respect to tissue composition, standards and feed values have not thus far benefited. Underfeeding growing animals leads to compensatory growth at the expense of the animals' own energy storage. These lean animals grow with a phenomenal efficiency on full feeding (J. T. Reid and White, 1977).

The relative accuracy of NE systems has been compared. The starch equivalent system (Section 24.7.4) and the California system (Section 24.7.2) may underestimate the value of forages in predicting gain. The error could be in the tabular values for digestibility or its conversion to NE, or in the requirements quoted for

maintenance and gain. The NRC tables may give values for forages that fit supplemented feedlot diets but overvalue forages when they are the main dietary ingredients. The problem here may also concern differing body compositions at low and high rates of gain; that is, the failure of NE gain to distinguish between growth and fattening, which might have different efficiencies. The problem may be related to the fiber requirement and its optimal level. If fiber is regarded as a required nutrient, then diets too low or too high in NDF can be regarded as unbalanced. All NE systems assume balanced diets.

24.4 Measuring Animal Efficiency

The simplest and oldest measurement of animal efficiency is the direct determination of body weight and composition following slaughter after carefully measured feeding for a specific period. Assessments of initial and final body weights and compositions have to be made. In this kind of measurement the costs of maintenance are obscure, as is energy lost as methane, unless some other animal experimentation and respiration measurements are conducted along with the feeding trial. In lactating animals the net amount of milk produced over a period of feeding can be measured, but measuring energy storage or tissue mobilization is a difficult and inconvenient problem. As a result, lactation balances have tended to be energy chamber balances, and gain has been measured by slaughter and analysis of the carcass. That is, fattening efficiency calculations have tended to depend on slaughter or other means of estimating body composition, while lactation efficiency values depend on respiration calorimetry, in which the total body balance of carbon, nitrogen, and oxygen can be determined. In production trials, measurement of feed utilization is often limited to digestibility. Energy losses in urine and methane production are usually estimated rather than directly determined. Ultimately, all estimates depend on validation by energy balance trials conducted in respiration chambers. The actual number of such total balances performed is limited because of the expense, labor, and technology required. The result is that prediction equations developed from the energy balance data are applied to various unmeasured situations. Unfortunately, the NRC does not distinguish measured from predicted values.

24.4.1 Respiration Calorimetry

Various physical devices and principles have been employed to determine the caloric balance of animals. Direct measurement requires the determination of heat production through some sensor device covering the

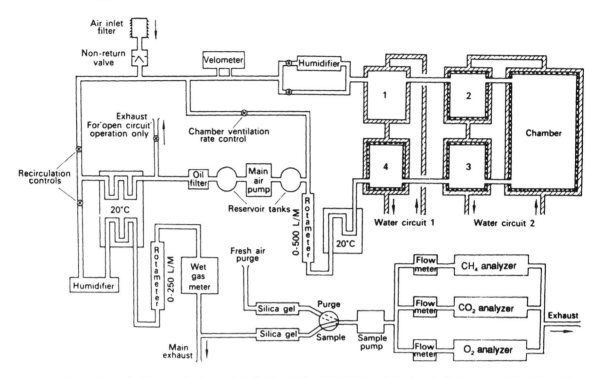

Figure 24.3. The ventilatory circuit for a gradient layer calorimeter (from Pullar, 1964). In this model heat production is measured directly by exchange with water. Exchanger units 1 and 2 equilibrate incoming air, and units 3 and 4, outgoing air. Other models rely on integrated networks of thermocouples to measure heat emission.

surface of the chamber or box in which the animal is housed (Figure 24.3). The system must be controlled at isothermal conditions to avoid stressing the animal and to alleviate the necessity of correcting heat capacities of water, air, and feed supplied to the animal. The original chamber constructed by Armsby at Pennsylvania State University was of this sort and used water circulated through the chamber walls to absorb the heat. Heat production was ascertained by integrating net temperature rise and volume of water circulated. Recent improvements utilize the technique of gradient layer calorimetry.

Measurement of total caloric balance is necessary to determine the chemically bound energy received in feed and voided in urine, feces, and methane. Consequently, the total gas exchange volume must be measured, including determination of oxygen consumption and CO_2 and methane production. The energy values of feed and feces are determined by bomb calorimetry. Urine is more difficult to burn in a bomb calorimeter and is often estimated by regression from nitrogen content or specific gravity. Ultimately, energy values of urine rest on bomb calorimetric measurement of dried urine samples.

24.4.2 Indirect Calorimetry

Energy balance can be estimated from the measured quantities of oxygen consumed and CO_2 produced.

The net oxidation of any dietary organic compound yields water and CO_2. Nitrogen is not oxidized but excreted as urea. Oxygen required for the oxidation of hydrogen lowers the respiratory quotient (RQ), which is the ratio of CO_2 produced to oxygen consumed. For pure carbohydrates this value is 1.0; the oxidation of fat produces a value of about 0.8. From this ratio it is possible to determine very accurately the caloric production from oxidation, because it is possible to use

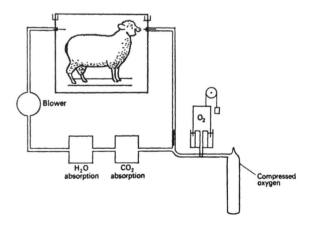

Figure 24.4. A closed-circuit respiration chamber (from Flatt, 1964). Heat production is calculated indirectly from respiratory quotients, the net oxygen used, and CO_2 and water produced. Oxygen is replenished by means of gas cylinders, and CO_2 and water are absorbed. Because of the difficulty of absorbing large amounts of CO_2 and water, this system is used mainly with smaller animals.

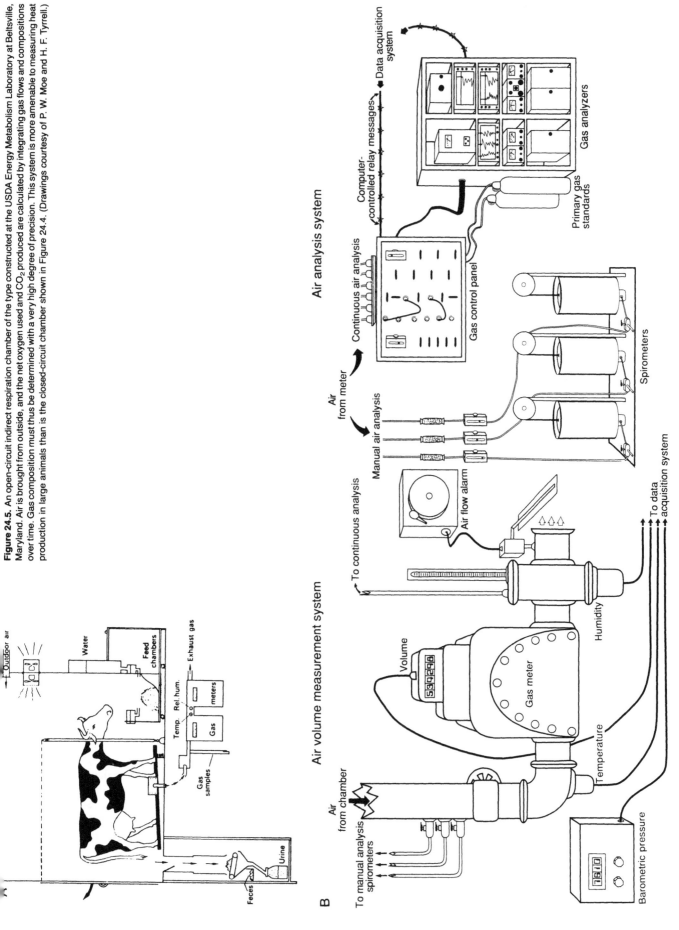

Figure 24.5. An open-circuit indirect respiration chamber of the type constructed at the USDA Energy Metabolism Laboratory at Beltsville, Maryland. Air is brought from outside, and the net oxygen used and CO_2 produced are calculated by integrating gas flows and compositions over time. Gas composition must thus be determined with a very high degree of precision. This system is more amenable to measuring heat production in large animals than is the closed-circuit chamber shown in Figure 24.4. (Drawings courtesy of P. W. Moe and H. F. Tyrrell.)

thermodynamic relationships to infer the carbon and hydrogen content of the substrate being oxidized from the RQ and its caloric content.

Two types of chamber systems are used to measure caloric balance by indirect means. In the closed system, air is cycled from the chamber through absorbers containing alkali to remove CO_2 and moisture while metered oxygen is supplied to the returning air (Figure 24.4). This system is often used for small animals. The open-circuit system is more practical for larger animals in view of the large amounts of absorbents necessary for recycling air in a closed system.

The open-circuit system utilizes outside air (Figure 24.5) and measures the metered flow of the exhaust, which is continually monitored for its CO_2, methane, oxygen, and moisture content. Integration of the values and compositions of outside air and the exhaust determines the respiration balance. The determinations of oxygen and CO_2 must be very precise because the differential composition between entering and exhaust air is not large.

Utilization of open-circuit indirect calorimetry has been limited for many years because of the large number of high-precision gas analyses that are required. The recent development of precision automatic sensing and integration devices in combination with computers, however, have made this type of calorimetry more feasible. Automatic sensing of CO_2 and methane in the inflow and outflow air with infrared absorbance, and of oxygen by paramagnetic resonance, is combined with automatic recording of gas volumes, composition, and computerization of data. This automation, while expensive, overcomes the serious labor problems that have historically hampered and limited energy balance measurements in large animals.

Open-circuit calorimetry has also opened the possibility of energy measurements on grazing animals. The necessary information on respiratory quotient and volume of expired air is provided by portable meters strapped to the animal. Face masks can be avoided by using a tracheal cannula (Figure 24.6), and the animal can feed normally. Intake information and fecal collection are also necessary. Open-circuit calorimetry proved that grazing animals require more energy for maintenance than stall-fed animals (Flatt et al., 1958).

Energy balance measurements lead to estimates of the energy stored or produced (NE) and the heat production (H), the sum of which must account for ME. Heat production is influenced by the character of the diet, the ME intake, and the size of the animal. The maintenance requirement is the amount of energy and feed required to run the animal machine and balance endogenous losses.

The concept of a variable maintenance requirement that responds to conditions requiring economy is consistent with natural biological function. For example, it

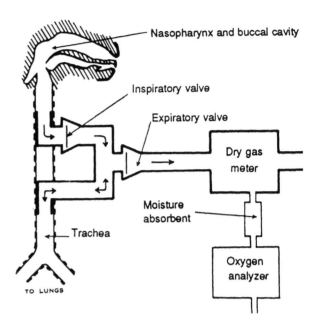

Figure 24.6. The tracheal cannula technique and gas metering system (from B. A. Young and Webster, 1963). The cannula is surgically inserted into the trachea. All inhaled air enters from the nostrils or mouth and all exhaled gas passes through the meter, which is usually strapped to the animal's back. A tube from the meter to a bag (not shown) also attached to the animal collects a small aliquot of the respiratory gases for analysis.

appears that deer exposed to very low temperatures in winter lower their metabolism and actually decrease their energy expenditure (Moen, 1973). Cattle and sheep are always treated according to the theory that increasing cold increases maintenance costs. There may indeed be differences in response among species.

Treating maintenance cost as a proportional function of metabolic size forces all variation in maintenance efficiency to become a part of the productive function. If true maintenance cost is larger than estimated, apparent net efficiency is increased. If maintenance cost is overestimated, net efficiency is underestimated.

24.4.3 Respiratory Quotients

The simplest way to obtain information on energy output is by using a face mask and a device for collecting exhaled air. Measurement of the respiratory quotient (CO_2-to-O_2 ratio) allows assessment of carbohydrate and fat utilization. The volumes of O_2 and CO_2 must be known to quantify net energy output. In ruminants, methane is formed from the reduction of CO_2, and the CO_2 output from the rumen dilutes that from animal respiration. These factors can cause significant errors because methane may represent 10% or more of the energy balance. The practice has been to estimate methane from the dietary intake based on regression equations. The accuracy of this could probably be improved if carbon balance equations were employed.

This would require sampling of the ruminal contents and measuring VFA ratios.

24.4.4 Double-labeled Water Techniques

Deuterium and [18]oxygen-labeled water have been used to estimate energy expenditure (Nagy, 1980), primarily in lizards. The principle depends on estimation of CO_2 production from the disproportionate amounts of deuterium and [18]oxygen in respired water. Assumptions are made that oxygen will appear in CO_2 and that all hydrogen is lost in water. It may be safe to ignore methane and hydrogen in carnivores, but not in herbivores.

There is no direct measurement of oxygen in this system, and so the respiratory quotient is unknown and is assumed not to vary. This leads to problems relative to nutritional status and fasting conditions, when fat becomes the main energy source with a corresponding drop in respiration quotient. Under these conditions energy expenditure is overestimated, which has been (probably erroneously) attributed to specific dynamic action (Seale, 1987). Energy expenditure is highly correlated with CO_2 production, however, and some correction can be made through calibration.

24.5 Energy Balance and Food Intake

The utilization of ME above maintenance is a linear function (Figure 24.7) in keeping with the law of conservation of energy. The realization of such linearity depends on the actual determined ME at each level of intake in order to adjust for digestibility depression. Energy systems that utilize constant estimates of digestibility tend to observe a curvilinear relation, as in Figure 24.2. Milk output in relation to incremental intake of estimated TDN tends to resemble a diminishing returns model, the differential being partly that of digestibility depression and partly the increasing deposition of adipose tissue at the expense of milk output with increasing feed intake.

Figure 24.8 shows proportional changes in balance resulting from increased intake. Fecal losses increase because of the greater rate of passage and escape of potentially available fiber to the feces, while urinary losses remain a relatively constant proportion. The losses of DE average about 5–12% in methane and 3–5% in urine. The heat increment is the largest loss, and it varies depending on the condition and functioning of the animal. Heat loss is about 20–30% in maintenance animals, 30% in lactating animals, and 42% in fattening animals. Fattening is a less efficient process than milk production.

The changing distributions of energy balance (Figure 24.8) depend on level of intake and create difficulty

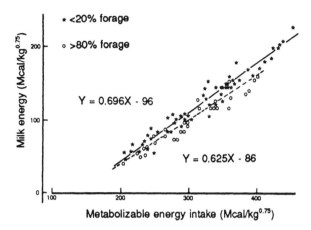

Figure 24.7. Relation between energy balance and intake of ME in dairy cattle (from Flatt and Coppock, 1965). The energy in produced milk is a part of the positive balance. Data are corrected for body tissue mobilization and deposition. Each point represents a digestion balance, so any effect of digestibility depression is removed. Data are essentially linear, although the efficiency slope is somewhat lower for animals on high-forage diets. See Figure 24.2 for a discussion of the problem of curvilinearity.

with the expression of NE, the constancy of which is clearly predicated on unitary input of DE. Actually, it is the DE that decreases with increasing intake, the drop being expressed in ME and NE (Tyrrell and Moe, 1975a). Equations that estimate NE and ME from TDN

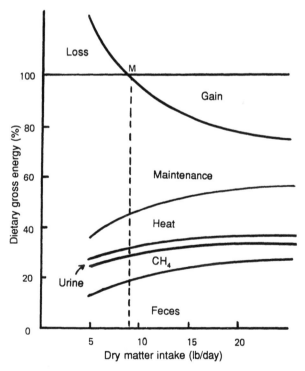

Figure 24.8. A view of the proportional changes in energy balance that accompany increasing intake (based on Mitchell and Hamilton, 1932). The proportion of energy lost in methane declines somewhat with increasing intake. Fecal losses increase through depression of digestibility.

utilize digestion values mostly determined near maintenance, while NE values for production are intended for ad libitum intakes. Newer values for dairy cattle reflect an adjustment for intake level (NRC, 1989); however, this is treated as a constant 4% per unit of maintenance intake for all feeds. The net discount is 8% at three times maintenance for the tabular values.[2] This is unrealistic because digestibility depression will vary depending on the competition between digestion and passage rates. There has been a tendency in the case of beef cattle to adjust the standard requirements rather than the feed values so that quoted values will agree with expected gain. In fact, it is the feed values themselves that are more in error than the unadjusted standards.

Feed interactions termed *associative effects* reflect the level-of-intake problem. The addition of concentrate to forage increases efficiency partly through the depression in methane, reduced rumination, and lowered heat increment. Associative effects occur to varying degrees depending on the combination of feeds and depend on a variety of factors. Energy in individual feeds may be used differently in metabolism. For example, Moe and Tyrrell (1975) found that animals fed corn grain stored 17% of the incremental feed net energy in milk and 83% in body tissue, while the proportions for animals fed beet pulp were 70% and 30%, respectively. Such differences may depend in part on fermentation products, particularly VFA ratios.

The classical view of fed energy values was that they were constants provided that the diet remained balanced with regard to other nutrients. Evidence of an associative effect might be taken as indicating a nutrient deficiency or imbalance. As H. H. Mitchell (1962, vol. 2:508) stated, "Except for differences in metabolizability, the energy value of all perfectly balanced rations is the same under the same conditions of feeding."

While there is no doubt that dietary imbalance does decrease energy efficiency, the modern view has moved away from this rigidity. Values of NE are now regarded as less certain and dependent on a variety of factors. There continues to be a problem of oversimplification in regard to prevailing concepts of NE that is represented in the tabular values for individual feeds. Actually, it is the composition and character of the total diet that determines animal response, and therefore NE. Expressing values for component feeds means that their combined values may be different from those observed when these feeds are fed as sole rations. This problem is apparent in the case of poor-quality forages, which, when added to a high-concentrate diet, may

appear to have a feeding value higher than their true energy content. The pragmatic approach is to assign tabular values that agree with the performance obtained by mixed diets, based on the argument that net energies should reflect optimal nutritional balances without deficiencies.

24.6 Factors Affecting Efficiency

The utilization of ME for production occurs at various efficiencies. These efficiencies represent an optimum, however, and the net efficiency may be influenced by the character of the total diet. Apart from the old question of dietary balance of essential nutrients, a number of other factors have received prominent discussion: (1) the comparative value of roughages and concentrates, (2) the influence of fiber or cell wall content, (3) the influence of eating and rumination time, and (4) the influence of rumen VFA ratios. These factors may also be discussed relative to the origin of the heat increment (Webster, 1978).

24.6.1 Roughages versus Concentrates

The origin of the movement to NE systems of calculating energy balance lies in the different efficiencies of roughages and concentrates. The Huffman-Moore controversy that led to the development of the NE systems now used in the United States involved this comparison. The problem remains to explain the high efficiency of digested energy in concentrates compared with roughages. Not only are roughages less efficient than concentrates, but within the roughage class itself poor-quality roughages are less efficient than higher-quality ones.

The reasons for the different efficiencies have been addressed by various theories, none of which has been adequately confronted by a comparison of the alternatives. Taking these theories in order of their historical presentation, they are (1) the level of fiber in the diet, (2) the ratio of propionate to acetate, and (3) the limitations of intake. The alternative to the second, which involves the negative association of propionate with methanogenesis, may be added to these. As in the case of other conflicting explanations of cause and effect, the actual reality may be a combination because it is not certain that the hypotheses are mutually exclusive.

24.6.2 The Negative Effect of Fiber

It was Kellner (1912) who first noted the nonlinearity between efficiency and fiber content, and he penalized the more fibrous feeds in calculating starch equivalents. The effect has been noted in the California

---

[2] See Section 25.6.4 for a definition and discussion of discounts of net energy.

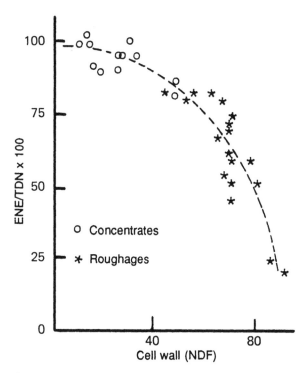

Figure 24.9. Relation between ratio of estimated net energy (ENE) to total digestible nutrients (TDN) and the cell wall content of concentrates and roughages (from Van Soest, 1971). The decline in efficiency is based on Morrison's ENE values and is in agreement with Meyer and Lofgreen (1959). Current NRC data cannot be used to derive such a relationship because most of those values are calculated from TDN by linear regression. The relation in this figure is consistent with the observation that heat production in animals fed coarse diets is related to eating and rumination time, both of which are closely related to the cell wall content of the diet. The decline in apparent efficiency also includes digestibility depression.

system (Meyer and Lofgreen, 1959) and in the ENE system, in which cell wall is curvilinearly associated with the NE-TDN ratio (Figure 24.9).

The question is, then, How does fiber have its effect? The answer is not simple, and a number of factors could be involved. It should be noted that in Figure 24.9 digestibility depression is obscured and would be larger for the higher-cell-wall forages. On the other hand, the low efficiency is to a considerable extent abolished by fine grinding or pelleting. The effect of grinding is to diminish rumination, which will save the animal some work of digestion, and to increase intake. Eating and rumination times are proportional to total cell wall intake and could be factors limiting voluntary intake under conditions where fill is limiting. Webster (1978) reported that when a sheep begins to eat, heat production increases 40–80%, the increase persisting for as long as 2 h. It is accompanied by increased heart rate. The heat production drops when the animal stops eating and seems to be related more to the time spent eating than to the amount eaten. The source of this heat

is obscure, however, because the physical work of eating and rumination does not account for the heat. Perhaps gastrointestinal motility is involved.

Rumination also contributes to heat increment. The increase in heat production when sheep begin to ruminate is about 1 kcal/min per kilogram of body weight. Webster (1978) estimated that cattle (400 kg) ruminating 8 h/day on a hay diet expend about 5 Mcal. Total grinding of the diet causes virtual cessation of rumination and lowers the heat increment. Heat production is partitioned between that associated with the digestive process and postabsorption heat production. Because grinding the diet also causes the VFA ratio to change, the rumination effect cannot be easily separated from the change in VFAs, which might affect caloric efficiency (MacRae and Lobley, 1986). Any theory regarding error in methane estimation or heat loss due to the work of rumination, peristalsis, and so on is based on consequences of effects of digestive processes, while the real effects of VFA ratios on efficiency are attributed solely to postabsorptive metabolism.

Fine grinding and reduction of fiber in the rations also change the VFA output of the rumen and the methane balance. The increase in the rumen propionate-to-acetate ratio and its effects on efficiency can be treated from the point of view of methane output or the postulated effect of acetate-propionate ratios on efficiency.

24.6.3 Positive Effects of Fiber

Fiber is probably essential for all herbivores, the best documented case being for dairy cattle. Inadequate levels of fiber are associated with low milk fat, rumen acidosis, and dietary inefficiency. Although high levels of dietary fiber are associated with high heat increment and a low conversion to net energy, the addition of forage to high-concentrate diets can improve feed conversion.

These observations led to D. R. Mertens's (1983) experiments relating NDF level in the diet to milk production, in which he found that forages of differing NDF content optimized milk production at differing optimum concentrate ratios (see Figure 7.16). The compositions of these total diets were near an average NDF content of 36%. Other experiments have found such an optimum at somewhat lower NDF content, but their results were obtained from cows at a higher production level. Forages of varying quality have elicited different levels of optimum production. The optimum occurs at the point of maximum feed intake (see Figure 21.5). There is no definitive or exact fiber requirement; it changes according to level of production (feed intake) and quality of the fiber.

Adequate fiber for dairy cattle and probably for most grazing ruminants results in the formation of an adequate rumen mat and also elicits the number of chews and ensalivation sufficient to provide adequate rumen buffering and motility. Net rumination is directly proportional to coarse NDF intake, and the result of such coarseness provides the mat, filtration, and rumination characteristics. Adequate coarse forage likely allows the finer concentrate-based NDF to become effective,[3] because concentrate-derived NDF can become occluded in the floating mat in the rumen. If there is inadequate coarse forage in the diet, the addition of extra coarse forage will overcome the problem.

Other factors modifying the optimum dietary level of fiber are the level of intake and body size. Higher milk production forces higher feed intake. Net rumen fill cannot compensate for the intake, so turnover increases. The result for net rumen composition is that relatively less coarse NDF is required for rumen function, causing the optimum to decline with increasing level of production. The optimum is less than 30% NDF at 40–50 kg milk/day, whereas Mertens's optimum (35–36% NDF) was obtained at about 20–25 kg milk/day. Lower levels of milk production and dietary intake probably allow the fiber-to-forage ratio to increase, which is consistent with the economics of feeding dairy cattle. Larger animals should need more fiber in their diets for optimal rumen function, since their rumens are less limiting relative to intake of dietary energy (Sections 4.6 and 4.7).

Some misunderstandings have resulted from Mertens's (1983) data. One is over the supposed constancy of the fiber requirement, which is, in fact, inconsistent for the above reasons. Many scientists have searched for the fiber measurement that will give the maximum correlation with dietary or production responses. Thus, for example, NDF is correlated with intake of forages while acid-detergent fiber (ADF) is best related to their digestibilities. The evaluation of any laboratory measurement such as fiber by statistical correlation is open to severe limitations (Chapters 6 and 25). Table 24.1 shows that using ADF to determine fiber recommendations (NRC, 1989) is of limited value. It is apparent that the NRC standard uses ADF based on alfalfa, with the result that the recommendations do not apply to corn silage, for example.

There have been criticisms of NDF because of its low correlation with animal response in practical diets, raising a call for a search for factors that will account for the aforementioned animal responses. Such measures would be evaluated by the size of their correlations with feed intake and production. These endeavors are futile because they violate principles of statistics and logic. If it be accepted that low fiber levels are limiting and high ones excessive relative to optimum

Table 24.1. NDF and ADF equivalence for alfalfa and corn silage based on NRC recommendations

Fiber source	NRC[a]	Alfalfa composition	Diet level[b]	Corn silage composition	Diet level[b]
NDF	28	47	60	47	60
ADF	21	35	60	28	75

Note: Values are presented as percentages of dry matter.
[a]Recommended minimum dietary level for cows in early lactation.
[b]Level necessary to provide fiber requirement from the forage sources.

animal performance, a positive relation with production will be expected at low fiber levels and a negative one at high levels of dietary fiber (see Figure 7.16). Near the optimum, where most dietary manipulations are conducted, the slope of the response approaches zero, so it will follow that all correlations of a response to any such measures of fiber will also approach zero and will be insignificant near the optimum.

24.6.4 Effective Fiber

The recognition of fiber as a required nutrient satisfies the condition stated by Mitchell that the optimum energy balance occurs when all nutrient limitations are satisfied. There is a remaining problem, however: quality of fiber. As I have pointed out, fiber quality affects the level of milk production even when fiber level is optimized. For example, optimizing concentrate supplementation to a high-quality forage results in a higher milk production than does supplementing a poorer-quality forage, which will require a higher concentrate-roughage ratio for optimization than a high-quality forage (see Figures 7.16 and 21.5).

Particle size presents another problem. Net NDFs of total mixed rations yield different optima when sources of differing fineness are fed (Briceno et al., 1987). The concept of effective fiber has been advanced to explain why fine particles do not promote rumination. Effective fiber is determined by discounting the NDF that passes a 1-mm screen. Such a definition is arbitrary in that when adequate coarse fiber is fed (i.e., that needed to form an effective rumen mat), the finer fiber becomes more effective because it is occluded by the filtration effects of the system. If inadequate coarse fiber is fed, the system fails because the finer fiber is ineffective even though fiber level is compositionally at the specified requirement. One solution is to set a required proportion of the dietary fiber that is above a specified size. Mertens (1983) suggested that 75% of ration NDF should come from coarse forages.

24.6.5 Methane Output

This factor has been relatively ignored as an explanation of feed efficiency, since methane loss is supposedly accounted for in the measurement of ME.

---

[3] Effective fiber has been somewhat arbitrarily defined as that NDF which does not pass through a millimeter screen (Mertens, 1985).

Since many ME values involve estimation of methane output from digestible carbohydrates and do not consider the negative association between methane production and propionate formation, however, one may wonder to what extent low efficiency associated with high acetate-propionate ratios is accounted for by methane production. Improved equations for estimating methane production indicate that the type of digestible carbohydrate (e.g., hemicellulose more than starch) promotes methane yield (Moe and Tyrrell, 1979). An alternative is to estimate methane output from VFA ratios in the rumen, using the carbon balance equations (Section 16.7.1).

The negative heat increment resulting from formate metabolism (J. T. Reid, 1962) could be related to methane production. Formate is known to be a precursor of methane and might serve to divert energy from methane synthesis. The excess of reduced enzymes from the carbon balance of the rumen might serve a beneficial purpose if a reductive substrate were provided, and might perhaps spare other energy liable to loss in methane synthesis. This is one sense in which the negative heat increment of formate might be understood. The energy-sparing action of formate likely lies in the rumen where it is metabolized and destroyed, and not in the tissue of the animal.

24.6.6 Acetate-Propionate Ratio

The concept that acetate is less efficiently utilized than propionate (Armstrong et al., 1957) has perhaps received more attention than any other hypothesis about energy efficiency. While substantial data indicate lower efficiency resulting from high acetate-producing diets (high forage), more recent studies indicate that it may be too simplistic to view the effect as due to the intrinsic metabolic efficiencies of acetate and propionate. For example, Tyrrell et al. (1979) found that the use of ME from acetate infused into a lactating cow on a hay diet was only 27% of the added energy, while that for a concentrate-fed cow was 73%. Thus it seems likely that the efficiency of acetate depends on the dietary situation and the proportions of glycogenic and lipogenic metabolites. A metabolic basis for the efficiency-promoting effects of propionate may be that propionate spares or supplies glucogenic amino acids, since the nonessential ones can be synthesized from propionate (MacRae and Lobley, 1986).

Experiments that found little difference between propionate and acetate efficiencies may be accounted for on the basis of the optimum ratio of metabolites. Furthermore, lower efficiency can be associated with the very high proportions of propionate associated with milk fat depression. In this situation increased tissue lipogenesis is promoted and occurs at a lower efficiency than lactation. Thus, it appears that the response of

efficiency to acetate-propionate ratios is not linear but is likely optimal at some middle range associated with adequate forage diets supplemented with substantial amounts of concentrate.

24.6.7 Ionophores and Energy

Ionophores interfere with cation transport and consequently the availability of hydrogen for methanogens. This forces carbon into an increased amount of propionate in the rumen fermentation, inhibits proteolysis, and sends a greater flow of undegraded feed protein to the lower digestive tract, even though microbial protein may be depressed (Section 16.9.2). The drop in net methane increases effective ME. Adjustments for animal performances are largely based on comparative feedlot data with steers (Fox et al., 1990).

24.6.8 Body Composition

Variability in the apparent efficiency of growth and weight gain involves the composition of the deposited tissues. Fat tissue has a much higher energy content than lean tissue, and mineralized bone contains very little energy. Further, body water and the gastrointestinal contents (particularly rumen fill) are not technically a part of the metabolic animal mass, even though the gut fermentation contributes to net metabolism.

The most accurate method of determining body composition involves slaughtering the animal and grinding up the empty carcass for laboratory analysis. This procedure has the disadvantage that composition becomes a single measurement at the final point of the animal's development. Direct determination of body composition by slaughter, grinding, and chemical analysis of the carcass is a laborious and expensive procedure, and alternative techniques are often utilized to estimate body composition. The favored procedure in the California NE system uses the specific gravity of the chilled empty carcass. Values are based on experimental determination of specific gravity and energy content and regression of the latter on the former. Fat becomes an increasing proportion of body composition as animals age, and this shift in composition must be accounted for in the analysis since fat is lighter than the other components of the body. Lower specific gravity is associated with higher energy content of tissue. Other techniques for estimating body composition include the use of dilution markers, ^{40}K, and sonic radiation. Most of these methods have been insufficiently accurate to serve as aids in NE measurements.

24.7 Systems for Net Energy

The surge of net energy measurements in the 1960s provided new information on which to base energy

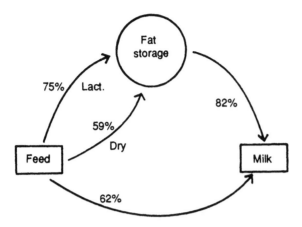

Figure 24.10. The efficiency of tissue storage of energy and its mobilization (data from Moe et al., 1971). Nonlactating animals store energy less efficiently than lactating animals. The overall efficiency of converting feed to milk is 62% (0.75 × U 0.82) when feed energy is stored during lactation and later mobilized. The corresponding value for energy stored during dry periods and later mobilized is only 48% (0.59 × 0.82).

calculations. In the United States the net energy system for lactating dairy cattle and the California system for beef cattle and sheep have replaced the older Morrison estimated net energies. In Germany there is continuing development of a net energy system replacing starch equivalents. In the United Kingdom, the ARC system is based on ME. Van der Honing and Alderman (1988) surveyed the various systems and their interrelationships.

The requirements for energy are better known than are the values of feeds (see Chapter 25), which are available only for domesticated cattle, sheep, goats, horses, rats, rabbits, cats, and dogs, so that wild animals, particularly zoo animals, generally are fed by extrapolation from the closest domesticated species. See Robbins, 1993, for a good discussion of the problems of wildlife nutrition.

24.7.1 The Beltsville System

The NE system developed at Beltsville describes energy relations for dairy animals, the value of feeds being expressed in units of milk energy that can be produced from them. Lactation involves much more energy flux than simple gain does. Milk production is often superimposed on the mobilization and loss of body tissue in high-producing animals. In lower-producing animals fat mobilization may be superimposed on tissue gain, which may include pregnancy and support of a fetus. All these functions are in addition to maintenance, and they operate at inherently different efficiencies. Gain is a less efficient process than tissue mobilization (average about 80% of lactation efficiency), and gain concurrent with milk production is more efficient than if the animal were not lactat-

ing (Figure 24.10). It is thus less efficient to mobilize gain stored in tissues during the dry period in the ensuing lactation than to convert feed directly.

One mathematical solution to this confused and variable situation would result in a profusion of NE units, one for each respective function. Instead, this problem is avoided by the use of a common currency into which the various biological functions are translated according to their respective efficiencies. Thus the cost of maintenance is expressed not as the actual heat production but in terms of the amount of milk energy equivalent to that heat production, taking into consideration the respective efficiencies of maintenance and milk production. Values in the NRC tables are set at zero tissue balance. The requirements of dry cows are treated separately from those of lactating animals. Efficiency of lactation, being higher than gain, is also closer to the apparent value for maintenance.

Energy mobilization and lactation efficiency are complicated problems in dairy cattle because of the difference in inherent efficiencies of energy storage and mobilization relative to tissue mobilization and lactation gain. In the Beltsville energy chambers, Flatt and Coppock (1965) used a high-producing cow (Lorna) to determine that while feed intake was about six times maintenance, energy expenditure was nine times maintenance for a period of 6 weeks during which milk production exceeded 45 kg/day and body weight remained relatively unchanged. These observations illustrate the inaccuracy of body weight measurements for assessing the value of tissue mobilization. Essentially, fat in adipose tissue was replaced by ruminal water over an extended period. Most of this water probably entered the rumen or other parts of the digestive tract, the gut expanding as intake rose and internal fat dissipated. As internal adipose tissue was lost, the volume of the rumen could expand, allowing feed intake to rise and ultimately come into balance with energy output in milk. The consequence of this mobilization is that efficiency of feed conversion is apparently greater in the earlier part of lactation (Figure 24.11), because the diet would be credited with the energy supplied by the mobilization.

Much of the apparent efficiency of milk production in practical terms varies according to genetic capacity, nutritional history, and stage of lactation in individual animals. Some animals tend to mobilize more body tissue in early lactation, while others, as observed in the Beltsville studies, are more nearly direct converters of feed energy with less obvious tissue exchange. The average dairy cow, however, loses tissue in early lactation that is replaced at a later stage. It is not possible to feed adequately at the early stage of lactation because the animal depends on energy mobilization.

Figure 24.12 shows that underfeeding during early

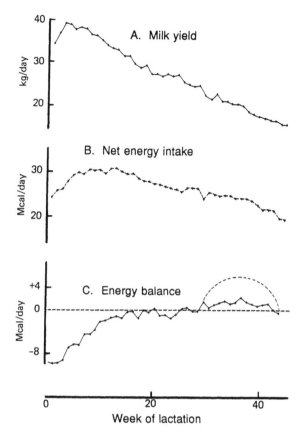

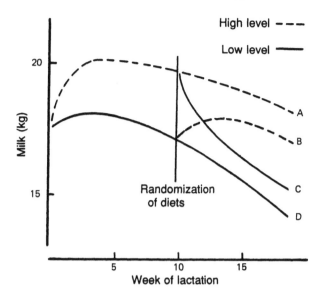

Figure 24.12. The influence of level of feeding on milk production in later lactation (data from Broster et al., 1969, and Moe and Tyrrell, 1975). At the tenth week of lactation all animals were randomly reassigned to either the high or the low plane of nutrition. The production of group B remained below that of control group A, although both received the same amount of feed; group C continued to produce at a higher level than group D, which received the same amount of feed as C after the changeover.

Figure 24.11. Relation among intake, milk yield, and net energy balance in high-producing dairy cows (from Bauman, 1979; see also Bauman and Currie, 1980). The data points are averaged from six cows that produced an average of 9,354 kg milk during 305-day lactations. Dashed line (right side, panel C) indicates recommended overfeeding during the last third of lactation to replenish body fat stores needed to support the next lactation. A major portion of the energy needed in early lactation is derived from tissue mobilization.

lactation reduces the milk output in the later part of lactation relative to cows maintained continuously on a high plane of nutrition. Refeeding underfed cows at the higher plane of nutrition will not restore the higher level of milk production, probably because hormonal regulation has relocated the energy set point. In this context it should be understood that the tabular values of net energy of lactation (NE_l) reflect feed utilization at zero tissue balance, since there is no available system that can entirely account for the tissue exchange.

Energy balance experiments indicate that lactating cows have limited ability to utilize body protein for milk production output. The ratio of mobilized tissue protein to fat is not greater than 0.15 (Moe and Tyrrell, 1975). The lactating cow therefore depends largely on rumen microbial protein output and dietary protein escaping the rumen fermentation. At lower levels of production the recycling urea pool and rumen synthesis are a more significant endogenous source of protein.

24.7.2 The California System

The NE values for maintenance and gain are based on a system developed in California, largely by the efforts of G. P. Lofgreen and W. N. Garrett (1968). In this system energy balances are based on slaughter experiments. Body composition is estimated sometimes by chemical analysis of the carcass but in most cases by the density of the empty carcass. The ME content of a total diet is partitioned into heat production and NE stored in body tissue. Two sets of trials with the same ration are conducted: one trial determines the ME at the maintenance level of intake, and the second uses animals fed ad libitum to determine body gain and composition. Energy in the carcass is subtracted from estimated ME consumed to estimate heat production and methane loss. In this calculation (often done with sheep) any difference in digestibility between the maintenance level and the unmeasured digestibility and ad libitum intake is also included in this difference. The result is apparent values of maintenance that are much larger than gain values.

Metabolizable energy is estimated from digestion trials in which DE is determined by bomb calorimetry of feed and feces, or in some cases by calculation from TDN. Urinary losses are measured, and methane production is often calculated from total apparent digestible carbohydrate using the method of Swift et al.

(1948). Estimates of body composition, fat, lean tissue, and bone are also needed (Section 24.6.8).

Originally, Lofgreen regressed NE on a logarithmic factor, F. The choice of a logarithmic function was apparently made in order to obtain a reasonable intercept of 77 kcal at zero intake for the regression of heat production on ME intake. This value is an estimate of fasting heat production (Lofgreen and Garrett, 1968). Metabolizable energy was estimated taking into account urinary loss and estimated methane production. Values of NE were calculated from ME, assuming a constant factor 0.82. The California system does not recognize the discordance exposed by the Beltsville experiment, in which $ME = 0.96 DE - 0.27$ (see Figure 25.3). More recently, the equations have been replaced by polynomial regressions (NRC, 1984), and the results appear to rank efficiency of gain higher than lactation (Van Soest et al., 1992b). This problem is discussed in Section 25.6.5.

One of the difficulties with the California system is the dual set of NE values it uses. Relative values of feed NE for maintenance and gain (NE_{m+g}) vary with the level of intake, although the values of NE_m and NE_g are assumed to be constant for a given diet, and NE_{m+g} is the weighted average of the contribution of the diet toward maintenance and gain. The system assumes a particular level of intake. If intake varies widely from the expected, a larger error occurs in the NE_{m+g} estimation. An alternative is offered in Section 25.6.5.

24.7.3 The Agricultural Research Council

The ARC (1980) in Great Britain has developed from K. L. Blaxter's work (1962) a new energy system based on ME. Efficiencies of specific animal functions are dealt with by the use of coefficients. The system has certain static rigidities. Change in digestibility with intake above maintenance is constant for all feeds, and the differing efficiencies of forages and concentrates are largely ignored, possibly because concentrates are fed at lower levels in Britain. Protein and microbial yields that escape from the rumen are treated as constants for a respective feedstuff. The estimates of feed energy value have been based on modified acid-detergent fiber (MADF), although there may be a move away from this and toward near infrared reflectance spectrophotometry. Neutral-detergent-fiber-based systems are still in the future.

24.7.4 Other European Systems

Historically, two kinds of systems based on comparative replacement values of feeds have been used in Europe: the Scandinavian feed unit system and the starch equivalent system, both of which are descen-

dants of the hay equivalents of Albrecht von Thaer. These are essentially NE systems because the productive value of feeds is measured according to their replacement value for a standard feed.

In the case of the Scandinavian feed unit system, the value of barley is taken as 100 and the relative quantity of any other feed required to replace the productive value of a unit of barley is taken as its feed unit equivalent. Like all NE systems the assumption is made that comparative rations are complete in all nutrient requirements other than the energy content being tested.

The starch equivalent is expressed as the NE value of feed in units relative to the NE of 1 kg of starch. This system, like TDN, converts the caloric equivalents of protein and fat into starch values, with the difference that their expression is in terms of productive value rather than in digestible nutrients. Starch equivalents were developed by Kellner (1912), who devised equations to estimate them from proximate composition.

The work of Blaxter (1986) and A. J. H. van Es (1978) has stimulated further revisions of feeding standards.[4] As a result, the older European systems have been replaced by systems that use ME, which is applied to all classes of ruminants. The push toward ME has been motivated at least partly by the desire for a unified system. Also, experts feel that NE and starch equivalent systems are to a degree nebulous because the heat increment cannot be described in terms of substance or entirely explainable processes, while the ME is based on an input and output of metabolized and unmetabolized substances that have energy values. Nevertheless, worldwide, most data are limited to apparent digestibilities, and even most ME values have to be estimated by regressions.

The problem posed by the adoption of ME is that the varying efficiencies of animal function—respectively, maintenance, gain, and lactation—result in ME not directly evaluating feedstuffs for their productive value, the reason that led to NE in the first place. Specific efficiency factors can be adopted for the respective animal function to solve this problem (Webster, 1978a, 1978b). The ME system thus resembles the Beltsville system in that there is only one unit of energy for expression of feed values, and the California system to the degree that separate efficiencies for maintenance and gain are derived from the ME value. Like the California system, digestibility is assumed to be constant and any depression becomes a part of the lower apparent efficiency of energy above maintenance.

All the existing European energy systems have been developed in northern Europe, where cool climates and growing conditions favor high-quality forages. Many of the environmental effects that would lead away from

---

[4] A detailed review of European systems is in van der Honing and Alderman, 1988.

proximate analyses and standardized digestion trials to include effects of intake on digestibility and problems of fiber quality have not been factors there. In southern Europe, where environments are more similar to those in the United States, U.S. systems tend to be used.

A more comprehensive development has occurred in France, where energy and protein systems (PDI) are competing with the most advanced mechanistic systems available in the United States (Daniel Sauvant pers. comm., 1991). The French systems have not been easily available outside France. The best discussion of the French systems is given in van der Honing and Alderman (1988).

25 Integrated Feeding Systems

The synthesis of a net feeding system that accounts for requirements, values, and quality of feeds relative to a net production output is the general goal of all net energy systems. Chapter 24 described the problems of measuring energy balance and deriving energy use efficiencies. These, joined with a protein evaluation system (Chapter 18), can now be combined with predicted digestibility and energy values to give an integrated feeding system. Such a system is presented in Sections 25.6.4–6.7. The digestibility estimate is the single largest error in most systems and is usually the least rational and least mechanistic operation. Yet mechanistic systems for predicting digestibility do exist. Here, in the final chapter, I will draw on the information presented in this book and introduce a net synthesis. This chapter also reviews the history of predictive systems and presents practical modeled systems and the philosophy of their use.

25.1 Predicting Feed Values

The prediction of feed values from compositional or in vitro information is a necessity because energy balance trials with animals are not possible on many by-product feeds; in addition, trials are expensive and unable to account for all the forages and major bulk ingredients included in ruminant diets. Solving the problems associated with predicting feed values from plant composition requires understanding plant chemistry and physiology, subjects that animal-oriented scientists who conduct feeding and balance trials do not always understand very well. Also, the time-honored system of empirical regression of a measurement such as fiber on digestibility or energy balance persists, even though there is no rational basis for making such predictions from regression analyses. Crude fiber has always presented problems—both in the methods for measuring it and in the statistical procedures used to analyze it. The advent of better fiber determination methods and associated information on plant growth, physiology, and composition allow more rational models to be designed. The methods have been

adopted, but new models have not been adopted, and meanwhile acid-detergent fiber (ADF) and neutral-detergent fiber (NDF) techniques are being misused. A part of this problem is that rapid and economical procedures for determining lignin have not been developed, yet all model systems for the mechanistic prediction of forage digestibility require a lignin determination.

25.1.1 Early Attempts at Feed Evaluation

The earliest notions regarding feed values were based on the quantities of different feeds required by livestock. These ideas were formulated into a coherent system (about 1790–1800) termed *hay equivalents*, whereby various feeds were quoted in amounts that could replace 100 pounds of meadow hay. The introduction of hay equivalents has been generally credited to Albrecht von Thaer, who wrote a four-volume treatise on agriculture (1809–12) that long remained a standard text. Tyler (1975), however, showed that many of the values quoted by Thaer (1809) were taken without citation from an Englishman named John Middleton.

Although many nutritionists have assumed that Thaer conducted feeding trials to obtain his hay values, there is no mention in his writings of his ever having conducted one. One of Thaer's own comments about feeding trials is worth quoting (Tyler's translation):

> I do not care tuppence for theoretical reasoning. Experience based on truly comparable experiments can alone decide. And these we do not have. I have indeed fed cooked and uncooked potatoes, but have not ventured to decide which was the more advantageous. In order to decide this, a dozen cows of completely equal condition must be placed in two different stalls and be fed half with cooked and half with uncooked potatoes, potatoes being their chief food and these should be weighed accurately as well as any additional food. Then the behavior of the cows, their milk production, addition of flesh and state of health in the following summer must be determined exactly. Thus, a fairly certain result would be obtained. However, such experi-

ments are not possible for private individuals and not even for our agricultural societies.

From this it appears that Thaer considered feeding trials too expensive.

Some nutritionists have supposed that Thaer used Heinrich Einhof's fiber values to evaluate feed value, considering the fiber fraction to represent the indigestible part of the feed, but no clear method for any such calculation is presented in his writings. He was critical of some of Einhof's fiber values as having given unreasonable results but may have used his personal judgment in altering values to what he thought they should be (as, more recently, Morrison is accused of having done). The origin of Thaer's hay values is unresolved. He obviously borrowed data from others, perhaps augmenting those values with his own observations on the approximate quantities of feeds required to maintain animals over the winter.

A major advance came with the experiments of Jean-Baptiste Boussingault (1845), who showed that feeds low in protein (nitrogen) do not support optimum growth. Accordingly, Boussingault developed a system of feed evaluation based on nitrogen content, although the system was limited by its failure to consider energy. Justus von Liebig also emphasized the protein content in foods as a basis for their evaluation, but he erroneously thought that protein was oxidized and wasted when work was performed. Liebig's errors resulted, as McCollum (1957) pointed out, from oversimplifying the conclusions of his experiment. Emil Wolff (1856) calculated hay equivalent values taking into account nitrogen and using fiber as an estimate of the indigestible portion of the feed. He also calculated feed composition on a proximate basis with the nitrogen-free extract (NFE) component obtained with the correction factor N × 6.38. Thus, for fifty years hay equivalent values were assigned to feeds on the basis of traditional feeding practices and, later, on the basis of calculations using what was then known about feed composition and its effect on animal performance.

25.1.2 Nomenclature

Chemistry as a science began to develop in the late eighteenth century, and biochemistry emerged in the late nineteenth century, along with the associated nutritional and animal sciences. As the fields emerged, they evolved their own terminologies, which have since changed and evolved so that one must read nineteenth-century literature, or even works published earlier in this century, very carefully. The words *protein* and *cellulose* date from 1830, *lignin* from 1854, and *hemicellulose* from about 1890. Unfortunately, the chemists who named hemicellulose thought it was a substance on its way to being converted to cellulose; hence

the literal meaning, "half cellulose." Although *hemicellulose* may not be the most accurate term, at this point there is no other word that distinguishes the insoluble lignified noncellulose carbohydrate from more available and relatively unlignified pectic material.

25.2 Digestion Trials

G. Haubner (1855) was the first to conduct digestion trials and to discover that fiber is, in fact, partly digestible. This discovery was amplified by Wilhelm Henneberg (1860), who, with his associate Friedrich Stohmann, conducted trials to determine the digestibilities of all the proximate constituents for a variety of feeds. This system of trials and analyses became widely adopted and has persisted to the present. Henneberg's work demolished the hay equivalent system, but his and his contemporaries' failure to cite earlier studies has hampered research into the animal nutrition literature before 1860. By 1860 the concept of digestible nutrients as a basis for evaluating individual feeds had developed; however, the lack of understanding of the relative nutritional values of the different proximate constituents prevented the conversion of this concept into a complete system of feed evaluation. Also, the field of animal nutrition was limited by the state of knowledge of basic biochemistry and physics.

By the late nineteenth century it was generally recognized that feed values and animal requirements could be expressed in terms of available energy and protein. The principle of the total digestible nutrient system—expressing feed constituents on an equivalent basis relative to energy content—emerged in Germany, perhaps in Liebig's university and from the use of Henneberg and Stohmann's tables (1860, 1864). Tables presenting "digestible nutrients in 100 lb" were first published by W. A. Henry (1898), and the term *total digestible nutrients* (TDN) was apparently first used by J. L. Hills (Hills et al., 1910). F. B. Morrison first used "total digestible nutrients" in his twentieth edition of *Feeds and Feeding* (1936). An advantage of the TDN system was that it used values that the average laboratory with ordinary equipment could measure. Although TDN was not adopted in Europe, a parallel development led to the starch equivalent system (Kellner, 1912). Both the starch equivalent and the Scandinavian feed unit retain an element of the old hay equivalent system, in which feeds are compared against a standard material—starch and barley, respectively—with a known value in terms of animal production.

There is still discussion concerning the means by which Morrison collected data and developed his tables. According to Moore et al. (1953), the estimated net energy (ENE) values were devised using various criteria and sources: (1) direct net energy determinations by Henry Armsby and others, (2) comparative results of feeding trials, (3) Fraps's production energy (PE) values, and (4) Morrison's own intelligent judgment. It is common knowledge among those who knew Morrison that he purged published data that were not in accord with his experience and made the feed values reflect practical conditions, an attitude not unlike that of Thaer more than a century earlier.

The necessity of good judgment in the interpretation of feed composition and evaluation data is becoming increasingly evident. The current emphasis on statistical analysis and the computer's ability to handle large masses of data have led to the philosophy that the datum is sacred, and that to purge or adjust values introduces subjectivity and bias. Mass collection and cataloging of worldwide data, however, has tended to obscure rather than to elucidate the value of various feeds for use under practical conditions (Chapter 2). The varying conditions and practices in feeding trials and analyses and the accuracy of digestion trial data at productive levels of feeding remain problems that future studies still have to resolve.

25.2.1 Estimated Net Energy

Estimated net energy is expressed in therms per 100 lb, a therm being equivalent to a megacalorie, the quantity of heat required to raise 1000 kg of water 1°C. Morrison's values for NE were derived from balance experiments in which each feed was compared with corn grain as a reference material. He augmented the data through interpolations based on his own experience with practical feeding situations. Although no theory of the time provided an understanding of the difference in efficiency between fattening and lactation, he provided special values for dairy cattle that were higher than those for beef animals and appear closer to actual present-day net energy of lactation (NE_l) values. This may be because the values were obtained under practical feeding conditions rather than by prediction from TDN values.

Moore et al. (1953) published equations for the conversion of TDN into ENE based on statistical analysis of data from Morrison and others. One difficulty with these regressions is that they are based indiscriminately on both determined and estimated values, since in Morrison's book the origin of individual numbers is not ascertainable. Values published by the National Research Council (NRC, 1970, 1971) under the guidance of Lorin Harris emphasized original observations and thus purged Morrison's values, although missing data were still filled in by regression calculations.

There are several defects of ENE as a unit for measuring efficiency. It is usually treated as a linear function of TDN, but it is, in fact, a system for adjusting TDN values to bring them into line with expected production. Another problem is that ENE as used is a NE

of production (NE_p) value largely without specification of an animal function. Nutritionists now know that the efficiency of maintenance differs, and also that lactation is a more efficient process than fattening or growth. This knowledge has contributed to the development of individual NE systems. As a result, ENE values have been abandoned and replaced by NE for maintenance (NE_m), NE for gain (NE_g), and NE for milk production (NE_l).

25.2.2 Units of Measurements

The units for expressing energy values have changed in the past and are still in a state of flux. In the United States, the NRC has adopted Mcal/kg in a move toward the metric system. In Europe, however, which already uses the metric system, a movement for further reduction in number of units favors the use of the joule as a common unit of energy. The joule does not provide any superior theoretical or practical accuracy, because all values are converted from calories to joules by the constant 4.184 (1 cal = 4.184 joules). While the NRC presently continues to use calories, the joule has been adopted by the American Institute of Nutrition, and most Europeans use joules.

25.3 Feed Analysis

The notion that feeds and food could be evaluated by chemical analysis emerged about 1800, along with attempts at evaluating feed, long before any such evaluations by balance trials, or indeed, any detailed knowledge of the chemical composition of plants. The origins of both idea and methods are obscure, but there seems to have been a consensus that woody fiber in plants was a priori indigestible, an attitude that prevailed until the work of Henneberg (1860) was published.

The origin of what is now called the crude fiber method is more complicated. Many modern texts erroneously credit the crude fiber method and the proximate system built around it to Henneberg. The most Henneberg could have contributed is standardized procedures that everyone followed. Henneberg himself wrote that he only adapted the methods of Einhof and gave them a new meaning. He spoke of the Einhof analysis as consisting of sequential extraction with ether, alcohol, water, dilute acid, and dilute alkali; however, even the most careful study of Einhof's extant writings reveals no such procedure. Moreover, as Tyler (1975) mentioned, no tradition of crude fiber analysis existed in the Moglin[1] laboratory after Einhof's death in 1808. It appears that Henneberg credited

---

[1] Moglin, Prussia, is the site of the first agricultural experiment station in Germany. The station was donated by King Friedrich Wilhelm III in 1804 to Albrecht von Thaer.

Table 25.1.   Early fiber values compared with values obtained with modern analyses

Source	Barley	Corn	Oats	Potatoes	Rye	Wheat	Meadow hay
Einhof, 1806	21.3	—	32.8	5.6	22	13.8	43
Gorham, 1820	—	3.3	—	—	—	—	—
Horsford, 1846	5.3	—	16.1	—	—	—	—
Wolff, 1856[a]	4.9	—	12–6	—	2.4	1.6	25–34
F. B. Morrison, 1956	6.0	2.4	13.5	2.2	2.7	2.9	26–34[b]
Van Soest, 1977[c]	21	12	31	4.7	22	14	40–56[d]

Note: All values are percentages of dry matter.
[a]Wolff's values are summaries of earlier literature.
[b]Mixed hay.
[c]Neutral-detergent fiber.
[d]Mixed clovers and grasses.

Einhof incorrectly, and the origin of the crude fiber method remains obscure.

The method Einhof used was essentially maceration with water after solvent extraction followed by sieving and filtration, as used by G. Pearson slightly earlier (Tyler, 1975). This method gave values much higher than the standard crude fiber method and much closer to those of neutral-detergent fiber (Table 25.1). The Pearson-Einhof method was used by August von Voelcker as late as the 1860s but was generally superseded in this period by a method using dilute acid and alkali (Way, 1853). In the judgment of the chemists of the time, crude fiber should be a "pure" fraction approximating the composition of cellulose. Corrections for nitrogen ($N \times 6.25$) and ash were often carefully performed, since the crude fiber was rarely ever free of those substances.

Wolff summarized (1856) hundreds of proximate analyses from various countries. He reported protein as $N \times 6.38$ and calculated NFE as dry matter less crude fiber, protein, ash, and ether extract; and he rejected fiber values obtained by the Pearson-Einhof method as being too high. Some idea of the reasons for favoring the acid-alkali method may be ascertained from the work of Eben Horsford (1846), who under Liebig's direction investigated the composition of cereals and found that exhaustively extracted fiber gave the elemental composition of pure carbohydrate. The concept of fiber as being pure cellulose may have originated about 1830 with Anselme Payen in France (Horsford, 1846). Insufficient extraction produced "impure" fiber residues. It may be surmised that the impurities were nitrogen and mineral matter, which are very difficult to completely remove from fiber preparations. It is clear that Horsford regarded acid extraction as optional, and it was used only as necessary to obtain a pure fiber. Other investigators used only alkali. Alkaline extraction may be the earliest form of the modern crude fiber method, for it was the method used by John Gorham of Harvard (1820)[2] to analyze Indian corn. It

may be that alkaline extraction (with or without acid preextraction), as well as ideas about the indigestibility of fiber, were borrowed from another field. Flax was, and is today, retted in a fermentation process to isolate fiber, but early chemists experimented with acid and alkaline treatments for more rapid processing. John Wilson (1853) mentioned that alkali was used to prepare fiber from flax experimentally as early as 1745.

The competition among fiber methods reflects the conflicting definitions of dietary fiber that continue to exist even today. The concept of fiber as a pure chemical entity survives in attempts to define fiber as cellulose and lignin, while the definition of dietary fiber as an unusable residue implies a biological unit of variable composition similar to that of Einhof's preparations. Einhof's concept of nutritive availability is essentially the same as that of the modern human nutritionist who attempts to measure fiber by in vitro procedures; that is, as the residue resistant to mammalian enzymes.

The defects of the crude fiber method and the proximate system were known to Henneberg (Van Soest, 1977), who noticed that the digestibility of crude fiber was higher than that of NFE in certain forages. He and his chemist, Stohmann, showed that lignin was an important component of the indigestible NFE. Although he was aware of the deficiencies of the crude fiber analysis, Henneberg was unable to suggest an improved method with the technology of his time. He did not, therefore, abandon crude fiber determinations. Imperfect as they were, they represented an important

---

[2] One of the curiosities of the development of scientific agriculture in the United States is that the implementation of the Land Grant Act (1862) led to separate agricultural colleges. Much of the research related to agriculture before 1860 was done at institutions such as Harvard, which after this period contributed comparatively little. This is in contrast with development in Britain and Europe, where most of the old institutions developed departments of agriculture. The result in America has been a peculiar separation of agriculture from the basic biological sciences with resulting problems of lower status associated with being part of newer institutions. This factor has impeded the introduction of basic concepts into applied agriculture (and ruminant nutrition) while also insulating the "Ivy" institutions from the physical problems of the agricultural sector. This remains an issue in contemporary agricultural research, as, for example, in attacks on the mode of funding government-supported (USDA) research.

characteristic of the physical nature of foods and feeds. Since Henneberg, no important advance in proximate feed analysis occurred until the 1960s, when renewed interest in fiber as a dietary component reopened the question of fiber definition and methods. The field is a peculiar one from the view of scientific history, since it has been marked by intransigence and the persistence of unresolved conflicting ideas for the past 150 years.

25.3.1 Modern Developments in Feed Analysis

In the 1930s there was a resurgence of interest in crude fiber determination but no successful replacement of the old method. Notable developments, however, were methods for analyzing cellulose, lignin, and normal-acid fiber. Because there was no comprehensive theory of feed fractionation and analysis, correlation with feed value was the only means of evaluating the validity of these chemical fractions as representing nutritionally homogeneous fractions. Since the apparent correlations of these chemical fractions with digestibility were often no better than those of crude fiber and the proximate analysis, little progress was made in improving the proximate system. The logical problems involved in an evaluation of such methods were discussed in Chapters 10 and 23.

Subsequent attacks on the problem involved a more comprehensive approach with the goal of accounting for most, if not all, of the feed dry matter. Notable is the work of Lauri Paloheimo and Irja Paloheimo (1949) in Finland, Blanche Gaillard (1962) in the Netherlands, and R. Waite and others in Britain. The problem with these comprehensive analyses is that, while the information they provide is excellent, the procedures are so detailed as to be economically noncompetitive with proximate analysis. This defect also is characteristic of component sugar analyses. The development of a more competitive methodology was achieved at the USDA laboratory in Beltsville, Maryland, where L. A. Moore supported projects on analytical methodology and net energy evaluation of feedstuffs. Under his sponsorship, the holocellulose system (Ely and Moore, 1955) and later the detergent system were developed.

The effects of dietary fiber in humans and nonruminant herbivores began receiving extensive attention in the mid-1970s, when studies revealed the negative epidemiological association between fiber and disease (Burkitt, 1973). The adaptation of the detergent system to human foods has required further refinements (Southgate, 1976a; J. B. Robertson and Van Soest, 1981; Van Soest et al., 1991). Other recent developments are near infrared reflectance spectrophotometry (NIRS), a nondestructive analysis requiring less than a minute that can search for and analyze any component having an NIRS spectrum, and nuclear magnetic resonance

(NMR), which may be more efficient than NIRS in detecting specific chemical structures but is at present limited by the expense and availability of instruments. Both NIRS and NMR are limited in their ability to detect minor components. These systems show the promise of providing all components needed for feed evaluation; however, they must first be calibrated against the wet laboratory procedures that describe the composition and nutritive value of feeds. The calibration is shifted by season, weather, plant species, and feed processing. These systems need to conform to the modeling principles outlined in Chapters 22 and 23.

The advance of nutritional science has been retarded by the retention of outmoded systems such as proximate analysis. In the future, no system that might replace proximate analysis should be allowed to attain such a permanent position. Practical feed analysis must keep pace with advances in feed chemistry.

25.4 Predicting Digestibility from Plant Composition

Much effort has been expended on devising regression equations to predict digestibility from forage composition. The unreliability of such equations is due to environmental and species variation, interactions among plant species in mixed forages, and the difficulty in accounting for basic cause-effect relationships between ruminant digestion and plant composition (Section 10.1.1). Although composition ultimately controls digestibility and availability of nutrients, forage components are not of equal value and do not form consistent associations. Some, such as lignin and cutin, have a primary effect of reducing the digestibility of cell wall, while others, such as cellulose and hemicellulose, have secondary effects that depend on their relationship with lignin. A third group, the cell contents, is entirely available and affects digestibility in an additive sense.

25.4.1 Environmental Interaction

Secondary factors relate to digestibility through the associations they form with the primary variables. Thus, for example, cellulose is related to digestibility through its association with the primary variable lignin. Fiber, which usually contains more cellulose than lignin, is similarly associated. Usually, high negative correlations between fiber and digestibility occur in first cuttings of temperate forage, in which lignin and cellulose are positively correlated, while fiber forms a much poorer association with lignin in later cuttings and in tropical forages. This latter effect is due to the lack of association between lignin and cellulose (Chapter 6). Statistically, combining forages from different

Table 25.2. Correlations (r) and standard deviations (s_b) for regressions of acid-detergent fiber on digestibilities of first and aftermath cuttings of forage

	First cut			Aftermath			Combined		
	n	r	s_b	n	r	s_b	n	r	s_b
West Virginia grasses	36	−0.89	3.0	23	−0.14	3.5	59	−0.81	3.3
Michigan grasses	7	−0.62	1.7	9	−0.80	3.3	16	−0.49	3.9
Missouri grasses	9	−0.41	4.2	14	−0.47	4.8	23	−0.56	5.0
West Virginia legumes	6	−0.87	2.7	5	−0.51	2.4	11	−0.75	3.0
Michigan legumes	5	−0.42	2.1	8	−0.93	1.5	13	−0.81	2.1
All grasses	52	−0.72	4.1	46	+0.05	5.7	98	−0.46	5.2
All legumes	11	−0.63	4.1	13	−0.47	6.8	24	−0.51	5.5
All forages	63	−0.70	4.1	59	−0.20	5.9	122	−0.47	5.3

Source: Van Soest et al., 1978a.

places or different times in the growing season increases the predictive error through environmental interaction (Table 25.2).

The reliability of regression systems depends on whether the standard forages on which the equation is founded reflect the balance of species and environmental interactions characteristic of the forages to be tested. Strict standards must be applied regarding the forage populations used to test the system. There should not be less than 20 forages of determined animal digestibility. Legumes and grasses should be equally represented and several species of each included. Reference forages should come from localities similar to those of the forages to be tested. Aftermath cuttings should be included as well as first cuttings. The reliability of the regression system should be tested on populations other than that from which the equation was derived but from similar localities. This allows the possibility of ascertaining two kinds of error: the standard error of an estimate and the bias of the system to over- or underestimate the correct value.

When prediction systems are tested against a properly selected group of forages, realistic estimates of the predictive errors are obtained (Table 25.3). Many prediction errors reported in the literature are too low because the authors failed to obtain a representative feed or forage population. Systems based on crude fiber and protein involve large errors because these methods do not assess the effect of environment. Protein is positively associated with digestibility even though it declines as the plant ages, but nitrogen fertilization increases crude protein content without greatly altering digestibility. Equations derived from the results of proximate analyses, while based on a large amount of data (Schneider, 1947), suffer from the limitations of the proximate system and the historical nature of much of the data. Also, cultivars, fertilization, and management practices have changed over the years. A bias in the TDN estimate can arise through the use of combined data from a variety of geographical regions. In the case of forages from the northeastern United States, there is a positive overestimate of three or four units of digestibility when the general crude fiber equations are used.

No individual fiber determination method gives satisfactory predictions over wide ranges of feeds and forages. Acid-detergent fiber is better than crude fiber, mainly because silica and lignin are recovered in the former. The advantage of ADF is apparent only when an adequate forage population is sampled to represent the interaction of plant species, maturity, and environmental effects. The error observed with forage populations balanced relative to the various factors affecting quality is much larger than that reported for data from cutting studies, which reflect principally age and maturity effects. The summative equations and in vitro digestion (Table 25.3) give better estimates for mixed forage populations.

The relationships of various compositional factors with digestibility and intake illustrate the contrasting

Table 25.3. Predictive errors associated with systems to estimate digestibility from composition

Method of estimation	Value predicted	Bias[a]	s_b[b]
		units of dig.	
Crude fiber	DDM[c]	—	11.0
Acid-detergent fiber	DDM	—	9.0
Equations based on crude fiber and protein[d]			
Legume and grass	TDN	+4.0	7.7
Mixed	TDN	+3.3	8.0
Summative equation[e]			
Unmodified	DDM	−1.0	6.1
With silica correction	DDM	+4.5	3.8
In vitro rumen digestibility[f]	DDM	+2.5	3.7
True digestibility[g]	DDM	+0.7	2.8

Source: Van Soest and Robertson, 1980.

Note: This evaluation used a balanced group of legumes and grasses that varied in geographic origin.

[a]Mean difference between predicted and observed values.

[b]Standard deviation from regression.

[c]Digestibility of DM.

[d]Equations from Adams et al., 1964, based on the system of Axellson.

[e]Equation from Goering and Van Soest, 1970.

[f]Tilley-Terry procedure.

[g]Modification of Tilley-Terry according to Goering and Van Soest, 1970.

Table 25.4. Correlations of various forage components with
voluntary intake and digestibility for 187 forages of diverse species

Component	Digestibility	Intake
Digestibility		
In vivo	—	+.61
In vitro[a]	+.80	+.47
Lignin	−.61	−.08
Acid-detergent fiber	−.75	−.61
Crude protein	+.44	+.56
Cellulose	−.56	−.75
Cell wall	−.45	−.76
Hemicellulose	−.12	−.58
Rate of digestion	+.53	+.44

Sources: Mertens, 1973; Van Soest et al., 1978a.

[a]Two-stage procedure of Tilley-Terry as modified by Goering and
Van Soest, 1970.

behaviors of feed components (Table 25.4). Lignin and
ADF are related more to digestibility than to intake,
while the reverse is characteristic of protein, cell wall,
cellulose, and hemicellulose. These correlations reflect the inherently different effects of feed components on the various physiologies of digestion, intake,
and feed efficiency. Although there is a significant positive association between digestibility and intake, there
are important exceptions, such as the case of fescue,
which contains alkaloids (Sections 12.7.3 and 13.7.2).

In vitro digestion data (Figure 8.4) can be used to
explain NDF's better relation to intake than to digestibility. When in vitro digestions are done at differing times of fermentation, the times at which the residues relate to intake and digestibility differ markedly.
The optimum time for association with intake is 6–12
h, while that for digestibility is 36 h. The rate of fermentation of cell contents is rapid and is largely complete by 6 h, while that for cell wall has just begun and
becomes asymptotic at 48–96 h. The optimum at 36 h
may reflect retention time in the rumen.

25.4.2 Contrast among Forage Species

Variation in the proportions of cell wall constituents
is quite apparent when one compares different plant
species (Sullivan, 1966), particularly legumes and
grasses (Table 25.5). Legumes generally show high
ratios of lignin to cellulose and low ratios of hemicellulose to cellulose. Also, they generally contain less
plant cell wall than grasses of the same digestibility, a
factor that compensates for the high lignification of the
legume cell wall.

The characteristic differences among plant species
with regard to the proportions of lignin, cellulose, and
hemicellulose result in different relationships between
lignin and digestibility (Figure 25.1). Legumes tend to
have about twice the lignin content of grasses with the
same digestibility. Sullivan (1959) showed significant
though small differences among grass species as well.

Lignification affects only the digestibility of cell wall;
the cell contents are completely available. When lignin
is expressed on a cell wall basis, legumes and grasses
form a single regression with indigestibility of NDF
(Figure 12.11).

Summative equations (Goering and Van Soest,
1970; Conrad et al., 1984) take advantage of this relationship to predict the digestibility of legume-grass
mixtures. Lignin can be expressed on a cell wall or
ADF basis with similar results. The use of ADF has
no biological basis but may reduce analytical error
through pairing of the lignin analysis with the residue
on which it is determined.

25.4.3 Misuse of Detergent Fiber Analysis

The use of ADF or NDF in direct regressions on
digestibility ignores the physiological basis that relates
fiber components with digestibility. Manipulation of
light and temperature during plant growth can create
both positive and negative associations with digestibility, or no correlations at all.

The digestion of all fiber fractions—NDF, ADF,
cellulose, and hemicellulose—is limited by lignification. Since lignification is widely variable, fiber fractions do not appear uniform within themselves. Statistically, the association of any fiber fraction with
digestibility depends on the degree to which it is associated with lignin. If the association between lignin
and fiber content is low (as in tropical grasses or fall
cuttings in temperate zones), fiber content will not be a
good predictor of digestibility.

Lignin itself is of reduced predictive value in treated
straws and in forages in which a soluble lignin component has been created (see Section 12.6). While the
formation of a soluble lignin component is associated
with a reduction in covalently bound lignin, a concurrent reduction in the net Klason value is not always
seen; indeed, the value can increase. The crude lignin
fraction accumulated in the Klason residue may contain a variety of components, of which cutin, Maillard
products, tannins, and plastic can be mentioned. All
are themselves indigestible, but their quantitative effects on digestibility differ. Adopting an ideal method
for determining true lignin will not necessarily improve matters because significant components affecting digestibility will then be ignored. The value of
lignin (particularly that obtained using the 72% sulfur
acid method) is somewhat limited because the quantitative influence of its subcomponents can vary. Nevertheless, use of lignin values in a summative system is
a vast improvement over simple regressions based on
fiber.

The recurring statement that better chemical methods need to be found for predicting digestibility and
nutritive value is misleading. Apart from the improve-

Table 25.5. Proportion of hemicellulose, cellulose, and lignin in the cell wall of forages

Forage	Common name	No. forages	CW	H (% CW)	C (% CW)	L (% CW)	H-C ratio
Legumes							
Medicago sativa	Alfalfa	14	51	22	56	21.7	0.39
	leaf	22	20	30	51	18.2	0.58
	stem	22	54	17	64	19.2	0.26
Trifolium pratense	Red clover	1	66	26	59	15.1	0.44
T. repens	Ladino clover	1	28	14	68	17.6	0.21
T. alexandrinum	Persian clover	1	53	16	65	18.9	0.25
Temperate grasses							
Bromus inermis	Brome	7	64	42	50	7.5	0.84
Dactylis glomerata	Orchardgrass	3	55	45	45	7.5	1.00
Festuca K-31	Fescue	4	54	43	46	7.2	0.93
Phalaris arundinacea	Reed canarygrass	2	62	40	51	7.5	0.78
P. arundinacea	Reed canarygrass	8	59	47	49	6.9	0.96
Phleum pratense	Timothy	2	57	45	49	4.9	0.92
Subtropical and tropical grasses							
Cenchrus	Kleingrass	2	66	53	37	7.7	1.43
	leaf	1	65	60	34	5.8	1.75
	stem	1	80	42	45	12.8	0.94
Cynodon dactylon	Bermudagrass	4	76	51	38	7.9	1.34
C. plectostachyus	Stargrass	10	68	44	45	10.0	0.98
	leaf	1	66	50	40	7.5	1.24
	stem	1	73	40	49	11.1	0.81
Digitaria decumbens	Pangola grass	11	68	38	48	11.1	0.79
	leaf	7	68	38	50	12.9	0.76
	stem	7	70	37	49	17.2	0.75
Panicum maximum	Guineagrass	2	66	37	50	8.0	0.74
Pennisetum purpureum	Napiergrass	4	63	34	51	10.4	0.67

Source: Van Soest, 1973b.
Note: CW = cell wall percentage of dry matter; H = hemicellulose; C = cellulose; L = lignin.

ment of methods for in vitro digestibility, chemical techniques better than those that already exist are most unlikely. This can be seen in the Lucas analysis and the nature of the summative components contributing to digestible and indigestible matter. Failure to assay for secondary limiting factors may increase error and promote overprediction of nutritive values. Certainly there are measurements that would improve the summative system, including crystallinity of starch and cellulosic carbohydrates and other related intrinsic fac-

tors influencing the rate of digestion, but none of these determinations is likely to involve a modification of fiber or lignin methodologies.

A proper enzymatic or microbial digestion system is sensitive to all intrinsic limiting factors in a feed, measured or unmeasured. The limitation of such in vitro systems is that they tell little about the actual chemical components involved. Thus, both in vitro and in vivo analyses are needed. Analyses provide descriptions of values of protein, fiber, fat, and other components that are important for feed formulation. Such inference is not produced by any of the measures of digestibility. As for fiber, descriptive use poses the question What is the right fiber value? and raises the issue of competitive assays for NDF, ADF, and crude fiber. Because the composition of crude fiber bears little relation to any original structural cell wall component, it can be ignored.

25.4.4 The Case for NDF over ADF

Neutral-detergent fiber is more closely associated with intake than digestibility, while ADF does not represent all the insoluble fiber that can provide scratch factor in the rumen, although it is usually better correlated with digestibility than NDF is. Because NDF represents the total insoluble matrix fiber it is better related to rumination, fill, passage, and feed intake and

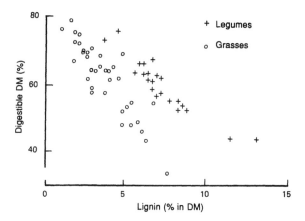

Figure 25.1. Dry matter digestibility and lignin content of temperate legumes and grasses (from Van Soest, 1964). The lignin content of legumes is about twice that of grasses with the same digestibility.

thus is more suitable for ration balancing. Furthermore, intake is more relevant to animal production than is digestibility.

25.5 Systems of Predicting Nutritive Value of Forages and Feeds

The overall conversion from composition to TDN, or from digestible energy (DE) to ME and NE, cannot be avoided in practical estimations of feed value, because frequently only compositional information is available. The consequence of this state of affairs is that the initial estimate of digestibility on which all further calculations are based constitutes by far the largest error in the overall estimation of NE. Unfortunately, the problems attendant on the estimates of digestibility often go unrecognized by those studying energy efficiencies.

Any error in estimating TDN or digestibility is passed along quantitatively to the final ME or NE estimate. Many NRC values of TDN have been estimated from proximate composition using equations similar to those developed by Schneider et al. (1951). Since such estimates are not distinguished from observed values in the tables, substantial error is likely, particularly with regard to tropical feeds.

The estimation of the productive value of any feedstuff involves several steps. The first is the determination of digestibility. This may be expressed as TDN or digestible organic matter (DOM). For many feeds and forages of low or nominal fiber content, this estimation of digestibility is potentially the largest error in estimation of NE. The second step is the conversion of TDN to an NE value relative to the type of production—whether lactation, gain, or maintenance. The final step is the discounting of energy relative to the level of intake above maintenance (see Section 25.6.4).

25.5.1 Predicting Digestibility from Equations

Although the use of prediction equations from the empirical regressions of digestibility on NDF and ADF has been criticized, such systems nevertheless represent the major current usage. Some prediction systems are better than others. Some are regional and thus achieve improved accuracy for local situations. P. Chandler (1990) published a survey of improved current equations in use and their variation when applied to a body of test forages of known value. The weakness of this evaluation is that it compares feeds by group—legumes, grasses, grains, and so on—rather than in multigroup associations. A major problem is that feeds from different plant families, which have characteristic lignin-to-cell-wall ratios, produce interactions when

regressions with fiber are used, and these interactions have not been evaluated.

Figure 25.2 illustrates the problems and bias that can result from environmental factors with data from Michigan corn grown in 1988 and 1989, which were, respectively, warm dry and normal wet years. Plants in 1988 were much less lignified and more digestible than 1989 plants, resulting in altered regressions between digestibility and ADF. The average lignin content of NDF was 6% with NDF digestibility of 50% in 1988, whereas in 1989 the lignin content of NDF was 7% with a digestibility of 42% (Allen and O'Neil, 1991). Slopes of the regressions of digestibility on ADF are similar for the two years but have quite different intercepts. When the data are combined, an interaction produces a steeper slope. The kind of interaction arises from geographical variation among microclimates.

As Figure 25.2 shows, forage plants behave differently under differing environmental circumstances. Corn silage is harvested at essentially maximum maturity, but its composition is profoundly affected by moisture, light, and net degree-days (Deinum, 1976). The ultimate lignification of leaf and stem is essentially determined by these factors. Lack of moisture was clearly more important than temperature in this instance. On the other hand, the degree of earing and formation of starch dilutes cell wall and is a major determinant of net fiber content. Dry weather often leads to poor grain yield, although this was not seen in the Michigan study. There was a lower crude protein content in the cooler year (Allen, 1992). Poor earing and grain yield tend to leave the photosynthetic energy as sugar in stalks and leaves, diluting fiber content but not yielding the expected net energy associated with a high grain yield.

25.5.2 Alternative Systems for Estimating Digestibility

Several alternative methods of estimating digestibility can be mentioned. They are discussed below in approximate order of decreasing accuracy.

Digestion balance or in vitro rumen digestion by the Tilley and Terry method or its modifications. Cellulase enzyme systems are of lesser value. Values of TDN can be estimated from digestible dry matter (DDM) by the equation:

$$TDN = DDM - \text{total ash} + \\ 1.25 \text{ ether extract} + 1.9 \qquad (25.1)$$

The error is not large for many feeds if the TDN and digestibility coefficients are assumed to be equal.

Moisture and ash are sufficiently variable in feeds to need to be individually determined. Values in pub-

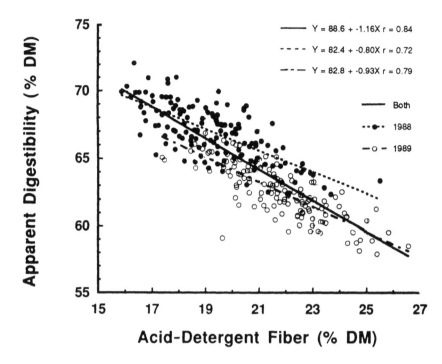

Figure 25.2. Relation between in vitro digestibility and acid-detergent fiber (ADF) in corn plants at harvest for the years 1988 and 1989 (courtesy of M. S. Allen, Michigan State University). The year 1988 was warm and dry with 2387 degree-days and 21 cm rain. In 1989, a wet year, there were 2072 degree-days and 41 cm rain.

lished tables are often unreliable because of differences in geography and environment. Grasses grown on soils with high levels of available silica have more ash and lower TDN than those grown on low-silica soils. If ash content is known, TDN should be adjusted with the equation:

$$\text{Adjusted TDN} = \text{table TDN} - \text{ash} (\%) + \text{table value for ash} \qquad (25.2)$$

Summative equations. Two summative systems exist: that of Goering and Van Soest (1970) and the newer one of Weiss et al. (1992). These are discussed in Section 25.5.3.

Estimation of TDN directly from tables. This is a relatively inaccurate procedure, especially for immature forages, but it may yield reasonable results for concentrates and by-product feeds if it is used carefully and with common sense. The procedure is to determine the NDF value of a specimen, find the same item, in Table A25.1, and add or subtract half the NDF difference between the two from TDN. For example, an alfalfa that contains 35% NDF (dry matter basis) is compared with early alfalfa, which has 41% NDF and a TDN of 67. Subtract 35 from 41, divide by 2, and add to 67 to give a TDN of 70. Always compare feeds with those of the nearest description. Feed comparisons should be undertaken only if the chemical compositions of the feeds to be compared are similar.

Acid-detergent fiber may be used in a similar way. A feed has 31% ADF, and tabular values are 29% ADF and TDN of 67. The difference is 2 units ADF. The full value of ADF should be used (a factor of 1.0). Thus, the TDN of the alfalfa is $67 - 2 = 65$. This procedure is much less accurate for forages and for feeds that are not well described than are equations 25.1 and 25.2.

Corn silage is a special case. The major factors influencing the quality of corn silage as it is growing are the proportion of grain to other plant matter and the conditions of growth. The corn plant is unique because it is harvested more or less at maximum maturity. Thus cumulative weather and genetic interactions are the prime cause of variability in composition and quality. Southern crops are likely to be more lignified than northern crops. Crops grown in drier summers will be less lignified and more digestible (Allen and O'Neil, 1991). There is also likely to be variation in net lignification among varieties. Values in the appendix tables (A25.1 and A25.2) are arranged according to North and South and are adjusted according to their grain content. Crops grown in dry years will be lower in both grain and lignin content and thus will approach higher-quality northern compositions. Net energy values may be adjusted for grain content. For each increase in 10% of grain, NE_g may be increased 0.5 Mcal and NE_l 0.6 Mcal. The grain content can be determined by manual separation and determination of the dry matter in the separated portions (Woody et al., 1984).

25.5.3 Summative Equations

Lignin would be a more accurate predictor of digestibility if species differences were adjusted for cell wall effects. Consequently, the summative equations

Table 25.6. Scheme for using the summative equation of Goering and Van Soest

Component	Analytical value[a]	Factor	Digestible amount[b]
Cellular contents	100-cell wall constituents	0.98	Add
Lignification of cell wall	Cell wall constituent analysis[c]		Add
Silica correction	SiO$_2$ analysis	3.0	Subtract
Heat-damage effect	Artifact lignin	1.0	Subtract
Estimated true DDM[d]			Sum
Metabolic fecal matter[e]			Subtract
Estimated apparent DDM			Difference

Source: Goering and Van Soest, 1970.

[a]All values must be expressed as percentage of whole dry matter.

[b]Analytical value is multiplied by factor to obtain the digestible amount, and then the process indicated in this column is performed.

[c]Estimated digestibility of cell wall from equation for (1) acid-detergent Klason lignin as percentage of ADF: Y = 147.3 − 78.9 log$_{10}$ ([L/ADF]100); or (2) permanganate lignin: Y = 180.8 − 96.6 log$_{10}$ ([L/ADF]100).

[d]DDM = digestible dry matter.

[e]See the discussion on variation of metabolic fecal matter in Chapter 18. The metabolic constant (M) for sheep is about 11.9; for cattle, 13.9; the regression for dairy cattle is the equation M = 36.57 − 0.257X, where X is estimated true digestibility.

Table 25.7. Summative system of Weiss et al. (1992) for calculating TDN

Cell contents	0.98 (100 − NDF$_{CP\text{-free}}$ − CP − EE) +
Protein	(exp$_{base10}$[−.012 ADICP])CP +
Lipid	2.25(EE − 1.5) +
Fiber	0.8(NDF$_{N\text{-free}}$ − LIG) (1 − [L/NDF$_{cp\text{-free}}$]$^{0.66}$ −
Metabolic	[5.5 + ash])

Note: CP = crude protein, EE = ether extract, ADICP = acid-detergent insoluble crude protein; L, LIG = lignin.

treat the unlignified cellular contents separately, assuming that mean true digestibility approaches 100%. Because of this treatment a metabolic loss has to be subtracted (see equation 22.25).

Several versions of summative systems are now available. The oldest is that of Goering and Van Soest (1970, Table 25.6). Others were designed by Conrad et al. (1984) and Weiss et al. (1992), and there is also a summative system for calculating digestibility in deer (Hanley et al., 1992; see Figure 13.5). The Van Soest system treats cell wall digestibility as a logarithmic function of the lignin content of ADF, while the systems of Conrad and Weiss use lignin content of NDF to the two-thirds power. Regressions of lignin on cell wall digestibility vary depending on the lignin method. Lignin values may require adjustment because of the effects of heating. Silica and tannin corrections may have to be applied; however, these also can be treated in a subtractive way in the summative model.

The summative system of Goering and Van Soest is shown in Table 25.6. The steps of a summative calculation involve the summation of digestible amounts from cellular contents and plant cell wall. The system of Weiss et al. (1992) adds subfractions for protein, lipid, and ash, as shown in Table 25.7.

Values for NDF, crude protein (CP), ether extract (EE), and ash are percentage of dry matter. NDF must be calculated nitrogen-free by subtracting neutral-detergent insoluble nitrogen (NDIN) multiplied by 6.25; ADICP is ADIN × 6.25 as percentage of dry matter. The Weiss equation sums the truly digestible contributions from cell contents, crude protein, ether extract, and lignified cell wall. It assumes that lignification affects only structural carbohydrate in NDF and limits digestibility by affecting surface area, hence

the two-thirds power function. Finally, a metabolic function is subtracted to obtain apparent digestibility.

Summative systems are superior to other regression systems when applied to a mixture of forages (legumes and grasses) but are less advantageous when applied within a single plant species. The equation will satisfactorily estimate the nutritive value of concentrates, something none of the other systems can do. In vitro rumen digestion estimated by the Tilley-Terry method or a modification of it predicts the digestibility of forages more accurately than the summative equation, but in vitro systems do not cope well with concentrates.

The problem with concentrates is that effects of lignification are restricted to only a small portion of the feed dry matter. Starch digestibility is affected by the level of intake, and starch can also depress fiber digestibility through inhibiting fermentation. These effects are difficult to predict either by regression equations or by an in vitro rumen digestion. Since concentrates are most often fed in combination with forage, interactions between forage and concentrate can be important. Fiber and starch are important factors in the depression in digestibility observed at high levels of intake, but predicting these effects requires modeling of rates of digestion and passage.

25.6 Adjustments to Digestibility Estimates

The conversion to NE can be considered in three steps: (1) the estimation of digestibility (TDN, DE), (2) the estimation of ME (outright or from TDN or DE), and (3) the estimation of NE (outright or from DE, TDN, or ME). Outright estimation requires consideration of composition and incorporates all the problems associated with estimating digestibility in addition to those encountered in converting to NE. Feed composition is likely to be an important factor influencing energy conversion from DE, TDN, or ME. The conversion equations published by the NRC do not consider compositional factors that modify the efficiency factors, particularly those involved with fiber and NDF. Also, the NRC equations for beef and dairy cattle are not mutually consistent.

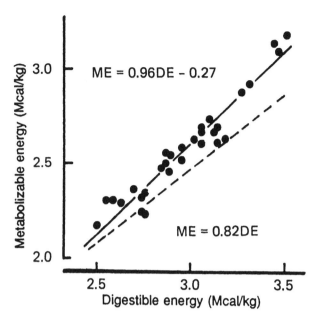

Figure 25.3. The relation of metabolizable energy (ME) to digestible energy (DE) (from Moe et al., 1972). The commonly assumed factor 0.82 (dashed line) underestimates true ME, which is relatively greater for more digestible feeds.

25.6.1 Conversion to TDN

The conversion of dry matter digestibility to any equivalent unit of energy involves adjustments for fat and ash content. Generally, the energy values of proteins and carbohydrates are equated, the higher energy content of protein being offset by the expenditure for nitrogen excretion. The ash correction is the most significant in the case of ruminant feeds, and digestible organic matter (DOM) is probably as satisfactory as TDN, since the amount of ether extract in many ruminant feeds is low and its composition is variable. Many of the TDN values used in the California NE system were calculated according to Lofgreen's equation (1953):

$$\text{TDN} = \text{digestible DM} + 1.25 \times$$
$$\text{digestible EE} - \text{total ash} \qquad (25.3)$$

The value 1.25 is used instead of 2.25 because a unit of digestible EE is included in the DM. This equation does not consider the variation in the energy equivalent of TDN, which is unavoidable if TDN is to be converted into calories.

25.6.2 Conversion to ME

It is commonly assumed that metabolizable energy is 82% of digestible energy; however, as might be suspected, this constancy is subject to revision. Regres-

sion of measured ME on DE shows a significant negative intercept (Figure 25.3) and a greater slope. The 0.82 factor tends to underestimate ME at higher digestible energies.

In many balance trials methane loss is calculated in the ME estimations, usually from the digestible carbohydrate. The assumption is that methane output is a constant proportion of the net rumen fermentation. No consideration is made of methane balance as a consequence of rumen VFA ratios. Fermentation balance can greatly alter the methane yield, and the escape of fermentable carbohydrate from the rumen would also contribute to deviation in predicted values. Most ME values have been calculated by regression from TDN or digestible energy (DE), often using the 82% factor. As a result, tabular values of ME have little advantage over TDN values.

25.6.3 Conversion to NE

The calculation of net energy values for feeds generally has been accomplished using regression equations derived from production trials. My opinion is that the dairy cattle values derived from actual respiration balances are more accurate than the beef cattle system. These values (edited by Paul Moe in NRC, 1971) show lactation and maintenance to be more efficient processes than gain. More recently produced values show gain to be more efficient than lactation (NRC, 1984). This is evidenced by a slope of greater than 1 when values of NE_g are plotted against NE_l, as shown in Figure 25.4.

Adjustments were made to the pre-1984 NRC beef equations to rectify the underestimated values for the energy content of concentrates; unfortunately, these modifications introduced inconsistencies between the beef and dairy systems. The excess slope in the newer beef equations may have resulted from the continued use of the 82% factor for calculating ME, whereas the dairy ME equation uses a slope of 0.96 (Moe et al., 1972). The dairy system includes an 8% discount for level of intake, whereas the NRC beef system has made no adjustment. The discount system uses the relationships developed from the earlier NRC publications in which the ratio of NE_g to NE_l is 0.78. This conversion factor is used throughout the discount system for beef cattle values (NRC, 1970; Moe et al., 1972).

The conversion of energy values to NE has practically involved conversion from TDN because few real ME values are available. The regressions and the practice have varied over the years, and current values for net energy of maintenance (NE_m) and NE_g for beef cattle are inconsistent relative to NE_l, as shown by the comparisons in Figure 25.4. The Cornell system, based on NRC values (1971), yields lower NE_g and

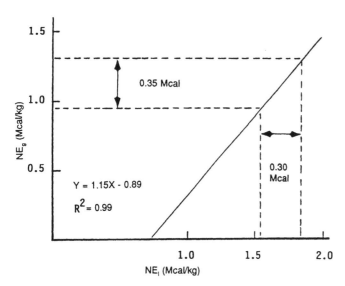

Figure 25.4. NRC values for NE_g versus NE_l (1984 and 1989). The slope of this regression, 1.15, is biologically unrealistic. Note that an increment of NE from added feed causes a larger increment in NE_g than in NE_l.

Table 25.8. Comparison of net energy regressions obtained from analyses of NRC values

Source	X	Y	Equation	No.
NRC 1971	NE_g	NE_l	Y = 0.78X + 0.31	25.4
	NE_m	NE_l	Y = 0.78X − 0.41	25.5
	NE_m	NE_g	Y = 1.01X − 0.71	25.6
NRC 1984, 1989	NE_g	NE_l	Y = 1.15X − 0.83	25.7
	NE_m	NE_l	Y = 1.33X − 0.47	25.8
	NE_m	NE_g	Y = 0.89X − 0.71	25.9

Sources: National Research Council 1971, 1984, 1989.

higher NE_m and insists on the slope of 0.78 as the relative efficiency of gain to lactation (Table 25.8). The lower efficiency of gain relative to lactation has been confirmed by many experiments. The 1984 NRC values allow gain to be more efficient than lactation, which is contrary to observed energy balance efficiencies. It is also not reasonable to expect lactation values to exceed maintenance values.

25.6.4 Discounted Values of Net Energy

Discounts of net energy arose because of the false assumption that digestibilities of feeds were invariable. In the case of beef cattle and sheep, digestibilities were measured in a representative subgroup of animals and applied to the calculation of efficiency of gain in the main group. For dairy cattle, digestibility measured at each level of milk production led to the disclosure of digestibility depression (Van Soest, 1973a; Tyrrell and Moe, 1975a). Most of the values for digestibility of feeds listed in the NRC tables were derived from a data base of sheep fed at maintenance level. The values of TDN at maintenance tend to be higher than the TDN values at productive levels of intake, so when the derived net energy equations were applied to these values, NE was overestimated. The responses of the beef (NRC, 1984) and dairy (NRC, 1989) committees differed. The beef committee chose to bury the difference by adjusting the equations, and the dairy committee introduced a factor to adjust for decline in digestibility with level of intake. Depression in digestibility is a more serious consideration for dairy cattle, in which intake varies greatly.

The depression in digestibility is largely accounted for by loss of potentially digestible NDF, and to a lesser extent by fecal starch, because these fractions are the slowest to degrade and therefore the most likely to escape the rumen undigested (Figure 21.17). Concentrates contribute more to the loss of digestible matter because concentrate particles are smaller, heavier, and less likely to be ruminated than their larger, lighter forage counterparts. The NRC has assumed that while depression in digestibility of individual feeds may vary, the average depression in an ordinary dairy diet is about 4% per unit of intake above maintenance and the differences among feeds cancel out because of the averaging process. Coppock (1987) pointed out that this may not be true.

By failing to consider variations among feeds in the decline in digestibility, the NRC has developed a less sensitive system for optimizing diets. A case in point is legumes and grasses. Alfalfa has always been a more productive forage per unit of TDN than most grasses. Alfalfa and clover elicit more production than does a grass of equal digestibility (Rattray and Joyce, 1974; D. J. Thomson, 1975). This difference arises from several causes: (1) the legume is usually consumed at a higher intake, at a higher ratio to maintenance; (2) the alfalfa suffers less from the drop in digestibility with level of intake (Van Soest, 1973a); and (3) legumes generally have less NDF than grasses. NDF pushes rumination (Welch and Smith, 1969), digestive work, and consequently heat increment (Webster, 1978a). Digestible NDF is also related to digestibility depression at increasing levels of intake. In the NRC publications, NE is predicted from TDN or DE, and all forages use the same equation.

Equations for calculating discounts and discounted values of NE are given in Table 25.9. Discount values have been computed, first, on the basis of in vivo data and, second, on the basis of in vitro digestion rates and available passage rates (Van Soest et al., 1979, 1984). Discounts are estimated from the integral function of escape at different levels of intake and expected rates of passage. The discount value represents the expected change in escape per unit of intake in units of maintenance. Discounts for feeds for which no laboratory or animal data are available are derived from equation

25.9. Discounts and calculation of NE$_l$ from TDN

	Source	Equation[a]	No.
...rsion from	NRC, 1984	$NE_{l-M} = 0.0266\ TDN - 0.12$	25.10
...d at 1 M	Van Soest[b]	$NE_{g-M} = 0.0207\ TDN - 0.50$	25.11
	Van Soest[b]	$NE_{m-M} = 0.0207\ TDN + 0.20$	25.12
	Weiss et al., 1992	$NE_{l-M} = 0.0228\ TDN_{3M} - 0.10$	25.13
...rsion with	Van Soest, 1971	$NE_{l-3M} = 0.01\ TDN_M \times$	25.14
...ount (3M)		$(2.86 - 35.5/[100 - NDF])$	
...unt (%)	Mertens, 1983	$D = 0.033 + 0.132\ (\%\ NDF) -$	25.15
		$0.033\ (\%\ TDN_M)$	
...onversion of	Van Soest, 1971	$NE_g = 0.78\ NE_l - 0.41$	25.16
...values		$NE_m = 0.78\ NE_l + 0.31$	25.17
		$NE_g = 1.01\ NE_m - 0.71$	25.18

...urces: National Research Council, 1970, 1971, 1984, 1989.

...te: Comparison of net energy regressions was obtained from regression an-... of NRC values (1970, 1984, 1989).

...ubscripts M and 3M specify level of intake in units of maintenance.

...quations derived by combining equations 25.4 and 25.10 or 25.4 and 25.11.

25.13 in Table 25.9 or by comparison with similar feeds for which there are data, using some common sense and practical judgment. Feeds that have a high cell wall content and a low degree of lignification usually have the largest discounts. Starch adds to the cell wall effect in cereal grains. The differential competition between digestion and passage rates may not be linear, so the change in digestibility is largest at between one and four times maintenance. Discounting above this level is problematic because few data are available. Theoretically, the rate of depression is proportional to the reciprocal of the square of intake (Van Soest et al., 1979). Discounts above five times maintenance intake may be reduced by one-half.

25.6.5 Calculating Discounts

The conversion from TDN to NE must be estimated for the specific animal production functions, whether lactation for dairy cattle or gain and maintenance for beef cattle. The values of NE$_l$ in Table A25.1 are based on equation 25.14 in Table 25.9. Alternatively, a combination of equations 25.10 and 25.15 may be used to calculate NE$_l$. These two alternatives are the only ones that recognize the different productive potentials of legumes and grasses because of their inherently different NDF contents. For this reason, equations 25.11 and 25.12 must be adjusted using equation 25.15 or discounts from Table A25.1. These combinations are alternatives to equation 25.14. Equations 25.10, 25.11, and 25.12 estimate NE relative to the intake of feed at which the TDN value was estimated. Therefore they need to be adjusted using discount values from Table A25.1. Equation 25.13 discounts with a flat 8% (4%/M) and should be used only on concentrates and high-quality forages, not on low-quality forages and high-NDF by-product feeds.

Gain values are at 2M in Table A25.1. The gain values may be calculated from NE$_m$ by subtracting 0.7

Mcal at 1M, 0.67 Mcal at 2M, and 0.63 Mcal at 3M. These values may be adjusted by the variable discounts in Table A25.1 as described in the example below. NE$_m$ is calculated from NE$_l$ using equation 25.11 and discounting upward to 2M. The discount factor is reduced by 0.78 in this calculation.

An example showing the use of the discounts is as follows: assuming a TDN (maintenance intake) of 60 and a discount factor of 9%, an NE$_{l-M}$ of 1.48 is obtained using equation 25.4. This value discounted to 3M is $1.48 \times (1 - 2 \times 0.09) = 1.21$ Mcal. For beef values, the value of NE$_m$ at three times maintenance intake (NE$_{m-3M}$) is calculated using equation 25.11: $0.78 \times 1.21 + .31 = 1.25$. This value is discounted upward $1.25\ (1 - 0.9 \times .78) = 1.34$ Mcal NE$_{m-2M}$. Subtracting $0.7 \times (1 - 0.09 \times 0.78) = 0.65$ Mcal yields an NE$_{g-2M}$ of 0.69 Mcal. The NDF and protein values for many commercial feedstuffs are sufficiently uniform that ration calculations can be based on tabular figures. Forages and silages are more variable and require individual laboratory analysis, as will feedstuffs that do not match descriptions in the table. Discount values for such feeds can be generated using equation 25.15.

Intake in units of maintenance is estimated by multiplying the dry matter intake of each dietary ingredient by its tabular value (Table A25.1) for NE$_g$ and summing the total daily intake in units of megacalories for a total net intake. This value is divided by $0.077 \times$ body weight $^{0.75}$ for NE$_g$, and $0.077/0.78 = 0.0987 \times$ body weight $^{0.75}$ for dairy cattle.

25.6.6 Validation

The NE$_g$ values in Table A25.1 and the associated equations (Table 25.9) have been validated with 91 observations of actual energy retained in growing cattle fed diets varying from all-corn silage to all corn (Figures 25.5 and 25.6). Nearly identical results were obtained with the 1984 NRC beef equations.

While the results for the NRC and Van Soest systems are comparable, the latter is built on a system in which efficiency of gain is 78% of lactation and the difference between NE$_g$ and NE$_m$ values is a relative constant, thus eliminating the need for two sets of NE values for gain and maintenance. The principal advantage of the Van Soest system is that only one set of net energy values (NE$_g$) is needed, since maintenance values (NE$_m$) are generated by subtracting a constant.

25.6.7 Discounts of Proteins

Microbial nitrogen normally exceeds escaping feed nitrogen by a factor of about 1.5 and is highly correlated with the escaping quantity provided adequate and appropriate energy is present (P. H. Robinson, 1983).

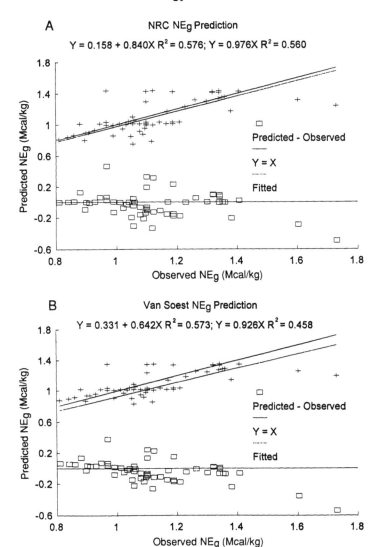

Figure 25.5. Comparison of NRC system (1984) for predicting net energy of gain (NE$_g$) in beef cattle (A) with (B) the system of Van Soest et al. (1992b). Lower line indicates the association of residual values. (Courtesy M. Barry and D. Fox, Cornell University, Ithaca, New York.)

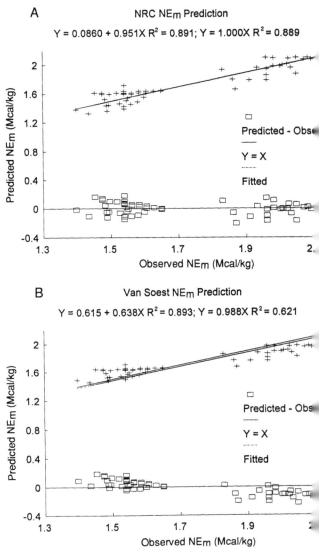

Figure 25.6. Comparison of NRC system (1984) for predicting net energy of maintenance (NE$_m$) in beef cattle (A) with (B) the system of Van Soest et al. (1992b). (Courtesy M. Barry and D. Fox, Cornell University, Ithaca, New York.)

The correlation between escape protein and potentially digestible organic matter is 0.74, and between microbial protein leaving the rumen and PDOM, 0.87 (Van Soest and Sniffen, 1984). These relationships were obtained in milk-producing dairy cattle over a wide range of intakes. The value of 1.5 is reduced to 1.05 (1.5 × 0.70) in the equations below to allow for a microbial crude protein digestibility of about 70%. The mechanistic justification for applying this association is that the great majority of efficient bacteria are attached to feed particles, the passage of which harvests rumen organisms for the lower tract. Thus microbial efficiency is tied to the escape of potentially digestible matter from the rumen. The dynamic forces driving microbial output have been underestimated by the NRC, which calculates microbial yield as a proportion of carbohydrates fermented. Microbes, like all animals, have maintenance requirements and respond to level of feeding (for microbes, this is the rate of fermentation). Microbial efficiency is also increased by rumen washout and therefore by feed intake (J. B. Russell et al., 1992). Of the respective carbohydrates, those in NDF or plant cell wall are the most efficient for protein yield. Therefore, forage quality is of prime importance. The combined increment of escape and microbial protein is highly correlated with discount factors provided rumen requirements for microbial growth are satisfied (Figure 18.14).

While the estimation of metabolizable protein (MP) is better obtained from the Cornell net carbohydrate and protein model, a calculation can be made from information provided here as indicated in the following equations:

$$ACP = TCP - (ADIN \times 6.25) + 1.0 \quad (25.19)$$

$$MP = EP + BP \quad (25.20)$$

$$EP = F_i \left(\frac{CP}{100}\right)(100 - DN_d - ADIN) \quad (25.21)$$

$$BP = 1.05EP + RCP \quad (25.22)$$

$$RCP = 0.70\left(\frac{4}{100F_m}\right)\left(\frac{4}{CP - 4}\right)$$

$$= \left(\frac{0.112F_i}{F_m(CP - 4)}\right) \quad (25.23)$$

$$\frac{MP_{kg}}{d} = 2.05\, F_i\left(\frac{CP}{100}\right)(100 - DN_d - ADIN) +$$

$$\left(\frac{0.112F_i}{F_m(CP - 4)}\right) \quad (25.24)$$

where EP is escaping (by-pass) protein, BP is microbial protein leaving the rumen, F_i is feed intake (kg/day), F_m is feed intake in units of maintenance, CP is crude protein percentage in the dry matter, DN_d is degradable nitrogen as a percentage of total CP nitrogen, and RCP is microbial protein derived from recycled nitrogen (Table A25.2).

Equation 25.19 shows the correction for unavailable protein: ADIN × 6.25 is subtracted from total crude protein (TCP). The adjusted crude protein (ACP) value that is compatible with ordinary crude protein values of undamaged feedstuffs is obtained by adding 1.0 to the difference TCP − ADIN, because this recognizes the average contribution of naturally lignified nitrogen (Figure 18.3). The ACP can be used in any equations involving calculation or balancing with CP. The addition of 1.0 adjusts for the normal level of ADIN in unheated feedstuffs. Feeds that have been derived from processes involving heat treatment or spontaneous heating are apt to contain a significant amount of nitrogen made unavailable through the Maillard reaction. Further study and analyses indicate that while ADIN has an apparent digestion coefficient, the utilization of the apparently digested fraction is very limited (Nakamura et al., 1991; Van Soest and Mason, 1991). The opinion put forth here is that ADIN may be quantitatively discounted from the measured ADIN multiplied by 6.25. Such feeds include distillers' grains, brewers' grains, dehydrated alfalfa meal, tomato pulp, haylages with more than 50% dry matter, and a number of other products. Tanniniferous feeds such as grape pomace and almond hulls also need to be discounted in the same manner.

Equation 25.20 indicate that metabolizable protein is the sum of escaping and microbial protein in the intestines. Equation 25.21 estimates escaping protein by subtracting discounted degradable nitrogen and ADIN from the true protein. It assumes that the availability of degradable protein for the rumen is decreased to the same degree as energy is discounted. Thus the discount increases the estimate of escaping protein from the rumen to the stomach and intestines with level of intake. This escape is at the expense of net energy lost from the rumen in the form of potentially digestible NDF. For example, by-product feeds such as distillers' grains and brewers' grains have relatively fine particle NDF prone to rumen escape, and therefore high discounts, which recognize the high potential of escape protein with these feeds. Escape of feed protein is inevitably associated with loss of net energy in the form of fermentable fiber because of the physical association of the insoluble protein with particles containing the NDF.

Equation 25.22 assumes that microbial protein escaping the rumen is from the microbes attached to escaping feed particles plus an additional increment (RCP) arising from microbes' use of recycled urea and salivary nitrogen. The efficiency of conversion of recycled endogenous nitrogen is influenced by feed intake and the crude protein level in the diet. Since the body pool of urea cannot expand in proportion to intake (because of homeostatic limits and because urinary loss is the means of relief), the value of recycled urea becomes diluted by feed intake and crude protein intake. Equation 25.23 estimates this contribution from recycled nitrogen and includes a function for the digestibility of bacteria (70%), the dilutions of the secreted pool in the digestive tract by feed intake ($4/100F_m$), and crude protein level in the diet [$4/(CP - 4)$]. The amount of this nitrogen as CP equivalent in the diet at maintenance is about 4% of intake, as evidenced by the fact that zero apparent digestibility occurs at about 4% of CP in the diet. From this level to about 8%, CP use of recycled nitrogen is 90–100%. Above this level the efficiency declines.

Figure 25.7 plots predicted crude protein equivalents (in units of crude protein in feed dry matter). The predicted values indicate that the crude protein level in the diet has a relatively larger effect than the level of feed intake in diluting the available recyclable nitrogen. The calculated efficiency of use of recycled nitrogen declines from 87% at 1M intake and 8% crude protein to 18% at 4M intake and 20% crude protein in the dry matter of the intake. The decline in value of recyclable nitrogen emphasizes the sensitivity of the high-producing dairy cow to inadequate levels of soluble protein in the rumen.

Equation 25.24 combines equations 25.21, 25.22, and 25.23 and estimates the net metabolizable protein relative to level of intake. This equation is not intended for use with diets having any of the following attributes:

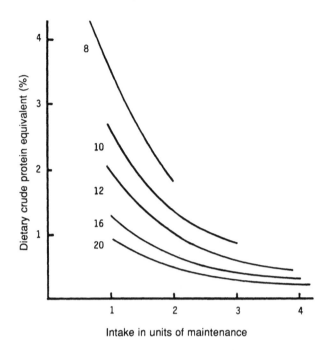

Figure 25.7. The calculated crude protein equivalent (as percentage of feed dry matter) of recycled nitrogen as related to dietary level of crude protein and feed intake. Isoclines are shown for diets containing 8, 10, 12, 16, and 20% crude protein. Values were calculated using a modification of equation 25.23.

CP less than 8%, deficient in NPN or in undegradable protein on typical feeds, or out of balance with requirements based on the Cornell net carbohydrate and protein system. Bacterial protein produced from urea included in the diet is not accounted for in these equations; equations that can address dietary urea and excesses of NPN or undegradable protein can be developed. Equation 25.24 assumes there are no deficiencies of fiber, energy, nitrogen fractions, or sulfur for microbial growth, and that there is no major amino acid imbalance or limitation because of poor carbohydrate quality. Feeding too little degradable nitrogen or feeding too much slow-fermenting carbohydrate will reduce microbial yield and result in an overestimation of bacterial protein. If soluble nitrogen or carbohydrates are limiting variables, the Cornell carbohydrate and protein model must be used (Fox et al., 1990, 1992).

An example of a ration balance using equation 25.24 and the energy discounts from Table A25.1 is shown in Table 25.10 for a cow yielding 120 lb/day of milk and eating at an apparent four times the maintenance level of intake. Estimates of bacterial protein are slightly higher than the value obtained by the Cornell carbohydrate and protein model, which is compensated by a slightly lower estimate of escaping protein. Note that the actual level of intake is 3.8M after discounts are

Table 25.10. Calculation of net energy and metabolizable protein using discounts for a 1400-lb (636-kg) cow producing 120 lb/day (54.5 kg) milk and eating at 4M[a]

	Alfalfa hay (E. bloom North)[a]	Corn grain, ground	Soybean meal (44)[a]	DDG + solubles[b]	Total diet
DM intake (kg/day)	12.7	11.4	2.3	2.7	29.1
TDN_M	62	87	81	88	75
NE_{I-3M}/kg[a]	1.42	2.04	1.78	1.58	—
Net intake	18.0	23.2	4.1	4.3	49.6[c]
Discount (%)[a]	3.5	3.3	5.1	13.5	—
NE_{I-4M}/kg	1.37	1.97	1.69	1.37	—
NE_I intake (Mcal/day)	17.4	22.5	3.9	3.7	47.5[c]
Crude protein[a]	19.0	10.0	49.0	29.5	18.8
ADIN[d]	11	5	2	21.0	10.9
DN_{3M}[d]	70	48	65	53	—
DN_{4M}[e]	68	46	62	46	—
EPN%[f]	21	49	36	33	32
EP% DMI[g]	4.0	4.9	17.6	9.7	5.96
EP (g/day)[h]	507	559	405	263	1734
BP (× 1.05) (g/day)[i]	—	—	—	—	1876
MP (sum) (g/day)	—	—	—	—	3610
MP est.[j]	—	—	—	—	3551

Source: Van Soest et al., 1992b.

[a]Values from Table A25.1 on a dry matter basis.

[b]DDG = dried distillers' grains.

[c]Approximate level of intake in units of maintenance is obtained by dividing $636^{0.75} \times 0.0987 = 12.50$ Mcal into estimated net intake of Mcal of 49.6 (line 4) = 3.97M. This has been rounded to 4.0 for this example. Note that the actual level of intake at 4M is 47.5/12.5 = 3.80M, after adjustment by the discount.

[d]Values from Table A25.2 as percentage of total crude protein.

[e]Degradable nitrogen using discount value from Table A25.1: multiply discount times table DN and subtract.

[f]Escaping protein nitrogen (EPN) as percentage of crude protein nitrogen obtained with the equation $100 - DN_{4M} - ADIN$.

[g](EPN × CP)/100.

[h]EP% × dry matter intake. Escaping protein (EP) = EPN × 6.25.

[i]Bacterial protein (BP) is estimated by multiplying 1.05 × EP%.

[j]The Cornell model predicted 1785 g EP and 1766 g BP and a total of 3551 g/day metabolizable protein (MP).

applied; however, the apparent intake of 4M must be used in order to obtain the discounted values.

Diets for ruminants are commonly balanced for NE and protein, with the assumption that NE promotes intake and production. This approach can lead to fiber-deficient diets with lowered rumen efficiencies, acidosis, and milk fat depression. When fiber is considered a dietary requirement and sufficient levels are provided, an adequate rumen mat forms and rumination is stimulated. As a result, the rumen is buffered by saliva and normal levels of digestibility are maintained. Animals on diets deficient in fiber show larger fecal losses of starch and other potentially digestible matter. The fiber value of a feed is related to its coarse NDF content, which establishes the rumen mat filtering system. This coarse fiber renders fine fiber more effective because it is captured through filtration, so estimates of effective NDF may be variable. Tabular values for NE apply only to diets optimally balanced for required nutrients. To optimize animal performance, requirements for fiber, protein (nitrogen), energy, minerals, and other cofactors must be satisfied.

25.7 Dynamic Systems and the Future

Tabular values of feeds quote static conditions, whereas the producing animal is a dynamic system. Computerized data accompanied by differential equations describing gastrointestinal and animal functions could render tables obsolete. Computers are capable of integrating the most complex mathematical functions, and it is no longer necessary to rely on inadequate and oversimplified static equations. The current status is an interim in the process of a revolution in the concepts of feed formulation and diet balancing for ruminants.

From the discussion of the variable utilization of the available energy in feeds and forages, it follows that feed values are not constant, but rather depend on retention time in the digestive tract, the latter being affected by the animal's species, size, and relative level of intake and metabolic function. Accounting for these variables requires models that reflect mechanistic principles of biochemistry and physics. Truly dynamic feed evaluation and ration calculation systems that take into account rate information on passage and digestion and integrate these rates at expected rumen turnovers relative to intake and production level are possible.

Systems founded on these principles include the Cornell carbohydrate and protein model (Fox et al., 1992), which deals with energy and protein; and the PDI system from Institut National de la Recherche Agronomique (INRA; Paris) and the model of Baldwin et al. (1987), which are primarily directed at protein. Rates of passage and escape apply equally to the starch and cell wall fractions of feeds and are used to estimate digestibility and ME in the Cornell model (Sniffen et al., 1992). These have been estimated as discounts to net energy in the Cornell system. A summative system for digestibility and net energy has been developed (Weiss et al., 1992; see Section 25.5.3). The detergent system and the validation of the summative model have made possible the mechanistic production of feed values, although this production has been delayed because lignin or alternative measures of cell wall digestibility are still needed. In addition, the competition of digestion and passage to determine the portion of available nutrients utilized or lost requires rate information that is only now becoming available. The concept has preceded the development. The Cornell system, now available on software (Fox et al., 1990), should replace all empirical statistical systems for feed evaluation. A software program is also available for the largely static NRC dairy cattle nutrient requirements (1989). The model continues to develop, and its present form will undoubtedly change.

A remaining problem has to do with amino acids and vitamins. When one compares the current state of understanding in monogastric nutrition with that of current understanding in ruminants with regard to amino acids and the B vitamins, ruminant nutrition indeed seems to be in a retarded state. But the retardation involves the enormous modifying factor of the rumen fermentation. Mechanistic models can be developed to predict rumen output for these dietary components and thus provide the requisite data to resolve questions about the ultimate net supply of amino acids and vitamins to the ruminant.

The guiding philosophy of this volume is the belief that science involves a mechanistic understanding of processes, the descriptions of which can be employed more reliably than descriptions based on stochastic or empirical models in which the causes and effects and mechanics are unclear. The problems of estimating digestibility from composition and the energy systems based on these estimates are important examples. The largest single issue in nutrition is the application of mechanistic principles to provide understanding.

Appendix: Tables of Feed Composition

Table A25.1.   Discounts of nutritive value and composition of ruminant feeds

Feed name	Int'l feed no.[a]	1M TDN (%)	Discount (%/M)	2M[b] NE_g[c]	3M[b] NE_l	Crude protein[d] (%)	NDF (%)	Lignin (%)	ADF (%)	EE (%)	Ash (%)
Alfalfa											
Meal, dehy. 15%	1-00-022	59	8.0	0.70	1.22	16.0[d]	50	13.0	36	2.5	10
Meal, dehy. 17%	1-00-023	61	8.0	0.74	1.26	17.0[d]	45	10.0	35	3.0	11
Hay, E. veg.	1-00-N	67	2.7	0.90	1.57	23.4	36	5.3	26	3.2	10
Hay, L. veg.	1-00-N	64	3.2	0.83	1.48	21.7	39	6.5	28	3.0	10
Hay, E. bl.	1-00-N	62	3.5	0.78	1.42	19.0	42	7.1	31	2.5	9
Hay, M. bl.	1-00-N	60	3.9	0.74	1.34	17.0	46	8.7	36	3.2	9
Hay, F. bl.	1-00-N	56	4.6	0.63	1.24	13.0	51	10.4	41	1.8	9
Hay, L. bl.	1-00-N	53	5.3	0.57	1.13	12.0	55	12.2	43	1.6	8
Hay, E. veg.	1-00-54-S	66	2.6	0.87	1.55	30.0	33	6.0	25	4.0	10
Hay, L. veg.	1-00-059-S	63	3.0	0.81	1.45	27.0	37	7.0	27	3.8	9
Hay, E. bl.	1-00-059-S	60	3.4	0.74	1.37	25.0	40	8.0	30	2.5	9
Hay, M. bl.	1-00-063-S	58	3.9	0.70	1.23	22.0	44	10.0	35	3.2	8
Hay, F. bl.	1-00-068-S	55	4.6	0.62	1.21	19.0	48	11.0	37	1.8	8
Hay, L. bl.	1-00-070-S	52	5.2	0.55	1.12	17.0	53	12.2	39	1.5	8
Hay, mature	1-00-71-S	50	6.0	0.48	1.02	14.0	58	14.4	44	1.3	7
Hay, seeded[a]	—	45	7.8	0.37	0.93	12.0	70	17.0	49	1.0	7
Hay, weathered[a]	—	48	6.0	0.37	0.97	10.0	58	15.0	45	—	8
Wilted silage											
E. bl. 35% DM	3-00-216	60	3.7	0.75	1.37	19.0[d]	43	10.0	33	3.2	9
M. bl. 38% DM	3-00-217	58	4.0	0.70	1.29	17.0[d]	47	11.0	35	3.1	9
F. bl. 40% DM	3-00-218	55	4.7	0.61	1.21	16.0[d]	51	12.0	42	2.7	8
Almond hulls											
Good quality[a]	—	60	3.1	0.75	1.37	0[d]	22	6.1	17	3.2	7
Medium quality	4-00-359	56	3.0	0.66	1.28	0[d]	31	10.0	27	3.0	7
Molded[b]	—	20	3.0	0	0.38	<0[d]	57	30.0	57	—	8
Bahiagrass, 30% DM	2-00-464	54	8.1	0.57	1.08	8.9	68	7.0	38	2.1	10
Hay	1-00-462	51	9.0	0.48	0.99	8.2	72	8.0	41	1.6	11
Bakery waste	4-00-466	89	2.0	1.36	2.15	9.0	18	1.0	13	12.7	5
Barley grain											
Heavy	4-00-549	84	4.3	1.21	1.92	13.0	19	2.0	7	2.1	3
Light[a]	—	77	4.3	1.11	1.82	14.0	28	2.9	10	2.3	4
Malt sprouts w/ hulls	4-00-545	71	8.6	0.89	1.45	28.1	46	3.0	18	1.4	7
Straw	1-00-498	49	8.5	0.45	0.97	4.3	80	11.0	59	1.9	7
Beet pulp, dehy.	4-00-669	74	3.0	1.03	1.73	7.7[d]	54	2.0	33	0.6	5
+Steffen's filtrate	4-00-675	66	3.0	0.87	1.53	10.0[d]	42	2.0	21	0.4	6
Bermudagrass, coastal hay, L. veg.	1-20-716	55	10.0	0.56	1.03	10.0	70	6.0	32	2.5	8
Bloodmeal	5-00-380	66	10.0	0.82	1.31	91.7	—	—	—	7.0	7
Brewers' grains											
Dehydrated	5-02-141	67	10.7	0.77	1.28	26.0[d]	46	6.0	24	7.2	4
Wet, 21% DM	5-02-142	70	10.7	0.82	1.34	26.0[d]	42	4.0	23	6.5	10
Brome											
Hay, Pre-bl.	1-00-887	57	6.8	0.65	1.18	10.5	65	5.0	38	2.6	10
Hay, L. bl.	1-00-888	55	7.9	0.62	1.10	7.4	72	8.0	44	2.3	9
Hay, smooth M. bl.	1-05-663	56	7.1	0.63	1.16	14.6	66	4.0	37	2.6	10
Hay, mature	1-00-944	52	8.2	0.54	1.02	5.8	71	8.0	45	3.0	8
Canarygrass, reed, hay	1-01-104	55	7.0	0.61	1.13	10.3	64	4.0	36	3.1	10
Carrot roots, fresh, 12% DM	2-01-146	84	0	1.26	2.11	9.9	9	0	8	1.4	8
Citrus pulp, dehy.	4-01-237	82	2.3	1.20	1.99	7.0[d]	23	3.0	22	3.7	7
Clover, ladino, hay	1-01-378	65	0.8	0.85	1.58	22.0	36	7.0	29	2.7	10
Clover, red, hay	1-01-415	55	5.1	0.62	1.19	16.0	56	10.0	41	2.8	9
Coconut meal[a]	—	64	8.0	0.74	1.33	21.5	56	10.0	—	—	—
Coffee grounds	1-01-576	50	0	0.55	1.21	0[d]	74	—	65	26.0	2
Hulls (bran)	1-01-577	0	0	0	0	—	93	20.0	71	1.3	5
Corn											
Bran[b]	—	83	18.0	1.04	1.34	11.0	60	1.0	16	8.2	3
Cobs, ground	1-02-782	50	10.0	0.49	0.94	3.2	90	7.0	35	0	2
Distillers' grain, dehy.	5-28-235	86	14.0	1.07	1.53	26.0[d]	50	5.0	17	9.8	4
Wet[a]	—	90	14.0	1.17	1.65	26.0[d]	40	4.0	15	9.9	4
Grains + solubles	5-02-843	88	13.5	1.12	1.58	29.5[d]	44	4.0	16	10.3	5
Solubles, dehy.	5-28-844	88	7.1	1.22	1.86	29.7[d]	23	1.0	7	9.2	8
Fodder	1-02-775	64	6.9	0.78	1.35	8.9	55	3.0	33	2.4	7
Gluten feed	5-28-243	83	13.5	1.06	1.49	25.6	45	1.0	12	7.5	5
Gluten meal	5-02-900	84	11.2	1.10	1.61	46.8	37	1.0	9	2.4	3

Table A25.1.   (Continued)

Feed name	Int'l feed no.[a]	1M TDN (%)	Discount (%/M)	2M[b] NE$_g$[c]	3M[b] NE$_l$	Crude protein[d] (%)	NDF (%)	Lignin (%)	ADF (%)	EE (%)	Ash (%)
Gluten meal, 60% CP	5-28-242	89	4.2	1.32	2.05	67.2	14	1.0	5	2.4	2
Grain cracked	4-21-017	82	5.0	1.15	1.84	10.0	9	0.2	3	4.3	2
Grain, flaked	4-20-224	88	3.0	1.32	2.10	10.0	9	0.2	3	4.3	2
Grain, ground	4-21-018	87	3.3	1.29	2.04	10.0	9	0.2	3	4.3	2
Grain, high moisture	4-20-771	88	3.0	1.32	2.10	10.6	9	0.2	3	4.3	2
Hominy	4-02-887	91	6.0	1.53	1.92	11.5	23	—	6	7.3	1
Stover	1-02-776	59	9.5	0.66	1.15	6.6	67	11.0	39	1.3	7
Whole ear (corn and cob)	4-28-238	82	6.0	1.17	1.81	9.0	28	2.0	12	3.7	2
Corn silage[a]	3-02-912										
45% grain	North	72	4.0	0.98	1.66	9.0	41	3.0	26	3.2	5
+ NPN	North	73	3.5	1.00	1.71	13.0	41	3.0	26	3.5	5
+ NPN + Ca	North	74	3.0	1.04	1.76	13.0	41	3.0	26	3.5	7
35% grain	North	69	4.5	0.92	1.57	8.6	46	4.0	28	2.6	7
25% grain	North	68	5.3	0.87	1.50	8.3	52	5.0	31	2.1	8
40% grain	South	66	4.0	0.85	1.50	9.2	45	4.0	27	3.1	4
+ NPN	South	67	3.5	0.89	1.56	13.2	45	4.0	27	3.1	4
+ NPN + Ca	South	68	3.0	0.93	1.61	13.0	45	4.0	27	3.1	6
25% grain	South	61	4.5	0.75	1.36	8.3	55	6.0	33	2.1	7
Stalkage	3-28-251	55	6.0	0.62	1.17	6.3	68	7.0	45	2.1	9
Immature (25% DM, no ears)	3-28-252	65	6.0	0.81	1.40	9.0	60	3.0	38	3.1	11
Cottonseed											
Hulls	1-01-599	45	10.0	0.37	0.84	4.1	90	24.0	73	1.7	4
Black whole seed	5-01-614	95	4.0	1.44	2.20	23.0	40	15.0	31	22.9	5
High lint[a]	—	90	5.0	1.32	2.03	20.0	47	16.0	37	20.0	4
Seed meal, mech. ex.	5-01-617	78	4.5	1.09	1.97	44.0	28	6.0	20	5.0	7
Solv. 41% CP	5-01-621	75	4.5	1.03	1.82	45.6	26	6.0	19	1.3	7
Solv. 43% CP	5-01-630	75	4.5	0.98	1.82	48.9	28	7.0	21	1.7	7
Feather meal	5-03-795	55[b]	10.0	0.60	1.07	78.0[d]	—	—	—	3.2	4
Fescue											
Alta, hay	1-05-684	55	7.6	0.60	1.12	10.2	70	6.5	41	2.2	10
K31, hay	1-09-187	60	6.3	0.71	1.29	16.4	63	4.0	35	6.1	9
Hay F. bl.	1-09-188	58	6.9	0.66	1.21	12.1	67	5.0	39	5.3	8
Mature	1-09-189	56	7.5	0.62	1.16	9.2	70	7.0	42	4.3	7
Meadow, hay	1-01-912	59	6.6	0.69	1.24	9.1	65	7.0	43	2.4	8
Fish meal	5-01-614	68	10.0	0.83	1.35	66.6	—	—	—	5.1	25
Grape pomace	1-02-208	33	10.0	0.12	0.57	0[d]	55	35.0	54	7.9	10
Linseed meal											
Mech. ex.	5-02-045	82	3.6	1.19	1.86	37.9	25	7.0	17	6.0	6
Meal solv. ex.	5-02-048	78	3.6	1.10	1.76	38.3	25	6.0	19	1.5	6
Lupins[a]	—	78	—	—	—	34.2	33	3.3	—	5.5	5.1
Meat and bone meal	5-00-388	60	10.0	0.69	1.19	50.0	—	—	—	11.8	38.1
Millet feed	4-03-120	84	13.0	1.01	1.53	12.9	—	4.0	17	3.9	3
Milo (see Sorghum)											
Molasses, beet, 79% DM[b]	4-00-668	68	0	0.94	1.70	3.0	0	0	0	0	11
Molasses, cane, 73% DM[b]	4-04-696	60	0	0.77	1.48	8.0	0	0	0	0	8
Napiergrass, fresh											
30 days 20% DM	2-03-158	55	8.1	0.59	1.10	8.7	70	10.0	45	3.0	9
60 days 23% DM	2-03-162	53	8.3	0.54	1.05	7.8	75	14.0	47	1.0	6
Needle and thread	2-03-199	49	9.0	0.45	0.94	4.1	83	6.0	43	5.4	20
Oats, grain, ground											
38 lb/bushel	4-03-309	77	3.7	1.10	1.77	13.3	32	3.0	16	5.4	3
32 lb/bushel	4-03-318	73	4.0	0.96	1.64	13.1	42	4.0	21	4.9	5
Hay	1-03-280	61	6.5	0.72	1.29	9.3	66	6.0	36	2.6	8
Hulls	1-03-289	35	10.0	0.24	0.62	3.9	78	8.0	42	1.8	7
Silage dough	3-04-396	57	8.0	0.59	1.17	10.0	56	9.0	32	4.2	7
Straw	1-03-283	50	9.0	0.48	0.97	4.4	70	14.0	47	2.2	8
Orchardgrass											
Hay, E. Bl.	1-03-425	65	5.5	0.82	1.42	10.2	57	4.0	34	2.8	10
Hay, L. bl.	1-03-428	54	7.2	0.59	1.11	8.4	65	5.0	38	3.4	9
Palm kernel oil meal[a]	—	57	7.2	0.66	1.20	1.5	69	14.0	—	—	—
Pangola grass, fresh, 21% DM	2-03-493	55	7.6	0.59	1.12	10.3	70	8.0	38	2.3	10
Pea, vines, silage, 25% DM	3-03-596	57	4.0	0.66	1.27	13.1	59	9.0	49	3.3	12
Peanut, hulls	1-08-028	22	3.4	0.00	0.43	7.8	74	23.0	65	2.0	4
Meal	5-03-649	83	2.5	1.30	1.98	52.0	14	—	6	6.3	6

(continued)

Table A25.1. (*Continued*)

Feed name	Int'l feed no.[a]	1M TDN (%)	Discount (%/M)	2M[b] NE_g[c]	3M[b] NE_l	Crude protein[d] (%)	NDF (%)	Lignin (%)	ADF (%)	EE (%)	Ash (%)
Pineapple, bran	4-03-722	68	18.0	0.70	1.04	4.6	73	7.0	37	1.5	3
Potato tubers	4-03-787	81	9.0	1.23	1.82	8.9	6	1.0	2	0.4	6
Poultry manure	5-14-015	58	3.5	0.22	0.81	28.2	38	2.0	14	2.4	22
Prickly pear, fresh, 17% DM	2-01-061	57	2.1	0.67	1.26	4.8	6	1.0	2	—	6
Rapeseed meal[a]	—	60	5.0	0.84	1.33	42.3	29	3.7	—	1.2	8
Red top, fresh, 29% DM	2-03-897	63	6.1	0.80	1.40	11.6	64	8.0	40	3.9	8
Rice											
Bran	4-03-928	70	6.6	0.92	1.55	13.0	25	3.3	14	15.1	10
Grain, ground	4-03-938	79	3.0	1.14	1.86	8.9	16	2.1	—	1.9	5
Grain, polished	4-03-932	89	2.7	1.34	2.12	8.6	—	—	1	0.8	1
Hulls	1-08-075	11	—	0	0.07	3.0	82	17.0	72	0.8	20
Russian thistle	1-08-497	50	6.4	0.51	1.04	11.2	64	11.0	44	3.0	15
Rye grain	4-04-047	86	2.5	1.29	2.08	13.8	19	1.0	—	1.7	2
Ryegrass, hay	1-04-077	64	3.8	0.84	1.45	8.6	41	2.0	30	2.2	10
Safflower, meal											
Mech. ex.	5-04-109	60	3.2	0.76	1.32	22.1	59	14.0	41	6.7	4
Solv. ex.	5-04-110	57	3.2	0.70	1.29	25.4	58	14.0	41	1.5	4
Sagebrush, big	2-07-992	50	2.2	0.54	1.16	9.3	42	12.0	30	11.0	7
Sorghum, grain, dry (milo)	4-08-140	76	6.4	1.08	1.80	12.4	23	1.4	5	3.1	2
Grain rolled[a]	—	82	6.4	1.14	1.78	12.4	23	1.4	5	3.1	2
Grain, steam flaked[a]	—	86	6.4	1.22	1.87	12.0	23	1.4	5	3.1	2
Silage, 30% DM	3-04-323	60	9.0	0.69	1.19	7.5	64	6.0	38	3.0	9
Soybean, hulls	1-04-560	80	18.0	0.93	1.29	12.1	67	2.0	50	2.1	5
Meal, 44	5-04-637	81	5.1	1.13	1.78	49.0	14	0.3	10	1.5	7
Meal, 49	5-04-612	81	5.1	1.13	1.78	55.0	8	0.2	6	1.0	7
Straw	1-04-567	42	—	0	0.40	5.2	70	16.0	54	1.5	6
Whole	5-04-610	94	4.0	1.42	2.21	42.8	13	0.2	9	18.8	6
Sudangrass, hay	1-04-480	56	7.1	0.63	1.16	11.3	66	4.0	39	1.8	10
Silage, 28% DM	3-04-499	55	8.0	0.61	1.10	10.8	71	5.0	42	2.8	10
Sunflower, seed meal	5-04-740	60	7.0	0.70	1.27	25.9	40	12.0	33	1.2	6
Tapioca[a]	—	84	4.0	1.15	1.95	3.1	8	—	—	0.8	3
Timothy											
Hay, L. veg.	1-04-881	70	4.7	0.96	1.56	14.0	55	3.0	30	3.0	8
Hay, E. bl.	1-04-882	64	5.7	0.80	1.39	11.0	61	4.0	34	2.9	7
Hay, M. bl.	1-04-883	59	6.8	0.69	1.24	9.0	67	5.0	36	2.7	7
Hay, L. bl.	1-04-885	56	7.3	0.63	1.16	8.0	68	6.0	40	2.6	7
Hay, milk st.	1-04-886	52	8.2	0.53	1.03	7.0	71	7.0	43	2.3	6
Hay, seed st.	1-04-888	47	8.9	0.33	0.79	6.0	72	9.0	45	2.0	6
Silage, L. bl. 42% DM	3-04-920	57	5.9	0.66	1.22	9.0[d]	65	7.0	41	3.1	7
Tomato pomace, dried	5-05-041	58	3.8	0.72	1.32	16.0[d]	55	11.0	50	10.3	8
Trefoil, birdft., hay	10-05-044	59	7.8	0.76	1.37	16.3	47	9.0	36	2.5	7
Turnip, fresh, 9% DM	4-05-067	85	0.5	1.23	2.00	11.8	44	1.0	34	1.9	2
Urea[a]	—	0	0	0	0	292.0	0	0	0	0	0
Vetch, hay	1-05-106	57	2.9	0.69	1.32	20.8	48	8.0	33	3.0	7
Wheat											
Bran	4-05-190	78	9.0	1.02	1.58	17.1	51	3.0	15	4.4	7
Grain, hard red spring	4-05-268	84	3.0	1.24	1.98	11.3	16	1.0	5	2.0	2
Grain, soft white	4-05-337	85	3.0	1.26	2.00	11.3	14	0.6	4	1.9	2
Hay, E. veg.	2-05-176	73	4.3	1.00	1.67	12.0	52	4.0	30	4.4	8
Hay, headed	1-05-172	58	7.0	0.67	1.21	8.5	68	7.0	41	2.2	7
Middlings	4-05-205	83	7.0	1.20	1.90	18.4	37	2.2	11	3.2	2
Silage dough[a]	—	59	7.0	0.66	1.25	8.1	52	7.7	43	3.0	8
Straw	1-05-175	44	11.6	0.33	0.78	3.6	85	14.0	54	1.8	8
Wheatgrass crest, hay	1-05-351	53	8.0	0.55	1.06	9.0	65	6.0	36	2.3	9
Whey, acid	4-08-134	78	0	1.13	1.95	14.2	—	—	—	0.7	10
Delact.	4-01-186	71	0	0.99	1.77	17.9	—	—	—	1.1	17
Winterfat	1-08-559	53	7.2	0.20	0.68	10.8	72	10.0	44	2.8	15

Source: Van Soest et al., 1992b.

Note: Definitions of abbreviations, variables, and international feed numbers are given in National Research Council, 1984.

[a]North-South designations according to Rohweder et al., 1978. There are no relevant international feed numbers for the north designation (and some other feedstuffs as well).

[b]All values are on a dry matter basis except as noted. Net energy (NE) values are in units of Mcal/kg. For conversion to air-dry basis use factor 0.90 or, preferably, a separately determined value for dry matter. Added values incorporated from Tamminga et al., 1990; van Straalen and Tamminga, 1990; Weisbjerg et al., 1991.

[c]Net energy of maintenance is estimated by adding 0.67 Mcal to tabular NE_g values. Note that this value is discounted to 2M. It may be adjusted upward for lower intakes or downward for higher intakes. Some NE_g values are somewhat lower than NRC values but balance a somewhat higher NE_m to give realistic production values. This was done to facilitate linear modeling of net energy values for beef cattle.

[d]Protein values need to be adjusted for protein bound in acid-detergent fiber (ADF). Obtain a laboratory analysis or use values from this table.

Table A25.2. Nitrogen and protein fractions in feedstuffs as percentage of total crude protein nitrogen

	Soluble nitrogen[a]	NPN[a]	NDIN[a]	ADIN[a]	B$_1$[a,b]	B$_2$[a,b]	B$_3$[a,b]	DN[a,c]
Alfalfa								
Meal, dehy.	28	28	25	17	0	47	8	52
Hay, L. veg.	30	28	15	10	2	55	5	74
Hay, E. bl.	29	27	18	11	2	53	7	70
Hay, M. bl.	28	26	25	14	2	47	11	66
Hay, F. bl.	27	25	29	16	2	44	13	63
Hay, L. bl.	26	24	33	18	2	41	15	59
Hay, mature	25	23	36	20	2	39	16	56
Hay, weathered	15	15	45	25	0	40	20	48
Wilt, silage								
E. bl., 35% DM	50	50	27	15	0	23	12	70
M. bl., 38% DM	45	45	32	18	0	23	14	56
F. bl., 40% DM	40	40	37	21	0	23	16	42
Bakery waste	40	30	6	3	10	54	3	74
Barley grain	17	5	8	5	12	75	3	66
Malt sprouts	48	40	27	4	8	25	23	62
Beet pulp, dehy.	27	26	53	11	1	20	42	43
Bloodmeal[d]	5	0	47[e]	1[e]	5	48	46	29
Brewers' grains								
Dehydrated	4	3	40	12	1	56	28	33
Wet, 21% DM	8	4	38	10	4	54	28	51
Coconut meal[f]	14	—	—	3	—	—	—	42
Corn								
Distillers' grain, dehy.								
Light	6	4	40	13	2	54	27	33
Intermediate	6	4	44	18	2	50	26	29
Dark	6	4	45	21	2	49	24	28
Very dark	6	4	47	36	2	47	11	27
Distillers' dehy. + solubles	19	17	62	21	2	19	41	53
Wet + solubles	25	17	55	12	8	20	43	55
Dried distillers' solubles	44	44	55	13	0	1	42	43
Gluten feed	49	49	8	2	0	43	6	74
Cracked grain	11	8	15	5	3	74	10	38
Gluten meal	4	3	11	2	1	85	9	45
Grain, flaked	8	6	15	5	2	77	10	48
Grain, ground	11	8	15	5	3	74	10	48
Grain, high moisture	40	40	15	5	0	44	9	57
Hominy	18	14	8	5	4	74	3	48
Whole ear (corn and cob)	16	11	18	3	5	66	15	48
High moisture	30	24	19	8	6	51	11	57
Corn silage								
45% grain	45	45	16	8	0	39	8	65
+NPN	63	63	15	5	0	22	10	74
35% grain	50	50	16	9	0	34	7	67
25% grain	55	55	16	9	0	29	7	70
Cottonseed								
Hull	20	10	13	13	10	67	0	61
Black, whole seed	40	1	6	6	39	54	0	73
Seed meal	20	8	10	8	12	70	2	57
Feathermeal[d]	9	8	50[e]	32[e]	1	41	18	31
Fish meal	12	0	1[e]	1[e]	12	87	0	22
Grass hay								
Cool season (C$_3$)[g]	50	50	30	15	0	20	15	80
Warm season (C$_4$)[g]	50	50	40	11	0	10	29	80
Linseed meal	20	10	10	2	10	70	8	65
Lupins[f]	26	—	—	—	—	—	—	83
Meat and bone meal	14	3	56	3	11	30	53	30
Molasses, beet, 70% DM	100	100	0	0	0	0	0	100
Molasses, cane, 73% DM	100	100	0	0	0	0	0	100
Oats, grain, ground	53	10	11	5	43	36	6	83
Hay	30	28	30	10	2	40	20	63
Silage, dough	50	50	30	10	0	20	20	77
Palm kernel meal[f]	9	—	—	7	—	—	—	46
Potato[f]	15	—	4	1	—	—	3	58
Peanut meal	33	9	10	1	24	57	9	80
Rapeseed meal	32	21	11	6	11	57	5	67
Rye, grain	53	10	7	4	43	40	3	81

(continued)

Table A25.2. (*Continued*)

	Soluble nitrogen[a]	NPN[a]	NDIN[a]	ADIN[a]	B_1[a,b]	B_2[a,b]	B_3[a,b]	DN[a,c]
Sorghum								
Grain, dry (milo)	12	4	10	5	8	78	5	47
Grain, high moist.	30	30	10	5	—	60	5	73
Silage, 30% DM (grain type)	45	45	50	5	0	5	45	71
Soybean								
Soy, hulls	18	13	20	14	5	62	6	58
Meal	20	11	5	2	9	75	3	65
Whole	44	10	4	3	34	52	1	74
Whole roast	6	6	24	7	0	71	16	30
Extruded	14	12	15	6	2	79	9	40
Sunflower, seed meal	30	11	8	5	19	62	3	—
Tapioca[f]	25	—	—	5	—	—	—	85
Timothy (*see* Grass)								
Tomato pomace, dried	—	—	—	20	—	—	—	—
Urea	100	100	0	0	0	0	0	100
Wheat								
Bran[f]	25	—	—	6	—	—	—	11
Grain, soft, white	30	22	4	2	8	66	2	73
Middlings	40	30	4	3	10	56	1	75
Silage (dough)	45	45	27	8	0	28	19	80
Whey, acid	100	—	0	0	—	—	—	100
Delact.	100	—	0	0	—	—	—	100

Source: Van Soest et al., 1992b.

Note: Definitions of abbreviations, variables, and international feed numbers are given in National Research Council, 1984.

[a]Percentage of total crude protein nitrogen: multiply these values for feed of nearest description by crude protein value in this table. Nitrogen fractions within forage species and described feeds do not vary greatly.

[b]The sum of B_1, B_2, and B_3 represents total available true protein (see Table 18.4 for definitions).

[c]Degradable nitrogen as percentage of total crude protein. Nitrogen is the sum of soluble nitrogen plus estimated degradable B_2 protein degrading in the rumen at three times maintenance. These values should be adjusted using discounts in Table A25.1 for level of intake. See Table 25.10 for an example of the calculation.

[d]Digestibility value based on Weisbjerg et al., 1991.

[e]NDF and ADF analyses were designed to be used on plant materials. Although there is no true NDF or ADF in animal products, an analytical value can be obtained that likely represents slower digesting protein fractions. Sulfites cannot be used in generating NDIN for the Cornell Net Carbohydrate and Protein Model (see Section 10.3.1).

[f]Data from Tamminga et al., 1990; van Straalen and Tamminga, 1990.

[g]Cool-season C_3 grasses include ryegrasses, orchardgrass, fescue, timothy, reed canarygrass; warm-season C_4 grasses include most tropical grasses, millets, bermudagrass, and sudangrass. Use these values for respective species in Table A25.1.

[h]Silages are quite variable, and analyses for nitrogen fractions are recommended.

References

Abdo, K.M., K.W. King, and R.W. Ensel. 1964. Protein quality of rumen microorganisms. J. Anim. Sci. 23:734.

Abou Akkada, A.R., D.A. Messmer, L.R. Fina, and E.E. Bartley. 1968. Distribution of 2-aminoethylphosphonic acid in some rumen microorganisms. J. Dairy Sci. 51:78.

Abrams, S.M., J.S. Shenk, M.O. Westerhaus, and F.E. Barton II. 1987. Determination of forage quality by near infrared reflectance spectroscopy: efficacy of broad-based calibration equations. J. Dairy Sci. 70:806.

Adams, R.S., J.H. Moore, E.M. Kesler, and G.Z. Stevens. 1964. New relationships for estimating TDN content of forages from chemical composition. J. Dairy Sci. 47:1461.

Aerts, J.V., J.L. de Boever, B.G. Cottyn, D.L. de Brabander, and F.X. Buysse. 1984. Comparative digestibility of feedstuffs by sheep and cows. Anim. Feed Sci. Technol. 12:47.

Aerts, J.V., D.L. de Brabander, B.G. Cottyn, F.X. Buysse, L.A. Carlier, and R.J. Moermans. 1978. Some remarks on the analytical procedure of Van Soest for the prediction of forage digestibility. Anim. Feed Sci. Technol. 3:309–322.

Agricultural and Food Research Council (AFRC), Technical Committee on Responses to Nutrients. 1991. Voluntary Intake of Cattle. Report no. 8. Nutr. Abstr. Rev. (Series B) 61:815–823.

Agriculture Research Council. 1980. The Nutrient Requirements of Ruminant Livestocks. Commonwealth Agricultural Bureau, Farnham Royal Slough.

Aiello, R.J., T.M. Kenna, and J.H. Herbein. 1984. Hepatic gluconeogenic and ketogenic interrelationships in the lactating cow. J. Dairy Sci. 67:1707.

Ainslie, S.J. 1991. Management systems for Holstein steers to utilize alfalfa, silage and improve carcass value. M.S. thesis, Cornell Univ., Ithaca, N.Y.

Akin, D.E. 1979. Comparison of bermudagrass and orchardgrass for differences in the mode of microbial attack of tissue types and with cell wall digestion. Proc. 15th Conf. on Rumen Function, Chicago, Nov. 28–29. Pp. 1–2.

Akin, D.E. 1989. Histological and physical factors affecting digestibility of forages. Agron. J. 81:17–25.

Akin, D.E., R.D. Hartley, W.H. Morrison III, and D.S. Himmelsbach. 1990. Diazonium compounds localize grass cell wall phenolics: relation to wall digestibility. Crop Sci. 30:985–989.

Albrecht, K.A., and R.E. Muck. 1991. Proteolysis in ensiled forage legumes that vary in tannin concentration. Crop Sci. 31:464–469.

Alexander, B.W., A.H. Gordon, J.A. Lomax, and A. Chesson. 1987. Composition and rumen degradability of straw from three varieties of oilseed rape before and after alkali, hydrothermal and oxidative treatment. J. Sci. Food Agric. 41:1–15.

Allen, M.S. 1982. Investigation into the use of rare earth elements as gastrointestinal markers. M.S. thesis, Cornell Univ., Ithaca, N.Y. P. 107.

Allen, M.S. 1992. Effect of plant population and maturity at harvest on corn forage quality. J. Anim. Sci. 70(Suppl. 1):191A.

Allen, M.S., and D.R. Mertens. 1988. Evaluating constraints on fiber digestion by rumen microbes. J. Nutr. 118:261–270.

Allen, M.S., D.R. Mertens, and C.J. Sniffen. 1981. A mathematical model of lag phase during ruminal fermentation. Abstr. Northeast American Dairy Science Assoc., American Society of Animal Science Meet. Univ. of Vermont, Burlington.

Allen, M.S., and K.A. O'Neil. 1991. Effect of environment on fiber components and fiber digestibility of corn forage. Proc. 21st Bienn. Conf. on Rumen Function, Chicago. P. 10.

Allen, M.S., J.B. Robertson, and P.J. Van Soest. 1984. A comparison of particle size methodologies and statistical treatments. Proc. Satellite Conf. of 6th Int. Symp. on Ruminant Physiology. P.M. Kennedy, ed. Can. Soc. Anim. Sci. Occas. Publ. no. 1. Edmonton, Alberta. Pp. 39–56.

Allison, M.J. 1978. The role of ruminal microbes in the metabolism of toxic constituents from plants. In: Effects of Poisonous Plants on Livestock. R.F. Keeler, K.R. Van Kampen, and L.F. James, eds. Academic Press, New York.

Allison, M.J., I.M. Robinson, R.W. Dougherty, and J.A. Bucklin. 1975. Grain overload in cattle and sheep; changes in microbial populations in the cecum and rumen. Am. J. Vet. Res. 36:181–185.

Allo, A.A., J.H. Oh, W.M. Longhurst, and G.E. Connelly. 1973. VFA production in the digestive systems of deer and sheep. J. Wildl. Manage. 37:202–211.

Altschul, A.M., ed. 1958. Processed Plant Protein Foodstuffs. Academic Press, New York.

Åman, P. 1993. Composition and structure of cell wall polysaccharides. In: Forage Cell Wall Structure and Digestibility. H.G. Jung, D.R. Buxton, R.D. Hatfield, and J. Ralph, eds. American Society of Agronomy, Madison, Wisc. Pp. 183–199.

American Society of Agricultural Engineers. 1961. Method of determining modulus of uniformity and modulus of fineness of ground feed. Agric. Eng. Yearbk. P. 126.

Ammerman, C.B., and R.D. Goodrich. 1983. Advances in mineral nutrition in ruminants. J. Anim. Sci. 57(Suppl. 2):519.

Andersen, R.A., and J.R. Todd. 1968. Estimation of total tobacco plant phenols by their bonding to polyvinylpyrrolidone. Tob. Sci. 12:107–111.

Anderson, J.W. 1985. Physiological and metabolic effects of dietary fiber. Fed. Proc. 44(14):2902–2905.

Anderson, R. 1985. Effect of prolonged wilting in poor conditions on the fermentation quality, metabolizability and net energy value of silage given to sheep. Anim. Feed Sci. Technol. 12:109.

Annison, E.F., and D.G. Armstrong. 1970. Volatile fatty acid metabolism and energy supply. In: Physiology of Digestion and Metabolism in the Ruminant. A.T. Phillipson, ed. Oriel Press, Newcastle upon Tyne, England. Pp. 422–437.

Anonymous. 1963. Protein foundation. N. Engl. J. Med. 269:1254.

Antoniou, T., and M. Hadjipanayiotou. 1985. The digestibility by sheep and goats of five roughages offered alone or with concentrates. J. Agric. Sci. 195:663–671.

Argenzio, R.A., J.E. Lowe, D.W. Pickard, and C.E. Stevens. 1974. Digesta passage and water exchange in the equine large intestine. Am. J. Physiol. 226:1035–1042.

Argenzio, R.A., and M. Southworth. 1974. Sites of organic acid production and absorption in gastrointestinal tract of the pig. Am. J. Physiol. 228:454.

Argenzio, R.A., and C.E. Stevens. 1984. The large bowel—a supplementary rumen? Proc. Nutr. Soc. 43:13–23.

Arman, P., and C.R. Field. 1973. Digestion in the hippopotamus. East Afr. Wildl. J. 11:9–17.

Arman, P., and D. Hopcraft. 1975. Nutritional studies on East African herbivores. I. Digestibilities of dry matter, crude fibre and crude protein in antelope, cattle and sheep. Br. J. Nutr. 33:255–264.

Armentano, L.E., and J.W. Young. 1983. Production and metabolism of volatile fatty acids, glucose and CO_2 in steers and the effects of monensin on volatile fatty acid kinetics. J. Nutr. 113:1265–1277.

Armstrong, D.G. 1972. Developments in cereal processing—ruminants. In: Cereal Processing and Digestion. Tech. Publ. U.S. Feed Grains Council, London Office. Pp. 9–37.

Armstrong, D.G. 1973. Factors affecting the protein value of fresh and conserved feeds. Proc. European Grasslands Fed., Uppsala.

Armstrong, D.G., K.L. Blaxter, and N.McC. Graham. 1957. The heat increments of mixtures of steam-volatile fatty acids in fasting sheep. Br. J. Nutr. 11:392–408.

Arnold, G.W. 1970. Regulation of food intake in grazing ruminants. In: Physiology of Digestion and Metabolism in the Ruminant. A.T. Phillipson, ed. Oriel Press, Newcastle upon Tyne, England. Pp. 264–276.

Arnold, G.W. 1985. Ingestive behavior. In: Ethology of Farm Animals. A.F. Fraser, ed. Elsevier, Amsterdam. P. 186.

Arnold, G.W., and M.L. Dudzinski. 1978. Ethology of the Free-ranging Animal. Elsevier, Amsterdam.

Arroyo-Aguilu, J.A., and J.L. Evans. 1972. Nutrient digestibility of low-fiber rations in the ruminant animal. J. Dairy Sci. 55:1266–1274.

Arroyo-Aguilu, J.A., S. Tessema, R.E. McDowell, P.J. Van Soest, A. Ramirez, and P.F. Randel. 1975. Chemical composition and in vitro digestibility of five heavily fertilized tropical grasses in Puerto Rico. J. Agric. Univ. P. R. 59:186–198.

Ash, R.W., and A. Dobson. 1963. The effect of absorption of the acidity of rumen contents. J. Physiol. 169:39–61.

Ashbell, G., and N. Lisker. 1988. Aerobic deterioration in maize silage stored in a bunker silo under farm conditions in a subtropical climate. J. Sci. Food Agric. 45:307–316.

Aspinall, G.O. 1965. Some recent developments in the chemistry of arabinogalactans. Int. Symp. Chemistry and Biochemistry of Lignin, Cellulose and Hemicellulose, Grenoble. Impr. Reunies de Chambery, Chambery.

Asquith, T.N., and L.G. Butler. 1985. Use of dye-labeled protein as spectrophotometric assay for protein precipitants such as tannin. J. Chem. Ecol. 11:1535–1544.

Aufrère, J. 1982. Etude de la prévision de la digestibilité des fourrages par une méthode enzymatique. Ann. Zootech. 31:111–130.

Austin, P.J., L.A. Suchar, C.T. Robbins, and A.E. Hagerman. 1989. Tannin-binding proteins in saliva of deer and their absence in saliva of sheep and cattle. J. Chem. Ecol. 15(4):1335–1347.

Bacon, C.W. 1988. Procedure for isolating the endophyte from tall fescue and screening isolates for ergot alkaloids. Appl. Environ. Microbiol. 54:2615–2618.

Bacon, C.W., P.C. Lyons, J.K. Porter, and J.D. Robbins. 1986. Ergot toxicity from endophyte-infected grasses: a review. Agron. J. 78:106–116.

Bacon, C.W., and M.R. Siegel. 1988. The endophyte of tall fescue. J. Prod. Agric. 1:45–55.

Bae, D.H. 1978. Study of efficiency of mastication and rumination in relation to amount of hay intake and body size in cattle. Ph.D. dissertation, Univ. of Vermont, Burlington.

Bae, D.H., B.E. Gilman, J.G. Welch, and R.H. Palmer. 1984. Quality of forage from Miscanthus sinensis. J. Dairy Sci. 67:630.

Bae, D.H., J.G. Welch, and A.M. Smith. 1979. Forage intake and rumination by sheep. J. Anim. Sci. 49:1292.

Baile, C.A., and F.C. Buonomo. 1987. Growth hormone-releasing factor effects on pituitary function, growth, and lactation. J. Dairy Sci. 70:467–473.

Baile, C.A., and M.A. Della-Fera. 1984. Peptidergic control of food intake in food-producing animals. Fed Proc. 43:2898–2902.

Bailey, C.B. 1958. The rate of secretion of mixed saliva in the cow. Proc. Nutr. Soc. 18:13.

Bailey, C.B. 1981. Silica metabolism and silica urolithiasis in ruminants: a review. Can J. Anim. Sci. 61:219–235.

Bailey, R.W. 1964. Oligosaccharides. Pergamon, Oxford. P. 179.

Bailey, R.W. 1973. Structural carbohydrates. In: Chemistry and Biochemistry of Herbage, vol. 1. G.W. Butler and R.W. Bailey, eds. Academic Press, London. Pp. 157–206.

Bailey, R.W., A. Chesson, and J.A. Munro. 1978. Plant cell wall fractionation and structural analysis. Am. J. Clin. Nutr. 31:577–581.

Bailey, R.W., and M. Ulyatt. 1970. Pasture quality and ruminant nutrition. II. Carbohydrate and lignin composition of detergent-extracted residues from pasture grasses and legumes. N. Engl. J. Agric. Res. 13:591–604.

Baird, D.G. 1982. Primary ketosis in the high producing dairy cow. Clinical and subclinical disorders, treatment, prevention, and outlook. J. Dairy Sci. 65:1.

Baker, G., L.H.P. Jones, and I.D. Wardrop. 1961. Oral phytoliths and mineral particles in the rumen of the sheep. Aust. J. Agric. Res. 12:462–472.

Bakker, R. 1986. The Dinosaur Heresies. Longman Scientific & Technical, Harlow Essex, England. P. 481.

Balasta, M.L.F.C., C.M. Perez, B.O. Juliano, D.B. Roxas, and C.P. Villareal. 1987. Effect of silica level on some properties of rice plants and grain. Presented at the 7th Workshop of the Australian Asian Fibrous Agricultural Residues Research Network, Philippines, April 1–3.

Balch, C.C. 1950. Factors affecting the utilization of food by dairy cows. 1. The rate of passage of food through the digestive tract. Br. J. Nutr. 4:361–388.

Balch, C.C. 1971. Proposal to use time spent chewing as an index of the extent to which diets for ruminants possess the physical property of fibrousness characteristic of roughages. Br. J. Nutr. 26:383–392.

Balch, C.C., P.A. Balch, S. Bartlett, M.P. Bertrum, V.W. Johnson, S.J. Rowland, and J. Turner. 1955. Studies of the secretion of milk of low fat content by cows on diets low in hay and high in concentrates. VI. The effect on the physical and biochemical processes of the reticulorumen. J. Dairy Res. 22:270.

Balch, C.C., and R.C. Campling. 1965. Rate of passage of digesta through the ruminant digestive tract. In: Physiology of Digestion in the Ruminant. R.W. Dougherty, ed. Butterworths, Washington, D.C. Pp. 108–123.

Baldwin, R.L. 1970. Energy metabolism in anaerobes. Am. J. Clin. Nutr. 23:1508–1518.

Baldwin, R.L., J. France, D.E. Beever, M. Gill, and J.H.M. Thornley. 1987. Metabolism of the lactating cow. Parts I, II, and III. J. Dairy Res. 54:77–145.

Ballard, F.J., R.W. Hanson, and D.S. Kronfeld. 1969. Gluconeogenesis and lipogenesis in tissue from ruminant and nonruminant animals. Fed. Proc. 28:218.

Banta, C.C., R.G. Warner, and J.B. Robertson. 1975. Protein nutrition of the golden hamster. J. Nutr. 105:38–45.

Barnes, R.F., and D.L. Gustine. 1973. Allelochemistry and forage crops. In: Anti-quality components of forages. Crop Science Society of America Meet., Miami Beach, Fla. Crop Science Society of America, Madison, Wisc., Pp. 1–13.

Barnes, R.F., and T.H. Taylor. 1985. Grassland agriculture and ecosystem concepts. In: Forages: The Science of Grassland Agriculture. 4th ed. M.E. Heath, R.F. Barnes, and D.S. Metcalfe, eds. Iowa State Univ. Press, Ames. Pp. 12–20.

Barry, T.N., and R.N. Andrews. 1973. Content and retention of sulphur in wool as affected by formaldehyde treatment of the diet, level of energy intake, and intraperitoneal supplementation with DL-methionine. N. Z. J. Agric. Res. 16:545–550.

Barry, T.N., P.F. Fennessy, and S.J. Duncan. 1973. Effect of formaldehyde treatment on the chemical composition and nutritive values of silage. N. Z. J. Agric. Res. 16:64–68.

Barry, T.N., and T.R. Manley. 1986. Interrelationships between the concentrations of total condensed tannin, free condensed tannin and lignin in Lotus sp. and their possible consequences in ruminant nutrition. J. Sci. Food Agric. 37:248–254.

Barry, T.N., T.R. Manley, and S.J. Duncan. 1986. The role of condensed tannins in the nutritional value of Lotus pedunculatus for sheep. 4. Sites of carbohydrate and protein digestion as influenced by dietary reactive tannin concentration. Br. J. Nutr. 55:123.

Bartiaux-Thill, N., and R. Oger. 1986. The indirect estimation of the digestibility in cattle of herbage from Belgian permanent pasture. Grass Forage Sci. 41:273–276.

Bartley, E.E., R.M. Meyer, and L.R. Fina. 1975. Feedlot or grain bloat. In: Digestion and Metabolism in the Ruminant. I.W. McDonald and A.C.I. Warner, eds. Univ. of New England Publ. Unit, Armidale, N.S.W., Australia. Pp. 551–562.

Barton, F.E. II, H.E. Amos, W.J. Albrecht, and D. Burdick. 1974. Treating peanut hulls to improve digestibility for ruminants. J. Anim. Sci. 38:860.

Bate-Smith, E.C. 1973. Haemanalysis of tannins: the concept of relative astringency. Phytochemistry 12:907–912.

Bate-Smith, E.C. 1977. Astringent tannins of Acer species. Phytochemistry 16:1421–1426.

Bath, I.H., and K.J. Hill. 1969. The lipolysis and hydrogenation of lipids in the digestive tract of the sheep. J. Agric. Sci. 68:139.

Bauchart, D., F. Legay-Carmier, M. Doreau, and B. Gaillard. 1990. Lipid metabolism of liquid-associated and solid-adherent bacteria in rumen contents of dairy cows offered lipid-supplemented diets. Br. J. Nutr. 63:563–578.

Bauchop, T., and R.W. Martucci. 1968. Ruminant-like digestion of the langur monkey. Science 161:698–699.

Bauman, D.E. 1974. Fat metabolism in ruminants. Proc. Cornell Nutrition Conf., Ithaca, N.Y. Pp. 69–73.

Bauman, D.E. 1979. Partitioning of nutrients in the high-producing dairy cow. Proc. Cornell Nutrition Conf., Ithaca, N.Y. Pp. 12–18.

Bauman, D.E. 1987. Bovine somatotropin in lactation: impact on management and feeding. Proc. Cornell Nutrition Conf., Ithaca, N.Y. P. 23.

Bauman, D.E., and W.B. Currie. 1980. Partitioning of nutrients during pregnancy and lactation: a review of mechanisms involving homeostasis and homeorhesis. J. Dairy Sci. 63:1514–1529.

Bauman, D.E., and C.L. Davis. 1975. Regulation of lipid metabolism. In: Digestion and Metabolism in the Ruminant. I.W. McDonald and A.C.I. Warner, eds. Univ. of New England Publ. Unit, Armidale, N.S.W., Australia. P. 496.

Bauman, D.E., C.L. Davis, and H.F. Buchholtz. 1971. Propionate production in the rumen of cows fed either a control or high-grain, low-fiber diet. J. Dairy Sci. 54:1282–1287.

Bauman, D.E., and S.N. McCutcheon. 1986. The effects of growth hormone and prolactin on metabolism. In: Control of Digestion and Metabolism in Ruminants. L.P. Milligan, W.L. Grovum, and A. Dobson, eds. Prentice-Hall, Englewood Cliffs, N.J. Pp. 436–455.

Baumgardt, B.R. 1970. Voluntary feed intake by ruminants: models and practical applications. Proc. Cornell Nutrition Conf. Ithaca, N.Y. P. 85.

Baxter, H.D., B.L. Bledsoe, M.J. Montgomery, and J.R. Owen. 1986. Comparison of alfalfa-orchardgrass hay stored in large round bales and conventional rectangular bales for lactating cows. J. Dairy Sci. 69:1854.

Beever, D.E., J.F. Coelho da Silva, H.H.D. Prescott, and D.G. Armstrong. 1972. The effect in sheep of physical form and stage of growth on the sites of digestion of a dried grass. Br. J. Nutr. 28:347–356.

Beever, D.E., and R.C. Siddons. 1986. Digestion and metabolism in the grazing ruminant. In: Control of Digestion and Metabolism of Ruminants. L.P. Milligan, W.L. Grovum, and A. Dobson, eds. Prentice-Hall, Englewood Cliffs, N.J. Pp. 479–497.

Beever, D.E., D.J. Thompson, and S.B. Cammell. 1976.

The digestion of frozen and dried grass by sheep. J. Agric. Sci. 86:443–452.

Beever, D.E., D.J. Thomson, S.B. Cammell, and D.G. Harrison. 1977. The digestion by sheep of silages made with and without the addition of formaldehyde. J. Agric. Sci. 88:61.

Beever, D.E., D.J. Thompson, and D.G. Harrison. 1974. Energy and protein transformation in the rumen and the absorption of nutrients by sheep on forage diets. Proc. 12th Grasslands Congr., Moscow. Vol. 3, pt. 1, pp. 56–62.

Ben-Ghedalia, D., G. Shefet, and Y. Drori. 1983. Chemical treatments for increasing the digestibility of cotton straw. I. Effect of ozone and sodium hydroxide treatments on rumen metabolism and on the digestibility of cell walls and organic matter. J. Agric. Sci. 100:393–400.

Bergen, W.G., E.H. Cash, and H.E. Henderson. 1974. Changes in nitrogenous compounds of the whole corn plant during ensiling and subsequent effects on dry matter intake by sheep. J. Anim. Sci. 39:629–637.

Bergen, W.G., D.B. Purser, and J.H. Cline. 1967. Enzymatic determination of the protein quality of individual rumen bacteria. J. Nutr. 92:357–364.

Bergen, W.G., D.B. Purser, and J.H. Cline. 1968. Effect of ration on the nutritive quality of rumen microbial protein. J. Anim. Sci. 27:1497.

Bergen, W.G., and M.T. Yokoyama. 1977. Productive limits to rumen fermentation. J. Anim. Sci. 46:573–584.

Berger, L., T. Klopfenstein, and R. Britton. 1979. Effect of sodium hydroxide on efficiency of rumen digestion. J. Anim. Sci. 49:1317–1323.

Bergman, E.N. 1975. Production and utilization of metabolites by the alimentary tract as measured in portal and hepatic blood. Proc. 4th Int. Symp. on Ruminant Physiology, Sydney, Australia. Pp. 292–305.

Bergman, E.N. 1990. Energy contributions of volatile fatty acids from the gastrointestinal tract in various species. Physiol. Rev. 70:567–590.

Bergman, E.N., and J.M. Pell. 1985. Integration of amino acid metabolism in the ruminant. In: Herbivore Nutrition in the Subtropics and Tropics. F.M.C. Gilchrist and R.J. Mackie, eds. The Science Press, Johannesburg, S. Afr. Pp. 613–628.

Bernal-Santos, G. 1989. Dynamics of rumen turnover in cows at various stages of lactation. Ph.D. dissertation, Cornell Univ., Ithaca, N.Y. P. 184.

Berry, C.S. 1986. Resistant starch: formation and measurement of starch that survives exhaustive digestion with amylolytic enzymes during the determination of dietary fibre. J. Cereal Sci. 4:301–314.

Bertone, A.L., P.J. Van Soest, D. Johnson, S.L. Ralston, and T.S. Stashak. 1989. Large intestinal capacity, retention times, and turnover rates of particulate ingesta associated with extensive large-colon resection in horses. Am. J. Vet. Res. 50:1621–1627.

Bertrand, R.P., and C. Demarquilly. 1986. Prediction of forage digestibility by principal component analysis of near infrared reflectance spectra. Anim. Feed Sci. Technol. 16:215–224.

Bickerstaffe, R., D.E. Noakes, and E.F. Annison. 1972. Quantitative aspects of fatty acid biohydrogenation, absorption and transfer into milk fat in the lactating goat, with special reference to the *cis*- and *trans*-isomers of octadecenoate and linoleate. Biochem. J. 130:607–617.

Bines, J.A., S. Suzuki, and C.C. Balch. 1969. The quantitative significance of long-term regulation of food intake in the cow. Br. J. Nutr. 23:695–704.

Bird, S.H., and R.A. Leng. 1978. The effects of defaunation of the rumen on the growth of cattle on low-protein high energy diets. Br. J. Nutr. 40:163–167.

Bittner, A.S., and J.C. Street. 1983. Monosaccharide composition of alcohol- and detergent-insoluble residues in maturing reed canarygrass leaves. J. Agric. Food Chem. 31:7–10.

Bjornhag, G. 1972. Separation and delay of contents in the rabbit colon. Swed. J. Agric. Res. 2:125–136.

Black, J.L., M. Gill, D.E. Beever, J.H.M. Thornley, and J.D. Oldham. 1987. Simulation of the metabolism of absorbed energy-yielding nutrients in young sheep: efficiency of utilization of acetate. J. Nutr. 117:105–115.

Black, J.L., and P.A. Kenney. 1984. Factors affecting diet selection by sheep. II. Height and density of pasture. Aust. J. Agric. Res. 35:565–578.

Blake, R.W., and A.A. Custodio. 1984. Feed efficiency: a composite trait of dairy cattle. J. Dairy Sci. 67:2075.

Blaxter, K.L. 1962. The Energy Metabolism of Ruminants. Hutchinson, London. P. 329.

Blaxter, K.L. 1986. Bioenergetics and growth: the whole and the parts. J. Anim. Sci. 63(Suppl. 2):1–10.

Block, E., L.D. Muller, and L.H. Kilmer. 1982. Brown midrib-3 versus normal corn plants (*Zea mays* L.) harvested as whole plant or stover and frozen fresh or preserved as silage for sheep. Can. J. Anim. Sci. 62:487.

Bodmer, R.E. 1990. Ungulate frugivores and the browser-grazer continuum. Oikos 57:319–325.

Bohn, P.J., and S.L. Fales. 1989. Cinnamic acid–carbohydrate esters: an evaluation of a model system. J. Sci. Food Agric. 48:1–7.

Bondi, A. 1958. Plant proteins. In: Processed Plant Protein Foodstuffs. A.M. Altschul, ed. Academic Press, New York. Pp. 43–63.

Bonner, J.M., and J.E. Varner. eds. 1965. The path of carbon in respiratory metabolism. In: Plant Biochemistry, J.M. Bonner and J.E. Varner, eds. Academic Press, New York.

Borneman, W.S., R.D. Hartley, W.H. Morrison, D.E. Akin, and L.G. Ljungdahl. 1990. Feruloyl and p-coumaroyl esterase from anaerobic fungi in relation to plant cell wall degradation. Appl. Microbiol. Biotechnol. 33:345–351.

Borneman-Storenkevitch, I.D., S.A. Borovick, and I.B. Borovsky. 1941. Rare earths in plants and soils. C.R. Acad. Sci. USSR 30:229.

Boussingault, J.B. 1845. Rural Economy in Its Relations with Chemistry, Physics and Meteorology (trans. of the original French edition of 1844 by G. Law). Bailliere, London.

Bowen, H.J.M. 1979. Environmental Chemistry of the Elements. Academic Press, New York.

Bowman, J.G.P. 1990. Strategies for overcoming endophyte-infected fescue. Ohio Beef Cattle Res. & Ind. Rept., Dept. Anim. Sci., Ohio State Univ. Pp. 155–159.

Brady, C.J. 1960. Redistribution of nitrogen in grass and leguminous fodder plants during wilting and ensilage. J. Sci. Food Agric. 11:276–284.

Bransby, D.I., B.E. Conrad, H.M. Dicks, and J.W. Drane. 1988. Justification for grazing intensity experiments: analyzing and interpreting grazing data. J. Range Manage. 41:274.

Bray, A.C., and A.R. Till. 1975. Metabolism of sulfur in the gastrointestinal tract. In: Digestion and Metabolism in the Ruminant. I.W. McDonald and A.C.I. Warner, eds. Univ. of New England Publ. Unit, Armidale, N.S.W., Australia. P. 243.

Breznak, J.A. 1982. Biochemical aspects of symbiosis between termites and their intestinal microbiota. In: Invertebrate-Microbial Interactions. Joint Symp. British Mycological Society and British Ecological Society, Univ. of Exeter. J.M. Anderson, A.D.M. Rayner, and D.W.H. Walton, eds. Cambridge Univ. Press, Cambridge. Pp. 173–204.

Breznak, J.A., and M.D. Kane. 1990. Microbial H_2/CO_2 acetogenesis in animal guts. Nature and nutritional significance. In 6th Int. Symp. on Microbial Growth. B.U. Bowien and J.R. Andreesen, eds. FEMS Microbiol. Rev. 54:369–371.

Breznak, J.A., and J.M. Switzer. 1986. Acetate synthesis from H_2 plus CO_2 by termite gut microbes. Appl. Environ. Microbiol. Pp. 52:623–630.

Briceno, J.V., H.H. Van Horn, B. Harris, Jr., and C.J. Wilcox. 1987. Effects of neutral detergent fiber and roughage source on dry matter intake and milk yield and composition of dairy cows. J. Dairy Sci. 70:298–308.

Briggs, P.K., J.P. Hogan, and R.L. Reid. 1957. The effect of volatile fatty acids, lactic acid, and ammonia on rumen pH in sheep. Aust. J. Agric. Res. 8:674–710.

Britton, R.A., T.J. Klopfenstein, R. Cleale, F. Foedeken, and V. Wilkerson. 1986. Methods of estimating heat damage in protein sources. Proc. Distillers' Feed Conf., Cincinnati. Distillers' Feed Research Council, Cincinnati. P. 67–74.

Broderick, G.A. 1978. In vitro procedures for estimating rates of ruminal protein degradation and proportions of protein escaping the rumen undegraded. J. Nutr. 108:181–190.

Brosh, A., I. Choshniak, A. Tadmor, and A. Shkolnik. 1986. Infrequent drinking, digestive efficiency and particle size of digesta in black Bedouin goats. J. Agric. Sci. 106:575.

Brosh, A., A. Shkolnik, and I. Choshniak. 1987. Effects of infrequent drinking on the nitrogen metabolism of Bedouin goats maintained on different diets. J. Agric. Sci. 109:165.

Brosh, A., I. Choshniak, A. Tadmer, and A. Shkolnik. 1988. Physico-chemical conditions in the rumens of Bedouin goats: effect of drinking, food quality and feeding time. J. Agric. Soc. Camb. 111:147–152.

Broster, W.H., V.J. Broster, and T. Smith. 1969. Effect on milk production of the level of feeding at two stages of the lactation. J. Agric. Sci. 72:229.

Bruckental, I., A.R. Lehrer, M. Weitz, J. Bernard, H. Kennit, and N. Kennit. 1987. Faecal output and estimated voluntary dry matter intake of grazing beef cows, relative to their live weight and to the digestibility of the pasture. Anim. Prod. 45:23.

Bryant, M.P. 1972. Commentary on the Hungate technique for culture of anaerobic bacteria. Am. J. Clin. Nutr. 25:1324.

Bryant, M.P. 1974. Nutritional features and ecology of predominant anaerobic bacteria of the intestinal tract. Am. J. Clin. Nutr. 27:1313.

Bryant, M.P. 1978. Cellulose digesting bacteria from human feces. Am. J. Clin. Nutr. 31(Suppl.):S113–S115.

Bryant, M.P. 1979. Microbial methane production—theoretical aspects. J. Anim. Sci. 48:193–201.

Bryant, M.P., and L.A. Burkey. 1953. Numbers and some predominant groups of bacteria in the rumen of cows fed different rations. J. Dairy Sci. 36:218.

Bryant, M.P., and M.J. Wolin. 1975. Rumen bacteria and their metabolic interactions. In: Proc. 1st Intersectional Congr. Int. Assoc. Microbiological Society, Vol. 2, Developmental Microbiology, Ecology. T. Hasegawa, ed. Science Council, Tokyo. Pp. 297–306.

Buchanan-Smith, J.G., and L.E. Phillip. 1986. Food intake in sheep following intraruminal infusion of extracts from lucerne silage with particular reference to organic acids and products of protein degradation. J. Agric. Sci. 106:611.

Bucher, A.C. 1984. A comparison of solvent systems for extraction of pectic substances from fruits and vegetables. M.S. thesis, Cornell Univ., Ithaca, N.Y.

Buchsbaum, R., J. Wilson, and I. Valiela. 1986. Digestibility of plant constituents by Canada geese and Atlantic brant. Ecology 67(2):386–393.

Bu'Lock, J.D., ed. 1987. Fermentation Kinetics and Modelling. Open Univ. Press, Milton Keynes.

Bunce, H.W.F. 1985. Fluoride in air, grass, and cattle. J. Dairy Sci. 68:1706.

Burkitt, D.P. 1973. Epidemiology of large bowel disease: the role of fiber. Proc. Nutr. Soc. 32:145–149.

Burns, R.E. 1971. Method for estimation of tannin in grain sorghum. Agron. J. 63:511–512.

Burroughs, W., D.K. Nelson, and D.R. Mertens. 1975. Protein physiology and its application in the lactating cow: the metabolizable protein feeding standard. J. Anim. Sci. 41:933–944.

Burrows, C.F., D.S. Kronfeld, C.A. Banta, and A.M. Merritt. 1982. Effects of fiber on digestibility and transit time in dogs. J. Nutr. 112:1726–1732.

Burton, G.W., and W.W. Hanna. 1985. Bermudagrass. In: Forages: The Science of Grassland Agriculture. 4th ed. M.E. Heath, R.F. Barnes, and D.S. Metcalfe, eds. Iowa State Univ. Press, Ames. P. 247.

Bush, L.P., and R.C. Buckner. 1973. Tall fescue toxicity. In: Anti-quality components of forages. A.G. Matches, ed. Crop Science Society of America, Madison, Wisc. Pp. 99–112.

Bush, L.P., C. Streeter, and R.C. Buckner. 1970. Perloline inhibition of in vitro ruminal cellulose digestion. Crop Sci. 10:108.

Butler, L.G. 1989. Effects of condensed tannin on animal nutrition. In: The Chemistry and Significance of Condensed Tannins. R.W. Hemingway and J.J. Karchesy, eds. Plenum Press, New York. Pp. 2401–2412.

Butterworth, M.H., and J.A. Diaz. 1970. Use of equations to predict the nutritive value of tropical grasses. J. Range Manage. 23:55–58.

Buxton, D.R., J.S. Hornstein, W.F. Wedin, and G.C. Marten. 1985. Forage quality in stratified canopies of alfalfa, birdsfoot trefoil, and red clover. Crop Sci. 25:273–279.

Calder, W.A. III. 1984. Size, Function and Life History. Harvard Univ. Press, Cambridge.

Cansunar, E., A.J. Richardson, G. Wallace, and C.S. Stewart. 1990. Effect of coumarin on glucose uptake by anaerobic rumen fungi in the presence and absence of *Methanobrevibacter smithii*. FEMS Microbiol. Lett. 70:157–160.

Capper, B.C. 1988. Genetic variation in the feeding value of cereal straw. Anim. Feed Sci. Technol. 21:127–140.

Care, A.D. 1988. A fresh look at hypomagnesaemia. Br. Vet. J. 144(1):3–4.

Carlisle, E.M. 1974. Silicon as an essential element. Fed. Proc. 33:1758.

Carlson, J.R., and R.G. Breeze. 1984. Ruminal metabolism of plant toxins with emphasis on indolic compounds. J. Anim. Sci. 58:1040.

Carpenter, J.R. 1976. Site of starch digestion and its relationship to milk fat depression in the lactating cow. Ph.D. dissertation, Cornell Univ., Ithaca, N.Y.

Casler, M.D., H. Talbert, A.K. Forney, N.J. Ehlke, and J.M. Reich. 1987. Genetic variation for rate of cell wall digestion and related traits in first cut smooth bromegrass. Crop Sci. 27:935–939.

Chalupa, W. 1972. Metabolic aspects of nonprotein nitrogen utilization in ruminant animals. Fed. Proc. 31:1152–1164.

Chalupa, W. 1977. Manipulating rumen fermentation. J. Anim. Sci. 46:585–599.

Chalupa, W., C.J. Sniffen, D.G. Fox, and P.J. Van Soest. 1991. Model generated protein degradation nutritional information. Proc. Cornell Nutrition Conf., Ithaca, N.Y. Pp. 44–51.

Chamberlain, D.G., and P.C. Thomas. 1979. Ruminal nitrogen metabolism and the passage of amino acids to the duodenum of sheep receiving diets containing hay and concentrates in various proportions. J. Sci. Food Agric. 30:677–686.

Chandler, J.A., W.J. Jewell, J.M. Gossett, P.J. Van Soest, and J.B. Robertson. 1980. Predicting methane fermentation biodegradability. Biotechnol. Bioeng. 10:93–107.

Chandler, P. 1990. Energy prediction of feeds by forage testing explored. Feedstuffs 62(36):12, 22.

Chandra, S., D.A. Prasad, and N. Krishna. 1985. Effect of sodium hydroxide treatment and/or extrusion cooking on the nutritive value of peanut hulls. Anim. Feed Sci. Technol. 12:187–194.

Chapin, F.S., J.D. McKendrick, and D.A. Johnson. 1986. Seasonal changes in carbon fractions in Alaskan tundra plants of differing growth form: implications for herbivory. J. Ecol. 74:707–731.

Charmley, E., and D.M. Veira. 1990. Inhibition of proteolysis at harvest using heat in alfalfa silages: effects on silage composition and digestion by sheep. J. Anim. Sci. 68:758–766.

Chaudhry, A.S., and A.J.F. Webster. 1993. The true digestibility and biological value for rats of undegraded dietary nitrogen in feeds for ruminants. Anim. Feed Sci. Technol. 42:209–221.

Cheeke, P.R. 1988. Toxicity and metabolism of pyrrolizidine alkaloids. J. Anim. Sci. 66:2343.

Chen, G., J.B. Russell, and C.J. Sniffen. 1987a. A procedure for measuring peptides in rumen fluid and evidence that peptide uptake can be a rate-limiting step in ruminal protein degradation. J. Dairy Sci. 70:1211–1219.

Chen, G., H.J. Strobel, J.B. Russell, and C.J. Sniffen. 1987b. Effect of hydrophobicity on utilization of peptides by ruminal bacteria in vitro. Appl. Environ. Microbiol. 53:2021–2025.

Chen, M., and M.J. Wolin. 1979. Effect of Monensin and Lasolacid-Sodium on the growth of methanogenic and rumen saccharolytic bacteria. Appl. Environ. Microbiol. 38:72–77.

Chenost, M. 1985. Estimation de la digestibilité de l'herbe ingéré au pâturage à partir de l'azote fécal et de quelques autres paramètres fécaux. Ann. Zootech. 34:205–228.

Chenost, M. 1986. Aspects méthodologiques de la prévision de la digestibilité de l'herbe pâturée par le mouton, les bovins et le cheval à partir de bols de l'oesophage et de diverses caractéristiques fécales. Ann. Zootech. 35:1–20.

Chenost, M., and W. Martin-Rosset. 1985. Comparaison entre espèces (mouton, cheval, bovin) de la digestibilité et des quantités ingérées des fourrages verts. Ann. Zootech. 34:291.

Cherney, J.H., J.D. Axtell, M.M. Hassen, and K.S. Anliker. 1988. Forage quality characterization of a chemically induced brown-midrib mutant in pearl millet. Crop. Sci. 28:783.

Cherney, J.H., D.J.R. Cherney, L.E. Sollenberger, J.A. Patterson, and K.V. Wood. 1990. Identification of 5-0-Caffeoylquinic acid in limpograss and its influence of fiber digestion. J. Agric. Food Chem. 38:2140.

Chesson, A. 1993. Mechanistic models of forage cell wall degradation. In: Forage Cell Wall Structure and Digestibility. H.G. Jung, D.R. Buxton, R.D. Hatfield, and J. Ralph, eds. American Society of Agronomy. Madison, Wisc. P. 347–376.

Chesson, A., and C.W. Forsberg. 1988. Polysaccharide degradation by rumen microorganisms. In: The Rumen Microbial Ecosystem. P.N. Hobson, ed. Elsevier Applied Science Publ., New York. Pp. 251–284.

Chesson, A., and J.A. Monro. 1982. Legume pectic substances and their degradation in the ovine rumen. J. Sci. Food Agric. 33:852–859.

Chibnall, A.C., and C.E. Glover. 1926. The extraction of sap from living leaves by means of compressed air. Ann. Bot. 40:491–497.

Christie, W.W. 1981. The effects of diet and other factors on the lipid composition of ruminant tissues and milk. In: Lipid Metabolism in Ruminant Animals. W.W. Christie, ed. Pergamon Press, New York. P. 193.

Church, D.C. 1975. Digestive physiology and nutrition of ruminants. In: Digestive Physiology, vol. 1. D.C. Church, ed. O & B Books, Corvallis, Ore.

Church, D.C., ed. 1988. The Ruminant Animal. Prentice-Hall, Englewood Cliffs, N.J. P. 564.

Cincotta, R.P., P.J. Van Soest, J.B. Robertson, C.M. Beall, and M.C. Goldstein. 1991. Foraging ecology of livestock on the Tibetan Changtang: a comparison of three adjacent grazing areas. Arct. Alp. Res. 23:149–161.

Clancy, M.J., P.J. Wangsness, and B.R. Baumgardt. 1976. Effect of conservation method on digestibility, nitrogen balance, and intake of alfalfa. J. Dairy Sci. 60:572–579.

Clancy, M.J., and R.K. Wilson. 1966. Development and application of a new chemical method for predicting the digestibility and intake of herbage samples. Proc. 10th Int. Grasslands Congr., Helsinki. Pp. 445–452.

Clar, U., H. Steingass, and K.H. Menke. 1988. Polyethylene and indigestible cell wall fractions according to Van Soest as markers for evaluation feed intake in sheep. (In German.) Arch. Tierernähr. 38:663.

Clark. J.H., H.R. Spires, and C.L. Davis. 1978. Uptake and metabolism of nitrogenous components by the lactating mammary gland. Fed. Proc. 37:1233.

Clemens, E.T., and C.E. Stevens. 1980. A comparison of gastrointestinal transit time in ten species of mammal. J. Agric. Sci. 94:735–737.

Clutton-Brock, T.H., and P.H. Harvey. 1983. The functional

significance of variation in body size among mammals. In: Advances in the Study of Mammalian Behavior. J.F. Eisenberg and D.G. Kleiman, eds. Spec. Publ. no. 7, American Society of Mammalogists. Pp. 632–658.

Cocimano, M.R., and R.A. Leng. 1967. Metabolism of urea in sheep. Br. J. Nutr. 21:353.

Coelho, M., F.G. Hembry, A.M. Saxton, and F.E. Barton. 1988. A comparison of microbial, enzymatic, chemical and near-infrared reflectance spectroscopy methods in forage evaluation. Anim. Feed Sci. Technol. 20:219–232.

Colburn, M.W., and J.L. Evans. 1968. Reference base, W^b, of growing steers determined by relating forage intake to body weight. J. Dairy Sci. 51:1073.

Collings, G.F., M.T. Yokoyama, and W.G. Bergen. 1978. Lignin as determined by oxidation with sodium chlorite and a comparison with permanganate lignin. J. Dairy Sci. 61:1156.

Collins, M. 1985a. Dry matter and nitrogen partitioning in mechanically dewatered alfalfa, red clover and birdsfoot trefoil. Agron. J. 77(6):923.

Collins, M. 1985b. Wetting effects on the yield and quality of legumes and legume-grass hays. Agron. J. 77:936.

Collins, M. 1988. Composition and fibre digestion in morphological components of an alfalfa-timothy sward. Anim. Feed Sci. Technol. 19:135–143.

Colucci, P.E. 1984. Comparative digestion and digesta kinetics in sheep and cattle. Ph.D. dissertation, Univ. of Guelph. P. 231.

Colucci, P.E., L.E. Chase, and P.J. Van Soest. 1982. Feed intake, apparent diet digestibility, and rate of particulate passage in dairy cattle. J. Dairy Sci. 65:1445–1456.

Combellas, J., E. González J., and R. Parra R. 1971. Composición y valor nutritivo de forrajes producidos en el trópico. I. Digestibilidad aparente y verdadera de las fracciones químicas. Agron. Trop. 6:483–494.

Comstock, J.M. 1986. Chemistry and Function of Pectins. M.L. Fishman and J.J. Jen, eds. ACS Symp. Ser. 310. American Chemical Society, Washington, D.C. P. 283.

Conklin, N.L. 1987. The potential nutritional value to cattle of some tropical browse species from Guanacaste, Costa Rica. Ph.D. dissertation, Cornell Univ., Ithaca, N.Y. 329 pp.

Conn, E.E. 1978. Cyanogenesis, the production of hydrogen cyanide, by plants. In: Effects of Poisonous Plants on Livestock. R.F. Keeler, K.R. Van Kampen, and L.F. James, eds. Academic Press, New York.

Conniffe, D., D. Browne, and M.J. Walshe. 1970. Experimental design for grazing trials. J. Agric. Sci. 74:339–342.

Connolly, J. 1976. Some comments on the shape of the gain-stocking rate curve. J. Agric. Sci. 86:103–109.

Conrad, H.R. 1966. Symposium on factors influencing the voluntary intake of herbage by ruminants: physiological and physical factors limiting feed intake. J. Anim. Sci. 25:227.

Conrad, H.R. 1983. Potassium-cation and anion balance in ruminant nutrition. In: Nutrition Institute: Minerals. National Feed Ingredient Assoc., West Des Moines, Iowa. Pp. 1–4.

Conrad, H.R., W.P. Weiss, W.O. Odwongo, and W.L. Shockey. 1984. Estimating net energy of lactation from components of cell solubles and cell walls. J. Dairy Sci. 67:427–436.

Cook, C.W., and J. Stubbendieck, eds. 1986. Range Re-

search: Basic Problems and Techniques. Society for Range Management, Denver, Colo.

Coors, J.G., C.C. Lowe, and R.P. Murphy. 1986. Selection for improved nutritional quality of alfalfa forage. Crop Sci. 26:843–848.

Coppock, C.E. 1986. Mineral utilization by the lactating cow—chlorine. J. Dairy Sci. 69:595.

Coppock, C.E. 1987. Supplying the energy and fiber needs of dairy cows from alternate feed sources. J. Dairy Sci. 70:1110–1119.

Coppock, C.E., J.K. Lanham, and J.I. Horner. 1987. A review of the nutritive value and utilization of whole cottonseed, cottonseed meal and associated by-products by dairy cattle. Anim. Feed Sci. Technol. 18:89–129.

Cordova, F.J., J.D. Wallace, and R.D. Pieper. 1978. Forage intake by grazing livestock: A review. J. Range Manage. 31:430–438.

Cotton, F.A., and G. Wilkinson. 1966. Advanced Inorganic Chemistry. Wiley, New York.

Couchman, J.F. 1959. Storage of hay. I. Effect of temperature on the "soluble" nitrogen, sugar and fat contents. J. Sci. Food Agric. 10:513–519.

Coughlan, M.P. 1991. Mechanisms of cellulose degradation by fungi and bacteria. Anim. Feed Sci. Technol. 32:77–100.

Coughlan, M.P., and L.G. Ljungdahl. 1988. Comparative biochemistry of fungal and bacterial cellulolytic enzyme systems. In: Biochemistry and Genetics of Cellulose Degradation. J.P. Aubert, P. Beguin, and J. Millet, eds. Academic Press, New York. Pp. 11–30.

Cowgill, U.M. 1973. Biogeochemistry of the rare-earth elements in aquatic macrophytes of Linsley Pond, North Brandford, Connecticut. Geochim. Cosmochim. Acta 37:2329.

Cowling, E.B., and W. Brown. 1969. Structural feature of cellulosic materials in relation to enzymatic hydrolysis. In: Cellulases and Their Application. E.T. Reese and G. Hajny, eds. Adv. Chem. Ser. 94:152–187.

Crampton, E.W., E. Donefer, and L.E. Lloyd. 1960. A nutritive index for forages. J. Anim. Sci. 19:538–544.

Crutzen, P.J., I. Aselmann, and W. Seiler. 1986. Methane production by domestic animals, wild ruminants, and other herbivorous fauna and humans. Tellus 38B:271.

Culvenor, C.C.J. 1987. Detrimental factors in pastures and forage. In: Temperate Pastures, Their Production, Use and Management. J.L. Wheeler, C.J. Pearson, and G.E. Robards, eds. CSIRO, Canberra. Pp. 435–445.

Cyr, N., R.M. Elofson, J.A. Ripmeester, and G.W. Mathison. 1988. Study of lignin in forages and wood by ^{13}C CP/MAS NMR; some evidence of polymerization and depolymerization. J. Agric. Food Chem. 36:1197–1201.

Czerkawski, J.W. 1967. The determination of lignin. Br. J. Nutr. 21:325–332.

Czerkawski, J.W. 1972. Fate of metabolic hydrogen in the rumen. Proc. Nutr. Soc. 31:141–146.

Czerkawski, J.W. 1984. Microbial fermentation in the rumen. Proc. Nutr. Soc. 43:101–118.

Czerkawski, J.W. 1986. An Introduction to Rumen Studies. Pergamon Press, New York. P. 236.

Czerkawski, J.W., and K.J. Cheng. 1988. Compartmentation in the rumen. In: The Rumen Microbial Ecosystem. P.N. Hobson, ed. Elsevier Applied Science Publ., New York. Pp. 361–386.

Damuth, J., and B.J. McFadden, eds. 1990. Body Size in

Mammalian Paleobiology: Estimation and Biological Implications. Cambridge Univ. Press, Cambridge.

Davis, C.L. 1967. Acetate production in the rumen of cows fed either control or low-fiber, high-grain diets. J. Dairy Sci. 50:1621–1625.

Davis, C.L., and R. Brown. 1970. Low-fat milk syndrome. In: Physiology of Digestion and Metabolism in the Ruminant. A.T. Phillipson, ed. Oriel Press, Newcastle upon Tyne, England. Pp. 545–565.

Dawson, R.M.C., and P. Kemp. 1970. Biodegradation of dietary fats in ruminants. In: Physiology of Digestion and Metabolism in the Ruminant. A.T. Phillipson, ed. Oriel Press, Newcastle upon Tyne, England. P. 504.

de Boer, G., J.J. Murphy, and J.J. Kennelly. 1987. Mobile nylon bag for estimating intestinal availability of rumen undegradable protein. J. Dairy Sci. 70:977.

De Boever, J.L., B.G. Cottyn, J.I. Andries, F.X. Buysse, and J.M. Vanacker. 1988. The use of a cellulase technique to predict digestibility, metabolizable and net energy of forages. Anim. Feed Sci. Technol. 19:247–260.

Dehority, B.A. 1993. Microbial ecology of cell wall fermentation. In: Forage Cell Wall Structure and Digestibility. H.G. Jung, D.R. Buxton, R.D. Hatfield, and J. Ralph, eds. American Society of Agronomy, Madison, Wisc. Pp. 377–395.

Deinum, B. 1976. Effect of age, leaf number and temperature on cell wall digestibility of maize. In: Carbohydrate Research in Plants and Animals. P.W. van Adrichem, ed. Landbouwhogeschool Misc. Pap. 12. Wageningen, Netherlands, P. 29.

Deinum, B. 1987. Genetic and environmental variation in digestibility of forage maize in Europe. Proc. of the Maize Conf. of Eucarpia, Nitra, Czechoslovakia, Sept. 7–17.

Deinum, B., and J.G.P. Dirven. 1975. Climate, nitrogen and grass. Comparison of yield and chemical composition of some temperate and tropical grass species grown at different temperatures. Neth. J. Agric. Sci. 23:69–82.

Deinum, B., and J.G.P. Dirven. 1976. Climate, nitrogen and grass. Comparison of production and chemical composition of Brachiaria ruziziensis and Setaria sphacelata grown at different temperatures. Neth. J. Agric. 24:67–78.

Deinum, B., A.J.H. van Es, and P.J. Van Soest. 1968. Climate, nitrogen and grass. 2. The influence of light intensity, temperature and nitrogen on in vivo digestibility of grass and the prediction of these effects from some chemical procedures. Neth. J. Agric. Sci. 16:217–233.

Deinum, B., and P.J. Van Soest. 1969. Prediction of forage digestibility from some laboratory procedures. Neth. J. Agric. Sci. 17:119–127.

Dekker, R.F.H. 1976. Hemicellulose degradation in the ruminant. In: Carbohydrate Research in Plants and Animals. P.J. Van Adrichem, ed. Landbouwhogeschool Misc. Pap. 12. Wageningen, Netherlands, P. 43.

Dell'Orto, V., E. Salimei, G. Savioni, A. Baldi, A. Lanzani, P. Bondioli, and L. Degano. 1990. Supplementation of diets for lactating cows with zinc as zinc sulfate and zinc soaps. J. Dairy Sci. 73 (Suppl. 1):165.

Demarquilly, C., J.M. Boissau, and G. Cuylle. 1965. Factors affecting the voluntary intake of green forage by sheep. Proc. 9th Int. Grasslands Congr., São Paulo, Brazil. Vol. 1, pp. 877–885.

Demment, M.W. 1982. The scaling of ruminoreticulum size with body weight in East African ungulates. Afr. J. Ecol. 20:43.

Demment, M.W., and G.B. Greenwood. 1988. Forage ingestion: effects of sward characteristics and body size. J. Anim. Sci. 66:2380.

Demment, M.W., and P.J. Van Soest. 1983. Body Size, Digestive Capacity, and Feeding Strategies of Herbivores. Winrock Int. Livestock Res. and Training, Ctr., Morrilton, Ark. P. 66.

Demment, M.W., and P.J. Van Soest. 1985. A nutritional explanation for body-size patterns of ruminant and nonruminant herbivores. Am. Nat. 125:641–672.

den Braver, E., and S. Eriksson. 1967. Determination of energy in grass hay by in vitro methods. Lantbrukshogsk. Ann. 33:751–765.

DeRuiter, J.M., and J.C. Burns. 1986. Rapid determination of cell wall monosaccharides in flaccidgrass. J. Agric. Food Chem. 34:780–786.

desBordes, C.K. 1981. Influence of specific gravity on rumination and passage of indigestible particles through the gastrointestinal tract. Ph.D. dissertation, Univ. of Vermont.

desBordes, C.K., and J.G. Welch. 1984. Influence of specific gravity on rumination and passage of indigestible particles. J. Anim. Sci. 59:470–475.

Devendra, C. 1978. The digestive efficiency of goats. World Rev. Anim. Prod. 14:9–22.

Dhanoa, M.S. 1988. On the analysis of dacron bag data for low degradability feeds. Grass Forage Sci. 43:441–444.

Dias–da Silva, A.A., and C.V.M. Guedes. 1990. Variability in the nutritive value of straw cultivars of wheat, rye and triticale and response to urea treatment. Anim. Feed Sci. Technol. 28:79–89.

Dias Filho, M.B., M. Corsi, and S. Cusato. 1992. Concentration, uptake and use efficiency of N, P and K in Panicum maximum Jacq. cv. tobiata under water stress. Pesqui. Agropecu. Bras. 27(3):381–387.

Dias Filho, M.B., M. Corsi, S. Cusato, and A. Pinheiro Camarao. 1991. In vitro organic matter digestibility and crude protein content of Panicum maximum Jacq. cv. tobiata under water stress. Pesqui. Agropecu. Bras. 26(10):1725–1729.

Dickenson, C.H., and G.J.F. Pugh, eds. 1974. Biology of Plant Decomposition, vol. 2. Academic Press, New York.

Dierenfeld, E.S. 1989. Vitamin E deficiency in zoo reptiles, birds, and ungulates. J. Zoo Wildl. Med. 20(1):3–11.

Dierenfeld, E.S., H.F. Hintz, J.B. Robertson, P.J. Van Soest, and O.T. Oftedal. 1982. Utilization of bamboo by the giant panda. J. Nutr. 112:636–641.

Dijkshoorn, W. 1973. Organic acids, and their role in ion uptake. In: Chemistry and Biochemistry of Herbage. G.W. Butler and R.W. Bailey, eds. Academic Press, New York. Pp. 163–188.

Dijkstra, J., H. Boer, J.V. Bruchem, M. Bruining, and S. Tamminga. 1993. Absorption of volatile fatty acids from the rumen of lactating dairy cows as influenced by volatile fatty acid concentration, pH and rumen liquid volume. Br. J. Nutr. 69:385–396.

Dinius, D.A., and B.R. Baumgardt. 1970. Regulation of food intake in ruminants. 6. Influence of caloric density of pelleted rations. J. Dairy Sci. 53:311–316.

Dixon, J.B., S.B. Weed, J.A. Kittrick, M.H. Milford, and J.L. White, eds. 1977. Minerals in Soil Environments. Soil Science Society of America, Madison, Wisc.

Dixon, R.M., J.J. Kennelly, and L.P. Milligan. 1983. Kinetics of [103]Ru phenanthroline and dysprosium particulate markers in the rumen of steers. Br. J. Nutr. 49:463.

Dobson, A. 1959. Active transport through the epithelium of the reticulo-rumen sac. J. Physiol. (Lond.) 146:235.

Dobson, A., A.F. Sellers, and V.H. Gatewood. 1976. Dependence of Cr-EDTA absorption from the rumen on luminal osmotic pressure. Am. J. Physiol. 231:1595–1600.

Dobson, D.E., E.M. Prager, and A.C. Wilson. 1984. Stomach lysozymes of ruminants. I. Distribution and catalytic properties. J. Biol. Chem. 259:11607–11625.

Dodson, P.M., J. Stocks, G. Holdsworth, and D.J. Galton. 1981. High-fibre and low-fat diets in diabetes mellitus. Br. J. Nutr. 46:289–294.

Dominguez-Bello, M.G., M. Lovera, P. Suarez, F. Michelangeli. 1993. Microbial digestive symbionts of the crop of the hoatzin (Opisthocomus hoazin): an avian foregut fermenter. Physiol. Zool. 66(3):374–383.

Donefer, E., E.W. Crampton, and L.E. Lloyd. 1966. The prediction of digestible energy intake potential (NVI) of forages using a simple in vitro technique. Proc. 10th Int. Grassland Congr., Helsinki. P. 442.

Doner, L.W. 1986. Analytical methods for determining pectin composition. In: Chemistry and Function of Pectins. M.L. Fishman and J.J. Jen, eds. American Chemical Society, Washington, D.C. Pp. 13–21.

Dowman, M.G., and F.C. Collins. 1982. The use of enzymes to predict the digestibility of animal feeds. J. Sci. Food Agric. 33:689–696.

Downey, G., P. Robert, D. Bertrand, and M.F. Devaux. 1987. Near infra-red analysis of grass silage by principal component analysis of transformed reflectance data. J. Sci. Food Agric. 41:219–230.

Drennen, T.E., and D. Chapman. 1992. Negotiating a response to climate change: role of biological emissions. Contemp. Policy Iss. 10:49–58.

Duke, G.E., E. Eccleston, S. Kirkwood, C.F. Louis, and H.P. Bedbury. 1984. Cellulose digestion by domestic turkeys fed low or high fiber diets. J. Nutr. 114:95–102.

Duncan, W.R.H., A.K. Lough, G.A. Garton, and P. Brooks. 1974a. Characterization of branched chain fatty acids from subcutaneous triaclyglycerols of barley-fed lambs. Lipids 9:669–673.

Duncan, W.R.H., E.R. Orskov, C. Fraser, and G.A. Garton. 1974b. Effect of processing of dietary barley and of supplementary cobalt and cyanocobalamin on the fatty acid composition of lamb triglycerides, with special reference to branched-chain components. Br. J. Nutr. 32:71–75.

Dworschak, E. 1980. Nonenzyme browning and its effect on protein nutrition. CRC Crit. Rev. Food Sci. Nutr. 13:1.

Eadie, J.M., and S.O. Mann. 1970. Development of the rumen microbial population: high starch diets and instability. In: Physiology of Digestion and Metabolism in the Ruminant. A.T. Phillipson, ed. Oriel Press, Newcastle upon Tyne, England. Pp. 335–347.

Eastwood, M.A., J.A. Robertson, W.G. Brydon, and D. MacDonald. 1983. Measurement of water-holding properties of fibre and their faecal bulking ability in man. Br. J. Nutr. 50:539.

Edwards, C.S. 1973. Determination of lignin and cellulose in forages by extraction with triethylene glycol. J. Sci. Food Agric. 24:381.

Eglinton, G., and R.J. Hamilton. 1967. Leaf epicuticular waxes. Science 156:1322–1332.

Ehle, F.R., F. Bas, B. Barno, R. Martin, and F. Leone. 1984. Particulate rumen turnover rate measurement as influenced by density of passage marker. J. Dairy Sci. 67:2910–2913.

Ehle, F.R., J.L. Jeraci, J.B. Robertson, and P.J. Van Soest. 1982. The influence of dietary fiber on digestibility, rate of passage and gastrointestinal fermentation in pigs. J. Anim. Sci. 55:1071–1081.

Ehle, F.R., and R.G. Warner. 1978. Nutritional implications of the hamster forestomach. J. Nutr. 108:1047–1053.

Einhof, H. 1806. Bemerkungen über die Nahrungsfähigkeit verschiedener vegetabilischer Produkte. Ann. Ackerbaues 4:627–659.

El Hag, G.A. 1976. A comparative study between desert goat and sheep efficiency of feed utilization. World Rev. Anim. Prod. 12:43–48.

Elliot, J.M. 1980. Propionate metabolism and vitamin B_{12}. In: Digestive Physiology and Metabolism in Ruminants. Y. Ruckebusch and P. Thivend, eds. MTP Press, Lancaster, England. P. 485.

Ellis, J.E., J.A. Wiens, C.F. Rodell, and J.C. Anway. 1976. A conceptual model of diet selection as an ecosystem process. J. Theor. Biol. 60:93–108.

Ellis, W.C., C. Lascano, and J.H. Matis. 1979. Sites contributing to compartmental flow to forage residues. Ann. Rech. Vet. 10:166–167.

Ellis, W.C., and H. Lippke. 1976. Nutritional values of forages. In: Grasses and Legumes in Texas—Development, Production and Utilization. E.C. Holt and R.D. Lewis, eds. Texas Agric. Exp. Sta. Res. Monog. RM6C. Pp. 26–66.

Elofson, R.M., J.A. Ripmaster, N. Cyr, L.P. Milligan, and G. Mathison. 1984. Nutritional evaluation of forages by high-resolution solid state ^{13}C-NMR. Can. J. Anim. Sci. 64:93.

Elsden, G.D., and G.H. Walker. 1942. Richmond's Dairy Chemistry. 4th ed. Charles Giffen and Co., London.

Ely, R.E., C.G. Melin, and L.A. Moore. 1956. Yields and protein content of holocellulose prepared from pepsin-treated froages. J. Dairy Sci. 39:1742.

Ely, R.E., and L.A. Moore. 1955. Holocellulose and the summative analysis of forages. J. Anim. Sci. 14:718–724.

England, P., and M. Gill. 1985. The effect of fish meal and sucrose supplementation on the voluntary intake of grass silage and live-weight gain of young cattle. Anim. Prod. 40:259–265.

Englyst, H.N., and J.H. Cummings. 1987. Resistant starch, a "new" food component: a classification of starch for nutritional purposes. In: Cereals in a European Context. Proc. 1st European Cong. of Food Science and Technology, Bournemouth, July 1986. Pp. 221–233.

Englyst, H.N., and J.H. Cummings. 1988. Improved method for measurement of dietary fiber as non-starch polysaccharides in plant foods. J. Assoc. Off. Anal. Chem. 71:808.

Englyst, H.N., and G.J. Hudson. 1987. Calorimetric method for routine measurement of dietary fibre as non-starch polysaccharides. A comparison with gas-liquid chromatography. Food Chem. 24:63–76.

Enzmann, J.W., R.D. Goodrich, and J.C. Meiski. 1969. Chemical composition and nutritive value of poplar bark. J. Anim. Sci. 29:653–660.

Erämetsä, O., M.L. Sihvonen, and A. Forssen. 1968. Rare earths in the human body. I. Yttrium. Ann. Med. Exp. Biol. Fenn. 46:179.

Eraso, F., and R.D. Hartley. 1990. Monomeric and dimeric phenolic constituents of plant cell walls—possible factors influencing wall biodegradability. J. Sci. Food Agric. 51:163–170.

Erdman, R.A., and L.W. Smith. 1985. Ytterbium binding among particle size fractions of forage cell walls. J. Dairy Sci. 68:3071.

Essig, H.W. 1988. Bloat. In: The Ruminant Animal: Digestive Physiology and Nutrition. Ed. D.C. Church. Prentice-Hall, Englewood Cliffs, N.J. P. 468.

Evans, E., and D.S. Miller. 1975. Bulking agents in the treatment of obesity. Nutr. Metab. 18:199–203.

Evans, T.R., and J.R. Wilson. 1984. Some responses of grasses to water stress and their implications for herbage quality and animal liveweight gain. In: The Impact of Climate on Grass Production and Quality. H. Riley and A.O. Skjelvag, eds. Norwegian State Agric. Res. St., Ås, Norway.

Evans, W.C. 1977. Biochemistry of the bacterial catabolism of aromatic compounds in anaerobic environments. Nature (Lond.) 270:17–22.

Fadel, J.G. 1984. Development of mathematical techniques to model ruminant fiber digestion systems. Ph.D. dissertation, Cornell Univ., Ithaca, N.Y. P. 128.

Fagerberg, B. 1988a. Phenological development in timothy, red clover and lucerne. Acta Agric. Scand. 38:159–170.

Fagerberg, B. 1988b. The change in nutritive value in timothy, red clover and lucerne in relation to phenological state, cutting time and weather conditions. Acta Agric. Scand. 38:347–362.

Fahey, G.C., Jr., L.D. Bourguin, E.C. Titgemeyer, and D.G. Atwell. 1992. Post-harvest treatment of fibrous feedstuffs to improve their nutritive value. In: Forage Cell Wall Structure and Digestibility. H.G. Jung, D.R. Buxton, R.D. Hatfield, and J. Ralph, eds. American Society of Agronomy, Madison, Wisc. P. 715–766.

Fahey, G.C., Jr., and H.G. Jung. 1983. Lignin as a marker in digestion studies: a review. J. Anim. Sci. 57:220–226.

Fahey, G.C. Jr., B.L. Miller, and H.W. Hadfield. 1979. Metabolic parameters affected by feeding various types of fiber to guinea pigs. J. Nutr. 109:77–83.

Faichney, G.J. 1975. The use of markers to partition digestion within the gastro-intestinal tract of ruminants. Proc. 4th Int. Symp. on Ruminant Physiology. I.W. McDonald and A.C.I. Warner, eds. Univ. of New England Publ. Unit. Armidale, N.S.W., Australia. Pp. 277–291.

Faichney, G. 1984. Application of the double marker method to the estimation of rumen particle fractions. Proc. Satellite Conf. 6th Int. Symp. on Ruminant Physiology. Techniques in Particle Size Analysis of Feed and Digesta in Ruminants. Ed. P.M. Kennedy. Can. Soc. Anim. Sci. Occas. Publ. no. 1. Edmonton, Alberta. P. 179.

Fairbairn, R., I. Alli, and B.E. Baker. 1988. Proteolysis associated with the ensiling of chopped alfalfa. J. Dairy Sci. 71:152–158.

Fales, S.L. 1985. Effects of temperature on fiber concentration, composition, and in vitro digestion kinetics of tall fescue. Agron. J. 78:963–966.

Farlow, J.O. 1987. Speculations about the diet and digestive physiology of herbivorous dinosaurs. Paleobiology 13(1):60–72.

Fauconneau, G. 1959. Les acides organiques des plantes fourrasores. II. Variations avec le stade de développement chez le dactygle et la fétuque des grès. Ann. Physiol. Veg. 2:181–189.

Fennessy, P.F., and T.N. Barry. 1973. Identification and measurement of 2,3-butanediol in silage. J. Sci. Food Agric. 24:643–648.

Fick, G.W., and S.C. Mueller. 1989. Alfalfa quality, maturity, and mean stage of development. Cornell Univ. Agron. Dept., Info. Bull. 217.

Fick, G.W., and D.W. Onstad. 1988. Statistical models for predicting alfalfa herbage quality from morphological or weather data. J. Prod. Agric. 1:160–166.

Flatt, W.P. 1964. Methods of calorimetry—indirect. In: The Science of Nutrition of Farm Livestock. International Encyclopedia of Food and Nutrition, vol. 17. D.P. Cuthbertson, ed. Pergamon Press, Oxford. Pp. 491–520.

Flatt, W.P., and C.E. Coppock. 1965. Physiological factors influencing the energy metabolism of ruminants. In: Physiology of Digestion of the Ruminant. R.W. Dougherty, ed. Butterworths, Washington, D.C. Pp. 240–253.

Flatt, W.P., D.R. Waldo, J.F. Sykes, and L.A. Moore. 1958. A proposed method for indirect calorimetry for energy metabolism studies with large animals under field conditions. Proc. 1st Symp. on Energy Metabolism, Copenhagen. Pp. 101–109.

Fleming, S.E., D. Marthinsen, and H. Kuhnlein. 1983. Colonic function and fermentation in men consuming high fiber diets. J. Nutr. 113:2535–2544.

Fonnesbeck, P.V. 1968. Digestion of soluble and fibrous carbohydrate of forage by horses. J. Anim. Sci. 27:1336–1344.

Foose, T. 1982. Trophic strategies of ruminant versus non-ruminant ungulates. Ph.D. dissertation, Univ. of Chicago.

Forbes, J.M. 1986. The Voluntary Food Intake of Farm Animals. Butterworths, London.

Ford, C.W., and R.D. Hartley. 1990. Cyclodimers of p-coumaric and ferulic acids in the cell walls of tropical grasses. J. Sci. Food Agric. 50:29–44.

Foster, L.A. 1988. Clinical ketosis. In: The Veterinary Clinics of North America. Metabolic Diseases of Ruminant Livestock, vol. 4, pt. 2. T.H. Herdt, ed. Harcourt Brace Jovanovich, New York. Pp. 253–267.

Fox, D.G., C.J. Sniffen, J.D. O'Connor, J.B. Russell, and P.J. Van Soest. 1990. The Cornell net carbohydrate and protein system for evaluating cattle diets. Search Agriculture. Cornell Univ. Agric. Exp. Stn. no. 34. Ithaca, N.Y.

Fox, D.G., C.J. Sniffen, J.D. O'Connor, J.B. Russell, and P.J. Van Soest. 1992. A net carbohydrate and protein system for evaluating cattle diets. III. Cattle requirements and diet adequacy. J. Anim. Sci. 70:3578–3596.

France, J., M.S. Dhanoa, R.C. Siddons, J.H.M. Thornley, and D.P. Poppi. 1988. Estimating the production of faeces by ruminants from faecal marker concentration curves. J. Theor. Biol. 135:383–391.

Francis, G.L., J.M. Gawthorne, and G.B. Storer. 1978. Factors affecting the activity of cellulases isolated from the rumen digesta of sheep. Appl. Environ. Microbiol. 36:643–649.

Francis-Smith, K., and D.G.M. Wood-Gush. 1977. Coprophagia as seen in thoroughbred foals. Equine Vet. J. 9:155–157.

Franklin, W.L. 1983. Contrasting socioecologies of South America's wild camelids: the vicuña and guanaco. In: Advances in the Study of Mammalian Behavior. J.F. Eisenberg and D.G. Kleiman, eds. Spec. Publ. no. 7, American Society of Mammalogists. Pp. 573–627.

Frear, D.E.H. 1950. Agricultural Chemistry, vol. 1. Van Nostrand, Toronto. Pp. 118–119.

Freeland, W.J., and D.H. Janzen. 1974. Strategies in herbivory by mammals: the role of plant secondary compounds. Am. Nat. 108:269–289.

French, A.D. 1973. Chemical and physical properties of starch. J. Anim. Sci. 37:1048–1061.

French, A.D. 1979. Allowed and preferred shapes of amylose. Baker's Dig. 1:39–46.

Frobish, R.A., and C.L. Davis. 1977. Effects of abomasal infusions of glucose and propionate on milk yield and composition. J. Dairy Sci. 60:204–209.

Fry, S.C. 1988. The Growing Plant Cell Wall: Chemical and Metabolic Analysis. Wiley, New York.

Fry, S.C., and J.G. Miller. 1989. Toward a working model of the growing cell wall. Phenolic cross-linking reactions in the primary cell walls of dicotyledons. In: Plant Cell Wall Polymers. Biogenesis and Biodegradation. N.G. Lewis and M.G. Paice, eds. ACS Symp. Ser. 399. American Chemical Society, Washington, D.C. Pp. 33–46.

Fukushima, R.S., B.A. Dehority, and S.C. Loerch. 1991. Modification of a colorimetric analysis for lignin and its use in studying the inhibitory effects of lignin on forage digestion by ruminal microorganisms. J. Anim. Sci. 69:295.

Gabrielsen, B.C. 1986. Evaluation of marketed cellulases for activity and capacity to degrade forage. Agron. J. 78:838–843.

Gaillard, B.D.E. 1958. A detailed summative analysis of the crude fibre and nitrogen-free extractives fraction of roughages. I. Proposed scheme of analysis. J. Sci. Food Agric. 9:170.

Gaillard, B.D.E. 1962. The relationship between cell wall constituents of roughages and the digestibility of the organic matter. J. Agric. Sci. 59:369.

Gaillard, B.D.E., and G.N. Richards. 1975. Presence of soluble lignin-carbohydrate complexes in the bovine rumen. Carbohydr. Res. 42:135–145.

Ganong, W.F. 1987. Review of Medical Physiology. Appleton & Lange, Norwalk, Conn.

Garton, G.A. 1964. Aspects of lipid metabolism in ruminants. In: Metabolism and Physiological Significance of Lipids. R.M.C. Dawson and D.M. Rhodes, eds. Wiley, New York. Pp. 335–349.

Garton, G.A. 1965. The digestion and assimilation of lipids. In: Physiology of Digestion in the Ruminant. R.W. Dougherty, ed. Butterworths, Washington, D.C. P. 390.

Garton, G.A. 1967. The digestion and absorption of lipids in ruminant animals. World Rev. Nutr. Diet. 7:225.

Garton, G.A., F.D. DeB. Hovell, and W.R.H. Duncan. 1972. Influence of dietary volatile fatty acids on the fatty-acid composition of lamb with special reference to the effect of propionate on the presence of branched chain components. Br. J. Nutr. 28:409–416.

Gee, J.M., R.M. Faulks, and I.T. Johnson. 1991. Physiological effects of retrograded, α-amylase-resistant cornstarch in rats. J. Nutr. 121:44–49.

Gengler, W.R., F.A. Martz, H.D. Johnson, G.F. Krause, and L. Hahn. 1970. Effect of temperature on food and water intake and rumen fermentation. J. Dairy Sci. 53:434–437.

Genthner, S.B.R., C.L. Davis, and M.P. Bryant. 1981. Features of rumen and sewage sludge strains of *Eubacterium limosum,* a methyl-utilizing and hydrogen–carbon dioxide-utilizing species. Appl. Environ. Microbiol. 42:12–19.

Gentry, A.W. 1978. Tragulidae and Camelidae. In: Evolution of African Mammals. V.J. Maglio and H.B.S. Cooke, eds. Harvard Univ. Press, London. Pp. 536–539.

Gentry, A.W., and J.J. Hooker. 1988. The phylogeny of the Artiodactyla. In: The Phylogeny and Classification of the Tetrapods. Vol. 2: Mammals. M.J. Benton, ed. Systematics Assoc. Spec. Vol. no. 35B. Clarendon Press, Oxford. Pp. 235–272.

Giesecke, D. 1970. Comparative microbiology of the alimentary tract. In: Physiology of Digestion and Metabolism in the Ruminant. A.T. Phillipson, ed. Oriel Press, Newcastle upon Tyne, England. Pp. 306–318.

Giesecke, D., and N.O. Van Gylswyck. 1975. A study of feeding types and certain rumen functions in six species of South African wild ruminants. J. Agric. Sci. 85:75–83.

Giger, S. 1985. Revue sur les méthodes de dosage de la lignine utilisées en alimentation animale. Ann. Zootech. 34:85–122.

Gill, M., A.J. Rook, and L.R.S. Thiago. 1988. Factors affecting the voluntary intake of roughages by the dairy cow. In: Nutrition and Lactation in the Dairy Cow. P.C. Garnsworthy, ed. Butterworths, London. Pp. 262–279.

Gill, S.S., H.R. Conrad, and J.W. Hibbs. 1969. Relative rate of in vitro cellulose disappearance as a possible estimator of digestible dry matter intake. J. Dairy Sci. 52:1687–1690.

Giner-Chavez, B., R.G. Warner, P.J. Van Soest, M.L. Thonney, and J.B. Robertson. 1990. Evaluation of internal markers for apparent dry matter digestibility in calves. (Abstr.) J. Anim. Sci. 68(Suppl. 1):583.

Glade, M.J. 1984. The influence of dietary fiber digestibility on the nitrogen requirements of mature horses. J. Anim. Sci. 58:638.

Gluckman, P.D., B.H. Breier, and S.R. Davis. 1987. Physiology of the somatotropic axis with particular reference to the ruminant. J. Dairy Sci. 70:442–466.

Godshalk, E.B., W.F. McClure, J.C. Burns, D.H. Timothy, and D.S. Fisher. 1988. Heritability of cell wall carbohydrates in switchgrass. Crop Sci. 28:736.

Goering, H.K. 1976. A laboratory assessment on the frequency of overheating in commercial dehydrated alfalfa samples. J. Anim. Sci. 43:869–872.

Goering, H.K., C.H. Gordon, R.W. Hemken, D.R. Waldo, P.J. Van Soest, and L.W. Smith. 1972. Analytical estimates of nitrogen digestibility in heat damaged forages. J. Dairy Sci. 55:1275.

Goering, H.K., and P.J. Van Soest. 1970. Forage fiber analyses (apparatus, reagents, procedures and some applications). USDA-ARS Agricultural Handbook 379. U.S. Government Printing Office, Washington, D.C.

Goering, H.K., P.J. Van Soest, and R.W. Hemken. 1973. Relative susceptibility of forages to heat damage as affected by moisture, temperature and pH. J. Dairy Sci. 56:137–143.

Goetsch, A.L., and F.N. Owens. 1985. Effects of sarsaponin on digestion and passage rates in cattle fed medium to low concentrate. J. Dairy Sci. 68:2377.

Goldschmidt, V.M. 1958. Geochemistry. A. Muir, ed. Clarendon Press, Oxford.

Goldstein, J.L., and T. Swain. 1963. Changes in tannins in ripening fruits. Phytochemistry 2:371–383.

Gómez, G., M. Valdivieso, D. De la Cuesta, and T.S. Salcedo. 1984. Effect of variety and plant age on the cyanide content of whole-root cassava chips and its reduction by sun-drying. Anim. Feed Sci. Technol. 11:57.

Gorham, J. 1820. Chemical analysis of Indian corn. N. Engl. J. Med. Surg. 4:320–328.

Gorosito, A.R., J.B. Russell, and P.J. Van Soest. 1985. Effect of carbon-4 and carbon-5 volatile fatty acids on digestion of plant cell wall in vitro. J. Dairy Sci. 68:840–847.

Goto, I., and D.J. Minson. 1977. Prediction of the dry matter digestibility of tropical grasses using a pepsin-cellulase assay. Anim. Feed Sci. Technol. 2:247–253.

Gottschalk, G. 1988. Cellulose degradation and the carbon cycle. In: Biochemistry and Genetics of Cellulose Degradation. J.P. Aubert, P. Beguin, and J. Miller, eds. Academic Press, New York. Pp. 3–8.

Gould, J.M. 1985. Enhanced polysaccharide recovery from agricultural residues and perennial grasses treated with alkaline hydrogen peroxide. Biotechnol. Bioeng. 27:893–896.

Grace, N.D., and D.R. Body. 1979. Dietary lipid and the absorption of Mg and Ca: effect of corn oil infused via the rumen on the absorption of Mg and Ca in sheep fed white clover. N. Z. J. Agric. Res. 22:405–410.

Grajal, A., S.D. Strahl, R. Parra, M.G. Dominguez, and A. Neher. 1989. Foregut fermentation in the hoatzin, a neotropical leaf-eating bird. Science 245:1236.

Grant, Richard J., V.F. Colenbrander, and D.R. Mertens. 1990. Milk fat depression in dairy cows: role of particle size of alfalfa hay. J. Dairy Sci. 73:1823–1833.

Grant, Richard J., and D.R. Mertens. 1992. Influence of buffer pH and raw corn starch addition on in vitro fiber digestion kinetics. J. Dairy Sci. 75:2762–2768.

Grant, Robert J. 1973. Digestibility of napier grass by Philippine cattle and water buffaloes, and in vitro digestibility and composition of Philippine feedstuffs. Ph.D. dissertation, Cornell Univ., Ithaca, N.Y.

Grant, Robert J., P.J. Van Soest, and R.E. McDowell. 1974. Influence of rumen fluid source and fermentation time on in vitro true dry matter digestibility. J. Dairy Sci. 57:1201–1205.

Grant, Robert J., P.J. Van Soest, R.E. McDowell, and C.B. Perez, Jr. 1974. Intake, digestibility and metabolic loss of napier grass by cattle and buffaloes when fed wilted, chopped and whole. J. Anim. Sci. 39:423–434.

Greenberg, N.A., and W.F. Shipe. 1979. Comparison of the abilities of trichloroacetic, picric, sulfosalicylic, and tungstic acids to precipitate protein hydrolysates and proteins. J. Food Sci. 44:735–737.

Grovum, W.L. 1984. The control of motility of the ruminoreticulum. In: Control of Digestion and Metabolism in Ruminants. L.P. Milligan, W.L. Grovum, and A. Dobson, eds. Prentice-Hall, Englewood Cliffs, N.J. Pp. 18–40.

Grovum, W.L. 1987. A new look at what is controlling feed intake. In: Symposium Proceedings: Feed Intake by Beef Cattle. F.N. Owens, ed. Oklahoma State Univ., Stillwater.

Grovum, W.L., and V.J. Williams. 1973. Rate of passage of digesta in sheep. IV. Passage of marker through the alimentary tract and the biological relevance of rate constants derived from the changes in concentration of marker in feces. Br. J. Nutr. 30:313–329.

Grovum, W.L., and V.J. Williams. 1977. Rate of passage of digesta in sheep. 6. The effect of level of food intake on mathematical predictions of the kinetics of digesta in the reticulorumen and intestines. Br. J. Nutr. 38:425–436.

Grubb, J.A., and B.A. Dehority. 1975. Effects of an abrupt change in ration from all roughage to high concentrate upon rumen microbial numbers in sheep. Appl. Microbiol. 30:404.

Grunes, D.L., and R.M. Welch. 1989. Plant contents of magnesium, calcium and potassium in relation to ruminant nutrition. J. Anim. Sci. 67:3485–3494.

Gupta, S.S., and T.P. Hilditch. 1951. The component acids and glycerides of a horse mesenteric fat. Biochem. J. 48:137.

Gutierrez-Vargas, R., J.A. Arroyo-Aguilu, and A. Ramirez-Ortiz. 1978. Voluntary intake, chemical composition and nutrient digestibility of Pangolagrass and stargrass hays. J. Agric. Univ. P. R. 62:389.

Hackenberger, M.K. 1987. Diet digestibilities and ingesta transit times of captive Asian (*Elephas maximus*) and African (*Loxodonta africana*) elephants. M.Sc. thesis, Univ. of Guelph.

Hagerman, A.E. 1987. Radial diffusion method for determining tannin in plant extracts. J. Chem. Ecol. 13:437.

Hagerman, A.E., and L.G. Butler. 1978. Protein precipitation method for the quantitative determination of tannins. J. Agric. Food Chem. 26:809–812.

Hagerman, A.E., and L.G. Butler. 1980. Determination of protein and tannin-protein precipitates. J. Agric. Food Chem. 28:944–947.

Hagerman, A.E., and L.G. Butler. 1989. Choosing appropriate methods and standards for assaying tannin. J. Chem. Ecol. 15:1795–1810.

Hagerman, A.E., C.T. Robbins, Y. Weerasuriya, T.C. Wilson, and C. McArthur. 1992. Tannin chemistry in relation to digestion. J. Range Manage. 45:57–62.

Hanley, T.A., C.T. Robbins, A.E. Hagerman, and C. McArthur. 1992. Predicting digestible protein and digestible dry matter in tannin-containing forages consumed by ruminants. Ecology 73(2):537–541.

Hansen, R.M., R.C. Clark, and W. Lawhorn. 1977. Foods of wild horses, deer and cattle in the Douglas Mountain area, Colorado. J. Range Manage. 30:116–118.

Harfoot, C.G. 1981. Lipid metabolism in the rumen. In: Lipid Metabolism in Ruminant Animals. W.W. Christie, ed. Pergamon Press, New York. P. 11.

Harkin, J.M. 1969. Methods of attacking the problems of lignin structure. Recent Adv. Phytochem. 2:35.

Harkin, J.M. 1973. Lignin. In: Chemistry and Biochemistry of Herbage, vol. 1. E.W. Butler and R.W. Bailey, eds. Academic Press, London. Pp. 323–373.

Harris, L.E., E.W. Crampton, A.D. Knight, and A. Denny. 1967. The collection and summarization of feed composition data. II. A proposed source form for collection of feed composition data. J. Anim. Sci. 26:97–105.

Harris, P.J., and R.D. Hartley. 1980. Phenolic constituents of the cell walls of monocotyledons. Biochem. Syst. Ecol. 8:153–160.

Harrison, D.G., D.E. Beever, D.J. Thomson, and D.F. Osbourn. 1973. The influence of diet upon the quantity and types of amino acids entering and leaving the small intestine of sheep. J. Agric. Sci. 81:391–401.

Harrison, J.H. 1989. Use of silage additives and their effect on animal productivity. Proc. Pacific Northwest Animal Nutrition Conf., Oct. 24–26, Boise, Idaho.

Hart, R.H. 1978. Stocking rate theory and its application to grazing on rangelands. In: Proc. 1st Int. Rangeland Congr. Denver Society for Range Management, Denver, Colo. Pp. 547–552.

Hart, R.H., J.W. Waggoner, Jr., T.G. Dunn, C.C. Kaltenbach, and L.D. Adams. 1988. Optimal stocking rate for cow-calf enterprises on native range and complementary improved pastures. J. Range Manage. 41:435.

Hartley, R.D. 1972. *p*-Coumaric and ferulic acid components of cell walls of ryegrass and their relationships with lignin and digestibility. J. Sci. Food Agric. 23:1347–1354.

Hartley, R.D. 1981. Chemical constitution, properties and processing of lignocellulosic wastes in relation to nutritional quality for animals. Agric. Environ. 6:87.

Hartley, R.D. 1983. Degradation of cell walls of forages by sequential treatment with sodium hydroxide and a commercial cellulase preparation. J. Sci. Food Agric. 34:29–36.

Hartley, R.D., and D.E. Akin. 1989. Effect of forage cell wall phenolic acids and derivatives on rumen microflora. J. Sci. Food Agric. 49:405–412.

Hartley, R.D., D.E. Akin, D.S. Himmelsbach, and D.C. Beach. 1990a. Microspectrophotometry of bermudagrass (*Cynodon dactylon*) cell walls in relation to lignification and wall biodegradability. J. Sci. Food Agric. 50:179–189.

Hartley, R.D., and P.J. Harris. 1981. Phenolic constituents of the cell walls of dicotyledons. Biochem. Syst. Ecol. 9:189–203.

Hartley, R.D., and E.C. Jones. 1976. Diferulic acid as a component of cell walls of *Lolium multiflorum*. Phytochemistry 15:1157–1160.

Hartley, R.D., and E.C. Jones. 1978. Phenolic components and degradability of the cell walls of the brown midrib mutant, bm3, of *Zea mays*. J. Sci. Food Agric. 29:777–782.

Hartley, R.D., E.C. Jones, and J.S. Fenlon. 1974. Prediction of the digestibility of forages by treatment of their cell walls with cellulolytic enzymes. J. Sci. Food Agric. 25:947–954.

Hartley, R.D., W.H. Morrison III, D.S. Himmelsbach, and W.S. Borneman. 1990b. Cross-linking of cell wall phenolic arabinoxylans in graminaceous plants. Phytochemistry 29:3705–3709.

Haslam, E. 1989. Plant Polyphenols. Vegetable Tannins Revisited. Cambridge Univ. Press, Cambridge. P. 230.

Hatfield, R.D. 1993. Cell wall matrix interactions and degradability. In: Forage Cell Wall Structure and Digestibility. H.G. Jung, D.R. Buxton, R.D. Hatfield, and J. Ralph, eds. American Society of Agronomy, Madison, Wisc. Pp. 285–314.

Haubner, G. 1855. Die Planzenfaser und ihre Verdaulichkeit. Z. Dtsch. Landwirtschaft. 6:177–182.

Hawkins, D.R., H.E. Henderson, and D.B. Purser. 1970. Effect of dry matter levels of alfalfa silage on intake and metabolism in the ruminant. J. Anim. Sci. 31:617–625.

Hayssen, V., A. Van Tienhoven, and A. Van Tienhoven. 1993. Asdell's Patterns of Mammalian Reproduction: A Compendium of Species-Specific Data. Cornell Univ. Press, Ithaca.

Healy, J.B., Jr., and L.Y. Young. 1979. Anaerobic biodegradation of eleven aromatic compounds to methane. Appl. Environ. Microbiol. 38:84–89.

Healy, J.B., Jr., L.Y. Young, and M. Reinhert. 1980. Methanogenic decomposition of ferulic acid, a model lignin derivative. Appl. Environ. Microbiol. 39:436–444.

Hegarty, M.P., and P.J. Peterson. 1973. Free amino acids, bound amino acids, amines and ureides. In: Chemistry and Biochemistry of Herbage, vol. 1. G.W. Butler and R.W. Bailey, eds. Academic Press, New York. Pp. 1–62.

Henneberg, W., and F. Stohmann. 1860, 1864. Begründung einer rationellen Fütterung der Wiederkäuer, vols. 1 and 2. Schwetschke und Söhne, Braunschweig.

Henry, W.A. 1898. Feeds and Feeding: A Handbook for the Student and Stockman. Madison, Wisc.

Hespell, R.B. 1987. Biotechnology and modifications of the rumen microbial ecosystem. Proc. Nutr. Soc. 46:407–413.

Hespell, R.B., and M.P. Bryant. 1979. Efficiency of rumen microbial growth: influence of some theoretical and experimental factors in Y_{ATP}. J. Anim. Sci. 49:1640–1659.

Hewitt, D., and J.E. Ford. 1982. Influence of tannins on the protein nutritional quality of food grains. Proc. Nutr. Soc. 41:7–18.

Hibberd, C.A., D.G. Wagner, R.L. Schemm, E.D. Mitchell, Jr., D.E. Weibel, and R.L. Hintz. 1982. Digestibility characteristics of isolated starch from sorghum and corn grain. J. Anim. Sci. 55:1490–1497.

Hill, G.M., P.R. Utley, and G.L. Newton. 1986. Digestibility and utilization of ammonia-treated and urea-supplemented peanut skin diets fed to cattle. J. Anim. Sci. 63:705.

Hill, K.J. 1961. Digestive secretions in the ruminant. In: Digestive Physiology and Nutrition of the Ruminant. D. Lewis, ed. Butterworths, London. Pp. 48–58.

Hills, J.L., C.H. Jones, and P.A. Benedict. 1910. Commercial feeding stuffs. Principles and practice of stock feeding. Vt. Agric. Exp. Stn. Bull. 152.

Himmelsbach, D.S., and F.E. Barton II. 1980. [13]C nuclear magnetic resonance of grass lignins. J. Agric. Food Chem. 28:1203–1208.

Himmelsbach, D.S., F.E. Barton II, and W.R. Windham. 1983. Comparison of carbohydrate, lignin, and protein ratios between grass species by cross polarization–magic angle spinning carbon-13 nuclear magnetic resonance. J. Agric. Food Chem. 31:401–404.

Hintz, H.F., R.A. Argenzio, and H.F. Schryver. 1971. Digestion coefficients, blood glucose levels and molar percentage of volatile acids in intestinal fluid of ponies fed varying forage-grain ratios. J. Anim. Sci. 33:992.

Hintz, H.F., H.F. Schryver, and M. Halbert. 1973. A note on the comparison of digestion by New World camels, sheep and ponies. Anim. Prod. 16:303–305.

Hintz, H.F., C.J. Sedgewick, and H.F. Schryver. 1976. Some observations on digestion of a pelleted diet by ruminants and non-ruminants. Int. Zoo. Yearb. 16:54–57.

Hirayama, K., S. Kawamura, T. Mitsuoka, and K. Tashiro. 1989. The faecal flora of the giant panda (*Ailuropoda melanoleuca*). J. Appl. Bacteriol. 67:411–415.

Hobson, P.N. 1988. The Rumen Microbial Ecosystem. Elsevier Applied Science Publ., New York. P. 527.

Hodge, J.E. 1953. Chemistry of browning reactions in model systems. J. Agric. Food Chem. 1:928–943.

Hodgson, J.F., R.M. Leach, Jr., and W.H. Allaway. 1962. Micronutrients in soils and plants in relation to animal nutrition. J. Agric. Food Chem. 10:171–174.

Hoernicke, H., W.F. Williams, D.R. Waldo, and W.P. Flatt. 1965. Composition and absorption of rumen gases and their importance for the accuracy of respiration trials with tracheostomized ruminants. Proc. 3d Symp. on Energy Metabolism. Academic Press, London. Pp. 165–178.

Hoffman, M., O. Steinhofel, and R. Fuchs. 1987. Studies of the digestibility of crude nutrients in horses. 2. Comparative studies of the digestive capacities of thoroughbred horses, ponies and wethers. Arch. Tierernähr. 37:351.

Hofmann, R.R. 1973. The Ruminant Stomach: Stomach Structure and Feeding Habits of East African Game Ruminants. East Afr. Lit. Bureau, Nairobi, Kenya. P. 354.

Hofmann, R.R. 1986. Morphophysiological evolutionary adaptations of the ruminant digestive system. In: Aspects of Digestive Physiology in Ruminants. Alan Dobson and Marjorie Dobson, eds. Cornell Univ. Press, Ithaca. P. 1.

Hofmann, R.R. 1988. Anatomy of the gastrointestinal tract. In: The Ruminant Animal. D.C. Church, ed. Prentice-Hall, Englewood Cliffs, N.J. P. 14.

Hofmann, R.R. 1989. Evolutionary steps of ecophysiological adaptation and diversification of ruminants: a comparative view of their digestive system. Oecologia 78:443–457.

Hofmann, R.R., and K. Nygren. 1992. Morphophysiological specialization and adaptation of the moose digestive system. Alces (Suppl. 1):91–100.

Hogan, J.P., P.A. Kenney, and R.H. Weston. 1987. Factors affecting the intake of feed by grazing animals. In: Temperate Pastures: Their Production, Use and Management. J.L. Wheeler, C.J. Pearson, and G.E. Robards, eds. CSIRO, Canberra. Pp. 317–327.

Hogan, J.P., and R.H. Weston. 1967. The digestion of chopped and ground roughages by sheep. II. The digestion of nitrogen and some carbohydrate fractions in the stomach and intestines. Aust. J. Agric. Res. 18:803.

Hogan, J.P., and R.H. Weston. 1970. Quantitative aspects of microbial protein synthesis in the rumen. In: Physiology of Digestion and Metabolism in the Ruminant. A.T. Phillipson, ed. Oriel Press, Newcastle upon Tyne, England. Pp. 474–485.

Hoglund, C.R. 1964. Comparative storage losses and feeding values of alfalfa and corn silage crops when harvested at different moisture levels and stored in gas-tight and conventional tower soils: an appraisal of research results. Michigan State Univ., Dept. of Agric. Econ. Mimeo 946.

Holdeman, L.V., I.J. Good, and W.E.C. Moore. 1976. Human fecal flora: variation in bacterial composition within individuals and a possible effect of emotional stress. Appl. Environ. Microbiol. 31:359–375.

Holechek, J.L., M. Vavra, and R.D. Pieper. 1982a. Botanical composition determination of range herbivore diets: a review. J. Range Manage. 35:309–315.

Holechek, J.L., M. Vavra, and R.D. Pieper. 1982b. Methods for determining the nutritive quality of range ruminant diets: a review. J. Anim. Sci. 54:364–376.

Holechek, J.L., H. Wofford, D. Arthun, M.L. Galyean, and J.D. Wallace. 1986. Evaluation of total fecal collection for measuring cattle forage intake. J. Range Manage. 39:2–4.

Holmgren, V.C. 1972. The other panda. Anim. Kingdom 75:6–10.

Hood, R.L., C.E. Allen, R.D. Goodrich, and J.C. Meiske. 1971. A rapid method for the direct chemical determination of water in fermented feeds. J. Anim. Sci. 33:1310–1314.

Hooper, A.P., and J.G. Welch. 1983. Chewing efficiency and body size of kid goats. J. Dairy Sci. 66:2551–2556.

Hooper, A.P., and J.G. Welch. 1985. Effects of particle size and forage composition on functional specific gravity. J. Dairy Sci. 68:1181.

Hoover, W.H., and S.D. Clarke. 1972. Fiber digestion in the beaver. J. Nutr. 102:9–15.

Hoover, W.H., B.A. Crooker, and C.J. Sniffen. 1976. Effects of differential solid-liquid removal rates on protozoa numbers in continuous cultures of rumen contents. J. Anim. Sci. 43:528.

Hoppe, P.P. 1977. Rumen fermentation and body weight in African ruminants. Proc. 13th Int. Congr. of Game Biologists. T.J. Peterle, ed. Wildlife Society, Washington, D.C. Pp. 141–150.

Hopps, H.C., E.M. Carlisle, J.A. McKeague, R. Siever, and P.J. Van Soest. 1977. Silicon. In: Geochemistry and the Environment. Vol. 2: The Relation of Other Selected Trace Elements to Health and Disease. E.T. Mertz, ed. National Academy of Sciences, Washington, D.C. P. 54.

Horecker, B.L., E. Stotz, and T.R. Hogness. 1939. The promotion effect of aluminum, chromium and the rare earths in the succinic dehydrogenase–cytochrome system. J. Biol. Chem. 128:251.

Horsford, E.N. 1846. Value of different kinds of vegetable food, based upon the amount of nitrogen. Philos. Mag. Ser. 3 29:365–397.

Horvath, P.J. 1981. The nutritional and ecological significance of acer-tannins and related polyphenols. M.S. thesis, Cornell Univ., Ithaca, N.Y. P. 138.

Houpt, T.R., and K.A. Houpt. 1968. Transfer of urea nitrogen across the rumen wall. Am. J. Physiol. 214:1296–1303.

Houpt, T.R., and K.A. Houpt. 1971. Nitrogen conservation by ponies fed a low-protein ration. Am. J. Vet. Res. 32(4):579–588.

Hsu, J.C., and M.H. Penner. 1989. Influence of cellulose structure on its digestibility in the rat. J. Nutr. 119:872–878.

Huber, T.L. 1976. Physiological effects of acidosis on feedlot cattle. J. Anim. Sci. 43:902–909.

Huffman, C.F., and C.W. Duncan. 1949. The nutritive value of alfalfa hay. III. Corn as a supplement to an all-alfalfa hay ration for milk production. J. Dairy Sci. 32:465–474.

Hughes, A.D. 1970. The non-protein nitrogen composition of grass silages. II. The changes occurring during the storage of silage. J. Agric. Sci. 75:421–431.

Hughes, A.D. 1971. The non-protein nitrogen composition of grass silage. III. The composition of spoilt silages. J. Agric. Sci. 76:329–336.

Hulse, J.H., E.M. Laing, and O.E. Pearson. 1980. Sorghum and the Millets: Their Composition and Nutritive Value. Academic Press, London.

Hume, I.D. 1974. Nitrogen and sulphur retention and fibre digestion by euros, red kangaroos and sheep. Aust. J. Zool. 22:13–23.

Hume, I.D. 1978. Evolution of the Macropodidae digestive system. Aust. Mammal. 2:37–41.

Hume, I.D. 1982. Digestive Physiology and Nutrition of Marsupials. Monographs on Marsupial Biology. Cambridge Univ. Press.

Hume, I.D., R.J. Moir, and M. Somers. 1970. Synthesis of microbial protein in the rumen. I. Influence of the level of nitrogen intake. Aust. J. Agric. Res. 21:283–296.

Hume, I.D., K.R. Morgan, and G.J. Kenagy. 1993. Digesta retention and digestive performance in sciurid and microtine rodents: effects of hindgut morphology and body size. Physiol. Zool. 66(3):396–411.

Hume, I.D., and A.C.I. Warner. 1980. The evolution of

fermentative digestion. In: Digestive Physiology and Metabolism in Ruminants. Y. Ruckebusch and P. Thivend, eds. M.T.P. Press, Lancaster, England. P. 665.

Hungate, R.E. 1950. The anaerobic, mesophilic, cellulolytic bacteria. Bacteriol. Rev. 14:1–49.

Hungate, R.E. 1966. The Rumen and Its Microbes. Academic Press, New York.

Hungate, R.E. 1979. Evolution of a microbial ecologist. Ann. Rev. Microbiol. 33:1–20.

Hungate, R.E., R.A. Mah, and M. Simesen. 1961. Rates of production of individual volatile fatty acids in the rumen of lactating cows. Appl. Microbiol. 9:554–561.

Hungate, R.E., G.D. Phillips, A. McGregor, D.P. Hungate, and H.K. Beuchner. 1959. Microbial fermentation in certain mammals. Science 130:1192–1194.

Hunter, R.A., and B.D. Siebert. 1985. Utilization of low-quality roughage by *Bos taurus* and *Bos indicus* cattle. 1. Rumen digestion. Br. J. Nutr. 53:637–648.

Huston, J.E., B.S. Rector, W.C. Ellis, and M.L. Allen. 1986. Dynamics of digestion in cattle, sheep, goats, and deer. J. Anim. Sci. 62:208–215.

Hutchinson, G.E. 1945. Aluminum in soils, plants and animals. Soil Sci. 60:29.

Iggo, A., and B.F. Leek. 1970. Sensory receptors in the ruminant stomach and their reflex effects. In: Physiology of Digestion and Metabolism in the Ruminant. A.T. Phillipson, ed. Oriel Press, Newcastle upon Tyne, England. Pp. 23–34.

Iiyama, K., T.B.T. Lam, P.J. Meikle, K. Ng, D.I. Rhodes, and B.A. Stone. 1993. Cell wall biosynthesis and its regulation. In: Forage Cell Wall Structure and Digestibility. H.G. Jung, D.R. Buxton, R.D. Hatfield, and J. Ralph, eds. American Society of Agronomy, Madison, Wisc. Pp. 621–683.

Iiyama, K., and A.F.A. Wallis. 1990. Determination of lignin in herbaceous plants by an improved acetyl bromide procedure. J. Sci. Food Agric. 51:145–162.

Iler, R.K. 1955. The Colloid Chemistry of Silica and Silicates. Cornell Univ. Press, Ithaca.

Illius, A.W., and I.J. Gordon. 1987. The allometry of food intake in grazing ruminants. J. Anim. Ecol. 56:989–999.

Illius, A.W., and I.J. Gordon. 1991. Prediction of intake and digestion in ruminants by a model of rumen kinetics integrating animal size and plant characteristics. J. Agric. Sci. 116:145–157.

Isaacson, H.R., F.C. Hinds, M.P. Bryant, and F.N. Owens. 1975. Efficiency of energy utilization by mixed bacteria in continuous culture. J. Dairy Sci. 58:1645–1659.

Ismail-Beigi, F., J.G. Reinhold, B. Faraji, and P. Abadi. 1977. Effects of cellulose added to diets of low and high fiber content upon the metabolism of calcium, magnesium, zinc and phosphorus by man. J. Nutr. 107:510–518.

Jackson, J.A., Jr., R.W. Hemken, J.A. Boling, R.J. Harmon, R.C. Buckner, and L.P. Bush. 1984. Loline alkaloids in tall fescue hay and seed and their relationship to summer fescue toxicosis in cattle. J. Dairy Sci. 67:104–109.

Jackson, M.G. 1977. Review article: the alkali treatment of straws. Anim. Feed Sci. Technol. 2:105.

Jacobson, D.R., H.H. VanHorn, and C.J. Sniffen. 1970. Lactating ruminants. Fed. Proc. 29:35.

Janis, C.M. 1976. The evolutionary strategy of the Equidae and the origins of rumen and cecal digestion. Evolution 30:757–774.

Janis, C.M. 1984. The use of fossil ungulate communities as indicators of climate and environment. In: Fossils and Climate. P. Brenchley, ed. John Wiley and Sons, New York. Pp. 85–104.

Janis, C.M. 1986. Evolution of horns and related structures in hoofed animals. Discovery 19(2):9–17.

Janis, C.M. 1987. Fossil ungulate mammals depicted on archeological artifacts. Cryptozoology 6:8–23.

Janis, C.M., and D. Ehrhardt. 1988. Correlation of relative muzzle width and relative incisor width with dietary preference in ungulates. Zool. J. Linn. Soc. 92:267–284.

Janis, C.M., and M. Fortelius. 1988. On the means whereby mammals achieve increased functional durability of their dentitions, with special reference to limiting factors. Biol. Rev. Camb. Philos. Soc. 63:197–230.

Janis, C.M., and K.M. Scott. 1987. The interrelationships of higher ruminant families with special emphasis on the members of the Cervoidea. Am. Mus. Novit. 2893:1–85.

Janis, C.M., and K.M. Scott. 1988. The phylogeny of the Ruminantia (Artiodactyla, Mammalia). In: The Phylogeny and Classification of the Tetrapods. Vol. 2: Mammals. M.J. Benton, ed. Systematics Assoc. Spec. Vol. no. 35B. Clarendon Press, Oxford. Pp. 273–282.

Janzen, D.H., M.W. Demment, and J.B. Robertson. 1985. How fast and why do germinating guanacaste seeds (*Enterolobium cyclocarpum*) die inside cows and horses? Biotropica 17:322–325.

Jarvis, M.C. 1984. Structure and properties of pectin gels in plant cell walls. Plant Cell Environ. 7:153.

Jenkins, D.J.A., A.L. Jenkins, T.M.S. Wolever, G.R. Collier, A. Venket Rao, and L.U. Thompson. 1987. Starchy foods and fiber: reduced rated of digestion and improved carbohydrate metabolism. Scand. J. Gastroenterol. 22:132.

Jenkins, D.J.A., A.R. Leeds, M.A. Gassull, B. Cochet, and K.G. Alberti. 1977. Decrease in postprandial insulin and glucose concentrations by guar and pectin. Ann. Intern. Med. 86:20.

Jenkins, T.C., and M.L. Thonney. 1988. Effect of propionate level in a volatile fatty acid salt mixture fed to lambs on weight gain, body composition and plasma metabolites. J. Anim. Sci. 66:1028–1035.

Jenny, B.F., and C.E. Polan. 1975. Postprandial blood glucose and insulin in cows fed high grain. J. Dairy Sci. 58:512–514.

Jeraci, J.L., and B.A. Lewis. 1989. Determination of soluble fiber components: $(1\rightarrow3; 1\rightarrow4)$-β-D-glucans and pectins. Anim. Feed Sci. Technol. 23:15–25.

Jeraci, J.L., J.B. Robertson, and P.J. Van Soest. 1980. Effects of inocula on feed classification. J. Anim. Sci. 51(Suppl.):372–373.

Johnson, D.E., M. Branine, and G.M. Ward. 1990. Methane emissions from livestock. Proc. American Feed Industry Assoc. Nutrition Symp. Pp. 33–55.

Johnson, D.E., T.M. Hill, and G.M. Ward. 1992. Methane emissions from cattle: global warming and management issues. Proc. Minnesota Nutrition Conf., St. Paul, Minn.

Johnson, L.R. 1987. Physiology of the Digestive Tract. 2 vols. Raven Press, New York.

Johnson, T.R., J.W. Thomas, C.A. Rotz, and M.B. Tesar. 1984. Drying rate of cut forages after spray treatments to hasten drying. J. Dairy Sci. 67:1745.

Johnson, W.L. 1966. The nutritive value of *Panicum maximum* (guinea grass) for cattle and water buffaloes in the tropics. Ph.D. dissertation, Cornell Univ., Ithaca, N.Y.

Jones, C.A. 1985. C$_4$ Grasses and Cereals, Growth, Development, and Stress Response. John Wiley & Sons, New York.

Jones, L.H.P. 1978. Mineral components of plant cell walls. Am. J. Clin. Nutr. 31(Suppl.):S94–S97.

Jones, L.H.P., and K.A. Handreck. 1967. Silica in soils, plant and animals. A review. Adv. Agron. 19:107.

Jones, R.J., and R.G. Megarrity. 1983. Comparative toxicity response of goats fed on *Leucaena leucocephala* in Australia and Hawaii. Aust. J. Agric. Res. 34:781–790.

Jones, W.T., R.B. Broadhurst, and J.W. Lyttleton. 1976. The condensed tannins of pasture legume species. Phytochemistry 9:1407–1409.

Jones, W.T., and J.W. Lyttleton. 1971. Bloat in cattle. 34. A survey of forage legumes that do and do not produce bloat. N. Z. J. Agric. Res. 14:101–107.

Jorgenson, N.A., and L.H. Schultz. 1965. Ration effects on rumen acids, ketogenesis and milk composition. II. Restricted roughage feeding. J. Dairy Sci. 48:1040–1045.

Jouany, J.P., D.I. Demeyer, and J. Grain. 1988. Effect of defaunating the rumen. Anim. Feed Sci. Technol. 21:229–266.

Jung, H.G. 1985. Inhibition of structural carbohydrate fermentation by forage phenolics. J. Sci. Food Agric. 36:74–80.

Jung, H.G. 1988. Inhibitory potential of phenolic-carbohydrate complexes released during ruminal fermentation. J. Agric. Food Chem. 36:782–788.

Jung, H.G., and D.A. Deetz. 1993. Cell wall lignification and degradability. In: Forage Cell Wall Structure and Digestibility. H.G. Jung, D.R. Buxton, R.D. Hatfield, and J. Ralph, eds. American Society of Agronomy, Madison, Wisc. Pp. 315–346.

Jung, H.G., and G.C. Fahey, Jr. 1983. Nutritional implications of phenolic monomers and lignin: a review. J. Anim. Sci. 57:206–219.

Jung, H.G., and K.P. Vogel. 1986. Influence of lignin on digestibility of forage cell wall material. J. Anim. Sci. 62:1703.

Kalu, B.A. 1976. Age and time of year effects on alfalfa quality and morphological components of three legumes. M.S. thesis, Cornell Univ., Ithaca, N.Y.

Kalu, B.A., and G.W. Fick. 1981. Quantifying morphological development of alfalfa for studies of herbage quality. Crop Sci. 21:267–271.

Kalu, B.A., and G.W. Fick. 1983. Morphological stage of development as a predictor of alfalfa herbage quality. Crop Sci. 23:1167–1172.

Kandylis, K. 1984. The role of sulphur in ruminant nutrition. A review. Livest. Prod. Sci. 11:611–624.

Kane, E.A., W.C. Jacobson, and P.M. Damewood, Jr. 1959. Effect of corn starch on digestibility of alfalfa hay. J. Dairy Sci. 42:849–855.

Kass, M., P.J. Van Soest, W.G. Pond, B.A. Lewis, and L.E. McDonald. 1980. Utilization of dietary fiber from alfalfa by growing swine. I. Apparent digestibility of diet components in specific segments of the gastrointestinal tract. J. Anim. Sci. 50:175–197.

Katz, I., and M. Keeny. 1966. Characterization of the octadecenoic acids in rumen digesta and rumen bacteria. J. Dairy Sci. 49:962.

Kautz, J.E., and G.M. Van Dyne. 1978. Comparative analyses of diets of bison, cattle, sheep, and pronghorn antelope on shortgrass prairie in northeastern Colorado. U.S.A. Proc. 1st Int. Rangelands Congr. D.N. Hyder, ed. Society for Range Management, Denver, Colo. P. 438.

Kay, R.N.B. 1960. The rate of flow and composition of various salivary secretions in sheep and calves. J. Physiol. 29:395–415.

Keeney, M. 1970. Lipid metabolism in the rumen. In: Physiology of Digestion and Metabolism in the Ruminant. A.T. Phillipson, ed. Oriel Press, Newcastle upon Tyne, England.

Kelly, N.C., P.C. Thomas, and D.G. Chamberlain. 1978. The digestion of dietary protein and synthesis of bacterial protein in the rumen in sheep given silages prepared with the addition of formic acid. Proc. Nutr. Soc. 37:34A.

Kellner, O. 1912. Die Ernährung der landwirtschaftlichen Nutztiere. Paul Parey, Berlin.

Kemp, P., and D.J. Lander. 1976. Inhibition of the biohydrogenation of dietary C$_{18}$ unsaturated fatty acids by rumen bacteria using some inhibitors of methanogenesis. Proc. Nutr. Soc. 35:31A–32A.

Kern, D.L., L.L. Slyter, J.M. Weaver, E.C. Leffel, and G. Samuelson. 1973. Pony caecum vs. steer rumen: the effect of oats and hay on the microbial system. J. Anim. Sci. 37:463–469.

Kern, D.L., L.L. Slyter, J.M. Weaver, and R.R. Oltjen. 1974. Ponies vs. steers: microbial and chemical characteristics of intestinal ingesta. J. Anim. Sci. 38:559–564.

Kertz, A.F., and J.P. Everett, Jr. 1975. Utilization of urea by lactating cows—an industry viewpoint. J. Anim. Sci. 41:945.

Ketelaars. J.J.M.H., and B.J. Tolkamp. 1992. Toward a new theory of feed intake regulation in ruminants. 1. Causes of differences in voluntary feed intake: critique of current views. Livest. Prod. Sci. 30:269–296.

Keys, J.E., Jr., and P.J. Van Soest. 1970. Digestibility of forages by the meadow vole (*Microtus pennsylvanicus*). J. Dairy Sci. 53:1502–1508.

Keys, J.E., Jr., P.J. Van Soest, and E.P. Young. 1969. Comparative study of the digestibility of forage cellulose and hemicellulose in ruminants and nonruminants. J. Anim. Sci. 29:11–15.

Keys, J.E., Jr., P.J. Van Soest, and E.P. Young. 1970. Effect of increasing dietary cell wall content on the digestibility of hemicellulose and cellulose in swine and rats. J. Anim. Sci. 31:172.

Kingsbury, J.M. 1978. Ecology of poisoning. In: Effects of Poisonous Plants on Livestock. R.F. Keeler, K.R. Van Kampen, and L.F. James, eds. Academic Press, New York.

Kirk, T.K., and R.L. Farrell. 1987. Enzymatic combustion: the microbial degradation of lignin. Annu. Rev. Microbiol. 41:465–505.

Kleiber, M. 1975. The Fire of Life, an Introduction to Animal Energetics. Rev. ed. R.E. Krieger Publ. Co., Huntington, N.Y. P. 38.

Klopfenstein, T., and R. Britton. 1987. Heat damage—real or artifact. Proc. Distillers' Feed Conf., Cincinnati. Pp. 84–86.

Kolattukudy, P.E. 1980. Biopolyester membranes of plants: cutin and suberin. Science 208:990–1000.

Koller, B.L., H.F. Hintz, J.B. Robertson, and P.J. Van Soest. 1978. Comparative cell wall and dry matter digestion in the cecum of the pony and the rumen of the cow using in vitro and nylon bag techniques. J. Anim. Sci. 47:209–215.

Kornegay, E.T. 1981. Soybean hull digestibility by sows and feeding value for growing-finishing swine. J. Anim. Sci. 53:138–145.

Korver, S. 1988. Genetic aspects of feed intake and feed efficiency in dairy cattle: a review. Livest. Prod. Sci. 20:1–14.

Kotb, A.R., and T.D. Luckey. 1972. Markers in nutrition. Nutr. Abstr. Rev. 42:813–845.

Krishnamoorthy, U., T.V. Muscato, C.J. Sniffen, and P.J. Van Soest. 1983a. Nitrogen fractions in selected feedstuffs. J. Dairy Sci. 65:217–255.

Krishnamoorthy, U., C.J. Sniffen, M.D. Stern, and P.J. Van Soest. 1983b. Evaluation of a mathematical model of rumen digestion and an in vitro simulation of rumen proteolysis to estimate the rumen-undegraded nitrogen content of feedstuffs. Br. J. Nutr. 50:555–568.

Kronfeld, D.S. 1970. Ketone body metabolism, its control, and its implications in pregnancy toxaemia, acetonaemia and feeding standards. In: Physiology of Digestion and Metabolism in the Ruminant. A.T. Phillipson, ed. Oriel Press, Newcastle upon Tyne, England. Pp. 566–583.

Kronfeld, D.S., and P.J. Van Soest. 1976. Carbohydrate nutrition. In: Comparative Animal Nutrition, vol. 1. M. Rechcigl. Jr., ed. Karger, Basel. Pp. 23–73.

Krumholz, L.R., and M.P. Bryant. 1986. *Eubacterium* sp. nov. requiring H_2 or formate to degrade gallate, pyrogallol, phloroglucinol or quercitin. Arch. Microbiol. 144:8–14.

Krumholz, L.R., R.L. Crawford, M.E. Hemling, and M.P. Bryant. 1986. A rumen bacterium degrading quercetin and trihydroxybenzenoids with concurrent use of formate or H_2. In: Plant Flavonoids in Biology and Medicine: Biochemical, Pharmacological, and Structure-Activity Relationships. Alan Liss, New York. Pp. 211–214.

Krysl, L.J., M.E. Hubbert, B.F. Sowell, G.E. Plumb, T.K. Jewett, M.A. Smith, and J.W. Waggoner. 1984. Horses and cattle grazing in the Wyoming red desert. I. Food habits and dietary overlap. J. Range Manage. 37(1):72.

Kuan, K.K., G. Stanogias, and A.C. Dunkin. 1983. The effect of proportion of cell-wall material from lucerne leaf meal on apparent digestibility, rate of passage and gut characteristics in pigs. Anim. Prod. 36:201.

Kudo, J., K.J. Cheng, M.R. Hanna, R.E. Howart, B.P. Goplen, and J.W. Costerton. 1985. Ruminal digestion of alfalfa strains selected for slow and fast initial rates of digestion. Can. J. Anim. Sci. 65:157.

Laby, R.H. 1975. Surface active agents in the rumen. In: Digestion and Metabolism in the Ruminant. I.W. McDonald and A.C.I. Warner, eds. Univ. of New England Publ. Unit, Armidale, N.S.W., Australia. Pp. 537–550.

Lajoie, S.F., S. Bank, T.L. Miller, and M.J. Wolin. 1988. Acetate production from hydrogen and [^{14}C]carbon dioxide by the microflora of human feces. Appl. Environ. Microbiol. 54:2723.

Lam, T.B.T., K. Iiyama, and B.A. Stone. 1990. Lignin in wheat internodes. Part 2. Alkaline nitrobenzene oxidation by wheat straw lignin and its fractions. J. Sci. Food Agric. 51:493–506.

Langer, P. 1987. Evolutionary patterns of Perissodactyla and Artiodactyla (Mammalia) with different types of digestion. Z. Zool. Syst. Evolutionsforsch. 25(3):212–236.

Langer, P. 1988. The Mammalian Herbivore Stomach: Comparative Anatomy Function and Evolution. G. Fischer, Stuttgart. P. 557.

Langlands, J.P. 1975. Techniques for estimating nutrient intake and its utilization by the grazing ruminant. In: Digestion and Metabolism in the Ruminant. I.W. McDonald and A.C.I. Warner, eds. Univ. of New England Publ. Unit. Armidale, N.S.W., Australia. Pp. 320–332.

Langston, C.W., H. Irwin, C.H. Gordon, C. Bouma, H.G. Wiseman, C.G. Melin, L.A. Moore, and J.R. McCalmont. 1958. Microbiology and Chemistry of Grass Silage. USDA Tech. Bull. 1187.

Lapierre, C., D. Jouin, and B. Monties. 1989. On the molecular origin of the alkali solubility of Gramineae lignins. Phytochemistry 28:1401–1403.

Larwence, A., F. Hammouda, and A. Salah. 1984. Valeur alimentaire des marcs de raisin. III. Rôle des tanins condensées dans la faible valeur nutritive des marcs de raisin chez le mouton: effet d'une addition de polyéthlyène glycol 4000. Ann. Zootech. 33:533–544.

Latham, M.J., B.E. Brooker, G.L. Pettipher, and P.J. Harris. 1978. *Ruminococcus flavefaciens* cell coat and adhesion to cotton cellulose and to cell walls in leaves of perennial ryegrass (*Lolium perenne*). Appl. Environ. Microbiol. 35:156–165.

Lau, M.M., and P.J. Van Soest. 1981. Titratable groups and soluble phenolics as indicators of the digestibility of chemically treated roughages. Anim. Feed Sci. Technol. 6:123.

Leat, W.M.F., and F.A. Harrison. 1975. Digestion, absorption and transport of lipids in the sheep. In: Digestion and Metabolism in the Ruminant. I.W. McDonald and A.C.I. Warner, eds. Univ. of New England Publ. Unit, Armidale, N.S.W., Australia. Pp. 481–495.

Leatherwood, J.M. 1969. Cellulase complex of ruminococcus and a new mechanism for cellulose degradation. In: Cellulases and Their Application. Adv. Chem. Ser. 95. American Chemical Society, Washington, D.C. Pp. 53–59.

Leclerc, F., and E. Lecrivin. 1985. Etude du compartiment d'ovine domestiques en élevage extensif sur le causse du larzac. In: Ethology of Farm Animals. A.F. Fraser, ed. Elsevier, Amsterdam.

Leek, B.F. 1983. Clinical diseases of the rumen: a physiologist's view. Vet. Rec. 113:10–14.

Lehmann, F. 1941. Die Lehre vom Ballast. Z. Tierernähr. Futtermittelkd. 5:155–173.

Leibensperger, R.Y., and R.E. Pitt. 1987. A model of clostridial dominance in ensilage. Grass Forage Sci. 42:297–317.

Leng, R.A. 1970. Formation and production of volatile fatty acids in the rumen. In: Physiology of Digestion and Metabolism in the Ruminant. A.T. Phillipson, ed. Oriel Press, Newcastle upon Tyne, England. Pp. 406–421.

Lengemann, F.W., and N.N. Allen. 1955. The development of rumen function in the dairy calf. 1. Some characteristics of the rumen contents of cattle of various ages. J. Dairy Sci. 38:651–656.

Lerner, J., E. Mathews, and I. Fung. 1988. Methane emission from animals: a global high resolution data base. Global Biogeochem. Cycles 2:139.

Lewis, B.A. 1978. Physical and biological properties of structural and other nondigestible carbohydrates. Am. J. Clin. Nutr. 31(Suppl.):S82–S85.

Lewis, D.H. 1980. Boron, lignification and the origin of vascular plants—a unified hypothesis. New Phytol. 84:209–299.

Lewis, N.G. 1988. Lignin biogenesis, biodegradation and

utilization. Bulletin de Liaison no. 14 du Groupe Polyphenols, Narbonne, France. Compte-rendu des Journées Internationales d'Etude et de l'Assemblée Générale. Université Brock, St. Catharines, Ont., 16–19 août 1988.

Lewis, S.M., D.P. Holzgraefe, L.L. Berger, G.C. Fahey, Jr., J.M. Gould, and G.F. Fanta. 1987. Alkaline hydrogen peroxide treatments of crop residues to increase ruminal dry matter disappearance in sacco. Anim. Feed Sci. Technol. 17:179–199.

Licitra, G., and P.J. Van Soest. 1991. Effect of laboratory handling on non-protein soluble nitrogen in cut Italian green forage at early stages of growth. J. Anim. Sci. 69(Suppl. 1):175.

Ling, J.R., and P.J. Buttery. 1978. The simultaneous use of ribonucleic acid, 35S, 1,6-diaminopimelic acid and 2-aminoethylphosphonic acid as markers of microbial nitrogen entering the duodenum of sheep. Br. J. Nutr. 39:165–179.

Lippke, H., W.C. Ellis, and B.F. Jacobs. 1986. Recovery of indigestible fiber from feces of sheep and cattle on forage diets. J. Dairy Sci. 69:403.

Liu, B.W.Y., and G.W. Fick. 1975. Yield and quality losses due to alfalfa weevil. Agron. J. 67:828.

Lofgreen, G.P. 1953. The estimation of total digestible nutrients from digestible organic matter. J. Anim. Sci. 12:359.

Lofgreen, G.P., and W.N. Garrett. 1968. A system for expressing net energy requirements and feed values for growing and finishing beef cattle. J. Anim. Sci. 27:793–806.

Lopez, J., N.A. Jorgensen, R.P. Niedermeier, and H.J. Larsen. 1970. Redistribution of nitrogen in urea-treated and soybean meal–treated corn silage. J. Dairy Sci. 53:1215–1224.

Lovern, J.A. 1965. Some analytical problems in the analysis of fish and fish products. J. Assoc. Off. Anal. Chem. 48:60–68.

Löwenthal, J. 1945. Method 18.42 In: Methods of Analysis. Association of Official Agricultural Chemists, Washington, D.C.

Lowry, J.B., and E.A. Sumpter. 1990. Problems with ytterbium precipitation as a method for determination of plant phenolics. J. Sci. Food Agric. 52:287.

Lucas, H.L. 1964. Stochastic elements in biological models; their sources and significance. In: Stochastic Models in Medicine and Biology. J. Gurland, ed. Univ. Wisconsin Press, Madison. P. 355.

Lucas, H.L., and W.W. Smart. 1959. Chemical composition and the digestibility of forages. Proc. 16th Pasture and Crop Improvement Conf., Mississippi State Univ., State College, Mississippi. Pp. 23–26.

Lucas, H.L., Jr., W.W.G. Smart, Jr., M.A. Cipolloni, and H.D. Gross. 1961. Relations between digestibility and composition of feeds and foods. S-45 Report, North Carolina State Coll.

Lyford, Jr., S.J. 1988. Growth and development of the ruminant digestive system. In: The Ruminant Animal. D.C. Church, ed. Prentice-Hall, Englewood Cliffs, N.J. P. 44.

Lyford, Jr., S.J., W.W.G. Smart, Jr., and G. Matrone. 1963. Digestibility of the alpha-cellulose and pentosan components of the cellulosic micelle of fescue and alfalfa. J. Nutr. 79:105–108.

MacRae, J.C., and G.E. Lobley. 1986. Interactions between energy and protein. In: Control of Digestion and Metabolism in Ruminants. L.P. Milligan, W.L. Grovum, and A. Dobson, eds. Prentice-Hall, Englewood Cliffs, N.J. Pp. 367–385.

Mader, T.L., R.G. Teeter, and G.W. Horn. 1984. Comparison of forage labeling techniques for conducting passage rate studies. J. Anim. Sci. 58:208.

Magnusson, G. 1963. The behavior of certain lanthanons in rats. Acta Pharmacol. Toxicol. 20(Suppl. 3):1–95.

Majak, W., and K.J. Cheng. 1987. Hydrolysis of the cyanogenic glycosides amygdalin, prunasin and linamarin by ruminal microorganisms. Can. J. Anim. Sci. 67:1133.

Makkar, H.P.S., R.K. Dawra, and B. Singh. 1987. Protein precipitation assay for quantitation of tannins: Determination of protein in tannin-protein complex. Anal. Biochem. 166:435–439.

Makkar, H.P.S., R.K. Dawra, and B. Singh. 1988a. Changes in tannin content, polymerization and protein precipitation capacity in oak (*Quercus incana*) leaves with maturity. J. Sci. Food Agric. 44:301.

Makkar, H.P.S., B. Singh, and R.K. Dawra. 1988b. Effect of tannin-rich leaves of oak on various microbial enzyme activities of the bovine rumen. Br. J. Nutr. 60:287–296.

Malechek, J.C., and D.F. Balph. 1987. Diet selection by grazing and browsing livestock. In: The Nutrition of Herbivores. J.B. Hacker and J.H. Ternouth, eds. Academic Press, New York. P. 121.

Males, J.R. 1987. Optimizing the utilization of cereal crop residues for beef cattle. J. Anim. Sci. 65:1124–1130.

Maloiy, G.M.O., and R.N.B. Kay. 1971. A comparison of digestion in red deer and sheep under controlled conditions. Q. J. Exp. Physiol. 56:257–266.

Mandels, M., and E.T. Reese. 1963. Inhibition of cellulases and beta-glucosidases. In: Advances in Enzymatic Hydrolysis of Cellulose and Related Materials. E.T. Reese, ed. MacMillan, New York. Pp. 115–158.

Mangan, J.L. 1988. Nutritional effects of tannins in animal feeds. Nutr. Res. Rev. 1:209–231.

Mann, M.E., R.D.H. Cohen, J.A. Kernan, H.H. Nicholson, D.A. Christensen, and M.E. Smart. 1988. The feeding value of ammoniated flax straw, wheat straw and wheat chaff for beef cattle. Anim. Feed Sci. Technol. 21:57–66.

Marks, D., J. Glyphis, and M. Leighton. 1987. Measurement of protein in tannin-protein precipitates using ninhydrin. J. Sci. Food Agric. 38:255–261.

Marshall, L.G., S.D. Webb, J.J. Sapkowsky, and D.M. Rupp. 1982. Mammalian evolution and the great American interchange. Science 215:1351–1357.

Marston, H.R. 1951. Energy transactions in homeothermic animals. Proc. Zool. Soc. N.S.W. 84:169.

Marten, G.C. 1985. Factors influencing feeding and value effective utilization of forages for animal production. In: Proc. 15th Int. Grasslands Congr. Kyoto, Japan. P. 89.

Marten, G.C., R.M. Jordan, and A.W. Hovin. 1981. Improved lamb performance associated with breeding for alkaloid reduction in reed canarygrass. Crop Sci. 21:295–298.

Martin, A.K. 1982. The origin of urinary aromatic compounds excreted by ruminants. 3. The metabolism of phenolic compounds to simple phenols. Br. J. Nutr. 48:497.

Martin, J.S., and M.M. Martin. 1983. Tannin assays in ecological studies. Precipitation of ribulose-1,5-bisphosphate carboxylase/oxygenase by tannic acid, quebracho, and oak leaf foliage extracts. J. Chem. Ecol. 9:285–294.

Martin, M.M. 1982. The role of ingested enzymes in the digestive processes of insects. In: Invertebrate-Microbial

Interactions. Joint Symp. British Mycological Society and British Ecological Society, Univ. of Exeter. J.M. Anderson, A.D.M. Rayner, and D.W.H. Walton, eds. Cambridge Univ. Press, Cambridge. Pp. 155–172.

Martin, M.M., and J.S. Martin. 1978. Cellulose digestion in the midgut of the fungus-growing termite *Macrotermes natalensis:* The role of acquired digestive enzymes. Science 199:1453–1455.

Martin, R.D., D.J. Chivers, A.M. MacLarnon, and C.M. Hladik. 1985. Gastrointestinal allometry in primates and other mammals. In: Size and Scaling in Primate Biology. W.L. Jungers, ed. Plenum Press, New York. Pp. 61–90.

Martinez, A., and D.C. Church. 1970. Effect of various mineral elements on in vitro rumen cellulose digestion. J. Anim. Sci. 36:982.

Mason, V.C. 1969. Some observations on the distribution and origin of nitrogen in sheep feces. J. Agric. Sci. 73:99–111.

Mason, V.C. 1979. The quantitative importance of bacterial residues in the non-dietary faecal nitrogen of sheep. I. Methodology studies. Z. Tierphysiol., Tierernähr. Futtermittelkd. 41:131.

Mason, V.C. 1984. Metabolism of nitrogenous compounds in the large gut. Proc. Nutr. Soc. 43:45–53.

Mason, V.C., J.E. Cook, M.S. Dhanoa, A.S. Keene, and R.D. Hartley. 1989. Chemical composition and nutritive value of sheep untreated and oven-ammoniated ryegrass hays prepared from crops on differing maturities. Anim. Feed Sci. Technol. 26:207–220.

Mason, V.C., J.E. Cook, M.S. Dhanoa, A.S. Keene, C.J. Hoadley, and R.D. Hartley. 1990. Chemical composition digestibility in vitro and biodegradability of grass hays oven-treated with different amounts of ammonia. Anim. Feed Sci. Technol. 29:237–249.

Mason, V.C., and J.H. Frederiksen. 1979. Partition of the nitrogen in sheep faeces with detergent solutions, and its application to the estimation of the true digestibility of dietary nitrogen and the excretion of non-dietary faecal nitrogen. Z. Tierphysiol., Tierernähr. Futtermittelkd. 41:121.

Masson, C., W. Alrahmoun, and J.L. Tisserand. 1986. Etude comparée de la quantité ingérée, de la digestibilité, de l'utilisation de l'azote, du temps moyen de rétention et du comportement alimentaire chez les jeunes caprins et ovins recevant différents régimes. Ann. Zootech. 35:49–60.

Mattos, W., and D.L. Palmquist. 1977. Biohydrogenation and availability of linoleic acid in lactating cows. J. Nutr. 107:1755–1761.

Mayer, F., Coughlan, M.P., Mori, Y., and Ljungdahl, L.G. 1987. Macromolecular organization of the cellulolytic enzyme complex of *Clostridium thermocellum* as revealed by electron microscopy. Appl. Environ. Microbiol. 53:2785–2792.

Mayes, R.W., C.S. Lamb, and P.M. Colgrove. 1986. The use of dosed and herbage *n*-alkanes as markers for the determination of herbage intake. J. Agric. Sci. 107:161.

McAllan, A.B. 1982. The fate of nucleic acids in ruminants. Proc. Nutr. Soc. 41:309–317.

McAllan, A.B., R. Knight, and J.D. Sutton. 1983. The effect of free and protected oils on the digestion of dietary carbohydrates between the mouth and duodenum of sheep. Br. J. Nutr. 49:433.

McBurney, M.I. 1985. Physicochemical and nutritive evaluation of chemically-treated feeds for ruminants. Ph.D. dissertation Cornell Univ., Ithaca, N.Y. P. 101.

McBurney, M.I., M.S. Allen, and P.J. Van Soest. 1986.

Praseodymium and copper cation-exchange capacities of neutral-detergent fibers relative to composition and fermentation kinetics. J. Sci. Food Agric. 37:666–672.

McBurney, M.I., P.J. Horvath, J.L. Jeraci, and P.J. Van Soest. 1985. Effect of in vitro fermentation using human fecal inoculum on the water-holding capacity of dietary fibre. Br. J. Nutr. 53:17–24.

McBurney, M.I., L.U. Thompson, D.J. Cuff, and C.J.A. Jenkins. 1988. Comparison of ileal effluents, dietary fibers, and whole foods in predicting the physiological importance of colonic fermentation. Am. J. Gastroenterol. 83:536–540.

McBurney, M.I., and P.J. Van Soest. 1984. Laboratory methods for estimation of digestibility and chemical and physical properties of fibrous roughages. Organisation for Economic Cooperation and Development, Paris. Pp. 1–21.

McBurney, M.I., P.J. Van Soest, and L.E. Chase. 1983. Cation exchange capacity of dietary fibers. J. Sci. Food Agric. 34:910–916.

McBurney, M.I., P.J. Van Soest, and J.L. Jeraci. 1987. Colonic carcinogenesis: the microbial feast or famine mechanism. Nutr. Cancer 10:23–28.

McCammon-Feldman, B. 1980. A critical analysis of tropical savanna forage consumption and utilization by goats. Ph.D. dissertation, Univ. of Illinois, Urbana-Champaign.

McCollum, E.V. 1957. A History of Nutrition. Houghton Mifflin, Boston.

McConnell, A.A., M.A. Eastwood, and W.D. Mitchell. 1974. Physical characteristics of vegetable foodstuffs that could influence bowel function. J. Sci. Food Agric. 25:1457.

McDonald, P. 1981. The Biochemistry of Silage. John Wiley & Sons, New York.

McDowell, G.H. 1983. Hormonal control of glucose homeostasis in ruminants. Proc. Nutr. Soc. 42:149.

McDowell, R.E. 1972. Improvement of Livestock Production in Warm Climates. W.H. Freeman, San Francisco.

McDowell, R.E. 1976. Importance of Ruminants of the World for Non-food Uses. Cornell Univ. Int. Agric. Mimeo 52.

McDowell, R.E. 1977. Ruminant Products: More Than Meat and Milk. Winrock Report, Winrock Int. Livestock Res. and Training Ctr., Morrilton, Ark.

McDowell, R.E. 1986. An animal science perspective on crop breeding and selection programmes for warm climates. Cornell Univ. Int. Agric. Mimeo 110.

McGavin, M., and C.W. Forsberg. 1989. Catalytic and substrate-binding domains of endoglucanase 2 from *Bacteroides succinogenes*. J. Bacteriol. 171:3310–3315.

McKersie, B.D. 1985. Effect of pH on protolysis in ensiled legume forage. Agron. J. 77:81–86.

McManus, W.R. 1981. Oesophageal fistulation technique as an aid to diet evaluation of grazing ruminant. In: Forage Evaluation: Concepts and Techniques. J.L. Wheeler and R.D. Mochrie, eds. American Forage and Grassland Council, Lexington, Ky. Pp. 249–260.

McNab, B.K. 1980. Food habits, energetics, and the population biology of mammals. Am. Nat. 116:106–124.

McNaughton, S.J. 1979. Tree grassland-herbivore dynamics. In: Serengeti Dynamics of an Ecosystem. A.R.E. Sinclair and M. Norton-Griffiths, eds. Univ. of Chicago Press, Chicago. Pp. 46–81.

McNeil, N.I., J.H. Cummings, and W.P.T. James. 1978.

Short chain fatty acid absorption by the human large intestine. Gut 19:819–824.

McQueen, R.W. 1974. Comparisons of plant cell wall degradation by fungal enzymes and ruminal micro-organisms in vitro. Ph.D. dissertation, Cornell Univ., Ithaca, N.Y.

McQueen, R.W., and P.J. Van Soest. 1975. Fungal cellulase and hemicellulase prediction of forage digestibility. J. Dairy Sci. 58:1482–1490.

Meara, M.L. 1955. Fats and other lipids. In: Moderne Methoden der Pflanzenanalyse, vol. 2. K. Paech and M.V. Tracey, eds. Springer Verlag, Berlin. Pp. 317–402.

Mehansho, H., D.K. Ann, L.G. Butler, J. Rogler, and D.M. Carlson. 1987. Induction of proline-rich proteins in hamster salivary glands by isoproterenol treatment and an unusual growth inhibition by tannins. J. Biol. Chem. 262(25):12344–12350.

Meijs, J.A.C. 1986. Concentrate supplementation of grazing diary cows. 2. Effect of concentrate composition on herbage intake and milk production. Grass Forage Sci. 41:229.

Menke, K.H., L. Raab, A. Salewski, H. Steinglass, D. Fritz, and W. Schneider. 1979. The estimation of the digestibility and metabolizable energy content of ruminant feedingstuffs from the gas production when they are incubated with rumen liquor in vitro. J. Agric. Sci. 93:217–222.

Menke, K.H., and H. Steingass. 1988. Estimation of the energetic feed value obtained from chemical analysis and in vitro gas production using rumen fluid. Anim. Res. Dev. 28:7.

Merchen, N.R., and L.D. Satter. 1983. Changes in nitrogenous compounds and sites of digestion of alfalfa harvested at different moisture contents. J. Dairy Sci. 66:789.

Merrill, A.L., and B.K. Watt. 1973. Energy value of foods—basis and derivation. U.S. Dept. Agric., Agric. Handb. 74 (rev. from 1955 ed.).

Mertens, D.R. 1973. Application of theoretical mathematical models to cell wall digestion and forage intake in ruminants. Ph.D. dissertation, Cornell Univ., Ithaca, N.Y.

Mertens, D.R. 1983. Using neutral detergent fiber to formulate dairy rations and estimate the net energy content of forages. Proc. Cornell Nutrition Conf., Ithaca, N.Y. Pp. 60–68.

Mertens, D.R. 1985. Factors influencing feed intake in lactating dairy cows: from theory to application using neutral detergent fiber. In: Proc. Georgia Nutrition Conf., Univ. of Georgia, Athens. Pp. 1–18.

Mertens, D.R. 1987. Predicting intake and digestibility using mathematical models of ruminal function. J. Anim. Sci. 64:1548.

Mertens, D.R., and L.O. Ely. 1979. A dynamic model of fiber digestion and passage in the ruminant for evaluating forage quality. J. Anim. Sci. 49:1085.

Mertens, D.R., and L.O. Ely. 1982. Relationship of rate and extent of digestion to forage utilization—a dynamic model evaluation. J. Anim. Sci. 54:895–905.

Mertz, W. 1981. The essential trace elements. Science 213:1332–1338.

Mertz, W. 1987. Trace Elements in Human and Animal Nutrition. 5th ed. Academic Press, New York.

Metson, A.J. 1978. Seasonal variations in chemical composition of pasture. I. Calcium, magnesium, potassium, sodium and phosphorus. J. Agric. Res. 21:341–353.

Meyer, J.H., and G.P. Lofgreen. 1959. Evaluation of alfalfa hay by chemical analyses. J. Anim. Sci. 18:1233–1242.

Miles, C.W., J.L. Kelsay, and N.P. Wong. 1988. Effect of dietary fiber on the metabolizable energy of human diets. J. Nutr. 118:1075–1081.

Milford, R., and D.J. Minson. 1965. Intake of tropical pasture species. Proc. 9th Int. Grasslands Congr., São Paulo, Brazil. Pp. 815–822.

Miller, B.L., G.C. Fahey, Jr., R.B. Rindsig, L.L. Berger, and W.G. Bottje. 1979. In vitro and in vivo evaluations of soybean residues ensiled with various additives. J. Anim. Sci. 49:1545.

Miller, R.E., R.H. Edwards, and G.E. Kohler. 1984. Measurement of cell rupture in macerated forages. J. Agric. Food Chem. 32:534–538.

Miller, R.W., R.W. Hemken, J.H. Vandersall, D.R. Waldo, M. Okamato, and L.A. Moore. 1965. Effect of feeding buffers to dairy cows grazing pearl millet or sudan grass. J. Dairy Sci. 48:1319.

Mills, C.F. 1987. Biochemical and physiological indicators of mineral status in animals: copper, cobalt and zinc. J. Anim. Sci. 65:1702.

Mills, S.E., D.C. Beitz, and J.W. Young. 1986a. Characterization of metabolic changes during a protocol for inducing lactation ketosis in dairy cows. J. Dairy Sci. 69:352.

Mills, S.E., D.C. Beitz, and J.W. Young. 1986b. Evidence for impaired metabolism in liver during induced lactation ketosis of dairy cows. J. Dairy Sci. 69:362.

Milne, A., M.K. Theodorou, M.G.C. Jordan, C. King-Spooner, and A.P.J. Trinci. 1989. Survival of anaerobic fungi in feces, in saliva, and in pure culture. Exp. Mycol. 13:27–37.

Milne, J.A., J.C. MacRae, A.M. Spence, and S. Wilson. 1978. A comparison of the voluntary intake and digestion of a range of forages at different times of the year by sheep and the red deer (Cervus elaphus). Br. J. Nutr. 40:347–358.

Milton, K., and M.W. Demment. 1988. Digestion and passage kinetics of chimpanzees fed high and low fiber diets and comparison with human data. J. Nutr. 118:1082–1088.

Milton, K., P.J. Van Soest, and J.B. Robertson. 1980. Digestive efficiencies of wild howler monkeys. Physiol. Zool. 53:402–409.

Minson, D.J. 1963. The effect of pelleting and wafering on the feeding value of roughage—a review. J. Br. Grassl. Soc. 18:39.

Minson, D.J. 1971. Influence of lignin and silicon on a summative system for assessing the organic matter digestibility of Panicum. Aust. J. Agric. Res. 22:589–598.

Minson, D.J. 1990. Forage in Ruminant Nutrition. Academic Press, New York.

Minson, D.J., and M.N. McLeod, 1970. The digestibility of temperate and tropical grasses. Proc. 11th Int. Grasslands Congr. Pp. 719–722.

Minson, D.J., J.C. Taylor, F.E. Alder, W.F. Raymond, J.E. Rudman, C. Line, and M.J. Head. 1960. A method for identifying the faeces produced by individual cattle or groups of cattle grazing together. J. Br. Grassl. Soc. 15:86–88.

Minson, D.J., and Wilson, J.R. 1980. Comparative digestibility of tropical and temperate forage—a contrast between grasses and legumes. J. Aust. Inst. Agric. Sci. 46:247—249.

Mitchell, H.H. 1962. Comparative Nutrition of Man and Domestic Animals, vols. 1 and 2. Academic Press, New York.

Mitchell, H.H., and T.S. Hamilton. 1932. The effect of the amount of feed consumed by cattle on the utilization of its energy content. J. Agric. Res. 45:163.

Mochi, U., and T.D. Carter. 1971. Hoofed Mammals of the World. Charles Scribner's Sons, New York.

Moe, P.W., W.P. Flatt, and H.F. Tyrrell. 1972. Net energy value of feeds for lactation. J. Dairy Sci. 55:945–958.

Moe, P.W., and H.F. Tyrrell. 1975. Symposium: production efficiency in the high producing cow. Efficiency of conversion of digested energy to milk. J. Dairy Sci. 58:602–610.

Moe, P.W., and H.F., Tyrrell. 1979. Methane production in dairy cows. J. Dairy Sci. 62:1583–1586.

Moe, P.W., H.F. Tyrrell, and W.P. Flatt. 1971. Loss of fat from dairy cows: energetics of body tissue mobilization. J. Dairy Sci. 54:548.

Moen, A.N. 1973. Wildlife Ecology. W.H. Freeman, San Francisco.

Moir, R.J. 1968. Ruminant digestion and evolution. In: Handbook of Physiology. Section G: Alimentary Canal. Vol. 5: Bile Digestion, Ruminal Physiology. C.F. Code, ed. American Physiological Society, Washington, D.C. Pp. 2673–2693.

Moody, E.G., P.J. Van Soest, R.E. McDowell, and G.L. Ford. 1971. Effect of high temperature and dietary fat on milk fatty acids. J. Dairy Sci. 54:10.

Moon, F.E., and A.K. Abou-Raja. 1952. The lignin fraction of animal feeding stuffs. II. Investigation of lignin determination procedures and development of a method for acid insoluble lignin. J. Sci. Food Agric. 3:407.

Moore, J.H., and R.C. Noble. 1975. Fetal and neonatal lipid metabolism. In: Digestion and Metabolism in the Ruminant. I.W. McDonald and A.C.I. Warner, eds. Univ. of New England Publ. Unit, Armidale, N.S.W., Australia. Pp. 465–480.

Moore, L.A. 1964. Nutritive value of forage as affected by physical form. I. General principles involved with ruminants and effect of feeding pelleted or wafered forage to dairy cattle. J. Anim. Sci. 23:230–238.

Moore, L.A., H.M. Irvin, and J.C. Shaw. 1953. Relationship between TDN and energy values of feeds. J. Dairy Sci. 36:93–97.

Moore, L.A., J.W. Thomas, and J.F. Sykes. 1960. The acceptability of grass/legume silage by dairy cattle. Proc. 8th Int. Grasslands Congr., Reading, England. Pp. 701–704.

Mora, M.I., G.C. Fahey, Jr., P.J. van der Aar, and L.L. Berger. 1983. Improving utilization of crop residues. Evaluation of two chemicals used to improve digestibility of crop residues. Anim. Feed Sci. Technol. 9:205–219.

Morley, F.H.W. 1987. Economics of pasture development and pasture research. In: Temperate Pastures: Their Production, Use and Management. J.L. Wheeler, C.J. Pearson, and G.E. Robards, eds. CSIRO, Canberra. Pp. 571–579.

Morris, E.J., and O.J. Cole. 1987. Relationship between cellulolytic activity and adhesion to cellulose in *Ruminococcus albus*. J. Gen. Microbiol. 133:1023–1032.

Morris, J.G. 1976. Oxygen and the obligate anaerobe. J. Appl. Bacteriol. 40:229–244.

Morris, J.N., J.W. Marr, and D.G. Clayton. 1977. Diet and heart: a postscript. Br. Med. J. 2:1307–1314.

Morrison, F.B. 1936. Feeds and Feeding. 20th ed. Morrison Publ. Co., Ithaca, N.Y.

Morrison, F.B. 1956. Feeds and Feeding. 22d ed. Morrison Publ. Co., Clinton, Iowa.

Morrison, I.M. 1972. Improvements in the acetyl bromide technique to determine lignin and digestibility and its application to legumes. J. Sci. Food Agric. 23:1463–1469.

Morrison, I.M. 1980. Hemicellulosic contamination of acid detergent residues and their replacement by cellulose residues in cell wall analysis. J. Sci. Food Agric. 31:639–645.

Morrison, I.M. 1988. Influence of chemical and biological pretreatments on the degradation of lignocellulosic material by biological systems. J. Sci. Food Agric. 42:295–304.

Mowat, D.N., R.S. Fulkerson, W.E. Tossell, and J.E. Winch. 1965. The in vitro digestibility and protein content of leaf and stem portions of forages. Can. J. Plant Sci. 45:321–331.

Muck, R.E. 1987. Dry matter level effects on alfalfa silage quality. I. Nitrogen transformations. Trans. ASAE 30(1):7–14.

Muck, R.E., and J.T. Dickerson. 1988. Storage temperature effects on proteolysis in alfalfa silage. Trans. ASAE 31(4):1005–1009.

Muecke, G.K., and P. Moller. 1988. The not-so-rare earths. Sci. Am. 258:72–77.

Mueller, S.C., and G.W. Fick. 1989. Converting alfalfa development measurements from mean stage by count to mean stage by weight. Crop Sci. 29:821–823.

Mueller-Harvey, I., R.D. Hartley, P.J. Harris, and E.H. Curzon. 1986. Linkage of p-coumaroyl and feruloyl groups to cell wall polysaccharides of barley straw. Carbohydr. Res. 148:71–85.

Mueller-Harvey, I., and A.B. McAllan. 1989. Tannins—their biochemistry and nutritional properties. In: Advances in Plant Cell Biochemistry and Biotechnology. I.M. Morrison, ed. JAI Press, London.

Mueller-Harvey, I., J.D. Reed, and R.D. Hartley. 1987. Characterization of phenolic compounds, including flavonoids and tannins, of ten Ethiopian browse species by high performance liquid chromatography. J. Sci. Food Agric. 39:1–14.

Mugerwa, J.S., and H.R. Conrad. 1975. Relationship of dietary nonprotein nitrogen to urea kinetics in dairy cows. J. Nutr. 101:1331–1342.

Munkenbeck, N.W. 1988. Comparison of methods of estimating palatability in dairy cattle. Ph.D. dissertation, Cornell Univ., Ithaca, N.Y.

Murphy, M. 1989. The influence of non-structural carbohydrates on rumen microbes and rumen metabolism in milk producing cows. Ph.D. dissertation, Swedish Univ. of Agr. Sci., Rapport 183. P. 154.

Murphy, M., P. Udén, D.L. Palmquist, and H. Wiktorsson. 1987. Rumen and total diet digestibilities in lactating cows fed diets containing full-fat rapeseed. J. Dairy Sci. 70:1572–1582.

Muscato, T.V., C.J. Sniffen, U. Krishnamoorthy, and P.J. Van Soest. 1983. Amino acid content of noncell and cell wall fractions in feedstuffs. J. Dairy Sci. 66:2198–2207.

Nafzaoui, A., and M. Vanbelle. 1986. Effects of feeding alkali-treated olive cake on intake, digestibility and rumen liquor parameters. Anim. Feed Sci. Technol. 14:139.

Nagy, K.A. 1980. CO_2 production in animals. Analysis of potential error in the doubly labeled water method. Am. J. Physiol. 238:466.

Nakamura, T., T. Klopfenstein, D. Gibb, and R. Britton. 1991. Growth efficiency and digestibility of heated protein. Beef Cattle Report. Univ. of Nebraska, Lincoln. Agric. Res. Div. Rep. MP-56. Pp. 36–41.

National Research Council. 1970. Nutrient Requirements of Beef Cattle. National Academy of Sciences, Washington, D.C.

National Research Council. 1971. Nutrient Requirements of Dairy Cattle. National Academy of Sciences, Washington, D.C.

National Research Council. 1972. Hazards of nitrate, nitrite, and nitrosamines to man and livestock. In: Accumulation of Nitrate. National Academy of Sciences, Washington, D.C. Pp. 46–75.

National Research Council. 1984. Nutrient Requirements of Beef Cattle. National Academy of Sciences, Washington, D.C.

National Research Council. 1987. Predicting Feed Intake of Food-producing Animals. National Academy of Sciences Press, Washington, D.C. P. 88.

National Research Council. 1989. Nutrient Requirements of Dairy Cattle. National Academy of Sciences, Washington, D.C.

Neilson, M.J., and G.N. Richards. 1978. The fate of the soluble lignin-carbohydrate complex produced in the bovine rumen. J. Sci. Food Agric. 29:573–579.

Nicholson, C.F. 1991. An optimization model of dual purpose cattle production in the humid lowlands of Venezuela. M.S. thesis, Cornell Univ., Ithaca, N.Y.

Nicolai, J.H., and W.E. Stewart. 1965. Relationship between forestomach and glycemia in ruminants. J. Dairy Sci. 48:56.

Nieduszynski, I., and R.H. Marchessault. 1971. Structure of B-D-(1–4)xylan hydrate. Nature 232:46–47.

Nielsen, F.H. 1991. Nutritional requirements for boron, silicon, vanadium, nickel, and arsenic: current knowledge and speculation. FASEB 5:2661–2668.

Nielsen, F.H., D.R. Myron, S.H. Givand, T.J. Zimmerman, and D.A. Ollerich. 1976. Nickel deficiency in rats. J. Nutr. 105:1620–1630.

Nielsen, F.H., H.T. Reno, L.O. Tiffin, and R.M. Welch. 1974. Nickel. In: Geochemistry and the Environment. E.T. Mertz, ed. National Academy of Sciences, Washington, D.C. P. 40.

Nikolic, J.A., and M. Jovanovic. 1986. Some properties of apple pomace ensiled with and without additives. Anim. Feed Sci. Technol. 15:57.

Nilsson, A., J.L. Hill, and H.L. Davies. 1967. An in vitro study of formonetin and biochanin A metabolism in rumen fluid from sheep. Biochim. Biophys. Acta 148:92.

Nocek, J.E. 1985. Evalution of specific variables affecting in situ estimates of ruminal dry matter and protein digestion. J. Anim. Sci. 60:1347.

Nocek, J.E., and J.E. English. 1986. In situ digestion kinetics: evaluation of rate determination procedures. J. Dairy Sci. 69:77.

Nocek, J.E., and J.B. Russell. 1988. Protein and energy as an integrated system. Relationship of ruminal protein and carbohydrate availability to microbial synthesis and milk production. J. Dairy Sci. 71:2070–2107.

Nolan, J.V. 1975. Quantitative models of nitrogen metabolism in sheep. In: Digestion and Metabolism in the Ruminant. I.W. McDonald and A.C.I. Warner, eds. Univ. of New England Publ. Unit, Armidale, N.S.W., Australia. Pp. 416–431.

Nolan, T., and J. Connolly. 1977. Mixed stocking by sheep and steers—a review. Herbage Abstr. 47:367–374.

Nordkvist, E., H. Graham, and P. Aman. 1989. Soluble lignin complexes isolated from wheat straw (Triticum arvense) and red clover (Trifolium pratense) stems by an invitro method. J. Sci. Food Agric. 48:311–321.

Norton, B.W. 1981. Differences between species in forage quality. In: Nutritional Limits to Animal Production from Pastures. J.B. Hacker, ed. Commonwealth Agricultural Bureau. Pp. 89–110.

Nygren, K., and R.R. Hofmann. 1990. Seasonal variations of food particle size in moose. Alces 26:44–50.

Nyman, M., T.F. Schweitzer, S. Tyren, S. Reimann, and N.-G. Asp. 1990. Fermentation of vegetable fiber in the digestive tract of rats and effects of fecal bulking and bile acid excretion. J. Nutr. 120:459–466.

O'Brien, S.J., W.G. Nash, D.E. Wildt, M.E. Bush, and R.E. Benveniste. 1985. A molecular solution to the riddle of the giant panda's phylogeny. Nature 317:140–144.

Odelson, D.A., and J.A. Breznak. 1983. Volatile fatty acid production by the hindgut microbiota of xylophagous termites. Appl. Environ. Microbiol. 45:1602–1613.

Oh, H.K., M.B. Jones, and W.M. Longhurst. 1968. Comparison of rumen microbial inhibition resulting from various essential oils isolated from relatively unpalatable plant species. Appl. Microbiol. 16:39–44.

Ohlsson, C., and W. Wedin. 1989. Phenological staging schemes for predicting red clover quality. Crop Sci. 29:416–420.

Ohshima, M., and P. McDonald. 1978. A review of the changes in nitrogenous compounds of herbage during ensilage. J. Sci. Food Agric. 29:497–505.

Oleszek, W. 1988. Solid-phase extraction-fractionation of alfalfa saponins. J. Sci. Food Agric. 44:43.

Olsson, K., P. Pernemalm, and O. Theander. 1978. Formation of aromatic compounds from carbohydrates. VII. Reaction of D-glucose and glycine in slightly acidic, aqueous solution. Acta Chem. Scand. 32:249–256.

Olubajo, F.O., P.J. Van Soest, and V.A. Oyenuga. 1974. Comparison and digestibility of four tropical grasses grown in Nigeria. J. Anim. Sci. 38:149–153.

Orpin, C.G., and K.N. Joblin. 1988. The rumen anaerobic fungi. In: The Rumen Microbial Ecosystem. P.N. Hobson, ed. Elsevier Applied Science Publ., New York. Pp. 21–76.

Ørskov, E.R., D. Benzie, and R.N.B. Kay. 1970. The effects of feeding procedure on closure of the oesophageal groove in young sheep. Br. J. Nutr. 24:785.

Ørskov, E.R., R.W. Mayes, and S.O. Mann. 1972. Postruminal digestion of sucrose in sheep. Br. J. Nutr. 28:425–432.

Ørskov, E.R., and I. McDonald. 1979. The estimation of protein degradability in the rumen from incubation measurements weighted according to rate of passage. J. Agric. Sci. 92:499–503.

Ørskov, E.R., G.W. Reid, and M. Kay. 1988. Prediction of intake by cattle from degradation characteristics of roughages. Anim. Prod. 46:29.

Osbourn, D.F., R.A. Terry, G.E. Outen, and S.B. Cammell. 1974. The significance of a determination of cell walls as the rational basis for the nutritive evaluation of forages. Proc. 12th Int. Grasslands Congr. Vol. 3, pp. 374–380.

Oscar, T.P., and J.W. Spears. 1988. Nickel-induced alterations of in vitro and in vivo ruminal fermentation. J. Anim. Sci. 66:198.

Owen, E., R.A. Wahed, R. Alimon, and W. El-Naiem. 1989. Strategies for feeding straw to small ruminants: upgrading or generous feeding to allow selective feeding. In: Overcoming Constraints to the Efficient Utilization of Agricultural By-products as Animal Feed. Proc. 4th Annu. Workshop of African Research Network for Agricultural By-products. A.N. Said and B.H. Dzowela, eds. International Livestock Center for Africa, Addis Ababa, Ethiopia. Pp. 1–21.

Owen-Smith, R.N. 1980. Factors influencing the transfer of plant products into large herbivore populations. In: Dynamic Changes in Savanna Ecosystems. B.J. Huntley and B.H. Walker, eds. Council for Scientific and Industrial Research, Pretoria.

Owen-Smith, R.N. 1987. Pleistocene extinctions: the pivotal role of megaherbivores. Paleobiology 13(3):351–362.

Owen-Smith, R.N. 1988. Megaherbivores. Cambridge Univ. Press, Cambridge.

Palmquist, D.L. 1972. Palmitic acid as a source of endogenous acetate and β-hydroxybutyrate in fed and fasted ruminants. J. Nutr. 102:1401–1406.

Palmquist, D.L. 1988. The feeding value of fats. In: Feed Science. E.R. Orskov, ed. Elsevier Science Publ., Amsterdam. Pp. 293–311.

Palmquist, D.L., C.L. Davis, R.E. Brown, and D.S. Sachan. 1969. Availability and metabolism of various substrates in ruminants. V. Entry rate into the body and incorporation into milk fat of D-β-hydroxybutyrate. J. Dairy Sci. 52:633–638.

Palmquist, D.L., T.C. Jenkins, and A.E. Joyner, Jr. 1986. Effect of dietary fat and calcium source on insoluble soap formation in the rumen. J. Dairy Sci. 69:1020.

Paloheimo, L. 1953. Some persistent misconceptions concerning the crude fiber and the nitrogen-free extract. Maataloustiet. Aikak. 25:16–22.

Paloheimo, L., and A. Mäkela. 1959. Further studies on the retention time of food in the digestive tract of cows. Suom. Maataloustiet. Seuran Julk. 94:15–39.

Paloheimo, L., and I. Paloheimo. 1949. On the estimation of the total of vegetable membrane substances. Maataloustiet. Aikak. 21:1.

Papadopoulos, Y.A., and B.D. McKersie. 1983. A comparison of protein degradation during wilting and ensiling of six forage species. Can. J. Plant Sci. 63:903–912.

Parks, T.G. 1973. The effects of low and high residue diets on the rate of transit and composition of the faeces. Proc. 4th Int. Symp. on Gastro-intestinal Motility. Mitchell Press, Vancouver. P. 369.

Parra, R. 1978. Comparison of foregut and hindgut fermentation in herbivores. In: The Ecology of Arboreal Folivores. G.G. Montgomery, ed. Smithsonian Institution Press, Washington, D.C. P. 205.

Parsons, A.J., I.R. Johnson, and A. Harvey. 1988. Use of a model to optimize the interaction between frequency and severity of intermittent defoliation and to provide a fundamental comparison of the continuous and intermittent defoliation of grass. Grass Forage Sci. 43:49–60.

Parsons, A.J., and P.D. Penning. 1988. The effect of the

duration of regrowth on photosynthesis, leaf death and the average rate of growth in a rotationally grazed sward. Grass Forage Sci. 43:15–27.

Peel, C.J., and D.E. Bauman. 1987. Somatotropin and lactation. J. Dairy Sci. 70:474–486.

Pell, A.N., and P. Schofield. 1993a. Microbial adhesion and degradation of plant cell walls. In: Forage Cell Wall Structure and Digestibility. H.G. Jung, D.R. Buxton, R.D. Hatfield, and J. Ralph, eds. American Society of Agronomy, Madison, Wisc. Pp. 397–424.

Pell, A.N., and P. Schofield. 1993b. Computerized monitoring of gas production to measure forage digestion in vitro. J. Dairy Sci. 76:1063–1073.

Penning, P.D. 1983. A technique to record automatically some aspects of grazing and ruminating behavior in sheep. Grass Forage Sci. 38:89–96.

Penning, P.D., and G.E. Hooper. 1985. An evaluation of the use of short-term weight changes in grazing sheep for estimating herbage intake. Grass Forage Sci. 40:79–84.

Perdock, H.B., and R.A. Leng. 1987. Hyperexcitability in cattle fed ammoniated roughages. Anim. Feed Sci. Technol. 17:121–144.

Peters, J.P., and J.M. Elliot. 1984. Endocrine changes with infusion of propionate in the dairy cow. J. Dairy Sci. 67:2455.

Peters, R.H. 1986. The Ecological Implications of Body Size. Cambridge Univ. Press, Cambridge. P. 329.

Phillipson, A.T. 1939. The movements of the pouches of the stomach of sheep. J. Exp. Physiol. 29:395–415.

Pichard, G.R. 1977. Forage nutritive value. Continuous and batch in vitro rumen fermentations and nitrogen solubility. Ph.D. dissertation, Cornell Univ., Ithaca, N.Y.

Pichard, G.R., and P.J. Van Soest. 1977. Protein solubility of ruminant feeds. Proc. Cornell Nutrition Conf., Ithaca, N.Y. Pp. 91–98.

Pirt, S.J. 1965. The maintenance energy of bacteria in growing cultures. Proc. R. Soc. Lond. B Biol. Sci. 163:224.

Pitt, R.E. 1990. Silage and Hay Preservation. Cornell Univ. Cooperative Extension NRAES-5. P. 53.

Pitt, R.E., and C.J. Sniffen. 1988. Silage inoculants. Cornell Univ. Agric. Eng. Ext. Bull. 452. Ithaca, N.Y. P. 11.

Pittman, D.A., S. Lakshmanan, and M.P. Bryant. 1967. Oligopeptide uptake by Bacteroides ruminicola. J. Bacteriol. 93:1499–1508.

Playne, M.J. 1978. Differences between cattle and sheep in their digestion and relative intake of a mature tropical grass hay. Anim. Feed Sci. Technol. 3:41–47.

Playne, M.J., and P. McDonald. 1966. The buffering constituents of herbage and of silage. J. Sci. Food Agric. 17:264–268.

Pond, K.R., W.C. Ellis, W.D. James, and A.G. Deswysen. 1985. Analysis of multiple markers in nutrition research. J. Dairy Sci. 68:745.

Pond, W.G., H.G. Jung, and V.H. Varel. 1988. Effect of dietary fiber on young adult genetically lean, obese and contemporary pigs: body weight, carcass measurements, organ weights and digesta content. J. Anim. Sci. 66:699–706.

Poos-Floyd, I.M., T.J. Klopfenstein, and R.A. Britton. 1985. Evaluation of laboratory techniques for predicting ruminal protein degradation. J. Dairy Sci. 68:829.

Porter, P.A. 1987. The acid detergent insoluble ash digestibility marker and its use in lactating dairy cows. Ph.D. dissertation, Cornell Univ., Ithaca, N.Y.

Price, J., M.A. Will, G. Paschaleris, and J.K. Chesters. 1988. Identification of thiomolybdates in digesta and plasma from sheep after administration of [99]Mo-labelled compounds into the rumen. Br. J. Nutr. 58:127–138.

Price, M.R.S. 1985. Game domestication for animal production in Kenya: the nutritional ecology of oryx, zebu cattle and sheep under free-range conditions. J. Agric. Sci. 104:375–382.

Prins, R.A., and D.A. Kreulen. 1991. Comparative aspects of plant cell wall digestion in insects. Anim. Feed Sci. Technol. 32:101–118.

Prins, R.A., T.P. Rooymand, M. Veldhuizen, M.A. Domhof, and W. Cline-Theil. 1983. Extent of plant cell wall digestion in several species of wild ruminants kept in the zoo. Zool. Gart. 53:393–403.

Prins, R.A., and L. Seekles. 1968. Effect of chloral hydrate on rumen metabolism. J. Dairy Sci. 51:883–887.

Provenza, F.D., and D.F. Balph. 1988. Development of dietary choice in livestock on rangelands and its implications for management. J. Anim. Sci. 66:2356–2368.

Pullar, J.D. 1964. Methods of calorimetry (A) direct. In: The Science of Nutrition of Farm Livestock. International Encyclopedia of Food and Nutrition, vol. 17, pt. 1. D.P. Cuthbertson, ed. Pergamon Press, London. Pp. 471–490.

Purser, D.B., and S.M. Buechler. 1966. Amino acid composition of rumen organisms. J. Dairy Sci. 49:81.

Putnam, P.A., and R.E. Davis. 1965. Postruminal fiber digestibility. J. Anim. Sci. 24:826–829.

Raleigh, R.J., R.J. Kartchner, and L.R. Rittenhouse. 1980. Chromic oxide in range nutrition studies. Oreg. Agric. Exp. Stn. Bull. 641.

Ralph, J., and R.F. Helm. 1993. Lignin structure and lignin-polysaccharide linkage. In: Forage Cell Wall Structure and Digestibility. H.G. Jung, D.R. Buxton, R.D. Hatfield, and J. Ralph, eds. American Society of Agronomy, Madison, Wisc. Pp. 201–246.

Rampino, M. 1989. Dinosaurs, comets and volcanoes. New Scientist, Feb. 18, pp. 54–58.

Rattray, P.V., and J.P. Joyce. 1974. Nutritive value of white clover and perennial ryegrass. 4. Utilization of dietary energy. N.Z. J. Agric. Res. 17:401–408.

Raven, J.A. 1983. The transport and function of silicon in plants. Biol. Rev. Camb. Philos. Soc. 58(2):179–207.

Raymond, W.F. 1969. The nutritive value of forage crops. Adv. Agron. 21:1–108.

Rebolé, A., and P. Alvira. 1986. Composition of vine branches with leaves of *Vitis vinifera* fresh and ensiled with different additives. Anim. Feed Sci. Technol. 16:89.

Rebolé, A., P. Alvira, and G. González. 1988. Digestibility in vivo of ensiled grapevines (branches and leaves): influence of the system of analysis in the detergent fibre scheme on the prediction of digestibility. Anim. Feed Sci. Technol. 22:173.

Reed, J.D. 1983. The nutritional ecology of game and cattle on a Kenyan ranch. Ph.D. dissertation, Cornell Univ., Ithaca, N.Y.

Reed, J.D. 1986. Relationships among soluble phenolics, insoluble proanthocyanidins and fiber in East African browse species. J. Range Manage. 39(1):5–7.

Reed, J.D. 1987. Phenolics, fiber, and fiber digestibility in bird resistant and non bird resistant sorghum grain. J. Agric. Food Chem. 35(4):461–464.

Reed, J.D., P.J. Horvath, M.S. Allen, and P.J. Van Soest. 1985. Gravimetric determination of soluble phenolics including tannins from leaves by precipitation with trivalent ytterbium. J. Sci. Food Agric. 36:255.

Reed, J.D., R.E. McDowell, P.J. Van Soest, and P.J. Horvath. 1982. Condensed tannins: a factor limiting the use of cassava forage. J. Sci. Food Agric. 33:213–220.

Reed, J.D., A. Tedla, and Y. Kebede. 1987. Phenolics, fibre and fibre digestibility in the crop residue from bird resistant and non–bird resistant sorghum varieties. J. Sci. Food Agric. 39:113.

Reeves, J.B., III. 1985. Lignin composition of chemically treated feeds as determined by nitrobenzene oxidation and its relationship to digestibility. J. Dairy Sci. 68:1976.

Regal, P.J. 1977. Ecology and evolution of flowering plant dominance. Science 196:622–629.

Reid, C.S.W., R.T.J. Clarke, F.R.M. Cockrem, W.T. Jones, J.T. McIntosh, and D.E. Wright. 1975. Physiological and genetical aspects of pasture (legume) bloat. In: Digestion and Metabolism in the Ruminant. I.W. McDonald and A.C.I. Warner, eds. Univ. of New England Publ. Unit, Armidale, N.S.W., Australia. Pp. 524–536.

Reid, C.S.W., and J.B. Cornwall. 1959. The mechanical activity of the reticulo-rumen of cattle. Proc. N.Z. Soc. Anim. Prod. 19:23.

Reid, J.T. 1961. Cut early for more milk. Hoard's Dairyman 106:417.

Reid, J.T. 1962. Energy values of feeds—past, present and future. Proc. Animal Nutrition Contributions to Modern Animal Agriculture. Spec. Bull., Cornell Univ., Ithaca, N.Y. Pp. 54–92.

Reid, J.T. 1970. Will meat, milk and egg production be possible in the future? Proc. Cornell Nutrition Conf., Ithaca, N.Y. P. 50.

Reid, J.T., and O.D. White. 1977. The phenomenon of compensatory growth. Proc. Cornell Nutrition Conf. for Feed Manufacturers, Ithaca, N.Y. Pp. 16–27.

Reid, R.L., B.S. Baker, and L.C. Vona. 1984. Effects of magnesium sulfate supplementation and fertilization on quality and mineral utilization of timothy hays by sheep. J. Anim. Sci. 59:1403.

Reid, R.L., and L.F. James. 1985. Forage-animal disorders. In: Forages: The Science of Grassland Agriculture. 4th ed. M.E. Heath, R.F. Barnes, and D.S. Metcalfe, eds. Iowa State Univ. Press, Ames. P. 430.

Reid, R.L., G.A. Jung, and W.V. Thayne. 1988. Relationships between nutritive quality and fiber components of cool season and warm season forages: a retrospective study. J. Anim. Sci. 66:1275–1291.

Reisenauer, H.M. 1976. Soil and plant testing in California. Univ. Calif. Div. Agric. Sci. Bull. 1879.

Renner, R. 1971. Carbohydrate requirements and availability for chickens. Proc. Cornell Nutrition Conf., Ithaca, N.Y. Pp. 79–85.

Rexen, F.P., P. Stigsen, and V.F. Kristensen. 1975. The effect of a new alkali technique on the nutritive value of straw. Proc. 9th Nutrition Conf. for Feed Manufacturers, Univ. of Nottingham, England, Jan. 5–7.

Rhoades, D.F., and R.G. Cates. 1976. Toward a general theory of plant antiherbivore chemistry. Recent Adv. Phytochem. 10:168–213.

Rice, R.W., J.G. Morris, B.T. Maeda, and R.L. Baldwin. 1974. Simulation of animal functions in models of production systems: ruminants on the range. Fed. Proc. 33:188–195.

Riewe, M.E. 1976. Principles of grazing management. In: Grasses and Legumes in Texas—Development, Production and Utilization. E.C. Holt and R.D. Lewis, eds. Texas Agr. Exp. Stn. Res. Monogr. RM6C. P. 169.

Riewe, M.E., and H. Lippke. 1970. Considerations in determining the digestibility of harvested forages. Proc. Nat. Conf. on Forage Evaluation and Utilization, Lincoln, Neb. P. F-1.

Robbins, C.T. 1973. The biological basis for the determination of carrying capacity. Ph.D. dissertation, Cornell Univ., Ithaca, N.Y.

Robbins, C.T. 1993. Wildlife Feeding and Nutrition. 2d ed. Academic Press, New York.

Robbins, C.T., A.E. Hagerman, P.J. Austin, C. McArthur, and T.A. Hanley. 1991. Variation in mammalian physiological responses to a condensed tannin and its ecological implications. J. Mammal. 72(3):480–486.

Robbins, C.T., P.J. Van Soest, W.W. Mautz, and A.N. Moen. 1975. Feed analysis and digestion with reference to white-tailed deer. J. Wildl. Manage. 39:67–79.

Robert, P., D. Bertrand, and C. Demarquilly. 1986. Prediction of forage digestibility by principal component analysis of near infrared reflectance spectra. Anim. Feed Sci. Technol. 16:215.

Roberts, M.S., and J.L. Gittleman. 1984. *Ailurus fulgens*. Mamm. Species 222:1–8.

Roberts, M.S., and D.S. Kessler. 1979. Reproduction in red pandas, *Ailurus fulgens* (Carnivora: Ailuropodidae). J. Zool., Lond. 188:235–249.

Robertshaw, D. 1984. The Evolution of Homeothermy: From Dinosaurs to Man. Spec. Report 329, Agric. Exp. Stn., Univ. of Missouri, Columbia.

Robertson, J.A., S.D. Murison, and A. Chesson. 1987. Estimation of the potential digestibility and rate of degradation of water-soluble dietary fiber in the pig cecum with a modified nylon bag technique. J. Nutr. 117:1402–1409.

Robertson, J.B., and P.J. Van Soest. 1975. A note on digestibility in sheep as influenced by level of intake. Anim. Prod. 21:89–92.

Robertson, J.B., and P.J. Van Soest. 1981. The detergent system of analysis and its application to human foods. In: The Analysis of Dietary Fiber in Food. W.P.T. James and O. Theander, eds. Marcel Dekker, New York. Pp. 123–158.

Robin, J.P., C. Mercier, F. Duprat, R. Charbonniere, and A. Guilbot. 1975. Amidons lintnérisés. Etudes chromatographique et enzymatique des résidus insolubles provenant de l'hydrolyse chlorhydrique d'amidons, de céréales, en particuliar de maïs cireux. Die Stärke 27:36–45.

Robinson, D.L., L.C. Kappal, and J.A. Boling. 1989. Management practices to overcome the incidence of grass tetany. J. Anim. Sci. 67:3470–3484.

Robinson, P.H. 1983. Development and initial testing of an in vivo system to estimate rumen and whole tract digestion in lactating dairy cows. Ph.D. dissertation, Cornell Univ., Ithaca, N.Y. P. 313.

Robinson, W.O. 1938. The agricultural significance of the minor elements. Am. Fert. 89:5.

Robinson, W.O., H. Bastron, and K.J. Murata. 1958. Biogeochemistry of the rare-earth elements with particular reference to hickory trees. Geochim. Cosmochim. Acta 14:55.

Rodrigue, C.B., and N.N. Allen. 1960. The effect of fine grinding of hay on ration digestibility, rate of passage, and fat content of milk. Can. J. Anim. Sci. 40:23–29.

Roe, D.A., K. Wrick, D. McLain, and P.J. Van Soest. 1978. Effects of dietary fiber sources on riboflavin absorption. Fed. Proc. 37:756.

Rogerson, A. 1958. Diet and partial digestion in secretions of the alimentary tract of the sheep. Br. J. Nutr. 12:164.

Rohweder, D.A., R.F. Barnes, and N. Jorgensen. 1978. Proposed hay grading standards based on laboratory analyses for evaluating quality. J. Anim. Sci. 47:747–759.

Rotz, C.A., J.R. Black, D.R. Mertens, and D.R. Buckmaster. 1989. DAFOSYM: A model of the dairy forage system. J. Prod. Agric. 2:83–91.

Roughan, P.G., and R. Holland. 1977. Predicting in vivo digestibilities of herbages by exhaustive enzymatic hydrolysis of cell walls. J. Sci. Food Agric. 28:1057–1064.

Roxas, D.B., L.S. Castillo, A. Obsioma, R.M. Lapitan, V.G. Momongan, and B.O. Juliano. 1984. Chemical composition and in vitro digestibility of straw from different varieties of rice. In: The Utilization of Agricultural Fibrous Residues as Animal Feeds. P.T. Doyle, ed. School of Agric. and Forest., Univ. of Melbourne, Parkville, Victoria, Australia. Pp. 39–46.

Ruckebusch, Y. 1988. Motility of the gastrointestinal tract. In: The Ruminant Animal. D.C. Church, ed. Prentice-Hall, Englewood Cliffs, N.J. P. 64.

Rücker, G., and O. Knabe. 1986. Dry matter loss and quality change due to the wilting of grass. 1. Dynamics of CO_2 production in wilting in swathes. Arch. Tierernähr. 36:749.

Russell, J.B. 1987. A proposed mechanism of Monensin action in inhibiting ruminal bacterial growth: effects on ion flux and proton motive force. J. Anim. Sci. 64:1519–1525.

Russell, J.B. 1989. What is tricarballylic acid? Proc. Cornell Nutrition Conf., Ithaca, N.Y. Pp. 22–24.

Russell, J.B., and N. Forsberg. 1986. Production of tricarballylic acid by rumen microorganisms and its potential toxicity in ruminant tissue metabolism. Br. J. Nutr. 56:153–162.

Russell, J.B., and R.B. Hespell. 1981. Microbial rumen fermentation. J. Dairy Sci. 64:1153–1169.

Russell, J.B., and S.A. Martin. 1984. Effects of various methane inhibitors on the fermentation of amino acids by mixed rumen microorganisms in vitro. J. Anim. Sci. 59:1329–1338.

Russell, J.B., and H.F. Mayland. 1987. Absorption of tricarballylic acid from the rumen of sheep and cattle fed forages containing *trans*-aconitic acid. J. Sci. Food Agric. 40:205–212.

Russell, J.B., J.D. O'Connor, D.G. Fox, P.J. Van Soest, and C.J. Sniffen. 1992. A net carbohydrate and protein system for evaluating cattle diets. I. Ruminal fermentation. J. Anim. Sci. 70:3551–3561.

Russell, J.B., C.J. Sniffen, and P.J. Van Soest. 1983. Effect of carbohydrate limitation on degradation and utilization of casein by mixed rumen bacteria. J. Dairy Sci. 66:763–775.

Russell, J.B., H.J. Strobel, and G. Chen. 1988. Enrichment and isolation of a ruminal bacterium with a very high specific activity of ammonia production. Appl. Environ. Microbiol. 54:872–877.

Russell, J.B., and P.J. Van Soest. 1984. In vitro ruminal fermentation of organic acids common in forage. Appl. Environ. Microbiol. 47:155–159.

Russell, J.B., and D.B. Wilson. 1988. Potential opportunities and problems for genetically altered rumen microorganisms. J. Nutr. 118:271.

Russell, J.D., A.R. Fraser, A.H. Gordon, and A. Chesson. 1988. Rumen digestion of untreated and alkali-treated cereal straws: a study by multiple internal reflectance infrared spectroscopy. J. Sci. Food Agric. 45:95–108.

Russell, J.R., A.W. Young, and N.A. Jorgensen. 1981. Effect of dietary corn starch intake on pancreatic amylase and intestinal maltase and pH in cattle. J. Anim. Sci. 52:1177–1182.

Russell, R.W., L. Moss, S.P. Schmidt, and J.W. Young. 1986. Effects of body size on kinetics of glucose metabolism and on nitrogen balance in growing cattle. J. Nutr. 116:2229.

Ryder, M.L., and S.K. Stephenson. 1968. Wool Growth. Academic Press, New York.

Saarinen, P., W. Jensen, and J. Alhojärvi. 1959. On the digestibility of high yield chemical pulp and its evaluation. Suom. Maataloustiet. Seuran Julk. Acta Agral. Fenn. 94:41.

Salton, M.R.J. 1960. The Bacterial Cell Wall. Elsevier, Amsterdam.

Salyers, A.A., M. O'Brien, and S.F. Kotarski. 1982. Utilization of chondroitin sulfate by *Bacteroides thetaiotaomicron* growing in carbohydrate-limited continuous culture. J. Bacteriol. 150:1008–1015.

Sanderson, M.A., and W.F. Wedin. 1988. Cell wall composition of alfalfa stems at similar morphological stages and chronological age during spring growth and summer regrowth. Crop Sci. 28:342–347.

Satter, L.D., and R.E. Roffler. 1975. Nitrogen requirement and utilization in dairy cattle. J. Dairy Sci. 58:1219–1237.

Satter, L.D., and L.L. Slyter. 1974. Effect of ammonia concentration on rumen microbial protein production in vitro. Br. J. Nutr. 32:199–208.

Sauvant, D., S. Giger, and D. Bertrand. 1985. Variations and precision of the in sacco dry matter digestion of concentrates and by-products. Anim. Feed Sci. Technol. 13:7.

Scaife, J.R., K.W.J. Wahle, and G.A. Garton. 1978. Utilization of methylmalonate for the synthesis of branched-chain fatty acids by preparations of chicken liver and sheep adipose tissue. Biochem. J. 176:799.

Schalk, A.F., and R.S. Amadon. 1928. Physiology of the ruminant stomach. North Dakota Agr. Exp. Stn. Bull. 216.

Schaller, G.B., Hu Jinchu, Pan Wenshi, and Zhn Jine. 1985. The Giant Pandas of Wolong. Univ. of Chicago Press, Chicago.

Schingoethe, D.J., F.M. Byers, and G.T. Schelling. 1988. Nutrient needs during critical periods of the life cycle. In: The Ruminant Animal, Digestive Physiology and Nutrition. D.C. Church, ed. Prentice-Hall, Englewood Cliffs, N.J. P. 421.

Schmidt-Nielsen, K. 1984. Scaling: Why Is Animal Size So Important? Cambridge Univ. Press, Cambridge. P. 239.

Schneider, B.H. 1947. Feeds of the World. Jarrett Printing Co., Charlestown, W. Va.

Schneider, B.H., H.L. Lucas, H.M. Pavlech, and M.A. Cipolloni. 1951. Estimation of the digestibility of feeds from their proximate composition. J. Anim. Sci. 10:706–713.

Schultz, J.C., I.T. Baldwin, and P.J. Nothnagle. 1981. He-

moglobin as a binding substrate in the quantitative analysis of plant tannins. J. Agric. Food Chem. 29:823–829.

Scott, B., T.G. Nasaij, and E.E. Bartley. 1979. Effect of Lasolacid or Monensin on lactic acid producing and using ruminal bacteria. Proc. 15th Conf. on Rumen Function. Nov. 28–29, Chicago.

Scott, D. 1986. Control of phosphorus balance in ruminants. In: Aspects of Digestive Physiology in Ruminants. Alan Dobson and Marjorie Dobson, eds. Cornell Univ. Press, Ithaca. P. 156.

Seale, D.B. 1987. Amphibia. In: Animal Energetics, vol. 2. T.J. Pandian and F.J. Vernberg, eds. Academic Press, New York. P. 468.

Sebastian, L., V.D. Mudgal, and P.G. Nair. 1970. Comparative efficiencies of milk production by Sahiwal cattle and Murrah buffalo. J. Anim. Sci. 30:253–256.

Seiden, R., and W.H. Pfander. 1957. The Handbook of Feedstuffs. Springer Publ. Co., New York.

Sellers, A.F., and C.E. Stevens. 1966. Motor functions of the ruminant forestomach. Physiol. Rev. 46:634–661.

Seoane, J.R., C.A. Baile, and F.H. Martin. 1972. Humoral factors modifying feeding behavior of sheep. Physiol. Behav. 8:993–995.

Sequeira, C.A. 1981. Effect of grinding on the in vitro digestibility of forages. M.S. thesis, Cornell Univ., Ithaca, N.Y. P. 126.

Setchell, K.D.R., and H. Adlercreutz. 1988. Mammalian lignans and phyto-oestrogens: recent studies on their formation, metabolism and biological role in health and disease. In: Role of the Gut Flora in Toxicity and Cancer. I.R. Rowland, ed. Academic Press, New York. Pp. 316–345.

Shaw, J.C. 1956. Ketosis in dairy cattle. A review. J. Dairy Sci. 39:402–434.

Shaw, J.C., and W.L. Ensor. 1959. Effect of feeding cod liver oil and unsaturated fatty acids on rumen volatile fatty acids and milk fat content. J. Dairy Sci. 6:461–465.

Shaw, J.C., W.L. Ensor, H.F. Tellechea, and S.D. Lee. 1960. Relation of diet to rumen volatile fatty acids, digestibility, efficiency of gain and degree of unsaturation of body fat in steers. J. Nutr. 71:203–208.

Shimojo, M., and I. Goto. 1989. Effect of sodium silicate on forage digestion with rumen fluid of goats or cellulase using culture solutions adjusted for pH. Anim. Feed Sci. Technol. 24:173–177.

Shorland, F.B. 1953. Animal fats: recent researches in the fats research laboratory. J. Sci. Food Agric. 4:498.

Short, H.L., D.E. Medin, and A.E. Anderson. 1975. Ruminoreticular characteristics of mule deer. J. Mammal. 46:196–199.

Shutt, D.A., and A.W.H. Braden. 1968. The significance of equol in relation to the eostrogenic responses in sheep ingesting clover with a high formononetin content. Aust. J. Agric. Res. 19:545–553.

Sibbald, A.R., T.J. Maxwell, and J. Eadie. 1979. A conceptual approach to the modelling of herbage intake by hill sheep. Agric. Syst. 4:119–134.

Simpson, G.G. 1945. The Principles of Classification and a Classification of Mammals. R. Tyler, ed. Bull. Am. Mus. Nat. Hist. 85:1.

Sinha, S. 1966. Complexes of the Rare Earths. Pergamon, New York.

Sisson, S., and J.D. Grossman. 1953. The Anatomy of the Domestic Animals. W.B. Saunders, Philadelphia.

Slade, L.M. 1970. Nitrogen metabolism in non-ruminant herbivores. Ph.D. dissertation, Univ. of California, Davis.

Slade, L.M., and D.W. Robinson. 1970. Nitrogen metabolism in nonruminant herbivores. 2. Comparative aspects of protein digestion. J. Anim. Sci. 3:761–763.

Slyter, L.L. 1976. Influence of acidosis on rumen function. J. Anim. Sci. 43:910–929.

Slyter, L.L., and P.A. Putnam. 1967. In vivo vs. in vitro continuous culture of ruminal microbial populations. J. Anim. Sci. 26:1421–1427.

Smart, W.W.G., Jr., T.A. Bell, N.W. Stanley, and W.A. Cope. 1961. Inhibition of rumen cellulase by an extract from sericea forage. J. Dairy Sci. 44:1945–1946.

Smith, D. 1973. The nonstructural carbohydrates. In: Chemistry and Biochemistry of Herbage, vol. 1. G.W. Butler and R.W. Bailey, eds. Academic Press, London. Pp. 106–156.

Smith, G.S. 1986. Gastrointestinal toxifications and detoxifications in ruminants in relation to resource management. In: Gastrointestinal Toxicology. K. Rozman and O. Hanninen, eds. Elsevier Science Publ., Amsterdam. Pp. 514–542.

Smith, G.S., and A.B. Nelson. 1975. Effects of sodium silicate added to rumen cultures on forage digestion with interactions of glucose, urea and minerals. J. Anim. Sci. 41:890–899.

Smith, G.S., A.B. Nelson, and E.J.A. Boggino. 1971. Digestibility of forages in vitro as affected by content of "silica." J. Anim. Sci. 33:466–471.

Smith, L.W. 1968. The influence of particle size and lignification upon rates of digestion and passage of uniformly labeled ^{14}C cell walls in the sheep. Ph.D. dissertation, Univ. of Maryland, College Park.

Smith, M.W. 1990. Ionophore-mineral interactions. Hoffmann–La Roche Symposium, Guelph, Canada, April 24. P. 20.

Smith, O.B., and A. Adegbola. 1985. Studies on the feeding value of agro-industrial by-products. II. Digestibility of cocoa-pod and cocoa-pod-based diets by ruminants. Anim. Feed Sci. Technol. 13:249.

Smith, R.H. 1975. Nitrogen metabolism in the rumen and the composition and nutritive value of nitrogen compounds entering the duodenum. In: Digestion and Metabolism in the Ruminant. I.W. McDonald and A.C.I. Warner, eds. Univ. of New England Publ. Unit, Armidale, N.S.W., Australia. P. 399.

Smith, R.H., and A.B. McAllan. 1973. Chemical composition of rumen bacteria. Proc. Nutr. Soc. 32:4, 9A–10A.

Sniffen, C.J., J.D. O'Connor, P.J. Van Soest, D.G. Fox, and J.B. Russell. 1992. A net carbohydrate and protein system for evaluating cattle diets. II. Carbohydrate and protein availability. J. Anim. Sci. 70:3562–3577.

Sniffen, C.J., and P.H. Robinson. 1984. Nutritional strategy. Can. J. Anim. Sci. 64:529–542.

Sniffen, C.J., J.B. Russell, and P.J. Van Soest. 1983. The influence of carbon source, nitrogen source and growth factors on rumen microbial growth. Proc. Cornell Nutrition Conf., Ithaca, N.Y. Pp. 26–33.

Soofi, R., G.C. Fahey, Jr., and L.L. Berger. 1982. In situ and in vivo digestibilities and nutrient intakes by sheep of alkali-treated soybean stover. J. Anim. Sci. 55:1206.

Southgate, D.A.T. 1969. Determination of carbohydrates in foods. II. Unavailable carbohydrates. J. Sci. Food Agric. 20:331–335.

Southgate, D.A.T. 1976a. Determination of Food Carbohydrates. Applied Science Publ., London. Pp. 1–178.

Southgate, D.A.T. 1976b. The chemistry of dietary fiber. In: Fiber and Human Nutrition. G. Spiller and R.J. Amen, eds. Plenum Press, New York. Pp. 31–36.

Southgate, D.A.T. 1981. Use of the Southgate method for unavailable carbohydrates in the measurement of dietary fiber. In: The Analysis of Dietary Fiber in Food. W.P.T. James and O. Theander, eds. Marcel Dekker, New York. Pp. 1–20.

Spears, J.W. 1984. Nickel as a "newer trace element" in the nutrition of domestic animals. J. Anim. Sci. 59:823–835.

Spencer, R.R., and G.W. Chapman, Jr. 1985. Surface wax of coastal bermuda grass. J. Agric. Food Chem. 22:654–656.

Stafford, H.A. 1989. The enzymology of proanthocyanidin biosynthesis. In: Chemistry and Significance of Condensed Tannins. R.W. Hemingway and J.J. Karchesy, eds. Plenum, New York. Pp. 47–70.

Stanton, T.L., R.W. Blake, R.L. Quaas, and L.D. Van Vleck. 1991. Response to selection of United States Holstein sires in Latin America. J. Dairy Sci. 74:651–664.

Steen, R.W.J. 1986. An evaluation of effluent from grass silage as a feed for beef cattle offered silage-based diets. Grass Forage Sci. 41:39–46.

Steinhour, W.D., and D.E. Bauman. 1986. Propionate metabolism: a new interpretation. In: Aspects of Digestive Physiology in Ruminants. Alan Dobson and Marjorie Dobson, eds. Cornell Univ. Press, Ithaca. P. 238.

Stephen, A.M., and J.H. Cummings. 1980. Mechanism of action of dietary fibre in the human colon. Nature 284:283–284.

Stern, M.D., and W.H. Hoover. 1979. Methods for determining and factors affecting rumen microbial protein synthesis: a review. J. Anim. Sci. 49:1590–1603.

Steven, D.H., and A.B. Marshall. 1970. Organization of rumen epithelium. In: Physiology of Digestion and Metabolism in the Ruminant. A.T. Phillipson, ed. Oriel Press, Newcastle upon Tyne, England. Pp. 80–100.

Stevens, C.E. 1978. Physiological implications of microbial digestion in the large intestine of mammals: relation to dietary factors. Am. J. Clin. Nutr. 31:5161–5168.

Stevens, C.E. 1980. The gastrointestinal tract of mammals: major variations. In: Comparative Physiology: Primitive Mammals. K. Schmidt-Nielsen and R. Taylor, eds. Cambridge Univ. Press, New York. Pp. 52–62.

Stevens, C.E. 1988. Comparative Physiology of the Vertebrate Digestive System. Cambridge Univ. Press, New York. P. 300.

Stevens, C.E., and A.F. Sellers. 1968. Rumination. In: Handbook of Physiology. Section 6: Alimentary Canal. C.F. Code, ed. American Physiological Society, Washington, D.C. Pp. 2699–2704.

Stevens, C.E., A.F. Sellers, and F.A. Spurrel. 1960. Function of the bovine omasum in ingesta transfer. Am. J. Physiol. 198:449.

Stevens, J.S., D.A. Levitsky, P.J. Van Soest, J.B. Robertson, H.J. Kalkwarf, and D.A. Roe. 1987a. Effect of psyllium gum and wheat bran on spontaneous energy intake. Am. J. Clin. Nutr. 46:812–817.

Stevens, J.S., P.J. Van Soest, J.B. Robertson, and D.A. Levitsky. 1987b. Mean transit time measurement by analysis of a single stool after ingestion of multicolored plastic pellets. Am. J. Clin. Nutr. 46:1048–1054.

Stewart, C.S., and M.P. Bryant. 1988. The rumen bacteria. In: The Rumen Microbial Ecosystem. P.N. Hobson, ed. Elsevier Applied Science Publ. New York. Pp. 21–76.

Stobbs, T.H. 1973a. The effect of plant structure on the intake of tropical pastures. I. Variation in the bite size of grazing cattle. Aust. J. Agric. Res. 24:809–819.

Stobbs, T.H. 1973b. The effect of plant structure on the intake of tropical pastures. II. Differences in sward structure, nutritive value, and bite size of animals grazing *Setaria anceps* and *Chloris gayana* at various stages of growth. Aust. J. Agric. Res. 24:821–829.

Stotzky, G. 1980. Surface interactions between clay minerals and microbes, viruses and soluble organics, and the probable importance of these interactions to the ecology of microbes in soil. In: Microbial Adhesion to Surfaces. Society Chemical Industry, London. Pp. 231–248.

Stout, P., J. Brownell, and R.G. Burau. 1967. Occurrences of *trans*-aconitate in range forage species. Agron. J. 59:21.

Streeter, C.L. 1969. A review of techniques used to estimate the in vivo digestibility of grazed forage. J. Anim. Sci. 29:757–768.

Strobel, H.J., and J.B. Russell. 1986. Effect of pH and energy spilling on bacterial protein synthesis by carbohydrate-limited cultures of mixed rumen bacteria. J. Dairy Sci. 69:2941–2947.

Struik, P.C., B. Deinum, and J.M.P. Hoefsloot. 1985. Effects of temperature during different stages of development on growth and digestibility of forage maize (*Zea mays* L.). Neth. J. Agric. Sci. 33:405–420.

Sudweeks, E.M., L.O. Ely, and L.R. Sisk. 1978. Effect of particle size of corn silage on digestibility and rumen fermentation. J. Dairy Sci. 62:292–296.

Sukhija, P.S., and D.L. Palmquist. 1988. Rapid method for determination of total fatty acid content and composition of feedstuffs and feces. J. Agric. Food Chem. 36:1202–1206.

Sullivan, J.T. 1959. A rapid method for the determination of acid-insoluble lignin in forages and its relation to digestibility. J. Anim. Sci. 18:1292.

Sullivan, J.T. 1966. Studies of the hemicelluloses of forage plants. J. Anim. Sci. 25:83–86.

Suttle, N.F. 1978. Effects of sulfur and molybdenum on the absorption of copper from forage crops by ruminants. Proc. Symp. Sulphur in Forages. J.C. Brogan, ed. An Foras Taluntais, Wexford, Ireland. Pp. 197–211.

Sutton, J.D., I.C. Hart, S.V. Morant, E. Schuller, and A.D. Simmonds. 1988. Feeding frequency for lactating cows: diurnal patterns of hormones and metabolites in peripheral blood in relation to milk-fat concentration. Br. J. Nutr. 60:265–274.

Swain, T. 1977. Secondary compounds as protective agents. Annu. Rev. Plant Physiol. 28:479–501.

Swain, T. 1979. Tannins and lignin. In: Herbivores: Their Interaction with Secondary Metabolites. G.A. Rosenthal, and D.H. Janzen, eds. Academic Press, New York. Pp. 657–682.

Swenson, M.J., and W.O. Reece, eds. 1993. Dukes' Physiology of Domestic Animals. 11th ed. Cornell Univ. Press, Ithaca.

Swift, R.W., and J.W. Bratzler. 1959. A comparison of the digestibility of forages by cattle and by sheep. Penn. Agric. Exp. Stn. Bull. 651. Pp. 1–5.

Swift, R.W., W.H. Bratzler, W.H. James, A.D. Tillman, and D.C. Meek. 1948. The effect of dietary fat on utilization of the energy and protein of rations by sheep. J. Anim. Sci. 7:475.

Tamminga, S. 1979. Protein degradation in the forestomachs of ruminants. J. Anim. Sci. 49:1615–1630.

Tamminga, S., A.M. van Vuuren, C.J. van der Koelen, R.S. Ketelaar, and P.L. van der Togt. 1990. Ruminal behavior of structural carbohydrates and crude protein from concentrate ingredients in dairy cows. Neth. J. Agric. Sci. 38:513–526.

Tang, Ren-Haun, Min-Min Su, Tong-Geng Cao, and Yuan-Fang Liu. 1985. Stimulation of proliferation of *Tetrahymena pyrifomis* by trace rare earths. Biol. Trace Element Res. 7:95.

Tangendjaja, B., J.B. Lowry, and R.B. Wills. 1985. Degradation of mimosine and hydroxy-4(TH)-pyridone (DHP) by Indonesian goats. Trop. Anim. Prod. 10:39.

Taylor, St. C.S., A.J. Moore, and R.B. Thiessen. 1986. Voluntary food intake relation to body weight among British breeds of cattle. Anim. Prod. 42:11.

Terrill, T.H., W.R. Windham, C.S. Hoveland, and H.E. Amos. 1989. Forage preservation method influences on tannin concentration, intake, and digestibility of *Sericea lespedeza* by sheep. Agron. J. 81:435–439.

Tessema, S. 1972. Nutritional value of some tropical grass species compared to some temperate grass species. Ph.D. dissertation, Cornell Univ. Ithaca, N.Y.

Thaer, Albrecht von. 1809. Grundsätze der rationellen Landwirtschaft, vol. 1. Realschulbuchhandlung, Berlin. Pp. 261–275.

Theander, O. 1976. The chemistry of dietary fibre. Marabou Symp., Sundyberg, Sweden, Suppl. 14:23–30.

Theander, O. 1980. Sugars in thermal processes. In: Carbohydrate Sweetners. P. Koivistoinen and I. Hyvonen, eds. Academic Press, New York. Pp. 185–199.

Theander, O., P. Aman, G.E. Miksche, and S. Yasuda. 1977. Carbohydrates, polyphenols, and lignin in seed hulls of different colors from turnip rapeseed. J. Agric. Food Chem. 25:270–273.

Theander, O., and E. Westerlund. 1993. Quantitative analysis of cell wall components. In: Forage Cell Wall Structure and Digestibility. H.G. Jung, D.R. Buxton, R.D. Hatfield, and J. Ralph, eds. American Society of Agronomy, Madison, Wisc. Pp. 83–104.

Theodorou, M.K., S.E. Lowe, and A.P.J. Trinci. 1988. The fermentative characteristics of anaerobic rumen fungi. Biosystems 21:371–376.

Theodorou, M.K., S.E. Lowe, and A.P.J. Trinci. 1992. Anaerobic fungi and the rumen ecosystem. Mycology Series, vol. 9. Marcel Dekker, New York. Pp. 43–72.

Theodorou, M.K., B.A. Williams, M.S. Dhanoa and A.B. McAllan. 1992b. A new laboratory procedure for estimating kinetic parameters associated with the digestibility of forages. Proc. Conf. Plant Cell Wall Digestion, U.S. Dairy Forage Research Center, Madison, Wisc. P. 20.

Theurer, C.B. 1970. Determination of botanical and chemical composition of the grazing animal's diet. Nat. Conf. Forage Quality Evaluation and Utilization. Nebraska Center for Continuing Education, Lincoln, Neb. Pp. J1–J17.

Theurer, C.B. 1986. Grain processing effects on starch utilization by ruminants. J. Anim. Sci. 63:1649–1662.

Thiago, L.R.L. de S., and R.C. Kellaway. 1982. Botanical

composition and extent of lignification affecting digestibility of wheat and oat straw and paspalum hay. Anim. Feed Sci. Technol. 7:71.

Thomas, B.A., and R.A. Spicer. 1986. The Evolution and Paleobiology of Land Plants. Croom Helm, Beckenham, Kent.

Thomas, C., and P.C. Thomas. 1985. Factors affecting the nutritive value of grass silages. In: Recent Advances in Animal Nutrition. W. Haresign and D.J.A. Cole, eds. Butterworths, London. Pp. 223–256.

Thomas, D., and R.P. de Andrade. 1986. The evaluation under grazing of legumes associated with *Andropogon gayanus* in a tropical savannah environment on the central plateau of Brazil. J. Agric. Sci. 107:37.

Thomas, J.W. 1978. Preservatives for conserved forage crops. J. Anim. Sci. 47:721–735.

Thomas, J.W., Y. Yu, T. Middleton, and C. Stallings. 1982. Estimations of protein damage. In: Protein Requirements for Cattle: Symposium. F.N. Owens, ed. MP109 Oklahoma State Univ., Stillwater. P. 81.

Thomas, P.C. 1973. Microbial protein synthesis. Proc. Nutr. Soc. 32:85–91.

Thomas, W.A. 1975. Accumulation of rare earths and circulation of cerium by mockernut hickory trees. Can. J. Bot. 53:1159.

Thompson, J.K., and R.W. Warren. 1979. Variations in composition of pasture herbage. Grass Forage Sci. 34:83.

Thomson, D.J. 1975. The nutritive value of red clover and perennial ryegrass harvested in the autumn. J. Br. Grassl. Soc. 30:89.

Thonney, M.L., R.W. Touchberry, R.D. Goodrich, and J.C. Meiske. 1976. Intraspecies relationship between fasting heat production and body weight: a re-evaluation of $W^{.75}$. J. Anim. Sci. 43:692–704.

Tilley, J.M.A., and R.A. Terry. 1963. A two-stage technique for the in vitro digestion of forage crops. J. Br. Grassl. Soc. 18:104–111.

Tolkamp, B.J., and J.J.M.H. Ketelaars. 1992. Toward a new theory of feed intake regulation in ruminants. 2. Costs and benefits of feed consumption: an optimization approach. Livest. Prod. Sci. 30:297–317.

Trowell, H. 1975. Dietary changes in modern times. In: Refined Carbohydrate Foods and Disease. D.P. Burkitt and H. Trowell, eds. Academic Press, London.

Tsuda, T., Y. Sasaki, and R. Kawashima, eds. 1991. Physiological Aspects of Digestion and Metabolism in Ruminants. Academic Press, New York.

Turner, A.W., and V.E. Hodgetts. 1955. Buffer systems in the rumen of the sheep. I. pH and bicarbonate concentration in relationship to $_pCO_2$. Aust. J. Agric. Res. 6:115–124.

Tyler, C. 1975. Albrecht Thaer's hay equivalents: fact or fiction? Nutr. Abstr. Rev. 45:1.

Tyrrell, H.F., and P.W. Moe. 1975a. Effect of intake on digestive efficiency. J. Dairy Sci. 58:602.

Tyrrell, H.F., and P.W. Moe. 1975b. Symposium: Production efficiency in the high producing cow. Effect on intake on digestive efficiency. J. Dairy Sci. 58:1151–1163.

Tyrrell, H.F., P.W. Moe, and W.P. Flatt. 1970. Influence of excess protein intake on energy metabolism of the dairy cow. Proc. 5th Symp. on Energy Metabolism, Vitznau, Switzerland. Pp. 69–72.

Tyrrell, H.F., and J.T. Reid. 1965. Prediction of the energy value of cow's milk. J. Dairy Sci. 48:1215–1223.

Tyrrell, H.F., P.J. Reynolds. and P.W. Moe. 1979. Effect of diet on partial efficiency of acetate use for body tissue synthesis by mature cattle. J. Anim. Sci. 48:598–606.

Udén, P. 1978. Comparable studies on rate of passage, particle size and rate of digestion in ruminants, equines, rabbits and man. Ph.D. dissertation, Cornell Univ., Ithaca, N.Y.

Udén, P. 1989. Comparative aspects of fibre fermentation in mammals. In: Proc. Int. Symp. on Comparative Aspects of the Physiology of Digestion in Ruminants and Hindgut Fermenters, Copenhagen, Denmark, July 6–8. E. Skadhauge and P. Norgaard, eds. Acta Vet. Scand. Suppl. Pp. 35–44.

Udén, P., P.E. Colucci, and P.J. Van Soest. 1980. Investigation of chromium, cerium and cobalt as markers in digesta. Rate of passage studies. J. Sci. Food Agric. 31:625–632.

Udén, P., R. Parra, and P.J. Van Soest. 1974. Factors influencing reliability of the nylon bag technique. J. Dairy Sci. 57:622.

Udén, P., T.R. Rounsaville, G.R. Wiggans, and P.J. Van Soest. 1982b. The measurement of liquid and solid digesta retention in ruminants, equines and rabbits given timothy (*Phleum pratense*) hay. Br. J. Nutr. 48:329–339.

Udén, P., and P.J. Van Soest. 1982a. Comparative digestion of timothy (*Phleum pratense*) fibre by ruminants, equines and rabbits. Br. J. Nutr. 47:267.

Udén, P., and P.J. Van Soest. 1982b. The determination of digesta particle size in some herbivores. Anim. Feed Sci. Technol. 7:35–44.

Underwood, E.J. 1977. Trace Elements in Human and Animal Nutrition. Academic Press, New York.

Utley, P.R., and R.E. Hellwig. 1985. Feeding value of peanut skins added to bermudagrass pellets and fed to growing beef calves. J. Anim. Sci. 60:329.

van der Honing, Y., and G. Alderman. 1988. Systems for energy evaluation of feeds and energy requirements for ruminants. Livest. Prod. Sci. 19:217–278.

Vandersall, J.H., R.W. Hemken, and J.E. Kunsman. 1964. The effect of ration on milk fat percentage and composition. Proc. Maryland Nutrition Conf. for Feed Manufacturers. Pp. 75–78.

van Dokkum, W., A. Wesstra, and F.A. Schippers. 1982. Physiological effects of fibre-rich types of bread. 1. The effect of dietary fibre from bread on the mineral balance of young men. Br. J. Nutr. 47:451.

van Es, A.J.H. 1978. Feed evaluation for ruminants. I. The systems in use from May 1977 onwards in the Netherlands. Livest. Prod. Sci. 5:331–346.

Van Hellen, R.W., and W.C. Ellis. 1973. Membranes for rumen in situ digestion techniques. J. Anim. Sci. 37:358.

Van Hellen, R.W., and W.C. Ellis. 1977. Sample container porosities for rumen in situ studies. J. Anim. Sci. 44:141–146.

Van Horn, H.H., B. Harris. Jr., M.J. Taylor, K.C. Bachman, and C.J. Wilcox. 1984. By-product feeds for lactating dairy cows: effects of cottonseed hulls, sunflower hulls, corrugated paper, peanut hulls, sugarcane bagasse, and whole cottonseed with additives of fat, sodium bicarbonate, and *Aspergillus oryzae* product on milk production. J. Dairy Sci. 67:2922.

Van Hoven, W., and D. Furstenburg. 1992. The use of purified condensed tannin as a reference in determining its influence on rumen fermentation. Comp. Biochem. Physiol. 101A:381–385.

Van Keulen, J., and B.A. Young. 1977. Evaluation of acid insoluble ash as a natural marker in mineral digestibility studies. J. Anim. Sci. 44:282.

van Milgen, J., L.L. Berger, and M.R. Murphy. 1992. Fractionation of substrate as an intrinsic characteristic of feedstuffs fed to ruminants. J. Dairy Sci. 75:124–131.

van Milgen, J., M.R. Murphy, and L.L. Berger. 1991. A compartmental model to analyze ruminal digestion. J. Dairy Sci. 74:2515–2529.

van Nevel, C.J., and D.I. Demeyer. 1979. Stoichiometry of carbohydrate fermentation and microbial growth efficiency in a continuous culture of mixed rumen bacteria. Eur. J. Appl. Microbiol. Biotech. 7:111–120.

van Nevel, C.J., and D.I. Demeyer. 1988. Manipulation of rumen fermentation. In: The Rumen Microbial Ecosystem. P.N. Hobson, ed. Elsevier, Amsterdam. Pp. 387–444.

Van Soest, P.J. 1955. Interrelationships between composition and physical condition of feeds, rumen fermentation products, blood constituents and milk fat of ruminants. Ph.D. dissertation, Univ. of Wisconsin, Madison.

Van Soest, P.J. 1963. Ruminant fat metabolism with particular reference to factors affecting low milk fat and feed efficiency. A review. J. Dairy Sci. 46:204–216.

Van Soest, P.J. 1964. Symposium on nutrition and forage and pastures: new chemical procedures for evaluating forages. J. Anim. Sci. 23:838–845.

Van Soest, P.J. 1965a. Symposium on factors influencing the voluntary intake of herbage by ruminants: voluntary intake in relation to chemical composition and digestibility. J. Anim. Sci. 24:834–843.

Van Soest, P.J. 1965b. Use of detergents in analysis of fibrous feeds. III. Study of effects of heating and drying on yield of fiber and lignin in forages. J. Assoc. Off. Anal. Chem. 48:785–790.

Van Soest, P.J. 1966. Forage intake in relation to chemical composition and digestibility: some new concepts. Proc. 23d Southern Pasture and Forage Crop Improvement Conf., Blacksburg, Va. Pp. 24–36.

Van Soest, P.J. 1967. Development of a comprehensive system of feed analyses and its application to forages. J. Anim. Sci. 26:119–128.

Van Soest, P.J. 1969a. Basis for relationship between nutritive value and chemically identifiable fractions. Natl. Conf. on Forage Quality Evaluation and Utilization, Lincoln, Nebr., Sept. 2–5.

Van Soest, P.J. 1969b. Chemical properties of fiber in concentrate feedstuffs. Proc. Cornell Nutrition Conf., Ithaca, N.Y. Pp. 17–21.

Van Soest, P.J. 1971. Estimations of nutritive value from laboratory analysis. Proc. Cornell Nutrition Conf., Ithaca, N.Y. Pp. 106–117.

Van Soest, P.J. 1973a. Revised estimates of the net energy values of feeds. Proc. Cornell Nutrition Conf., Ithaca, N.Y. Pp. 11–23.

Van Soest, P.J. 1973b. The uniformity and nutritive availability of cellulose. Fed. Proc. 32:1804–1808.

Van Soest, P.J. 1975. Physico-chemical aspects of fibre digestion. In: Digestion and Metabolism in the Ruminant. I.W. McDonald and A.C.I. Warner, eds. Univ. of New England Publ. Unit, Armidale, N.S.W., Australia. Pp. 352–365.

Van Soest, P.J. 1977. Plant fiber and its role in herbivore nutrition. Cornell Vet. 67:307–326.

Van Soest, P.J. 1978. Dietary fibers: their definition and nutritional properties. Am. J. Clin. Nutr. 31(Suppl.):S12–S20.

Van Soest, P.J. 1981. Limiting factors in plant residues of low biodegradability. Agric. Environ. 6:135–143.

Van Soest, P.J. 1985. Comparative fiber requirements of ruminants and nonruminants. Proc. Cornell Nutrition Conf., Ithaca, N.Y. Pp. 52–60.

Van Soest, P.J. 1988a. Fibre in the diet. In: Comparative Nutrition. K. Blaxter and I. Macdonald, eds. John Libbey, London. P. 215.

Van Soest, P.J. 1988b. A comparison of grazing and browsing ruminants in the use of feed resources. In: Increasing Small Ruminant Productivity in Semi-arid Areas. E.F. Thomson and F.S. Thomson, eds. International Center for Agricultural Research in the Dry Areas. Printed in the Netherlands. Pp. 67–79.

Van Soest, P.J. 1989. On the digestibility of bound N in the distillers grains: a reanalysis. Proc. Cornell Nutrition Conf., Ithaca, N.Y. Pp. 127–135.

Van Soest, P.J. 1993. Physicochemical properties of fiber. In: Fibers. G. Csomis, J. Kusche, and S. Meryn, eds. Springer Verlag, Berlin. Pp. 22–38.

Van Soest, P.J., N.L. Conklin, and P.J. Horvath. 1987. Tannins in foods and feeds. Proc. Cornell Nutrition Conf., Ithaca, N.Y. Pp. 115–122.

Van Soest, P.J., J.G. Fadel, and C.J. Sniffen. 1979. Discount factors for energy and protein in ruminant feeds. Proc. Cornell Nutrition Conf., Ithaca, N.Y. P. 63.

Van Soest, P.J., T. Foose, and J.B. Robertson. 1983a. Comparative digestive capacities of herbivorous animals. Proc. Cornell Nutrition Conf., Ithaca, N.Y. Pp. 51–59.

Van Soest, P.J., D.G.Fox, D.R. Mertens, and C.J. Sniffen. 1984. Discounts for net energy and protein—fourth revision. Proc. Cornell Nutrition Conf., Ithaca, N.Y. Pp. 121–136.

Van Soest, P.J., J. France, and R.C. Siddons. 1992a. On the steady state turnover of compartments in the ruminant gastro-intestinal tract. J. Theor. Biol. 159:135–145.

Van Soest, P.J., J.L. Jeraci, T. Foose, K. Wrick, and F. Ehle. 1982. Comparative fermentation of fibre in man and other animals. In: Fibre in Human and Animal Nutrition. G. Wallace and L. Bell, eds. Int. Symp. Dietary Fibre, Massey Univ., Palmerston North, New Zealand. Bull. 20. Pp. 75–80.

Van Soest, P.J., and L.H.P. Jones. 1968. Effect of silica in forages upon digestibility. J. Dairy Sci. 51:1644–1648.

Van Soest, P.J., and L.H.P. Jones. 1988. Analysis and classification of dietary fibre. In: Trace Element Analytical Chemistry in Medicine and Biology. P. Bratter and P. Schramel, eds. Walter de Gruyter, Berlin. Pp. 351–370.

Van Soest, P.J., and V.C. Mason. 1991. The influence of the Maillard reaction upon the nutritive value of fibrous feeds. Anim. Feed Sci. Technol. 32:45–53.

Van Soest, P.J., and R.W. McQueen. 1973. The chemistry and estimation of fibre. Proc. Nutr. Soc. 32:123–130.

Van Soest, P.J., and D.R. Mertens. 1974. Composition and nutritive characteristics of low quality cellulosic wastes. Fed. Proc. 33:1942–1944.

Van Soest, P.J., D.R. Mertens, and B. Deinum. 1978a. Preharvest factors influencing quality of conserved forage. J. Anim. Sci. 47:712–720.

Van Soest, P.J., and J.B. Robertson. 1976a. Composition and nutritive value of uncommon feedstuffs. Proc. Cornell Nutrition Conf., Ithaca, N.Y. Pp. 102–111.

Van Soest, P.J., and J.B. Robertson. 1976b. Chemical and physical properties of dietary fibre. Proc. Miles Symp. of the Nutrition Society of Canada. Halifax, Nova Scotia. Pp. 13–25.

Van Soest, P.J., and J.B. Robertson. 1977. What is fibre and fibre in food? Nutr. Rev. 35:12–22.

Van Soest, P.J., and J.B. Robertson. 1980. Systems of analysis for evaluating fibrous feeds. In: Standardization of Analytical Methodology for Feeds. W.J. Pidgen, C.C. Balch, and M. Graham, eds. IDRC-134e. Int. Dev. Res. Centre, Ottawa, Canada. Pp. 49–60.

Van Soest, P.J., and J.B. Robertson. 1985. Analysis of forages and fibrous foods. Lab Manual for Animal Science 613. Dept. Animal Sci., Cornell University, Ithaca, N.Y.

Van Soest, P.J., J.B. Robertson, and B.A. Lewis. 1991. Methods for dietary fiber, neutral detergent fiber, and non-starch polysaccharides in relation to animal nutrition. J. Dairy Sci. 74:3583–3597.

Van Soest, P.J., J.B. Robertson, D.A. Roe, J. Rivers, B.A. Lewis, and L.R. Hackler. 1978b. The role of dietary fiber in human nutrition. Proc. Cornell Nutrition Conf., Ithaca, N.Y. Pp. 5–12.

Van Soest, P.J., M.B. Rymph, and D.J. Fox. 1992b. Discounts for net energy and protein—fifth revision. Proc. Cornell Nutrition Conf., Ithaca, N.Y. Pp. 40–68.

Van Soest, P.J., and C.J. Sniffen. 1984. Nitrogen fractions in NDF and ADF. Proc. Distillers' Feed Conf., Cincinnati. Pp. 73–82.

Van Soest, P.J., C.J. Sniffen, and M.S. Allen. 1986. Rumen dynamics. In: Aspects of Digestive Physiology in Ruminants. A. Dobson and M.J. Dobson, Cornell Univ. Press, Ithaca. Pp. 21–42.

Van Soest, P.J., P. Udén, and K.F. Wrick. 1983b. Critique and evaluation of markers for use in nutrition of humans and farm and laboratory animals. Nutr. Rep. Int. 27:17–28.

Van Soest, P.J., and R.H. Wine. 1967. Use of detergents in the analysis of fibrous feeds. IV. Determination of plant cell wall constituents. J. Assoc. Off. Anal. Chem. 50:50–55.

Van Soest, P.J., and R.H. Wine. 1968. Determination of lignin and cellulose in acid-detergent fiber with permanganate. J. Assoc. Off. Anal. Chem. 51:780–785.

Van Soest, P.J., R.H. Wine, and L.A. Moore. 1966. Estimation of the true digestibility of forages by the in vitro digestion of cell walls. Proc. 10th Int. Grasslands Congr., Helsinki. Pp. 438–441.

van Straalen, W.M., and S. Tamminga. 1990. Protein degradation of ruminant diets. In: Feedstuff Evaluation. J. Wiseman and D.J.A. Cole, eds. Butterworths, London. Pp. 55–72.

Varel, V.H., H.G. Jung, and W.G. Pond. 1988. Effects of dietary fiber on young adult genetically lean, obese and contemporary pigs: rate of passage, digestibility and microbiological data. J. Anim. Sci. 66:707–712.

Varvikko, T. 1986. Microbially corrected amino acid composition of rumen-undegraded feed protein and amino acid degradability in the rumen of feeds enclosed in nylon bags. Br. J. Nutr. 56:131.

Veira, D.M. 1986. The role of ciliate protozoa in nutrition of the ruminant. J. Anim. Sci. 63:1547–1560.

Vennesland, B., P.A. Castric, E.E. Conn, L.P. Solomonson, M. Volini, and J. Westley. 1982. Cyanide metabolism. Fed. Proc. 41(10):2639–2648.

Vercellotti, J.R., A.A. Salyers, W.S. Bullard, and T.D. Wilkins. 1975. Breakdown of mucin and plant polysaccharides in the human colon. Can. J. Biochem. 55:1190–1196.

Verite, R. 1980. Appreciation of the nitrogen value of feeds for ruminants. In: Standardization of Analytical Methodology for Feeds. W.J. Pigden, C.C. Balch, and M. Graham, eds. IDRC-134e. Int. Dev. Res. Centre, Ottawa. Pp. 87–96.

Vickery, R.C. 1953. Chemistry of the Lanthanons. Butterworths, London.

Vidal, H.M., D.E. Hogue, J.M. Elliot, and E.F. Walker. 1969. Digesta of sheep fed different hay-grain ratios. J. Anim. Sci. 29:62–68.

Virtanen, A.I. 1933. The A.I.V. method of preserving fresh fodder. Emp. J. Exp. Agric. 1:143–155.

Virtanen, A.I. 1966. Milk production of cows on protein-free feed. Science 153:1603–1604.

Visek, W.J. 1978. Diet and cell growth modulation by ammonia. Am. J. Clin. Nutr. 31(Suppl.):S216–S220.

Visek, W.J., and J.B. Robertson. 1973. Dried brewer's grains in dog diets. Proc. Cornell Nutrition Conf., Ithaca, N.Y. Pp. 40–49.

Vogels, G.D., W.F. Hoppe, and C.K. Stumm. 1980. Association of methanogenic bacteria with rumen ciliates. Appl. Environ. Microbiol. 40:608–613.

Volenec, J.J., J.H. Cherney, and K.D. Johnson. 1987. Yield components, plant morphology, and forage quality of alfalfa as influenced by plant population. Crop Sci. 27:321–326.

Voorhies, M.R., and J.R. Thomasson. 1979. Fossil grass anthoecia within Miocene rhinoceros skeletons: diet in an extinct species. Science 206:331–333.

Wagner, D.G., and J.K. Loosli. 1967. Studies on the energy requirements of high-producing dairy cows. Cornell Univ. Agric. Exp. Stn. Memoir no. 400.

Wahle, K.W.J., W.R.H. Duncan, and G.A. Garton. 1979. Propionate metabolism in different species of ruminants. Ann. Rech. Vet. 10:362–364.

Wahle, K.W.J., and C.T. Livesay. 1985. The effect of Monensin supplementation of dried grass or barley diets on aspects of propionate metabolism, insulin secretion and lipogenesis in the sheep. J. Sci. Food Agric. 36:227–236.

Wahle, K.W.J., and S.M. Paterson. 1979. The utilization of methylmalonyl-CoA for branched-chain fatty acid synthesis by preparations from bovine (Bos taurus) adipose tissue. Am. J. Biochem. 10:433–437.

Waite, R., and A.R.N. Gorrod. 1959. The comprehensive analysis of grasses. J. Sci. Food Agric. 10:317.

Waldo, D.R. 1970. Factors influencing the voluntary intake of forages. Proc. Natl. Conf. Forage Quality Evaluation and Utilization, Lincoln, Neb. P. E-1.

Waldo, D.R., C.E. Coppock, L.A. Moore, and J.F. Sykes. 1961. Estimates of the net energy maintenance requirement of grazing dairy cattle. Abstr. 5th Int. Congr. Nutrition, Washington, D.C. Vol. 20, p. 8.

Waldo, D.R., and N.A. Jorgensen. 1981. Forages for high animal production: nutritional factors and effects of conservation. J. Dairy Sci. 64:1207.

Waldo, D.R., and L.H. Schultz. 1956. Lactic acid production in the rumen. J. Dairy Sci. 39:1453–1460.

Waldo, D.R., L.W. Smith, and E.L. Cox. 1972. Model of

cellulose disappearance from the rumen. J. Dairy Sci. 55:125–129.

Waller, S.S., and J.K. Lewis. 1979. Occurrence of C_3 and C_4 photosynthetic pathways in North American grasses. J. Range Manage. 32:12–28.

Wanderley, R.C., C.B. Theurer, S. Rahnema, and T.H. Noon. 1985. Automated long-term total collection versus indicator method to estimate duodenal digesta flow in cattle. J. Anim. Sci. 61:1550.

Wardrop, I.D., and J.B. Combe. 1960. The post-natal growth of the visceral organs of the lamb. I. The growth of the visceral organs of the grazing lamb from birth to sixteen weeks of age. J. Agri. Sci. 54:140–143.

Warner, A.C.I. 1956. Criteria for establishing the validity of in vitro studies with rumen microorganisms in so-called artificial rumen systems. J. Gen. Microbiol. 14:733–748.

Warner, A.C.I. 1969. Binding of the ^{51}Cr complex of ethylenediamine tetra-acetic acid to particulate matter in the rumen. Vet. Record (April 26):441–442.

Warner, R.G., and W.P. Flatt. 1965. Anatomical developments of the ruminant stomach. In: Physiology of Digestion in the Ruminant. R.W. Dougherty, R.S. Allen, W. Burroughs, N.L. Jacobson, and A.D. McGilliard, eds. Butterworths, Washington, D.C. Pp. 24–38.

Waters, C.J., M.A. Kitcherside, and A.J.F. Webster. 1992. Problems associated with estimating the digestibility of undegraded dietary nitrogen from acid-detergent insoluble nitrogen. Anim. Feed Sci. Technol. 39:279–291.

Way, J.T. 1853. On the relative nutritive and fattening properties of different natural and artificial grasses. J. Royal Agric. Soc. Engl. 14:171.

Webster, A.J.F. 1978a. Prediction of the energy requirements for growth in beef cattle. World Rev. Nutr. Diet. 30:189–226.

Webster, A.J.F. 1978b. Energy metabolism and requirements. In: Digestive Physiology and Nutrition of Ruminants, vol. 2, 2d ed. D.C. Church, ed. O&B Books, Corvallis, Ore. Pp. 210–229.

Wedig, C.L., E.H. Jaster, and K.J. Moore. 1988. Effect of brown midrib and normal genotypes of sorghum × sudangrass on ruminal fluid and particulate rate of passage from the rumen and extent of digestion at various sites along the gastrointestinal tract in sheep. J. Anim. Sci. 66:539.

Weisbjerg, M.R., T. Hvelplund, and D.L. Palmquist. 1991. Ruminal, intestinal, and total digestibilities of diets supplemented with fat and hydrolyzed feather meal/blood meal. J. Dairy Sci. 74(Suppl. 1):258.

Weiss, W.P., H.R. Conrad, and W.L. Shockey. 1986. Digestibility of nitrogen in heat damaged alfalfa. J. Dairy Sci. 69:2658.

Weiss, W.P., H.R. Conrad, and N.R. St. Pierre. 1992. A theoretically-based model for predicting total digestible nutrient values of forages and concentrates. Anim. Feed Sci. Technol. 39:95–110.

Welch, J.G. 1967. Appetite control in sheep by indigestible fibers. J. Anim. Sci. 26:849–854.

Welch, J.G. 1982. Rumination, particle size and passage from the rumen. J. Anim. Sci. 54:885.

Welch, J.G., and A.M. Smith. 1969. Effect of varying amounts of forage intake on rumination. J. Anim. Sci. 28:827–830.

Welch, J.G., and A.M. Smith. 1971. Physical stimulation of rumination activity. J. Anim. Sci. 33:1118–1123.

Weller, R.A. 1957. The amino acid composition of hydrolysates of microbial preparations from the rumen of sheep. Aust. J. Biol. Sci. 10:384–389.

Weller, R.A., F.V. Gray, and A.F. Pilgrim. 1958. The conversion of plant nitrogen to microbial nitrogen in the rumen of the sheep. Br. J. Nutr. 12:421.

Weller, R.A., and A.F. Pilgrim. 1974. Passage of protozoa and volatile fatty acids from the rumen of the sheep and from a continuous in vitro fermentation system. Br. J. Nutr. 32:341.

Wheeler, J.L. 1987. Pastures and pasture research in southern Australia. In: Temperate Pastures: Their Production, Use and Management. J.L. Wheeler, C.J. Pearson, and G.E. Robards, eds. CSIRO, Canberra. Pp. 3–31.

Wheeler, W.E. 1977. Buffers in dairy cattle feeds. Proc. Cornell Nutrition Conf., Ithaca, N.Y. Pp. 53–58.

White, D.H., and F.H.W. Morley. 1977. Estimation of optimal stocking rate of merino sheep. Agric. Syst. 2:289–304.

Whitehead, D.C., K.M. Goulden, and R.D. Hartley. 1985. The distribution of nutrient elements in cell wall and other fractions of the herbage of some grasses and legumes. J. Sci. Food Agric. 36:311–318.

Whyte, R.O. 1962. The myth of tropical grasslands. Trop. Agric. 39:1–12.

Wieghart, M., R. Slepetis, J.M. Elliot, and D.F. Smith. 1986. Glucose absorption and hepatic gluconeogenesis in dairy cows fed diets varying in forage content. J. Nutr. 116:839.

Wilkie, K.C.B. 1979. The hemicelluloses of grasses and cereals. Adv. Carbohydr. Chem. Biochem. 36:215–264.

Wilkins, R.J. 1974. The nutritive value of silages. Proc. Univ. of Nottingham Nutrition Conf. no. 8. H. Swan and D. Lewis, eds. Butterworths, London. Pp. 167–189.

Williams, R.S., and W.H. Olmstead. 1936. The effect of cellulose, hemicellulose and lignin on the weight of the stool: a contribution to the study of laxation in man. J. Nutr. 11:433–449.

Wilson, J. 1853. Flax, its treatment, agricultural and technical. J. Royal Agric. Soc. 14:187–210.

Wilson, J.R. 1981. Environmental and nutritional factors affecting herbage quality. In: Nutritional Limits to Animal Production from Pastures. Proc. Int. Symp., St. Lucia, Queensland, Australia. J.B. Hacker, ed. Commonwealth Agricultural Bureau. Pp. 111–131.

Wilson, J.R. 1983. Effects of water stress on in vitro dry matter digestibility and chemical composition of herbage of tropical pasture species. Aust. J. Agric. Res. 34(4):377–390.

Wilson, J.R. 1993. Organization of forage plant tissues. In: Forage Cell Wall Structure and Digestibility. H.G. Jung, D.R. Buxton, R.D. Hatfield, and J. Ralph, eds. American Society of Agronomy, Madison, Wisc. Pp. 1–32.

Wilson, J.R., and J.B. Hacker. 1987. Comparative digestibility and anatomy of some sympatric C_3 and C_4 arid zone grasses. Aust. J. Agric. Res. 38:287.

Wiseman, J. ed. 1984. Fats in Animal Nutrition. Butterworths, Boston.

Woese, C.R. 1987. Bacterial evolution. Microbiol. Rev. 51:221–271.

Wofford, H., J.L. Holechek, M.L. Galyean, J.D. Wallace and M. Cardenas. 1985. Evaluation of fecal indices to predict cattle diet quality. J. Range Manage. 38:450.

Wohlt, J.E., J.F. Fiallo, and M.E. Miller. 1981. Composition of by-products of the essential-oil industry and their potential as feeds for ruminants. Anim. Feed Sci. Technol. 6:115–121.

Wolff, E. 1856. Die naturgesetzlichen Grundlagen des Ackerbaues. 3d ed. Otto Weigand, Leipzig. Pp. 953–954.

Wolin, M.J. 1960. A theoretical rumen fermentation balance. J. Dairy Sci. 40:1452.

Wolin, M.J. 1975. Interactions between the bacterial species of the rumen. In: Digestion and Metabolism in the Ruminant. I.W. McDonald and A.C.I. Warner, eds. Univ. of New England Publ. Unit, Armidale, N.S.W. Pp. 134–148.

Wong, E. 1973. Plant phenolics. In: Chemistry and Biochemistry of Herbage, vol. 1. G.W. Butler and R.W. Bailey, eds. Academic Press, London. Pp. 265–322.

Wood, H.G. 1991. Life with CO or CO_2 and H_2 as a source of carbon and energy. FASEB 5:156–163.

Wood, P.J. 1990. Physicochemical properties and physiological effects of the $(1\to3)$ $(1\to4)$-β-D-glucan from oats. In: New Developments in Dietary Fiber, Physiological, Physicochemical, and Analytical Aspects. I. Furda and C.J. Brine, eds. Plenum Press, New York. Pp. 119–134.

Wood, T.M. 1989. Mechanisms of cellulose degradation by enzymes from aerobic and anaerobic fungi. In: Enzyme Systems in Lignocellulose Degradation. M.P. Coughlan, ed. Elsevier Applied Science, London. Pp. 17–35.

Woodwell, G.M., R.H. Whittaker, W.A. Reiners, G.E. Likens, C.C. Delwiche, and D.B. Botkin. 1978. The biota and the world carbon budget. Science 199:141–146.

Woody, H.D., D.G. Fox, and J.R. Black. 1984. Effect of diet grain content on performance of growing and finishing cattle. J. Anim. Sci. 57:710.

Woolford, M.K. 1975a. Microbiological screening of the straight chain fatty acids (C1–C12) as potential silage additives. J. Sci. Food Agric. 26:219–228.

Woolford, M.K. 1975b. Microbiological screening of food preservatives, cold sterilants and specific antimicrobial agents as potential silage additives. J. Sci. Food Agric. 26:229–237.

Wrenn, T.R., J. Bitman, R.A. Waterman, J.R. Weyant, D.L. Wood, L.L. Strozinski, and N.W. Hooven, Jr. 1978. Feeding protected and unprotected tallow to lactating cows. J. Dairy Sci. 61:49–58.

Wu, Z., and D.L. Palmquist. 1991. Synthesis and biohydrogenation of fatty acids by ruminal microorganisms in vitro. J. Dairy Sci. 74:3035–3046.

Wyburn, R.S. 1980. The mixing and propulsion of the stomach contents of ruminants. In: Digestive Physiology and Metabolism of Ruminants. Y. Ruckebusch and P. Thivend, eds. MTP Press, Lancaster, England. Pp. 35–52.

Young, B.A., and M.E.P. Webster. 1963. A technique for the estimation of energy expenditure in sheep. Aust. J. Agric. Res. 14:867–873.

Young, J.W., J.J. Veenhuizen, J.K. Drackley, and T.R. Smith. 1990. New insights into lactation ketosis and fatty liver. Proc. Cornell Nutrition Conf., Ithaca, N.Y. P. 60.

Yu, Y., and J.W. Thomas. 1976. Estimation of the extent of heat damage in alfalfa haylage by laboratory measurement. J. Anim. Sci. 42:766–774.

Zadrazil, F., G.C. Galletti, R. Piccaglia, G. Chiavari, and O. Francioso. 1991. Influence of oxygen and carbon dioxide on cell wall degradation by white-rot fungi. Anim. Feed Sci. Technol. 32:137–142.

Zadrazil, F., and P. Reiniger, eds. 1988. Treatment of Lignocellulosics with White Rot Fungi. Elsevier Applied Science, London.

Zeikus, J.G. 1980. Fate of lignin and related aromatic substrates in anaerobic environments. In: Biodegradation: Microbiology, Chemistry and Potential Applications, vol. 1. T.K. Kirk, T. Higuchi, and H. Chang, eds. CRC Press, Boca Raton, Fla. Pp. 101–109.

Zemmelink, G. 1986. The effects of hot climate on feed quality and intake. World Rev. Anim. Prod. 22:83.

Author Index

Subject Index

www.ingramcontent.com/pod-product-compliance
Ingram Content Group UK Ltd.
Pitfield, Milton Keynes, MK11 3LW, UK
UKHW011455280225
455513UK00020B/52